Applied Mathematical Sciences

Volume 207

The mathematization of all sciences, the fading of traditional scientific boundaries, the impact of computer technology, the growing importance of computer modeling and the necessity of scientific planning all create the need both in education and research for books that are introductory to and abreast of these developments. The purpose of this series is to provide such books, suitable for the user of mathematics, the mathematician interested in applications, and the student scientist. In particular, this series will provide an outlet for topics of immediate interest because of the novelty of its treatment of an application or of mathematics being applied or lying close to applications. These books should be accessible to readers versed in mathematics or science and engineering, and will feature a lively tutorial style, a focus on topics of current interest, and present clear exposition of broad appeal. A compliment to the Applied Mathematical Sciences series is the Texts in Applied Mathematics series, which publishes textbooks suitable for advanced undergraduate and beginning graduate courses.

More information about this series at http://www.springer.com/series/34

Andrea Manzoni • Alfio Quarteroni • Sandro Salsa

Optimal Control of Partial Differential Equations

Analysis, Approximation, and Applications

Springer

Andrea Manzoni
MOX, Department of Mathematics
Politecnico di Milano
Milano, Italy

Alfio Quarteroni
MOX, Department of Mathematics
Politecnico di Milano
Milano, Italy

Sandro Salsa
Department of Mathematics
Politecnico di Milano
Milano, Italy

ISSN 0066-5452 ISSN 2196-968X (electronic)
Applied Mathematical Sciences
ISBN 978-3-030-77228-4 ISBN 978-3-030-77226-0 (eBook)
https://doi.org/10.1007/978-3-030-77226-0

Mathematics Subject Classification (2020): 35-XX, 44-XX, 65N-XX

This Springer imprint is published by the registered company Springer Nature Switzerland AG
The registered company address is: Gewerbestrasse 11, 6330 Cham, Switzerland

Preface

This is a book on Optimal Control Problems (OCPs): how to formulate them, how to set up a suitable mathematical framework for their analysis, how to approximate them numerically, how to use them to address both academic and real life applications.

An OCP consists in searching for a *control function* u and a corresponding *state function* $y = y(u)$ yielding the optimum (either maximum or minimum) of a quantity, called *cost* or *objective function*, that depends on both u and $y = y(u)$. This latter is the solution of an equation, the so-called state (or governing) equation. Typically, u represents a set of data for the state equation to be properly chosen in order to achieve the optimum cost.

An OCP can therefore be regarded as an optimization problem subject to the constraint expressed by the state equation; further additional constraints, both on the control function and the state function, may also apply.

OCPs arise in a broad variety of situations. Depending upon the nature of the state equation, we talk about OCPs for algebraic systems, ordinary differential equations (ODEs) or partial differential equations (PDEs). In this book we will restrict to state equations that are well posed (i.e. they admit a unique solution that depends continuously on the problems' data). However, OCPs can be formulated as well for ill posed state equations. Their analysis is more involved; the interested reader can find examples in, e.g., [96].

The literature is paved by an almost uncountable number of excellent textbooks and monographs in the field of optimal control problems, see, e.g., [177, 178, 148, 151, 263, 44]. In this book we primarily deal with OCPs governed by PDEs. Our aim is to cover the whole range going from the set up and the theoretical analysis of the OCP, to the derivation of the system of optimality conditions, the proposition of suitable numerical methods, their analysis, and the application to a wide class of problems of practical relevance. This broad span represents one of the distinguishing features of our book. We cover the basic types of governing equations – algebraic systems, linear ODEs and linear PDEs of elliptic and parabolic type. Hyperbolic (wave) equations require special attention – indeed, the controllability of hyperbolic equations and systems is more challenging because of the lack of dissipation mechanisms – and we only address them occasionally by presenting an example and few exercises, referring to the pertinent literature; see, e.g., [72]. Moreover, we treat OCPs that are relevant in the field of engineering and applied sciences. Among those, we mention OCPs for advection-diffusion equations, Navier-Stokes equations governing the motion of incompressible viscous fluids, the monodomain electrophysiology problem for the cardiac electrical potential, shape optimization, and parameter identification problems. Depending upon the case, the cost function to be minimized may represent energy dissipation in a mechanical system, drag forces acting on an airfoil, the maximum time for defibrillation in a cardiac tissue affected by arrhythmias, etc.

A quick outline of the book is reported below. Chapter 1 is a self-contained introduction which addresses a handful of meaningful and representative OCPs, and presents an overview of the mathematical issues arising in the field of optimal control problems for PDEs. Although these are not trivial concepts, we have made an effort to render this subject accessible to a large audience, including readers who are not necessarily acquainted with the mathematical intricacies of PDEs.

Part I, from Chapter 2 to 4, deals with optimization in finite dimensional spaces as well as with optimal control problems for algebraic systems and (linear) ordinary differential equations. After the problem setting, we carry out the well posedness analysis and derive first-order optimality conditions for both free (unconstrained) and constrained optimal control problems. We then review some of the most widely used algorithms for the numerical computation of functional minimizers, and discuss the case of state equations governed by algebraic equations or by a Cauchy problem for ordinary differential equations. This part has been conceived to be accessible to students with a basic knowledge of Calculus and Linear Algebra. Given the introductory character of Chapters 2 and 3, exercises are proposed from Chapter 4 on.

Part II, from Chapter 5 to 8, represents the core of the book. We first address OCPs for linear elliptic PDEs with quadratic cost functionals, and their numerical approximation. We cover both constrained and (free) unconstrained OCPs, with algebraic constraints on the control. State constrained OCPs (i.e. with constraints on the solution of the state equation) are mathematically more challenging and will be treated only marginally. Then we turn to OCPs for initial-boundary value problems of parabolic type and their numerical approximation. Different types of controls are considered and first order optimality conditions are derived. This part is suitable for advanced undergraduate programs in mathematics, applied mathematics and engineering; in principle, it can be directly accessed without passing through Part I. For pedagogical convenience, our presentation follows an inductive approach: starting from simple problems that allow a *hands-on* treatment, the reader is progressively led to a general framework, suitable to face broader classes of problems.

Part III, from Chapter 9 to 11, is aimed at covering advanced OCPs involving nonlinear PDEs and non-quadratic cost functions. We start by providing a general framework for nonlinear state equations, with an introductory preview on the optimization in Banach and Hilbert spaces. We then derive first order optimality conditions for OCPs with constraints on the control functions. The analysis of this type of problems requires some basic concepts of differential calculus in Banach spaces that are provided to the reader's benefit. The two more classical approaches for the numerical approximation of these OCPs are presented: iterative methods, and all-at-once methods. We then turn to more advanced applications involving OCPs for Navier-Stokes equations (both steady and unsteady) in fluid dynamics and a nonlinear diffusion reaction system that models the propagation of the electric potential through the myocardium (the heart's muscle). Our final application concerns shape optimization problems. Because of the mathematical intricacy and the vastity of this subject, we confine ourselves to a quick introduction to shape functionals and shape derivatives, to the derivation of first-order optimality conditions, and to a quick presentation of the numerical methods for shape optimization. Two classical applicative examples conclude this part: the minimization of the compliance of an elastic structure, and the drag minimization for a steady Navier-Stokes flow. This part requires a more in-depth mathematical knowledge in order to be fully appreciated. However, by surfing on the more intricate mathematical concepts, all the readers can find precise and constructive guidelines to set up and solve OCPs in the context of more complex applications.

For the reader's convenience, those sections that address the more subtle mathematical concepts (with full proofs) are labelled by a $*$. The writing has been conceived in such a way that skipping the $*$ parts would not compromise the reading and the understanding of all of the practical aspects necessary to implement OCPs. However, we hope that readers more passionate with Math will find the $*$ parts stimulating.

The book also features two Appendices. Appendix A is aimed at providing a wealth of tools that are essential to carry out a rigorous theoretical analysis of OCPs. These include tools of Functional Analysis and functional spaces that are essential to formulate and analyze PDEs. Appendix B offers instead a preview on the numerical approximation of PDEs by the classical Galerkin methods. We hope that readers unacquainted with these subjects could find them useful before facing some of the Chapters of this book.

While writing this book, we benefitted from comments, insights and suggestions of many colleagues and students. In particular, the authors would like to thank Luca Dedè, Federico Negri and Stefano Pagani for many fruitful discussions on numerical methods for optimal control problems. Among the students who have sat through our courses on Optimal Control of PDEs, we would like to thank Ludovica Cicci, Niccolò Dal Santo, Monica Huynh, Matteo Salvador, and Carlo Sinigaglia. In addition, special thanks are due to Francesca Bonadei and Francesca Ferrari of Springer Italia for their invaluable help in the preparation of the manuscript.

Lausanne, Switzerland
Milan, Italy
November 2021

Andrea Manzoni
Alfio Quarteroni
Sandro Salsa

List of Algorithms

Contents

About the Authors

Andrea Manzoni, PhD is an associate professor of numerical analysis at Politecnico of Milan. He is the author of 2 books and of approximately 50 papers. In 2012, he won the ECCOMAS Award for the best PhD thesis in Europe about computational methods in applied sciences and engineering and the Biannual SIMAI prize (Italian Society of Applied and Industrial Mathematics) in 2017. His research interests include the development of reduced-order modelling techniques for PDEs, PDE-constrained optimization, uncertainty quantification, computational statistics, and machine/deep learning.

Alfio Quarteroni is a professor of numerical analysis at Politecnico di Milano and professor emeritus at EPFL, Lausanne. He is the author of 25 books, editor of 12 books, and author of about 400 papers. He is the recipient of two ERC Advanced Grants. He is a member of the Italian Academy of Science, the European Academy of Science, Academia Europaea, and the Lisbon Academy of Science. His research group at EPFL has carried out the mathematical simulation for the Alinghi sailing boat, the winner of two editions (2003 and 2007) of America's Cup. His research interests include mathematical modeling and its applications at large.

Sandro Salsa is professor emeritus at the Politecnico di Milano, where he has been one of the main founders of the educational program in mathematical engineering. His research interest ranges over diverse aspects of nonlinear, nonlocal, singular or degenerate elliptic and parabolic partial differential equations, with particular emphasis on free boundary problems. He is author of 13 books and several papers with the most prestigious scientific mathematical journals.

About the Authors

Chapter 1
Introduction: Representative Examples, Mathematical Structure

The purpose of this chapter is to introduce a general framework for optimal control problems (OCPs) governed by partial differential equations (PDEs), as well as a few intuitive examples and some significant cases. In Sect. 1.1 we present some preliminary concepts common to many different OCPs and provide the basic ingredients useful to formulate the examples that are introduced in Sect. 1.2. In Sects. 1.3–1.7 we deal with more advanced problems in engineering and applied sciences. In Sect. 1.8 we prepare the general setting and notations that will be used throughout the book for the formulation and the analysis of OCPs. More general PDE-constrained optimization problems, such as shape optimization and parameter estimation problems, are introduced in Sects. 1.9–1.10. Finally, we sketch some general theoretical features and numerical challenges shared by these problems in Sect. 1.11 and 1.12.

1.1 Optimal Control Problems Governed by PDEs

Problems arising from applied sciences are often modeled by PDEs which depend on a set of *input data*, such as physical or material coefficients, boundary and initial conditions, source terms, as well as on the *geometrical configuration* describing the domain where the problem is formulated, which can be considered as *input* itself. Often, a problem governed by PDEs – usually denoted as *state system* – needs to be controlled or optimized by acting on (one, or more of) these input variables. This is a difficult mathematical task that may pose severe computational challenges.

Solving a *forward problem* (or *direct problem*) consists in computing the solution of a given PDE (the state system) and, possibly, some *output* of interest. Instead, solving an *optimal control problem* governed by a PDE consists of minimizing (or maximizing) a physical quantity, called the *cost function* and depending on the PDE solution itself, by acting on some suitable *control variables*; these latter are some of the *data* required by the PDE.

In an *optimal design* (or *shape optimization*) problem we look for that specific shape of the geometrical domain of the PDE that minimizes (or maximizes) a suitable cost function.

Other problems that can often be cast in the form of PDE-constrained optimization are *parameter estimation* and *data assimilation* problems. In the former case, we aim at identifying some *input data* (the parameters) which are unknown, or only partially known, by relying on some observations (or measures) of a given *output* of interest. In the latter case, a numerical model of a given system, usually affected by uncertainty, is improved by incorporating system observations.

The main focus of this book is on optimal control problems for PDEs, whose formulation, analysis and numerical approximation are treated in details in Chapters 5–10.

A. Manzoni et al., *Optimal Control of Partial Differential Equations*,
Applied Mathematical Sciences 207, https://doi.org/10.1007/978-3-030-77226-0_1

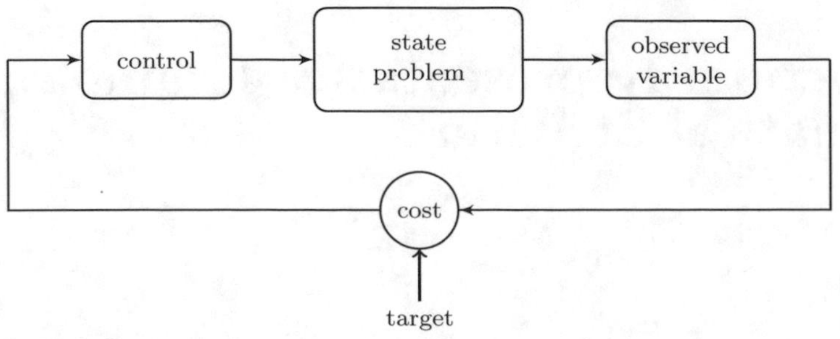

Fig. 1.1 Structure of an unconstrained optimal control problem

Optimal control problems for PDEs are characterized by four main ingredients: a state problem, a control function, a cost function and, possibly, some constraints (see Fig. 1.1).

The *state* problem can be either an algebraic system, an initial-value problem for ordinary differential equations, or a boundary-value problem for partial differential equations. Its solution, denoted by y, depends on a variable u representing the *control* that can be exerted on the system. In this way, the state problem associates to every control a state solution $y = y(u)$.

The goal is to find the control u in such a way that an *observed* variable z, also called *observation* – a function of u through y – is as near as possible to a desired value z_d, the so-called *target*.

The state problem is said to be *controllable* if a control u exists such that the observed variable z matches *exactly* the desired value z_d. Achieving exact controllability is neither easy nor (always) possible[1] and depends on the class of systems under consideration.

For instance, controllability of heat and wave equations poses different challenges because of their different behavior with respect to time reversal. If we wish to control the state of the system at a given time, the heat equation is not exactly controllable for any time, due to its regularizing effect; on the contrary, the time needed to control waves needs to be large enough due to their finite speed of propagation. See, in this respect, [278], as well as [175, 168].

The difficulties mentioned above motivate replacing the problem of controllability by one of *optimization*, in which case we do not require the output variable (or observation) z to be exactly equal to the target z_d, rather the *distance* between z and z_d (in a suitable sense) to be minimized. Control and optimization problems are thus two intimately related concepts, as we will see later on. In this respect, another essential ingredient of an optimal control problem is the *cost* function, which depends on the control function u and on the state solution $y = y(u)$, and has to be minimized, reflecting the goal of the problem. Depending on the context, the cost function may represent a suitable distance, a function of the state solution having a physical meaning, such as energy, flux, forces and so on.

A further (optional) ingredient is given by constraints that may act on the control u and/or on the state y (respectively, control constraints or state constraints); they restrict the set of admissible controls or prescribe desired properties of the state solution.

In the following section we introduce two simple examples, while more advanced problems are described in Sections 1.3–1.7. A more precise mathematical setting for OCPs is provided later in Sect. 1.8.

---

[1] Take for instance the simple case in which the state problem is the linear algebraic system $A\mathbf{y} = \mathbf{b}$, where A is a given $n \times n$ non singular matrix and $\mathbf{b}$ a given vector of $\mathbb{R}^n$. Assume moreover that the observation is represented by one solution component, say the first one, and the control be one of the components of the right hand side, say the last one. The question therefore reads: "Find $u \in \mathbb{R}$ such that the solution of the linear system $A\mathbf{y} = \mathbf{b} + (0, \ldots, 0, u)^\top$ satisfy $y_1 = z_1$", being z_1 a given value. In general, this problem admits no solution.

1.2 An Intuitive Example: Optimal Control for Heat Transfer

We now introduce some OCPs related to *heat transfer*. An example is provided by a body occupying a region (the domain) $\Omega \subset \mathbb{R}^3$ that has to be heated (or cooled). In this case, the state y represents its temperature, whereas the state system is modeled by the (stationary) heat equation. To heat or cool the system, we apply to the volume a heat source u. The latter represents the control function which has to be chosen so that on one side the corresponding temperature distribution $y = y(u)$ in the domain be as near as possible to a given target temperature z_d, and on the other side the effort in heating (or cooling) the system is as low as possible. Clearly, the state y, the control u and the target z_d are all functions of the spatial coordinates $\mathbf{x} \in \Omega \subset \mathbb{R}^3$.

This problem can be modeled by seeking for the solution of the following *minimization* problem for the cost function $J(u)$,

$$J(u) = \frac{1}{2}\int_{\Omega} (y(u) - z_d)^2 d\mathbf{x} + \frac{\alpha}{2}\int_{\Omega} u^2 d\mathbf{x} \; \rightarrow \; \min \tag{1.1}$$

where $y = y(u)$ is the solution of the following *state* problem

$$\begin{cases} -\Delta y = f + u & \text{in } \Omega \\ y = g_D & \text{on } \Gamma = \partial\Omega, \end{cases} \tag{1.2}$$

given by a Dirichlet boundary problem for the stationary heat equation.

Notation (1.1), $J(u) \; \rightarrow \; \min$, is an abridge notation that stands for: *we look for u that minimizes $J(u)$ over all possible control functions that belong to a suitable functional space.* The latter will depend on the problem at hand and will be made precise from case to case.

Here the temperature g_D on the boundary Γ of the domain and the source term f in Ω are prescribed. Thus, the first term appearing in the cost[2] function (1.1) represents a suitable *distance* between the target z_d and the temperature $y = y(u)$, whereas the second term can be thought as the heating cost, i.e. the cost paid to enable the action of the control. In this respect, $\alpha \geq 0$ is a parameter that can be used like a tuning unitary cost. From a mathematical point of view, we will see in the following chapters that this term sometimes ensures the well-posedness of the optimal control problem, providing some regularization property on its solution.

In the case we are interested in heating or cooling only a portion of the region Ω, i.e. by controlling the heat source in a subdomain $\Omega_c \subset \Omega$, u can be replaced by $u\chi_{\Omega_c}(\mathbf{x})$, being $\chi_D(\mathbf{x})$ the indicator function of the set D,

$$\chi_D(\mathbf{x}) = \begin{cases} 1 & \mathbf{x} \in D \\ 0 & \mathbf{x} \notin D. \end{cases} \tag{1.3}$$

Alternatively, we might be interested to reach a desired temperature just in a portion $\Omega_{\text{obs}} \subset \Omega$ of the region, so that instead of (1.1) we should minimize

$$J(u) = \frac{1}{2}\int_{\Omega_{\text{obs}}} (y(u) - z_d)^2 d\mathbf{x} + \frac{\alpha}{2}\int_{\Omega} u^2 d\mathbf{x} \; \rightarrow \; \min . \tag{1.4}$$

If the target is to reach a desired thermal flux across Γ, (1.4) is replaced by

$$J(u) = \int_{\Gamma} \left(\frac{\partial y(u)}{\partial n} - z_{d_\Gamma} \right)^2 d\sigma + \frac{\alpha}{2}\int_{\Omega} u^2 d\mathbf{x} \; \rightarrow \; \min \tag{1.5}$$

[2] J is in fact a functional of u. For this reason, we can equivalently use the expressions *cost function* or *cost functional.*

provided $y(u)$ is sufficiently regular. Here $\partial y/\partial n = \nabla y \cdot \mathbf{n}$ is the normal derivative of u, i.e. the derivative of u along the normal direction $\mathbf{n}$ to the boundary (directed outward). For the *heat equation*, this latter provides the outer thermal flux across the boundary.

In all these cases we say that we are dealing with a *distributed control* problem; the observation of the system can be distributed, when it takes place in the domain Ω or in a subdomain Ω_{obs}, as in (1.1) or (1.10), or at the boundary, like for the cost functional (1.10).

An important class is that of the *feedback controls*, so-called because of their direct dependence on the state variable. A typical example is represented by a thermostat which can switch on and off. This is a *piecewise constant* control function taking only the values 0 or 1, that measures the temperature in the region to be cooled or heated, compares that value with a desired temperature value z_d, and acts (through the control u) on the heating plant according to the sign of the difference $y(u) - z_d$, also called the *feedback error*. In this way, we may think that it turns heating on when $y(u) < z_d$, whereas it turns it off when $y(u) > z_d$.

In a similar way, we might control the thermal flux on a portion Γ_N of the boundary, so that the state equation can be formulated as follows:

$$\begin{cases} -\Delta y = f & \text{in } \Omega \\ y = g_D & \text{on } \Gamma_D \\ \dfrac{\partial y}{\partial n} = \gamma(u - y) & \text{on } \Gamma_N, \end{cases} \tag{1.6}$$

where Γ_D and Γ_N provide a disjoint partition of Γ. Here we look for

$$J(u) = \frac{1}{2}\int_{\Omega_{\text{obs}}} (y(u) - z_d)^2 d\mathbf{x} + \frac{\alpha}{2}\int_{\Gamma_N} u^2 d\sigma \ \to \ \min \tag{1.7}$$

by replacing the second integral in (1.4) with a boundary integral over Γ_N.

Alternatively, we shall seek

$$J(u) = \int_{\Gamma_D} \left(\frac{\partial y(u)}{\partial n} - z_{d_\Gamma}\right)^2 d\sigma + \frac{\alpha}{2}\int_{\Gamma_N} u^2 d\sigma \ \to \ \min \tag{1.8}$$

if we want the temperature to be close to a desired distribution over a portion of the boundary (disjoint from the one where we act through the control). In this case, we suppose that the surroundings are kept at temperature u and assume that the inward heat flux from Γ_N, affected by our control variable, depends linearly on the difference $u - y$, with $\gamma > 0$ (radiation[3] or Robin condition). In the case where $\partial u/\partial n = u$ on Γ_N, we directly control the heat flux across Γ_N (Neumann condition).

Alternatively, we can directly control the temperature on a portion Γ_D of the boundary, that is the Dirichlet datum of the state problem:

$$\begin{cases} -\Delta y = f & \text{in } \Omega \\ y = u & \text{on } \Gamma_D \\ \dfrac{\partial y}{\partial n} = g_N & \text{on } \Gamma_N \end{cases} \tag{1.9}$$

by seeking u like in (1.1), (1.4) or, otherwise,

$$J(u) = \int_{\Gamma_N} (y(u) - z_{d_\Gamma})^2 \, d\sigma + \frac{\alpha}{2}\int_{\Gamma_D} u^2 d\sigma \ \to \ \min. \tag{1.10}$$

[3] This formula is based on Newton's law of cooling, for which the heat loss from the surface of a body is a linear function of the temperature drop from the surroudings to the surface.

In all these cases we may need to impose some constraints on the control function. Typically, by requiring that either

$$u \leq u_{\max}$$

or

$$u \in [u_{\min}, u_{\max}],$$

we model some restrictions in the control heating capacity. This is a constrained OCP, with constraints on the control functions. A more complex case (that we only marginally consider, see Sect. 5.12) is that of *state constraints*, that is, constraints on the state variable, e.g. under the form

$$y > 0 \qquad \text{over } \;\Omega_{con} \subseteq \Omega.$$

We conclude by mentioning *shape optimization* problems, where the shape of the domain Ω itself acts as a control variable. For instance, if we minimize the cost functional

$$J(\Omega) = \int_{\Omega} (u(\Omega) - z_d)^2 d\mathbf{x} \;\rightarrow\; \min$$

among a family of admissible shapes, we look for that shape allowing to approach as much as possible the desired temperature distribution z_d in the domain.

The list of applications involving OCPs in engineering, applied sciences and industry is endless. In the following sections we present some complex cases, much closer to real-life applications than those described until now. We deal with physical models and OCPs of increasing difficulty, highlighting for each case the additional mathematical ingredients required to formulate the problem.

1.3 Control of Pollutant Emissions from Chimneys

Our first application deals with the control of emissions released by industrial chimneys into the atmosphere. In this case, the spreading of the pollutant can be modeled by an advection-diffusion-reaction equation, while the emission we want to control is described by a source term (either distributed or pointwise).

Indeed, the dispersion into the atmosphere of pollutants (like sulfur dioxide, SO_2, or carbon dioxide, CO_2) is a physical process depending on many factors, such as meteorological situation, air composition and pollutants species. With some simplifications, we may assume that at urban scales the concentration behavior can be modeled by stationary advection-diffusion equations, neglecting (without significantly lowering the model accuracy) the reactive terms:

$$\begin{cases} -\mathrm{div}(\nu \nabla y) + \mathbf{b} \cdot \nabla y = u & \text{in } \Omega \subset \mathbb{R}^2 \\ y = 0 & \text{on } \Gamma_D \\ \nu \dfrac{\partial y}{\partial n} = 0 & \text{on } \Gamma_N. \end{cases} \tag{1.11}$$

Here $y = y(u)$ is the pollutant concentration, u is a control function representing, for instance, the pollutant emission rate, $\nu = \nu(\mathbf{x})$ is the (turbulent) diffusion coefficient, whereas $\mathbf{b} = \mathbf{b}(\mathbf{x})$ is a given velocity field representing the wind speed. Molecular diffusivity can be neglected.

In this case, the state system is represented by a boundary value problem for an elliptic PDE. For simplicity we have imposed homogeneous Dirichlet conditions over the inflow boundary $\Gamma_D = \{\mathbf{x} \in \partial\Omega : \mathbf{b}(\mathbf{x}) \cdot \mathbf{n}(\mathbf{x}) < 0\}$, where $\mathbf{n}(\mathbf{x})$ is the normal (outward) unit vector, and a homogeneous Neumann condition on the outflow boundary $\Gamma_N = \partial\Omega \backslash \Gamma_D$. This latter choice is dictated by numerical reasons, since the far region downfield the chimneys has been truncated by considering a bounded domain Ω (see Fig. 1.2).

The system (1.11) models the diffusion and transport of pollutant at the emission height in a plane $\mathbf{x} = (x_1, x_2)$ parallel to soil, at height, say, $x_3 = H$, assuming that the tallness of every emission plant equals H. Since a complete model should take into account 3D phenomena related with air motion, following [78] we can adopt a quasi-3D model by considering the pollutant concentration $\tilde{y}(x_1, x_2, x_3) = \pi(x_1, x_2, x_3)y(x_1, x_2)$ in a parallelepipedal domain, where

$$\pi(x_1, x_2, x_3) = \exp\left(-\frac{1}{2}\left(\frac{x_3 - H}{\sigma_3}\right)^2\right) + \exp\left(-\frac{1}{2}\left(\frac{x_3 + H}{\sigma_3}\right)^2\right)$$

takes into account the distribution of the concentration with respect to the vertical coordinate x_3. The function π mimics a Gaussian function which, in case of constant wind field, models the pollutant concentration in a parallelepipedal domain as being emitted by a unique source. We denote by σ_3 the vertical dispersion coefficient, which accounts for atmospherical stability and soil orography.

Our target consists in regulating the emission rates from N chimneys in order to keep the pollutant concentration below a desired threshold in an observation area Ω_{obs}, say a town. If we denote by D_i the spatial region representing the i-th chimney and u_i its emission rate (normalized with respect to an appropriate emission volume), then an *admissible control* takes the form

$$u(\mathbf{x}) = \sum_{i=1}^{N} u_i \, \chi_{D_i}(\mathbf{x}), \qquad u_i \geq 0.$$

Let us introduce the observed variable

$$Cy(u) = \pi(x_1, x_2, 0)y(u) = 2\exp\left(-\frac{1}{2}\left(\frac{H}{\sigma_3}\right)^2\right) y(u),$$

which represents the projection of pollutant concentration at soil. Thus, our optimal control problem is to solve the minimization problem

$$J(u) = \int_{\Omega_{obs}} (Cy(u) - z_d)^2 d\mathbf{x} \; \rightarrow \; \min \tag{1.12}$$

over the set of admissible controls, where z_d is the desired pollutant concentration at soil (limited to the observation area Ω_{obs}).

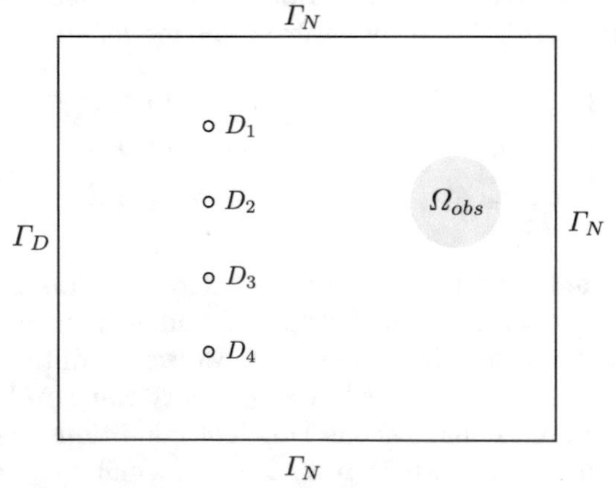

Fig. 1.2 Domain for emission control problem in air

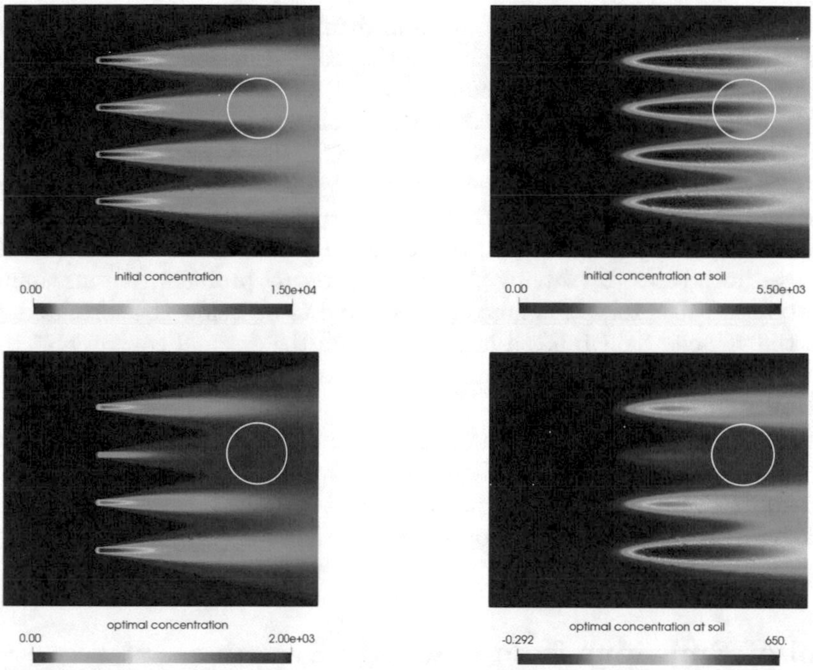

Fig. 1.3 Control of pollutant emissions from chimneys (case of horizontal velocity field). Top: pollutant initial concentrations at effective height $z = H$ (left) and at soil, $z = 0$ (right). Bottom: solution of the optimal control problem, in terms of pollutant concentrations at effective height $z = H$ (left) and at soil, $z = 0$

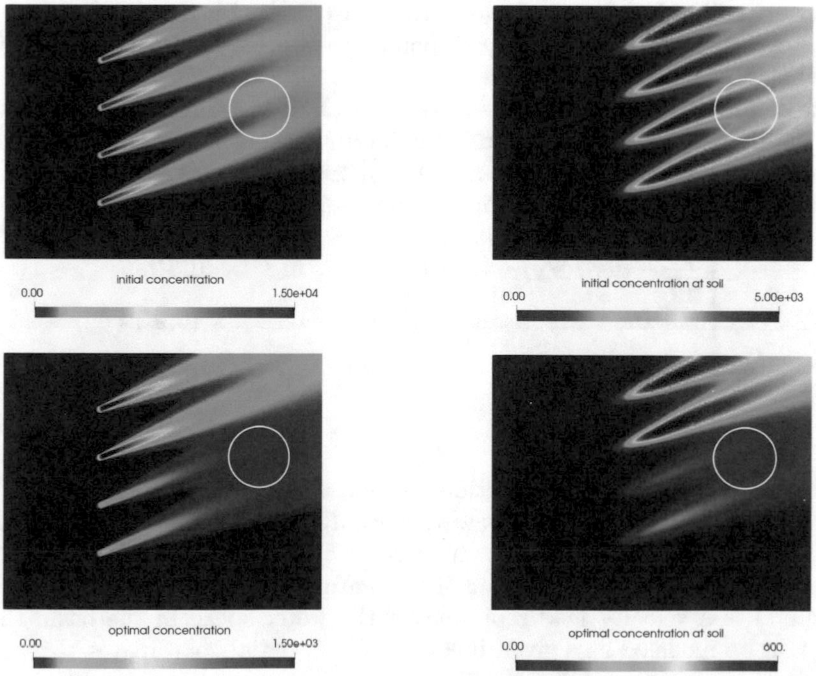

Fig. 1.4 Control of pollutant emissions from chimneys (case of oblique velocity field). Top: pollutant initial concentrations at effective height $z = H$ (left) and at soil, $z = 0$ (right). Bottom: solution of the optimal control problem, in terms of pollutant concentrations at effective height $z = H$ (left) and at soil, $z = 0$

In particular, the emission areas χ_{D_i} are approximately at the same distance from the observation area Ω_{obs}, so that we can consider the same diffusion coefficient $\nu = \sigma_{12}^2 V/2r$, where σ_{12} is the horizontal dispersion coefficient.

For the sake of illustration, in Figs. 1.3–1.4 the solution of the optimal control problem (1.11)-(1.12) in the case of $N = 4$ chimneys is displayed. In this particular case, $z_d = 100\mu g/m^3$ is chosen as target concentration level (the attention level is $125\mu g/m^3$, while the alarm level is $250\mu g/m^3$), as well as a maximum emission rate $u_{max} = 800g/s$ at height $H = 100m$, and two different scenarios regarding the velocity field: either $\mathbf{b} = (2.5, 0)m/s$ or $\mathbf{b} = (2.5, 1)m/s$. We point out that, in both cases, by solving the optimal control problem, the maximum value of SO_2 concentration in the observation area Ω_{obs} is found to be below the threshold. Obviously, a difference in the velocity field $\mathbf{b}$ has a huge impact on the optimal control able to reach the target, as we see by comparing Figs. 1.3 and 1.4. See, e.g., [78] for more details on the model and its numerical solution.

With respect to the heat transfer examples of the previous section, the optimal control problem here considered stands on a different state problem (advection-diffusion equation), features a distributed observation and a localized control function, which may even reduced to *pointwise* controls in the limit of the control areas $|\Omega_i| \to 0$.

1.4 Control of Emissions from a Sewage System

If either the rate of pollutant release or the advection field vary in time, the model introduced in the previous section is no longer appropriate. Let us consider, for example, the pollutant released by a sewage system (after going through a purification plant) into a water body near a town. As before, our goal is to regulate the emission rate from the drainage (and its location, in a more general case) in order to keep the pollutant concentration below a desired level in an observation area Ω_{obs} located, say, along the coast, right by the town. Other examples are given, e.g., by a plant in a river or by shipping tanks, with the goal of keeping pollutant concentration below a desired threshold in a specific area.

For the sake of simplicity, we model the water basin by a two-dimensional domain $\Omega \subset \mathbb{R}^2$, neglecting its depth (otherwise said, assuming pollutant to surf on the water). In such a case, the evolution of the pollutant concentration $y(t) = y(t; u)$ over a time interval $(0, T)$, $T < +\infty$, can be described by an unsteady advection-diffusion equation:

$$\begin{cases} \dfrac{\partial y}{\partial t} - \operatorname{div}(\nu \nabla y) + \mathbf{b} \cdot \nabla y = u & \text{in } \Omega \times (0,T) \\ y = 0 & \text{on } \Gamma_D \times (0,T) \\ \nu \dfrac{\partial y}{\partial n} = 0 & \text{on } \Gamma_N \times (0,T) \\ y(0) = y_0 & \text{on } \Omega \end{cases} \tag{1.13}$$

where, in addition to the boundary conditions (now defined for any $t \in (0, T)$), we also need to specify an initial condition in Ω. In this way, the state problem is an initial/boundary value problem for a parabolic PDE. Here ν denotes the mass diffusivity (of the pollutant substances) in water, which can be considered constant if the water temperature is about constant. The vector $\mathbf{b} = \mathbf{b}(\mathbf{x}, t)$ is a velocity field representing the water speed in the basin that, in the simplest case, can be modeled as a given function of the spatial coordinates.

In more realistic models, the transport field $\mathbf{b}$ could be provided by the solution of an *external* fluid dynamics problem, such as the (steady or unsteady) Navier-Stokes equations, or the *shallow water* equations (see [2]). Alternatively, and similarly to what we did in the previous example, we can model the water speed with a Gaussian-like function, see [29].

According to the above assumptions, the same setting of the previous example can be adapted to the current situation. In particular, an admissible control takes the form

$$u(\mathbf{x},t) = U(t)\,\chi_D(\mathbf{x}), \qquad U(t) \geq 0,$$

where $\chi_D(\mathbf{x})$ denotes the characteristic function of the spatial region D from which the pollutant is released and $U(t)$ its emission rate, a function of time. Thus, we can consider the following minimization problem,

$$J(u) = \int_0^T \int_{\Omega_{obs}} (Cy(\mathbf{x},t;u) - z_d(\mathbf{x},t))^2 d\mathbf{x}\, dt \ \to \min$$

over the set of admissible controls, where z_d is the desired pollutant concentration in the observation area Ω_{obs}, and the observed variable

$$Cy(\mathbf{x},t;u) = \chi_{\Omega_{obs}}(\mathbf{x})y(\mathbf{x},t;u),$$

is defined as the restriction of y to Ω_{obs}. Alternatively, we might consider

$$Cy(\mathbf{x},t;u) = \chi_{\Omega_{obs}}(\mathbf{x})y(\mathbf{x},T;u),$$

for which the corresponding minimization problem becomes

$$J(u) = \int_{\Omega_{obs}} (Cy(\mathbf{x},T;u) - z_d(\mathbf{x},T))^2 d\mathbf{x} \ \to \min$$

over the set of admissible controls. In this case, the goal is to reach a desired target at a given final time, where the dynamics of the system is fully developed.

1.5 Optimal Electrical Defibrillation of Cardiac Tissue

The active mechanical behavior of the heart muscle, which allows to pump blood into the vascular network, depends on the electrical excitation of the muscle, which goes under the name of *electrical activation* [68]. At each heart beat, the electrical signal spreads throughout the ventricular chambers, whose mechanical contraction induces blood ejection. A similar behavior occurs within the atria, whose relaxation lets the blood flow into the heart [221, 219]. In the healthy case, the electrical activation of the heart is an extremely organized (and efficient) process. However, disturbances in the formation and/or propagation of electrical signals may induce reentrant activation patterns which lead to tachycardia, that is, a noticeable increase in the heart's activation rate. In the worst case, this may turn to an even less organized activation pattern, the so-called *fibrillation*.

The only reliable therapy to terminate (otherwise lethal) fibrillations and restore a healthy cardiac rhythm is *electrical defibrillation*, consisting in delivering a strong electrical shock by injecting external currents through a set of electrodes. This restores a spatially uniform activation pattern, recovering an extracellular potential distribution which dampens voltage gradients. Electrical defibrillation is nowadays achieved by implanting a suitable device (the so-called cardioverter defibrillator) that can monitor heart rhythm and then deliver electrical discharges when needed.

In this case, the state problem is represented by the so-called *monodomain model*, a nonlinear unsteady diffusion-reaction equation for the transmembrane potential v (describing the macroscopic electrical activity in the heart tissue) coupled with an ODE describing the evolution of a *gating* or *recovery* variable w; see, e.g., [68, 220, 259]. Denoting by $\Omega \subset \mathbb{R}^d$, $d = 2, 3$, a portion of (isolated) cardiac tissue, the monodomain problem reads as follows:

$$\begin{cases} \dfrac{\partial v}{\partial t} - \mathrm{div}(\sigma \nabla v) + I_{ion}(v, w) = I_e & \text{in } \Omega \times (0, T) \\ \dfrac{\partial w}{\partial t} = g(v, w) & \text{in } \Omega \times (0, T) \\ \dfrac{\partial v}{\partial n} = 0 & \text{on } \partial\Omega \times (0, T) \\ v(0) = v_0, \qquad w(0) = w_0 & \text{in } \Omega, \end{cases} \tag{1.14}$$

where $y = (v, w)$ plays the role of state variable, $\sigma : \Omega \to \mathbb{R}^{d \times d}$ denotes the intracellular conductivity tensor, I_{ion} is the current density flowing through ionic channels across the membrane, and I_e is an extracellular current density stimulus, representing the defibrillation action, which plays the role of control function.

The ionic activity is usually modeled through a system of ODEs; here we consider the simplest case involving just a single ionic current variable w, and a single ODE. A very popular choice is the so-called Fitzhugh-Nagumo model, where we set:

$$\begin{aligned} I_{ion}(v, w) &= \eta_0 v \left(1 - \frac{v}{v_{th}}\right)\left(1 - \frac{v}{v_{pk}}\right) + \eta_1 v w, \\ g(v, w) &= \eta_2 \left(\frac{v}{v_{pk}} - \eta_3 w\right). \end{aligned} \tag{1.15}$$

Here η_j, $j = 0, \dots, 3$ are positive real coefficients, $v_{th} > 0$ is a threshold potential and v_{pk}, $v_{pk} > v_{th}$, is the peak potential [68]. In (1.14) we apply homogeneous Neumann conditions because we assume that both intracellular and extracellular domains are electrically isolated; the initial value of the transmembrane potential and ion current variables are prescribed to, say, constant values v_0 and w_0, respectively. See e.g [45, 259, 68] for more on the subject.

To perform an *optimal* electrical defibrillation, we aim at determining an optimal extracellular current I_e, which depends on the defibrillation pulse to be controlled, such that the electrical activity in the tissue – described by the transmembrane potential v – is damped in an optimal way. The current is induced by external devices consisting in electrode plates Ω_j, $j = 1, \dots, N$, placed in the tissue; to each plate Ω_j is associated an independent pulse $u_j = u_j(t)$ so that I_e can be modeled as

$$I_e(\mathbf{x}, t) = \sum_{j=1}^{N} u_j(t) \chi_{\Omega_j}(\mathbf{x}) \xi_{[0, t_f]}(t),$$

where t_f is the duration of the electrical pulse; χ_{Ω_j} and $\xi_{[0,t_f]}$ denote instead the characteristic functions of Ω_j and $[0, t_f]$, respectively. Hence, the control function is given by the vector $\mathbf{u}(t) = (u_1(t), \dots, u_N(t))$.

Distinguishing features of defibrillation procedures are *(i)* the duration of the electrical pulse, *(ii)* the distribution of the electric potential at the end of a defibrillation, which is indeed a relevant factor determining the failure or the success of the procedure, and *(iii)* the strength of the applied current.

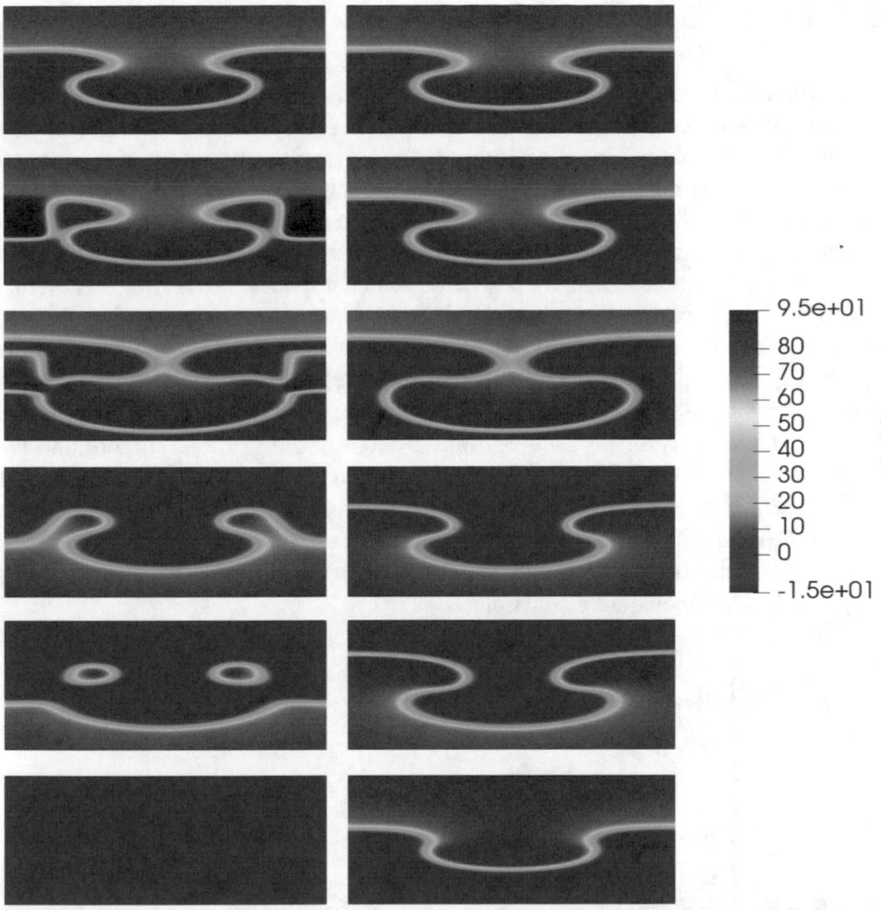

Fig. 1.5 From top to bottom: transmembrane potential at different time instants with optimal electrical defibrillation (left) and without electrical defibrillation (right)

Following [62, 163, 161], we consider the minimization problem

$$J(u) = \int_0^{t_f} \left(\kappa + \frac{\alpha}{2}|\mathbf{u}(t)|^2\right) dt + \frac{\mu}{2}\int_\Omega |v(\mathbf{x}, t_f)|^2 d\mathbf{x} \to \ \min$$

subject to the state system (1.14) and the constraints

$$|u_j(t)| \leq M_j, \quad j = 1, \ldots, N, \quad 0 \leq t \leq t_f.$$

The first term in the cost functional takes into account the duration of the electrical pulse, the second one reflects the need to maintain the electrical shock as small as possible, whereas the third one enforces the requirement of a more uniform potential distribution after the shock; κ, α, μ are positive constants. The electrical potential without electrical defibrillation and in presence of an optimal electrical defibrillation, is shown in Fig. ??. We highlight the role of the optimal control, which stabilizes the reentry wave so that at the end of the selected time window $[0, t_f)$ the electrical potential is uniform and no longer shows the typical pattern characterizing reentry waves as in the case of the top, left panel. We will analyze this problem in Sect. 10.2.

1.6 Optimal Flow Control for Drag Reduction

Another situation where optimal control comes naturally into play is when one looks for the reduction of resistance on bodies immersed in a fluid. In several fields, such as ship design and aeronautics engineering, this is aimed at increasing the performance of a given system.

We consider the problem of reducing the viscous drag of a body immersed in a fluid and moving with respect to a fluid with a prescribed relative velocity (a generalization to optimal shape design will be presented in Sect. 1.7). In fact, drag or frictional forces increase if the boundary layer, in this case modeled by the fluid velocity field located in a thin layer close to the surface of the body, becomes turbulent. We aim at reducing the viscous drag by controlling the velocity of the fluid over a portion $\Gamma_c \subset \Gamma_B$ of the boundary of the body, representing some holes (see Fig. 1.6): positive velocities yield a blowing effect, that drives away the fluid from the body, whereas negative velocities give a sucking effect, and the fluid is aspirated in.

We suppose that our body occupies a volume B with boundary Γ_B, surrounded by a fluid, filling a domain $\Omega = D \setminus B$; here $D \subset \mathbb{R}^d$ $(d = 2, 3)$ denotes a large box around the body. Then $\partial\Omega = \Gamma_B \cup \partial D$. The goal is to minimize the drag coefficient by regulating a suitable flow $\boldsymbol{u}$ across a portion Γ_c of the boundary Γ_B. For the sake of simplicity, we consider a steady fluid flow, denoting by $\boldsymbol{v}$ and π its velocity and pressure, respectively. The state system is given by the following steady Navier-Stokes equations:

$$\begin{cases} -\mathrm{div}\mathbf{T}(\boldsymbol{v}, \pi) + \rho(\boldsymbol{v} \cdot \nabla)\boldsymbol{v} = \mathbf{0} & \text{in } \Omega \\ \mathrm{div}\boldsymbol{v} = 0 & \text{in } \Omega \\ \boldsymbol{v} = \boldsymbol{v}_\infty & \text{on } \Gamma_{in} \\ \boldsymbol{v} = \mathbf{0} & \text{on } \Gamma_B \setminus \Gamma_c \\ \boldsymbol{v} = \boldsymbol{u} & \text{on } \Gamma_c \\ \boldsymbol{v} \cdot \mathbf{n} = 0, \qquad (\mathbf{T}(\boldsymbol{v}, \pi)\mathbf{n}) \cdot \mathbf{t} = 0 & \text{on } \Gamma_w \\ \mathbf{T}(\boldsymbol{v}, \pi)\mathbf{n} = \mathbf{0} & \text{on } \Gamma_{out} \end{cases} \tag{1.16}$$

where ρ is the fluid density and $\boldsymbol{v}_\infty$ represents the incoming flow far from the body (if we think the body steady and the fluid flowing around it); conversely, $-\boldsymbol{v}_\infty$ is the velocity of the body moving in the fluid; $\mathbf{n}$ and $\mathbf{t}$ are the (outward directed) normal and the tangent unit vectors on the boundary $\partial\Omega$. Note that both the state $\mathbf{y} = (\boldsymbol{v}, \pi)$ and the control function $\boldsymbol{u}$ are vector functions.

In the case of a Newtonian fluid, the (Cauchy) stress tensor $\mathbf{T}(\boldsymbol{v}, \pi)$ depends linearly on the symmetric part of the velocity gradient, that is

$$\mathbf{T}(\boldsymbol{v}, \pi) = -\pi\mathbf{I} + 2\mu\boldsymbol{\sigma}(\boldsymbol{v}), \qquad \boldsymbol{\sigma}(\boldsymbol{v}) = \frac{1}{2}(\nabla\boldsymbol{v} + (\nabla\boldsymbol{v})^T),$$

where μ is the dynamic viscosity of the fluid.

The drag force (which represents the force component in the direction of the flow velocity) on a body in relative motion with respect to a fluid can be computed from the stress tensor as

$$F_D = -\int_{\Gamma_B} (\mathbf{T}(\boldsymbol{v}, \pi)\mathbf{n}) \cdot \mathbf{e}_\infty \, d\sigma.$$

Note that by inserting a minus sign we conform to the convention that the force is positive when acting on the fluid. The *drag coefficient* C_D is a dimensionless quantity commonly used to quantify the resistance of an object in a fluid environment, and is given, for the two-dimensional case at hand, by

$$C_D = \frac{F_D}{q_\infty\, d} = \frac{F_D}{\frac{1}{2}\rho V_\infty^2\, d}$$

where d is a characteristic dimension of the body, $\mathbf{e}_\infty$ is the unit vector directed as the incoming flow $\boldsymbol{v}_\infty = V_\infty \mathbf{e}_\infty$, and $q_\infty = \frac{1}{2}\rho V_\infty^2$. Note that in a more general three-dimensional case d would be replaced by a characteristic area A of the body.

Our purpose is thus to solve the optimization problem

$$J(\mathbf{u}) = C_D(\boldsymbol{v}(\boldsymbol{u}), \pi(\boldsymbol{u})) \;\rightarrow\; \min$$

over a suitable set of admissible controls on Γ_c, where

$$C_D(\boldsymbol{v}, \pi) = -\frac{1}{q_\infty} \int_{\Gamma_B} (\mathbf{T}(\boldsymbol{v}, \pi)\mathbf{n}) \cdot \mathbf{e}_\infty \, d\sigma \tag{1.17}$$

is the *drag coefficient* and $(\boldsymbol{v}, \pi) = (\boldsymbol{v}(\mathbf{u}), \pi(\mathbf{u}))$ the solution of the state system (1.16).

According to the classification above, this is a *boundary control* problem, with boundary observation, where the observed variable is given by

$$C(\boldsymbol{v}, \pi) = -\frac{1}{q_\infty d}(\mathbf{T}(\boldsymbol{v}, \pi)\mathbf{n}) \cdot \mathbf{e}_\infty \Big|_{\Gamma_B}.$$

Evaluating the cost functional (1.17) can be troublesome from a numerical point of view; for this reason, when solving numerically this problem, it is preferable to convert it into an equivalent functional involving a distributed observation.

Another option is to solve the optimization problem

$$J(\mathbf{u}) = 2\mu \int_\Omega |\boldsymbol{\sigma}(\boldsymbol{v}(\mathbf{u}))|^2 \, d\mathbf{x} = 2\mu \int_\Omega \boldsymbol{\sigma}(\boldsymbol{v}(\mathbf{u})) : \boldsymbol{\sigma}(\boldsymbol{v}(\mathbf{u}))\, d\mathbf{x} \;\rightarrow\; \min, \tag{1.18}$$

where $\nabla \boldsymbol{v} : \nabla \boldsymbol{w} = \sum_{i,j=1}^d \frac{\partial v_i}{\partial x_j}\frac{\partial w_i}{\partial x_j}$. In this case, J is given by the *dissipation function*, representing the rate at which heat energy is conducted into the fluid. Note that if Dirichlet conditions are assigned on the whole boundary and a constant velocity field is imposed far from the body, (1.18) is equivalent to (1.17) (see Sect. 11.7.2).

The solutions (velocity, pressure, and streamline velocity) corresponding to the problem without control and that with optimal control on the velocity are plotted in Fig. 1.7. In particular, a comparison between the streamlines highlights the effects of the optimization in reducing the wake extension. An analysis of this problem is presented in Sect. 10.1.

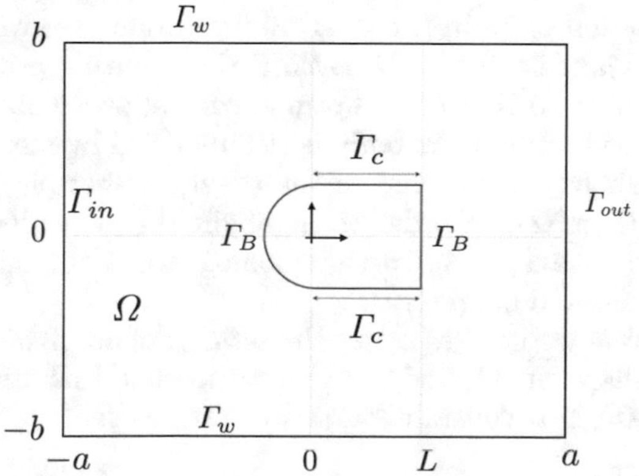

Fig. 1.6 Domain for velocity control problem

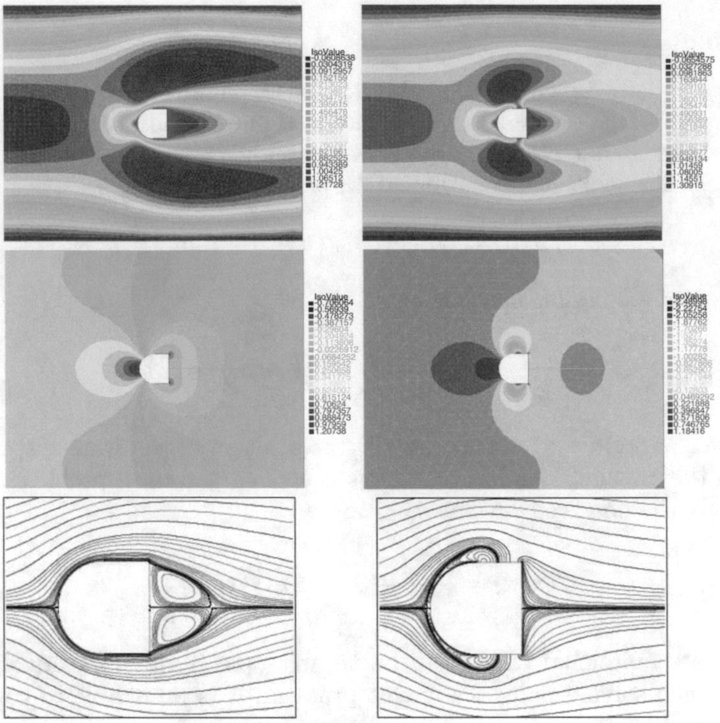

Fig. 1.7 Velocity (magnitude), pressure and streamlines of the velocity field in the case without velocity control (left) and in the optimal control case (right)

1.7 Optimal Shape Design for Drag Reduction

The drag minimization problem, formulated as an optimal control problem in Sect. 1.6, can be alternatively reformulated as a *shape optimization* problem: in this case, we look for the geometrical shape of the body that minimizes the drag forces acting on its surface. This is a common problem in aerodynamics, where the optimal design of an airfoil (e.g., the wing cross section) can yield remarkable fuel savings and global improvement of aircraft performance, but also in hydrodynamics, where for instance an optimal design of ship hulls yields substantial reduction in power requirements.

In this context, the control variable is the shape of the domain itself or, more precisely, its boundary or else a portion of it. In the drag minimization case we look for optimal shape change of the boundary Γ_B of the body B. Even if the resulting problem is harder to solve, its formulation is very similar to that of an optimal control problem.
In particular, if the flow field around the body is still modeled by means of the steady Navier-Stokes equations (1.16) already considered in the previous example, we denote this time its solution by $(\boldsymbol{v}, \pi) = (\boldsymbol{v}(\Omega), \pi(\Omega))$ in order to highlight the dependence on the shape of the domain Ω. Moreover, we have $\Gamma_c = \emptyset$, so that homogeneous Dirichlet boundary conditions $\boldsymbol{v} = \mathbf{0}$ are imposed on the whole boundary Γ_B.

To formulate our problem we need to select the set $\mathcal{O}_{ad}$ of admissible shapes. Since we only admit variations of the shape of the body B and we assume that its volume is constant (so that the volume of $\Omega = D \setminus B$ is constant, too), we can choose

$$\mathcal{O}_{ad} = \left\{ \Omega \in \mathcal{O} \,:\, \partial\Omega \setminus \Gamma_B \text{ is fixed}, \; |\Omega| = \int_\Omega d\mathbf{x} = m_0 \right\},$$

where $\mathcal{O}$ is a suitable set of smooth shapes. In this way, we end up with the following shape optimization problem with boundary observation

$$J(\Omega) = c_D(\boldsymbol{v}(\Omega), \pi(\Omega)) \;\rightarrow\; \min \tag{1.19}$$

over the set $\mathcal{O}_{ad}$.

By considering the same setting of the previous example, through the shape optimization problem (1.19), we obtain a reduction of drag coefficient of about 15%; initial and optimal shapes, as well as velocity and pressure fields in both cases, are reported in Fig. 1.8. With respect to the boundary velocity control addressed in Sect. 1.6, the reduction of the wake extension at the rear of the body is now even more evident, thus making shape optimization more effective in order to design the shape of aerodynamic/hydrodynamic bodies. This problem is considered in Sect. 11.7.

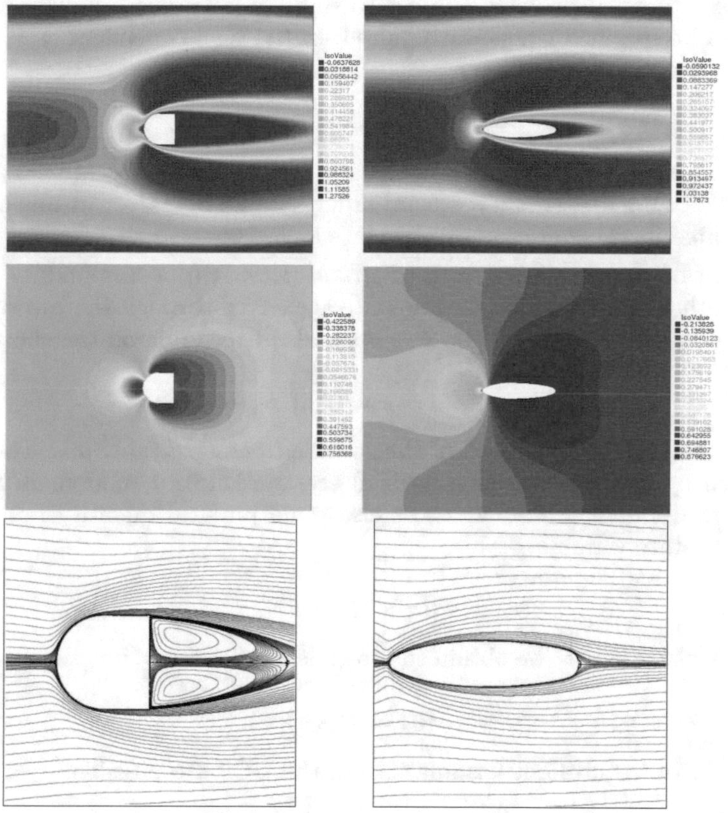

Fig. 1.8 Velocity (magnitude), pressure and streamlines of the velocity field in the case without velocity control (left) and in the optimal control case (right).

1.8 A General Setting for OCPs

From the previous sections we have learned that for a given *state* or *controlled* system, the goal of an *optimal control* problem is to find the best *control input* such that the observation of some output of interest (depending on the solution of the state system) reaches a prescribed target, in a suitable sense.

In abstract form, such a problem is characterized by:

- the *control function* u, that belongs to a functional vector space $\mathcal{U}$, the *control space*. In general, a control is subject to suitable constraints, usually characterized through a set of equalities or inequalities, that define a subset $\mathcal{U}_{ad} \subseteq \mathcal{U}$ of admissible controls;
- the *state* of the system (or state variable) $y = y(u)$, depending on the control u and belonging to a suitable functional space V. It satisfies the *state equation*, given in our case by either a boundary value problem for an elliptic PDE or an initial-boundary value problem for an evolution PDE. In the former case, the state equation writes

$$G(y,u) = Ay - f - Bu = 0, \tag{1.20}$$

where A is a second-order, elliptic partial differential operator (linear or not). Equation (1.20) models a physical problem subject to suitable boundary conditions; additional assumptions will be specified whenever required. In the time-dependent case, the state equation takes the form:

$$G(y,u) = \frac{\partial y}{\partial t} + Ay - f - Bu = 0, \qquad t \in (0,T) \tag{1.21}$$

where $T \in (0, +\infty)$ and A is still a second-order, elliptic operator. In this case, we need to impose a suitable initial condition on the state at $t = 0$;
- the control can be, e.g., a source term (*distributed control*) or a boundary data (*boundary control*). In both cases, its action on the state is exerted through the *control operator* B;
- the *observation function*, denoted by $z = z(u)$, also depending on the control u through y and a suitable operator $C : V \to \mathcal{Z}$,

$$z = Cy. \tag{1.22}$$

This function belongs to a space $\mathcal{Z}$ of the so-called *observed functions*. The observation is typically given by the state variable itself, its restriction to a subdomain $\Omega_{\text{obs}} \subset \Omega$ or to a portion of the boundary $\Gamma_{\text{obs}} \subset \Gamma$, or else some physical quantities depending on the gradient of the state variable;
- the *cost* (or *objective*) *functional*

$$\tilde{J} : V \times \mathcal{U} \to \mathbb{R}.$$

Inserting the state $y = y(u)$ we obtain the so-called *reduced cost* functional $J : \mathcal{U} \to \mathbb{R}$, that is

$$J(u) = \tilde{J}(y(u), u). \tag{1.23}$$

In this framework, two equivalent formulations of the OCP are possible:

1. *full space approach* (*minimization with respect to* (y, u))

$$\begin{gathered} \tilde{J}(y,u) \to \min \\ \text{subject to} \\ G(y,u) = 0, \quad y \in V, \quad u \in \mathcal{U}_{ad}. \end{gathered} \tag{1.24}$$

A minimizing pair $(\hat{u}, \hat{y})$ satisfying (1.24) is called *optimal control* and *optimal state*, respectively. Note that the state problem $G(y,u) = 0$ plays the role of an *equality constraint*, and that y and u are regarded as two independent variables.
Moreover, we may also require that the state variable fulfills additional *constraints* – also referred to as *state constraints* – that is, $y \in V_{ad} \subset V$.

2. *reduced space approach* (*minimization with respect to* u). If the state system can be used to express the state variable y as a function of u, that is, $y = y(u)$, then problem (1.24) reduces to

$$J(u) = \tilde{J}(y(u), u) \to \min \quad \text{subject to} \quad u \in \mathcal{U}_{ad}. \tag{1.25}$$

From a computational standpoint, it can be advantageous to treat y and u as independent variables; this approach stands at the basis of the well-known *Lagrange multiplier* method, which will be extensively used throughout the book.

Remark 1.1. Sometimes, (1.24) is also called *Simultaneous Analysis and Design* formulation of an OCP, while (1.25) is called *Nested Analysis and Design* formulation of an OCP; see, e.g., [13]. •

A typical expression of the cost functional is

$$\tilde{J}(y, u) = P(y) + Q(u) = \frac{1}{2}\|Cy - z_d\|^2_{\mathcal{Z}} + \frac{\alpha}{2}\|u\|^2_{\mathcal{U}}, \qquad \alpha \geq 0.$$

Depending on the observation, P may assume several forms, such as

$$P(y) = \int_{\Omega_{obs}} (y - z_d)^2 d\mathbf{x}, \qquad P(y) = \int_{\Omega} |\nabla(y - z_d)|^2 d\mathbf{x},$$

corresponding to the choice $Cy = u\chi_{\Omega_{obs}}$ or $Cy = \nabla y$, respectively, as distributed observation functions, or

$$P(y) = \int_{\Gamma_{obs}} (y - z_d)^2 d\sigma, \qquad P(y) = \int_{\partial\Omega} \left(\frac{\partial y}{\partial n} - z_d\right)^2 d\sigma,$$

corresponding to the choice $Cy = y|_{\Gamma_{obs}}$ i.e. the trace of the state on a portion $\Gamma_{obs} \subseteq \partial\Omega$, or the trace of its normal derivative $Cy = \partial y/\partial n$, called boundary observation function.

For time-dependent PDEs, observations can be made over the whole time interval,

$$P(y) = \int_0^T \int_{\Omega_{obs}} (y(\mathbf{x}, t) - z_d(\mathbf{x}, t))^2 d\mathbf{x}dt, \quad \text{or} \quad P(y) = \int_0^T \int_{\partial\Omega} \left(\frac{\partial y}{\partial n}(\sigma, t) - z_d(\mathbf{x}, t)\right)^2 d\sigma dt,$$

or restricted to the final time $t = T$ only,

$$P(y) = \int_{\Omega_{obs}} (y(\mathbf{x}, T) - z_d(\mathbf{x}))^2 d\mathbf{x}, \quad \text{or} \quad P(y) = \int_{\Omega} |\nabla(y(\mathbf{x}, T) - z_d(\mathbf{x}))|^2 d\mathbf{x}.$$

The choice of Q can be inspired by different aims, for instance minimizing a suitable norm of the control function, introducing a regularization term for the minimization problem, etc. Specific examples will be introduced later on.

1.9 Shape Optimization Problems

The framework introduced above proves very useful in other contexts as well. With no pretention to be exhaustive, we consider in this book the extension to shape optimization problems.

In a *shape optimization* problem, the goal is, say, to minimize an objective function which involves the solution of a state system, defined over a geometrical domain of interest, by controlling the shape of the domain itself. As in the optimal control case, the state system can be modeled by either a boundary value problem for a stationary PDE, or an initial/boundary value problem for a time-dependent PDE. For the sake of simplicity, we focus on the former,

$$A(\Omega)y(\Omega) = f(\Omega), \tag{1.26}$$

where we highlight the dependence of both the partial differential operator and the right-hand side on the shape of the domain Ω. The latter belongs to a space $\mathcal{O}_{ad} \subseteq \mathcal{O}$ of admissible shapes that plays in the current context the role of control space. From now on, we identify the shape of the domain with the domain itself, for the sake of simplicity.

Also the state space $V = V(\Omega)$ may depend on the domain Ω, for instance in the case where non-homogeneous Dirichlet conditions are imposed on a portion of the boundary $\partial\Omega$. Here the objective is given by a cost functional $J(\Omega)$, also called *shape functional*, defined on $\mathcal{O}_{ad}$. Similarly to (1.25), a shape optimization problem can be written as

$$\begin{gathered} J(\Omega) = \tilde{J}(y(\Omega), \Omega) \rightarrow \min \\ \text{subject to} \\ \Omega \in \mathcal{O}_{ad}. \end{gathered} \tag{1.27}$$

A minimizing shape $\hat{\Omega}$ is called *optimal shape*. Also in a shape optimization problem the objective function can be expressed as the sum of two terms

$$\tilde{J}(y, \Omega) = P(y) + Q(\Omega) = \frac{1}{2}\|Cy - z_d\|^2_{\mathcal{Z}} + Q(\Omega).$$

The observation function P obeys to problem-dependent criteria as in the optimal control case, whereas Q may depend on shape features such as the volume (or area) of Ω, its perimeter, etc.

Shape optimization problems under the form (1.27) can be regarded as a particular case of a larger class of problems, usually denoted as *optimal design* problems in *structural optimization*. In this context, some data of the mathematical model describing the behavior of a structure are considered to be *design parameters* (or control variables), by means of which a structure is tuned until a desired property is achieved. Depending on the nature of the design parameters, three different approaches can be envisaged:

(i) *sizing or parametric optimization*: design parameters reflect material properties of the structure, or simple geometrical features, such as typical *sizes*; the control variables enter into the coefficients of the PDE. In this case we deal with constrained optimization problems in $\mathbb{R}^p$, being p the number of the design parameters;

(ii) *(geometrical) shape optimization* : the position of the boundary of the shape under control is optimized without varying its topology;

(iii) *topological optimization*: the topology of a structure, as well as its shape, is optimized by allowing for example the inclusion of holes.

Topology optimization problems require deeper mathematical tools than those introduced in this book; the interested reader may refer, e.g., to [27, 135]; in Chapter 11 we will focus instead on (simple) geometrical shape optimization problems (case *(ii)*).

1.10 Parameter Estimation Problems

Another class of problems that can be cast in the family of OCPs are parameter estimation problems. Indeed, when setting a mathematical model to describe complex phenomena, not all model inputs may be directly measurable. For example, in the case of the monodomain problem (1.14)–(1.15), the electrical conductivity function σ can be hard to assess for a given individual. On the other hand, estimating this quantity is essential for the model. This raise the issue of *parameter estimation*: the missing parameters are typically recovered starting from measurements concerning a specific quantity that can be related to the solution of the state problem. In this book we limit ourselves to underline some analogies between parameter estimation and optimal control, and to show some examples.

The so-called *variational approach* recasts parameter estimation in the framework of PDE-constrained optimization, by considering the equations governing the problem at hand as state system and the discrepancy between the observation of the state and the measured data as cost functional to be minimized. Exploiting a mathematical analogy, the parameters to be estimated can be regarded as if they were control variables in the case of OCPs; be aware, however, that these are model parameters that escape the possibility of being controlled!

In abstract mathematical terms, we can formulate the parameter estimation problem as follows. Assume that the state problem reads

$$A(\boldsymbol{\mu})y(\boldsymbol{\mu}) = f(\boldsymbol{\mu}), \tag{1.28}$$

where now we have highlighted the dependence on the missing parameter $\boldsymbol{\mu} \in \mathcal{P} \subset \mathbb{R}^p$ (here assumed to be a vector). The available measurements z_d can be exploited to estimate $\boldsymbol{\mu}$ by minimizing the error (in the least squares sense) between z_d itself and the observation $z(\boldsymbol{\mu}) = Cy(\boldsymbol{\mu})$,

$$J(\boldsymbol{\mu}) = \frac{1}{2}\|Cy(\boldsymbol{\mu}) - z_d\|^2_{\mathcal{Z}} + \frac{\alpha}{2}Q(\boldsymbol{\mu}) \ \rightarrow \ \min \tag{1.29}$$

where $Q : \mathcal{P} \rightarrow \mathbb{R}^+$ typically plays the role of regularization term. For instance, the choice

$$Q(\boldsymbol{\mu}) = \|\boldsymbol{\mu} - \boldsymbol{\mu}_p\|^2_{\mathcal{P}}, \tag{1.30}$$

where $\boldsymbol{\mu}_p \in \mathcal{P}$ represents the *a priori* guess on $\boldsymbol{\mu}$, yields the celebrated Tikhonov regularization (see e.g. [91]).

It is worth mentioning that measurements are typically affected by noise and/or uncertainties. This might call for the set up of an approach aimed at recovering the way uncertainties are propagated into the model and eventually affect the estimates. We can for instance look for confidence bands about the estimated quantities – and not only point estimates, as done with the variational approach – and characterize the statistical distribution of the unknown parameters, by formulating a suitable *statistical inverse problem* in, e.g., a Bayesian setting [154, 260, 256]. More in general, statistical inversion recasts the parameter estimation problem in a larger space of probability distributions [154]. This strategy provides a better characterization of the prior information brought, e.g., by the regularization terms in (1.30), under the form of a prior probability density function of the unknown inputs. For more on this topic, see, e.g., [256, 257].

Another (indeed, very relevant) case where optimal control theory can be exploited is that of *(variational) data assimilation*, where experimental measurements of a quantity modeled by the state system can be incorporated (or *assimilated*) into the (computational) model in order to better characterize model inputs that are unavailable [14, 171].

1.11 Theoretical Issues

The analysis of optimal control problems and more general PDE-constrained optimization problems presents several challenges. A classical approach is that of J.L. Lions [177], see Chapters 5 and 9. Another (more general) approach follows the Lagrangian formalism: the OCP is regarded as a constrained minimization problem for which a Lagrangian functional is introduced. The associated Lagrange multiplier is denoted as *adjoint state* or *co-state*. In this framework, should it exist, the optimum is a saddle point of the Lagrangian functional.

In this section we highlight the most relevant issues addressed in the following chapters.

1. *Well-posedness.* We first need to study the existence of an optimal control and, possibly, its uniqueness. At this stage, we can rely on well-posedness results for the state problem (given for example by the Lax-Milgram theorem in the simplest case of linear, elliptic PDEs), and on suitable assumptions on both the cost functional and the space of admissible controls.
2. *Optimality conditions.* Once the well-posedness of the problem has been stated, we need to derive a system of equations capable to characterize its solution. First-order optimality conditions express necessary conditions to be satisfied by the solution. They usually involve the state problem, the so-called *adjoint* problem and a condition on the control function involving the expression of the *gradient* of the cost functional.
   For instance, a distributed control problem like (1.1), with state system described by an elliptic PDE like (1.2), leads to the optimality system:
$$\begin{cases} -\Delta y = f + u & \text{in } \Omega & \text{state problem} \\ y = g_D & \text{on } \partial\Omega & \\ -\Delta p = y - z_d & \text{in } \Omega & \text{adjoint problem} \\ p = 0 & \text{on } \partial\Omega & \\ p + \alpha u = 0 & & \text{optimality condition.} \end{cases} \tag{1.31}$$
   Its derivation in more general cases is provided in Chapters 5 and 7.
3. *Sensitivity analysis.* Once the optimal control problem has been solved, we might be interested to study the way the optimal value of the cost functional varies with respect to control perturbations. This is called *sensitivity analysis* and requires the characterization of the gradient of the cost functional.

We will investigate these issues in Chapters 5,7 and 9. For a gentle warm up, we start with optimization of finite dimensional OCPs in Chapter 4.

1.12 Numerical Approximation of an OCP

Once the solution has been characterized by a system of optimality conditions (OC), the corresponding coupled system of PDEs has to be approximated numerically. This poses several challenges from a numerical point of view.

In this section we anticipate some general ideas related to the numerical solution of this class of problems; algorithms will be extensively analyzed in Chapters 6, 8, and 9.6.

For the sake of simplicity, let us consider a general PDE-constrained optimization problem, including all the cases listed in the previous sections (the only exception is given by shape optimization problems, which feature several differences, and will be addressed later on), under the following form:

$$\tilde{J}(y,u) \to \min \quad \text{subject to} \quad G(y,u) = 0, \quad y \in V, \quad u \in \mathcal{U}_{ad} \tag{1.32}$$

For the numerical solution of this problem, two different paradigms can be adopted:

- *optimize-then-discretize* (OtD): we first write down the optimality system at the continuous level and then we solve it numerically;
- *discretize-then-optimize* (DtO): we first discretize the state problem and the expression of the cost functional, and then look for a discrete control, so that the optimality conditions are stated directly at the finite dimensional level.

In general, these two strategies do not necessarily yield the same result. In any case, *discretization schemes* and *optimization methods* are fundamental constituents of the numerical solution of PDE-constrained optimization problems.
The goal of this section is to highlight the main sources of complexity and/or difficulty entailed by the numerical solution of this class of problems:

1. *Discretization of the state problem (and its adjoint)*: the choice of a suitable discretization method is the first step af any numerical approximation of a PDE-constrained optimization problem. In this book we will focus on the Galerkin-Finite Element Method (FEM); however, this choice is not restrictive, the whole construction remains valid for other discretization techniques as well. In particular, the choice of discrete spaces, numerical solvers for linear systems, preconditioning techniques, etc., is usually dictated by the very nature of the state problem. Since the numerical solution of the state and adjoint problems has to be performed many times when solving iteratively a PDE-constrained optimization problem, using efficient numerical PDE solvers is mandatory.
Moreover, very often also the control function is a vector of (control) parameters. In this case the optimization can be performed with respect to a finite-dimensional variable. The smaller the dimension p of the control parameter vector, the more effective the discrete representation. Note that, after (spatial) discretization, the size of the discrete control function might be at most equal to the one of the discrete state variable in the case of distributed controls. For instance, dealing with control parameters is often preferable in the case of shape optimization problems to avoid boundary deformations and possible remeshing [134, 200].
2. *Numerical Optimization*: by formally eliminating the PDE constraint using the solution operator $u \to y(u) \, : \, G(y(u),u) = 0$, we can keep only the control variable as optimization variable in (1.32). The resulting problem is usually solved by *iterative methods* for the minimization of the reduced cost functional $J(u) = \tilde{J}(y(u),u)$ over the control space. This class of methods is typically referred to as *iterative approach* for the solution of the OCP. For instance, if $\mathcal{U}_{ad} = \mathcal{U}$, we can exploit standard nonlinear optimization algorithms (such as descent methods or trust region methods, see Chap. 3) in a straightforward way. These algorithms require the *repeated solution* of the state and the adjoint problems, as well as the sequential updating of the control (or optimization) variables, until a suitable convergence criterion is achieved. Indeed, this requires many numerical evaluations of the PDEs and of quantities related with the state and the adjoint solutions, such as the gradient or the Hessian of the cost functional.

3. *Computational complexity*: it emerges, from the considerations above, that PDE-constrained optimization problems are computationally demanding, because of their intrinsic iterative nature. Essential features related to the state problem, the control variables and the observation operator may greatly affect the rate of convergence of the optimization procedure, and therefore the global computational cost. Alternatively, at least in some particular cases and under suitable assumptions, the coupled system of the optimality conditions can be solved as a whole. For instance, in the case of linear-quadratic problems in absence of further control constraints, the set of optimality conditions yields a linear system with a *saddle-point* structure, for which effective numerical methods have to be used. This approach is usually referred to as *all-at-once method* [226, 241, 141] and can be regarded as an alternative to the iterative approach discussed above. For more general OCPs, methods such as *active set* or *sequential quadratic programming* [208, 148] entail a sequence of saddle point problems of similar nature.

In Chapters 6 and 8 we introduce the main ideas underlying these numerical strategies, as well as some of the most popular optimization methods. We consider first the simpler (yet significant and non-trivial) case of OCPs with quadratic cost functional and linear state equations. In fact, when tackling more difficult PDE-constrained optimization problems, usual optimization techniques require the solution of simpler linear-quadratic problems, making indispensable the development of efficient methods for these latter problems.

Shape optimization problems feature additional difficulties also from the numerical standpoint. For instance, they require further effort for representing and deforming efficiently the shape of the underlying geometry, as well as for evaluating shape derivatives or shape-dependent quantities. As we will see in Chap. 11, one can benefit from representing shapes by means of a (possibly small) number of design parameters, hence adopting suitable shape parametrization techniques. See, e.g., [134, 200] for a survey of numerical techniques in shape optimization.

The numerical solution of parameter estimation problems features further difficulties because of their ill-posedness, which often demand for special care in the choice of computational techniques, regularization parameter selections, and so on. We remark that also a statistical approach to uncertainty in inverse problems requires in any case the repeated evaluation of outputs (depending on the state solution) over large sample sets, for instance in order to compute sample statistics such as expectations, variances, and higher order moments. Useful references for the numerical solution of inverse problems are, e.g., [268, 154, 260].

Part I
A Preview on Optimization and Control in Finite Dimensions

Chapter 2
Prelude on Optimization: The Finite Dimensional Case

As pointed out in Sect. 1.1, optimization is a crucial ingredient of control problems. In this chapter we review the optimization of functions in finite dimensional spaces. This sets the ground for the investigation of the infinite dimensional case, the natural ambient for the OCPs governed by PDEs. In Sect. 2.1 we set up the problem and we analyze it. We address free and constrained optimization of functions in $\mathbb{R}^n$ in Sects. 2.2 and 2.3, respectively. In particular, we introduce the Lagrange multipliers and the Karush-Kuhn-Tucker methods for treating the case of equality constraints and inequality constraints, respectively. These ideas will then be generalized to the case of infinite dimensional spaces in Chapter 9. Some notions of practical interest when dealing with Euclidean spaces, such as vector/matrix norms, scalar products and basic issues related with matrix operations, are reviewed in Sect. B.1. In particular, for the scalar product in $\mathbb{R}^n$, we use the notations $\mathbf{x} \cdot \mathbf{y}$ or $(\mathbf{x}, \mathbf{y})$.

2.1 Problem Setting and Analysis

We want to minimize an *objective* (or *cost*) function $J : \mathbb{R}^n \to \mathbb{R}$, over a set $K \subseteq \mathbb{R}^n$, that is

$$J(\mathbf{x}) \;\to\; \min_{\mathbf{x}\in K}. \tag{2.1}$$

A point where the minimum of J is achieved is called a *minimizer* or *minimization point* (or also *optimal decision*). Note that $\max J = -\min(-J)$, so that we can restrict ourselves to minimization problems. Problem (2.1) is:

- a *free optimization* problem when $K = \mathbb{R}^n$ or, more in general, K is *open*. Solving a free optimization problem consists therefore in seeking minimizers that are *interior points* of the set K;
- a *constrained optimization* problem if $K \subset \mathbb{R}^n$ is not an open set. In many relevant cases $K \subset \mathbb{R}^n$ is defined by means of *equality* or *inequality constraints* under the following form

$$\begin{aligned} K = \{\mathbf{x} \in \mathbb{R}^n \;:\; & \varphi_j(\mathbf{x}) = b_j \;\; \forall\, j = 1, \ldots, m, \\ & \varphi_i(\mathbf{x}) \geq b_i \;\; \forall\, i = m+1, \ldots, m+p\}, \quad m < n. \end{aligned} \tag{2.2}$$

When the search is over the whole domain K, (2.1) is a *global* minimization problem. Optimization algorithms are generally tailored for *local* minimizers, that is, for points $\hat{\mathbf{x}}$ for which there exists a ball $B_r(\hat{\mathbf{x}})$ such that

$$J(\hat{\mathbf{x}}) \leq J(\mathbf{x}) \qquad \forall\, \mathbf{x} \in K \cap B_r(\hat{\mathbf{x}}).$$

A. Manzoni et al., *Optimal Control of Partial Differential Equations*,
Applied Mathematical Sciences 207, https://doi.org/10.1007/978-3-030-77226-0_2

Clearly, a local minimizer $\hat{\mathbf{x}}$ is a global minimizer in $K \cap B_r(\hat{\mathbf{x}})$. In the sequel, the expression *minimum* or *minimizer* will always refer to *global* minimization.

2.1.1 Well Posedness Analysis

The first issue we address is the well-posedness of problem (2.1). Let K be a *closed* set. Here are the most common properties that guarantee the *existence* of a minimum:

(*a*) J is *lower semicontinuous* in K. This means that, for any $\hat{\mathbf{x}} \in K$,

$$J(\hat{\mathbf{x}}) \leq \lim \inf_{\mathbf{x} \to \hat{\mathbf{x}}} J(\mathbf{x}).$$

Equivalently, the lower-level sets

$$E_c = \{\mathbf{x} \in K \; : \; J(\mathbf{x}) \leq c\}, \qquad c \in \mathbb{R} \tag{2.3}$$

are closed (see Fig. 2.1).

(*b*) K is *bounded*; otherwise, we require that the following *coercivity at* ∞ condition holds

$$J(\mathbf{x}) \to +\infty \text{ as } |\mathbf{x}| \to +\infty. \tag{2.4}$$

If K is bounded, since it is also closed, it is compact. Instead, if K is unbounded, condition (2.4) reduces the search for the minimum to a compact set. Indeed, if $\mathbf{z} \in K$ and $c = J(\mathbf{z})$, the lower-level set (2.3) is non-empty, closed and bounded, that is compact, and

$$\inf_{\mathbf{x} \in K} J(\mathbf{x}) = \inf_{\mathbf{x} \in E_c} J(\mathbf{x}).$$

To seek the minimum of J it is therefore sufficient to consider the restriction of J to E_c.

The above conditions ensure the existence of a minimum for problem (2.1) thanks to the Weierstrass Theorem, see e.g. [234]. Let us examine the key steps of its proof, emphasizing the role of the conditions (*a*) and (*b*). As we have seen, we can assume that K is compact.

1. *Construction of a minimizing sequence.* Let $\lambda = \inf_{\mathbf{x} \in K} J(\mathbf{x})$ (finite or $-\infty$). By definition of *infimum*, there exists a sequence $\{\mathbf{x}_k\} \subset K$ such that

$$J(\mathbf{x}_k) \to \lambda \quad \text{as } k \to \infty.$$

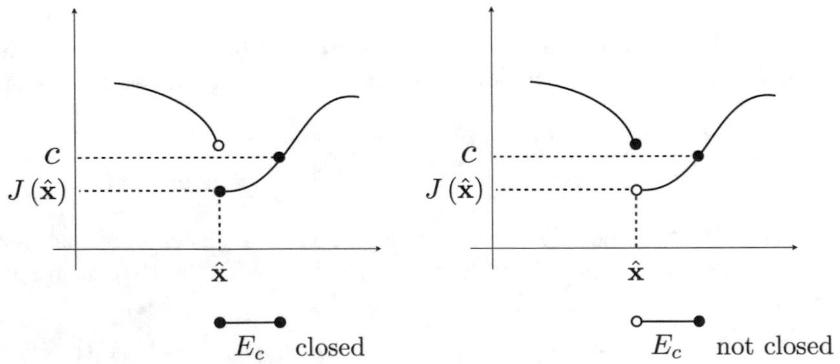

Fig. 2.1 A lower semicontinuous function (left) and a function which is not lower semicontinuous (right)

2. *Extraction of a convergent subsequence.* Since K is compact, a subsequence $\{\mathbf{x}_{k_j}\}$ exists such that $\mathbf{x}_{k_j} \to \hat{\mathbf{x}}$, $j \to \infty$, with $\hat{\mathbf{x}} \in K$.
3. *Use of the lower semicontinuity of J.* Since

$$J(\hat{\mathbf{x}}) \leq \lim \inf_{j \to +\infty} J(\mathbf{x}_{k_j}) = \lambda$$

we deduce that $J(\hat{\mathbf{x}}) = \lambda$ (in particular, $\lambda > -\infty$) and that $\hat{\mathbf{x}}$ is a minimizer.

Clearly, infinitely many minimization points might exist. Simple conditions ensuring *uniqueness* are the *convexity* of the set K and the *strict convexity* of the cost function J.

Definition 2.1. A set $K \subseteq \mathbb{R}^n$ is *convex* if, for all $\mathbf{x}, \mathbf{y} \in K$ and all $t \in [0, 1]$, the point $(1-t)\mathbf{x} + t\mathbf{y}$ is in K, that is, every point on the line segment between $\mathbf{x}$ and $\mathbf{y}$ belongs to K.

Definition 2.2. Let $K \subseteq \mathbb{R}^n$ be convex. A function $J : K \to \mathbb{R}$ is convex (resp. strictly convex) if, for any $\mathbf{x}, \mathbf{y} \in K$, $\mathbf{x} \neq \mathbf{y}$,

$$J(t\mathbf{x} + (1-t)\mathbf{y}) \leq tJ(\mathbf{x}) + (1-t)J(\mathbf{y}) \qquad (\text{resp. } <). \tag{2.5}$$

Geometrically, convexity means that the graph of J lies above its tangent plane at any point. Strict convexity rules out *flat* parts of the graph (see Fig. 2.2, left). When J is differentiable[1] in K, (2.5) is equivalent to

$$J(\mathbf{x}) - J(\mathbf{y}) \geq \nabla J(\mathbf{y}) \cdot (\mathbf{x} - \mathbf{y}) \qquad (\text{resp. } >),$$

Let now K be convex and J be strictly convex. Should there exist two minimizers $\hat{\mathbf{x}}_1$ and $\hat{\mathbf{x}}_2$ on K, we would end up with

$$\lambda \leq J\left(\frac{1}{2}\hat{\mathbf{x}}_1 + \frac{1}{2}\hat{\mathbf{x}}_2\right) < \frac{1}{2}J(\hat{\mathbf{x}}_1) + \frac{1}{2}J(\hat{\mathbf{x}}_2) = \lambda$$

which is impossible. Summarizing, we have shown the following theorem:

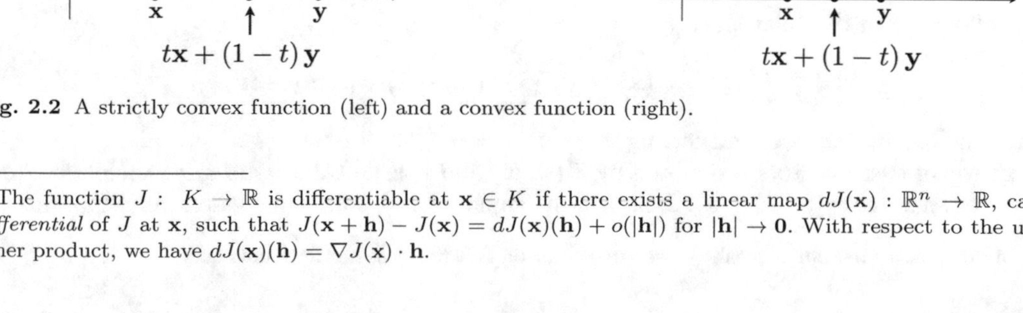

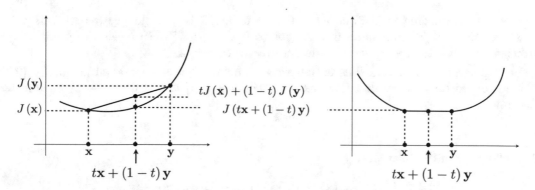

Fig. 2.2 A strictly convex function (left) and a convex function (right).

[1] The function $J : K \to \mathbb{R}$ is differentiable at $\mathbf{x} \in K$ if there exists a linear map $dJ(\mathbf{x}) : \mathbb{R}^n \to \mathbb{R}$, called *differential* of J at $\mathbf{x}$, such that $J(\mathbf{x} + \mathbf{h}) - J(\mathbf{x}) = dJ(\mathbf{x})(\mathbf{h}) + o(|\mathbf{h}|)$ for $|\mathbf{h}| \to \mathbf{0}$. With respect to the usual inner product, we have $dJ(\mathbf{x})(\mathbf{h}) = \nabla J(\mathbf{x}) \cdot \mathbf{h}$.

Theorem 2.1 (Weierstrass). *Let $K \subseteq \mathbb{R}^n$ be a closed, non-empty set. Assume that $J : K \to \mathbb{R}$ is a lower semicontinuous function on K and coercive at ∞ if K is unbounded. Then there exists $\hat{\mathbf{x}} \in K$ such that*

$$J(\hat{\mathbf{x}}) = \min_{\mathbf{x} \in K} J(\mathbf{x}). \tag{2.6}$$

Moreover, if K is convex and J is strictly convex, the minimizer is unique.

2.1.2 Convexity, Optimality Conditions, and Admissible Directions

Let $\hat{\mathbf{x}}$ be a minimizer for J. If K is convex and J is differentiable, it is easy to derive a first order optimality condition. As

$$J(\hat{\mathbf{x}} + t(\mathbf{x} - \hat{\mathbf{x}})) - J(\hat{\mathbf{x}}) \geq 0$$

for any $\mathbf{x} \in K$ and $0 < t \leq 1$, dividing by t, letting $t \to 0$ and denoting by $dJ(\hat{\mathbf{x}})$ the differential of J at $\hat{\mathbf{x}}$, we find

$$dJ(\hat{\mathbf{x}})(\mathbf{x} - \hat{\mathbf{x}}) = \nabla J(\hat{\mathbf{x}}) \cdot (\mathbf{x} - \hat{\mathbf{x}}) \geq 0 \qquad \forall\, \mathbf{x} \in K. \tag{2.7}$$

This is a *variational inequality* and constitutes a first order necessary optimality condition. This condition becomes also sufficient if J is convex, since

$$J(\mathbf{x}) - J(\hat{\mathbf{x}}) \geq \nabla J(\hat{\mathbf{x}}) \cdot (\mathbf{x} - \hat{\mathbf{x}}) \qquad \forall\, \mathbf{x} \in K; \tag{2.8}$$

if (2.7) holds $\hat{\mathbf{x}}$ is a minimizer. Thus, the following result (also referred to as *Fermat theorem*) holds:

Theorem 2.2. *Let K be a convex, non-empty set and $\hat{\mathbf{x}}$ a minimizer for J. If J is differentiable at $\hat{\mathbf{x}}$, then (2.7) holds. Conversely, if J is convex in K and (2.7) holds, then $\hat{\mathbf{x}}$ is a minimizer. Moreover, if J is strictly convex, $\hat{\mathbf{x}}$ is the unique minimizer.*

Remark 2.1. According to (2.7) either $\nabla J(\hat{\mathbf{x}}) = \mathbf{0}$ (i.e., $\hat{\mathbf{x}}$ is a stationary (or critical) point for J), for instance when $\hat{\mathbf{x}}$ is an interior point of K, or the $\nabla J(\hat{\mathbf{x}})$ meets every admissible *direction*[2] $\mathbf{h} = \mathbf{x} - \hat{\mathbf{x}}$, $\mathbf{x} \in K$, at an angle not greater than $\pi/2$. •

For a general set K and a differentiable cost function J, the variational inequality (2.7) holds for all vectors $\mathbf{h}$ such that there exists a sequence $\{\mathbf{x}_k\} \subset K$ with $\mathbf{x}_k \to \hat{\mathbf{x}}$, $\mathbf{x}_k \neq \hat{\mathbf{x}}$, for all k, and

$$\frac{\mathbf{x}_k - \hat{\mathbf{x}}}{|\mathbf{x}_k - \hat{\mathbf{x}}|} \to \frac{\mathbf{h}}{|\mathbf{h}|}.$$

This follows from the relation

$$0 \geq J(\mathbf{x}_k) - J(\hat{\mathbf{x}}) = \nabla J(\hat{\mathbf{x}}) \cdot (\mathbf{x}_k - \hat{\mathbf{x}}) + o(|\mathbf{x}_k - \hat{\mathbf{x}}|),$$

upon dividing by $|\mathbf{x}_k - \hat{\mathbf{x}}|$ and letting $\mathbf{x}_k \to \hat{\mathbf{x}}$.

The set of these vectors is denoted by $T(\hat{\mathbf{x}}, K)$ and is called the set of *admissible directions* or the *tangent cone* at $\hat{\mathbf{x}} \in K$. It is a non empty closed cone that essentially contains all

[2] From now on, a *direction* is meant to be any vector on $\mathbb{R}^n$, not necessarily of unit length.

the vectors "tangent" to K or pointing towards the interior of K. For instance, in the case represented on the left of Fig. 2.3, $T(\hat{\mathbf{x}}, K)$ coincides with the set $\{\mathbf{h} \in \mathbb{R}^2 \ : h_1 \geq 0, \ h_2 \geq 0\}$. Note that $T(\hat{\mathbf{x}}, K)$ is not necessarily convex as in Fig. 2.3, right).

The variational inequality

$$\nabla J(\hat{\mathbf{x}}) \cdot \mathbf{h} \geq 0 \qquad \forall \, \mathbf{h} \in T(\hat{\mathbf{x}}, K) \tag{2.9}$$

expresses the fact that, near $\hat{\mathbf{x}}$, J cannot decrease along a direction $\mathbf{h}$ in the tangent cone. As we will see in the next sections, a basic question in optimization is how to turn (2.9) into a more practical condition.

2.2 Free (Unconstrained) Optimization

For the reader's convenience, we review few additional elementary facts about free optimization. According to Remark 2.1, when $\hat{\mathbf{x}}$ is a minimizer belonging to the interior of K, a *first-order* optimality condition is

$$\nabla J(\hat{\mathbf{x}}) = \mathbf{0}.$$

This is a necessary condition that becomes sufficient when both K and J are convex. Besides, in this case any local minimizer is also a global one, as $J(\mathbf{x}) \geq J(\hat{\mathbf{x}}) \ \forall \mathbf{x} \in K$ thanks to (2.8). If $J \in C^2(K)$, the nature of a critical point $\hat{\mathbf{x}}$ can be analyzed using the Taylor development

$$J(\hat{\mathbf{x}} + \mathbf{h}) - J(\hat{\mathbf{x}}) = \frac{1}{2} \sum_{i,j=1}^{n} (D^2 J)_{ij}(\hat{\mathbf{x}}) h_i h_j + o(|\mathbf{h}|^2) \tag{2.10}$$

where

$$(D^2 J)_{ij}(\mathbf{x}) = \frac{\partial^2 J}{\partial x_i \partial x_j}(\mathbf{x}), \qquad i, j = 1, \ldots, n$$

are the elements of the symmetric Hessian matrix $D^2 J(\mathbf{x})$ of J. If $D^2 J(\hat{\mathbf{x}})$ is positive definite, then $\hat{\mathbf{x}}$ is a local minimizer. The following result holds.

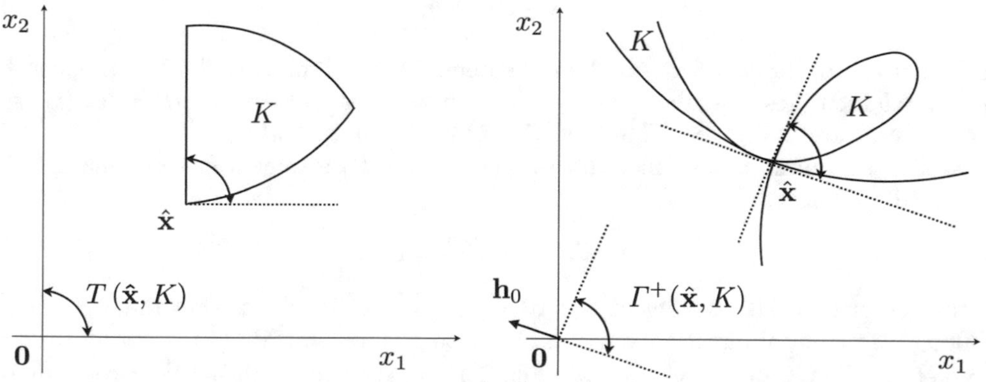

Fig. 2.3 Left: the tangent cone $T(\hat{\mathbf{x}}, K)$ is convex. Right: $T(\hat{\mathbf{x}}, K)$ is nonconvex, since $T(\hat{\mathbf{x}}, K) = \Gamma^+(\hat{\mathbf{x}}, K) \cup \mathbf{h}_0$. The notation $\Gamma^+(\hat{\mathbf{x}}, K)$ is used here to denote the directions $\mathbf{h}$ such that $\nabla\varphi_1(\hat{\mathbf{x}}) \cdot \mathbf{h} \geq 0$ and $\nabla\varphi_2(\hat{\mathbf{x}}) \cdot \mathbf{h} \geq 0$, and will be clear in Sect. 2.3.2.

Theorem 2.3. *Let K be open and $J : K \to \mathbb{R}$ be a differentiable function. If $\hat{\mathbf{x}}$ is a (local) minimizer of J, then $\hat{\mathbf{x}}$ is a stationary point for J, that is $\nabla J(\hat{\mathbf{x}}) = \mathbf{0}$. Furthermore, if $J \in C^2(K)$, then $D^2 J(\hat{\mathbf{x}})$ is positive semidefinite. On the other hand, if $J \in C^2(K)$, $\hat{\mathbf{x}}$ is a stationary point for J and $D^2 J(\hat{\mathbf{x}})$ is a positive definite matrix, then $\hat{\mathbf{x}}$ is a local minimizer for J.*

The interested reader can find further and more general results, e.g., in [36]. We devote Sects. 3.1-3.2 to the most popular algorithms for free optimization; they can also be exploited to build algorithms suitable for constrained problems.

2.3 Constrained Optimization

As we have already mentioned, in many relevant cases, the set K is characterized by equality constraints or inequality constraints. In this section we sketch the most important ideas related to the methods of *Lagrange multipliers* and *Karush-Kuhn-Tucker multipliers.* We refer the reader to [36, 55, 184] for further details and the proofs of the results we introduce.

2.3.1 Lagrange Multipliers: Equality Constraints

Consider first the following two-dimensional case ($n = 2$)

$$\begin{cases} J(x_1, x_2) \ \to \ \min \\ \quad \text{subject to} \\ \varphi(x_1, x_2) = b. \end{cases} \tag{2.11}$$

We assume that $J : \mathbb{R}^2 \to \mathbb{R}$ and $\varphi : \mathbb{R}^2 \to \mathbb{R}$ are continuously differentiable functions. The second equation represents the constraint. Suppose $\hat{\mathbf{x}} = (\hat{x}_1, \hat{x}_2)$ is a minimizer for (2.11) and assume the following, key *regularity condition*[3]

$$\nabla \varphi(\hat{\mathbf{x}}) \neq \mathbf{0}. \tag{2.12}$$

Then, thanks to the *implicit function* theorem, in a neighborhood of $\hat{\mathbf{x}}$ the constraint $\varphi(x_1, x_2) = b$ identifies a regular curve γ, that can be parametrized as $\mathbf{x}(t) = (x_1(t), x_2(t))$, with $t \in (-\delta, \delta)$ and $\mathbf{x}(0) = \hat{\mathbf{x}}$. Note that $\nabla \varphi(\hat{\mathbf{x}})$ is normal to γ at $\hat{\mathbf{x}}$.

Since $\hat{\mathbf{x}}$ is a minimizer, the function $t \mapsto g(t) = J(\mathbf{x}(t))$ attains a free minimum at $t = 0$, so that $g'(0) = 0$ or, equivalently,

$$\nabla J(\hat{\mathbf{x}}) \cdot \dot{\mathbf{x}}(0) = 0. \tag{2.13}$$

Now, the vector $\mathbf{h} = \dot{\mathbf{x}}(0)$ is tangent to γ at $\hat{\mathbf{x}}$, so that (2.13) tells us that the derivative of J at $\hat{\mathbf{x}}$ along a tangential direction vanishes. We also deduce that $\nabla J(\hat{\mathbf{x}})$ is normal to γ at $\hat{\mathbf{x}}$, and therefore it is parallel to $\nabla \varphi(\hat{\mathbf{x}})$ (see Fig. 2.4). This means that, for some real number $\hat{\lambda}$, we can write

$$\nabla J(\hat{\mathbf{x}}) = \hat{\lambda} \nabla \varphi(\hat{\mathbf{x}}). \tag{2.14}$$

[3] In general, we say that a point $\mathbf{x} \in E_b := \{\mathbf{x} \in \mathbb{R}^n : \varphi(\mathbf{x}) = b\}$ is *regular* if (2.12) holds.

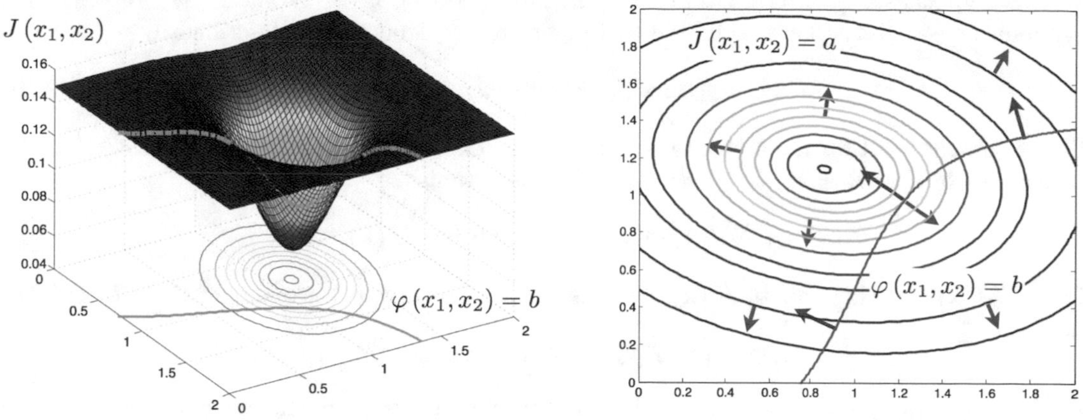

Fig. 2.4 A graphical sketch of the Lagrange multipliers method in the case $n = 2$: both the contour lines of J and the constraint $\varphi = b$ are shown. At the minimizer $\hat{\mathbf{x}}$ the gradients ∇J and $\nabla \varphi$ are parallel

In summary we obtain the following first-order necessary optimality condition: under the above hypotheses, there exists $\hat{\lambda} \in \mathbb{R}$, called *Lagrange multiplier*, such that (2.14) holds.

Introducing the Lagrangian function

$$\mathcal{L}(\mathbf{x},\lambda) = J(\mathbf{x}) + \lambda(b - \varphi(\mathbf{x})),$$

we can recover the above optimality conditions for problem (2.11) by requiring the vanishing of the gradient of $\mathcal{L}$, that is:

$$\begin{cases} \dfrac{\partial \mathcal{L}}{\partial x_1}(\hat{\mathbf{x}}, \hat{\lambda}) = 0 \\ \dfrac{\partial \mathcal{L}}{\partial x_2}(\hat{\mathbf{x}}, \hat{\lambda}) = 0 \\ \dfrac{\partial \mathcal{L}}{\partial \lambda}(\hat{\mathbf{x}}, \hat{\lambda}) = 0. \end{cases}$$

Note that the third equation reproduces the constraint.

In the general case, K is expressed by a set of equality constraints

$$K = \{\mathbf{x} : \varphi_j(\mathbf{x}) = b_j,\ j = 1, \ldots, m\}$$

where $\varphi_j : \mathbb{R}^n \to \mathbb{R}$, $j = 1, \ldots, m$, with $m < n$, are m continuously differentiable functions and $\mathbf{b} = (b_1, \ldots, b_m)$. Let $J : \mathbb{R}^n \to \mathbb{R}$ be a differentiable cost function and consider the problem:

$$\begin{cases} J(\mathbf{x}) \ \to \ \min \\ \text{subject to} \\ \mathbf{x} \in K. \end{cases} \tag{2.15}$$

Let $\hat{\mathbf{x}}$ be a minimizer and assume that the vectors $\nabla \varphi_j(\hat{\mathbf{x}})$, $j = 1, \ldots, m$ are *linearly independent* or, equivalently, by setting $\boldsymbol{\varphi} = (\varphi_1, \ldots, \varphi_m)$, that the Jacobian $\nabla \boldsymbol{\varphi} : \mathbb{R}^n \to \mathbb{R}^{m \times n}$ of $\boldsymbol{\varphi}$, defined by

$$(\nabla \boldsymbol{\varphi}(\mathbf{x}))_{ij} = \frac{\partial \varphi_i(\mathbf{x})}{\partial x_j}, \qquad i = 1, \ldots, m\ ,\ j - 1, \ldots, n$$

is a *surjective* map.

This hypothesis guarantees that in a neighborhood of $\hat{\mathbf{x}}$, K geometrically represents a hypersurface of dimension $n-m$, whose tangent hyperplane at $\hat{\mathbf{x}}$ is described by

$$\Pi\left(\hat{\mathbf{x}}\right)=\left\{\mathbf{h}:\nabla\varphi_j\left(\hat{\mathbf{x}}\right)\cdot\mathbf{h}=0,\quad j=1,\ldots,m\right\}. \tag{2.16}$$

The argument used for the above two-dimensional problem can be extended to prove that the derivatives of J at $\hat{\mathbf{x}}$ along *any* tangential direction vanish,

$$\nabla J(\hat{\mathbf{x}})\cdot\mathbf{h}=0 \qquad \text{for all } \mathbf{h}\in\Pi\left(\hat{\mathbf{x}}\right).$$

Thus, $\nabla J(\hat{\mathbf{x}})$ is orthogonal to $\Pi\left(\hat{\mathbf{x}}\right)$. Thanks to (2.16), every vector orthogonal to $\Pi\left(\hat{\mathbf{x}}\right)$ takes the form

$$\lambda_1\nabla\varphi_1\left(\hat{\mathbf{x}}\right)+\lambda_2\nabla\varphi_2\left(\hat{\mathbf{x}}\right)+\ldots+\lambda_m\nabla\varphi_m\left(\hat{\mathbf{x}}\right)$$

for some scalars $\lambda_j\in\mathbb{R}$, $j=1,\ldots,m$. We are therefore led to state the following result, whose proof can be found, for instance, in [208, Theorem 12.1]:

Theorem 2.4. *Under the above hypotheses, there exist m scalars $\hat{\lambda}_j\in\mathbb{R}$, $j=1,\ldots,m$, called (Lagrange) multipliers, such that*

$$\nabla J(\hat{\mathbf{x}})=\hat{\lambda}_1\nabla\varphi_1\left(\hat{\mathbf{x}}\right)+\hat{\lambda}_2\nabla\varphi_2\left(\hat{\mathbf{x}}\right)+\ldots+\hat{\lambda}_m\nabla\varphi_m\left(\hat{\mathbf{x}}\right). \tag{2.17}$$

By denoting $\boldsymbol{\lambda}=(\lambda_1,\ldots,\lambda_m)$ and introducing the Lagrangian function

$$\mathcal{L}\left(\mathbf{x},\boldsymbol{\lambda}\right)=J\left(\mathbf{x}\right)+\sum_{j=1}^{m}\lambda_j(b_j-\varphi_j\left(\mathbf{x}\right)), \tag{2.18}$$

the optimality condition (2.17) can be written in the form

$$\nabla_{\mathbf{x}}\mathcal{L}(\hat{\mathbf{x}},\hat{\boldsymbol{\lambda}})=\mathbf{0}$$

while the constraints are obtained from the equation

$$\nabla_{\boldsymbol{\lambda}}\mathcal{L}(\hat{\mathbf{x}},\hat{\boldsymbol{\lambda}})=\mathbf{0}.$$

Remark 2.2. The multipliers $\hat{\lambda}_j$ play a role in the *sensitivity analysis*. Indeed, let $\hat{\mathbf{x}}\left(\mathbf{b}\right)$ be a minimizer for (2.15). If $\hat{\mathbf{x}}\left(\mathbf{b}\right)$ is a differentiable function of $\mathbf{b}$, the optimal value

$$m\left(\mathbf{b}\right)=J\left(\hat{\mathbf{x}}\left(\mathbf{b}\right)\right)$$

is differentiable, too, and it satisfies

$$\frac{\partial m}{\partial b_j}=\hat{\lambda}_j, \qquad j=1,\ldots,m. \tag{2.19}$$

In other words, $\hat{\lambda}_j$ represents the sensitivity (measured by the partial derivative $\partial m/\partial b_j$) of the optimal value $m(\mathbf{b})$ to the variations of the level b_j of the j-th constraint. •

Remark 2.3. If the vectors $\nabla\varphi_j\left(\hat{\mathbf{x}}\right)$, $j=1,\ldots,m$, are linearly independent, we say that $\hat{\mathbf{x}}$ is a *regular point*, or that the constraints are *regular* at $\hat{\mathbf{x}}$. Clearly, this is only a sufficient condition for the existence of multipliers, as the following example shows. •

Example 2.1. In the case of the following problem:

$$\begin{cases} J(x_1, x_2, x_3) = x_2^2 \ \rightarrow \ \min \\ \text{subject to} \\ \varphi_1(x_1, x_2, x_3) = x_1 + x_2 + x_3 = 0 \\ \varphi_2(x_1, x_2, x_3) = x_1^2 - x_2^2 = 0 \end{cases} \tag{2.20}$$

$\hat{\mathbf{x}} = (0,0,0)$ is a solution, $\nabla\varphi_1(\hat{\mathbf{x}}) = (1,1,1)$ and $\nabla\varphi_2(\hat{\mathbf{x}}) = (0,0,0)$ are not independent. However, (2.17) is verified by choosing $\lambda_1 = 0$ and $\lambda_2 \in \mathbb{R}$ arbitrary.
On the other hand, for the following problem:

$$\begin{cases} J(x_1, x_2) = x_1^2 + x_2^2 \ \rightarrow \ \min \\ \text{subject to} \\ \varphi(x_1, x_2) = (x_1 - 1)^3 - x_2^2 = 0 \end{cases} \tag{2.21}$$

the solution $\hat{\mathbf{x}} = (1,0)$ is the point on the curve $\varphi_1(x_1, x_2) = 0$ closest to the origin (see Fig. 2.5). Here $\nabla J(\hat{\mathbf{x}}) = (2,0)$ and $\nabla\varphi(\hat{\mathbf{x}}) = (0,0)$, hence $\hat{\mathbf{x}}$ is not a regular point and no multiplier exists. •

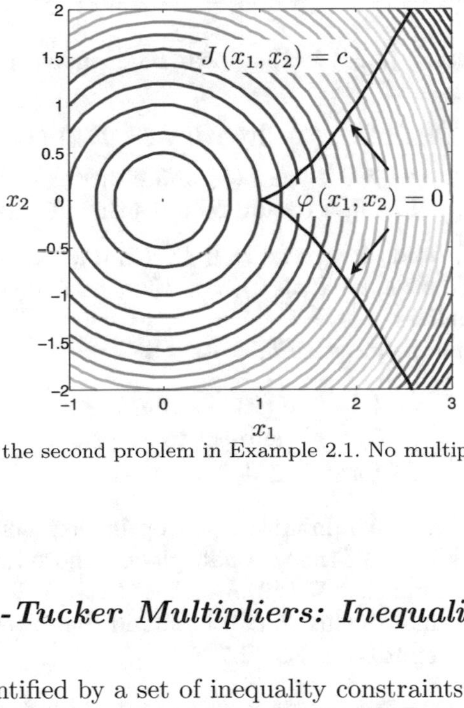

Fig. 2.5 A graphical sketch of the second problem in Example 2.1. No multiplier exists at the minimizer $\hat{\mathbf{x}}$

2.3.2 Karush-Kuhn-Tucker Multipliers: Inequality Constraints

Assume now that K is identified by a set of inequality constraints ($m = 0$ in (2.2)). Under a suitable regularity condition of the constraints – the so-called *constraints qualification* (CQ) condition – the variational inequality (2.9) can be expressed more conveniently in terms of multipliers.

Let $J : \mathbb{R}^n \rightarrow \mathbb{R}$ and $\varphi_i : \mathbb{R}^n \rightarrow \mathbb{R}$, $i = 1, \ldots, p$ be continuously differentiable functions and define

$$K = \{\mathbf{x} : \varphi_i(\mathbf{x}) \geq b_i, \ i = 1, \ldots, p\}.$$

If $\hat{\mathbf{x}}$ is a minimizer for

$$\begin{cases} J(\mathbf{x}) \ \rightarrow \ \min \\ \text{subject to} \\ \mathbf{x} \in K \end{cases} \tag{2.22}$$

from (2.9) we know that

$$\nabla J(\hat{\mathbf{x}}) \cdot \mathbf{h} \geq 0 \qquad \forall\, \mathbf{h} \in T\,(\hat{\mathbf{x}}, K)\,, \tag{2.23}$$

where $T(\hat{\mathbf{x}}, K)$ is the *tangent cone at* $\hat{\mathbf{x}}$ of K. As in the case of equality constraints, we express (2.23) in terms of the gradients $\nabla\varphi_i\,(\hat{\mathbf{x}})$ of the constraints. Let

$$I\,(\hat{\mathbf{x}}) = \{i = 1, \ldots, p\,:\,\varphi_i\,(\hat{\mathbf{x}}) = b_i\}$$

be the set of *active constraints at* $\hat{\mathbf{x}}$, and define the (positive) *polar cone* of K at $\hat{\mathbf{x}}$ as

$$\Gamma^+\,(\hat{\mathbf{x}}, K) = \{\mathbf{h} : \nabla\varphi_i\,(\hat{\mathbf{x}}) \cdot \mathbf{h} \geq 0,\; i \in I\,(\hat{\mathbf{x}})\}\,.$$

If $\nabla\varphi_i\,(\hat{\mathbf{x}})$ does not vanish, it points towards the interior of K, and the polar cone $\Gamma^+\,(\hat{\mathbf{x}}, K)$ collects the vectors $\mathbf{h}$ meeting all the $\nabla\varphi_i\,(\hat{\mathbf{x}})$ at an angle not greater than $\pi/2$, see Fig. 2.3.

Definition 2.3. We say that the *constraints qualification* (CQ) condition is satisfied at $\hat{\mathbf{x}}$ if the following relation holds:

$$\Gamma^+\,(\hat{\mathbf{x}}, K) = T\,(\hat{\mathbf{x}}, K)\,. \tag{2.24}$$

A simple condition ensuring (2.24) is the linear independence of the vectors $\nabla\varphi_i\,(\hat{\mathbf{x}})\,, i \in I\,(\hat{\mathbf{x}})$. Alternative conditions are:

- there exists $\mathbf{h}^* \in \mathbb{R}^n$ such that $\nabla\varphi_i\,(\hat{\mathbf{x}}) \cdot \mathbf{h}^* > 0\;\forall\, i \in I\,(\hat{\mathbf{x}})$ (*Cottle condition*);
- there exists $\bar{\mathbf{x}}$ such that $\varphi_i\,(\bar{\mathbf{x}}) > 0\;\forall\, j \in I\,(\hat{\mathbf{x}})$ and either φ_i is convex or $\varphi_i\,(\bar{\mathbf{x}}) \geq 0$ and φ_i is affine (*Slater condition*). The first option requires that K has non empty interior.

Thanks to (2.24), (2.23) holds for every $\mathbf{h} \in \Gamma^+\,(\hat{\mathbf{x}}, K)$, and implies that $\nabla J(\hat{\mathbf{x}})$ can be expressed as a linear combination of $\nabla\varphi_i\,(\hat{\mathbf{x}})$, $i \in I(\hat{\mathbf{x}})$, with *nonnegative coefficients*, called *Karush-Kuhn-Tucker multipliers.*

First consider the case $n = p = 2$, with $b_1 = b_2 = 0$

$$\begin{cases} J\,(\mathbf{x}) \;\to \min \\ \text{subject to} \\ \varphi_1(\mathbf{x}) \geq 0, \quad \varphi_2(\mathbf{x}) \geq 0\,. \end{cases} \tag{2.25}$$

Referring to Fig. 2.6, if $\hat{\mathbf{x}}$ is a minimizer, the optimality condition (2.23) implies that $\nabla J(\hat{\mathbf{x}})$ belongs to the intersection of the two half planes whose interior normals are $\nabla\varphi_1\,(\hat{\mathbf{x}})$ and $\nabla\varphi_2\,(\hat{\mathbf{x}})$. We easily conclude that $\nabla J\,(\hat{\mathbf{x}}) = \lambda_1\nabla\varphi_1\,(\hat{\mathbf{x}}) + \lambda_2\nabla\varphi_2\,(\hat{\mathbf{x}})$, with $\lambda_1 > 0, \lambda_2 > 0$.

In the general case, the same conclusion can be found thanks to the Farkas lemma, whose geometrical intepretation is reported in Fig. 2.7:

Lemma 2.1 (Farkas). *Let* $\mathbf{w}_1, \ldots, \mathbf{w}_p, \mathbf{v} \in \mathbb{R}^n$ *be a set of vectors in* $\mathbb{R}^n$. *The following inclusion*

$$\{\mathbf{h} : \mathbf{w}_i \cdot \mathbf{h} \geq 0,\; i = 1, \ldots, p\} \subseteq \{\mathbf{h} : \mathbf{v} \cdot \mathbf{h} \geq 0,\; i = 1, \ldots, p\} \tag{2.26}$$

holds if and only if there exist nonnegative λ_i, $i = 1, \ldots, p$, *such that*

$$\mathbf{v} = \sum_{i=1}^{p} \lambda_i \mathbf{w}_i.$$

By choosing $\mathbf{v} = \nabla J(\hat{\mathbf{x}})$ and $\mathbf{w}_i = \nabla\varphi_i\,(\hat{\mathbf{x}})$, $i = 1, \ldots, p$ in Farkas Lemma 2.1, we deduce the following fundamental result:

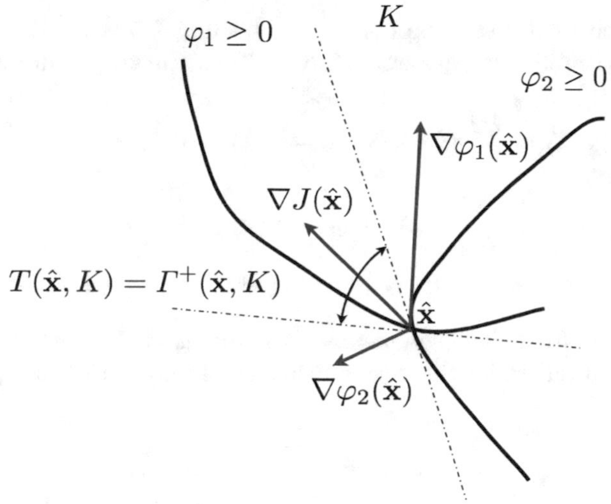

Fig. 2.6 A graphical sketch of the Karush-Kuhn-Tucker multipliers method in the case $n = p = 2$.

Theorem 2.5 (Karush, 1939; Kuhn-Tucker, 1951). *Let* $\hat{\mathbf{x}}$ *be a minimizer for problem* (2.22) *and suppose that the constraint qualification (CQ) condition is satisfied at* $\hat{\mathbf{x}}$. *Then, there exist* $\hat{\lambda}_i \geq 0$, $i \in I(\hat{\mathbf{x}})$, *called* Karush-Kuhn-Tucker (KKT) multipliers, *such that*

$$\nabla J(\hat{\mathbf{x}}) = \sum_{i \in I(\hat{\mathbf{x}})} \hat{\lambda}_i \nabla \varphi_i(\hat{\mathbf{x}}), \qquad \hat{\lambda}_i \geq 0. \tag{2.27}$$

We point out that only the active constraints at $\hat{\mathbf{x}}$ appear in formula (2.27). Since $\hat{\mathbf{x}}$ is not a priori known, it is more convenient to rewrite (2.27) in terms of the Lagrangian function, defined as

$$\mathcal{L}(\mathbf{x}, \boldsymbol{\lambda}) = J(\mathbf{x}) + \sum_{i=1}^{p} \lambda_i (b_i - \varphi_i(\mathbf{x})), \tag{2.28}$$

where $\boldsymbol{\lambda} = (\lambda_1, \dots, \lambda_p)$ is the vector of KKT multipliers, and all the constraints are considered.

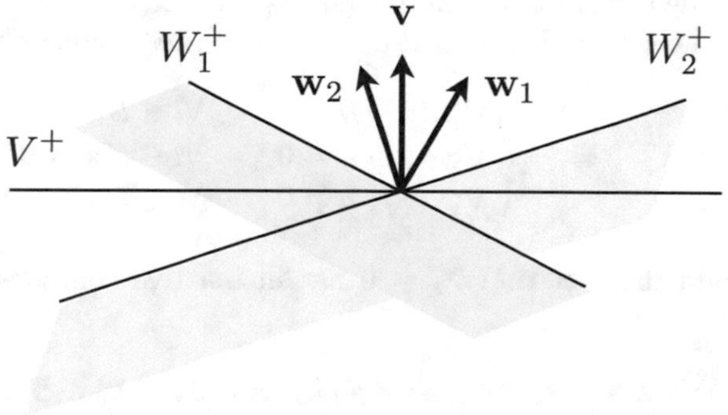

Fig. 2.7 Geometrical interpretation of Farkas lemma: if W_i^+ is the half space of the vectors forming with $\mathbf{w}_i$ an acute angle, the set at the left-hand side of (2.26) is given by $\bigcap_{i=1}^{p} W_i^+$. Since $W_1^+ \cap W_2^+ \subseteq V^+$, $\mathbf{v}$ is a linear combination of $\mathbf{w}_1$ and $\mathbf{w}_2$ with nonnegative coefficients.

The optimal necessary condition expressed in Theorem 2.5 is equivalent to require that $(\hat{\mathbf{x}}, \hat{\boldsymbol{\lambda}})$ is a solution of the following system (*Karush-Kuhn-Tucker* conditions):

$$\frac{\partial \mathcal{L}}{\partial x_k}(\hat{\mathbf{x}}, \hat{\boldsymbol{\lambda}}) = \frac{\partial J}{\partial x_k}(\hat{\mathbf{x}}) - \sum_{i=1}^{p} \hat{\lambda}_i \frac{\partial \varphi_i}{\partial x_k}(\hat{\mathbf{x}}) = 0 \quad k = 1, \quad , n \tag{2.29}$$

$$\frac{\partial \mathcal{L}}{\partial \lambda_i}(\hat{\mathbf{x}}, \hat{\boldsymbol{\lambda}}) = b_i - \varphi_i(\hat{\mathbf{x}}) \leq 0 \qquad i = 1, \ldots, p \tag{2.30}$$

$$\hat{\lambda}_i \geq 0, \qquad \hat{\lambda}_i \left(b_i - \varphi_i\left(\hat{\mathbf{x}}\right)\right) = 0 \qquad i = 1, \ldots, p. \tag{2.31}$$

Equations (2.31) are also called *complementary equations.* Note that, if $i \notin I(\hat{\mathbf{x}})$, from (2.31) we get $\lambda_i = 0$, so that only the active constraints play a role in equations (2.29) and we recover equation (2.27).

2.3.3 Second Order Conditions

Second order conditions are also available. We treat both m equality and p inequality constraints at the same time, and assume that both the cost function and the constraints are twice continuously differentiable. The first-order optimality conditions are therefore:

$$\begin{aligned} &\nabla_{\mathbf{x}} \mathcal{L}(\hat{\mathbf{x}}, \hat{\boldsymbol{\lambda}}) = \mathbf{0} && \\ &\varphi_j(\hat{\mathbf{x}}) = b_j, && \forall\, j \in E = \{1, \ldots, m\} \\ &\varphi_{m+i}(\hat{\mathbf{x}}) \geq b_{m+i}, && \forall\, m+i \in I = \{m+1, \ldots, m+p\} \\ &\hat{\lambda}_i \geq 0, \quad \hat{\lambda}_i(b_i - \varphi_i(\hat{\mathbf{x}})) = 0, && \forall\, i \in I; \end{aligned} \tag{2.32}$$

moreover, denote by $I(\hat{\mathbf{x}}) \subseteq I$ the set of active (inequality) constraints at $\hat{\mathbf{x}}$. In this case, the *polar cone* of K at $\hat{\mathbf{x}}$ is given by

$$\Gamma^+(\hat{\mathbf{x}}, K) = \{\mathbf{h} : \nabla \varphi_j(\hat{\mathbf{x}}) \cdot \mathbf{h} = 0 \ \forall\, j \in E, \quad \nabla \varphi_i(\hat{\mathbf{x}}) \cdot \mathbf{h} \geq 0 \ \forall\, i \in I(\hat{\mathbf{x}})\}.$$

Second order conditions involve the action of the Hessian of $\mathcal{L}$ along directions tangential to the constraints. Given the polar cone $\Gamma^+(\hat{\mathbf{x}}, K)$ and a pair $(\hat{\mathbf{x}}, \hat{\boldsymbol{\lambda}})$ satisfying (2.32), we introduce the *critical* cone $C(\hat{\mathbf{x}}, \hat{\boldsymbol{\lambda}}, K)$ of those directions belonging to $\Gamma^+(\hat{\mathbf{x}}, K)$ for which first order conditions do not provide any information on the nature of $\hat{\mathbf{x}}$.

Precisely, the critical cone can be defined as the set of those directions $\mathbf{h}$ such that:

$$\mathbf{h} \in C(\hat{\mathbf{x}}, \hat{\boldsymbol{\lambda}}, K) \quad \Leftrightarrow \quad \begin{cases} \nabla \varphi_i(\hat{\mathbf{x}}) \cdot \mathbf{h} = 0 & \forall i \in E, \\ \nabla \varphi_i(\hat{\mathbf{x}}) \cdot \mathbf{h} = 0 & \forall i \in I(\hat{\mathbf{x}}) \text{ with } \hat{\lambda}_i > 0, \\ \nabla \varphi_i(\hat{\mathbf{x}}) \cdot \mathbf{h} \geq 0 & \forall i \in I(\hat{\mathbf{x}}) \text{ with } \hat{\lambda}_i = 0. \end{cases} \tag{2.33}$$

In fact, from (2.33) and the fact that $\hat{\lambda}_i = 0$ for all inactive components $i \in I \setminus I(\hat{\mathbf{x}})$, we immediately obtain that

$$\mathbf{h} \in C(\hat{\mathbf{x}}, \hat{\boldsymbol{\lambda}}, K) \quad \Rightarrow \quad \hat{\lambda}_i \nabla \varphi_i(\hat{\mathbf{x}}) \cdot \mathbf{h} = 0 \qquad \forall i \in E \cup I;$$

hence, from $(2.32)_1$ and the definition (2.28) of the Lagrangian function, we find that

$$\mathbf{h} \in C(\hat{\mathbf{x}}, \hat{\boldsymbol{\lambda}}, K) \quad \Rightarrow \quad \nabla J(\hat{\mathbf{x}}) \cdot \mathbf{h} = \sum_{i \in E \cup I} \hat{\lambda}_i \nabla \varphi_i(\hat{\mathbf{x}}) \cdot \mathbf{h} = 0.$$

It follows that $C(\hat{\mathbf{x}}, \hat{\boldsymbol{\lambda}}, K)$ is exactly the subset of $\Gamma^+(\hat{\mathbf{x}}, K)$ we are interested in. A necessary condition for a local minimizer at which the CQ condition is fulfilled reads as follows:

Proposition 2.1. *Let* $(\hat{\mathbf{x}}, \hat{\lambda})$ *satisfy system* (2.32)*. Assume that the CQ condition is satisfied at* $\hat{\mathbf{x}}$*. Then*

$$D^2_{\mathbf{xx}}\mathcal{L}(\hat{\mathbf{x}}, \hat{\boldsymbol{\lambda}})\mathbf{h} \cdot \mathbf{h} \geq 0 \qquad \forall\, \mathbf{h} \in C(\hat{\mathbf{x}}, \hat{\boldsymbol{\lambda}}, K).$$

More significant is the following sufficient condition for a strict local minimizer:

Proposition 2.2. *Suppose that for some point* $\hat{\mathbf{x}} \in K$ *there is a multiplier vector* $\hat{\boldsymbol{\lambda}}$ *such that the first-order conditions* (2.32) *are satisfied. Suppose also that*

$$D^2_{\mathbf{xx}}\mathcal{L}(\hat{\mathbf{x}}, \hat{\boldsymbol{\lambda}})\mathbf{h} \cdot \mathbf{h} > 0 \qquad \forall\, \mathbf{h} \in C(\hat{\mathbf{x}}, \hat{\boldsymbol{\lambda}}, K),\ \mathbf{h} \neq \mathbf{0}.$$

Then $\hat{\mathbf{x}}$ *is a strict local minimizer.*

Note that the CQ condition is not needed in Proposition 2.2, since we are already assuming the existence of the multipliers.

Chapter 3
Algorithms for Numerical Optimization

We consider the optimization of a cost function $J : \mathbb{R}^n \to \mathbb{R}$. Numerical methods for optimization problems are based on *iterative* algorithms which compute a finite sequence of points $\{\mathbf{x}_k\}_{k\geq 1}$ approaching a local optimizer. For a comprehensive survey on optimization algorithms see, e.g., [208, 80, 36]. From a numerical standpoint, we can make a general distinction between optimization in $\mathbb{R}^n$ (*free optimization*) and optimization over a closed set $K \subset \mathbb{R}^n$ characterized by either equality and/or inequality constraints (*constrained optimization*). We first illustrate two families of numerical algorithms for solving free optimization problems: *descent methods* (Sect. 3.1) and *trust-region* methods (Sect. 3.2). Then, in Sects. 3.3, 3.4 and 3.5 we describe some widely used algorithms for solving constrained optimization problems, such as the *penalty* method, the *augmented Lagrangian* method and the more general *sequential quadratic programming* method. As already noticed in Sect. 2.1, we can restrict our analysis to minimization problems without loss of generality. From now on, the word *optimization* should therefore understood as *minimization.*

Most algorithms for nonlinear optimization require the knowledge of the derivatives of the cost function J. However, if J is not differentiable, or if the computation of its derivatives is nontrivial, we can rely on *derivative free* methods. The latter exploit either the comparison among function evaluations in different directions at each step, or low-order local approximants of J in order to assess its local behavior and localize the minimizer; see for instance [270, 273, 71, 167]. A very popular derivative-free method is the method of Nelder and Mead [206] (see also [166] for the analysis of its convergence properties). A widespread approach makes use of finite difference approximations of partial derivatives of J; see e.g. [208].

More efficient methods are usually employed in the solution of large-scale problems, like those arising from the discretization of OCPs for PDEs. For any given starting point $\mathbf{x}_0$, these methods build up a sequence

$$\{\mathbf{x}_k\}_{k\geq 1} \quad \text{such that} \quad J(\mathbf{x}_{k+1}) < J(\mathbf{x}_k)$$

iteratively, until a suitable stopping criterion is fulfilled. Several strategies allow to move from the current iterate $\mathbf{x}_k$ to the new one $\mathbf{x}_{k+1}$:

- *descent* or *line-search* methods: the algorithm selects a direction $\mathbf{d}_k$ and then computes a step length to move from the current iterate $\mathbf{x}_k$ along that direction to find a new iterate such that $J(\mathbf{x}_{k+1}) < J(\mathbf{x}_k)$;
- *trust region* methods: the information about the cost function J is exploited to construct a model function which, in a given region (the so-called *trust region*) near the current iterate $\mathbf{x}_k$, resembles the actual cost function J.

Both strategies enable a substantial simplification of the original problem (2.1). In the former, at each iteration problem (2.1) is reduced to a one dimensional problem, as $\mathbf{x}_{k+1}$ is

A. Manzoni et al., *Optimal Control of Partial Differential Equations*,
Applied Mathematical Sciences 207, https://doi.org/10.1007/978-3-030-77226-0_3

obtained by restricting the function J along the direction $\mathbf{d}_k$, and then minimizing a single variable function alongwith. In the latter, instead, the cost function is replaced by a model function – for instance, a quadratic function – for which the search of a local minimizer is easier, even if the dimension of the problem is not reduced. For large problems (say $n > 10^3$) trust region methods may become computationally unaffordable unless an efficient implementation is carried out.

For the sake of notation, we highlight that in this chapter, as well as in those regarding the numerical approximation of OCPs, the *gradient* is understood (as all the other vectors) as column vector, to make notation more closely related to the one adopted when implementing numerical methods.

3.1 Free Minimization by Descent Methods

In this section we describe the most popular *descent* methods, for which the minimization problem can be stated as follows: given an initial guess $\mathbf{x}_0$, the method consists in constructing iteratively a sequence $\{\mathbf{x}_k\}$ such that

$$\mathbf{x}_{k+1} = \mathbf{x}_k + \tau_k \mathbf{d}_k \qquad k = 1, 2, \dots \tag{3.1}$$

where $\mathbf{d}_k$ represents a *descent direction*, that is a vector that satisfies:

$$\begin{aligned} &\mathbf{d}_k^\top \nabla J(\mathbf{x}_k) < 0 \qquad && \text{if } \nabla J(\mathbf{x}_k) \neq \mathbf{0} \\ &\mathbf{d}_k = \mathbf{0} && \text{if } \nabla J(\mathbf{x}_k) = \mathbf{0}, \end{aligned} \tag{3.2}$$

and τ_k is a *step length* (possibly not too small, for the efficiency of the method) that can be chosen in a variety of ways, as we discuss later. Descent directions (3.2) are called in this way since $\nabla J(\mathbf{x}_k)$ gives in $\mathbb{R}^n$ the direction of maximum increase of J starting from $\mathbf{x}_k$; $\mathbf{d}_k^\top \nabla J(\mathbf{x}_k)$ is the directional derivative of J along $\mathbf{d}_k$. Formula (3.1) yields a descent method if the vectors $\mathbf{d}_k$ are chosen according to (3.2). Thus, by requiring $(3.2)_1$, we step in the direction opposite to the gradient, that is (in principle) towards a minimum of the function. At every iteration, descent methods require the evaluation of the cost function and its gradient. The evaluation of the latter, in particular, may be quite involved. A pseudocode implementing a descent method is reported in Algorithm 3.1.

Algorithm 3.1 Descent method

Input: Maximum iterations number $k_{\max}$, stopping tolerance ε, starting point $\mathbf{x}_0$
Output: Minimizer $\hat{\mathbf{x}}$

set $k = 0$, $err = \varepsilon + 1$
while $err > \varepsilon$ and $k < k_{max}$
    compute the search direction $\mathbf{d}_k$
    compute the step length τ_k with a backtracking routine (see Section 3.1.2)
    set $\mathbf{x}_{k+1} = \mathbf{x}_k + \tau_k \mathbf{d}_k$
    set $err = |\nabla J(\mathbf{x}_k)|$
        (alternatively, set $err = |J(\mathbf{x}_{k+1}) - J(\mathbf{x}_k)|$, or $err = |\mathbf{x}_{k+1} - \mathbf{x}_k|$)
    set $k = k + 1$
end while
set $\hat{\mathbf{x}} = \mathbf{x}_{k-1}$

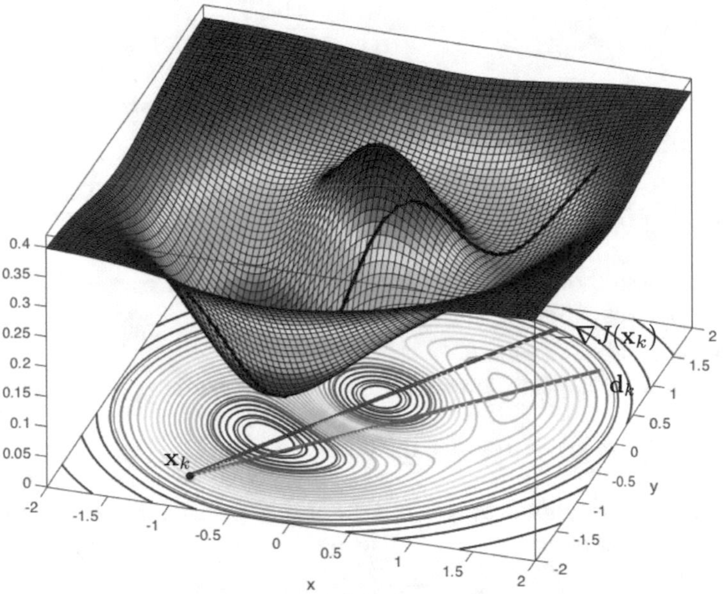

Fig. 3.1 Contour lines of a function, its restriction along a descent direction $\mathbf{d}_k$ and the direction of its gradient evaluated at $\mathbf{x}_k$

Typically the search direction $\mathbf{d}_k$ could depend on both the gradient and the Hessian of the objective function J; several algorithms are available with different choices of $\mathbf{d}_k$, each of them bearing advantages and disadvantages in terms of computational and storage costs. An example of descent direction is shown in Fig. 3.1. Once $\mathbf{d}_k$ has been chosen, the optimal step length τ_k, which ensures the maximum (negative) variation of J along $\mathbf{d}_k$, has to be computed.

3.1.1 Choice of descent directions

The most popular strategies to choose a descent direction are recalled below.

- *Steepest-descent method*
  The steepest descent method (also-called gradient method) makes use of the following direction
  $$\mathbf{d}_k = -\nabla J(\mathbf{x}_k) \tag{3.3}$$
  at every step. This method requires the calculation of the gradient $\nabla J(\mathbf{x}_k)$, but not of the Hessian. While it is globally convergent (that is, it converges for any choice of the initial guess $\mathbf{x}_0$), it converges only linearly (see Section 3.1.3).
- *Conjugate gradient method*
  The conjugate gradient (CG) method improves (in principle) the convergence of the gradient method by moving along the directions:
  $$\begin{aligned} \mathbf{d}_0 &= -\nabla J(\mathbf{x}_0), \\ \mathbf{d}_k &= -\nabla J(\mathbf{x}_k) + \beta_k \mathbf{d}_{k-1}, \quad k > 0 \end{aligned} \tag{3.4}$$
  where $\beta_k \in \mathbb{R}$ is to be determined by requiring that $\mathbf{d}_{k-1}$ and $\mathbf{d}_k$ be conjugated with respect to the Hessian matrix $\mathbf{H}_k = D^2 J(\mathbf{x}_k)$, that is, $\mathbf{d}_k^\top \mathbf{H}_k \mathbf{d}_{k-1} = 0$. When J is quadratic, $\mathbf{H}_k = \mathbf{H}$ is constant, and in this case β_k is given by

$$\beta_k = \frac{\nabla J(\mathbf{x}_k)^\top \mathbf{H}\mathbf{d}_{k-1}}{\mathbf{d}_{k-1}^\top \mathbf{H}\mathbf{d}_{k-1}}. \tag{3.5}$$

When J is a general nonlinear function, different choices of β_k are used; examples are given by the Fletcher-Reeves and the Polak-Ribière formulae

$$\beta_k^{FR} = \frac{\|\nabla J(\mathbf{x}_k)\|^2}{\|\nabla J(\mathbf{x}_{k-1})\|^2}, \qquad \beta_k^{PR} = \frac{\nabla J(\mathbf{x}_k)^\top (\nabla J(\mathbf{x}_k) - \nabla J(\mathbf{x}_{k-1}))}{\|\nabla J(\mathbf{x}_{k-1})\|^2}. \tag{3.6}$$

If J is quadratic, these choices coincide with (3.5); other choices are available in literature, see for instance [95, 208, 258].

- *Newton and quasi-Newton methods*
The idea of the Newton method is to minimize the quadratic approximation

$$m(\mathbf{d}) = J(\mathbf{x}_k) + \mathbf{d}^\top \nabla J(\mathbf{x}_k) + \frac{1}{2}\mathbf{d}^\top D^2 J(\mathbf{x}_k)\mathbf{d} \tag{3.7}$$

of J in a suitable neighborhood of $\mathbf{x}_k$. If $D^2 J(\mathbf{x}_k)$ is positive definite, the vector $\mathbf{d}_k$ that minimizes $m(\mathbf{d})$ is given by

$$\mathbf{d}_k = -(D^2 J(\mathbf{x}_k))^{-1} \nabla J(\mathbf{x}_k), \tag{3.8}$$

so that the descent direction $\mathbf{d}_k$ is the solution of the linear system

$$D^2 J(\mathbf{x}_k)\mathbf{d}_k = -\nabla J(\mathbf{x}_k). \tag{3.9}$$

Newton methods converge quadratically but only locally (i.e., provided the initial guess is chosen within a suitable neighborhood of the solution), see Section 3.1.3. Moreover, at each iteration they require the computation of the Hessian $\mathbf{H}_k = D^2 J(\mathbf{x}_k)$ and therefore the solution of the linear system (3.9) may become computationally prohibitive when n is very large. For instance, in the case of OCPs for PDEs, methods requiring the evaluation of the Hessian of the cost functional may become computationally demanding for large scale problems.
Quasi-Newton methods employ a suitable approximation $\mathbf{B}_k$ of $D^2 J(\mathbf{x}_k)$ in equation (3.9), and provide reliable surrogates of the Newton method in optimal control problems. The associated search direction is therefore obtained as

$$\mathbf{d}_k = -\mathbf{B}_k^{-1} \nabla J(\mathbf{x}_k). \tag{3.10}$$

that is, as the solution of the linear system

$$\mathbf{B}_k \mathbf{d}_k = -\nabla J(\mathbf{x}_k). \tag{3.11}$$

Matrix $\mathbf{B}_k$ is required to:

  - fulfill the *secant equation*

$$\mathbf{B}_{k+1}(\mathbf{x}_{k+1} - \mathbf{x}_k) = \nabla J(\mathbf{x}_{k+1}) - \nabla J(\mathbf{x}_k);$$

  - be symmetric (since $D^2 J(\mathbf{x})$ is symmetric) and positive definite;
  - provide a good approximation of $D^2 J(\hat{\mathbf{x}})$ along $\mathbf{d}_k$, more precisely,

$$\lim_{k\to\infty} \frac{|(\mathbf{B}_k - D^2 J(\hat{\mathbf{x}}))\mathbf{d}_k|}{|\mathbf{d}_k|} = 0. \tag{3.12}$$

A popular strategy to update $\mathbf{B}_k$ is the BFGS recursive method (from Broyden-Fletcher-Goldfarb-Shanno), according to which

$$\mathbf{B}_{k+1} = \mathbf{B}_k - \frac{\mathbf{B}_k \mathbf{s}_k \mathbf{s}_k^\top \mathbf{B}_k}{\mathbf{s}_k^\top \mathbf{B}_k \mathbf{s}_k} + \frac{\mathbf{y}_k \mathbf{y}_k^\top}{\mathbf{y}_k^\top \mathbf{s}_k}, \tag{3.13}$$

where $\mathbf{s}_k = \mathbf{x}_{k+1} - \mathbf{x}_k$ and $\mathbf{y}_k = \nabla J(\mathbf{x}_{k+1}) - \nabla J(\mathbf{x}_k)$; $\mathbf{B}_0$ can either be an approximation of $D^2 J(\mathbf{x}_0)$, or a scaled identity matrix. The BFGS method is reported in Algorithm 3.2. In practical implementations of quasi-Newton methods, sometimes we directly solve (3.10) by replacing the exact inverse $\mathbf{B}_k^{-1}$ with an approximation $\mathbf{C}_k$ based on the recursive formula

$$\mathbf{C}_{k+1} = (\mathbf{I} - \rho_k \mathbf{s}_k \mathbf{y}_k^\top) \mathbf{C}_k (I - \rho_k \mathbf{y}_k \mathbf{s}_k^\top) + \rho_k \mathbf{s}_k \mathbf{s}_k^\top, \qquad \rho_k = \frac{1}{\mathbf{y}_k^\top \mathbf{s}_k}, \tag{3.14}$$

and setting[1] $\mathbf{d}_k = -\mathbf{C}_k \nabla J(\mathbf{x}_k)$.

Algorithm 3.2 BFGS method

Input: Maximum number of iterations $k_{\max}$, stopping tolerance ε, starting point $\mathbf{x}_0$, initial inverse Hessian approximation $\mathbf{C}_0$

Output: Minimizer $\hat{\mathbf{x}}$

```
set k = 0, err = ε + 1
while err > ε and k < k_max
    compute d_k = −C_k ∇J(x_k);
    compute step length τ_k with a line-search routine (see Section 3.1.2)
    set x_{k+1} = x_k + τ_k d_k
    compute s_k = x_{k+1} − x_k, y_k = ∇J(x_{k+1}) − ∇J(x_k),
    update C_k by computing C_{k+1} through formula (3.14)
    set err = |∇J(x_k)|
        (alternatively, set err = |J(x_{k+1}) − J(x_k)|, or err = |x_{k+1} − x_k|)
    set k = k + 1.
end while
set x̂ = x_{k−1}
```

An example of the iterates generated by steepest descent, CG, Newton and quasi-Newton methods is provided by Example 3.1 and represented in Fig. 3.2, whereas the descent directions leading to these methods are summarized in Table 3.1.

steepest descent	$\mathbf{d}_k = -\nabla J(\mathbf{x}_k)$
conjugate gradient	$\mathbf{d}_k = -\nabla J(\mathbf{x}_k) + \beta_k \mathbf{d}_{k-1}$
Newton	$\mathbf{d}_k = -(D^2 J(\mathbf{x}_k))^{-1} \nabla J(\mathbf{x}_k)$
quasi-Newton	$\mathbf{d}_k = -\mathbf{B}_k^{-1} \nabla J(\mathbf{x}_k)$

Table 3.1 Choice of descent directions

We remark that although the steepest descent (or gradient) directions (3.3) verify the conditions (3.2) for any $k \geq 0$, the choices (3.8) and (3.10) yield descent directions only when $D^2 J(\mathbf{x}_k)$ and $\mathbf{B}_k$, respectively, are positive definite matrices. The directions (3.4) employed in the CG method are descent directions under suitable conditions, as detailed in Section 3.1.3.

[1] At each iteration, this approach requires to store both $\mathbf{C}_{k-1}$ and $\mathbf{C}_k$, a task which could be troublesome for large scale optimization problems. The *limited-memory BFGS method* circumvents this difficulty, by storing, instead of a fully dense $n \times n$ approximation matrix, only a few vectors of length n; see [208, Section 7.2].

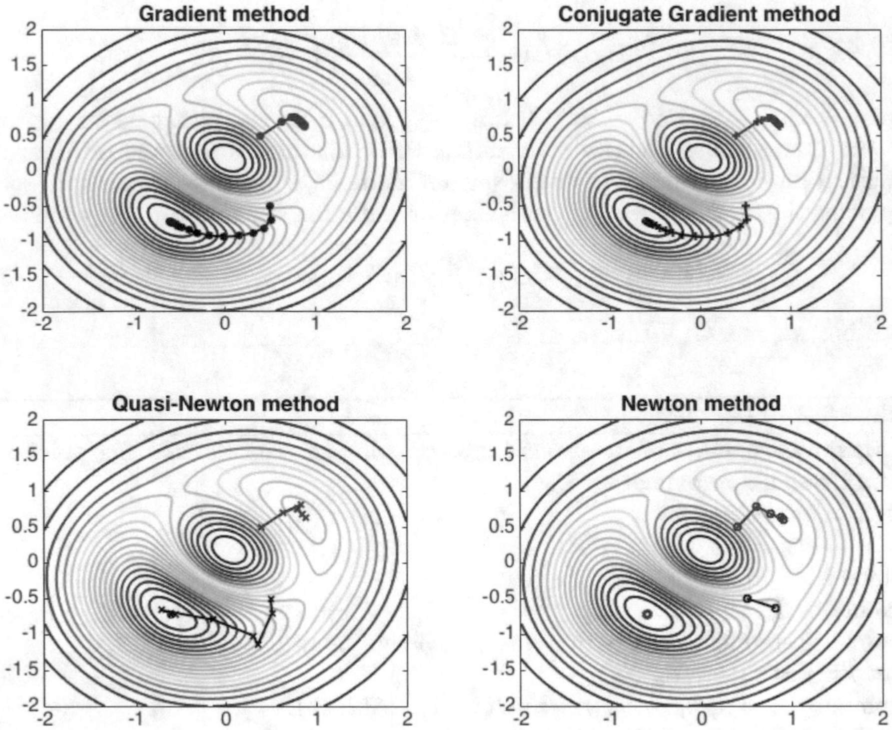

Fig. 3.2 Minimization of the function (3.15), whose global minimization point is $\hat{\mathbf{x}} \approx (-0.5954, -0.7161)$. Together with the contour plot of f, the iterations of the steepest descent and the CG method (top), quasi-Newton and Newton methods (bottom) are reported, obtained by taking $\mathbf{x}_0 = (0.5, -0.5)$ (in black) and $\mathbf{x}_0 = (0.4, 0.5)$ (in red)

Example 3.1. The function

$$f(\mathbf{x}) = \frac{2}{5} - \frac{1}{10}(5x_1^2 + 5x_2^2 + 3x_1x_2 - x_1 - 2x_2)e^{-(x_1^2+x_2^2)} \tag{3.15}$$

has a global minimizer $\hat{\mathbf{x}} \approx (-0.5954, -0.7161)$. For the approximation of $\hat{\mathbf{x}}$, we exploit a descent method. First, we set $\mathbf{x}_0 = (0.5, -0.5)$, a tolerance $\varepsilon = 10^{-5}$ and a maximum number of iterations $k_{max} = 200$. We compare the steepest descent method, the CG method (with Polak-Ribière formula for evaluating β_k at each step), the quasi-Newton method and the Newton method. The first three methods converge to the global minimizer, requiring 11, 29 and 36 iterations, respectively. In the case of the Newton method, after 200 iterations the stopping test would not be fulfilled, and the Newton method does not converge (recall that the Newton method is only locally convergent). Choosing a different initial guess $\mathbf{x}_0 = (0.4, 0.5)$, again with $\varepsilon = 10^{-5}$ and $k_{max} = 200$, all the methods converge to a local minimizer $\mathbf{x} = (0.8873, 0.6395)$. This example highlights several features: different methods can have a different convergence behavior, some converge, other may not converge for the same initial guess. When they converge, convergence rates can be different. Finally, the limit may sometimes be the global minimizer, most often however it coincides with a local minimizer. Contour plots of the function (3.15) and the sequences generated by descent methods with the two initial guesses above are reported in Fig. 3.2. •

3.1.2 Step Length Evaluation and Inexact Line-Search

Once the direction $\mathbf{d}_k$ has been chosen, we need to compute a step length τ_k. The most natural choice is to seek the minimum of the restriction of J along $\mathbf{d}_k$, that is

$$\tau_k = \arg\min_{\tau \in \mathbb{R}} J(\mathbf{x}_k + \tau \mathbf{d}_k). \tag{3.16}$$

When the cost function J is quadratic, i.e. of the form

$$J(\mathbf{x}) = \frac{1}{2}\mathbf{x}^\top \mathbf{Q}\mathbf{x} - \mathbf{b}^\top \mathbf{x}, \tag{3.17}$$

for a suitable matrix $\mathbf{Q}$ and a vector $\mathbf{b}$, one can easily perform an exact line-search: the solution of problem (3.16) can be found by requiring that the derivative of $J(\mathbf{x}_k + \tau \mathbf{d}_k)$ with respect to τ vanishes. In this case, we find

$$\tau_k = \frac{\mathbf{g}_k^\top \mathbf{d}_k}{\mathbf{d}_k^\top \mathbf{Q}\mathbf{d}_k}, \qquad \mathbf{g}_k = -\nabla J(\mathbf{x}_k) = \mathbf{b} - \mathbf{Q}\mathbf{x}_k. \tag{3.18}$$

If J is not a quadratic function, solving (3.16) exactly shall require an iterative method and will be therefore rather involved. A popular alternative is to rely on an *inexact line-search* procedure, based for instance on the following *Wolfe conditions*: a step size $\tau_k > 0$ is accepted if

$$\begin{aligned} J(\mathbf{x}_k + \tau_k \mathbf{d}_k) &\le J(\mathbf{x}_k) + \sigma \tau_k \mathbf{d}_k^\top \nabla J(\mathbf{x}_k), \\ \mathbf{d}_k^\top \nabla J(\mathbf{x}_k + \tau_k \mathbf{d}_k) &\ge \delta \mathbf{d}_k^\top \nabla J(\mathbf{x}_k) \end{aligned} \tag{3.19}$$

where $0 < \sigma < \delta < 1$ are given constants. The first condition, also known as *Armijo rule*, assures the cost variations to be proportional to both τ_k and the directional derivative $\mathbf{d}_k^\top \nabla J(\mathbf{x}_k)$. The second condition ensures instead that the directional derivative of J along $\mathbf{d}_k$ at $\mathbf{x}_k + \tau_k \mathbf{d}_k$ is larger than δ times the directional derivative of J along $\mathbf{d}_k$ at $\mathbf{x}_k$. In other words, $\mathbf{x}_k + \tau_k \mathbf{d}_k$ is a good candidate if at that point the cost function decreases less than at $\mathbf{x}_k$: this also ensures that the step length is not too small when starting from a point where J has a very small directional derivative. More restrictive conditions are the so-called *strong Wolfe conditions*, obtained by replacing $(3.19)_2$ by the following one (see [208, Lemma 3.1]):

$$|\mathbf{d}_k^\top \nabla J(\mathbf{x}_k + \tau_k \mathbf{d}_k)| \ge -\delta \mathbf{d}_k^\top \nabla J(\mathbf{x}_k). \tag{3.20}$$

The *backtracking* algorithm provides a simple strategy to find a step length τ_k fulfilling the Wolfe conditions (3.19): an initial value $\tau = 1$ is set and then progressively reduced every time by a given factor ρ (typically, $\rho \in [1/10, 1/2]$) until the first condition of (3.19) is fulfilled. The second condition in (3.19) is usually not checked because the *backtracking* algorithm ensures that the resulting steps do not become too small. A simple implementation is provided in Algorithm 3.3. Often, a quadratic or cubic interpolation is used to model the behavior of J along the direction $\mathbf{d}_k$, so that $\mathbf{x}_{k+1}$ results as the minimization point of the interpolating polynomial. This is the so-called *quadratic* or *cubic line-search* [208].

3.1.3 Convergence of Descent Methods

In this section we recall some convergence results for the descent methods introduced so far. For a sequence $\{\mathbf{x}_k\}_{k \ge 0}$ converging to $\hat{\mathbf{x}}$, that is

$$\lim_{k \to \infty} |\mathbf{x}_k - \hat{\mathbf{x}}| = 0,$$

Algorithm 3.3 Backtracking algorithm

Input: descent direction $\mathbf{d}_k$, current iterate $\mathbf{x}_k$, $\sigma \in (0,1)$, $\rho \in [1/10, 1/2)$
Output: step length τ_k

```
set τ = 1, k = 0
while J(x_k + τ d_k) > J(x_k) + σ τ d_k^T ∇J(x_k)
    set τ = ρ τ
end while
set τ_k = τ.
```

we say that its *rate of convergence* is equal to p (≥ 1) if there is a positive constant C such that

$$|\mathbf{x}_{k+1} - \hat{\mathbf{x}}| \leq C |\mathbf{x}_k - \hat{\mathbf{x}}|^p$$

for all k larger than some $\bar{k} \geq 0$. When $p = 1$ the sequence is linearly convergent; in that case the constant C is required to be smaller than 1.

• Let us consider the steepest descent method (3.1)-(3.3); as we have already remarked, the choice (3.3) always yields descent directions. If $J \in C^2(\mathbb{R}^n)$ is bounded from below and the step lengths satisfy the Wolfe conditions (3.19), the steepest descent method is globally convergent. In the case of a quadratic cost function (3.17) and exact line searches, the method is linearly convergent[2] (see for instance [184]).

• Let us now turn to the CG method (3.1)-(3.4) and consider a cost function $J \in C^1(\mathbb{R}^n)$, with Lipschitz continuous gradient. By assuming that the step lengths τ_k fulfill the strong Wolfe conditions $(3.19)_1$-(3.20) with $0 < \sigma < \delta < 1/2$ and the initial guess $\mathbf{x}_0$ is such that the set $A = \{\mathbf{x} : J(\mathbf{x}) \leq J(\mathbf{x}_0)\}$ is bounded, then the CG method (3.1)-(3.4) with the Fletcher-Reeves choice for $\beta_k = \beta_k^{FR}$ is globally convergent to a stationary point for J (see for instance [208, 258] for the proof). Note that instead of $\delta < 1$ as in $(3.19)_1$-(3.20), here we have required that $\delta < 1/2$. It is possible to show (see for instance [208]) that under this condition we have

$$\nabla J(\mathbf{x}_k)^\top \mathbf{d}_k = -|\nabla J(\mathbf{x}_k)|^2 + \beta_k^{FR} \nabla J(\mathbf{x}_k)^\top \mathbf{d}_{k-1} < 0.$$

Thus, any line search procedure fulfilling the strong Wolfe conditions (with $\delta < 1/2$) ensures that the directions $\{\mathbf{d}_k\}_{k\geq 1}$ are descent directions[3] for J. A similar result holds for the CG method with the Polak-Ribière choice $\beta_k = \beta_k^{PR}$, provided that we consider $\beta_k^{PR+} = \max\{\beta_k^{PR}, 0\}$ instead of β_k^{PR} and the step lengths τ_k according to a modified version of the strong Wolfe conditions. See for instance [207, 208, 258] for a more detailed analysis.

• Let us now consider the Newton method (3.1)-(3.8) with step lengths fulfilling the Wolfe conditions (3.19) and $J \in C^2(\mathbb{R}^n)$. If the Hessian matrices $\mathbf{H}_k = D^2 J(\mathbf{x}_k)$ are positive definite[4] for any $k \geq 0$ and their condition numbers are uniformly bounded from above,

$$\exists\, M > 0 \text{ such that } \kappa(\mathbf{H}_k) \leq M \qquad \forall k \geq 0, \tag{3.21}$$

[2] This result agrees with the classical convergence result for the gradient method when regarded as iterative method for the solution of linear systems; see for instance [222, Theorem 4.10]. In fact, solving a free optimization problem with a quadratic cost function by means of the gradient method is equivalent to the solution of the linear system arising from the first-order necessary optimality condition. Nevertheless, the convergence of the steepest descent method is the same for any nonlinear objective function; see, e.g., [208].

[3] We recall that the CG method in exact arithmetic will return the exact solution in at most n iterations in the case of a quadratic cost function (3.17), exact line searches yielding (3.18) and the choice (3.5) for the parameter β_k (thus, when regarded as iterative method for the solution of linear systems). Choosing conjugate directions thus improves substantially the convergence properties of the gradient (or steepest descent) method. See for instance [222] for a comprehensive analysis.

[4] When this condition is not met, the Hessian matrix could be suitably modified, see, e.g., [223, Sect. 7.5].

then the sequence $\{\mathbf{x}_k\}_{k\geq 0}$ generated according to (3.1)-(3.8) converges to a stationary point $\hat{\mathbf{x}}$ for J. Moreover, by considering step lengths $\tau_k = 1$ for all k larger than some $\bar{k} \geq 0$, that is, when we are sufficiently close to $\hat{\mathbf{x}}$, the Newton method converges quadratically (i.e. with rate $p = 2$); see for instance [208, Theorem 3.2].

• Finally, in the case of quasi-Newton methods (3.1)-(3.10) we can prove that if $J \in C^2(\mathbb{R}^n)$ is bounded from below and the matrices $\mathbf{B}_k$ are positive definite for any $k \geq 0$ and fulfill the relation (3.21) where $\mathbf{H}_k$ is replaced by $\mathbf{B}_k$, then the BFGS method converges to a minimizer for J superlinearly (i.e. with rate $p \in (1, 2)$); see for instance [80, 208]. We remark that the BFGS matrices (3.13) are symmetric and positive definite provided that $\mathbf{y}_k^\top \mathbf{s}_k > 0$, and this latter condition is guaranteed if the step lengths fulfill the Wolfe conditions (3.19).

As a general principle, achieving a rapid convergence in optimization algorithms is often in contrast with the need of ensuring a global convergence. For instance, the steepest descent method is globally convergent but is usually quite slow, whereas the Newton method converges rapidly provided $\mathbf{x}_0$ is close enough to a local minimizer.

3.2 Free optimization by trust region methods

The *trust region* method exploits an approximate quadratic model of the cost function to generate steps towards the minimization point, as done in (3.7). Precisely, we select a ball $B_{\delta_k}(\mathbf{x}_k)$ inside which we *trust* the quadratic model to be a suitable representation of the cost function. Then, we set $\mathbf{x}_{k+1} = \mathbf{x}_k + \mathbf{p}_k$, where the step $\mathbf{p}_k$ is determined by seeking a minimizer of the quadratic model in the trust region $B_{\delta_k}(\mathbf{x}_k)$, so that the direction and the step length can be chosen simultaneously. The size δ_k of the trust region is crucial for the convergence of the algorithm. In case a step is considered not acceptable, we reduce the size of the trust region and seek a new minimizer; instead, if the quadratic model accurately predicts the behavior of the cost function, the size of the trust region may be increased to enable longer steps.

More precisely, at each iteration k, we start from a *trusted* value $\delta_k > 0$ and build a quadratic model

$$m_k(\mathbf{p}) = J(\mathbf{x}_k) + \mathbf{p}^\top \nabla J(\mathbf{x}_k) + \frac{1}{2}\mathbf{p}^\top \mathbf{H}_k \mathbf{p}$$

for the cost function J; here $\mathbf{H}_k$ may denote either the Hessian $D^2 J(\mathbf{x}_k)$ of J at $\mathbf{x}_k$, or a suitable symmetric approximation (as in quasi Newton methods). At each step, we look for a solution $\mathbf{p}_k$ of the subproblem

$$m_k(\mathbf{p}) \to \min \qquad \text{subject to} \quad |\mathbf{p}| \leq \delta_k. \tag{3.22}$$

If the solution $\mathbf{p}_k$, obtained with the quadratic model, is feasible also for the original minimization problem (see below), we define $\mathbf{x}_{k+1} = \mathbf{x}_k + \mathbf{p}_k$, then we update the quadratic model, possibly modify the trust region, and iterate in the same way. Otherwise, we reduce the size of the trust region and we solve again (3.22).

When $\mathbf{H}_k$ is positive definite and $|\mathbf{H}_k^{-1}\nabla J(\mathbf{x}_k)| < \delta_k$, the solution of (3.22) lies in the interior of the trust region, thus, it is a critical point of m_k, that is

$$\mathbf{p}_k = -\mathbf{H}_k^{-1}\nabla J(\mathbf{x}_k).$$

Otherwise, we solve (see Fig. 3.3)

$$m_k(\mathbf{p}) \to \min \qquad \text{subject to} \quad |\mathbf{p}| = \delta_k. \tag{3.23}$$

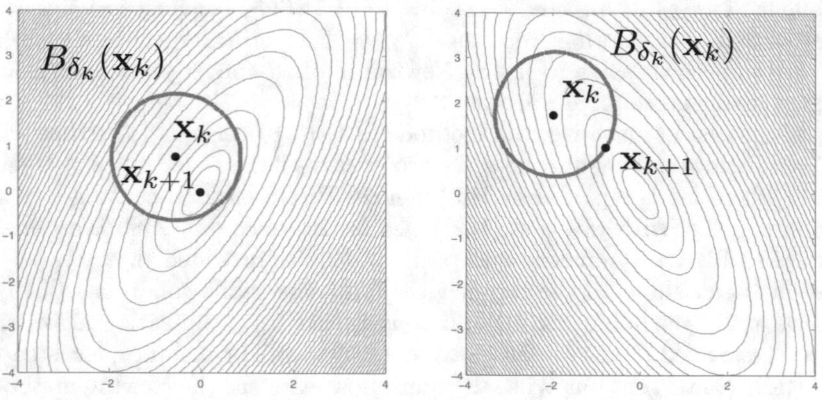

Fig. 3.3 A quadratic model with positive definite Hessian whose minimum lies inside (left) or outside (right) the trust region $B_{\delta_k}(\mathbf{x}_k)$.

For this constrained minimization problem, we seek a critical point of the Lagrangian function

$$\mathcal{L}_k(\mathbf{p}, \lambda) = m_k(\mathbf{p}) + \frac{\lambda}{2}(\delta_k^2 - |\mathbf{p}|^2).$$

Precisely, we look for a vector $\mathbf{p}_k$ and a scalar multiplier[5] $\lambda^k < 0$ such that

$$\begin{cases} (\mathbf{H}_k - \lambda_k \mathbf{I})\mathbf{p}_k = -\nabla J(\mathbf{x}_k), \\ \delta_k^2 - |\mathbf{p}_k|^2 = 0, \\ \mathbf{H}_k - \lambda_k \mathbf{I} \text{ is positive semidefinite.} \end{cases} \tag{3.24}$$

We still need to ensure that the solution $\mathbf{p}_k$ to the subproblem (3.22) is *feasible* for the original minimization problem, and that the size δ_k of the trust region is well-chosen. The criterion usually employed to assess the quality of $\mathbf{p}_k$ is based on the agreement between the variation of J and that of the quadratic model m_k, evaluated at the points $\mathbf{x}_k$ and $\mathbf{x}_k + \mathbf{p}_k$. More precisely, we evaluate the ratio between the actual reduction and that predicted by the quadratic model,

$$\rho_k = \frac{J(\mathbf{x}_k) - J(\mathbf{x}_k + \mathbf{p}_k)}{m_k(\mathbf{0}) - m_k(\mathbf{p}_k)}: \tag{3.25}$$

- if ρ_k is close to 1, there is good agreement between m_k and J, thus we can accept $\mathbf{p}_k$ and define $\mathbf{x}_{k+1} = \mathbf{x}_k + \mathbf{p}_k$. If the minimizer lies on the boundary of the trust region, we expand the trust region for the next iteration;
- if $\rho_k \approx 0$ or $\rho_k < 0$, $\mathbf{p}_k$ is rejected, we shrink the trust region by reducing its radius ρ_k and seek a new $\mathbf{p}_k$ by solving (3.22) again;
- finally, if $\rho_k > 0$ is significantly smaller than 1, we keep the same trust region.

An implementation of the trust region method is reported in Algorithm 3.4. $\delta > 0$ denotes the maximum radius, δ_0 $(0 < \delta_0 < \delta)$ the initial radius, η_1, η_2 $(0 < \eta_1 < \eta_2 < 1)$ two real parameters for the update of the trust region, and μ $(0 \leq \mu < \eta_1)$ a real parameter for assessing the feasibility of the solution; see [208, Section 4.2] for possible choice of δ, η_1, η_2 and μ.

[5] The Lagrange multiplier λ_k is negative since the quadratic model $m_k(\mathbf{p})$ is a convex function. In this case, the size of its level sets increases along the direction of its gradient, and the minimization point is the first point where a level set is tangent to the spherical surface representing the constraint. At that point, the gradient of

---

Algorithm 3.4 Trust region method

Input: initial guess $\mathbf{x}_0$, $\delta > 0$, $0 < \eta_1 < \eta_2 < 1$, $0 \le \mu < \eta_1$
Output: Minimizer $\hat{\mathbf{x}}$

1: **for** $k = 0, 1, 2, \ldots$
2:      compute $J(\mathbf{x}_k)$, $\nabla J(\mathbf{x}_k)$, $\mathbf{H}_k$ and solve problem (3.22)
3:      compute ρ_k from (3.25)
4:      **if** $\rho_k > \mu$
5:          set $\mathbf{x}_{k+1} = \mathbf{x}_k + \mathbf{p}_k$
6:      **else**
7:          set $\mathbf{x}_{k+1} = \mathbf{x}_k$
8:      **end if**
9:      **if** $\rho_k < \eta_1$
10:        set $\delta_{k+1} = \gamma_1 \delta_k$
11:      **else**
12:          **if** $\eta_1 \le \rho_k \le \eta_2$
13:             set $\delta_{k+1} = \delta_k$
14:          **else**
15:             **if** $\rho_k > \eta_2$ and $|\mathbf{p}_k| = \delta_k$
16:                set $\delta_{k+1} = \min(\gamma_2 \delta_k, \delta)$
17:             **end if**
18:          **end if**
19:      **end if**
20: **end for**
21: set $\hat{\mathbf{x}} = \mathbf{x}_k$

A final remark: if $J \in C^2(\mathbb{R}^n)$ is bounded from below, the norms $\|\mathbf{H}_k\|$ are uniformly bounded for any $k \ge 0$, and the chosen steps yield a sufficient reduction in the quadratic model m_k, then the sequence $\{\mathbf{x}_k\}_{k \ge 0}$ obtained through the trust region algorithm converges to a critical point of J. See for instance [247, 70] for the convergence analysis.

3.3 Constrained Optimization by Projection Methods

Let us now focus on the solution of constrained optimization problems, for which a huge variety of methods can be employed, depending on the type of constraints and/or the nature of the cost function to be minimized. Here we describe the most relevant techniques which prove to be useful in the design of numerical algorithms to solve OCPs for PDEs. We refer to [36, 208, 258] for further techniques and details.

A first approach consists in adapting a descent method to the constrained case

$$\begin{cases} J(\mathbf{x}) \to \min \\ \quad \text{subject to} \\ \quad\quad \mathbf{x} \in K. \end{cases} \tag{3.26}$$

Since we minimize over a closed set $K \subset \mathbb{R}^n$, there is no guarantee that the iterates $\{\mathbf{x}_k\}_{k \ge 0}$ obtained by formula (3.1) belong to K. To enforce the condition $\mathbf{x}_k \in K$, after evaluating each

---

the constraint is orthogonal to the spherical surface, pointing outward from it (see Fig. 3.3), so that the two gradients are parallel but point in opposite directions.

iterate we apply the projection operator

$$Pr_K : \mathbb{R}^n \to K, \qquad \mathbf{x} \mapsto Pr_K(\mathbf{x}) : \quad Pr_K(\mathbf{x}) = \arg\min_{\mathbf{y}\in K} |\mathbf{x}-\mathbf{y}|^2.$$

This yields e.g. to the so-called *projected steepest descent* method if the descent direction is given by (3.3),

$$\mathbf{x}_{k+1} = Pr_K(\mathbf{x}_k - \tau_k \nabla J(\mathbf{x}_k)), \qquad k \geq 0.$$

Note that in this case the search for the step length τ_k is slightly more tricky than in the unconstrained case, since the function $\phi(\tau) = J(Pr_K(\mathbf{x}_k - \tau\nabla J(\mathbf{x}_k))$ is not necessarily differentiable. Evaluating the projection operator for an arbitrary set K is in general rather involved. However, there are situations where the projection can be easily obtained, as the ones that follow.

Example 3.2. For instance, if

$$K = \{\mathbf{x} :\ x_i \geq a_i,\ i = 1,\dots,n\} \quad \text{or} \quad K = \{\mathbf{x} :\ x_j \leq b_j,\ j = 1,\dots,n\}$$

that is, K is characterized by a set of inequality constraints on the $\mathbf{x}$ components, the vector $\mathbf{y} = Pr_K(\mathbf{x})$ reads

$$y_i = \max(x_i, a_i) \quad \text{or} \quad y_i = \min(x_i, b_i), \quad i = 1,\dots,n.$$

In the case of *box constraints* K reads

$$K = \{\mathbf{x} :\ a_j \leq x_j \leq b_j,\ j = 1,\dots,n\}. \tag{3.27}$$

Then the components of $\mathbf{y} = Pr_K(\mathbf{x})$ are

$$y_j = \min\{b_j, \max\{a_j, x_j\}\}, \qquad j = 1,\dots,n. \tag{3.28}$$

With an abuse of notation, we can more easily express (3.27) as $K = \{\mathbf{x} \in \mathbb{R}^n :\ \mathbf{a} \leq \mathbf{x} \leq \mathbf{b}\}$. •

Example 3.3. Let $c \in \mathbb{R}$ be a given constant and consider the set

$$K = \{\mathbf{x} \in \mathbb{R}^n \ :\ \frac{1}{n}\sum_{j=1}^{n} x_j = c\} \subset \mathbb{R}^n.$$

In this case $\mathbf{y} = Pr_K(\mathbf{x})$ is such that $(\mathbf{y}-\mathbf{x})\cdot(\mathbf{s}-\mathbf{y}) \geq 0\ \forall \mathbf{s} \in K$, and its components are given by

$$y_j = x_j - \frac{1}{n}\sum_{j=1}^{n}(x_j - c), \qquad j = 1,\dots,n. \tag{3.29}$$

•

Example 3.4. If $K = \{\mathbf{x} \in \mathbb{R}^n \ :\ \mathbf{x} \in \overline{B(\mathbf{x}_0, R)}\}$ where $B(\mathbf{x}_0, R)$ is the ball of center $\mathbf{x}_0$ and radius R, $\mathbf{y} = Pr_K(\mathbf{x})$ is given by

$$\mathbf{y} = \begin{cases} \mathbf{x}, & \mathbf{x} \in K \\ \mathbf{x}_0 + R\dfrac{\mathbf{x}-\mathbf{x}_0}{|\mathbf{x}-\mathbf{x}_0|}, & \mathbf{x} \notin K. \end{cases}$$ •

Trust-region methods can also be extended to problems involving box constraints. The main difference with the unconstrained case is that, instead of problem (3.22), we now seek the solution of the following subproblem at each step:

$$\begin{cases} m_k(\mathbf{p}) \ \to\ \min \\ \text{subject to} \\ |\mathbf{p}| \leq \delta_k, \quad \mathbf{a} \leq \mathbf{x_k} + \mathbf{p} \leq \mathbf{b} \end{cases}$$

where the inequalities for $\mathbf{x}_k + \mathbf{p}$ have to be intended componentwise.

The following sections are devoted instead to more general numerical methods for constrained optimization (see Fig. 3.4). The table below is a preview of the different situations and associated numerical methods that will be addressed.

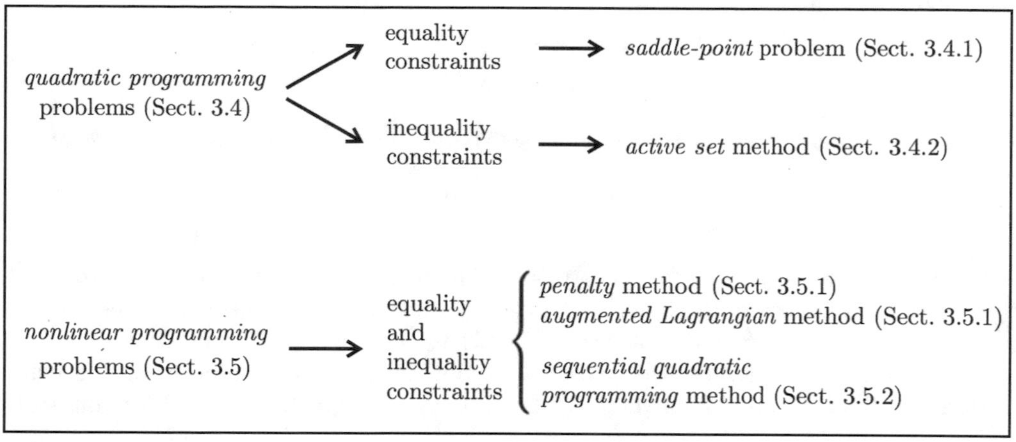

Fig. 3.4 A preview of the methods presented in Sections 3.4 and 3.5

3.4 Constrained Optimization for Quadratic Programming Problems

Consider a *quadratic programming* (QP) problem of the form:

$$\begin{cases} J(\mathbf{x}) = \dfrac{1}{2}\mathbf{x}^\top \mathbf{G}\mathbf{x} - \mathbf{x}^\top \mathbf{c} \ \rightarrow \ \min \\ \qquad\qquad \text{subject to} \\ \mathbf{a}_i^\top \mathbf{x} = b_i, \qquad i \in E \\ \mathbf{a}_i^\top \mathbf{x} \geq b_i, \qquad i \in I, \end{cases} \tag{3.30}$$

where $\mathbf{G} \in \mathbb{R}^{n\times n}$ is a symmetric positive (semi)definite matrix, E and I are finite sets of indices and $\mathbf{c}$, $\{\mathbf{a}_i\}_i$ are vectors in $\mathbb{R}^n$. Apart from its own interest, this kind of problems arises in many algorithms for nonlinear programming problems, such as in the case of *sequential quadratic programming* (see Sect. 3.5.2), where a sequence of quadratic subproblems need to be solved.

3.4.1 Equality Constraints: a Saddle-Point Problem

We start by considering the case of equality constraints:

$$\begin{cases} J(\mathbf{x}) = \dfrac{1}{2}\mathbf{x}^\top \mathbf{G}\mathbf{x} - \mathbf{x}^\top \mathbf{c} \ \rightarrow \ \min \\ \qquad\qquad \text{subject to} \\ \qquad\qquad \mathbf{A}\mathbf{x} = \mathbf{b}, \end{cases} \tag{3.31}$$

where $\mathbf{A} \in \mathbb{R}^{m\times n}$ (with $m < n$) is a given matrix.

By applying the Lagrange multiplier method described in Sect. 2.3.1 with

$$\mathcal{L}(\mathbf{x}, \boldsymbol{\lambda}) = J(\mathbf{x}) - \boldsymbol{\lambda}^\top(\mathbf{b} - \mathbf{A}\mathbf{x}),$$

the first-order necessary conditions state that, if $\hat{\mathbf{x}}$ is a solution of (3.31), there exists a vector of Lagrange multipliers $\hat{\boldsymbol{\lambda}}$ such that

$$\begin{pmatrix} \mathbf{G} & \mathbf{A}^\top \\ \mathbf{A} & 0 \end{pmatrix} \begin{pmatrix} \hat{\mathbf{x}} \\ \hat{\boldsymbol{\lambda}} \end{pmatrix} = \begin{pmatrix} \mathbf{c} \\ \mathbf{b} \end{pmatrix}. \tag{3.32}$$

This is a *saddle point system*; thanks to a general result of linear algebra, the *saddle point matrix*

$$\mathcal{K} = \begin{pmatrix} \mathbf{G} & \mathbf{A}^\top \\ \mathbf{A} & 0 \end{pmatrix} \tag{3.33}$$

is nonsingular if and only if $\mathbf{G} = \mathbf{G}^T$ is positive semidefinite, $\mathbf{A}$ has full row rank and $ker(\mathbf{G}) \cap ker(\mathbf{A}) = \{0\}$. In this case, provided that $\mathbf{G}$ is (symmetric and) positive definite, the matrix $\mathbf{S} = \mathbf{A}\mathbf{G}^{-1}\mathbf{A}^\top$ (named *Schur complement* of $\mathbf{G}$) is (symmetric and) positive definite, too. Under the assumptions above, the vector $\hat{\mathbf{x}}$ is the unique global solution of the equality-constrained QP (3.31). See Appendix B.1.3 for further properties of saddle point systems.

Solving a QP with only equality constraints is thus equivalent to solve the linear system (3.32). The saddle-point matrix $\mathcal{K}$ is in general indefinite (with both positive and negative eigenvalues). Depending on the size and the ill-conditioning of this matrix, the system (3.33) can be solved according to different strategies: either by direct LU block factorization methods or by iterative Krylov methods, such as the generalized minimal residual method (GMRES). In the former case, some blocks will involve the Schur complement matrix $\mathbf{S}$; this requires using iterative procedures with suitable preconditioners; see e.g. [30] or [4, 226]. Note also that, in order to preserve the symmetry, a symmetric indefinite factorization under the form $\mathbf{P}^T\mathcal{K}\mathbf{P} = \mathbf{L}\mathbf{B}\mathbf{L}^T$, where $\mathbf{P}$ is a permutation matrix, is often preferable with respect to the classical Gaussian elimination method with partial pivoting. In the latter case, the global system needs to be preconditioned, however the choice of the preconditioner depends on the specific problem at hand; see Sect. 6.6.

3.4.2 Inequality Constraints: Active Set Method

In this section we describe the *active set* method for the solution of quadratic problems under the form (3.30). Of course, this is not the only available technique to solve such a problem. For instance, a well-known alternative is given by the *interior-point* method. However, the *active set* method is more intuitive and allows to take advantage of the techniques for equality-constrained QPs already discussed. We start by observing that, if the active set $I(\hat{\mathbf{x}})$ defined as

$$I(\hat{\mathbf{x}}) = \{j \in E \cup I \ : \ \mathbf{a}_j^\top \hat{\mathbf{x}} = b_j\}$$

were known, (3.30) would reduce to an equality-constrained QP,

$$\begin{cases} J(\mathbf{x}) = \dfrac{1}{2}\mathbf{x}^\top \mathbf{G}\mathbf{x} + \mathbf{x}^\top \mathbf{c} \ \rightarrow \ \min \\ \qquad\qquad \text{subject to} \\ \mathbf{a}_i^\top \mathbf{x} = b_i, \qquad i \in I(\hat{\mathbf{x}}), \end{cases}$$

as those addressed in the previous section. Unfortunately, the determination of the active set represents the main challenge of inequality-constrained QPs.

The goal of an active set method is to construct a sequence of active sets, by starting from an initial guess and using repeatedly the information provided by gradients and Lagrange multipliers to adding/dropping indices to the current active set, until some optimality condition is achieved. The set W_k of the indices corresponding to all the active constraints at step k is called *working set*. In particular, we require that the gradients of the constraints in the working set are linearly independent.

To turn this into practice, given $\mathbf{x}_k$ and W_k, first of all we check if $\mathbf{x}_k$ minimizes J in the subspace defined by the current working set. If not, we solve an equality-constrained QP problem, where we consider just the constraints corresponding to the working set, which are all treated as equalities:

$$\begin{cases} q_k(\mathbf{p}) = \dfrac{1}{2}\mathbf{p}^\top \mathbf{G}\mathbf{p} + \mathbf{g}_k^\top \mathbf{p} \ \rightarrow \ \min \\ \qquad\qquad \text{subject to} \\ \mathbf{a}_i^\top \mathbf{p} = 0, \qquad i \in W_k, \end{cases} \tag{3.34}$$

where $\mathbf{p} = \mathbf{x} - \mathbf{x}_k$, $\mathbf{g}_k = \mathbf{G}\mathbf{x}_k + \mathbf{c}$. We remark that

$$J(\mathbf{x}) = J(\mathbf{x}_k + \mathbf{p}) = \frac{1}{2}\mathbf{p}^\top \mathbf{G}\mathbf{p} + \mathbf{g}_k^\top \mathbf{p} + \mathbf{h}_k = q_k(\mathbf{p}) + \mathbf{h}_k.$$

Note that $\mathbf{h}_k = \frac{1}{2}\mathbf{x}_k^\top \mathbf{G}\mathbf{x}_k + \mathbf{c}^\top \mathbf{x}_k$ is independent of $\mathbf{p}$, and it can be dropped from the cost function in (3.34). Let $\mathbf{p}_k$ be the solution of this QP subproblem, which can be computed following the procedure of Section 3.4.1, as $\mathbf{G}$ is positive definite.

Moreover, since $\mathbf{a}_i^\top \mathbf{p}_k = 0$ for any $i \in W_k$, we have

$$\mathbf{a}_i^\top(\mathbf{x}_k + \alpha \mathbf{p}_k) = \mathbf{a}_i^\top \mathbf{x}_k = b_i, \qquad \forall\, \alpha \in \mathbb{R}.$$

Since the constraints in the working set W_k are fulfilled at $\mathbf{x}_k$, they are also fulfilled at $\mathbf{x}_k + \alpha\mathbf{p}_k$, for any $\alpha \in \mathbb{R}$: thus, $\mathbf{p}_k$ is a feasible direction to move along.

We still have to select a step length along the direction $\mathbf{p}_k$. Usually, we set $\alpha = 1$ if $\mathbf{x}_k + \mathbf{p}_k$ is still feasible with respect to any constraint, otherwise we set

$$\mathbf{x}_{k+1} = \mathbf{x}_k + \alpha_k \mathbf{p}_k,$$

where the step length $\alpha_k \in [0, 1]$ is selected as the largest value for which every constraint is fulfilled. As already seen, the constraints $i \in W_k$ are satisfied regardless of the choice of α_k, so that we can derive an explicit expression of the step size by considering only the constraints $i \notin W_k$. We remark that:

- if $\mathbf{a}_i^\top \mathbf{p}_k \geq 0$ for some $i \notin W_k$, we have

$$\mathbf{a}_i^\top(\mathbf{x}_k + \alpha_k \mathbf{p}_k) \geq \mathbf{a}_i^\top \mathbf{x}_k \geq \mathbf{b}_i \qquad \forall \alpha_k \geq 0.$$

  In this case, the i-th constraint is fulfilled for any nonnegative step size;
- if instead $\mathbf{a}_i^\top \mathbf{p}_k < 0$ for some $i \notin W_k$, we have that $\mathbf{a}_i^\top(\mathbf{x}_k + \alpha_k \mathbf{p}_k) \geq \mathbf{b}_i$ only if

$$0 \leq \alpha_k \leq \frac{\mathbf{b}_i - \mathbf{a}_i^\top \mathbf{x}_k}{\mathbf{a}_i^\top \mathbf{p}_k}.$$

Thus, we can select

$$\alpha_k = \min\left(1, \min_{i \notin W_k, \mathbf{a}_i^\top \mathbf{p}_k < 0} \frac{b_i - \mathbf{a}_i^\top \mathbf{x}_k}{\mathbf{a}_i^\top \mathbf{p}_k}\right). \tag{3.35}$$

The constraint i allowing to reach the minimum in the formula (3.35) is called *blocking constraint*: if $\alpha_k < 1$, the step along the direction $\mathbf{p}_k$ is blocked by some constraint which does not belong to W_k, a new working set W_{k+1} is formed by adding one of the blocking constraints to W_k. Instead, if $\alpha_k = 1$ and any new constraint is active at $\mathbf{x}_k + \alpha\mathbf{p}_k$, there are no blocking constraints on the current iteration, so that $\mathbf{x}_{k+1} = \mathbf{x}_k + \mathbf{p}_k$ and $W_{k+1} = W_k$.

Then, we iterate by adding one constraint at a time to the working set until we reach a minimization point $\hat{\mathbf{x}}$ of J over the current working set $\hat{W}$; $\hat{\mathbf{x}}$ is such that $\mathbf{p} = \mathbf{0}$ is a solution to (3.34). In this case, the optimality conditions give

$$\sum_{i\in\hat{W}} \mathbf{a}_i\hat{\lambda}_i = \mathbf{g} = \mathbf{G}\hat{\mathbf{x}} + \mathbf{c} \tag{3.36}$$

for some Lagrange multipliers $\hat{\lambda}_i$, $i \in \hat{W}$. Thus, the pair $(\hat{\mathbf{x}}, \hat{\boldsymbol{\lambda}})$ satisfies the KKT conditions if we set to zero the multipliers corresponding to the inequality constraints that are not in the working set, that is $\hat{\lambda}_i = 0$ for $i \notin \hat{W}$ and $\hat{\lambda}_i \geq 0$ for $i \in \hat{W} \cap I$.

Instead, if one (or more) multiplier $\hat{\lambda}_i$ for $i \in \hat{W} \cup I$ is negative, the last KKT condition (2.31) is not fulfilled and the cost function q_k can be decreased by dropping one of these constraints: in this case we remove from the working set an index i for which $\hat{\lambda}_i < 0$ and solve a new subproblem (3.34) for the new step. It can be proved that this strategy yields an admissible direction with respect to the dropped constraint, which is also a descent direction if second order sufficient conditions for the resulting problem are fulfilled (see Sect. 2.3.3).

We remark that although any index i for which $\hat{\lambda}_i < 0$ usually generates such a direction, in practice the one that is chosen corresponds to the negative multiplier with the largest absolute value. This is motivated by the results of the sensitivity analysis of Sect. 2.3.1: when a constraint is removed, the rate of decrease of the cost function is proportional to the magnitude of the multiplier associated to that constraint. To conclude, we have that any $\mathbf{p}_k \neq \mathbf{0}$ which solves (3.34) and satisfies second-order sufficient optimality conditions for the current working set is a descent direction for J. See [208] for the proof.

The whole procedure is reported in Algorithm 3.5. For strictly convex QPs, this algorithm achieves the solution $\hat{\mathbf{x}}$ in a finite number of iterations.

Algorithm 3.5 Active Set method

Input: given $\mathbf{x}_0$, W_0,
for $k = 0, 1, 2, \ldots$
    solve problem (3.33) to find $\mathbf{p}_k$;
    if $p_k = 0$
        compute multipliers $\hat{\lambda}_i$ satisfying (3.36) with $\hat{W} = W_k$;
        if $\hat{\lambda}_i \geq 0 \; \forall \, i \in W_k \cap I$
            set $\hat{\mathbf{x}} = \mathbf{x}_k$ and **exit**
        else
            set $j = \arg\min_{j\in W_k \cap I} \hat{\lambda}_j$;
            set $\mathbf{x}_{k+1} = \mathbf{x}_k$; $W_{k+1} = W_k$;
        end if
    else
        compute α_k from (3.35);
        set $\mathbf{x}_{k+1} = \mathbf{x}_k + \alpha_k\mathbf{p}_k$;
        if $\alpha_k < 1$ (there are blocking constraints)
            obtain W_{k+1} by adding one of the blocking constraints to W_k;
        else
            $W_{k+1} = W_k$;
        end if
    end if
end for

Remark 3.1. For constrained optimization problems involving box constraints (see Sect. 3.3), applying an active set method could be troublesome since the working set would change rather slowly; indeed, at most one constraint is added to (or dropped from) the working set at each step. This can represent a serious drawback when dealing with large-scale optimization problems, if the initial working set is far from the one corresponding to the exact solution. •

3.5 Constrained Optimization for More General Problems

In this section we present some of the most popular techniques for the solution of a constrained minimization problem. A first strategy consists in restating it as a free optimization problem, a second option consists instead of solving *local* quadratic approximations in the frame of quadratic programming problems.

3.5.1 Penalty and Augmented Lagrangian Methods

Penalty methods transform the constrained problem (for simplicity, we pose $b_i = 0$ for any $i \in E \cup I$)

$$\begin{cases} J(\mathbf{x}) \rightarrow \min \\ \text{subject to} \\ \varphi_i(\mathbf{x}) = 0, \qquad i \in E \\ \varphi_i(\mathbf{x}) \geq 0, \qquad i \in I \end{cases} \tag{3.37}$$

into an unconstrained one, by replacing J with the new cost function (the so-called *penalty* function)

$$P(\mathbf{x}, \mu) = J(\mathbf{x}) + \frac{\mu}{2} \sum_{i \in E} \varphi_i^2(\mathbf{x}) + \frac{\mu}{2} \sum_{i \in I} (\max\{-\varphi_i(\mathbf{x}), 0\})^2, \tag{3.38}$$

where $\mu > 0$ is a penalty parameter. If the constraints are not satisfied at $\mathbf{x}$, the sums appearing in (3.38) measure the distance of $\mathbf{x}$ from the set K of the constraints. The larger μ, the more the constraints violation is penalized.

If $\hat{\mathbf{x}}$ is a solution of (3.37), then $\hat{\mathbf{x}}$ is a minimizer of $P(\mathbf{x}, \mu)$. On the other hand, under suitable assumptions of regularity for J and $\{\varphi_i\}_i$, $i \in E \cup I$, the minimizer $\hat{\mathbf{x}} = \hat{\mathbf{x}}(\mu)$ of $P(\mathbf{x}, \mu)$ is such that

$$\lim_{\mu \to \infty} \hat{\mathbf{x}}(\mu) = \hat{\mathbf{x}}.$$

Hence $\hat{\mathbf{x}}(\mu)$ is a good approximation of the solution $\hat{\mathbf{x}}$ for $\mu \gg 1$. However, large values of μ make the minimization problem for $P(\mathbf{x}, \mu)$ ill-conditioned, so that a better strategy is to consider a monotonically increasing sequence of penalty parameters $\{\mu_k\}_{k \geq 0}$ and to compute, for each μ_k, an approximation $\hat{\mathbf{x}}_k$ given by

$$\hat{\mathbf{x}}_k = \arg\min_{\mathbf{x} \in \mathbb{R}^n} P(\mathbf{x}, \mu_k), \tag{3.39}$$

by exploiting one of the free optimization methods of Sect. 3.1.

The sequence μ_k should grow rather quickly, otherwise the convergence of the method would be too slow. A common choice is to start with a moderate value of μ_0 and then set $\mu_{k+1} = \beta \mu_k$ for $k > 0$, where β is an integer, typically between 4 and 10, to be chosen according to the difficulty occurred while solving (3.39) at the previous step (see [36]). More precisely, if (3.39) has required many iterations at step k, μ_{k+1} will be slightly larger than μ_k, otherwise larger increments can be considered. Furthermore, at step k we initialize the new iteration by the

Algorithm 3.6 Penalty method

Input: $\mu_0 > 0$, $\{\varepsilon_k\}_{k\geq 0}$ such that $\varepsilon_k > 0$, $\varepsilon_k \to 0$ for $k \to \infty$, $\bar{\varepsilon} > 0$, initial guess $\mathbf{x}_0^s$,

```
for k = 0, 1, 2, ...
    Find an approximate minimization point x_k of P(·, μ_k), starting at x_k^s,
    with tolerance ε_k on the stopping criterion;
    if k ≥ 1 and |∇_x P(x_k, μ_k)| ≤ ε̄
        set x̂ = x_k and exit;
    else
        set x_k^s = x_{k-1}
        choose β and compute μ_{k+1} = βμ_k > μ_k
    end if
end for
```

solution $\mathbf{x}_k^s = \mathbf{x}_{k-1}$ of (3.39) computed at the previous iteration. The whole procedure is reported in Algorithm 3.6.

By assuming J and $\{\varphi_i\}_{i\in E}$ to be regular, if $\mu_k \to \infty$ and $\varepsilon_k \to 0$, it is proved in [208, Theorem 17.2] that if a limit point $\hat{\mathbf{x}}$ is feasible and the constraint gradients $\nabla\varphi_i(\hat{\mathbf{x}})$ are linearly independent, then $\hat{\mathbf{x}}$ is a KKT point for problem (3.37). Further details can be found in [36].

A variant of the penalty method makes use, instead of $P(\mathbf{x}, \mu)$, of the *augmented Lagrangian* function

$$\mathcal{L}_A(\mathbf{x}, \boldsymbol{\lambda}, \mu) = J(\mathbf{x}) - \sum_{i\in E} \lambda_i \varphi_i(\mathbf{x}) + \frac{\mu}{2} \sum_{i\in I} \varphi_i^2(\mathbf{x}).$$

For the sake of simplicity, we consider only equality constraints, that is $I = \emptyset$ in (3.37). In this case, we build a sequence $\{\mu_k\}_{k\geq 0}$, $\mu_k \to \infty$, such that the corresponding sequence $(\mathbf{x}_k, \boldsymbol{\lambda}_k)$ converges to a point which fulfills the optimality conditions for the Lagrangian function $\mathcal{L}(\mathbf{x}, \boldsymbol{\lambda}) = J(\mathbf{x}) - \sum_{i\in E} \lambda_i \varphi_i(\mathbf{x})$. At the k-th iteration, given the multipliers vector $\boldsymbol{\lambda}_k = \{(\lambda_i)_k\}_{i\in E}$, we need to compute an approximate solution $\mathbf{x}_k$ to the problem

$$\min_{\mathbf{x}\in\mathbb{R}^n} \mathcal{L}_A(\mathbf{x}, \boldsymbol{\lambda}_k, \mu_k)$$

and then update $\boldsymbol{\lambda}_k$. For the update, a strategy consists of imposing that $\nabla_{\mathbf{x}}\mathcal{L}_A(\mathbf{x}_k\boldsymbol{\lambda}_k, \mu_k) \approx 0$, since we aim at converging to a critical point of $\mathcal{L}_A$. Since

$$\nabla_{\mathbf{x}}\mathcal{L}_A(\mathbf{x}_k, \boldsymbol{\lambda}_k, \mu_k) = \nabla J(\mathbf{x}_k) - \sum_{i\in E} (\lambda_i)_k - \mu_k\varphi_i(\mathbf{x}_k))\nabla\varphi_i(\mathbf{x}_k)$$

and thanks to the property that if $(\hat{\mathbf{x}}, \hat{\boldsymbol{\lambda}})$ is a critical point of $\mathcal{L}$, then from (2.17)

$$\nabla_{\mathbf{x}}\mathcal{L}(\hat{\mathbf{x}}, \hat{\boldsymbol{\lambda}}) = \nabla J(\hat{\mathbf{x}}) - \sum_{i\in E} \hat{\lambda}_i \nabla\varphi_i(\hat{\mathbf{x}}) = \mathbf{0},$$

by direct comparison we obtain $(\lambda_i)_k - \mu_k\varphi_i(\mathbf{x}_k) \approx \hat{\lambda}_i$, and the following update rule

$$(\lambda_i)_{k+1} = (\lambda_i)_k - \mu_k\varphi_i(\mathbf{x}_k). \tag{3.40}$$

The update of the parameter μ_k follows instead the same rules introduced for the penalty method. The whole procedure is reported in Algorithm 3.7.

Algorithm 3.7 Augmented Lagrangian method

Input: $\mu_0 > 0$, $\{\varepsilon_k\}_{k\geq 0}$ such that $\varepsilon_k > 0$, $\varepsilon_k \to 0$ for $k \to \infty$, $\bar{\varepsilon} > 0$, initial guess $\mathbf{x}_0^s$,

for $k = 0, 1, 2, \ldots$
  Find an approximate minimization point $\mathbf{x}_k$ of $\mathcal{L}_A(\cdot, \boldsymbol{\lambda}_k, \mu_k)$, starting at $\mathbf{x}_k^s$,
  with tolerance ε_k on the stopping criterion;
  if $k \geq 1$ and $|\nabla_{\mathbf{x}}\mathcal{L}_A(\mathbf{x}_k, \boldsymbol{\lambda}_k, \mu_k)| \leq \bar{\varepsilon}$
    set $\hat{\mathbf{x}} = \mathbf{x}_k$ and **exit**;
  else
    set $(\lambda_i)_{k+1} = (\lambda_i)_k - \mu_k \varphi_i(\mathbf{x}_k)$
    set $\mathbf{x}_k^s = \mathbf{x}_{k-1}$ and choose β and compute $\mu_{k+1} = \beta\mu_k > \mu_k$;
  end if
end for

If we know the exact Lagrange multiplier vector $\hat{\boldsymbol{\lambda}}$, then the solution $\hat{\mathbf{x}}$ of (3.37) (in the case $I = \emptyset$) is a minimizer for $\mathcal{L}_A(\mathbf{x}, \hat{\boldsymbol{\lambda}}, \mu)$ for all μ sufficiently large. Not only, we can obtain a reasonable estimate of $\hat{\mathbf{x}}$ by minimizing $\mathcal{L}_A(\mathbf{x}, \hat{\boldsymbol{\lambda}}, \mu)$ even though we only know a reasonable estimate of $\hat{\boldsymbol{\lambda}}$, and μ is not particularly large. See for instance [208, Theorem 17.6].

3.5.2 Sequential Quadratic Programming

Sequential quadratic programming (SQP) represents one of the most effective strategies for solving general (nonlinear) programming problems. SQP is an iterative procedure which generates, at each step, an optimization subproblem in which a quadratic model of the objective is minimized, subject to a linearization of the constraints. At the simplest stage, this technique can be regarded as the application of the Newton method (for the solution of nonlinear equations) to the system of KKT optimality conditions (2.29)–(2.31) obtained for the given constrained optimization problem.

We start by considering a local version of the SQP algorithm for the problem

$$\begin{cases} J(\mathbf{x}) \to \min \\ \text{subject to} \\ \varphi_i(\mathbf{x}) = b_i, \qquad 1 \leq i \leq m \end{cases} \tag{3.41}$$

for some real b_i's, and denote by $\mathbf{A}(\mathbf{x})^\top = [\nabla\varphi_1(\mathbf{x}), \ \ldots, \ \nabla\varphi_m(\mathbf{x})]$ the transpose[6] of the Jacobian matrix of the constraints. We introduce the Lagrangian functional as

$$\mathcal{L}(\mathbf{x}, \boldsymbol{\lambda}) = J(\mathbf{x}) - \boldsymbol{\lambda}^\top (\mathbf{b} - \boldsymbol{\varphi}(\mathbf{x})),$$

where $\boldsymbol{\varphi}(\mathbf{x}) = [\varphi_1(\mathbf{x}), \ \ldots, \ \varphi_m(\mathbf{x})]^\top$. The first-order optimality conditions for (3.41) can be written as a system of $n+m$ (nonlinear) equations in the $n+m$ unknowns $(\mathbf{x}, \boldsymbol{\lambda})$:

$$\mathcal{F}(\mathbf{x}, \boldsymbol{\lambda}) = \begin{bmatrix} \nabla J(\mathbf{x}) + \mathbf{A}(\mathbf{x})^\top \boldsymbol{\lambda} \\ \boldsymbol{\varphi}(\mathbf{x}) - \mathbf{b} \end{bmatrix} = \mathbf{0}. \tag{3.42}$$

Any solution $(\hat{\mathbf{x}}, \hat{\boldsymbol{\lambda}})$ of (3.41) for which the constraint qualification (CQ) condition – see Sect. 2.3.2 – holds (that is, for which the matrix $\mathbf{A}(\mathbf{x})^\top$ has full rank) satisfies (3.42).

[6] By defining $\mathbf{A}(\mathbf{x})^\top$ in this way, $\mathbf{A}(\mathbf{x})$ is indeed the Jacobian matrix coherently with the definition introduced in Sect. 2.3.1, in which gradients were considered (as commonly done in Analysis) as *row* vectors.

Applying Newton's method (for nonlinear equations) to system (3.42) gives the so-called *Lagrange-Newton method.* Starting from an initial guess $(\mathbf{x}_0, \boldsymbol{\lambda}_0)$, we generate the sequence $\{(\mathbf{x}_k, \boldsymbol{\lambda}_k)\}_{k\geq 0}$ as

$$\begin{bmatrix} \mathbf{x}_{k+1} \\ \boldsymbol{\lambda}_{k+1} \end{bmatrix} = \begin{bmatrix} \mathbf{x}_k \\ \boldsymbol{\lambda}_k \end{bmatrix} + \begin{bmatrix} \delta\mathbf{x} \\ \delta\boldsymbol{\lambda} \end{bmatrix} \tag{3.43}$$

where the increment $(\delta\mathbf{x}, \delta\boldsymbol{\lambda})^\top$ solves the following Newton-KKT system

$$\begin{bmatrix} D^2_{\mathbf{xx}}\mathcal{L}(\mathbf{x}_k, \boldsymbol{\lambda}_k) & \mathbf{A}(\mathbf{x}_k)^\top \\ \mathbf{A}(\mathbf{x}_k) & \mathbf{0} \end{bmatrix} \begin{bmatrix} \delta\mathbf{x} \\ \delta\boldsymbol{\lambda} \end{bmatrix} = - \begin{bmatrix} \nabla J(\mathbf{x}_k) + \mathbf{A}(\mathbf{x}_k)^\top \boldsymbol{\lambda}_k \\ \varphi(\mathbf{x}_k) - \mathbf{b} \end{bmatrix}. \tag{3.44}$$

This requires the calculation of the Jacobian of $\mathcal{F}(\mathbf{x}, \boldsymbol{\lambda})$ – that is, the Hessian of the Lagrangian $\mathcal{L}(\mathbf{x}, \boldsymbol{\lambda})$ – with respect to $\mathbf{x}$ and $\boldsymbol{\lambda}$, which is a nonsingular matrix if:

- the constraint Jacobian $\mathbf{A}(\mathbf{x}_k)$ has full row rank;
- the matrix $D^2_{\mathbf{xx}}\mathcal{L}(\mathbf{x}_k, \boldsymbol{\lambda}_k)$ is positive definite on the tangent space of the constraints, that is, $\mathbf{d}^\top D^2_{\mathbf{xx}}\mathcal{L}(\mathbf{x}_k, \boldsymbol{\lambda}_k)\mathbf{d} > 0$ for all $\mathbf{d} \neq \mathbf{0}$ such that $\mathbf{A}(\mathbf{x}_k)\mathbf{d} = \mathbf{0}$.

The second condition above essentially matches the second-order sufficient condition of Proposition (2.1). We remark that the same system can be obtained if, at the iterate $(\mathbf{x}_k, \boldsymbol{\lambda}_k)$, we approximate (3.41) by means of the following equality-constrained quadratic program[7]:

$$\begin{cases} m_k(\mathbf{p}) = J(\mathbf{x}_k) + \nabla J(\mathbf{x}_k)^\top \mathbf{p} + \dfrac{1}{2}\mathbf{p}^\top D^2_{\mathbf{xx}}\mathcal{L}(\mathbf{x}_k, \boldsymbol{\lambda}_k)\mathbf{p} \rightarrow \min \\ \text{subject to} \\ \mathbf{A}(\mathbf{x}_k)^\top \mathbf{p} + \varphi(\mathbf{x}_k) - \mathbf{b} = \mathbf{0}. \end{cases} \tag{3.45}$$

Under the above assumptions, the unique solution $(\mathbf{p}_k, \mathbf{l}_k)$ of (3.45) satisfies the following first-order conditions:

$$\begin{cases} D^2_{\mathbf{xx}}\mathcal{L}(\mathbf{x}_k, \boldsymbol{\lambda}_k)\mathbf{p}_k + \nabla J(\mathbf{x}_k) + \mathbf{A}(\mathbf{x}_k)^\top \mathbf{l}_k = \mathbf{0} \\ \mathbf{A}(\mathbf{x}_k)^\top \mathbf{p}_k + \varphi(\mathbf{x}_k) - \mathbf{b} = \mathbf{0} \end{cases} \tag{3.46}$$

and is strictly related to the solution of the Newton-KKT system (3.44); in fact, by summing $\mathbf{A}(\mathbf{x}_k)^\top \boldsymbol{\lambda}_k$ to both sides of the first equation in (3.44), we obtain

$$\begin{bmatrix} D^2_{\mathbf{xx}}\mathcal{L}(\mathbf{x}_k, \boldsymbol{\lambda}_k) & \mathbf{A}(\mathbf{x}_k)^\top \\ \mathbf{A}(\mathbf{x}_k) & \mathbf{0} \end{bmatrix} \begin{bmatrix} \delta\mathbf{x} \\ \boldsymbol{\lambda}_{k+1} \end{bmatrix} = \begin{bmatrix} -\nabla J(\mathbf{x}_k) \\ -\varphi(\mathbf{x}_k) \end{bmatrix} \tag{3.47}$$

so that $\mathbf{l}_k = \boldsymbol{\lambda}_{k+1} =$ and $\mathbf{p}_k = \delta\mathbf{x}$ solve (3.44). The new iterate $(\mathbf{x}_{k+1}, \boldsymbol{\lambda}_{k+1})$ can therefore be viewed either as the solution of the quadratic program (3.45) or as $(\mathbf{x}_{k+1}, \boldsymbol{\lambda}_{k+1}) = (\mathbf{x}_k + \delta\mathbf{x}, \boldsymbol{\lambda}_k + \delta\boldsymbol{\lambda})$, $(\delta\mathbf{x}, \delta\boldsymbol{\lambda}_{\boldsymbol{\lambda}})$ being the solution to the k-th iterative step (3.44) of Newton iterations applied to the first order optimality system (3.42). The Lagrange-Newton method (which is also called *local SQP method*) for solving (3.41) is implemented in Algorithm 3.8.

In the case of a general nonlinear programming problem under the form (3.37), we can linearize both equality and inequality constraints to get

$$\begin{cases} m_k(\mathbf{p}) = J(\mathbf{x}_k) + \nabla J(\mathbf{x}_k)^\top \mathbf{p} + \dfrac{1}{2}\mathbf{p}^\top D^2_{\mathbf{xx}}\mathcal{L}(\mathbf{x}_k, \boldsymbol{\lambda}_k)\mathbf{p} \rightarrow \min \\ \text{subject to} \\ \nabla\varphi_i(\mathbf{x}_k)^\top \mathbf{p} + \varphi_i(\mathbf{x}_k) = 0, \qquad i \in E \\ \nabla\varphi_i(\mathbf{x}_k)^\top \mathbf{p} + \varphi_i(\mathbf{x}_k) \geq 0, \qquad i \in I \end{cases} \tag{3.48}$$

[7] Since $\nabla J(\mathbf{x}_k)^\top \mathbf{p} = \nabla_x \mathcal{L}(\mathbf{x}_k, \boldsymbol{\lambda}_k)^\top \mathbf{p}$, $m_k(\mathbf{p})$ can be seen as a quadratic approximation of $\mathcal{L}$ at $\mathbf{x}_k$.

Algorithm 3.8 Local Sequential Quadratic Programming (SQP) method

---

Input: Initial guess $(\mathbf{x}_0, \boldsymbol{\lambda}_0)$,

set $k = 0$;

repeat

  evaluate $J(\mathbf{x}_k)$, $\nabla J(\mathbf{x}_k)$, $D^2_{\mathbf{xx}}\mathcal{L}(\mathbf{x}_k, \boldsymbol{\lambda}_k)$, $\varphi(\mathbf{x}_k)$, $\mathbf{A}(\mathbf{x}_k)$;

  solve the quadratic program (3.45) to obtain $(\mathbf{p}_k, \mathbf{l}_k)$;

  set $\mathbf{x}_{k+1} = \mathbf{x}_k + \mathbf{p}_k$ and $\boldsymbol{\lambda}_{k+1} = \mathbf{l}_k$;

until a convergence criterion is fulfilled.

---

in analogy to (3.45). By using the active set method described in Sect. 3.4.2, the new iterate $(\mathbf{x}_k + \mathbf{p}_k, \boldsymbol{\lambda}_{k+1})$ yields the solution $\mathbf{p}_k$ and the corresponding Lagrange multiplier of (3.48).

The local (quadratic) convergence of the SQP method follows from the application of Newton's method to the nonlinear system given by the KKT conditions (3.46).

However, a SQP method must be able to converge also when the initial guess is far from the (unknown) solution, or in the case of nonconvex problems. In this respect, two alternative strategies, the *trust-region SQP* or the *line-search SQP* methods can be used in order to properly adjust the step length at each iteration.

Here we only describe the latter case because it is more intuitive and simple to implement; further details about trust-region SQP and other possible approaches can be found for instance in [208]. In the line-search SQP case:

- to decide whether to accept or reject a step, we could control its size as in the line search method for unconstrained optimization. In this respect, we can introduce a *merit function* $\Phi(\mathbf{x}; \mu)$ (as we already did, for instance, in the case of the augmented Lagrangian function) and accept a step only if it yields a sufficient decrease in the merit function. In analogy with the Armijo rule (3.19) the step length $\alpha > 0$ is required to be small enough to satisfy the inequality
$$\Phi(\mathbf{x} + \alpha\mathbf{p}; \mu) \leq \Phi(\mathbf{x}; \mu) + \eta\alpha D(\Phi(\mathbf{x}; \mu); \mathbf{p})$$
for some $\eta \in (0, 1)$; $D(\Phi(\mathbf{x}; \mu); \mathbf{p})$ is the directional derivative of $\Phi(\mathbf{x}; \mu)$ in the direction $\mathbf{p}$;
- if the quadratic subproblem (3.48) is convex, we can use an *active-set* method to solve it at each iteration, by initializing the working set for each QP subproblem with the final active set from the previous SQP iteration;
- since the Hessian of the Lagrangian function is not often easily computable, in (3.48) it is replaced by a quasi-Newton approximation, obtained for instance with the BFGS formula (3.13).

By using these recipes in Algorithm (3.8), we obtain a *line-search SQP* (or *global SQP*) method for solving nonlinear programming problems such as (3.37). We report it in Algorithm 3.9.

In practice, SQP methods often manage to converge to a solution from a remote initial guess. Nevertheless, classical global convergence results can be stated by requiring several restrictive (and often very difficult to verify) assumptions.

Instead, local convergence is guaranteed under more natural hypotheses, such as the CQ condition (2.24), given that both J and the constraints are twice differentiable with Lipschitz continuous second derivatives at $(\hat{\mathbf{x}}, \hat{\boldsymbol{\lambda}})$.

Under these assumptions, provided that $(\mathbf{x}_0, \boldsymbol{\lambda}_0)$ is sufficiently close to $(\hat{\mathbf{x}}, \hat{\boldsymbol{\lambda}})$, the sequence $\{\mathbf{x}_k, \boldsymbol{\lambda}_k\}_{k>0}$ generated by the SQP algorithm converges quadratically to $(\hat{\mathbf{x}}, \hat{\boldsymbol{\lambda}})$ when an exact Hessian matrix is used, similarly to what happens with the Newton method in the case of unconstrained minimization problems.

Algorithm 3.9 Global Sequential Quadratic Programming (SQP) method

---

Input: Initial guess $(\mathbf{x}_0, \boldsymbol{\lambda}_0)$, $\eta \in (0, 0.5)$, $\tau, \rho \in (0, 1)$,
 1: evaluate $J(\mathbf{x}_0)$, $\nabla J(\mathbf{x}_0)$, $\boldsymbol{\varphi}(\mathbf{x}_0)$, $\mathbf{A}(\mathbf{x}_0)$;
 2: choose an initial Hessian approximation B_0 (if a quasi-Newton approximation is used),
    otherwise compute $D^2\mathcal{L}(\mathbf{x}_0, \boldsymbol{\lambda}_k)$;
 3: set $k = 0$;
 4: **repeat**
 5:     solve (3.48) to obtain $(\mathbf{p}_k, \hat{\boldsymbol{\lambda}})$;
 6:     set $\delta\boldsymbol{\lambda} = \hat{\boldsymbol{\lambda}} - \boldsymbol{\lambda}_k$;
 7:     choose μ_k such that $D(\Phi(\mathbf{x}_k; \mu_k); \mathbf{p}_k)$ is *sufficiently* negative;
 8:     set $\alpha_k = 1$;
 9:     **while** $\Phi(\mathbf{x}_k + \alpha_k \mathbf{p}_k; \mu) > \Phi(\mathbf{x}_k; \mu_k) + \eta\alpha D(\Phi(\mathbf{x}_k; \mu_k); \mathbf{p}_k)$
10:         reset $\alpha_k \leftarrow \tau_\alpha \alpha_k$, for some $\tau_\alpha \in (0, \tau)$;
11:     **end while**
12:     set $\mathbf{x}_{k+1} = \mathbf{x}_k + \alpha_k \mathbf{p}_k$ and $\boldsymbol{\lambda}_{k+1} = \boldsymbol{\lambda}_k + \alpha_k \delta\boldsymbol{\lambda}$;
13:     evaluate $J(\mathbf{x}_{k+1})$, $\nabla J(\mathbf{x}_{k+1})$, $\boldsymbol{\varphi}(\mathbf{x}_{k+1})$, $\mathbf{A}(\mathbf{x}_{k+1})$ (and possibly $D^2\mathcal{L}(\mathbf{x}_{k+1}, \boldsymbol{\lambda}_{k+1})$);
14:     **if** a quasi-Newton approximation is used
15:         set $\mathbf{s}_k = \alpha_k \mathbf{p}_k$, $\mathbf{y}_k = \nabla_{\mathbf{x}}\mathcal{L}(\mathbf{x}_{k+1}, \boldsymbol{\lambda}_{k+1}) - \nabla_{\mathbf{x}}\mathcal{L}(\mathbf{x}_k, \boldsymbol{\lambda}_{k+1})$;
16:         compute B_{k+1} by updating B_k with the BFGS formula (3.13);
17:     **end if**
18: **until** a convergence criterion is fulfilled.

On the other hand, if quasi-Newton approximate Hessians are used, $\{\mathbf{x}_k, \boldsymbol{\lambda}_k\}_{k>0}$ converges superlinearly (although not quadratically) to the solution under some assuptions similar to those introduced in Sect. 3.1.3.

In conclusion, SQP methods generalize to the case of constrained minimization problems the Newton and quasi-Newton methods for unconstrained minimization problems, and feature the same convergence properties.

Chapter 4
Prelude on Control: The Case of Algebraic and ODE Systems

In this chapter we describe two important classes of finite-dimensional OCPs, governed by either a linear system of algebraic equations, or a linear ODE system. This choice is motivated by several reasons. These problems serve as a *warm up* to develop the analysis and derive the system of optimality conditions by exploiting the results of Chap. 2. Moreover, when treating the numerical discretization of OCPs governed by PDEs we naturally land on finite dimensional problems. Last, but not least, these two classes of problems can be relevant *per se*. For instance, control problems governed by ODEs arise in several areas, e.g. finance and economics, where a vast literature exists.

4.1 Algebraic Optimal Control Problems

In this section we consider an OCP governed by a linear system of algebraic equations. Let us denote by $\mathcal{U}_{ad}$ a convex closed nonempty subset of $\mathcal{U} = \mathbb{R}^q$, a control vector $\mathbf{u} \in \mathcal{U}_{ad}$ and a state vector $\mathbf{y} \in V = \mathbb{R}^N$, with $q \leq N$. V denotes the state space, $\mathcal{U}$ the control space, $\mathcal{U}_{ad}$ the space of admissible controls. The state equation is given by the algebraic system

$$\mathbf{A}\mathbf{y} = \mathbf{f} + \mathbf{B}\mathbf{u} \tag{4.1}$$

where $\mathbf{A} \in \mathbb{R}^{N\times N}$ is a nonsingular matrix, $\mathbf{f} \in \mathbb{R}^N$ and $\mathbf{B} \in \mathbb{R}^{N\times q}$ is a matrix of rank q. We consider the minimization problem over $V \times \mathcal{U}_{ad}$,

$$\tilde{J}(\mathbf{y}, \mathbf{u}) = \frac{1}{2}|\mathbf{y} - \mathbf{z}_d|^2 + \frac{\alpha}{2}|\mathbf{u}|^2 \;\rightarrow\; \min \qquad (\alpha \geq 0). \tag{4.2}$$

Since $\mathbf{A}$ is non-singular, from (4.1) we obtain

$$\mathbf{y}(\mathbf{u}) = \mathbf{A}^{-1}(\mathbf{f} - \mathbf{B}\mathbf{u}); \tag{4.3}$$

by inserting (4.3) in (4.2), we can introduce the so-called *reduced cost functional* $J : \mathcal{U} \rightarrow \mathbb{R}$ and formulate the OCP as an optimization problem for J over $\mathcal{U}_{ad}$,

$$J(\mathbf{u}) = \tilde{J}(\mathbf{y}(\mathbf{u}), \mathbf{u}) \rightarrow \min\,. \tag{4.4}$$

In the following subsections, we discuss the well-posedness of problem (4.4), by recalling the theoretical results introduced in Sect. 2.1.

A. Manzoni et al., *Optimal Control of Partial Differential Equations*,
Applied Mathematical Sciences 207, https://doi.org/10.1007/978-3-030-77226-0_4

4.1.1 Existence and Uniqueness of the Solution

In order to prove existence of a solution to (4.4), we use the Weierstrass Theorem 2.1. To prove the uniqueness of the solution, we show that J is strictly convex.

The reduced cost functional J is clearly continuous since it is a composition of linear and quadratic functions of $\mathbf{u}$. Thus, since $\mathcal{U}_{ad}$ is closed, for the existence of a minimizer it is sufficient that $\mathcal{U}_{ad}$ be bounded, or to check condition (2.4), the coercivity at ∞.

Let $\mathbf{S} = \mathbf{A}^{-1}\mathbf{B} \in \mathbb{R}^{N\times q}$ be the so-called *solution matrix*; since $\mathbf{B}$ has rank q, $\mathbf{S}$ has rank q, too, and its singular values $\sigma_1 \geq \ldots \geq \sigma_q > 0$ are such that

$$\sigma_q^2|\mathbf{v}|^2 \leq (\mathbf{Sv}, \mathbf{Sv}) \leq \sigma_1^2|\mathbf{v}|^2 \qquad \forall \mathbf{v} \in \mathbb{R}^q. \tag{4.5}$$

Now, we can write

$$\begin{aligned} J(\mathbf{u}) &= \frac{1}{2}\left|\mathbf{A}^{-1}(\mathbf{f}+\mathbf{Bu}) - \mathbf{z}_d\right|^2 + \frac{\alpha}{2}|\mathbf{u}|^2 = \\ &= \frac{1}{2}\left|\mathbf{A}^{-1}\mathbf{Bu}\right|^2 - \left(\mathbf{A}^{-1}\mathbf{Bu}, \mathbf{A}^{-1}\mathbf{f} - \mathbf{z}_d\right) + \frac{1}{2}\left|\mathbf{A}^{-1}\mathbf{f} - \mathbf{z}_d\right|^2 + \frac{\alpha}{2}|\mathbf{u}|^2. \end{aligned}$$

By setting

$$((\mathbf{u}, \mathbf{v})) = \left(\mathbf{A}^{-1}\mathbf{Bu}, \mathbf{A}^{-1}\mathbf{Bv}\right) + \frac{\alpha}{2}(\mathbf{u}, \mathbf{v}) = (\mathbf{Su}, \mathbf{Sv}) + \frac{\alpha}{2}(\mathbf{u}, \mathbf{v}) \tag{4.6}$$

and

$$L\mathbf{u} = \left(\mathbf{A}^{-1}\mathbf{Bu},\ \mathbf{A}^{-1}\mathbf{f} - \mathbf{z}_d\right) = \left(\mathbf{Su}, \mathbf{A}^{-1}\mathbf{f} - \mathbf{z}_d\right),$$

we can write

$$J(\mathbf{u}) = \frac{1}{2}((\mathbf{u}, \mathbf{u})) - L\mathbf{u} + c$$

with $c = \frac{1}{2}\left|\mathbf{A}^{-1}\mathbf{f} - \mathbf{z}_d\right|^2$. We have

$$\left(\sigma_q^2 + \frac{\alpha}{2}\right)|\mathbf{u}|^2 \leq ((\mathbf{u}, \mathbf{u})) \leq \left(\sigma_1^2 + \frac{\alpha}{2}\right)|\mathbf{u}|^2 \qquad \forall \mathbf{u} \in \mathbb{R}^q \tag{4.7}$$

thanks to (4.5), and

$$|L\mathbf{u}| \leq |\mathbf{Su}|\left|\mathbf{A}^{-1}\mathbf{f} - \mathbf{z}_d\right| \leq \sigma_1\sqrt{2c}\,|\mathbf{u}|. \tag{4.8}$$

Therefore

$$J(\mathbf{u}) \geq \frac{1}{2}\left(\sigma_q^2 + \frac{\alpha}{2}\right)|\mathbf{u}|^2 - \sigma_1\sqrt{2c}\,|\mathbf{u}| + c \to +\infty \quad \text{if } |\mathbf{u}| \to \infty \tag{4.9}$$

which expresses the coercivity at ∞ of J. Thus, there exists (at least) one minimizer. Thanks to (4.7), J is also *strictly convex* (see Exercise 1), so that the minimizer is unique.

Note that, if $\alpha = 0$, the matrix $\mathbf{B}$ must have full rank in order for (4.7) to hold, and thus $((\mathbf{u}, \mathbf{u}))$ to be positive.

Remark 4.1. Since $((\mathbf{u}, \mathbf{w})) = ((\mathbf{w}, \mathbf{u}))$, $((\cdot\,, \cdot))$ is a scalar product in $\mathcal{U}$. More generally, $((\mathbf{u}, \mathbf{w}))$ is a bilinear, symmetric, coercive form; see Sect. A.4. •

Remark 4.2. In more general cases, (4.2) is replaced by

$$\tilde{J}(\mathbf{y}, \mathbf{u}) = P(\mathbf{y}) + Q(\mathbf{u}) \tag{4.10}$$

where $P : V \to \mathbb{R}$ and $Q : \mathcal{U} \to \mathbb{R}$.

For instance,

$$P(\mathbf{y}) = \frac{1}{2}|\mathbf{Cy} - \mathbf{z}_d|^2 \tag{4.11}$$

where $\mathbf{C} \in \mathbb{R}^{m \times N}$ and $\mathbf{z} = \mathbf{Cy} \in \mathcal{Z} = \mathbb{R}^m$ is the observation, and

$$Q(\mathbf{u}) = \frac{\alpha}{2}(\mathbf{Nu}, \mathbf{u}),$$

where $\mathbf{N} \in \mathbb{R}^{q \times q}$ is a given symmetric and positive definite matrix. For the cost function

$$J(\mathbf{u}) = P(\mathbf{y}(\mathbf{u})) + Q(\mathbf{u}) = \frac{1}{2}|\mathbf{Cy}(\mathbf{u}) - \mathbf{z}_d|^2 + \frac{\alpha}{2}(\mathbf{Nu}, \mathbf{u}),$$

with $\mathbf{N} \in \mathbb{R}^{q \times q}$ being a symmetric positive definite matrix, the same conclusion as above holds provided $\alpha > 0$ or the matrix $\mathbf{CS}$ has full rank. •

4.1.2 Optimality conditions

Due to the strict convexity of J, the optimality condition (2.7)

$$dJ(\hat{\mathbf{u}})(\mathbf{v} - \hat{\mathbf{u}}) = (\nabla J(\hat{\mathbf{u}}), \mathbf{v} - \hat{\mathbf{u}}) \geq 0 \qquad \forall \mathbf{v} \in \mathcal{U}_{ad}, \tag{4.12}$$

is both necessary and sufficient for $\hat{\mathbf{u}}$ to be optimal.

We first consider the case of the cost functional (4.2). The Jacobian matrix of $\mathbf{y} = \mathbf{y}(\hat{\mathbf{u}})$ is

$$\nabla \mathbf{y}(\mathbf{u}) = \mathbf{A}^{-1}\mathbf{B},$$

hence we obtain (chain rule)

$$\begin{aligned} &\left(\mathbf{y}(\hat{\mathbf{u}}) - \mathbf{z}_d, \mathbf{A}^{-1}\mathbf{B}(\mathbf{v} - \hat{\mathbf{u}})\right) + \alpha(\hat{\mathbf{u}}, \mathbf{v} - \hat{\mathbf{u}}) \\ &\quad = (\mathbf{y}(\hat{\mathbf{u}}) - \mathbf{z}_d, \mathbf{y}(\mathbf{v}) - \mathbf{y}(\hat{\mathbf{u}})) + \alpha(\hat{\mathbf{u}}, \mathbf{v} - \hat{\mathbf{u}}) \geq 0. \qquad \forall \mathbf{v} \in \mathcal{U}_{ad}. \end{aligned} \tag{4.13}$$

The last expression of the variational inequality is of limited practical interest because of the presence of $\mathbf{y}(\mathbf{v}) - \mathbf{y}(\hat{\mathbf{u}})$. On the other hand, the inequality in the second line requires the inverse matrix of $\mathbf{A}$.

A more efficient way to proceed benefits from the use of an auxiliary variable, the so-called *adjoint state* (or *Lagrange multiplier*), $\hat{\mathbf{p}} = \hat{\mathbf{p}}(\hat{\mathbf{u}})$, which is the solution of the following *adjoint problem*

$$\mathbf{A}^\top \hat{\mathbf{p}} = \mathbf{y}(\hat{\mathbf{u}}) - \mathbf{z}_d.$$

Then, noting that

$$\mathbf{A}(\mathbf{y}(\mathbf{v}) - \mathbf{y}(\hat{\mathbf{u}})) = \mathbf{B}(\mathbf{v} - \hat{\mathbf{u}}),$$

the variational inequality (4.13) becomes

$$\begin{aligned} 0 &\leq (\mathbf{y}(\hat{\mathbf{u}}) - \mathbf{z}_d, \mathbf{y}(\mathbf{v}) - \mathbf{y}(\hat{\mathbf{u}})) + \alpha(\hat{\mathbf{u}}, \mathbf{v} - \hat{\mathbf{u}}) \\ &= (\mathbf{A}^\top \hat{\mathbf{p}}, \mathbf{y}(\mathbf{v}) - \mathbf{y}(\hat{\mathbf{u}})) + \alpha(\hat{\mathbf{u}}, \mathbf{v} - \hat{\mathbf{u}}) \\ &= (\hat{\mathbf{p}}, \mathbf{B}(\mathbf{v} - \hat{\mathbf{u}})) + \alpha(\hat{\mathbf{u}}, \mathbf{v} - \hat{\mathbf{u}}) = \left(\mathbf{B}^\top \hat{\mathbf{p}}, \mathbf{v} - \hat{\mathbf{u}}\right) + \alpha(\hat{\mathbf{u}}, \mathbf{v} - \hat{\mathbf{u}}). \end{aligned} \tag{4.14}$$

We thus reach the following conclusion:

Theorem 4.1. *Assume that* $\mathbf{A} \in \mathbb{R}^{N\times N}$ *is nonsingular and either* $\alpha > 0$ *or* $\mathbf{B} \in \mathbb{R}^{N\times q}$ *has full rank. Let us consider the following OCP:*

$$\begin{cases} J(\mathbf{u}) = \frac{1}{2}|\mathbf{y}(\mathbf{u}) - \mathbf{z}_d|^2 + \frac{\alpha}{2}|\mathbf{u}|^2 \to \min \\ \qquad \textit{subject to} \\ \qquad \mathbf{u} \in \mathcal{U}_{ad}. \end{cases} \tag{4.15}$$

If $\hat{\mathbf{u}}$ *is the minimizer and* $\hat{\mathbf{y}} = \mathbf{y}(\hat{\mathbf{u}})$ *is the corresponding optimal state, there exists* $\hat{\mathbf{p}} = \mathbf{p}(\hat{\mathbf{u}})$ *such that the following system of first-order necessary optimality conditions holds:*

$$\begin{cases} \mathbf{A}\hat{\mathbf{y}} = \mathbf{f} + \mathbf{B}\hat{\mathbf{u}} \\ \mathbf{A}^\top\hat{\mathbf{p}} = \hat{\mathbf{y}} - \mathbf{z}_d \\ (\mathbf{B}^\top\hat{\mathbf{p}} + \alpha\hat{\mathbf{u}}, \mathbf{v} - \hat{\mathbf{u}}) \geq 0 \qquad \forall\, \mathbf{v} \in \mathcal{U}_{ad}. \end{cases} \tag{4.16}$$

Conversely, if $\hat{\mathbf{y}}$, $\hat{\mathbf{p}}$ *and* $\hat{\mathbf{u}}$ *are solutions to (4.16), then* $\hat{\mathbf{u}}$ *is the optimal control for* (4.15), *with associated optimal state* $\hat{\mathbf{y}}$.

The control problem[1] is said to be *unconstrained* if $\mathcal{U}_{ad} = \mathcal{U}$, and *constrained* if $\mathcal{U}_{ad} \subsetneq \mathcal{U}$. In this latter case, the control is subject to a set of constraints, usually defined through a set of equalities or inequalities. Note that in the unconstrained case $\mathcal{U}_{ad} = \mathcal{U} = \mathbb{R}^q$, $\mathbf{v} - \hat{\mathbf{u}}$ can be any vector in $\mathbb{R}^q$. Therefore the variational inequality (4.14) reduces to the equation

$$\mathbf{B}^\top\hat{\mathbf{p}} + \alpha\hat{\mathbf{u}} = \mathbf{0}. \tag{4.17}$$

Then, (4.16) can be written under the form of a block linear system, as follows:

$$\begin{bmatrix} \mathbf{I} & 0 & -\mathbf{A}^\top \\ 0 & \alpha\mathbf{I} & \mathbf{B}^\top \\ -\mathbf{A} & \mathbf{B} & 0 \end{bmatrix} \begin{bmatrix} \mathbf{y} \\ \mathbf{u} \\ \mathbf{p} \end{bmatrix} = \begin{bmatrix} \mathbf{z}_d \\ \mathbf{0} \\ -\mathbf{f} \end{bmatrix}. \tag{4.18}$$

This system features a *saddle point* structure. In fact, provided we group the state and the control vector in a unique variable $\mathbf{w} = (\mathbf{y}, \mathbf{u})^\top \in \mathbb{R}^{N+q}$ and define

$$\mathbf{G} = \begin{bmatrix} \mathbf{I} & 0 \\ 0 & \alpha\mathbf{I} \end{bmatrix}, \qquad \mathbf{E} = \begin{bmatrix} -\mathbf{A} & \mathbf{B} \end{bmatrix}, \qquad \mathbf{b} = \begin{bmatrix} 0 \\ -\mathbf{f} \end{bmatrix},$$

we can rewrite (4.25) as

$$\begin{bmatrix} \mathbf{G} & \mathbf{E}^\top \\ \mathbf{E} & 0 \end{bmatrix} \begin{bmatrix} \mathbf{w} \\ \mathbf{p} \end{bmatrix} = \begin{bmatrix} \mathbf{z}_d \\ \mathbf{b} \end{bmatrix}. \tag{4.19}$$

We will often deal with this kind of systems in the following chapters. In Chapter 6 several preconditioning techniques are shown to make the application of iterative solvers feasible in the case of large-scale systems under this form. Here we only point out that (4.19) is a saddle-point system provided we regard state and control vectors as a unique variable. Moreover, if $\alpha \neq 0$ the matrix $\mathbf{G}$ has full rank and system (4.19) admits a unique solution. Several other spectral properties of saddle-point systems are recalled in sect. B.1.3.

In the unconstrained case, if $\alpha > 0$ we obtain from (4.17) that

$$\hat{\mathbf{u}} = -\frac{1}{\alpha}\mathbf{B}^\top\hat{\mathbf{p}}. \tag{4.20}$$

[1] The state equation always plays the role of constraint, as it expresses a link between the state and the control; however, we usually refer to constrained OCPs when the control and/or the state are subject to a set of additional constraints.

Then, from the state and adjoint equations, we derive the following system:

$$\begin{cases} \mathbf{A}\hat{\mathbf{y}} = \mathbf{f} - \dfrac{1}{\alpha}\mathbf{B}\mathbf{B}^\top\hat{\mathbf{p}} \\ \mathbf{A}^\top\hat{\mathbf{p}} = \hat{\mathbf{y}} - \mathbf{z}_d. \end{cases}$$

Solving this optimality system yields the optimal state vector and the corresponding adjoint state; then, using (4.20) we can retrieve the optimal control vector.

4.1.3 Gradient, Sensitivity and Minimum Principle

In this section we focus on $\nabla J(\mathbf{u})$, on its evaluation and its usage for the sensitivity analysis.

First of all, we highlight the important role played by the adjoint state on the solution *sensitivity.* Taking first the case $\alpha = 0$, the third equation of (4.16) gives

$$(\mathbf{B}^\top\hat{\mathbf{p}}, \mathbf{v}) \geq (\mathbf{B}^\top\hat{\mathbf{p}}, \hat{\mathbf{u}}),$$

that is

$$(\mathbf{B}^\top\hat{\mathbf{p}}, \hat{\mathbf{u}}) = \min_{\mathbf{v}\in\mathcal{U}_{ad}} (\mathbf{B}^\top\hat{\mathbf{p}}, \mathbf{v}).$$

This last relation is also known as *minimum principle.* On the other hand, from (4.13) we have that if $\hat{\mathbf{y}}$, $\hat{\mathbf{u}}$ and $\hat{\mathbf{p}}$ solve (4.16), then

$$dJ(\hat{\mathbf{u}})(\mathbf{v} - \hat{\mathbf{u}}) = (\mathbf{B}^\top\hat{\mathbf{p}}, \mathbf{v} - \hat{\mathbf{u}}),$$

which is nonnegative thanks to the minimum principle. From (4.13) we also find[2]

$$dJ(\hat{\mathbf{u}})(\mathbf{v} - \hat{\mathbf{u}}) = (\nabla J(\hat{\mathbf{u}}), \mathbf{v} - \hat{\mathbf{u}}) = (\mathbf{B}^\top\hat{\mathbf{p}}, \mathbf{v} - \hat{\mathbf{u}}) \qquad \forall \mathbf{v} \in \mathcal{U}_{ad},$$

which gives the following expression for the (*reduced*) *gradient* of J

$$\nabla J(\hat{\mathbf{u}}) = \mathbf{B}^\top\mathbf{p}(\hat{\mathbf{u}}). \tag{4.21}$$

Actually, formula (4.21) holds for any $\mathbf{u}$. Hence, $\mathbf{B}^\top\mathbf{p}(\mathbf{u})$ denotes the direction of *steepest ascent* of J starting from a given value $\mathbf{u}$ of the control vector. This expression, sometimes called *sensitivity,* plays a key role in several algorithms for the solution of an OCP as shown, e.g., in Sect. 6.3.

In the case $\alpha \neq 0$, we have instead

$$\nabla J(\mathbf{u}) = \mathbf{B}^\top\mathbf{p} + \alpha\mathbf{u}.$$

For a general cost functional under the form $J(\mathbf{u}) = \tilde{J}(\mathbf{y}(\mathbf{u}), \mathbf{u})$ and a state problem like (4.1), the adjoint problem reads

$$\mathbf{A}^\top\mathbf{p} = (\nabla_\mathbf{y}\tilde{J}(\mathbf{y}, \mathbf{u}))^\top \tag{4.22}$$

and the (reduced) gradient is

$$\nabla J(\mathbf{u}) = \mathbf{B}^\top\mathbf{p} + \nabla_\mathbf{u}\tilde{J}(\mathbf{y}, \mathbf{u}). \tag{4.23}$$

[2] Recall that $dJ(\mathbf{u}) \in \mathbb{R}^{q\times q}$ is the differential of J at $\mathbf{u}$, and that for any $\mathbf{v} \in \mathbb{R}^q$, $dJ(\mathbf{u})(\mathbf{h}) = (\nabla J(\mathbf{u}), \mathbf{h})$, being $\nabla J(\mathbf{u}) \in \mathbb{R}^q$ the gradient of J at $\mathbf{u}$.

Consider for instance the case

$$\tilde{J}(\mathbf{y}, \mathbf{u}) = \frac{1}{2}|\mathbf{C}\mathbf{y} - \mathbf{z}_d|^2_{\mathbf{M}} + \frac{\alpha}{2}|\mathbf{u}|^2_{\mathbf{N}},$$

where $\alpha > 0$, $\mathbf{C} \in \mathbb{R}^{m\times N}$, $\mathbf{M} \in \mathbb{R}^{m\times m}$ and $\mathbf{N} \in \mathbb{R}^{q\times q}$ are two symmetric and positive definite matrices, and

$$|\mathbf{z}|_{\mathbf{M}} = (\mathbf{M}\mathbf{z}, \mathbf{z})^{1/2}, \qquad |\mathbf{u}|_{\mathbf{N}} = (\mathbf{N}\mathbf{u}, \mathbf{u})^{1/2} \qquad \forall \mathbf{z} \in \mathbb{R}^m, \forall \mathbf{u} \in \mathbb{R}^q.$$

Then, we find

$$\nabla_{\mathbf{y}}\tilde{J}(\mathbf{y}, \mathbf{u}) = \mathbf{C}^\top\mathbf{M}(\mathbf{C}\mathbf{y} - \mathbf{z}_d), \qquad \nabla_{\mathbf{u}}\tilde{J}(\mathbf{y}, \mathbf{u}) = \alpha\mathbf{N}\mathbf{u}$$

and the optimality conditions read as follows (see Exercise 2 for a detailed derivation):

$$\begin{cases} \mathbf{A}\hat{\mathbf{y}} = \mathbf{f} + \mathbf{B}\hat{\mathbf{u}} \\ \mathbf{A}^\top\hat{\mathbf{p}} = \mathbf{C}^\top\mathbf{M}(\mathbf{C}\hat{\mathbf{y}} - \mathbf{z}_d) \\ \left(\mathbf{B}^\top\hat{\mathbf{p}} + \alpha\mathbf{N}\hat{\mathbf{u}}, \mathbf{v} - \hat{\mathbf{u}}\right) \geq 0 \qquad \forall\, \mathbf{v} \in \mathcal{U}_{ad}. \end{cases} \tag{4.24}$$

In the unconstrained case ($\mathcal{U}_{ad} = \mathcal{U}$) we obtain $\mathbf{B}^\top\hat{\mathbf{p}} + \alpha N\hat{\mathbf{u}} = \mathbf{0}$ so that the system (4.24) now can be written in the form

$$\begin{bmatrix} \mathbf{C}^\top\mathbf{M}\mathbf{C} & 0 & -\mathbf{A}^\top \\ 0 & \alpha\mathbf{N} & \mathbf{B}^\top \\ -\mathbf{A} & \mathbf{B} & 0 \end{bmatrix} \begin{bmatrix} \mathbf{y} \\ \mathbf{u} \\ \mathbf{p} \end{bmatrix} = \begin{bmatrix} \mathbf{C}^\top\mathbf{M}\mathbf{z}_d \\ \mathbf{0} \\ -\mathbf{f} \end{bmatrix}. \tag{4.25}$$

4.1.4 Direct vs. Adjoint Approach

So far we have considered the *adjoint approach* to compute the gradient of the cost functional, for each $\mathbf{u} \in \mathbb{R}^q$. This approach requires to solve the state problem to compute $\mathbf{y}$ and then the adjoint problem to compute $\mathbf{p}$, in order to evaluate the expression of the gradient in (4.23).

Alternatively, the *direct approach* provides the expression of the gradient without requiring the solution of the adjoint problem. However, $q + 1$ problems have to be solved in this case, thus making this approach computationally attractive only if the dimension q of the control variable is small. This may occur, for instance, if the control variable is parametrized, that is, it is defined as a function of (possibly few) *design parameters.* In literature the direct approach is also called *sensitivity* approach, since the q additional problems to be solved are meant to characterize the sensitivity of J with respect to each control variable. We only describe the direct approach in the algebraic case, while we refer to, e.g., [120, 153] for the case of PDE-constrained optimization problems. A possible way to speed up the evaluation of the derivative of a functional depending on the solution of a PDE is proposed, e.g., in [116], where it is applied to both a parameter identification problem for a Navier-Stokes flow and an OCP for a diffusion-reaction equation.

To compare the two approaches, let us consider the problem

$$\tilde{J}(\mathbf{y}, \mathbf{u}) \to \min \quad \text{subject to} \qquad \mathbf{g}\,(\mathbf{y}, \mathbf{u}) = \mathbf{0}, \quad \mathbf{y} \in V, \mathbf{u} \in \mathcal{U}.$$

Assuming that the Jacobian matrix $\nabla_{\mathbf{y}}\mathbf{g}(\mathbf{y}, \mathbf{u})$ is invertible, the state problem $\mathbf{g}(\mathbf{y}, \mathbf{u}) = \mathbf{0}$ implicitly provides $\mathbf{y} = \mathbf{y}(\mathbf{u})$, and the gradient of $J(\mathbf{u}) = \tilde{J}(\mathbf{y}(\mathbf{u}), \mathbf{u})$ can be obtained by the *chain rule*

$$\nabla J(\mathbf{u}) = \nabla_{\mathbf{u}}\tilde{J}(\mathbf{y}(\mathbf{u}), \mathbf{u}) + \nabla_{\mathbf{y}}\tilde{J}(\mathbf{y}(\mathbf{u}), \mathbf{u})\nabla_{\mathbf{u}}\mathbf{y}(\mathbf{u}), \tag{4.26}$$

with

$$\nabla_{\mathbf{u}}\tilde{J}(\mathbf{y}(\mathbf{u}),\mathbf{u}) = \left(\frac{\partial \tilde{J}}{\partial u_1}(\mathbf{y}(\mathbf{u}),\mathbf{u}),\ldots,\frac{\partial \tilde{J}}{\partial u_q}(\mathbf{y}(\mathbf{u}),\mathbf{u})\right) \in \mathbb{R}^{1\times q},$$

$$\nabla_{\mathbf{y}}\tilde{J}(\mathbf{y}(\mathbf{u}),\mathbf{u}) = \left(\frac{\partial \tilde{J}}{\partial y_1}(\mathbf{y}(\mathbf{u}),\mathbf{u}),\ldots,\frac{\partial \tilde{J}}{\partial y_N}(\mathbf{y}(\mathbf{u}),\mathbf{u})\right) \in \mathbb{R}^{1\times N}.$$

In equation (4.26)$\nabla_{\mathbf{u}}\mathbf{y}(\mathbf{u}) \in \mathbb{R}^{N\times q}$ is the *sensitivity matrix*, whose components represent the sensitivities of the state with respect to the control variables

$$(\nabla_{\mathbf{u}}\mathbf{y}(\mathbf{u}))_{ij} = \frac{\partial y_i}{\partial u_j}(\mathbf{u}), \qquad i=1,\ldots,N,\ j=1,\ldots,q.$$

To obtain it, we derive the expression $\mathbf{g}(\mathbf{y}(\mathbf{u}),\mathbf{u}) = \mathbf{0}$ with respect to $\mathbf{u}$, thus finding

$$\mathbf{0} = \nabla_{\mathbf{y}}\mathbf{g}(\mathbf{y}(\mathbf{u}),\mathbf{u})\nabla_{\mathbf{u}}\mathbf{y}(\mathbf{u}) + \nabla_{\mathbf{u}}\mathbf{g}(\mathbf{y}(\mathbf{u}),\mathbf{u}). \tag{4.27}$$

Substituting (4.27) in (4.26) gives

$$\nabla J(\mathbf{u}) = \nabla_{\mathbf{u}}\tilde{J}(\mathbf{y}(\mathbf{u}),\mathbf{u}) - \nabla_{\mathbf{y}}\tilde{J}(\mathbf{y}(\mathbf{u}),\mathbf{u})\left(\nabla_{\mathbf{y}}\mathbf{g}(\mathbf{y}(\mathbf{u}),\mathbf{u})\right)^{-1}\nabla_{\mathbf{u}}\mathbf{g}(\mathbf{y}(\mathbf{u}),\mathbf{u}). \tag{4.28}$$

At this point we have two ways to compute the gradient, depending upon the way the second term on the right-hand side of (4.28) is treated:

- in the *direct approach*, we solve for $\nabla_{\mathbf{u}}\mathbf{y}(\mathbf{u})$ in (4.27) – that is, we compute the q vectors $\partial\mathbf{y}/\partial u_j$, $j=1,\ldots,q$, solutions of the q (possibly, nonlinear) systems

$$\nabla_{\mathbf{y}}\mathbf{g}(\mathbf{y}(\mathbf{u}),\mathbf{u})\frac{\partial \mathbf{y}}{\partial u_j}(\mathbf{u}) = -\frac{\partial \mathbf{g}}{\partial u_j}(\mathbf{y}(\mathbf{u}),\mathbf{u})$$

  and evaluate the gradient by (4.26);
- in the *adjoint approach* we solve for $\mathbf{p}$ in

$$\left(\nabla_{\mathbf{y}}\mathbf{g}(\mathbf{y}(\mathbf{u}),\mathbf{u})\right)^\top \mathbf{p} = \left(\nabla_{\mathbf{y}}\tilde{J}(\mathbf{y}(\mathbf{u}),\mathbf{u})\right)^\top$$

  and then, using (4.28), we evaluate the gradient as

$$\nabla J(\mathbf{u}) = \nabla_{\mathbf{u}}\tilde{J}(\mathbf{y}(\mathbf{u}),\mathbf{u}) - \mathbf{p}^\top\nabla_{\mathbf{u}}\mathbf{g}(\mathbf{y}(\mathbf{u}),\mathbf{u}) \in \mathbb{R}^{1\times q}.$$

The size of $\nabla_{\mathbf{y}}\mathbf{g}(\mathbf{y},\mathbf{u})$ is $N\times N$; this makes the direct approach much more expensive than the adjoint one since we need to solve a $N\times N$ system q times – as involved by (4.27) – to evaluate $\nabla J(\mathbf{u})$. For this reason, the direct approach is usually out of reach from a computational standpoint; recall that the gradient of the cost function must be evaluated at every iteration when using, e.g., a descent method. On the other hand, to evaluate $\nabla J(\mathbf{u})$ the adjoint approach requires the solution of a $N\times N$ system only once; however, the adjoint solver may be rather involved to implement when algebraic systems arise from the discretization of a nonlinear or time-dependent OCP. In this latter case, provided q is very small, the direct approach is a viable alternative to get $\nabla J(\mathbf{u})$.

Remark 4.3. In the case of a linear state system, $\mathbf{g}(\mathbf{y},\mathbf{u}) = \mathbf{f}+\mathbf{B}\mathbf{u}-\mathbf{A}\mathbf{y}$, so that $\nabla_{\mathbf{y}}\mathbf{g}(\mathbf{y},\mathbf{u}) = -\mathbf{A}$, $\nabla_{\mathbf{u}}\mathbf{g}(\mathbf{y},\mathbf{u}) = \mathbf{B} = (\mathbf{b}_1,\ldots,\mathbf{b}_q) \in \mathbb{R}^{N\times q}$. Problem (4.27) yields the linear systems

$$\mathbf{A}\frac{\partial \mathbf{y}}{\partial u_i} - \mathbf{b}_i = \mathbf{0}, \qquad i=1,\ldots,q.$$

The adjoint problem takes instead the form (see (4.22))

$$\mathbf{A}^\top\mathbf{p} = -\left(\nabla_{\mathbf{y}}\tilde{J}(\mathbf{y},\mathbf{u})\right)^\top,$$

so that the gradient becomes

$$\nabla J(\mathbf{u}) = \nabla_{\mathbf{u}} \tilde{J}(\mathbf{y}(\mathbf{u}), \mathbf{u}) - \mathbf{B}^\top \mathbf{p} \in \mathbb{R}^{1\times q}. \qquad \bullet$$

Both these approaches fit the so-called *reduced space approach* (also denoted as *Nested Analysis and Design* paradigm), for which the solver of the state system is nested within the optimization solver. The former must supply the latter with updated state variables which fulfill the state problem and corresponding sensitivity information (e.g., the gradient of the cost functional). The optimization solver then uses the information to generate the next value of the control variables, according, e.g., to a *descent method* (see Sect. 3.1); the updated control is passed as input to the PDE solver at the next step, until convergence. This approach is also referred to as *iterative* or *black-box* method; we will discuss it in detail in Chapter 6. Note that the state problem is implicitly fulfilled at each step, and that the optimization solver only needs to store the control variables.

4.2 Formulation as a Constrained Optimization Problem

In this section we formulate the algebraic OCP (4.1)–(4.4) as a constrained optimization problem, by taking advantage of the approaches of Sect. 2.3, based on Lagrange multipliers and Karush-Kuhn-Tucker multipliers.

4.2.1 Lagrange Multipliers

The algebraic OCP (4.1)–(4.2) can be rewritten as

$$\min_{\mathbf{y}\in V, \mathbf{u}\in \mathcal{U}_{ad}} \tilde{J}(\mathbf{y}, \mathbf{u}) \qquad \text{subject to} \qquad \mathbf{g}(\mathbf{y}, \mathbf{u}) = \mathbf{0}, \tag{4.29}$$

with

$$\mathbf{g}(\mathbf{y}, \mathbf{u}) = \mathbf{f} + \mathbf{B}\mathbf{u} - \mathbf{A}\mathbf{y}. \tag{4.30}$$

To address the equality constraint $\mathbf{g}(\mathbf{y}, \mathbf{u}) = \mathbf{0}$, we introduce the Lagrangian $\mathcal{L} : \mathbb{R}^{2N+q} \to \mathbb{R}$,

$$\mathcal{L}(\mathbf{y}, \mathbf{u}, \mathbf{p}) = \tilde{J}(\mathbf{y}, \mathbf{u}) + (\mathbf{p}, \mathbf{g}(\mathbf{y}, \mathbf{u})) = \frac{1}{2}|\mathbf{y} - \mathbf{z}_d|^2 + \frac{\alpha}{2}|\mathbf{u}|^2 + (\mathbf{p}, \mathbf{f} + \mathbf{B}\mathbf{u} - \mathbf{A}\mathbf{y}),$$

where $\mathbf{p} \in \mathbb{R}^N$ plays the role of *Lagrange multiplier*. The corresponding system of (first-order, necessary) optimality conditions (4.16) reads:

$$\begin{cases} \nabla_{\mathbf{p}} \mathcal{L}(\hat{\mathbf{y}}, \hat{\mathbf{u}}, \hat{\mathbf{p}}) = \mathbf{0} \\ \nabla_{\mathbf{y}} \mathcal{L}(\hat{\mathbf{y}}, \hat{\mathbf{u}}, \hat{\mathbf{p}}) = \mathbf{0} \\ (\nabla_{\mathbf{u}} \mathcal{L}(\hat{\mathbf{y}}, \hat{\mathbf{u}}, \hat{\mathbf{p}}), (\mathbf{v} - \hat{\mathbf{u}})) \geq 0 \qquad \text{for all } \mathbf{v} \in \mathcal{U}_{ad}. \end{cases}$$

We remark that $\hat{\mathbf{p}}$ is in fact the adjoint variable, i.e. the solution of the adjoint equation (4.22). In other words, the state equation can be easily recovered by differentiating the Lagrangian function with respect to the adjoint variable $\mathbf{p}$ and requiring it to be zero, that is

$$\mathbf{A}\hat{\mathbf{y}} = \mathbf{f} + \mathbf{B}\hat{\mathbf{u}}.$$

In the same way, by differentiating the Lagrangian with respect to the state variable $\mathbf{y}$, we easily obtain the adjoint equation

$$\mathbf{A}^\top \hat{\mathbf{p}} = -(\hat{\mathbf{y}} - \mathbf{z}_d).$$

Finally, the variational inequality follows from differentiation of the Lagrangian with respect to the control variable $\mathbf{u}$

$$(\alpha\hat{\mathbf{u}} - \mathbf{B}^\top \hat{\mathbf{p}}, \mathbf{v} - \hat{\mathbf{u}}) \geq 0 \qquad \forall\, \mathbf{v} \in \mathcal{U}_{ad}.$$

Note that we still have to account for the constraint $\mathbf{v} \in \mathcal{U}_{ad}$ (we will return on this point in the following subsection). As we will see in the next chapters, deriving the optimality conditions of an OCP (and, more generally, a PDE-constrained optimization problem) by introducing Lagrange multipliers and the corresponding Lagrangian function is a general approach, valid for any cost functional and any state equation, provided suitable assumptions on regularity/differentiability of the involved expressions are made. Formulating the OCP (4.29) as a constrained optimization problem fits the so-called *full space approach* (also denoted as *Simultaneous Design and Analysis* approach), in which the constraints provided by the set of equations corresponding to the state system are explicitly specified.

Following the notation introduced in Sect. 4.1.4 and extending the conclusions we obtained so far to the case of a more general OCP (*without control contraints*), the first-order optimality conditions state that there exists $\hat{\mathbf{p}}$ such that the system

$$\begin{cases} (\nabla_{\mathbf{y}} \tilde{J}(\hat{\mathbf{y}}, \hat{\mathbf{u}}))^\top - (\nabla_{\mathbf{y}} \mathbf{g}(\hat{\mathbf{y}}, \hat{\mathbf{u}}))^\top \hat{\mathbf{p}} = \mathbf{0} \\ (\nabla_{\mathbf{u}} \tilde{J}(\hat{\mathbf{y}}, \hat{\mathbf{u}})))^\top - (\nabla_{\mathbf{u}} \mathbf{g}(\hat{\mathbf{y}}, \hat{\mathbf{u}}))^\top \hat{\mathbf{p}} = \mathbf{0} \\ \mathbf{g}(\hat{\mathbf{y}}, \hat{\mathbf{u}}) = \mathbf{0} \end{cases} \tag{4.31}$$

is fulfilled at the optimum $(\hat{\mathbf{y}}, \hat{\mathbf{u}}, \hat{\mathbf{p}})$; note the formal analogy with the system (3.42).

If $\tilde{J}(\mathbf{y}, \mathbf{u})$ is quadratic and $\mathbf{g}(\mathbf{y}, \mathbf{u})$ is linear with respect to both the state $\mathbf{y}$ and the control $\mathbf{u}$, problem (4.29) is nothing but a *quadratic programming problem*. Solving the linear system (4.31) provides the optimal state, control and adjoint variables. In the case the OCP stems from the discretization of a PDE-constrained optimization, the system (4.31) has very large dimension; moreover, in the linear-quadratic case, its *saddle-point* structure calls for efficient preconditioners; see Sect. 6.6 for a detailed discussion.

If the state system is nonlinear, then the equations in (4.31) are linear in the multiplier $\mathbf{p}$ but nonlinear in the optimization variables $\mathbf{y}$ and $\mathbf{u}$. By using the standard Newton method, we look for the search direction $\delta\mathbf{x} = (\delta\mathbf{y}, \delta\mathbf{u})$ and the multiplier update $\delta\mathbf{p}$, which fulfill, at each step $k = 0, 1, \ldots$, the following linear system:

$$\begin{bmatrix} H_{\mathbf{yy}}(\mathbf{y}_k, \mathbf{u}_k, \mathbf{p}_k) & H_{\mathbf{yu}}(\mathbf{y}_k, \mathbf{u}_k, \mathbf{p}_k) & (\nabla_{\mathbf{y}} \mathbf{g}(\mathbf{y}_k, \mathbf{u}_k))^\top \\ H_{\mathbf{uy}}(\mathbf{y}_k, \mathbf{u}_k, \mathbf{p}_k) & H_{\mathbf{uu}}(\mathbf{y}_k, \mathbf{u}_k, \mathbf{p}_k) & (\nabla_{\mathbf{u}} \mathbf{g}(\mathbf{y}_k, \mathbf{u}_k))^\top \\ \nabla_{\mathbf{y}} \mathbf{g}(\mathbf{y}_k, \mathbf{u}_k) & \nabla_{\mathbf{u}} \mathbf{g}(\mathbf{y}_k, \mathbf{u}_k) & 0 \end{bmatrix} \begin{bmatrix} \delta\mathbf{y} \\ \delta\mathbf{u} \\ \delta\mathbf{p} \end{bmatrix}$$
$$= - \begin{bmatrix} (\nabla_{\mathbf{y}} \tilde{J}(\mathbf{y}_k, \mathbf{u}_k))^\top - (\nabla_{\mathbf{y}} \mathbf{g}(\mathbf{y}_k, \mathbf{u}_k))^\top \mathbf{p}_k \\ (\nabla_{\mathbf{u}} \tilde{J}(\mathbf{y}_k, \mathbf{u}_k))^\top - (\nabla_{\mathbf{u}} \mathbf{g}(\mathbf{y}_k, \mathbf{u}_k))^\top \mathbf{p}_k \\ \mathbf{g}(\mathbf{y}_k, \mathbf{u}_k) \end{bmatrix}, \tag{4.32}$$

where the blocks

$$\mathbf{H}_{\mathbf{yy}}(\mathbf{y}, \mathbf{u}, \mathbf{p}) = D^2_{\mathbf{yy}} \tilde{J}(\mathbf{y}, \mathbf{u}) + \sum_{i=1}^N p_i\, D^2_{\mathbf{yy}} \mathbf{g}_i(\mathbf{y}, \mathbf{u}),$$

$$\mathbf{H}_{\mathbf{uu}}(\mathbf{y}, \mathbf{u}, \mathbf{p}) = D^2_{\mathbf{uu}} \tilde{J}(\mathbf{y}, \mathbf{u}) + \sum_{i=1}^N p_i\, D^2_{\mathbf{uu}} \mathbf{g}_i(\mathbf{y}, \mathbf{u}),$$

$$\mathbf{H}_{\mathbf{yu}}(\mathbf{y},\mathbf{u},\mathbf{p}) = D^2_{\mathbf{yu}}\tilde{J}(\mathbf{y},\mathbf{u}) + \sum_{i=1}^{N} p_i\, D^2_{\mathbf{yu}}\mathbf{g}_i(\mathbf{y},\mathbf{u}),$$

$$\mathbf{H}_{\mathbf{uy}}(\mathbf{y},\mathbf{u},\mathbf{p}) = D^2_{\mathbf{uy}}\tilde{J}(\mathbf{y},\mathbf{u}) + \sum_{i=1}^{N} p_i\, D^2_{\mathbf{uy}}\mathbf{g}_i(\mathbf{y},\mathbf{u}),$$

correspond to the Hessian of the Lagrangian function; then, we set

$$\begin{bmatrix}\mathbf{y}_{k+1}\\ \mathbf{u}_{k+1}\\ \mathbf{p}_{k+1}\end{bmatrix} = \begin{bmatrix}\mathbf{y}_{k}\\ \mathbf{u}_{k}\\ \mathbf{p}_{k}\end{bmatrix} + \begin{bmatrix}\delta\mathbf{y}\\ \delta\mathbf{u}\\ \delta\mathbf{p}\end{bmatrix}.$$

Note the formal analogy between (4.32) and (3.44). System (4.32) corresponds to the (necessary and sufficient) optimality conditions for the linear-quadratic problem

$$\begin{aligned}&\frac{1}{2}\begin{bmatrix}\delta\mathbf{y}\\ \delta\mathbf{u}\end{bmatrix}^\top \mathbf{H}(\mathbf{y}_k,\mathbf{u}_k,\mathbf{p}_k)\begin{bmatrix}\delta\mathbf{y}\\ \delta\mathbf{u}\end{bmatrix} + \begin{bmatrix}\nabla_{\mathbf{y}}\mathbf{g}(\mathbf{y}_k,\mathbf{u}_k)\\ \nabla_{\mathbf{u}}\mathbf{g}(\mathbf{y}_k,\mathbf{u}_k)\end{bmatrix}^\top\begin{bmatrix}\delta\mathbf{y}\\ \delta\mathbf{u}\end{bmatrix} \to \min\\ &\text{subject to}\\ &\nabla_{\mathbf{y}}\mathbf{g}(\mathbf{y}_k,\mathbf{u}_k)\delta\mathbf{y} + \nabla_{\mathbf{u}}\mathbf{g}(\hat{\mathbf{y}}_k,\hat{\mathbf{u}}_k)\delta\mathbf{u} + \mathbf{g}(\hat{\mathbf{y}}_k,\hat{\mathbf{u}}_k) = \mathbf{0}\end{aligned} \tag{4.33}$$

where

$$\mathbf{H}(\mathbf{y}_k,\mathbf{u}_k,\mathbf{p}_k) = \begin{bmatrix}\mathbf{H}_{\mathbf{yy}}(\mathbf{y}_k,\mathbf{u}_k,\mathbf{p}_k) & \mathbf{H}_{\mathbf{yu}}(\mathbf{y}_k,\mathbf{u}_k,\mathbf{p}_k)\\ \mathbf{H}_{\mathbf{uy}}(\mathbf{y}_k,\mathbf{u}_k,\mathbf{p}_k) & \mathbf{H}_{\mathbf{uu}}(\mathbf{y}_k,\mathbf{u}_k,\mathbf{p}_k)\end{bmatrix}.$$

Solving a sequence of quadratic programs under the form (4.33) yields the so-called Sequential Quadratic Programming (SQP) method, which indeed is equivalent to use a Newton method applied on the optimality system. For this reason, SQP is also referred to as *Lagrange-Newton method.* The SQP method provides a locally quadratic convergent method for finding stationary points of constrained optimization problems; this technique will be exploited for the solution of OCPs involving nonlinear PDEs, see Sect. 9.6.2.

The approach described above to find the optimal solution of (4.29) is an instance of *all-at-once* or *one-shot* method, extensively analyzed in Sect. 9.6.2. Note that in this case solving system (4.31) requires to store all the variables, not just the control one.

4.2.2 Control Constraints: Karush-Kuhn-Tucker Multipliers

So far, we have exploited the Lagrangian function to eliminate the equality constraints expressed by the state equation. However, a similar approach can be used also to deal with the constraint $\mathbf{u} \in \mathcal{U}_{ad}$, provided $\mathcal{U}_{ad}$ is identified by a set of inequality constraints.

For the sake of analysis, here we consider the case – yet, of practical relevance – of admissible controls $\mathcal{U}_{ad}$ defined through a set of *box constraints*, under the form

$$\mathcal{U}_{ad} = \{\mathbf{u} \in \mathbb{R}^q \,:\, u_{a,i} \le u_i \le u_{b,i},\ i = 1,\dots,q\} \subset \mathcal{U}. \tag{4.34}$$

If we consider the minimization of the cost functional (4.2), the variational inequality expressing the *minimum principle* in (4.16) can be rewritten as

$$(\mathbf{B}^\top\hat{\mathbf{p}} + \alpha\hat{\mathbf{u}}, \mathbf{v}) \ge (\mathbf{B}^\top\hat{\mathbf{p}} + \alpha\hat{\mathbf{u}}, \hat{\mathbf{u}}) \qquad \forall\, \mathbf{v} \in \mathcal{U}_{ad}, \tag{4.35}$$

see Sect. 4.1.3. Hence, we can recognize that $\hat{\mathbf{u}}$ minimizes over $\mathcal{U}_{ad}$ the function

$$\mathbf{v} \mapsto (\nabla_{\mathbf{u}}\mathcal{L}(\hat{\mathbf{y}},\hat{\mathbf{u}},\hat{\mathbf{p}}), \mathbf{v}) = (\mathbf{B}^\top\hat{\mathbf{p}} + \alpha\hat{\mathbf{u}}, \mathbf{v}).$$

Thanks to (4.34), the minimization can be done componentwise. Then (4.35) is equivalent to

$$\left(\mathbf{B}^\top \hat{\mathbf{p}} + \alpha \hat{\mathbf{u}}\right)_i \hat{u}_i = \min_{u_{a,i} \le u_i \le u_{b,i}} \left(\mathbf{B}^\top \hat{\mathbf{p}} + \alpha \hat{\mathbf{u}}\right)_i u_i, \qquad i = 1, \dots, q.$$

From this observation, it follows that

$$\hat{u}_i = \begin{cases} u_{a,i} & \text{if } (\mathbf{B}^\top \hat{\mathbf{p}} + \alpha \hat{\mathbf{u}})_i > 0 \\ \tilde{u}_i \in [u_{a,i}, u_{b,i}] & \text{if } (\mathbf{B}^\top \hat{\mathbf{p}} + \alpha \hat{\mathbf{u}})_i = 0 \\ u_{b,i} & \text{if } (\mathbf{B}^\top \hat{\mathbf{p}} + \alpha \hat{\mathbf{u}})_i < 0. \end{cases}$$

From the minimum principle we cannot recover any direct information on the value of those components u_i associated to an index i for which $(\mathbf{B}^\top \hat{\mathbf{p}} + \alpha \hat{\mathbf{u}})_i = 0$. However, in this case

$$\hat{u}_i = -\frac{1}{\alpha} (\mathbf{B}^\top \hat{\mathbf{p}})_i.$$

Furthermore, if $\mathbf{B}^\top \hat{\mathbf{p}} + \alpha \hat{\mathbf{u}} \neq \mathbf{0}$, then each component of $\hat{\mathbf{u}}$ is either equal to $u_{a,i}$ or to $u_{b,i}$: in this case, $\hat{\mathbf{u}}$ is called *bang-bang* control. We note that (4.35) yields

$$\hat{u}_i = \min \left\{ u_{b,i}, \max \left\{ u_{a,i}, -\frac{1}{\alpha} (\mathbf{B}^\top \hat{\mathbf{p}})_i \right\} \right\},$$

that is to say,

$$\hat{\mathbf{u}} = Pr_{\mathcal{U}_{ad}} \left(-\frac{1}{\alpha} \mathbf{B}^\top \hat{\mathbf{p}} \right),$$

where $Pr_{\mathcal{U}_{ad}}(\mathbf{x})$ is the projection[3] of $\mathbf{x} \in \mathbb{R}^q$ over the n-dimensional hypercube $\mathcal{U}_{ad}$.

The minimum principle (4.35) can be more conveniently expressed in terms of Karush-Kuhn-Tucker (KKT) multipliers. Let us introduce the Lagrangian

$$\mathcal{L}(\mathbf{y}, \mathbf{u}, \mathbf{p}) = \tilde{J}(\mathbf{y}, \mathbf{u}) + (\mathbf{p}, \mathbf{g}(\mathbf{y}, \mathbf{u})) + (\mathbf{u} - \mathbf{u}_a, \boldsymbol{\lambda}_a) + (\mathbf{u}_b - \mathbf{u}, \boldsymbol{\lambda}_b) \tag{4.36}$$

similarly to (2.28); $\boldsymbol{\lambda}_a$ and $\boldsymbol{\lambda}_b$ are the Lagrange multipliers associated to the inequality constraints $\mathbf{u} - \mathbf{u}_a \ge \mathbf{0}$ and $\mathbf{u}_b - \mathbf{u} \ge \mathbf{0}$, respectively. Here we denote by $\mathbf{u}_a$, $\mathbf{u}_b$ the vectors whose components are given by $(\mathbf{u}_a)_i = u_{a,i}$, $(\mathbf{u}_b)_i = u_{b,i}$, so that the box constraints can be expressed as $\mathbf{u}_a \le \mathbf{u} \le \mathbf{u}_b$.

According to the results of Sect. 2.3.2, if $\hat{\mathbf{u}}$ is an optimal control for the problem (4.29), (4.30),(4.34), with associated state $\hat{\mathbf{y}}$, there exist KKT multipliers $\hat{\mathbf{p}} \in \mathbb{R}^N$ and $\hat{\boldsymbol{\lambda}}_a, \hat{\boldsymbol{\lambda}}_b \in \mathbb{R}^q$ not both vanishing, such that the following KKT optimality conditions hold:

$$\nabla_{\mathbf{p}} \mathcal{L}(\hat{\mathbf{y}}, \hat{\mathbf{u}}, \hat{\mathbf{p}}, \boldsymbol{\lambda}_\mathbf{a}, \boldsymbol{\lambda}_\mathbf{b}) = \mathbf{0} \tag{4.37}$$
$$\nabla_{\mathbf{y}} \mathcal{L}(\hat{\mathbf{y}}, \hat{\mathbf{u}}, \hat{\mathbf{p}}, \boldsymbol{\lambda}_\mathbf{a}, \boldsymbol{\lambda}_\mathbf{b}) = \mathbf{0} \tag{4.38}$$
$$\nabla_{\mathbf{u}} \mathcal{L}(\hat{\mathbf{y}}, \hat{\mathbf{u}}, \hat{\mathbf{p}}, \boldsymbol{\lambda}_\mathbf{a}, \boldsymbol{\lambda}_\mathbf{b}) = \mathbf{0} \tag{4.39}$$
$$\nabla_{\boldsymbol{\lambda}_a} \mathcal{L}(\hat{\mathbf{y}}, \hat{\mathbf{u}}, \hat{\mathbf{p}}, \boldsymbol{\lambda}_\mathbf{a}, \boldsymbol{\lambda}_\mathbf{b}) \le \mathbf{0} \tag{4.40}$$
$$\nabla_{\boldsymbol{\lambda}_b} \mathcal{L}(\hat{\mathbf{y}}, \hat{\mathbf{u}}, \hat{\mathbf{p}}, \boldsymbol{\lambda}_\mathbf{a}, \boldsymbol{\lambda}_\mathbf{b}) \le \mathbf{0} \tag{4.41}$$
$$\hat{\boldsymbol{\lambda}}_a \ge \mathbf{0}, \quad \hat{\boldsymbol{\lambda}}_b \ge \mathbf{0}, \qquad (\mathbf{u} - \mathbf{u}_a, \boldsymbol{\lambda}_a) = 0, \quad (\mathbf{u}_b - \mathbf{u}, \boldsymbol{\lambda}_b) = 0. \tag{4.42}$$

As in the case of Lagrange multpliers, the adjoint equation can be recovered from the vanishing of the derivative of the Lagrangian functional $\mathcal{L}$ with respect to $\mathbf{y}$. In the same way, the state equation corresponds to the vanishing of the derivative of $\mathcal{L}$ with respect to $\mathbf{p}$ (see equations (4.37) and (4.38), respectively).

[3] The projection of $\mathbf{x} \in \mathbb{R}^q$ over the box $Q = \Pi_{j=1}^q [a_j, b_j]$ is given by the vector $\mathbf{y} = Pr_Q(\mathbf{x})$, whose components are $y_j = \min\{b_j, \max\{a_j, x_j\}\}, j = 1, \dots, q$.

Equation (4.39) gives instead

$$\nabla_{\mathbf{u}}\tilde{J}(\hat{\mathbf{y}},\hat{\mathbf{u}})+\mathbf{B}^\top\hat{\mathbf{p}}-\hat{\boldsymbol{\lambda}}_a+\hat{\boldsymbol{\lambda}}_b=\mathbf{0}. \tag{4.43}$$

Equations (4.42) are also called *complementary equations.*

From (4.40)–(4.42) we can express the multipliers $\hat{\boldsymbol{\lambda}}_a$, $\hat{\boldsymbol{\lambda}}_b$ in terms of $\nabla_{\mathbf{u}}\tilde{J}(\hat{\mathbf{y}},\hat{\mathbf{u}})+\mathbf{B}^\top\hat{\mathbf{p}}$. Indeed, (4.40)–(4.41) yield

$$\nabla_{\boldsymbol{\lambda}_a}\mathcal{L}(\hat{\mathbf{y}},\hat{\mathbf{u}},\hat{\mathbf{p}},\boldsymbol{\lambda}_{\mathbf{a}},\boldsymbol{\lambda}_{\mathbf{b}})=\mathbf{u}-\mathbf{u}_a\geq\mathbf{0},$$
$$\nabla_{\boldsymbol{\lambda}_b}\mathcal{L}(\hat{\mathbf{y}},\hat{\mathbf{u}},\hat{\mathbf{p}},\boldsymbol{\lambda}_{\mathbf{a}},\boldsymbol{\lambda}_{\mathbf{b}})=\mathbf{u}_b-\mathbf{u}\geq\mathbf{0}$$

that is, the expression of the bound constraints. Then:

- for the complementary equations, either

  $$\hat{\lambda}_{b,i}=0 \quad\text{when}\quad \hat{u}_{b,i}-u_i>0,$$

  so that $\hat{\lambda}_{a,i}=(\nabla_{\mathbf{u}}\tilde{J}(\hat{\mathbf{y}},\hat{\mathbf{u}})+\mathbf{B}^\top\hat{\mathbf{p}})_i\geq 0$, or

  $$\hat{\lambda}_{b,i}>0 \quad\text{when}\quad \hat{u}_{b,i}-u_i=0,$$

  so that necessarily, $\hat{\lambda}_{a,i}=0$ and then $\hat{\boldsymbol{\lambda}}_{b,i}=0-(\nabla_{\mathbf{u}}\tilde{J}(\hat{\mathbf{y}},\hat{\mathbf{u}})+\mathbf{B}^\top\hat{\mathbf{p}})_i>0$;
- similarly, either

  $$\hat{\lambda}_{a,i}=0 \quad\text{when}\quad u_i-\hat{u}_{a,i}>0,$$

  so that $\hat{\lambda}_{b,i}=-(\nabla_{\mathbf{u}}\tilde{J}(\hat{\mathbf{y}},\hat{\mathbf{u}})+\mathbf{B}^\top\hat{\mathbf{p}})_i\geq 0$, or

  $$\hat{\lambda}_{a,i}>0 \quad\text{when}\quad u_i-\hat{u}_{b,i}=0,$$

  so that necessarily, $\hat{\lambda}_{b,i}=0$ and then $\hat{\boldsymbol{\lambda}}_{a,i}=0+(\nabla_{\mathbf{u}}\tilde{J}(\hat{\mathbf{y}},\hat{\mathbf{u}})+\mathbf{B}^\top\hat{\mathbf{p}})_i>0$.

In conclusion, we obtain

$$\hat{\boldsymbol{\lambda}}_a=\left(\nabla_{\mathbf{u}}\tilde{J}(\hat{\mathbf{y}},\hat{\mathbf{u}})+\mathbf{B}^\top\hat{\mathbf{p}}\right)^+,\quad \hat{\boldsymbol{\lambda}}_b=\left(\nabla_{\mathbf{u}}\tilde{J}(\hat{\mathbf{y}},\hat{\mathbf{u}})+\mathbf{B}^\top\hat{\mathbf{p}}\right)^-$$

where $(\mathbf{x})^+$ and $(\mathbf{x})^-$ denote the positive and negative part of $\mathbf{x}$, respectively.

A possible strategy to solve the KKT system (4.37)–(4.42) exploits the so-called *primal-dual active set* method, which indeed generates a sequence of problems, each one characterized by a set of active and inactive control constraints, and without inequality constraints. This technique will be exploited for the solution of constrained linear-quadratic OCPs, see Sect. 6.8.

4.3 Control Problems Governed by ODEs

So far we have introduced the reader to the basic concepts and optimality conditions in a stationary decision problem, emphasizing the double nature of *optimization* and *control.* Now we consider a class of optimal control problems, still in a finite dimensional setting, with both state and control variables evolving in time.

More specifically, we deal with control problems governed by a linear autonomous ordinary differential (ODE) system. A complete analysis and, consequently, the numerical approximation of optimal control problems governed by ODEs would require mathematical tools from dynamic optimization, calculus of variation and optimal control theory which deserve an entire book. For this reason, we only introduce some basic notions, to highlight useful facts in view of the treatment of OCPs involving time-dependent PDEs. Those readers interested to deepen the analysis of OCPs for ODEs can refer, e.g., to [265, 186, 155, 216].

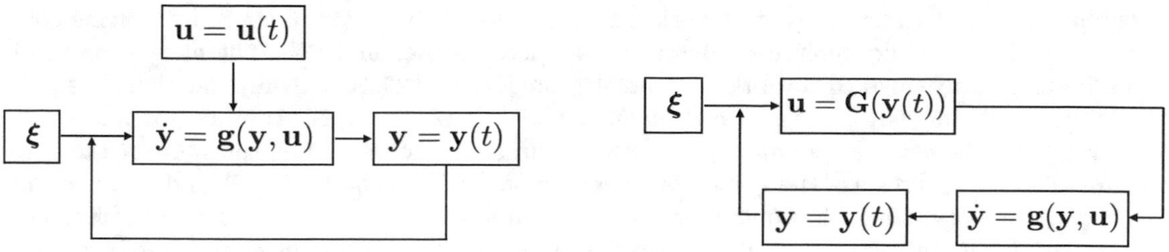

Fig. 4.1 Left: open loop control; right: closed loop (or feedback) control

In our presentation the *state variable,* describing the system evolution, is a vector valued function $\mathbf{y} : [0, T] \to \mathbb{R}^n$ subject to a *control process* (or *dynamics*) given by the ODE system

$$\dot{\mathbf{y}} = \mathbf{g}(\mathbf{y}, \mathbf{u}) \tag{4.44}$$

with initial and final conditions

$$\mathbf{y}(0) = \boldsymbol{\xi}, \qquad \mathbf{y}(T) \in \mathbf{Y}_T, \tag{4.45}$$

where $\mathbf{Y}_T \subseteq \mathbb{R}^n$ is the *final target set.* If $\mathbf{Y}_T = \mathbb{R}^n$ the end point is free. Also note that the final time T could be either fixed or unknown. Indeed, an important example where T is unknown is the so called *minimum time problem,* see Sect. 4.5.

A solution of (4.44), (4.45), also called *path* or *trajectory,* is influenced by the function $\mathbf{u} : [0, T] \to \mathbb{R}^m$ that acts as a *control* on the dynamics for $\mathbf{y}$. A control is in general subject to constraints, e.g. $\mathbf{u}(t) \in U$, for every $t \in [0, T]$, where $U \subseteq \mathbb{R}^m$ is called *control region.*

The set of controls satisfying the constraints and such that there is a unique corresponding trajectory respecting the initial and final conditions constitutes the class $\mathcal{C}$ of the *admissible controls.* A control can influence the dynamics in two ways, as shown in Figure 4.1:

- in an *open loop* control scheme, at each time a control is chosen independently of the current state;
- in a *closed loop* (also called *feedback*) control scheme, $\mathbf{u}(t)$ has a functional dependence on $\mathbf{y}(t)$ and the dynamics can be written in the form (*control synthesis*)

$$\mathbf{y} = \mathbf{g}(\mathbf{y}, \mathbf{G}(\mathbf{y})).$$

A feedback control works more efficiently since at each time the optimal control is chosen accordingly with the knowledge of the state at the same time.

The problem we have to face consists in selecting an admissible control and the corresponding trajectory in order to *minimize* (or maximize) a *cost* (or *objective*) *functional* of the form

$$J(\mathbf{u}) = \psi(\mathbf{y}(T)) + \int_0^T f(\mathbf{y}(s), \mathbf{u}(s))\, ds. \tag{4.46}$$

The term $\psi(\mathbf{y}(T))$ is called *terminal payoff.* An admissible control that minimizes (or maximizes) the cost is called *optimal control* and the corresponding trajectory an *optimal path.*

From now on we consider the case where the state equation is given by a linear system of ODEs. We address different cases of cost functional to be minimized: the scalar product of the state solution with a vector of (given) constant coefficients; the time to reach a desired state (involving the relevant concepts of *controllability* and *observability*); quadratic functional, leading to a special case of *Riccati equation.*

Remark 4.4. The description of numerical methods for the solution of control problems governed by ODEs would deserve (at least) a complete chapter and does not fit the target of this book, mainly devoted to optimal control problems for PDEs. We only mention that it is possible to distinguish between *direct methods* and *indirect methods*. Another possible strategy relies on *dynamic programming* [185]. An indirect method (such as the *indirect shooting method*) seeks a solution of the system of necessary optimality conditions [54]. A direct method first discretizes (in time) the control problem, then applies nonlinear programming techniques to the resulting finite-dimensional optimization problem, using only control and state variables as optimization variables [261, 37]. We also highlight that, in the case of indirect methods, the state and adjoint equations are integrated forward and backward in time, respectively, from their initial or final conditions. In particular, the adjoint equations are stable provided that the state equations are themselves stable [56]; integrating the adjoint equations backward in time then requires to *reconstruct* the state information at each time instant. This can be pursued, for instance, by storing the state variables at each time step during the forward integration, and then interpolating these values during the backward integration. A simple example in this respect is proposed in Sect. 4.7. •

4.4 Linear Payoff, Free End Point

In this simple case it is possible to convey the essential features of time-dependent control problems in an elementary way and to unveil some underlying geometrical aspects.

Given a non-zero vector $\boldsymbol{\eta} \in \mathbb{R}^n$ and the linear function $\psi(\mathbf{y}) = \boldsymbol{\eta} \cdot \mathbf{y}$, we consider the following linear control problem,

$$\psi(\mathbf{y}(T)) = \boldsymbol{\eta} \cdot \mathbf{y}(T) \to \min$$

for a given time $T > 0$, *subject to*

$$\begin{cases} \dot{\mathbf{y}}(t) = \mathbf{A}\mathbf{y}(t) + \mathbf{B}\mathbf{u}(t), & 0 < t < T \\ \mathbf{y}(0) = \boldsymbol{\xi} \end{cases} \tag{4.47}$$

where $\mathbf{A} \in \mathbb{R}^{n \times n}$ and $\mathbf{B} \in \mathbb{R}^{n \times m}$ are two constant matrices; note that $\mathbf{y}(T)$ is free. An *admissible control* is a bounded, piecewise continuous function $\mathbf{u} : [0, T] \to \mathbb{R}^m$. By a simple rescaling and change of variables, we may assume without loss of generality that $|\mathbf{u}(t)| \leq 1$ for every $t \in [0, T]$. Recall that a function $\mathbf{u}$ is *piecewise continuous* if every component u_j, $j = 1, \ldots, m$, is continuous at every $t \geq 0$, except for a *finite* number of jump discontinuities[4], say at $t_1, \ldots, t_N \in (0, T)$, depending in general on u_j. We denote by $\mathcal{C}_T$ the class of admissible controls; note that $\mathcal{C}_T$ is a convex set.

By a *solution* of (4.47) we mean a *continuous* function $\mathbf{y}(t) = \mathbf{y}(t; \boldsymbol{\xi}, \mathbf{u}) : [0, T] \to \mathbb{R}^n$, differentiable at each point of continuity of $\mathbf{u}$, satisfying the ODE at these points. Clearly, as dictated by the equation itself, some component of $\mathbf{y}$ may have a corner at some of the points $t_1, \ldots, t_N$. The solution of (4.47) can be represented in closed form by means of *the variation of constants formula*. In fact

Proposition 4.1. *For every* $\mathbf{u} \in \mathcal{C}_T$, *the initial value problem (4.47) has a unique solution given by*

$$\mathbf{y}(t; \boldsymbol{\xi}, \mathbf{u}) = e^{t\mathbf{A}}\boldsymbol{\xi} + \int_0^t e^{(t-s)\mathbf{A}}\mathbf{B}\mathbf{u}(s)\, ds \qquad t \in [0, T]. \tag{4.48}$$

[4] A discontinuity at $t = 0$ or $t = T$ can be eliminated without loss of generality redefining $u(0) = u(0^+)$ and $u(T) = u(T^-)$.

The exponential $e^{t\mathbf{A}}$ is given by the formula[5]

$$e^{t\mathbf{A}} = \sum_{k=0}^{\infty} \frac{\mathbf{A}^k}{n!} = \mathbf{I} + \mathbf{A} + \frac{\mathbf{A}^2}{2} + \cdots + \frac{\mathbf{A}^k}{k!} + \cdots . \tag{4.49}$$

If $\mathbf{u}^* \in \mathcal{C}_T$ and $\mathbf{y}^*(t) = \mathbf{y}(t; \boldsymbol{\xi}, \mathbf{u}^*)$ is the corresponding trajectory, our goal is to derive *necessary* and/or *sufficient conditions* for $\mathbf{u}^*$and $\mathbf{y}^*$ to be optimal.

We first give a geometric characterization of $\mathbf{y}^*(T)$. Let us introduce the set

$$\mathcal{R}(t; \boldsymbol{\xi}) = \{\mathbf{y}(t; \boldsymbol{\xi}, \mathbf{u}) : \mathbf{u} \in \mathcal{C}_T\}$$

of all states that can be reached at time t starting from $\boldsymbol{\xi}$, using all the admissible controls. We call $\mathcal{R}(t; \boldsymbol{\xi})$ the *reachable set.* We have the following

Lemma 4.1. *For every $t \in [0, T]$, $\mathcal{R}(t; \boldsymbol{\xi})$ is a bounded, convex set and $\mathbf{y}^*(T) \in \partial\mathcal{R}(T; \boldsymbol{\xi})$.*

Proof. To prove that $\mathcal{R}(t; \boldsymbol{\xi})$ is a bounded set, observe that, since $|\mathbf{u}(t)| \le 1$,

$$\begin{aligned}|\mathbf{y}(t; \boldsymbol{\xi}, \mathbf{u})| &\le \left|e^{t\mathbf{A}}\boldsymbol{\xi}\right| + \left|\int_0^t e^{(t-s)\mathbf{A}}\mathbf{B}\mathbf{u}ds\right| \le \left|e^{t\mathbf{A}}\boldsymbol{\xi}\right| + \int_0^T \left|e^{(t-s)\mathbf{A}}\mathbf{B}\mathbf{u}(s)\right| ds \\ &\le \max_{t\in[0,T]} \left|e^{t\mathbf{A}}\boldsymbol{\xi}\right| + \max_{t\in[0,T]} \left\|e^{t\mathbf{A}}\mathbf{B}\right\| T \le (|\boldsymbol{\xi}| + \|\mathbf{B}\| T) \max_{t\in[0,T]} \left\|e^{t\mathbf{A}}\right\|.\end{aligned}$$

Thus, for every $t \in [0, T]$, $\mathcal{R}(t; \boldsymbol{\xi})$ is contained in the ball B_R, centered at the origin with radius

$$R = (|\boldsymbol{\xi}| + \|\mathbf{B}\| T) \max_{t\in[0,T]} \left\|e^{t\mathbf{A}}\right\|. \tag{4.50}$$

Therefore $\mathcal{R}(t; \boldsymbol{\xi})$ is bounded, actually equi-bounded with respect to $t \in [0, T]$, since R is independent of t.

Let now $\mathbf{u}_1, \mathbf{u}_2 \in \mathcal{C}_T$ and $\mathbf{y}(t; \boldsymbol{\xi}, \mathbf{u}_1), \mathbf{y}(t; \boldsymbol{\xi}, \mathbf{u}_2) \in \mathcal{R}(t; \boldsymbol{\xi})$. Since $\mathcal{C}_T$ is a convex set, for all $\lambda \in (0, 1)$, $\mathbf{u}_\lambda = \lambda\mathbf{u}_1 + (1-\lambda)\mathbf{u}_2 \in \mathcal{C}_T$, and, because of the linearity of (4.47),

$$\mathbf{y}(t; \boldsymbol{\xi}, \mathbf{u}_\lambda) = \lambda\mathbf{y}(t; \boldsymbol{\xi}, \mathbf{u}_1) + (1-\lambda)\mathbf{y}(t; \xi, \mathbf{u}_2).$$

Therefore $\mathcal{R}(t; \boldsymbol{\xi})$ is convex. Finally, $\mathbf{y}^*(T)$ optimal means

$$\boldsymbol{\eta} \cdot (\mathbf{y} - \mathbf{y}^*(T)) \ge 0 \qquad \forall \mathbf{y} \in \mathcal{R}(T; \boldsymbol{\xi}). \tag{4.51}$$

If $\mathbf{y}^*(T)$ were an interior point of $\mathcal{R}(T; \boldsymbol{\xi})$, for $s > 0$, small, we would have

$$\mathbf{y} = \mathbf{y}^*(T) - s\boldsymbol{\eta} \in \mathcal{R}(T; \boldsymbol{\xi})$$

and

$$\boldsymbol{\eta} \cdot (\mathbf{y} - \mathbf{y}^*(T)) = -s|\boldsymbol{\eta}|^2 < 0,$$

which contradicts (4.51). □

If $K \subset \mathbb{R}^n$ is a bounded convex set, we say that a *non-zero* vector $\boldsymbol{\zeta}$ is an *interior normal to a hyperplane supporting K at $\mathbf{z} \in \partial K$*, if

$$\boldsymbol{\zeta} \cdot (\mathbf{y} - \mathbf{z}) \ge 0 \qquad \forall \mathbf{y} \in K.$$

Thus, from (4.51) and Lemma 4.1, we conclude that (see Figure 4.2)

Proposition 4.2. *$\boldsymbol{\eta}$ is an interior normal to a hyperplane supporting $\mathcal{R}(T; \boldsymbol{\xi})$ at $\mathbf{y}^*(T)$.*

We want to turn (4.51) into an operative condition able to select the optimal control $\mathbf{u}^*$.

---

[5] Recall that, since $\left\|\mathbf{A}^k\right\| \le \|\mathbf{A}\|^k$, the numerical series $\sum \frac{\|\mathbf{A}\|^k}{k!}$ is always convergent, hence the series in (4.49) is well defined for every square matrix $\mathbf{A}$.

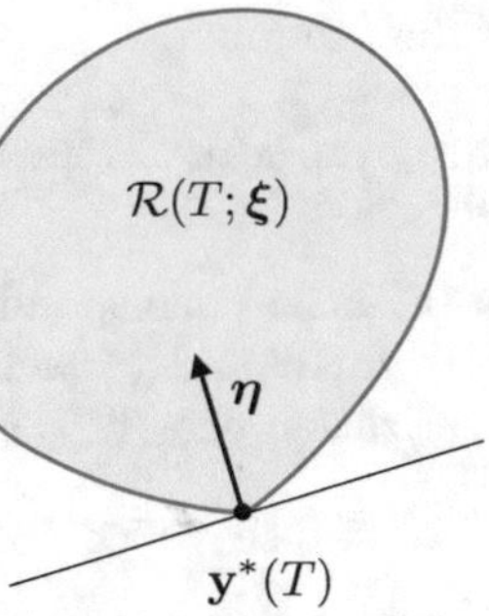

Fig. 4.2 $\boldsymbol{\eta}$ is interior normal to a supporting hyperplane for $\mathcal{R}(T;\mathbf{x})$ at $\mathbf{y}^*(T)$

Observe that (4.51) is equivalent to

$$\int_0^T \boldsymbol{\eta}\cdot e^{(T-s)\mathbf{A}}\mathbf{B}\mathbf{u}^*(s)\,ds \le \int_0^T \boldsymbol{\eta}\cdot e^{(T-s)\mathbf{A}}\mathbf{B}\mathbf{u}(s)\,ds \qquad \forall \mathbf{u}\in\mathcal{C}_T. \tag{4.52}$$

Then, introduce the *adjoint state* (multiplier)

$$\mathbf{p}^*(t) = e^{(T-t)\mathbf{A}^\top}\boldsymbol{\eta} \tag{4.53}$$

which is the unique solution of the following *adjoint problem*

$$\begin{cases} \dot{\mathbf{p}}^*(t) = -\mathbf{A}^\top\mathbf{p}^*(t), & 0<t<T \\ \mathbf{p}^*(T) = \boldsymbol{\eta}. \end{cases} \tag{4.54}$$

Note that in problem (4.54) time runs backward. The final condition at time T is also called *transversality condition*[6]. Then (4.52) reads

$$\int_0^T \mathbf{B}^\top\mathbf{p}^*(s)\cdot\mathbf{u}^*(s)\,ds \le \int_0^T \mathbf{B}^\top\mathbf{p}^*(s)\cdot\mathbf{u}(s)\,ds \qquad \forall \mathbf{u}\in\mathcal{C}_T. \tag{4.55}$$

Inequality (4.55) is a special case of the so called *minimum principle*, a cornerstone of the whole theory. We have proved the following

Theorem 4.2. *Let $\mathbf{u}^*\in\mathcal{C}_T$ and $\mathbf{y}^*(t)=\mathbf{y}^*(t;\xi,\mathbf{u}^*)$, $0\le t\le T$. Let $\mathbf{p}^*(t)$ be the solution of the adjoint problem (4.54). Then $\mathbf{u}^*$ is optimal if and only if (4.55) holds or, equivalently,*

$$\forall\mathbf{v},\ |\mathbf{v}|\le 1:\ \mathbf{B}^\top\mathbf{p}^*(t)\cdot\mathbf{u}^*(t)\le\mathbf{B}^\top\mathbf{p}^*(t)\cdot\mathbf{v} \tag{4.56}$$

for every $t\in[0,T]$, except for a finite number of points.

Proof. Clearly, (4.56) implies (4.55). Vice versa, if (4.56) is not true, there exists $\bar{\mathbf{v}}$, $|\bar{\mathbf{v}}|\le 1$, such that

$$\mathbf{B}^\top\mathbf{p}^*(t)\cdot\mathbf{u}^*(t) > \mathbf{B}^\top\mathbf{p}^*(t)\cdot\bar{\mathbf{v}} \tag{4.57}$$

at infinitely many points $t\in[0,T]$ and, in particular, at some continuity point s of $\mathbf{B}^\top\mathbf{p}^*(t)\cdot\mathbf{u}^*(t)$. Thus, (4.57) holds for each $t\in I_\varepsilon=(s-\varepsilon,s+\varepsilon)$, for some $\varepsilon>0$. Define

$$\tilde{\mathbf{v}}(s)=\begin{cases}\bar{\mathbf{v}} & \text{in } I_\varepsilon\\ \mathbf{u}^*(s) & \text{in } (0,T)\setminus I_\varepsilon.\end{cases}$$

Then $|\tilde{\mathbf{v}}(s)|\le 1$ and inserting $\mathbf{u}=\tilde{\mathbf{v}}$ into (4.55), we get a contradiction. □

[6] This is a terminology borrowed from the classical Calculus of Variation.

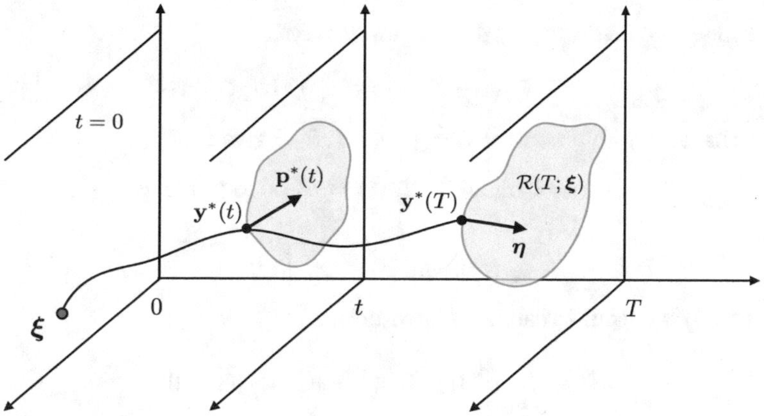

Fig. 4.3 Transporting $\boldsymbol{\eta}$ back in time along the optimal path $\mathbf{y}^*$

Let us define the *Hamiltonian function*

$$H(\mathbf{y}, \mathbf{u}, \mathbf{p}) = \mathbf{p} \cdot (\mathbf{A}\mathbf{y} + \mathbf{B}\mathbf{u}). \tag{4.58}$$

$H(\mathbf{y}^*, \mathbf{u}^*, \mathbf{p}^*)$ is constant over $[0, T]$ (see Exercise 5). As a consequence, the function

$$t \mapsto \mathbf{B}^\top \mathbf{p}^*(t) \cdot \mathbf{u}^*(t)$$

is continuous over $[0, T]$ and the pointwise minimum principle (4.56) holds for every $t \in [0, T]$.
Using (4.55) we can prove the following result (see Exercise 6)

Corollary 4.1. *For every* $t \in [0, T]$, $\mathbf{y}^*(t) \in \partial\mathcal{R}(t; \xi)$ *and*

$$\mathbf{p}^*(t) \cdot (\mathbf{y} - \mathbf{y}^*(t)) \geq 0 \qquad \forall \mathbf{y} \in \mathcal{R}(t; \boldsymbol{\xi}). \tag{4.59}$$

The geometrical meaning of (4.59) is the following. If $\mathbf{u}^*$ is optimal, for every t, $\mathbf{y}^*(t)$ lies on the boundary of the reachable set $\mathcal{R}(t; \boldsymbol{\xi})$ and $\mathbf{p}^*(t)$ is an *interior normal* to a hyperplane supporting $\mathcal{R}(t; \xi)$ at $\mathbf{y}^*(t)$. This normal is obtained by transporting $\mathbf{p}^*(T) = \boldsymbol{\eta}$ backwards in time through the adjoint equation along the optimal trajectory (see Fig. 4.3).

4.4.1 Uniqueness of Optimal Control. Normal Systems

Let us now explore the uniqueness of an optimal control. A consequence of (4.55) is that an optimal control $\mathbf{u}^* = (u_1^*, \ldots, u_m^*)$ is given by

$$u_j^*(t) = -\text{sgn}\left[\mathbf{p}^*(t) \cdot \mathbf{b}_j\right], \qquad j = 1, 2, \ldots, m, \qquad t \in [0, T] \tag{4.60}$$

where $\mathbf{b}_j$ is the jth column of $\mathbf{B}$ and

$$\text{sgn}(y) = \begin{cases} -1 & \text{if } y < 0 \\ +1 & \text{if } y > 0 \\ c \in [-1, 1] & \text{if } y = 0. \end{cases}$$

Indeed, if for some j we had $u_j^*(t) < 1$ on some interval

$$E_j^- \subseteq \{t \in [0,T] : \mathbf{p}^*(t) \cdot \mathbf{b}_j < 0\},$$

we could select the admissible control $\tilde{\mathbf{u}} = (\tilde{u}_1, \ldots, \tilde{u}_m)$ given by

$$\tilde{u}_j(t) = 1 \text{ on } E_j^-, \quad \tilde{u}_j(t) = u_j^*(t) \text{ outside } E_j^-$$

and, for $k \neq j$,

$$\tilde{u}_k(t) = u_k^*(t) \qquad \text{in } [0,T].$$

Inserting $\tilde{\mathbf{u}}$ in (4.55) we would reach a contradiction

$$0 > \int_{E_j} \mathbf{p}^*(t) \cdot \mathbf{b}_j \left(1 - u_j^*(s)\right) ds \geq 0.$$

Thus, $u_j^*(t) = 1$ on $\{t \in [0,T] : \mathbf{p}^*(t) \cdot \mathbf{b}_j < 0\}$. Similarly we deduce that $u_j^*(t) = -1$ on the set

$$\{t \in [0,T] : \mathbf{p}^*(t) \cdot \mathbf{b}_j > 0\}.$$

Clearly, since we are assuming $\boldsymbol{\eta} \neq \mathbf{0}$, the analytic function

$$t \longmapsto \mathbf{p}^*(t) \cdot \mathbf{b}_j = e^{(T-t)\mathbf{A}^\top} \boldsymbol{\eta} \cdot \mathbf{b}_j = e^{T\mathbf{A}^\top} \boldsymbol{\eta} \cdot e^{-t\mathbf{A}} \mathbf{b}_j \tag{4.61}$$

is either identically zero in $[0,T]$ or it vanishes at most at a finite number of points[7].

Then formula (4.60) uniquely determines the optimal control if and only if the following condition holds:

$$\text{for every } j = 1, \ldots, m, \text{ the function (4.61) is not identically zero.} \tag{4.62}$$

Thus if (4.62) holds, the optimal control $\mathbf{u}^*$ assumes only the values ± 1, except at a finite number of points, corresponding to the zeros of (4.61). In this case, we say that $\mathbf{u}^*$ is a *bang-bang* control, with a finite number of *switching points* (the zeros of (4.61)).

We summarize our conclusions in the following

Proposition 4.3. *If condition* (4.62) *holds, there is a unique optimal control* $\mathbf{u}^*$ *given by*

$$u_j^*(t) = -sgn\left[\mathbf{p}^*(t) \cdot \mathbf{b}_j\right], \qquad j = 1, 2, \ldots, m, \ t \in [0,T].$$

$\mathbf{u}^*$ *is a bang-bang control with at most a finite number of switching points.*

When condition (4.62) holds for every $\boldsymbol{\eta} \neq \mathbf{0}$, we say that the control system is *normal*, according to the following

Definition 4.1. We say that the control system $\dot{\mathbf{y}}(t) = \mathbf{A}\mathbf{y}(t) + \mathbf{B}\mathbf{u}(t)$ is normal if, for every $\boldsymbol{\nu} \in \mathbb{R}^n, \boldsymbol{\nu} \neq 0$, and every $j = 1, \ldots, m$, the (analytic) functions

$$t \longmapsto \boldsymbol{\nu} \cdot e^{-t\mathbf{A}} \mathbf{b}_j \tag{4.63}$$

can vanish in $[0,T]$ only at a finite number of points.

It is useful to have simple tests for normality. The simplest one is expressed in the following

[7] Recall that $e^{(T-t)\mathbf{A}} = e^{T\mathbf{A}} e^{-t\mathbf{A}}$.

Theorem 4.3. *The control system $\dot{\mathbf{y}}(t) = \mathbf{A}\mathbf{y}(t) + \mathbf{B}\mathbf{u}(t)$ is normal if and only if for every $j = 1, \ldots, m$ the vectors*

$$\mathbf{b}_j,\ \mathbf{A}\mathbf{b}_j,\ \mathbf{A}^2\mathbf{b}_j, \ldots, \mathbf{A}^{n-1}\mathbf{b}_j \tag{4.64}$$

are linearly independent.

Proof. Assume that the vectors (4.64) are independent and that the control system is *not* normal. Then there exists a non-zero vector $\boldsymbol{\nu}$ such that

$$\boldsymbol{\nu} \cdot e^{-t\mathbf{A}}\mathbf{b}_j \equiv 0 \tag{4.65}$$

for some j. Letting $t = 0$ we get $\boldsymbol{\nu} \cdot \mathbf{b}_j = 0$. Differentiating (4.65) we have

$$\boldsymbol{\nu} \cdot e^{-t\mathbf{A}} \left(-\mathbf{A}\right) \mathbf{b}_j \equiv 0$$

and for $t = 0$ we obtain $\boldsymbol{\nu} \cdot \mathbf{A}\mathbf{b}_j = 0$.

Differentiating (4.65) $n - 1$ times and letting $t = 0$ we get

$$\boldsymbol{\nu} \cdot \left(-\mathbf{A}\right)^k \mathbf{b}_j = 0, \qquad k = 0, \ldots, n-1$$

which implies that the vectors (4.64) must be linearly dependent, that is, we get a contradiction.

Assume now that, for some j, the vectors (4.64) are linearly dependent. Then

$$\boldsymbol{\nu} \cdot \mathbf{A}^k\mathbf{b}_j = 0, \qquad k = 0, \ldots, n-1 \tag{4.66}$$

for some non-zero vector $\boldsymbol{\nu}$. Set

$$h\left(t\right) = \boldsymbol{\nu} \cdot e^{-t\mathbf{A}}\mathbf{b}_j.$$

We show that $h\left(t\right) \equiv 0$. For that, by setting $D = d/dt$ and $D^k = d^k/dt^k$, we have

$$D^k h\left(t\right) = \boldsymbol{\nu} \cdot \left(-\mathbf{A}\right)^k e^{-t\mathbf{A}}\mathbf{b}_j. \tag{4.67}$$

Let $\sigma\left(\lambda\right) = \det\left(-\mathbf{A} - \lambda\mathbf{I}\right) = \sum_{k=0}^{n} a_k\lambda^k$ be the characteristic polynomial of the matrix $-\mathbf{A}$. From Caley-Hamilton Theorem (see Sect. B.1.1), we deduce that $\sigma\left(-\mathbf{A}\right) = \mathbf{O}$, where $\mathbf{O}$ denotes the zero matrix.

On the other hand, applying the nth–order differential operator $\sigma\left(D\right) = \sum_{k=0}^{n} a_k D^k$ to h, and recalling that $D = d/dt$, we have, from (4.67),

$$\sigma\left(D\right) h\left(t\right) = \boldsymbol{\nu} \cdot \sigma\left(D\right) e^{-t\mathbf{A}}\mathbf{b}_j = \boldsymbol{\nu} \cdot \sigma\left(-\mathbf{A}\right) e^{-t\mathbf{A}}\mathbf{b}_j = 0. \tag{4.68}$$

Finally, (4.66) gives

$$D^k h\left(0\right) = 0, \qquad k = 0, \ldots, n-1. \tag{4.69}$$

Now, (4.68) is a linear homogeneous ODE of order n for h whereas (4.69) provides a set of n homogeneous initial conditions for h and its derivatives up to order $n - 1$. Thus, we must have $h\left(t\right) \equiv 0$, which contradicts the fact that the control system $\dot{\mathbf{y}} = \mathbf{A}\mathbf{y} + \mathbf{B}\mathbf{u}$ is normal. □

4.5 Minimum Time Problems

4.5.1 Controllability

An important problem is to drive some state variable $\mathbf{y} = \mathbf{y}\left(t\right)$ from a given initial state $\boldsymbol{\xi}$ to a given target state $\mathbf{y}_{tar}$ in *minimum time.* Precisely, let us consider the problem

$$T \to \min$$

under the dynamics

$$\begin{cases} \dot{\mathbf{y}}\left(t\right) = \mathbf{A}\mathbf{y}\left(t\right) + \mathbf{B}\mathbf{u}\left(t\right), & 0 < t < T \\ \mathbf{y}\left(0\right) = \boldsymbol{\xi},\ \mathbf{y}\left(T\right) = \mathbf{y}_{tar}. \end{cases} \tag{4.70}$$

This problem is often called *controllability problem.* Since the final time is unknown, an admissible control is a piecewise continuous function $\mathbf{u} : [0, +\infty) \to \mathbb{R}^m$, with $\left|\mathbf{u}\left(t\right)\right| \leq 1$ for all $t \geq 0$. We denote by $\mathcal{C}_\infty$ this class of controls.

Before deriving the optimality conditions, some remarks are in order. First of all, dealing with a given final state introduces additional difficulties with respect to the free end-point case. The first basic issue is to check if $\mathbf{y}_{tar}$ is reachable in a finite time. This is not always the case, as the next example shows, even if we take the unconstrained class of admissible control $\mathcal{C}$, given by the set of all piecewise continuous functions $\mathbf{u} : [0, +\infty) \to \mathbb{R}^m$.

Example 4.1. Suppose we want to use the dynamics

$$\dot{\mathbf{y}}(t) = \mathbf{A}\mathbf{y}(t) + \mathbf{B}\mathbf{u}(t)$$

with

$$\mathbf{A} = \begin{bmatrix} -2 & 1 \\ 0 & 1 \end{bmatrix} \quad \text{and} \quad \mathbf{B} = \begin{bmatrix} 1 & 0 \\ 1 & 0 \end{bmatrix}$$

to drive the state $\mathbf{y} = (y_1, y_2)$ from the origin $\boldsymbol{\xi} = \mathbf{0}$ to some target point $\mathbf{y}_{tar}$ in minimum time T. We know that the set of reachable points from $\boldsymbol{\xi} = \mathbf{0}$ in some time T is given by

$$\mathcal{R}(T; \mathbf{0}) = \{\mathbf{y}(T; \mathbf{0}, \mathbf{u}) : \mathbf{u} \in \mathcal{C}\}$$

and that

$$\mathbf{y}(T; \mathbf{0}, \mathbf{u}) = \int_0^T e^{(T-s)\mathbf{A}} \mathbf{B}\mathbf{u}(s)\, ds.$$

The question is: can we always find $\mathbf{u} = (u_1, u_2)$ and a finite T such that $\mathbf{y}(T; \mathbf{0}, \mathbf{u}) = \mathbf{y}_{tar}$? Let us compute the matrix $e^{\tau\mathbf{A}}\mathbf{B}$. We have

$$e^{\tau\mathbf{A}}\mathbf{B} = \left[I + \tau\mathbf{A} + \frac{\tau^2}{2!}\mathbf{A}^2 + \cdots + \frac{\tau^k}{k!}\mathbf{A}^k + \cdots \right] \mathbf{B}. \tag{4.71}$$

By Caley-Hamilton Theorem, every square matrix is a solution of its own characteristic equation. Since the characteristic equation of $\mathbf{A}$ is given by

$$\det(\mathbf{A} - \lambda\mathbf{I}) = (-2 - \lambda)(1 - \lambda) = \lambda^2 + \lambda - 1 = 0,$$

we have that $\mathbf{A}^2 = -\mathbf{A} + \mathbf{I}$ and, by induction, for every $k > 1$,

$$\mathbf{A}^k = c_k I + c_{k-1}\mathbf{A}$$

for suitable real numbers c_k, c_{k-1}, so that

$$\mathbf{A}^k\mathbf{B} = c_k\mathbf{B} + c_{k-1}\mathbf{A}\mathbf{B} = c_k \begin{bmatrix} 1 & 0 \\ 1 & 0 \end{bmatrix} + c_{k-1} \begin{bmatrix} -1 & 0 \\ -1 & 0 \end{bmatrix}$$

Observe now that for any vector of the form

$$\boldsymbol{\nu} = c \begin{bmatrix} 1 \\ -1 \end{bmatrix}, \qquad c \in \mathbb{R},$$

we have

$$\boldsymbol{\nu}^\top\mathbf{B} = \mathbf{0}, \quad \boldsymbol{\nu}^\top\mathbf{A}\mathbf{B} = \mathbf{0}.$$

Then $\boldsymbol{\nu}^\top\mathbf{A}^k\mathbf{B} = \mathbf{0}$ for every $k \geq 0$, which in turn implies $\boldsymbol{\nu}^\top e^{\tau\mathbf{A}}\mathbf{B} = \mathbf{0}$ for every $\tau \geq 0$. Hence

$$\boldsymbol{\nu} \cdot \int_0^t e^{(t-s)\mathbf{A}}\mathbf{B}\mathbf{u}(s)\, ds = 0 \qquad \text{for every } t \geq 0$$

and every $\mathbf{u} \in \mathcal{C}$. In conclusion, $\mathcal{R}(t; \mathbf{0})$ is contained in the hyperplane orthogonal to $\boldsymbol{\nu}$. Therefore, if $\boldsymbol{\nu} \cdot \mathbf{y}_{tar} \neq 0$, there is no way to steer $\mathbf{y}$ from the origin to $\mathbf{y}_{tar}$. •

The above example leads to the concept of *controllability*. For the time being, we keep the class $\mathcal{C}$ of admissible controls.

Definition 4.2. The autonomous control system

$$\dot{\mathbf{y}}(t) = \mathbf{A}\mathbf{y}(t) + \mathbf{B}\mathbf{u}(t), \qquad t > 0$$

is called $\mathcal{C}-$ controllable if for any given pair of points $\boldsymbol{\xi}$ and $\mathbf{y}_{tar}$, there exists an admissible control $\mathbf{u} \in \mathcal{C}$ that steers $\mathbf{y}$ from $\boldsymbol{\xi}$ to $\mathbf{y}_{tar}$ in finite time.

According to Definition 4.2, the control process in Example 4.1 is *not* $\mathcal{C}-$controllable: let us explore the reason. A closer look at the example reveals that the key point lies in the possibility to exhibit a vector which is orthogonal to *all* the columns of the matrix

$$(\mathbf{B}, \mathbf{AB}) = \begin{bmatrix} 1 & 0 & -1 & 0 \\ 1 & 0 & -1 & 0 \end{bmatrix}.$$

This is possible because the matrix $(\mathbf{B}, \mathbf{AB})$ has rank $1 < 2$. The same argument clearly works in dimension n with reference to the $n \times nm$ matrix

$$\left(\mathbf{B}, \mathbf{AB}, \mathbf{A}^2\mathbf{B}, \ldots, \mathbf{A}^{n-1}\mathbf{B}\right). \tag{4.72}$$

In fact, if the columns of the matrix (4.72) are linearly dependent, that is

$$\text{rank}\left(\mathbf{B}, \mathbf{AB}, \mathbf{A}^2\mathbf{B}, \ldots, \mathbf{A}^{n-1}\mathbf{B}\right) < n,$$

it is always possible to find a vector $\boldsymbol{\nu} \in \mathbb{R}^n$ such that $\mathcal{R}(t; \mathbf{0})$ is contained in the hyperplane orthogonal to $\boldsymbol{\nu}$, so that the system is not controllable.

Conversely, if

$$\text{rank}\left(\mathbf{B}, \mathbf{AB}, \mathbf{A}^2\mathbf{B}, \ldots, \mathbf{A}^{n-1}\mathbf{B}\right) = n, \tag{4.73}$$

the system is controllable, otherwise there would exist a non-zero vector $\boldsymbol{\nu}$ such that, say,

$$\boldsymbol{\nu} \cdot \int_0^1 e^{(1-s)\mathbf{A}}\mathbf{B}\mathbf{u}(s)\, ds = \mathbf{0}$$

for all $\mathbf{u} \in \mathcal{C}$. This implies[8], for the column vector $\boldsymbol{\nu}$,

$$\boldsymbol{\nu}^\top e^{(1-s)\mathbf{A}}\mathbf{B} = \mathbf{0} \text{ for every } s \in [0, 1]. \tag{4.74}$$

Letting $s = 1$, we find $\boldsymbol{\nu}^\top\mathbf{B} = \mathbf{0}$. Differentiating (4.74) $n - 1$ times and setting $s = 1$, we obtain

$$\boldsymbol{\nu}^\top\mathbf{AB} = \boldsymbol{\nu}^\top\mathbf{A}^2\mathbf{B} = \cdots = \boldsymbol{\nu}^\top\mathbf{A}^{n-1}\mathbf{B} = \mathbf{0},$$

which contradicts the maximum rank condition (4.73). We summarize the above conclusions in the following

Theorem 4.4. *The linear autonomous control system*

$$\dot{\mathbf{y}}(t) = \mathbf{A}\mathbf{y}(t) + \mathbf{B}\mathbf{u}(t), \qquad t > 0$$

is $\mathcal{C}-$ *controllable if and only if*

$$\textit{rank}\left(\mathbf{B}, \mathbf{AB}, \mathbf{A}^2\mathbf{B}, \ldots, \mathbf{A}^{n-1}\mathbf{B}\right) = n. \tag{4.75}$$

[8] In fact, for any given $g \in C([a, b])$, if $\int_a^b g(t)\, u(t) = 0$ for every $u \in C([a, b])$ then $g = 0$ in $[a, b]$.

Note that (4.75) is independent of T. The following consequence is also immediate.

Corollary 4.2. *If the system is normal, then it is* $\mathcal{C}-$*controllable.*

Note that the converse is true only for $n = 1$. Take for instance $n = m$, $\mathbf{A} = \mathbf{I}$ and any $\mathbf{B}$ non singular. Then, clearly, $\text{rank}\big(\mathbf{B}, \mathbf{AB}, \mathbf{A}^2\mathbf{B}, \ldots, \mathbf{A}^{n-1}\mathbf{B}\big) = \text{rank}(\mathbf{B}, \mathbf{B}, \mathbf{B}, \ldots, \mathbf{B}) = n$ but the system is not normal.

4.5.2 Observability

Controllability of a system is related to *observability*. Consider the linear ODE system

$$\begin{cases} \dot{\mathbf{y}}\,(t) = \mathbf{A}\mathbf{y}(t), & t > 0 \\ \mathbf{y}\,(0) = \boldsymbol{\xi}. \end{cases} \tag{4.76}$$

Assume that we can only observe some linear function of the solution

$$\mathbf{z}\,(t) = \mathbf{C}\mathbf{y}(t) \tag{4.77}$$

where $\mathbf{C} \in \mathbb{R}^{q \times n}$ is a constant matrix, so that $\mathbf{z}\,(t) \in \mathbb{R}^q$. We wonder *if the solution* $\mathbf{y}\,(t)$ *can be uniquely reconstructed for each* $t \geq 0$*, from the sole knowledge of* $\mathbf{z}\,(t)$ *on some interval* $[0, T]$. The interesting case is $q < n$, corresponding to reconstructing an $n-$dimensional vector $\mathbf{y}$ from a lower dimensional observed vector $\mathbf{z}$. Since $\mathbf{y}\,(t)$ is uniquely determined in any interval by the knowledge of $\boldsymbol{\xi}$, the problem reduces to determine $\boldsymbol{\xi}$ uniquely. When this is possible for *any* $\boldsymbol{\xi}$ we say that the system (4.76), (4.77) is *observable*. The following theorem shows the connection between observability and controllability.

Theorem 4.5. *The system (4.76), (4.77) is observable if and only if the control system*

$$\dot{\mathbf{x}}\,(t) = \mathbf{A}^\top \mathbf{x}\,(t) + \mathbf{C}^\top \mathbf{u}\,(t)\,, \qquad t > 0 \tag{4.78}$$

is $\mathcal{C}-$*controllable, that is if and only if*

$$\text{rank}\left(\mathbf{C}^\top, \mathbf{A}^\top\mathbf{C}^\top, (\mathbf{A}^\top)^2\mathbf{C}^\top, \ldots, (\mathbf{A}^\top)^{n-1}\mathbf{C}^\top\right) = n. \tag{4.79}$$

Proof. The solution of (4.76) is given by $\mathbf{y}\,(t) = e^{\mathbf{A}t}\boldsymbol{\xi}$, hence

$$\mathbf{z}\,(t) = \mathbf{C}e^{\mathbf{A}t}\boldsymbol{\xi}, \qquad t \geq 0. \tag{4.80}$$

Suppose the system (4.76), (4.77) is not observable. This means that there are $\boldsymbol{\xi}_1$, $\boldsymbol{\xi}_2$, $\boldsymbol{\xi}_1 \neq \boldsymbol{\xi}_2$ such that

$$\mathbf{C}e^{\mathbf{A}t}\boldsymbol{\xi}_1 = \mathbf{C}e^{\mathbf{A}t}\boldsymbol{\xi}_2 \qquad \forall t \geq 0$$

or, setting $\boldsymbol{\xi}_0 = \boldsymbol{\xi}_1 - \boldsymbol{\xi}_2$,

$$\mathbf{C}e^{\mathbf{A}t}\boldsymbol{\xi}_0 = \mathbf{0} \qquad \forall t \geq 0. \tag{4.81}$$

Differentiating (4.81) $n-1$ times with respect to t and setting $t = 0$, we obtain

$$\mathbf{C}\boldsymbol{\xi}_0 = \mathbf{CA}\boldsymbol{\xi}_0 = \cdots = \mathbf{CA}^{n-1}\boldsymbol{\xi}_0 = \mathbf{0}$$

or, equivalently,

$$\boldsymbol{\xi}_0^\top\mathbf{C}^\top = \boldsymbol{\xi}_0^\top\mathbf{A}^\top\mathbf{C}^\top = \ldots = \boldsymbol{\xi}_0^\top(\mathbf{A}^\top)^{n-1}\mathbf{C}^\top = \mathbf{0},$$

which contradicts the rank condition (4.79). Assume now that

$$\text{rank}\left(\mathbf{C}^\top, \mathbf{A}^\top\mathbf{C}^\top, (\mathbf{A}^\top)^2\mathbf{C}^\top, \ldots, (\mathbf{A}^\top)^{n-1}\mathbf{C}^\top\right) < n.$$

Then there exists a non-zero vector $\boldsymbol{\xi}_0$ such that

$$\mathbf{C}\boldsymbol{\xi}_0 = \mathbf{CA}\boldsymbol{\xi}_0 = \cdots = \mathbf{CA}^{n-1}\boldsymbol{\xi}_0 = \mathbf{0}. \tag{4.82}$$

Caley-Hamilton Theorem gives

$$\mathbf{A}^k = c_{k-1}I + c_{k-2}\mathbf{A} + c_{k-2}\mathbf{A}^2 + \cdots + c_0\mathbf{A}^{k-1} \qquad \forall k \geq 1$$

for suitable real numbers $c_0, \ldots, c_{k-1}$. Using (4.82) we infer $\mathbf{C}\mathbf{A}^k\boldsymbol{\xi}_0 = \mathbf{0}$ $\forall k \geq 1$ and in turn $\mathbf{C}e^{\mathbf{A}t}\boldsymbol{\xi}_0 = \mathbf{0}$ for every $t \geq 0$. Therefore

$$\mathbf{C}\mathbf{y}(t) = \mathbf{C}e^{\mathbf{A}t}\boldsymbol{\xi}_0 = \mathbf{0} \qquad \forall t \geq 0,$$

hence system (4.76), (4.77) is not observable. □

4.5.3 Optimality conditions

We now return to our class $\mathcal{C}_\infty$ of admissible controls. The control constraint $|\mathbf{u}(t)| \leq 1$ clearly limits the system controllability. Typically, given $\mathbf{y}_{tar}$, there will be only a subset $E \subset \mathbb{R}^n$ of initial points (the so called *controllability region*) that can be steered to $\mathbf{y}_{tar}$, as shown in Example 4.1. In this section we assume that an optimal control $\mathbf{u}^*$ exists, with optimal time T^*. To find a characterization of $\mathbf{u}^*$, we may re-interpret our mimimum time problem in terms of the reachable sets $\mathcal{R}(\boldsymbol{\xi}; t)$ introduced in Sect. 4.4.

Then $\mathbf{y}(T^*; \boldsymbol{\xi}, \mathbf{u}^*) = \mathbf{y}_{tar} \in \mathcal{R}(\boldsymbol{\xi}; T^*)$ and T^* is the smallest time t for which $\mathbf{y}_{tar} \in \mathcal{R}(\boldsymbol{\xi}; t)$, otherwise we should have $\mathbf{y}(\tau; \boldsymbol{\xi}, \mathbf{u}^*) = \mathbf{y}_{tar}$ for some time $\tau < T^*$. However, since T^* is unknown, it is convenient to write

$$\mathbf{y}(t; \boldsymbol{\xi}, \mathbf{u}) = e^{t\mathbf{A}}\boldsymbol{\xi} + \int_0^t e^{(t-s)\mathbf{A}}\mathbf{B}\mathbf{u}(s)\, ds = e^{t\mathbf{A}}\left[\boldsymbol{\xi} + \int_0^t e^{-s\mathbf{A}}\mathbf{B}\mathbf{u}(s)\, ds\right]$$

and set

$$\mathbf{z}(t) = \mathbf{z}(t; \mathbf{u}) = \int_0^t e^{-s\mathbf{A}}\mathbf{B}\mathbf{u}(s)\, ds, \qquad \mathbf{z}_{tar} = e^{-T^*\mathbf{A}}\mathbf{y}_{tar} - \boldsymbol{\xi}.$$

If we define

$$Z(t) = \{\mathbf{z}(t; \mathbf{u}) \; : \; \mathbf{u} \in \mathcal{C}_\infty\},$$

then

$$Z(t) = e^{-t\mathbf{A}}\mathcal{R}(\boldsymbol{\xi}; t) - \boldsymbol{\xi}$$

and the condition $\mathbf{y}_{tar} \in \mathcal{R}(\boldsymbol{\xi}; T^*)$ is equivalent to $\mathbf{z}_{tar} \in Z(T^*)$.

Since, for every $t \geq 0$, $\mathcal{R}(\boldsymbol{\xi}; t)$ is bounded and convex, so is $Z(t)$. The set $Z(t)$ grows with time; indeed, $Z(0) = \mathbf{0}$ and $Z(t_1) \subset Z(t_2)$ for $t_1 < t_2$. Thus T^* coincides with the first time $Z(t)$ intersects $\mathbf{z}_{tar}$. By the continuity of the trajectories, it must be $\mathbf{z}_{tar} \in \partial Z(T^*)$.

From convex analysis we infer the existence of a *non-zero* vector $\boldsymbol{\eta}^*$ which is the *interior normal* to a hyperplane supporting $Z(T^*)$ at $\mathbf{y}_{tar}$, that is

$$\boldsymbol{\eta}^* \cdot (\mathbf{z} - \mathbf{z}_{tar}) \geq 0 \qquad \forall \mathbf{z} \in Z(T^*). \tag{4.83}$$

This inequality translates into the following version of the minimum principle,

$$\boldsymbol{\eta}^* \cdot \int_0^{T^*} e^{-s\mathbf{A}}\mathbf{B}(\mathbf{u}(s) - \mathbf{u}^*(s))\, ds \geq 0.$$

Note now that $\mathbf{p}^*(t) = e^{-t\mathbf{A}^\top}\boldsymbol{\eta}^*$ is the solution of the adjoint problem

$$\begin{cases} \dot{\mathbf{p}}^*(t) = -\mathbf{A}^\top\mathbf{p}^*(t), & 0 < t < T^* \\ \mathbf{p}^*(0) = \boldsymbol{\eta}^*. \end{cases} \tag{4.84}$$

Note that here both T^* and $\boldsymbol{\eta}^*$ are unknowns. Reasoning as in Sect. 4.4, we deduce the following

Theorem 4.6. *Let* $\mathbf{u}^*$ *be an optimal control steering* $\boldsymbol{\xi}$ *to* $\mathbf{y}_{tar}$ *in minimum time* T^*. *Then there exists a non-zero vector* $\boldsymbol{\eta}^*$ *such that*

$$\int_0^{T^*} \mathbf{B}^\top \mathbf{p}^*(s) \cdot (\mathbf{u}(s) - \mathbf{u}^*(s))\, ds \geq 0 \qquad \forall \mathbf{u} \in \mathcal{C}_\infty \tag{4.85}$$

where $\mathbf{p}^*(t)$ *is the solution of the adjoint problem (4.84). If the control system is normal, the optimal control is a unique bang-bang control, given by*

$$u_j^*(t) = -sgn(\mathbf{p}^*(t) \cdot \mathbf{b}_j), \qquad j = 1, 2, \ldots, m, \quad t \in [0, T^*],$$

with at most a finite number of switching points.

We apply the above theorem to a classical example.

Example 4.2 (Control under a repulsive force). Consider the problem

$$T \to \min$$

under

$$\begin{cases} \ddot{y}(t) = k^2 y(t) + u(t), & t \in (0, T) \\ y(0) = y_0,\ \dot{y}(0) = v_0, & y(T) = \dot{y}(T) = 0 \end{cases} \tag{4.86}$$

with $k \neq 0$ and $|u| \leq 1$.

The above problem describes a unit mass particle, initially located at y_0, with velocity v_0, in motion under the action of a repulsive force $k^2 y$ and a control force u. Our goal is to select u in order to steer the particle to the origin with zero velocity, in minimum time.

Setting $y_1 = y, y_2 = \dot{y}$, we write (4.86) in the form (here $n = 2, m = 1$)

$$\begin{cases} \dot{y}_1 = y_2 \\ \dot{y}_2 = k^2 y_1 + u \end{cases} \tag{4.87}$$

or

$$\begin{bmatrix} \dot{y}_1 \\ \dot{y}_2 \end{bmatrix} = \begin{bmatrix} 0 & 1 \\ k^2 & 0 \end{bmatrix} \begin{bmatrix} y_1 \\ y_2 \end{bmatrix} + \begin{bmatrix} 0 \\ 1 \end{bmatrix} u$$

with

$$\begin{bmatrix} y_1(0) \\ y_2(0) \end{bmatrix} = \begin{bmatrix} y_0 \\ v_0 \end{bmatrix}, \begin{bmatrix} y_1(T) \\ y_2(T) \end{bmatrix} = \begin{bmatrix} 0 \\ 0 \end{bmatrix}.$$

Notice that

$$\mathbf{b} = \begin{bmatrix} 0 \\ 1 \end{bmatrix} \qquad \text{and} \qquad \mathbf{Ab} = \begin{bmatrix} 1 \\ 0 \end{bmatrix}$$

so that the system is normal and therefore $\mathcal{C}-$controllable.

The adjoint equations are:

$$\begin{cases} \dot{p}_1 = -k^2 p_2 \\ \dot{p}_2 = -p_1. \end{cases}$$

from which we find

$$p_1(t) = -\alpha\beta \cosh(k + \beta t),\ p_2(t) = \alpha \sinh(k + \beta t), \qquad \alpha, \beta \in \mathbb{R}.$$

The minimum principle gives

$$u^*(t) = -\text{sgn}[\mathbf{p}^*(t) \cdot \mathbf{b}] = -\text{sgn}[p_2(t)] = -\text{sgn}[\alpha \sinh(k + \beta t)].$$

This proves that u^* is indeed a bang-bang control with at most one switching time t_s, corresponding to the zero of $\sinh(k + \beta t)$.

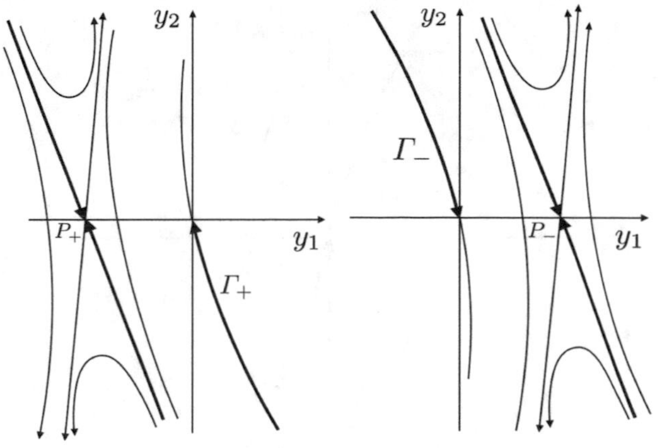

Fig. 4.4 Phase portrait for the systems S_+, S_-

To proceed, we examine the trajectories of the systems obtained by setting $u = 1$ and $u = -1$ in (4.87)

$$(S_+)\begin{cases}\dot{y}_1 = y_2\\ \dot{y}_2 = k^2 y_1 + 1\end{cases} \quad\text{and}\quad (S_-)\begin{cases}\dot{y}_1 = y_2\\ \dot{y}_2 = k^2 y_1 - 1.\end{cases}$$

The two systems have the equilibria $P_+ = \left(-1/k^2, 0\right)$ and $P_- = \left(1/k^2, 0\right)$, respectively, obtained by solving the algebraic systems

$$\begin{cases}y_2 = 0\\ k^2 y_1 + 1 = 0\end{cases} \quad\text{and}\quad \begin{cases}y_2 = 0\\ k^2 y_1 - 1 = 0.\end{cases}$$

Since $\det A = -k^2 < 0$, both equilibria are *saddle points*, with unstable manifolds given, respectively, by the straight lines

$$y_2 = k\left(y_1 + 1/k^2\right), \quad y_2 = k\left(y_1 - 1/k^2\right)$$

and the stable manifolds given, respectively, by the straight lines

$$y_2 = -k\left(y_1 + 1/k^2\right), \quad y_2 = -k\left(y_1 - 1/k^2\right).$$

The phase portrait of systems S_+ and S_- is shown in Figure 4.4. If the initial point (y_0, v_0) belongs to one of the two orbits Γ_+ or Γ_-, then clearly these orbits coincide with the optimal path, with optimal control $u^* = 1$ and $u^* = -1$, respectively. After a finite time T^* (the optimal time) these orbits reach the origin. If (y_0, v_0) does not belong to Γ_+ or Γ_- we need to distinguish two cases. Consider the strip

$$E = \left\{(y_1, y_2) : -k\left(y_1 + 1/k^2\right) < y_2 < -k\left(y_1 - 1/k^2\right)\right\}$$

between the two stable manifolds of systems S_+ and S_-.

If $(y_0, v_0) \notin E$, by looking at the phase portraits we see that there is no possibility to reach the origin using one or more orbits of either system.

If $(y_0, v_0) \in E$, the optimal trajectory is constructed in the following way. There are only two possibilities:

a) (y_0, v_0) belongs to an orbit O_+ of S_+. Then we choose $u^* = 1$ and move along O_+ until we intersect Γ_- at some *switching time* t_s. At that time we choose $u^* = -1$ and move along Γ_- until we reach the origin at time T^* (the optimal time);

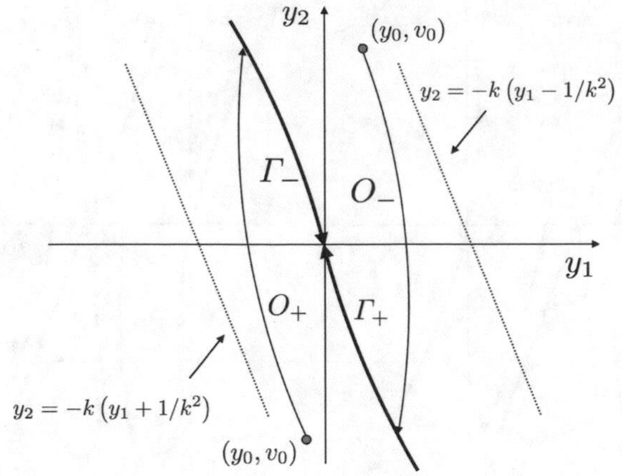

Fig. 4.5 Optimal paths and controllability strips for the control under repulsive force problem

b) (y_0, v_0) belongs to an orbit O_- of S_-. Then we choose $u^* = -1$ and move along O_- until we intersect Γ_+ at some *switching time* t_s. At that time we choose $u^* = 1$ and move along Γ_+ until we reach the origin at time T^* (the optimal time).

The two optimal paths are shown in figure 4.5. The strip E is called *controllability region* and its size depends on k for a given constraint on u. Indeed, k^2 represents the intensity of the repulsive force: a larger k corresponds to a smaller width of E.

It turns out that u^* can be put into feedback form. Indeed, let $y_2 = w(y_1)$ be the function whose graph is given by $\Gamma = \Gamma_+ \cup \Gamma_-$. This graph is called *switching locus.* Define now

$$G(y_1, y_2) = \begin{cases} -1 & \text{on } \Gamma_- \text{ and } y_2 > w(y_1) \\ 1 & \text{on } \Gamma_+ \text{ and } y_2 < w(y_1) \end{cases}$$

and $G(0,0) = 0$. Then we can write u^* in the feedback form

$$u^*(t) = G(y_1^*(t), y_2^*(t)), \tag{4.88}$$

that is, the optimal trajectory is a solution of the system

$$\begin{cases} \dot{y}_1 = -y_2 \\ \dot{y}_2 = -k^2 y_1 + G(y_1, y_2) \end{cases}$$

with $(y_0, v_0) \in E$. Note that (4.88) allows us to choose the value of u^* at time t by only looking at the position on the trajectory. •

4.6 Quadratic Cost

In this last section we consider quadratic cost functionals of the form

$$J(\mathbf{u}) = \psi(\mathbf{y}(T)) + \int_0^T [f(\mathbf{y}(s)) + q(\mathbf{u}(s))]\, ds \tag{4.89}$$

where

$$\psi(\mathbf{y}) = \frac{1}{2}\mathbf{S}\mathbf{y} \cdot \mathbf{y}, \quad f(\mathbf{y}) = \frac{1}{2}\mathbf{P}\mathbf{y} \cdot \mathbf{y}, \quad q(\mathbf{u}) = \frac{1}{2}\mathbf{Q}\mathbf{u} \cdot \mathbf{u};$$

$\mathbf{S}, \mathbf{P}$ are $n \times n$ positive symmetric semidefinite matrices while $\mathbf{Q}$ is a $m \times m$ symmetric positive definite matrix.

We want to solve the quadratic control problem, also referred to as *linear-quadratic regulator* (LQR)

$$J(\mathbf{u}) \to \min \tag{4.90}$$

under the usual dynamics

$$\begin{cases} \dot{\mathbf{y}}(t) = \mathbf{A}\mathbf{y}(t) + \mathbf{B}\mathbf{u}(t), & 0 < t < T \\ \mathbf{y}(0) = \boldsymbol{\xi} \end{cases} \tag{4.91}$$

for a given fixed $T > 0$, and the class $\mathcal{C}_T$ of piecewise continuous admissible controls. Existence and uniqueness of an optimal control $\mathbf{u}^*$ will follow from the general theory in Chapter 5.

4.6.1 First-order conditions

For nonlinear cost functionals, the elementary approach followed thus far in order to find first order necessary and sufficient optimality conditions does not work any longer.

To mimic what we did in Sect. 2.1.2, we suppose that

$$J(\mathbf{u}^*) \leq J(\mathbf{u}) \qquad \forall \mathbf{u} \in \mathcal{C}_T$$

and that $\mathbf{y}^*(t) = \mathbf{y}^*(t; \xi, \mathbf{u}^*)$, $0 \leq t \leq T$, is the corresponding optimal trajectory.

We need a notion of differential (or derivative) of the cost functional. Let $\mathbf{v} \in \mathcal{C}_T$, $\theta \neq 0$,

$$\mathbf{u}_\theta = \mathbf{u}^* + \theta \mathbf{v} \in \mathcal{C}_T$$

and let $\mathbf{y}_\theta$ be the corresponding trajectory. We define *the differential of* J *at* $\mathbf{u}^*$ *along the direction* $\mathbf{v}$ by the formula

$$dJ(\mathbf{u}^*)\mathbf{v} = \lim_{\theta \to 0} \frac{J(\mathbf{u}^* + \theta \mathbf{v}) - J(\mathbf{u}^*)}{\theta}.$$

Lemma 4.2. *For the quadratic cost functional (4.89), we have that*

$$dJ(\mathbf{u}^*)\mathbf{v} = \mathbf{S}\mathbf{y}^*(T) \cdot \mathbf{z}(T) + \int_0^T [\mathbf{P}\mathbf{y}^*(s) \cdot \mathbf{z}(s) + \mathbf{Q}\mathbf{u}^*(s) \cdot \mathbf{v}(s)]\, ds, \tag{4.92}$$

where

$$\mathbf{z}(t) = \int_0^t e^{(T-s)\mathbf{A}} \mathbf{B}\mathbf{v}(s)\, ds.$$

Proof. We have

$$J(\mathbf{u}_\theta) - J(\mathbf{u}^*) = \psi(\mathbf{y}_\theta(T)) - \psi(\mathbf{y}^*(T)) + \int_0^T [f(\mathbf{y}_\theta(s)) - f(\mathbf{y}^*(s)) + q(\mathbf{u}_\theta(s)) - q(\mathbf{u}^*(s))]\, ds. \tag{4.93}$$

Observe that

$$\mathbf{y}_\theta(t) - \mathbf{y}^*(t) = \theta \int_0^t e^{(T-s)\mathbf{A}} \mathbf{B}\mathbf{v}(s)\, ds = \theta \mathbf{z}(t)$$

and

$$\begin{aligned} \psi(\mathbf{y}_\theta(T)) - \psi(\mathbf{y}^*(T)) &= \frac{1}{2} S\mathbf{y}_\theta(T) \cdot \mathbf{y}_\theta(T) - \frac{1}{2} S\mathbf{y}^*(T) \cdot \mathbf{y}^*(T) \\ &= \frac{1}{2} S(\mathbf{y}_\theta(T) - \mathbf{y}^*(T)) \cdot (\mathbf{y}_\theta(T) - \mathbf{y}^*(T)) + S\mathbf{y}^*(T) \cdot (\mathbf{y}_\theta(T) - \mathbf{y}^*(T)) \\ &= \theta S\mathbf{y}^*(T) \cdot \mathbf{z}(T) + \theta^2 \psi(\mathbf{z}(T)). \end{aligned} \tag{4.94}$$

Similarly,

$$f(\mathbf{y}_\theta) - f(\mathbf{y}^*) = \theta P\mathbf{y}^* \cdot \mathbf{z} + \theta^2 f(\mathbf{z}),$$
$$q(\mathbf{u}_\theta) - q(\mathbf{u}^*) = \theta Q\mathbf{u}^* \cdot \mathbf{v} + \theta^2 q(\mathbf{v}).$$

Inserting the above expressions into (4.93), dividing by $\theta \neq 0$ and letting $\theta \to 0$, we get (4.92). □

The formula (4.92) can be put in a more reveiling (and useful) form, by writing the cost functional in the augmented form below, after using the state equation, and having introduced a *multiplier* (or *adjoint* state) $\mathbf{p}$, to be chosen later,

$$J(\mathbf{u}) = \psi(\mathbf{y}(T)) + \int_0^T [f(\mathbf{y}(s)) + q(\mathbf{u}(s))]\, ds + \int_0^T \mathbf{p}(s) \cdot [A\mathbf{y}(s) + B\mathbf{u}(s) - \mathbf{y}(s)]\, ds.$$

Note that, indeed, the second integral is zero. Integrating by parts the last term we get

$$J(\mathbf{u}) = \psi(\mathbf{y}(T)) + \int_0^T [f(\mathbf{y}(s)) + q(\mathbf{u}(s))]\, ds$$
$$+ \int_0^T \{\mathbf{p}(s) \cdot [\mathbf{A}\mathbf{y}(s) + \mathbf{B}\mathbf{u}(s)] + \dot{\mathbf{p}}(s) \cdot \mathbf{y}(s)\}\, ds - \mathbf{p}(T) \cdot \mathbf{y}(T) + \mathbf{p}(0) \cdot \boldsymbol{\xi}.$$

Using the augmented form of J, we deduce the following expression for $dJ(\mathbf{u}^*)\mathbf{v}$:

$$dJ(\mathbf{u}^*)\mathbf{v} = \mathbf{S}\mathbf{y}^*(T) \cdot \mathbf{z}(T) + \int_0^T [\mathbf{P}\mathbf{y}^*(s) \cdot \mathbf{z}(s) + \mathbf{Q}\mathbf{u}^*(s) \cdot \mathbf{v}(s)]\, ds$$
$$+ \int_0^T \left[\mathbf{A}^\top \mathbf{p}(s) \cdot \mathbf{z}(s) + \mathbf{B}^\top \mathbf{p}(s) \cdot \mathbf{v}(s) + \dot{\mathbf{p}}(s) \cdot \mathbf{z}(s)\right] ds - \mathbf{p}^\top(T) \cdot \mathbf{z}(T), \tag{4.95}$$

valid for every $\mathbf{v} \in \mathcal{C}_T$.

We now choose $\mathbf{p} = \mathbf{p}^*$, where $\mathbf{p}^*$ is the solution of the *adjoint* final-value problem

$$\begin{cases} \dot{\mathbf{p}}^*(t) = -\mathbf{A}^\top \mathbf{p}^*(t) - \mathbf{P}\mathbf{y}^*(t), & 0 < t < T \\ \mathbf{p}^*(T) = \mathbf{S}\mathbf{y}^*(T). \end{cases} \tag{4.96}$$

In this way we get rid of the terms containing $\mathbf{z}$ and (4.95) becomes

$$dJ(\mathbf{u}^*)\mathbf{v} = \int_0^T \left[\mathbf{Q}\mathbf{u}^*(s) + \mathbf{B}^\top \mathbf{p}^*(s)\right] \cdot \mathbf{v}(s)\, ds \qquad \forall \mathbf{u}, \mathbf{v} \in \mathcal{C}_T. \tag{4.97}$$

Remark 4.5. Since formula (4.97) expresses the linear functional $\mathbf{v} \mapsto dJ(\mathbf{u}^*)\mathbf{v}$ as an inner product (in $L^2(0,T)$), we coherently define the *gradient of* J at $\mathbf{u}^*$ by the formula

$$\nabla J(\mathbf{u}^*) = Q\mathbf{u}^* + B^\top \mathbf{p}^*. \tag{4.98}$$

We will see some numerical implication of this formula in Sect. 4.7. •

We now go back to our OCP. Choosing $\mathbf{v} = \mathbf{u} - \mathbf{u}^*$ and $\theta \in (0,1]$, by the convexity of $\mathcal{C}_T$, $\mathbf{u}_\theta \in \mathcal{C}_T$ and we can wrtite

$$J(\mathbf{u}^* + \theta(\mathbf{u} - \mathbf{u}^*)) - J(\mathbf{u}^*) \geq 0.$$

Dividing by θ and letting $\theta \to 0$ we end up with the following *variational inequality*:

$$dJ(\mathbf{u}^*)(\mathbf{u} - \mathbf{u}^*) = \int_0^T \left[\mathbf{Q}\mathbf{u}^*(s) + \mathbf{B}^\top \mathbf{p}^*(s)\right] \cdot (\mathbf{u}(s) - \mathbf{u}^*(s))ds \geq 0 \qquad \forall \mathbf{u} \in \mathcal{C}_T. \tag{4.99}$$

Reasoning as in Sect. 4.4, (4.99) implies that for every $\mathbf{u} \in \mathbb{R}^m$, $|\mathbf{u}| \leq 1$,

$$\nabla J(\mathbf{u}^*)(s) \cdot (\mathbf{u} - \mathbf{u}^*(s)) = (\mathbf{Q}\mathbf{u}^*(s) + \mathbf{B}^\top \mathbf{p}^*(s)) \cdot (\mathbf{u}(s) - \mathbf{u}^*(s)) \geq 0 \tag{4.100}$$

(*minimum principle*) for all $s \in [0, T]$ except at the discontinuity points of $\mathbf{u}^*$.

Since $\mathbf{Q}$ is positive, we recognize from (4.97) or (4.100) that $\mathbf{u}^*$ is the projection of $-\mathbf{Q}^{-1}\mathbf{B}^\top \mathbf{p}^*$ onto $\mathcal{C}_T$, that is,

$$\mathbf{u}^* = Pr_{\mathcal{C}_T}\left(-\mathbf{Q}^{-1}\mathbf{B}^\top \mathbf{p}^*\right),$$

according to what we have seen in Sect. 4.2.2. We summarize the above results in the following

Theorem 4.7. *Let* $\mathbf{u}^* \in \mathcal{C}_T$ *and* $\mathbf{y}^*(t) = \mathbf{y}^*(t; \boldsymbol{\xi}, \mathbf{u}^*)$, $0 \leq t \leq T$. *Let* $\mathbf{p}^*(t)$ *be the solution of the adjoint problem* (4.96). *Then* $\mathbf{u}^*$ *is optimal if and only if*

$$\int_0^T \left[\mathbf{Q}\mathbf{u}^*(s) + \mathbf{B}^\top \mathbf{p}^*(s)\right] \cdot (\mathbf{u}(s) - \mathbf{u}^*(s))\, ds \geq 0 \qquad \forall \mathbf{u} \in \mathcal{C}_T. \tag{4.101}$$

As a consequence, the minimum principle (4.100) holds for all $s \in [0, T]$ *except at the discontinuity points of* $\mathbf{u}^*$ *and*

$$\mathbf{u}^* = Pr_{\mathcal{C}_T}\left(-\mathbf{Q}^{-1}\mathbf{B}^\top \mathbf{p}^*\right). \tag{4.102}$$

4.6.2 The Riccati equation

In this section we show that through the solution of a final value problem for a particular matrix equation (the so called *Riccati equation*) the optimal control that minimizes the quadratic cost function (4.89) can be expressed in feedback form, realizing in this way a control synthesis. In turn, this gives complete information on the optimal cost value.

For simplicity we remove the control restriction $|\mathbf{u}(t)| \leq 1$ in $[0, T]$. In this case Theorem 4.7 still holds except for formula (4.102). Let t be a continuity point of $\mathbf{u}$. Then, the minimum principle

$$\left[\mathbf{Q}\mathbf{u}^*(t) + \mathbf{B}^\top \mathbf{p}^*(t)\right] \cdot (\mathbf{u} - \mathbf{u}^*(t)) \geq 0 \tag{4.103}$$

holds for every $\mathbf{u} \in \mathbb{R}^m$. Since $\mathbf{u}$ is arbitrary, it must be

$$\mathbf{Q}\mathbf{u}^*(t) + \mathbf{B}^\top \mathbf{p}^*(t) = \mathbf{0} \tag{4.104}$$

whence

$$\mathbf{u}^*(t) = -\mathbf{Q}^{-1}\mathbf{B}^\top \mathbf{p}^*(t). \tag{4.105}$$

Since $\mathbf{p}^*$ is continuous in $[0, T]$, this formula implies that also $\mathbf{u}^*$ is continuous in $[0, T]$. In order to express $\mathbf{u}^*$ in terms of $\mathbf{y}^*$, let us write

$$\mathbf{p}^*(t) = \mathbf{R}(t)\, \mathbf{y}^*(t) \tag{4.106}$$

where $\mathbf{R}$ is the so-called *Riccati matrix*, that will be determined from the dynamics and the adjoint equation, independently of the initial condition $\boldsymbol{\xi}$. We have, from (4.105) and (4.106),

$$\dot{\mathbf{y}}(t) = \mathbf{A}\mathbf{y}^*(t) + \mathbf{B}\mathbf{u}^*(t) = \mathbf{A}\mathbf{y}^*(t) - \mathbf{B}\mathbf{Q}^{-1}\mathbf{B}^\top \mathbf{R}(t)\, \mathbf{y}^*(t). \tag{4.107}$$

On the other hand, differentiating (4.106) we get

$$\dot{\mathbf{p}}^*(t) = \dot{\mathbf{R}}(t)\, \mathbf{y}^*(t) + \mathbf{R}(t)\, \dot{\mathbf{y}}^*(t) = [\dot{\mathbf{R}}(t) + \mathbf{R}(t)\, \mathbf{A} - \mathbf{B}\mathbf{Q}^{-1}\mathbf{B}^\top \mathbf{R}(t)]\mathbf{y}^*(t).$$

Finally, using (4.96),

$$\dot{\mathbf{p}}^*(t) = \left[-\mathbf{A}^\top \mathbf{R}(t) - \mathbf{P}\right]\mathbf{y}^*(t)$$

and comparing the last two expressions for $\dot{\mathbf{p}}$, we deduce for $\mathbf{R}$ the *Riccati (matrix) equation*

$$\dot{\mathbf{R}}(t) = -\mathbf{R}(t)\mathbf{A} - \mathbf{A}^\top \mathbf{R}(t) + \mathbf{R}(t)\mathbf{B}\mathbf{Q}^{-1}\mathbf{B}^\top \mathbf{R}(t) - \mathbf{P}. \tag{4.108}$$

Since $\mathbf{p}^*(T) = \mathbf{S}\mathbf{y}^*(T)$, to (4.108) we can associate the *final* condition

$$\mathbf{R}(T) = \mathbf{S}. \tag{4.109}$$

The final value problem (4.108), (4.109) always admits a regular solution $\mathbf{R}^*$ defined in a neighborhood $[t_0, T]$, $t_0 \geq 0$, uniquely determined as a symmetric and positive semidefinite matrix, since $\mathbf{S}$ meets these assumptions. It turns out that $\mathbf{R}^*$ is defined in the whole interval $[0, T]$ (see, e.g., [172, Sect. 3.3]). The symmetry of $\mathbf{R}^*$ implies, in particular, that to solve problem (4.108), (4.109) only $n(n+1)/2$ rather than n^2 equations have to be solved.

Note that $\mathbf{R}(t)$ can be found without knowing either $\mathbf{y}(t)$ or $\mathbf{u}(t)$; this allows to *(i)* first solve a final-value problem for $\mathbf{R}(t)$ and then *(ii)* solve an initial-value problem for $\mathbf{y}(\mathbf{t})$, once the optimal control has been expressed through (4.105) and (4.106). We summarize the above results in the following

Theorem 4.8. *There exists a unique symmetric and positive semidefinite solution* $\mathbf{R}^*(t)$ *of problem (4.108)–(4.109) in* $[0, T]$. *The quadratic cost control problem (4.90), (4.91) admits a unique solution* $\mathbf{u}^*$ *given by*

$$\mathbf{u}^*(t) = -\mathbf{Q}^{-1}\mathbf{B}^\top \mathbf{R}^*(t)\mathbf{y}^*(t)$$

where $\mathbf{y}^*$ *is the solution of the linear ODE system*

$$\dot{\mathbf{y}}^*(t) = \left[\mathbf{A} - \mathbf{B}\mathbf{Q}^{-1}\mathbf{B}^\top \mathbf{R}^*(t)\right]\mathbf{y}^*(t), \qquad \mathbf{y}^*(0) = \boldsymbol{\xi}.$$

Moreover, the minimum cost value is given by

$$J(\mathbf{u}^*) = \frac{1}{2}\mathbf{R}^*(0)\boldsymbol{\xi}\cdot\boldsymbol{\xi}. \tag{4.110}$$

Proof. We have only to check formula (4.110). Since $\mathbf{u}^* = -\mathbf{Q}^{-1}\mathbf{B}^\top\mathbf{R}^*\mathbf{y}^*$,

$$J(\mathbf{u}^*) = \frac{1}{2}\mathbf{S}\mathbf{y}^*(T)\cdot\mathbf{y}^*(T) + \frac{1}{2}\int_0^T \left[\mathbf{P} + \mathbf{R}^*(s)\mathbf{B}\mathbf{Q}^{-1}\mathbf{B}^\top\mathbf{R}^*(s)\right]\mathbf{y}^*(s)\cdot\mathbf{y}^*(s)\,ds.$$

From the Riccati equation,

$$\mathbf{P} = -\dot{\mathbf{R}}^*(s) - \mathbf{R}^*(s)\mathbf{A} - \mathbf{A}^\top\mathbf{R}^*(s) + \mathbf{R}^*(s)\mathbf{B}\mathbf{Q}^{-1}\mathbf{B}^\top\mathbf{R}^*(s)$$

so that

$$\begin{aligned} J(\mathbf{u}^*) &= \frac{1}{2}\mathbf{S}\mathbf{y}^*(T)\cdot\mathbf{y}^*(T) \\ &\quad + \frac{1}{2}\int_0^T \left[2\mathbf{R}^*(s)\mathbf{B}\mathbf{Q}^{-1}\mathbf{B}^\top\mathbf{R}^*(s) - \mathbf{R}^*(s)\mathbf{A} - \mathbf{A}^\top\mathbf{R}^*(s) - \dot{\mathbf{R}}^*(s)\right]\mathbf{y}^*(s)\cdot\mathbf{y}^*(s)\,ds. \end{aligned}$$

Thanks to (4.107) the expression under the integral coincides with

$$-\frac{d}{dt}\left[\mathbf{R}^*(s)\mathbf{y}^*(s)\cdot\mathbf{y}^*(s)\right].$$

Thus, we conclude the proof by noting that

$$J(\mathbf{u}^*) = \frac{1}{2}\mathbf{S}\mathbf{y}^*(T)\cdot\mathbf{y}^*(T) - \frac{1}{2}\int_0^T \frac{d}{dt}\left[\mathbf{R}^*(s)\mathbf{y}^*(s)\cdot\mathbf{y}^*(s)\right]ds = \frac{1}{2}\mathbf{R}^*(0)\boldsymbol{\xi}\cdot\boldsymbol{\xi}.$$

□

Remark 4.6. We can recover the maximum principle by introducing the *Hamiltonian function*

$$H(\mathbf{y},\mathbf{p},\mathbf{u})=\frac{1}{2}(\mathbf{Py}\cdot\mathbf{y}+\mathbf{Qu}\cdot\mathbf{u})+\mathbf{p}\cdot(\mathbf{Ay}+\mathbf{Bu}). \tag{4.111}$$

Then equation (4.104) corresponds to

$$\nabla_{\mathbf{u}}H(\mathbf{y}^*(s),\mathbf{p}^*(s),\mathbf{u}^*(s))=\mathbf{Qu}^*(s)+\mathbf{B}^\top\mathbf{p}^*(s)=\mathbf{0}. \tag{4.112}$$

Thanks to the convexity of the function

$$\mathbf{u}\to\frac{1}{2}\mathbf{Qu}\cdot\mathbf{u}+\mathbf{p}\cdot\mathbf{Bu},$$

(4.112) is equivalent to

$$H(\mathbf{y}^*(s),\mathbf{p}^*(s),\mathbf{u})\geq H(\mathbf{y}^*(s),\mathbf{p}^*(s),\mathbf{u}^*(s))\qquad\forall\mathbf{u}\in\mathbb{R}^n \tag{4.113}$$

which in turn is the usual form to express the minimum principle. Moreover, observe that

$$\dot{\mathbf{p}}^*(t)=-\nabla_{\mathbf{y}}H(\mathbf{y}^*(t),\mathbf{p}^*(t),\mathbf{u}^*(t))$$

corresponds to the *adjoint equation.* •

Remark 4.7. To solve the Riccati equation (4.108), it could be useful to consider the matrix factorization

$$\mathbf{R}(t)=\mathbf{Y}(t)\,\mathbf{X}^{-1}(t).$$

Then, $\mathbf{R}(t)$ solves the Riccati system (4.108)–(4.109) if and only if the matrices $\mathbf{X}$ and $\mathbf{Y}$ solve the following *linear* ODE system

$$\begin{bmatrix}\dot{\mathbf{X}}(t)\\ \dot{\mathbf{Y}}(t)\end{bmatrix}=\begin{bmatrix}\mathbf{A} & -\mathbf{BQ}^{-1}\mathbf{B}^\top\\ \mathbf{P} & -\mathbf{A}^\top\end{bmatrix}\begin{bmatrix}\mathbf{X}(t)\\ \mathbf{Y}(t)\end{bmatrix},\qquad 0\leq t\leq T,$$

with $\mathbf{X}^{-1}(T)\mathbf{Y}(T)=\mathbf{S}$. •

Example 4.3. Let us solve the one-dimensional problem

$$x^2(T)+\frac{1}{2}\int_0^T\left[x^2(s)+u^2(s)\right]ds\to\min$$

under the dynamics

$$\dot{x}=-x+\sqrt{3}u,\quad x(0)=\xi.$$

The Riccati equation for the scalar function $R=R(t)$ is

$$\dot{R}(t)=2R(t)+3R^2(t)-1 \tag{4.114}$$

with the final condition

$$R(T)=2. \tag{4.115}$$

The constant $k=-1$ is a solution of (4.114). Setting

$$R=-1+\frac{1}{Y}$$

we see that Y solves the linear equation

$$\dot{Y}=4Y-3$$

with $Y(T)=1/3$. We find

$$Y(t)=\frac{3}{4}-\frac{5}{12}\exp\{-4(T-t)\}.$$

Thus

$$R(t) = -1 + \left[\frac{3}{4} - \frac{5}{12}\exp\{-4(T-t)\}\right]^{-1}.$$

The optimal trajectory is the solution of the initial value problem

$$\dot{x}(t) = [-1 - 3R(t)]\, x(t), \qquad x(0) = \xi$$

while the optimal control is given by

$$u(t) = -\sqrt{3}R(t)\, x(t).$$

Finally, the optimal cost value is

$$\frac{1}{2}R(0)\,\xi^2 = \frac{1}{2}\left[-1 + \left(\frac{3}{4} - \frac{5}{12}e^{-4T}\right)\right]^{-1}\xi^2.$$

•

4.6.3 The Algebraic Riccati Equation

In the special case $T = \infty$ and $\psi = 0$, we want to minimize the cost functional

$$J(\mathbf{u}) = \frac{1}{2}\int_0^{\infty} [\mathbf{P}\mathbf{y}\cdot\mathbf{y} + \mathbf{Q}\mathbf{u}\cdot\mathbf{u}]\, ds \tag{4.116}$$

under the usual dynamics (4.91); this problem is sometimes referred to as *time-invariant* LQR. Since there is no terminal payoff, there are no constraints on the final value $\mathbf{p}(T)$ or, equivalently, on $\mathbf{R}(T)$. We can thus seek for a constant matrix $\mathbf{R}$ fulfilling the Riccati matrix equation (4.108), that is,

$$\mathbf{R}\mathbf{A} + \mathbf{A}^{\top}\mathbf{R} - \mathbf{R}\mathbf{B}\mathbf{Q}^{-1}\mathbf{B}^{\top}\mathbf{R} + \mathbf{P} = \mathbf{0}. \tag{4.117}$$

This equation is also referred to as the *algebraic Riccati (matrix) equation.* Since $\mathbf{Q}$ is positive definite, we can express it as $\mathbf{Q} = \mathbf{H}^{\top}\mathbf{H}$, so that the cost function becomes

$$J(\mathbf{u}) = \frac{1}{2}\int_0^{\infty} \left[|\mathbf{H}\mathbf{y}|^2 + \mathbf{Q}\mathbf{u}\cdot\mathbf{u}\right] dt.$$

Moreover, let us assume that the pair $(\mathbf{A}, \mathbf{H})$ is observable. It is well-known that equation (4.117) admits a variety of solutions: it may have no solution at all or, if it does have one, it can admit both real and complex solutions, some of them being Hermitian or symmetric matrices; there can be even infinitely many solutions. Due to the underlying physical problem, however, only non-negative solutions are of interest to us. Therefore, we are mainly concerned with the existence and uniqueness of such solution.

We also assume that $(\mathbf{A}, \mathbf{B})$ is controllable, so that the state can be steered to the origin in finite time and, due to the observability assumption, equilibrium at the origin is the only solution corresponding to the vanishing of the cost function. Hence, the solution of the optimal control problem converges to the origin, so that $\lim_{t\to\infty}\mathbf{y}(t) = \mathbf{0}$. In other words, the optimal control required for moving from any state to the origin can be found by applying a feedback law as in (4.118) and letting the system evolve in closed loop. Precisely, the following result holds (see, e.g., [21, Sect. 3.5] for the proof and [15, Chap. 6] for further details.

Theorem 4.9. *Consider the quadratic cost control problem (4.108)-(4.109) with* $T = +\infty$ *and* $J(\mathbf{u})$ *given by* (4.116). *Assume that* $(\mathbf{A}, \mathbf{B})$ *is controllable and that* $(\mathbf{A}, \mathbf{H})$ *is observable. The optimal control is given by*

$$\mathbf{u}^* = -\mathbf{Q}^{-1}\mathbf{B}^\top \mathbf{R}^* \mathbf{y}^* \tag{4.118}$$

where $\mathbf{R}^*$ *is the unique (symmetric) positive definite solution of the algebraic Riccati equation* (4.117) *and* $\mathbf{y}^*$ *is the solution of the linear ODE system*

$$\dot{\mathbf{y}}^*(t) = \left[\mathbf{A} - \mathbf{B}\mathbf{Q}^{-1}\mathbf{B}^\top\mathbf{R}^*\right]\mathbf{y}^*(t), \qquad \mathbf{y}^*(0) = \boldsymbol{\xi}.$$

Moreover, the minimum cost value is given by

$$J(\mathbf{u}^*) = \frac{1}{2}\mathbf{R}^*\boldsymbol{\xi}\cdot\boldsymbol{\xi}. \tag{4.119}$$

Example 4.4. Consider the mechanical system shown in Figure 4.6, where $d_1, \dots, d_4$ denote displacements from an equilibrium position, and u_1, u_2, u_3 represent control forces acting on the masses. For the sake of simplicity, we can take masses and spring (stiffness) constants equal to 1, that is, $m_1 = \dots = m_4 = 1$, $k_1 = \dots = k_4 = 1$. By applying the Newton law, we can derive the equations of the motion for the system, which read as follows:

$$\begin{aligned}
m_1\ddot{d}_1 &= -k_1 d_1 + k_2(d_2 - d_1) + u_1 = -2d_1 + d_2 + u_1 \\
m_2\ddot{d}_2 &= -k_2(d_2 - d_1) + k_3(d_3 - d_2) - u_1 = d_1 - 2d_2 + d_3 - u_1 - u_3 \\
m_3\ddot{d}_3 &= -k_3(d_3 - d_2) + k_4(d_4 - d_3) + u_2 = d_2 - 2d_3 + d_4 + u_2 \\
m_4\ddot{d}_4 &= -k_4(d_4 - d_3) - u_2 = d_3 - d_4 - u_2
\end{aligned}$$

Introducing four new variables, $d_5 = \dot{d}_1, \dots, d_8 = \dot{d}_4$ and setting $\mathbf{y} = (d_1, \dots, d_8) \in \mathbb{R}^8$ we can rewrite the system above as a (coupled) system of first-order linear ODEs under the form (4.109) with

$$\mathbf{A} = \begin{bmatrix} 0 & 0 & 0 & 0 & 1 & 0 & 0 & 0 \\ 0 & 0 & 0 & 0 & 0 & 1 & 0 & 0 \\ 0 & 0 & 0 & 0 & 0 & 0 & 1 & 0 \\ 0 & 0 & 0 & 0 & 0 & 0 & 0 & 1 \\ -2 & 1 & 0 & 0 & 0 & 0 & 0 & 0 \\ 1 & -2 & 1 & 0 & 0 & 0 & 0 & 0 \\ 0 & 1 & -2 & 1 & 0 & 0 & 0 & 0 \\ 0 & 0 & 1 & -1 & 0 & 0 & 0 & 0 \end{bmatrix}, \qquad \mathbf{B} = \begin{bmatrix} 0 & 0 & 0 \\ 0 & 0 & 0 \\ 0 & 0 & 0 \\ 0 & 0 & 0 \\ 1 & 0 & 0 \\ -1 & 0 & -1 \\ 0 & 1 & 0 \\ 0 & -1 & 0 \end{bmatrix}.$$

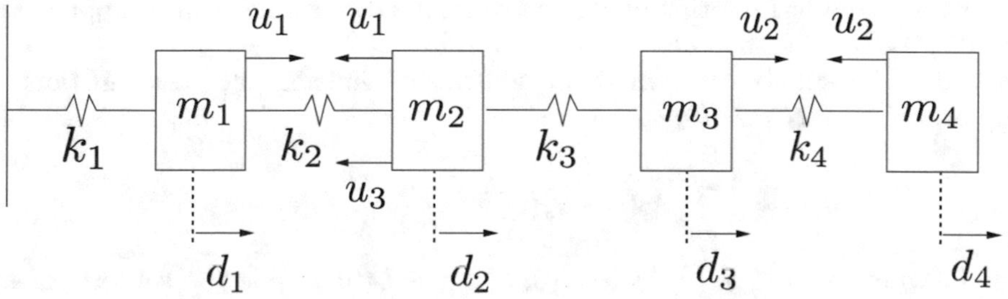

Fig. 4.6 A mechanical system made by four masses and four springs

We want to determine the optimal control forces so that the cost functional

$$J(\mathbf{u}) = \frac{1}{2}\int_0^{\infty} \left[|\mathbf{y}(t)|^2 + |\mathbf{u}(t)|^2\right] ds \tag{4.120}$$

is minimized. Note that (4.120) is nothing but (4.116) once we have set

$$\mathbf{P} = \begin{bmatrix} \mathbf{I}_4 & 0 \\ 0 & 0 \end{bmatrix}, \qquad \mathbf{Q} = \mathbf{I}_3$$

where $\mathbf{I}_d \in \mathbb{R}^d$ is the $d \times d$ identity matrix. Using an initial condition

$$\mathbf{y}(0) = \boldsymbol{\xi} = (0, 0, 0, 1, 0, 0, 0, 0),$$

we can determine the displacements for the open loop system (that is, with $\mathbf{u}(t) = \mathbf{0}$) and the closed loop system with optimal feedback control (4.118). To this aim, we can use the Matlab command care

```
[R,E,K] = care(A, B, P, Q)
```

to determine the Riccati matrix $\mathbf{R}$ and the gain matrix $\mathbf{K} = -\mathbf{Q}^{-1}\mathbf{B}^\top\mathbf{R}$ or, alternatively, the Matlab command lqr

```
[K,R,E] = lqr(A, B, P, Q)
```

The (optimal) gain matrix $\mathbf{K} = -\mathbf{Q}^{-1}\mathbf{B}^\top\mathbf{R}$ results as follows,

$$\mathbf{K} = \begin{bmatrix} -0.1579 & 0.3393 & 0.0447 & 0.1948 & -0.3975 & 0.4758 & 0.5069 & 0.5787 \\ 0.1754 & 0.1755 & -0.1997 & 0.3923 & -0.0514 & 0.0203 & 0.0234 & 1.1090 \\ 0.1458 & 0.7680 & 0.2497 & 0.1537 & 0.6683 & 1.1442 & 0.8593 & 0.8796 \end{bmatrix}.$$

Note that Matlab returns as K the negative of the optimal gain matrix in the notation we have adopted. The solution computed with the Matlab command ode23 (implementing an explicit, single-step Runge-Kutta method [246]) for both the open loop system and the closed loop system are reported in Figure 4.7. •

4.7 Hints on Numerical Approximation

We close this chapter by illustrating a possible strategy to approximate the solution of an OCP governed by the linear ODE (4.91) and a quadratic cost functional under the form (4.89). Rather than providing a detailed numerical treatment, we simply highlight a possible strategy for the numerical solution of OCPs involving ODEs, which can be applied also in the case of PDEs, as we will see in Chapters 6 and 8.

We consider the simple case where we aim at minimizing the cost functional $J(u) = \tilde{J}(y(u), u)$ over the set $\mathcal{U} = L^2(0,T)$, where

$$\tilde{J}(y, u) = \frac{1}{2}\int_0^T |y(t) - z_d(t)|^2 dt + \frac{\alpha}{2}\int_0^T |u(t) - u_d(t)|^2 dt, \tag{4.121}$$

with $\alpha > 0$, where $y(t) \in V = \left\{y \in L^2(0,T) : \dot{y} \in L^2(0,T)\right\}$ is the solution of the linear Cauchy problem

$$\begin{cases} \dot{y}(t) = ay(t) + u(t), & 0 < t < T \\ y(0) = y_0. \end{cases} \tag{4.122}$$

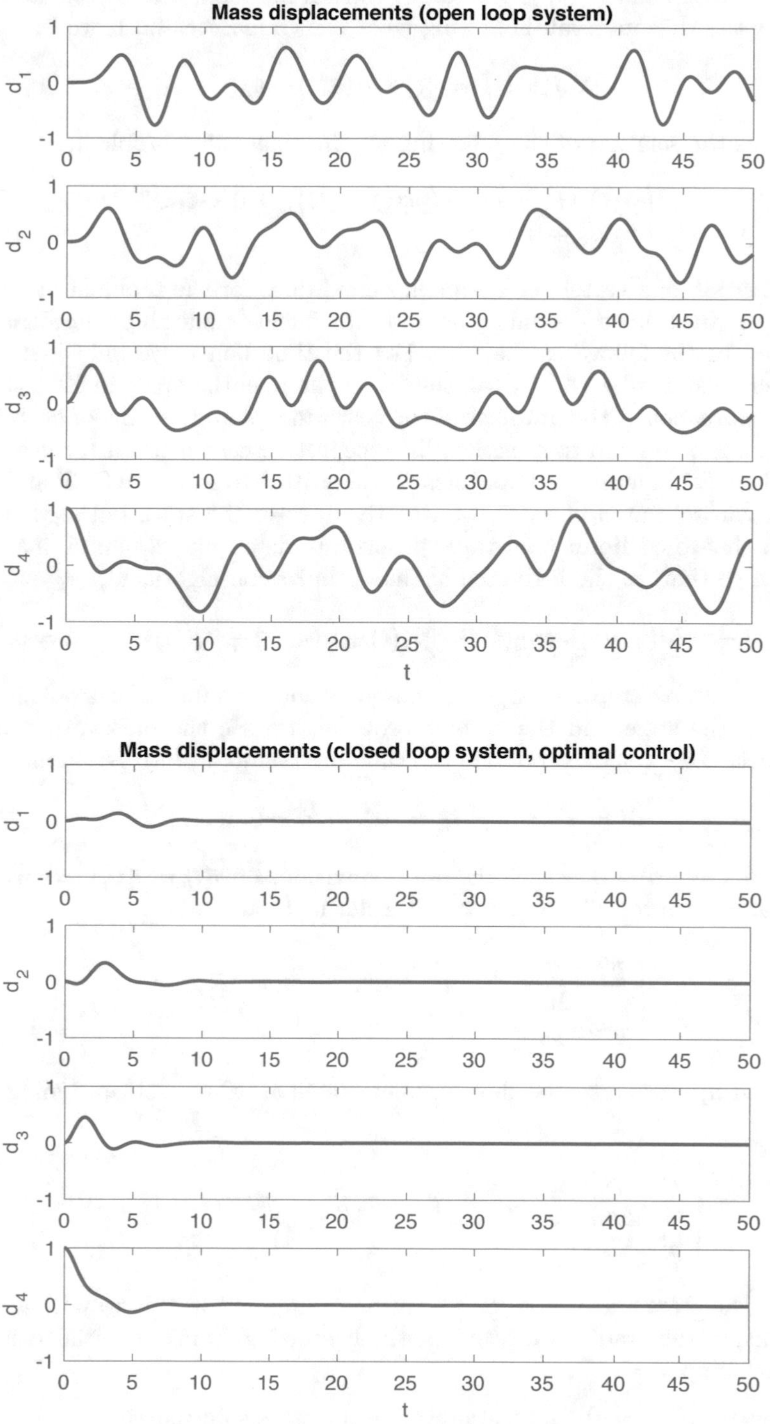

Fig. 4.7 Mass displacements for the open loop system (top) and the closed loop system (bottom)

Here $z_d(t)$ is the target and $u_d(t)$ is the target control function, that is, the desired profile of the control over the time interval. Following what shown in Sect. 4.6.1, we can deduce that

$$\nabla J(u)(t) = p(t) + \alpha(u(t) - u_d(t))$$

where $p(t) \in V$ is the solution of the following adjoint final-value problem,

$$\begin{cases} \dot{p}\,(t) = -ap(t) - (y(t) - z_d(t)), & 0 < t < T \\ p\,(T) = 0. \end{cases} \tag{4.123}$$

A rigorous derivation of a system of optimality conditions for the problem at hand, adopting the standpoint of constrained optimization and introducing a suitable Lagrangian functional, will be provided in the following chapters. For the time being, we only highlight that the minimum principle stated by (4.97), together with the remarks made in Sect. 4.1.3, allow us to identify the expression of the gradient of the cost functional, in terms of both the state and the adjoint variables. We can now exploit the steepest descent method (or gradient method) introduced in Sect. 3.1 as numerical optimization algorithm to solve the OCP at hand. Starting from an initial control function $u_0(t)$, we iteratively solve the state initial-value problem to determine $y(t)$, then the adjoint final value problem to determine $z(t)$, and finally update the control function so that, at the k-th step of the optimization algorithm,

$$u_{k+1}(t) = u_k(t) - \tau\nabla J(u_k)(t) = u_k(t) - \tau(p_k(t) + \alpha(u_k(t) - u_d(t))), \qquad k = 0, 1, \ldots \tag{4.124}$$

where $\tau > 0$ is a suitable step length, here chosen as constant for the sake of simplicity.

To solve both the state and the adjoint problem, we use the backward (implicit) Euler method (see Sect. B.3). Let us subdivide the time interval $[0, T]$ into N subintervals, so that

$$\Delta t = T/N, \qquad t_n = n\Delta t, \quad n = 0, 1, \ldots, N$$

and denote by y^n, u^n, $n = 0, \ldots, N$ the approximations of $y(t_n)$, $u(t_n)$, respectively. Given the (approximated) control u^n, the backward Euler method reads:

$$\begin{cases} \dfrac{y^n - y^{n-1}}{\Delta t} = ay^n + u^n, & n = 1, \ldots, N, \\ y^0 = y_0. \end{cases} \tag{4.125}$$

Similarly, we can approximate the dual problem (a final-value problem, going backward in time) as follows:

$$\begin{cases} \dfrac{p^n - p^{n+1}}{\Delta t} = ap^n + (y^n - z_d^n), & n = N-1, \ldots, 0 \\ p^N = 0. \end{cases} \tag{4.126}$$

We can exploit the state and the adjoint solutions computed in this way in an optimization loop, implementing the gradient descent method: starting from an initial control u_0^n, $n = 0, \ldots, N$, at each step $k = 1, \ldots$:

- given the control u_k^n, $n = 0, \ldots, N$, determine the state solution y_k^n, $n = 0, \ldots, N$ through (4.125) with u_k^n in place of u^n, and y_k^n in place of y^n;
- determine the adjoint solution z_k^n, $n = N, \ldots, 0$ through (4.126) with z_k^n in place of z^n, and y_k^n in place of y^n;
- update the control according to the discrete counterpart of (4.124), that is,

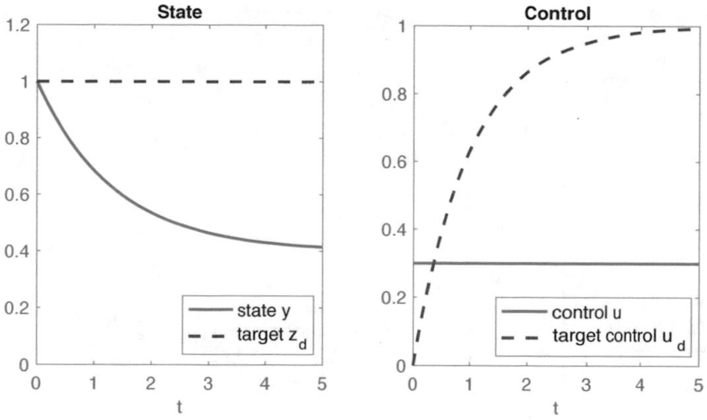

Fig. 4.8 State solution (left) and control (right) at the initial step $k = 0$ of the optimization procedure. The target z_d and the target control u_d are also reported (dotted lines)

$$u_{k+1}^n = u_k^n - \tau(p_k^n + \alpha(u_k^n - u_d^n)), \quad n = 0, \ldots, N;$$

- iterate until a suitable stopping criterion is fulfilled. For instance, the optimization loop can be stopped as soon as the difference between two successive iterates is smaller than a given tolerance $\varepsilon > 0$, that is,

$$|J(u_{k+1}) - J(u_k)| < \varepsilon,$$

or

$$\left(\sum_{n=0}^{N} |p_k^n + \alpha(u_k^n - u_d^n)|^2 \right)^{1/2} < \varepsilon.$$

Example 4.5. We now report a test case solved using this procedure. We consider $T = 5$, $a = -0.75$, $y_0 = 1$ and $\alpha = 0.1$. Moreover, we set

$$z_d = 1, \qquad u_d = 1 - e^{-t}.$$

We discretize both the state and the adjoint problems introducing a uniform partition of the time interval $[0, 5]$, with subintervals of length $\Delta t = 0.0125$. As initial control, we set a constant function $u_0(t) = 3$; we then consider the descent method discussed above, by setting a constant step length $\tau = 1$ and a tolerance $\varepsilon = 10^{-6}$ on the stopping criterion based on the difference between two successive iterates. For the case at hand, the cost functional is approximated by the (composite) trapezoidal formula, resulting in

$$J_k = \frac{1}{2}\Delta t \left[\frac{1}{2}(y_k^N - z_d^N)^2 + \sum_{n=1}^{N-1} (y_k^n - z_d^n)^2 \right];$$

note that $y_k^0 - z_d^0 = y_0 - 1 = 0$. Starting from an initial value of the cost functional of $J_0 = 5.497 \cdot 10^{-1}$, after K = 113 steps the optimization loop stops with $J_{113} = 2.353 \cdot 10^{-3}$, having reached $|J(u_{113}) - J(u_{112})| = 9.735 \cdot 10^{-7} < \varepsilon$. The state and the control at the beginning of the optimization procedure are reported in Fig. 4.8, whereas the optimal state, optimal control and corresponding adjoint solution are reported in Fig. 4.9.

•

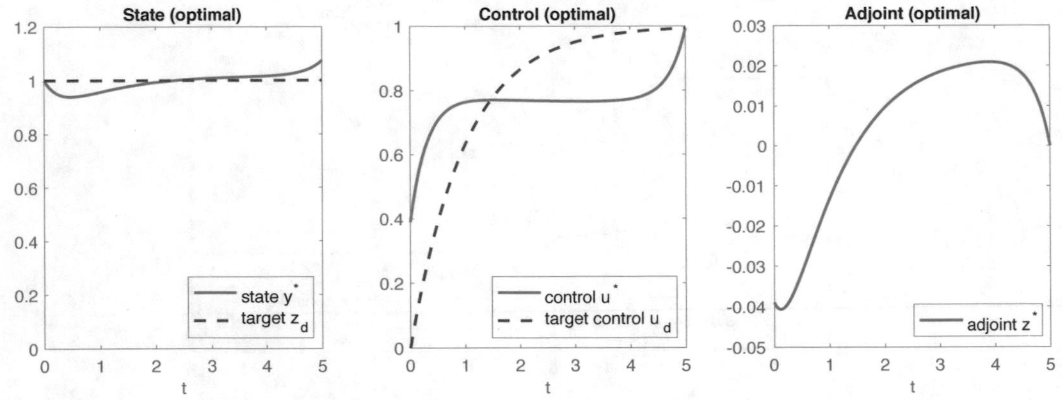

Fig. 4.9 Optimal state solution (left), optimal control (center) and corresponding adjoint solution (right) at the end of the optimization procedure

4.8 Exercises

1. Let $(\!(\mathbf{u},\mathbf{v})\!)$ be as in (4.6). Deduce from (4.7) that $(\!(\mathbf{u},\mathbf{u})\!)$ is strictly convex.
2. Let
$$\tilde{J}(\mathbf{y},\mathbf{u}) = \frac{1}{2}|\mathbf{C}\mathbf{y} - \mathbf{z}_d|^2_{\mathbf{M}} + \frac{\alpha}{2}|\mathbf{u}|^2_{\mathbf{N}}, \qquad \alpha > 0.$$
Show that
$$(\nabla_{\mathbf{y}}\tilde{J}(\mathbf{y},\mathbf{u}),\mathbf{z}) = (\mathbf{M}(\mathbf{C}\mathbf{y} - \mathbf{z}_d), \mathbf{C}\mathbf{z})$$
and therefore that
$$\nabla_{\mathbf{y}}\tilde{J}(\mathbf{y},\mathbf{u}) = \mathbf{C}^\top\mathbf{M}(\mathbf{C}\mathbf{y} - \mathbf{z}_d);$$
finally, deduce the optimality conditions (4.24).
3. Solve the problem
$$x_2 \to \min$$
subject to
$$\begin{cases} \dot{x}_1 = -x_2, & t \in (0,1) \\ \dot{x}_2 = -x_1 + u \end{cases}$$
with $x_1(0) = 0, x_2(0) = 10$ and $|u| \leq 1$.
4. Solve the problem
$$\int_0^T \left[px^2(s) + u^2(s)\right] \to \min$$
with $p > 0$, subject to
$$\begin{cases} \dot{x} = ax + bu, & t \in (0,T) \\ x(0) = 1. \end{cases}$$
5. Show that the Hamiltonian function (4.58)
$$t \mapsto H^*(t) = H(\mathbf{y}^*(t), \mathbf{u}^*(t), \mathbf{p}^*(t)) = \mathbf{p}^*(t) \cdot (\mathbf{A}\mathbf{y}^*(t) + \mathbf{B}\mathbf{u}^*(t))$$
is constant on $[0,T]$.

[*Hint.* a) First show that $t \mapsto \mathbf{p}^*(t) \cdot \mathbf{B}\mathbf{u}^*(t)$ is continuous on $[0, T]$. Let t be a point of discontinuity for $\mathbf{u}^*$. Use (4.56) to write, for h positive and small,

$$\mathbf{p}^*(t+h) \cdot \mathbf{B}\mathbf{u}^*(t+h) \le \mathbf{p}^*(t+h) \cdot \mathbf{B}\mathbf{u}^*(t-h)$$

and

$$\mathbf{p}^*(t-h) \cdot \mathbf{B}\mathbf{u}^*(t-h) \le \mathbf{p}^*(t-h) \cdot \mathbf{B}\mathbf{u}^*(t+h).$$

Then let $h \to 0$ to get

$$\mathbf{p}^*(t) \cdot \mathbf{B}\mathbf{u}^*(t_+) \le \mathbf{p}^*(t) \cdot \mathbf{B}\mathbf{u}^*(t_-)$$

and

$$\mathbf{p}^*(t) \cdot \mathbf{B}\mathbf{u}^*(t_-) \le \mathbf{p}^*(t) \cdot \mathbf{B}\mathbf{u}^*(t_+).$$

b) Show that $\dot{H}^*(t)$ exists and it is zero on $(0, T)$. We have, using (4.56)

$$\begin{aligned} &H^*(t+h) - H^*(t) \\ &= \mathbf{p}^*(t+h) \cdot (\mathbf{A}\mathbf{y}^*(t+h) + \mathbf{B}\mathbf{u}^*(t+h)) - \mathbf{p}^*(t) \cdot (\mathbf{A}\mathbf{y}^*(t) + \mathbf{B}\mathbf{u}^*(t)) \\ &= [\mathbf{p}^*(t+h) - \mathbf{p}^*(t)] \cdot \mathbf{A}\mathbf{y}^*(t+h) + \mathbf{p}^*(t) \cdot [\mathbf{A}\mathbf{y}^*(t+h) - \mathbf{A}\mathbf{y}^*(t)] + \\ &\quad \underbrace{+\mathbf{p}^*(t+h) \cdot [\mathbf{B}\mathbf{u}^*(t+h) - \mathbf{B}\mathbf{u}^*(t)]}_{\le 0} + [\mathbf{p}^*(t+h) - \mathbf{p}^*(t)] \cdot \mathbf{B}\mathbf{u}^*(t). \end{aligned}$$

Dividing by $h > 0$ and letting $h \to 0^+$ we get

$$\begin{aligned} \limsup_{h \to 0^+} \tfrac{1}{h} [H^*(t+h) - H^*(t)] &\le \dot{\mathbf{p}}^*(t) \cdot [\mathbf{A}\mathbf{y}^*(t) + \mathbf{B}\mathbf{u}^*(t)] + \mathbf{p}^*(t) \cdot \mathbf{A}\dot{\mathbf{y}}^*(t) \\ &= \dot{\mathbf{p}}^*(t) \cdot \dot{\mathbf{y}}^*(t) + \mathbf{p}^*(t) \cdot \mathbf{A}\dot{\mathbf{y}}^*(t) \\ &= -\mathbf{A}^\top \mathbf{p}^*(t) \cdot \dot{\mathbf{y}}^*(t) + \mathbf{p}^*(t) \cdot \mathbf{A}\dot{\mathbf{y}}^*(t) = 0. \end{aligned}$$

Similarly, dividing by $h < 0$ and letting $h \to 0^-$ one gets

$$\liminf_{h \to 0^-} \tfrac{1}{h} [H^*(t+h) - H^*(t)] \ge 0$$

and therefore $\dot{H}^*(t)$ exists and it is zero in $(0, T)$.]

6. Prove Corollary 4.1.
[*Hint:* From (4.56) we have, for every $s \in (0, T)$ and every admissible $\mathbf{u}$

$$\boldsymbol{\eta} \cdot e^{(T-s)\mathbf{A}} [\mathbf{B}\mathbf{u}^*(s) - \mathbf{B}\mathbf{u}(s)] \le 0$$

or

$$e^{(T-t)\mathbf{A}^T} \boldsymbol{\eta} \cdot e^{(t-s)\mathbf{A}} [\mathbf{B}\mathbf{u}^*(s) - \mathbf{B}\mathbf{u}(s)] \le 0.$$

Integrate over $(0, t)$, recalling (4.48) and (4.53).]

7. Consider the problem

$$G(\mathbf{y}(T)) \to \min$$

for a given fixed time T, subject to

$$\begin{cases} \dot{\mathbf{y}}(t) = \mathbf{A}\mathbf{y}(t) + \mathbf{B}\mathbf{u}(t), & 0 < t < T \\ \mathbf{y}(0) = \boldsymbol{\xi} \end{cases}$$

with $|\mathbf{u}| \le 1$. Prove that, if we assume that $G : \mathbb{R} \to \mathbb{R}$ is a differentiable convex function, then $\mathbf{u}^*$ is optimal if and only if:
a) $\mathbf{y}^*(T) \in \partial R(t; \boldsymbol{\xi})$, and
b) either $\boldsymbol{\eta} = \nabla G(\mathbf{y}^*(T)) = \mathbf{0}$ or

$$\boldsymbol{\eta} \cdot (\mathbf{y} - \mathbf{y}^*(T)) \ge 0 \quad \forall \mathbf{y} \in \mathcal{R}(t; \boldsymbol{\xi}).$$

Write the adjoint problem and deduce a generalization of the minimum principle of Theorem 4.2.

[*Hint:* For point a), write $G(\mathbf{y}) - G(\mathbf{y}^*(T)) = \nabla G(\mathbf{y}^*(T)) \cdot (\mathbf{y} - \mathbf{y}^*(T)) + o(\mathbf{y} - \mathbf{y}^*(T))$. For point b), from convexity, $G(\mathbf{y}) - G(\mathbf{y}^*(T)) \geq \nabla G(\mathbf{y}^*(T)) \cdot (\mathbf{y} - \mathbf{y}^*(T))$.]

8. Consider the following state system

$$\dot{\mathbf{y}}(t) = \begin{bmatrix} 0 & 1 \\ 0 & 0 \end{bmatrix} \mathbf{y}(t) + \begin{bmatrix} 0 \\ 1 \end{bmatrix} \mathbf{u}(t), \qquad t > 0.$$

Determine the optimal control in terms of the feedback law $\mathbf{u}^*(t) = -\mathbf{K}\mathbf{y}^*(t)$ by solving problem (4.117)–(4.118) with

$$\mathbf{P} = \begin{bmatrix} p^2 & 0 \\ 0 & 0 \end{bmatrix}, \qquad \mathbf{Q} = 1,$$

and $p > 0$. [*Hint:* first solve the algebraic Riccati equation (4.117), in which

$$\mathbf{A} = \begin{bmatrix} 0 & 1 \\ 0 & 0 \end{bmatrix}, \qquad \mathbf{B} = \begin{bmatrix} 0 \\ 1 \end{bmatrix};$$

then find $\mathbf{u}^*(t) = -\mathbf{Q}^{-1}\mathbf{B}^\top\mathbf{R}\mathbf{y}^*(t)$ where $\mathbf{Q}^{-1}\mathbf{B}^\top\mathbf{R} = \mathbf{K} = [1/p \quad \sqrt{2/p}]$.]

Part II
Linear-Quadratic Optimal Control Problems

Chapter 5
Quadratic control problems governed by linear elliptic PDEs

The examples we have chosen in Chapter 1 give a flavor of the wide range of problems that can be tackled within the optimal control framework. In this chapter we focus on the case of linear-quadratic OCPs, namely OCPs involving a quadratic cost functional and a state problem given by a linear elliptic PDE. We consider both unconstrained and constrained problems. In Sect. 5.1 we analyze an unconstrained linear-quadratic OCP governed by an advection-diffusion equation. We prove its well-posedness and obtain a system of optimality conditions, following an *adjoint-based* approach. We also recover the optimality conditions by the Lagrange multipliers method. The constrained case is considered in Sect. 5.2. In Sect. 5.3 we provide a general framework for linear-quadratic OCPs governed by abstract variational problems in presence of *control constraints*; we also recall the main results on the well-posedness of boundary value problems in Sect. 5.4. Next, in Sects. 5.5–5.11 we apply the above analysis to OCPs governed by general, second order boundary value problems for elliptic equations in divergence form, with different kinds of distributed/boundary controls, more general observations of the state variable (i.e., over a subregion or on a portion of the boundary) and various kinds of quadratic cost functionals. A simple case of OCP involving *state constraints* is also considered in Sect. 5.12. Finally, OCPs governed by a Stokes system are presented in Sect. 5.13. For readers not well-acquainted with the analysis of PDEs, the suitable tools for abstract variational formulations, Sobolev spaces, etc., are provided in Appendix A.

5.1 Optimal Heat Source (1): an Unconstrained Case

Our initial illustrative example concerns the heat control problem of Sect. 1.2. In particular, we want to reach a given temperature distribution z_d over a *bounded* domain $\Omega \subset \mathbb{R}^d$, by controlling a distributed heat source u (the *control* variable) and seeking the solution of the following *minimization* problem

$$\tilde{J}(y, u) = \frac{1}{2} \int_{\Omega} (y - z_d)^2 d\mathbf{x} + \frac{\beta}{2} \int_{\Omega} u^2 d\mathbf{x} \ \rightarrow \ \min \tag{5.1}$$

with $\beta > 0$, where $y = y(u)$ (the *state* variable) is the solution of the following *state* problem

$$\begin{cases} \mathcal{E}y = -\Delta y + \mathbf{b} \cdot \nabla y = f + u & \text{in } \Omega \\ y = 0 & \text{on } \Gamma = \partial\Omega. \end{cases} \tag{5.2}$$

A. Manzoni et al., *Optimal Control of Partial Differential Equations*,
Applied Mathematical Sciences 207, https://doi.org/10.1007/978-3-030-77226-0_5

Here $\mathbf{b} \in C^1\left(\bar{\Omega};\mathbb{R}^d\right)$ is a given velocity field with $\text{div}\mathbf{b} = 0$ in Ω, $f \in L^2(\Omega)$ is a source term, while $\mathbf{b} \cdot \nabla y$ represents a drift or transport term. If $\mathbf{b} = \mathbf{0}$, we find the simpler diffusion equation (1.2) of Sect. 1.2. Since u appears as a distributed source in the state equation, this is a distributed control problem.

$\tilde{J}(y,u)$ is the *cost functional*. The second term in (5.1), called *penalization term*, besides helping in the well-posedness analysis, prevents using "too large" controls in the minimization of $\tilde{J}$.

The main issues to be addressed concern existence and, possibly, uniqueness of an optimal pair $(\hat{y},\hat{u})$, and the derivation of necessary and/or sufficient optimality conditions.

5.1.1 Analysis of the State Problem

For the well-posedness analysis of problem (5.2) it is convenient to turn it into an equivalent integral form, the so-called *weak formulation*. We proceed along the following steps:

1. select a space of *smooth test functions* (or *test space*) compatible with the boundary conditions, multiply the differential equation by a test function and integrate over the domain Ω;
2. assume that all the data are smooth. Integrating by parts the divergence terms (here only the diffusion term $\Delta y = \text{div}\nabla y$) and using the boundary conditions, we obtain an *integral equation*, valid for every test function. This is a *weak or variational formulation*;
3. check that the original formulation can be recovered by integrating by parts in reverse order. This means that, for *smooth* solutions and data, the two formulations are equivalent (coherence principle);
4. extend the variational formulation to the state space V (usually a *Sobolev space*), represented by the closure of the test space in a suitable norm, and interpret it as an *abstract variational problem* in V (see Sect. A.4).

The main tool ensuring that the abstract variational problem admits a (unique) solution is the Lax-Milgram Theorem A.6. We now implement the above strategy to achieve a weak formulation of (5.2).

1. We select $C_0^1(\Omega)$ as the space of test functions, having continuous first derivatives and vanishing outside a bounded closed subset of Ω. In particular, *these test functions vanish in a neighborhood of* Γ. We multiply the differential equation by $\varphi \in C_0^1(\Omega)$ and integrate over Ω. We get
$$\int_\Omega \{-\Delta y + \mathbf{b}\cdot\nabla y - f - u\}\,\varphi\, d\mathbf{x} = 0. \tag{5.3}$$
2. We integrate by parts the first term, obtaining, since $\varphi = 0$ on Γ,
$$-\int_\Omega \Delta y\,\varphi\, d\mathbf{x} = \int_\Omega \nabla y\cdot\nabla\varphi\, d\mathbf{x} + \int_\Gamma \nabla y\cdot\mathbf{n}\,\varphi\, d\sigma = \int_\Omega \nabla y\cdot\nabla\varphi\, d\mathbf{x},$$
where $\mathbf{n}$ is the outward unit normal to Γ. Then, from (5.3), we get the following integral equation,
$$\int_\Omega \{\nabla y\cdot\nabla\varphi + (\mathbf{b}\cdot\nabla y)\,\varphi\}\, d\mathbf{x} = \int_\Omega (f+u)\,\varphi\, d\mathbf{x} \qquad \forall\varphi\in C_0^1(\Omega). \tag{5.4}$$
A solution y of (5.4) is called *weak* or *variational* solution to problem (5.2). Thus (5.2) implies (5.4).

3. On the other hand, assume (5.4) is true. Integrating by parts in the reverse order, we return to (5.3), which entails $-\Delta y + \mathbf{b} \cdot \nabla y - f - u = 0$ in Ω, due to the arbitrariness of φ (see Lemma A.4). Thus, *for smooth solutions, (5.2) and (5.4) are equivalent.*
4. Observe that (5.4) only involves first order derivatives of the solution and of the test function. Then, we extend the variational formulation to $V = H_0^1(\Omega)$, the closure of $C_0^1(\Omega)$ in the norm $\|u\|_{H_0^1(\Omega)} = \|\nabla u\|_{L^2(\Omega)}$ (see Sect. A.5) and we arrive to the following final weak formulation:

Find $y \in V = H_0^1(\Omega)$ such that

$$\int_{\Omega} \{\nabla y \cdot \nabla \varphi + (\mathbf{b} \cdot \nabla y) \varphi\} \, d\mathbf{x} = \int_{\Omega} (f + u)\varphi \, d\mathbf{x} \qquad \forall \varphi \in V. \tag{5.5}$$

Introducing the bilinear form

$$a(y, \varphi) = \int_{\Omega} \{\nabla y \cdot \nabla \varphi + (\mathbf{b} \cdot \nabla y) \varphi\} \, d\mathbf{x} \tag{5.6}$$

and the linear functional

$$F\varphi = \int_{\Omega} (f + u)\varphi \, d\mathbf{x},$$

equation (5.5) corresponds to the following abstract variational problem:

Find $y \in V$ such that

$$a(y, \varphi) = F\varphi \qquad \forall \varphi \in V. \tag{5.7}$$

The well-posedness of this problem follows from the Lax-Milgram Theorem.

Proposition 5.1. *For each $u \in L^2(\Omega)$, problem (5.5) has a unique solution $y = y(u) \in H_0^1(\Omega)$. Moreover*

$$\|\nabla y\|_{L^2(\Omega)} \leq C_P \left\{ \|f\|_{L^2(\Omega)} + \|u\|_{L^2(\Omega)} \right\} \tag{5.8}$$

where $C_P > 0$ is the Poincaré constant defined in (A.37).

Proof. We check that the hypotheses of the Lax-Milgram Theorem hold with $V = H_0^1(\Omega)$. Recall that in $H_0^1(\Omega)$ the Poincaré inequality $\|u\|_{L^2(\Omega)} \leq C_P \|\nabla u\|_{L^2(\Omega)}$ holds (see Theorem A.10), for some (Poincaré) constant C_P, depending on d and the diameter of Ω.

Continuity of $a(\cdot, \cdot)$. The Schwarz and Poincaré inequalities yield (with $|\mathbf{b}|_\infty = \max_{\bar{\Omega}} |\mathbf{b}|$):

$$|a(y, \varphi)| \leq \|\nabla y\|_{L^2(\Omega)} \|\nabla \varphi\|_{L^2(\Omega)} + |\mathbf{b}|_\infty \|\nabla y\|_{L^2(\Omega)} \|\varphi\|_{L^2(\Omega)} \leq \left(1 + C_P |\mathbf{b}|_\infty\right) \|\nabla y\|_{L^2(\Omega)} \|\nabla \varphi\|_{L^2(\Omega)}.$$

Coercivity of $a(\cdot, \cdot)$. We have

$$a(y, y) = \int_{\Omega} |\nabla y|^2 \, d\mathbf{x} + \int_{\Omega} (\mathbf{b} \cdot \nabla y) \, y d\mathbf{x} = \int_{\Omega} |\nabla y|^2 \, d\mathbf{x} = \|\nabla y\|^2_{L^2(\Omega)}.$$

Indeed, since $y = 0$ on Γ and $\operatorname{div}\mathbf{b} = 0$, via an integration by parts, we find

$$\int_{\Omega} (\mathbf{b} \cdot \nabla y) \, y d\mathbf{x} = \frac{1}{2} \int_{\Omega} \mathbf{b} \cdot \nabla (y^2) d\mathbf{x} = \frac{1}{2} \int_{\Gamma} (\mathbf{b} \cdot \mathbf{n}) y^2 d\sigma - \frac{1}{2} \int_{\Omega} \operatorname{div}\mathbf{b} \, y^2 d\mathbf{x} = 0.$$

Continuity of $F(\cdot)$. The Schwarz and Poincaré inequalities give

$$|F\varphi| = \left| \int_{\Omega} (f + u)\varphi d\mathbf{x} \right| \leq \|(f + u)\|_{L^2(\Omega)} \|\varphi\|_{L^2(\Omega)} \leq C_P \left(\|f\|_{L^2(\Omega)} + \|u\|_{L^2(\Omega)} \right) \|\nabla \varphi\|_{L^2(\Omega)}.$$

Hence $L \in H^{-1}(\Omega)$ (the dual space of $H_0^1(\Omega)$) and $\|L\|_{H^{-1}(\Omega)} \leq C_P(\|f\|_{L^2(\Omega)} + \|u\|_{L^2(\Omega)})$. □

5.1.2 Existence and Uniqueness of an Optimal Pair. A First Optimality Condition

Given a control $u \in L^2(\Omega)$, the state problem (5.5) has a unique weak solution $y(u) \in H_0^1(\Omega)$ that we may substitute into the analytical expression of $\tilde{J}$ to get the *reduced cost functional* (depending only on u)

$$J(u) = \tilde{J}(y(u), u) = \frac{1}{2}\int_\Omega (y(u) - z_d)^2 \, d\mathbf{x} + \frac{\beta}{2}\int_\Omega u^2 d\mathbf{x}. \tag{5.9}$$

Thus, our minimization problem (5.1) becomes

$$J(u) \to \min \tag{5.10}$$

over the space $\mathcal{U} = L^2(\Omega)$. Once an optimal control $\hat{u}$ is known, the corresponding optimal state is $\hat{y} = y(\hat{u})$. This is the *reduced space* formulation of the OCP.

To derive optimality conditions it is convenient to rewrite $J(u)$ as

$$J(u) = \frac{1}{2}q(u, u) - Lu + c \tag{5.11}$$

where:

- q is a nonnegative, bilinear, *symmetric, continuous* and *coercive* form in $\mathcal{U} = L^2(\Omega)$;
- L is a *linear, continuous functional* in $\mathcal{U} = L^2(\Omega)$;
- c is a constant (irrelevant in the optimization).

Under these assumptions, we can derive necessary and sufficient first order optimality conditions, as done in the finite dimensional case. We will see below how to cast the cost functional in (5.9) in the form (5.11).

Let $\varepsilon \neq 0$, $v \in L^2(\Omega)$ and compare the values of J at $\hat{u}$ and $\hat{u} + \varepsilon v$:

$$\begin{aligned} J(\hat{u} + \varepsilon v) - J(\hat{u}) &= \left[\frac{1}{2}q(\hat{u} + \varepsilon v, \hat{u} + \varepsilon v) - L(\hat{u} + \varepsilon v) + c\right] - \left[\frac{1}{2}q(\hat{u}, \hat{u}) - L\hat{u} + c\right] \\ &= \varepsilon\left[q(\hat{u}, v) - Lv\right] + \frac{1}{2}\varepsilon^2 q(v, v), \end{aligned} \tag{5.12}$$

where we have used the bilinearity of q and the linearity of L. Dividing by ε we get

$$\frac{J(\hat{u} + \varepsilon v) - J(\hat{u})}{\varepsilon} = q(\hat{u}, v) - Lv + \frac{1}{2}\varepsilon q(v, v). \tag{5.13}$$

The left hand side is the differential quotient $[\Phi(\varepsilon) - \Phi(0)]/\varepsilon$ of the real variable function $\Phi(\varepsilon) = J(\hat{u} + \varepsilon v)$.

Now, if $\hat{u}$ is a minimizer, then $J(\hat{u} + \varepsilon v) - J(\hat{u}) \geq 0$ for any ε and v. Then, letting $\varepsilon \to 0^+$, we obtain

$$0 \leq \Phi'(0) = \frac{d}{d\varepsilon}J(\hat{u} + \varepsilon v)\bigg|_{\varepsilon=0} = q(\hat{u}, v) - Lv$$

while, letting $\varepsilon \to 0^-$, we get

$$0 \geq \Phi'(0) = \frac{d}{d\varepsilon}J(\hat{u} + \varepsilon v)\bigg|_{\varepsilon=0} = q(\hat{u}, v) - Lv.$$

Therefore it must be $q(\hat{u}, v) - Lv = 0$ for all $v \in L^2(\Omega)$.

Conversely, if $q(\hat{u}, v) - Lv = 0$ for all $v \in L^2(\Omega)$, then from (5.12)

$$J(\hat{u} + \varepsilon v) - J(\hat{u}) = \frac{1}{2}\varepsilon^2 q(v, v) \geq 0 \qquad \forall v \in L^2(\Omega),$$

whence $\hat{u}$ is a minimizer. Thus, our first order optimality condition reads

$$\left.\frac{d}{d\varepsilon}J(\hat{u}+\varepsilon v)\right|_{\varepsilon=0} = q\left(\hat{u}, v\right) - Lv = 0 \qquad \forall v \in L^2(\Omega). \tag{5.14}$$

Since $v \longmapsto q\left(\hat{u}, v\right) - Lv$ is a linear continuous functional on $L^2(\Omega)$, the left hand side of equation (5.14) defines an element of the dual space of $L^2(\Omega)$, called the *derivative*[1] of J *at* $\hat{u}$. More generally, we can define the derivative of J at any $u \in L^2(\Omega)$; for it we use the symbol $J'\left(u\right)$ and its action *along the direction* v is precisely given by the formula

$$J'\left(u\right)v = \left.\frac{d}{d\varepsilon}J(u+\varepsilon v)\right|_{\varepsilon=0} = q\left(u, v\right) - Lv \qquad \forall v \in L^2\left(\Omega\right). \tag{5.15}$$

In this way, equation (5.14) can be written in the more revealing form

$$J'\left(\hat{u}\right)v = 0 \qquad \forall v \in L^2(\Omega) \tag{5.16}$$

and takes the name of *Euler equation.*

Still we have to check existence and uniqueness of a minimizer. However, under our hypotheses on q and L, thanks to the Lax-Milgram Theorem, the Euler equation (5.16) has a unique solution $\hat{u} \in L^2\left(\Omega\right)$ and this must be the *unique minimizer*, with corresponding *unique optimal state* $\hat{y} = y\left(\hat{u}\right)$.

We now apply the above strategy to our OCP, by writing our functional (5.9) in the form (5.11). Let $y_0(u)$ be the (weak) solution of problem (5.2) with $f = 0$, that is, $\mathcal{E}y_0(u) = u$, $y_0(u) = 0$ on Γ. Note that $y_0(u) = y(u) - y(0)$ and that the map $u \mapsto y_0(u)$ is *linear.*

With these notations $J(u)$ can be written in the form (5.11) where

$$q(u, v) = \int_\Omega y_0(u)y_0(v)\,d\mathbf{x} + \beta\int_\Omega uv\,d\mathbf{x} \tag{5.17}$$

is a bilinear form,

$$Lu = \int_\Omega y_0(u)(z_d - y(0))\,d\mathbf{x} \tag{5.18}$$

is a linear functional in $L^2(\Omega)$ and

$$c = \frac{1}{2}\int_\Omega (y(0) - z_d)^2\,d\mathbf{x}$$

is a constant.

Moreover, q is symmetric (obvious), continuous and coercive in $L^2(\Omega)$. In fact, from (5.8) with $f = 0$, using the Schwarz and Poincaré inequalities,

$$\begin{aligned}|q(u, v)| &\le \|y_0(u)\|_{L^2(\Omega)}\,\|y_0(v)\|_{L^2(\Omega)} \\ &\le C_P^2\,\|\nabla y_0(u)\|_{L^2(\Omega)}\,\|\nabla y_0(v)\|_{L^2(\Omega)} \le C_P^4\,\|u\|_{L^2(\Omega)}\,\|v\|_{L^2(\Omega)}\end{aligned}$$

which gives the continuity of q. The $L^2(\Omega)$-coercivity of q follows from

$$q(u, u) = \int_\Omega (y_0(u))^2 d\mathbf{x} + \beta\int_\Omega u^2\,d\mathbf{x} \ge \beta\,\|u\|^2_{L^2(\Omega)},$$

[1] This derivative is a particular case of Gâteaux derivative, defined in Sect. A.7. In Calculus of Variation, J' is called *first variation* of J and denoted by δJ.

since $\beta > 0$. In the same way,

$$|Lv| \le \|z_d - y(0)\|_{L^2(\Omega)} \|y_0(v)\|_{L^2(\Omega)} \le C_P^2 \|z_d - y(0)\|_{L^2(\Omega)} \|v\|_{L^2(\Omega)},$$

which shows that L is continuous in $L^2(\Omega)$.

Moreover, recalling that $y_0(u) = y(u) - y(0)$, from (5.15) we have for all $u, v \in L^2(\Omega)$,

$$\begin{aligned} J'(u)v &= \int_\Omega y_0(u)y_0(v)\, d\mathbf{x} + \beta \int_\Omega uv\, d\mathbf{x} - \int_\Omega (z_d - y(0))y_0(v)\, d\mathbf{x} \\ &= \int_\Omega (y(u) - z_d)\, y_0(v) d\mathbf{x} + \beta \int_\Omega uv\, d\mathbf{x}. \end{aligned}$$

Then, we infer existence and uniqueness of an optimal control $\hat{u} \in L^2(\Omega)$, characterized by the Euler equation (with $\hat{y} = y(\hat{u})$)

$$J'(\hat{u})v = \int_\Omega (\hat{y} - z_d)\, y_0(v) d\mathbf{x} + \beta \int_\Omega \hat{u}v\, d\mathbf{x} = 0 \qquad \forall v \in L^2(\Omega). \tag{5.19}$$

We summarize everything in the following Theorem.

Theorem 5.1. *There exists a unique optimal control $\hat{u} \in L^2(\Omega)$, solution of the OCP (5.1) and (5.2). Moreover, $\hat{u}$ is optimal if and only if the Euler equation (5.19) holds.*

Remark 5.1. We used the full coercivity of $q(\cdot,\cdot)$ only to prove existence and uniqueness of the optimal control. A closer inspection reveals that to derive the optimality condition (5.19) it is enough to assume that q is nonnegative, that is $q(u,u) \ge 0\ \forall u \in L^2(\Omega)$. •

5.1.3 Use of the Adjoint State

Although the Euler equation (5.19) provides a characterization of the optimal control $\hat{u}$, it is not suitable for its actual computation. Indeed, due to the presence of the factor $y_0(v) = y(v) - y(0)$ in the analytical expression of the derivative of J, to compute $J'(\hat{u})v$ one needs to solve the state equation for both $\hat{u}$ and v.

Moreover, formula (5.19) does not offer a readable expression for the gradient of J at $\hat{u}$, $\nabla J(\hat{u})$, which is the Riesz element associated to the functional $J'(\hat{u})$, see Sect. A.2.2. $\nabla J(\hat{u})$ is an element of the control space and corresponds to the direction of maximum increase (or ascent) for J. This motivates the use of $\nabla J(\hat{u})$ in constructing approximation algorithms based e.g. on the steepest descent method (see Sect. 6.3).

Our next task is to obtain a formula for $J'(\bar{u})v$, at any point $\bar{u}$, that does not involve $y_0(v)$ and that allows us to easily compute $\nabla J(\bar{u})$. The idea is to use the state equation introducing as a test function a *multiplier* $p \in H_0^1(\Omega)$, to be chosen suitably later on, and write $J(u)$ in the augmented form

$$J(u) = \frac{1}{2}\int_\Omega (y(u) - z_d)^2\, d\mathbf{x} + \frac{\beta}{2}\int_\Omega u^2 d\mathbf{x} + \int_\Omega \{(f+u)p - \nabla y(u)\cdot\nabla p - (\mathbf{b}\cdot\nabla y(u))p\}\, d\mathbf{x}. \tag{5.20}$$

In fact, we have just added a term which is zero for all u, thanks to (5.5). Consequently, recalling that $y(u) = y_0(u) + y(0)$, with $y_0(u)$ linear, the linear part Lu in (5.18) now becomes

$$\tilde{L}u = \int_\Omega y_0(u)(z_d - y(0))\, d\mathbf{x} + \int_\Omega \{up - \nabla y_0(u)\cdot\nabla p - (\mathbf{b}\cdot\nabla y_0(u))\, p\}\, d\mathbf{x}$$

and $\tilde{L}$ is still a linear, continuous functional in $L^2(\Omega)$. Therefore (5.15) yields the following formula for the derivative at a point $\bar{u}$: for all $v \in L^2(\Omega)$,

$$J'(\bar{u})\,v = \int_\Omega (y(\bar{u}) - z_d)\, y_0(v)\, d\mathbf{x} + \int_\Omega (p + \beta\bar{u}) v d\mathbf{x} - \int_\Omega \{\nabla y_0(v) \cdot \nabla p + (\mathbf{b} \cdot \nabla y_0(v))\, p\}\, d\mathbf{x}.$$

We now choose the multiplier in order to get rid of the terms containing $y_0(v)$. For that, consider the following *adjoint* problem:

given $\bar{u}$, *find* $p \in V = H_0^1(\Omega)$ *such that*

$$\int_\Omega \{\nabla\psi \cdot \nabla p + (\mathbf{b} \cdot \nabla\psi)\, p\}\, d\mathbf{x} = \int_\Omega (y(\bar{u}) - z_d)\, \psi\, d\mathbf{x} \qquad \forall\psi \in V. \tag{5.21}$$

Observe that the bilinear form on the left hand side of (5.21),

$$a^*(p, \psi) = \int_\Omega \{\nabla\psi \cdot \nabla p + (\mathbf{b} \cdot \nabla\psi)\, p\}\, d\mathbf{x},$$

is precisely the *adjoint* of the bilinear form a in (5.6), given by

$$a^*(p, \psi) = a(\psi, p) \qquad \forall p, \psi \in V,$$

obtained by interchanging the arguments.

An integration by parts reveals that (5.21) represents the weak formulation of the problem

$$\begin{cases} -\Delta p - \operatorname{div}(\mathbf{b}p) = -\Delta p - \mathbf{b} \cdot \nabla p = y(\bar{u}) - z_d & \text{in } \Omega \\ p = 0 & \text{on } \Gamma, \end{cases} \tag{5.22}$$

as $\operatorname{div}\mathbf{b} = 0$.

Since a^* enjoys the same properties of a and $y(u) - z_d \in L^2(\Omega)$, by the Lax-Milgram Theorem the adjoint problem (5.21) has a unique solution $\bar{p} = p(\bar{u}) \in H_0^1(\Omega)$, called *adjoint state* or *multiplier*. Using (5.21), we deduce the following formula for the derivative of J

$$J'(\bar{u})\,v = \int_\Omega (\bar{p} + \beta\bar{u}) v\, d\mathbf{x}. \tag{5.23}$$

Note that (5.23) expresses the action of $J'(\bar{u})$ on an element $v \in L^2(\Omega)$ as the inner product of $\bar{p} + \beta\bar{u}$ and v. This precisely[2] means that

$$\nabla J(\bar{u}) = \bar{p} + \beta\bar{u}.$$

The way the multiplier $\bar{p}$ enters into the expressions of $J'(\bar{p})$ and $\nabla J(\bar{p})$ highlights that its role is not merely to get rid of uncomfortable terms.

In particular, if $\bar{u} = \hat{u}$, the Euler equation (5.19) takes the form

$$J'(\hat{u})\,v = \int_\Omega (\hat{p} + \beta\hat{u}) v\, d\mathbf{x} = 0 \qquad \forall v \in L^2(\Omega) \tag{5.24}$$

which gives immediately

$$\hat{p} + \beta\hat{u} = 0 \quad \text{or} \quad \hat{u} = -\frac{1}{\beta}\hat{p} \qquad \text{a.e. in } \Omega. \tag{5.25}$$

In fact, $\hat{u}$ enjoys a better regularity than just belonging to $\mathcal{U}_{ad}$, as from (5.25) $\hat{u} \in H_0^1(\Omega)$. However, this extra regularity of the optimal control is not true in general, as we well see later.

[2] More rigorously, $\nabla J(u) = Ep(u) + \beta u$, where E is the embedding operator of $H_0^1(\Omega)$ into $L^2(\Omega)$.

Summarizing, we have proven the following result.

Proposition 5.2. *The control* $\hat{u}$ *and the state* $\hat{y} = y(\hat{u})$ *are optimal for the OCP* (5.1)–(5.2) *if and only if, together with the adjoint state (or multiplier)* $\hat{p}$, *they fulfill the following system of optimality conditions:*

$$\begin{cases} \mathcal{E}\hat{y} = -\Delta\hat{y} + \mathbf{b}\cdot\nabla\hat{y} = f + \hat{u} & \text{in } \Omega, \quad \hat{y} = 0 \ \text{ on } \Gamma \\ \mathcal{E}^*\hat{p} = -\Delta\hat{p} - \mathbf{b}\cdot\nabla\hat{p} = \hat{y} - z_d & \text{in } \Omega, \quad \hat{p} = 0 \ \text{ on } \Gamma \\ \hat{p} + \beta\hat{u} = 0 & \text{a.e. in } \Omega. \end{cases}$$

5.1.4 The Lagrange Multipliers Approach

The weak form of the state and adjoint equations can also be obtained after introducing the *Lagrangian* $\mathcal{L} = \mathcal{L}(y, u, p)$, given by

$$\begin{aligned} \mathcal{L}(y,u,p) &= \tilde{J}(y,u) - a(y,p) + (f+u,p)_{L^2(\Omega)} \\ &= \frac{1}{2}\|y - z_d\|^2_{L^2(\Omega)} + \frac{\beta}{2}\|u\|^2_{L^2(\Omega)} - a(y,p) + (f+u,p)_{L^2(\Omega)}. \end{aligned} \tag{5.26}$$

In (5.26) y, u, p have to be considered as *independent variables.* Notice that $\mathcal{L}$ is linear in p. Proceeding as in the definition of J', we can define the partial derivatives of $\mathcal{L}$ with respect to p, y, u as follows:

$$\begin{aligned} \mathcal{L}'_p(\hat{y},\hat{u},\hat{p})\,\varphi &= \frac{d}{d\varepsilon}\mathcal{L}(\hat{y},\hat{u},\hat{p}+\varepsilon\varphi)\Big|_{\varepsilon=0} && \forall\varphi \in V \\ \mathcal{L}'_y(\hat{y},\hat{u},\hat{p})\,\psi &= \frac{d}{d\varepsilon}\mathcal{L}(\hat{y}+\varepsilon\psi,\hat{u},\hat{p})\Big|_{\varepsilon=0} && \forall\psi \in V \\ \mathcal{L}'_u(\hat{y},\hat{u},\hat{p})\,v &= \frac{d}{d\varepsilon}\mathcal{L}(\hat{y},\hat{u}+\varepsilon v,\hat{p})\Big|_{\varepsilon=0} && \forall v \in \mathcal{U} = L^2(\Omega). \end{aligned}$$

We have

$$\mathcal{L}'_p(\hat{y},\hat{u},\hat{p})\,\varphi = -a(\hat{y},\varphi) + (f+\hat{u},\varphi)_{L^2(\Omega)} = 0 \qquad \forall\varphi \in V$$

that corresponds to the state equation. Moreover

$$\begin{aligned} \mathcal{L}'_y(\hat{y},\hat{u},\hat{p})\,\psi &= \tilde{J}'_y(\hat{y},\hat{u})\,\psi - a(\psi,\hat{p}) \\ &= (\hat{y} - z_d,\psi)_{L^2(\Omega)} - a^*(\hat{p},\psi) = 0 \qquad \forall\psi \in V \end{aligned} \tag{5.27}$$

corresponds to the adjoint equation, while

$$\begin{aligned} \mathcal{L}'_u(\hat{y},\hat{u},\hat{p})\,v &= \tilde{J}'_u(\hat{y},\hat{u})\,v + (v,\hat{p})_{L^2(\Omega)} \\ &= (v,\beta\hat{u})_{L^2(\Omega)} + (v,\hat{p})_{L^2(\Omega)} = 0 \qquad \forall v \in L^2(\Omega) \end{aligned}$$

represents the Euler equation.

5.2 Optimal Heat Source (2): a Box Constrained Case

Suppose now that in the optimal heat source problem of Sect. 5.1, the heat source control u is more realistically subject to some pointwise limitations, under the form of *box constraints*. Precisely, define a set of admissible controls as

$$\mathcal{U}_{ad} = \left\{u \in L^2(\Omega) : a \le u(\mathbf{x}) \le b, \text{ a.e. in } \Omega\right\}. \tag{5.28}$$

where a, b are either constants or $L^\infty(\Omega)$ functions. Observe that $\mathcal{U}_{ad}$ is a bounded, closed and convex subset of $L^2(\Omega)$. Indeed, if $u \in \mathcal{U}_{ad}$, then $\|u\|_{L^2(\Omega)} \le \max\{\|a\|_{L^\infty(\Omega)}, \|b\|_{L^\infty(\Omega)}\} |\Omega|^{1/2}$ and therefore $\mathcal{U}_{ad}$ is bounded. Moreover, if $u, v \in \mathcal{U}_{ad}$, any linear convex combination of u and v, $tu + (1-t)v$, $0 \le t \le 1$, belongs to $\mathcal{U}_{ad}$. Finally, if $a \le u_k(\mathbf{x}) \le b$ a.e. in Ω and $u_k \to u$ in $L^2(\Omega)$ then $a \le u(\mathbf{x}) \le b$ a.e. in Ω.

5.2.1 Optimality Conditions

Putting aside for the time being the existence/uniqueness issues, let us focus on optimality conditions. Also we allow $\beta = 0$ in the expression of the cost functional. As in the unconstrained case, we compare the value of J at $\hat{u}$ and at another admissible point, given by the convex combination

$$tv + (1-t)\hat{u} = \hat{u} + t(v - \hat{u}), \qquad 0 \le t \le 1$$

where $v \in \mathcal{U}_{ad}$. From (5.12) we have

$$J(\hat{u} + t(v - \hat{u})) - J(\hat{u}) = t\left[q(\hat{u}, v - \hat{u}) - L(v - \hat{u})\right] + \frac{1}{2}t^2 q(v - \hat{u}, v - \hat{u}).$$

If $\hat{u}$ is a minimizer, then the left hand side is nonnegative and, since this time we can divide by $t > 0$ only, letting $t \to 0^+$, we get the *variational inequality*

$$J'(\hat{u})(v - \hat{u}) = q(\hat{u}, v - \hat{u}) - L(v - \hat{u}) \ge 0 \qquad \forall v \in \mathcal{U}_{ad}. \tag{5.29}$$

On the other hand, if (5.29) holds, then

$$J(\hat{u} + t(v - \hat{u})) - J(\hat{u}) \ge \frac{1}{2}t^2 q(v - \hat{u}, v - \hat{u}) \ge 0 \qquad \forall v \in \mathcal{U}_{ad}$$

and therefore $\hat{u}$ is a minimizer.

Thus, in our control constrained case, a first necessary and sufficient optimality condition is given by the variational inequality (5.29). Writing it in explicit form, and noticing that $y_0(v - \hat{u}) = y(v) - y(\hat{u})$, we get the following

Proposition 5.3. *The control $\hat{u} \in \mathcal{U}_{ad}$ is optimal for the OCP (5.1) and (5.2), with $u \in \mathcal{U}_{ad}$, if and only if the following variational inequality holds*

$$J'(\hat{u})(v - \hat{u}) = \int_\Omega (\hat{y} - z_d)(y(v) - y(\hat{u}))\, d\mathbf{x} + \beta \int_\Omega \hat{u}(v - \hat{u})\, d\mathbf{x} \ge 0 \qquad \forall v \in \mathcal{U}_{ad}. \tag{5.30}$$

Introducing the multiplier $\hat{p}$ and the adjoint problem (5.21), we can use formula (5.23) for the derivative of J to obtain a simplified version of the variational inequality (5.30), namely

$$\int_\Omega (\hat{p} + \beta\hat{u})(v - \hat{u})\, d\mathbf{x} \ge 0 \qquad \forall v \in \mathcal{U}_{ad}. \tag{5.31}$$

In terms of the Lagrangian function, (5.31) reads

$$\mathcal{L}'_u(\hat{y}, \hat{u}, \hat{p})(v - \hat{u}) \geq 0 \qquad \forall v \in \mathcal{U}_{ad}$$

or

$$\mathcal{L}'_u(\hat{y}, \hat{u}, \hat{p})\,\hat{u} = \min_{v \in \mathcal{U}_{ad}} \mathcal{L}'_u(\hat{y}, \hat{u}, \hat{p})\,v.$$

For this reason, (5.31) is sometimes called *minimum principle.*

Still we have to check existence/uniqueness of a minimizer. According to Proposition 5.3, both existence and uniqueness follows if we prove that the variational inequality

$$q(\hat{u}, v - \hat{u}) - L(v - \hat{u}) \geq 0 \qquad \forall v \in \mathcal{U}_{ad} \tag{5.32}$$

has a unique solution $\hat{u} \in \mathcal{U}_{ad}$. Since $\mathcal{U}_{ad}$ is a closed convex set, this is a consequence of Theorem 5.5, proven in Sect. 5.3.5.

5.2.2 Projections onto a Closed Convex Set of a Hilbert Space

In the case $\beta > 0$ the variational inequality (5.31) has an important geometric interpretation in terms of projection over a closed convex set. Under this assumption, we can put $z = -\beta^{-1}\hat{p}$ and write (5.31) in the form

$$(\hat{u} - z, v - \hat{u})_{L^2(\Omega)} \geq 0 \qquad \forall v \in \mathcal{U}_{ad}. \tag{5.33}$$

The (general) theorem below characterizes the solution $\hat{u}$ of (5.33) as the element of $\mathcal{U}_{ad}$ which realizes the distance $\|\hat{u} - z\|$ from $\mathcal{U}_{ad}$, that is

$$\|\hat{u} - z\| = \operatorname{dist}(\mathcal{U}_{ad}, z) = \inf_{u \in \mathcal{U}_{ad}} \|u - z\|.$$

Such element is called the *projection* of z onto $\mathcal{U}_{ad}$, which we denote by

$$\hat{u} = Pr_{\mathcal{U}_{ad}}(z).$$

Theorem 5.2. *Let H be a Hilbert space and $K \subseteq H$ be a closed, convex set. For every $z \in H$ there exists a unique projection $Pr_K(z)$. Moreover $\hat{u} = Pr_K(z)$ if and only if $\hat{u} \in K$ and*

$$(\hat{u} - z, v - \hat{u}) \geq 0 \qquad \forall v \in K \tag{5.34}$$

or, equivalently,

$$(\hat{u} - z, \hat{u}) = \min_{v \in K} (\hat{u} - z, v). \tag{5.35}$$

Proof. Let $d = \inf_{u \in K} \|u - z\|$ and $\{u_n\} \subset K$ be a minimizing sequence, i.e. $\|u_n - z\| \to d$ as $n \to \infty$. As K is convex, $\frac{1}{2}(u_m + u_n) \in K$ and, from the parallelogram law, we can write

$$\|u_n - u_m\|^2 = 2\|u_n - z\|^2 + 2\|u_m - z\|^2 - 4\left\|z - \tfrac{1}{2}(u_m + u_n)\right\|^2 \leq 2\|u_n - z\|^2 + 2\|u_m - z\|^2 - 4d^2.$$

Letting $m, n \to \infty$ we deduce $\|u_n - u_m\| \to 0$ and therefore $\{u_n\}$ is a Cauchy sequence. Since H is complete, u_n converges to some $\hat{u} \in K$ since K is closed. On the other hand $\|u_n - z\| \to \|\hat{u} - z\|$ by the norm continuity, whence $\|\hat{u} - z\| = d$ and $\hat{u} = Pr_K(z)$.

To show uniqueness, assume by contradiction that $\|\hat{u} - z\| = \|\hat{v} - z\| = d$. Using again the parallelogram law, we have

$$\|\hat{u} - \hat{v}\|^2 = 2\|\hat{u} - z\|^2 + 2\|\hat{v} - z\|^2 - 4\left\|z - \tfrac{1}{2}(\hat{u} + \hat{v})\right\|^2 \le 2d^2 + 2d^2 - 4d^2 = 0$$

whence $\hat{u} = \hat{v}$.

To prove (5.34), fix $v \in K$. Assume $\hat{u} = Pr_K(z)$. By the convexity of K, for any $t \in [0,1]$, $tv + (1-t)\hat{u} \in K$ so that we can write

$$\|z - \hat{u}\|^2 \le \|z - [tv + (1-t)\hat{u}]\|^2 = \|(z - \hat{u}) - t(v - \hat{u})]\|^2 = \|z - \hat{u}\|^2 - 2t(z - \hat{u}, v - \hat{u}) + t^2\|v - \hat{u}\|^2.$$

This implies $0 \le 2(\hat{u} - z, v - \hat{u}) + t\|v - \hat{u}\|^2$ and letting $t \to 0$ we obtain (5.34). Conversely, if $\hat{u} \in K$ and (5.34) holds, for every $v \in K$, we have

$$\|z - v\|^2 = \|z - \hat{u} + \hat{u} - v\|^2 = \|z - \hat{u}\|^2 + 2(z - \hat{u}, v - \hat{u}) + \|v - \hat{u}\|^2 \ge \|z - \hat{u}\|^2$$

which shows that $\hat{u} = Pr_K(z)$. □

For the case at hand, $H = \mathcal{U}$ and $K = \mathcal{U}_{ad}$.

Corollary 5.1. *The projection map $Pr_K : H \to K$, $z \longmapsto Pr_K(z)$, is*

(a) strictly monotone, that is

$$(Pr_K(z) - Pr_K(w), z - w) > 0 \quad \forall w, z \in H,\ z \neq w;$$

(b) non-expansive, that is

$$\|Pr_K(w) - Pr_K(z)\| \le \|w - z\| \quad \forall w, z \in H. \tag{5.36}$$

Proof. (a) If $w, z \in H$, we have for all $v \in K$

$$(Pr_K(w) - w, v - Pr_K(w)) \ge 0 \tag{5.37}$$
$$(Pr_K(z) - z, v - Pr_K(z)) \ge 0. \tag{5.38}$$

Taking $v = Pr_K(w)$ in (5.38) and $v = Pr_K(z)$ in (5.37), we obtain

$$(Pr_K(w) - w, Pr_K(z) - Pr_K(w)) \ge 0$$
$$(Pr_K(z) - z, Pr_K(z) - Pr_K(w)) \le 0.$$

Subtracting these two inequalities and rearranging the terms, we deduce

$$\|Pr_K(z) - Pr_K(w)\|^2 \le (Pr_K(z) - Pr_K(w), z - w) \quad \forall w, z \in H. \tag{5.39}$$

Hence Pr_K is a *strictly monotone* operator.

(b) By Cauchy-Schwarz inequality,

$$(Pr_K(z) - Pr_K(w), z - w) \le \|Pr_K(z) - Pr_K(w)\|\,\|z - w\|;$$

(5.36) then follows by (5.39) simplifying by $\|Pr_K(z) - Pr_K(w)\|$. □

5.2.3 Karush-Kuhn-Tucker Conditions

In the case of box constraints, explicit pointwise equivalent versions of the variational inequality (5.31) are available.

Lemma 5.1. *Let $\mathcal{U}_{ad}$ be as in (5.28), $\beta \geq 0$. Then, if $\hat{u} \in \mathcal{U}_{ad}$, the variational inequality (5.31) is equivalent to each one of the following pointwise conditions:*

(i) for all $v \in \mathcal{U}_{ad}$,

$$[\hat{p}(\mathbf{x}) + \beta\hat{u}(\mathbf{x})][v(\mathbf{x}) - \hat{u}(\mathbf{x})] \geq 0 \quad \text{a.e. in } \Omega; \tag{5.40}$$

(ii) for a.e. $\mathbf{x}$ in Ω,

$$\hat{u}(\mathbf{x}) = \begin{cases} a & \text{if } \hat{p}(\mathbf{x}) + \beta\hat{u}(\mathbf{x}) > 0 \\ b & \text{if } \hat{p}(\mathbf{x}) + \beta\hat{u}(\mathbf{x}) < 0 \end{cases} \tag{5.41}$$

and $\hat{p}(\mathbf{x}) + \beta\hat{u}(\mathbf{x}) = 0$ everywhere else;

(iii) if $\beta > 0$, for a.e. $\mathbf{x}$ in Ω,

$$\hat{u}(\mathbf{x}) = \min\left\{b, \max\left\{a, -\beta^{-1}\hat{p}(\mathbf{x})\right\}\right\} = \max\left\{a, \min\left\{b, -\beta^{-1}\hat{p}(\mathbf{x})\right\}\right\}. \tag{5.42}$$

Remark 5.2. When $\beta > 0$, (5.42) provides an explicit expression of the projection operator, analogous to that already found in (3.28) in the case of a finite dimensional space. Moreover, if $\hat{p} \in H^1(\Omega)$, (5.42) indicates that $\hat{u}$ not only belongs to $L^2(\Omega)$ but also to $H^1(\Omega)$.

If $\beta = 0$, (5.41) gives

$$\hat{u}(\mathbf{x}) = \begin{cases} a & \text{if } \hat{p}(\mathbf{x}) > 0 \\ \in [a, b] & \text{if } \hat{p}(\mathbf{x}) = 0 \\ b & \text{if } \hat{p}(\mathbf{x}) < 0. \end{cases}$$

In particular, if $a = -1, b = 1$,

$$\hat{u}(\mathbf{x}) = -\operatorname{sign} \hat{p}(\mathbf{x}).$$

We see that where $\hat{p}(\mathbf{x}) = 0$ no information on $\hat{u}$ is available. When it is possible to prove that $\hat{p}(\mathbf{x}) \neq 0$ a.e. in Ω, then $\hat{u}$ assumes only the extreme values a or b a.e. in Ω. In this case $\hat{u}$ is called a *bang-bang* control. A simple situation in which this happens is described below.

Assume for simplicity that $\mathbf{b} = \mathbf{0}$ and that our equation makes sense pointwise. Then if $-\Delta z_d - f$ is either $> a$ or $< b$, then $\hat{p}(\mathbf{x}) \neq 0$ a.e. in Ω. In fact, if $\hat{p}(\mathbf{x}) = 0$ on a set $E \subset \Omega$ of positive measure, we infer from the adjoint equation that on this set $\hat{y} = z_d$ on E. Then, from the state equation we deduce that $\hat{u} + f = -\Delta\hat{y} = -\Delta z_d$ on E. In this set,

$$\hat{u} = -\Delta z_d - f.$$

Therefore, since $a \leq \hat{u} \leq b$, we reach a contradiction. •

Proof. Clearly (i) implies (5.31) just by integration of (5.40) over Ω. On the other hand, assume by contradiction that (5.31) holds and there exists $w \in \mathcal{U}_{ad}$ such that

$$[\hat{p}(\mathbf{x}) + \beta\hat{u}(\mathbf{x})][w(\mathbf{x}) - \hat{u}(\mathbf{x})] < 0 \tag{5.43}$$

in a set A of positive measure. Define

$$\bar{v}(\mathbf{x}) = \begin{cases} w(\mathbf{x}) & \text{in } A \\ \hat{u}(\mathbf{x}) & \text{in } \Omega \backslash A. \end{cases}$$

Then $\bar{v} \in \mathcal{U}_{ad}$ and inserting it into (5.33) we get the contradiction

$$\int_\Omega (\hat{p} + \beta\hat{u})(\bar{v} - \hat{u})\, d\mathbf{x} = \int_A [\hat{p}(\mathbf{x}) + \beta\hat{u}(\mathbf{x})][w(\mathbf{x}) - \hat{u}(\mathbf{x})]\, d\mathbf{x} < 0.$$

Thus (i) is equivalent to (5.31). A direct check shows that (i) is equivalent to (ii).

Let $\beta > 0$; we check that (ii) is equivalent to (iii). Assume that (ii) holds. If $\hat{p}(\mathbf{x}) + \beta\hat{u}(\mathbf{x}) > 0$, and therefore $\hat{u}(\mathbf{x}) = a$, then $\max\left\{a, -\beta^{-1}\hat{p}(\mathbf{x})\right\} = a$ and (5.42) is true. If $\hat{p}(\mathbf{x}) + \beta\hat{u}(\mathbf{x}) < 0$, and therefore $\hat{u}(\mathbf{x}) = b$, then $\min\left\{b, -\beta^{-1}\hat{p}(\mathbf{x})\right\} = b$ and again (5.42) is true.

Conversely, assume that (iii) holds. If $\min\left\{b, \max\left\{a, -\beta^{-1}\hat{p}(\mathbf{x})\right\}\right\} = a$, then $-\beta^{-1}\hat{p}(\mathbf{x}) \leq a = \hat{u}(\mathbf{x})$ while if $\min\left\{b, \max\left\{a, -\beta^{-1}\hat{p}(\mathbf{x})\right\}\right\} = -\beta^{-1}\hat{p}(\mathbf{x})$ then $a \leq -\beta^{-1}\hat{p}(\mathbf{x}) \leq b$ and therefore $-\beta^{-1}\hat{p}(\mathbf{x}) = \hat{u}(\mathbf{x})$.

Finally, if $\min\left\{b, \max\left\{a, -\beta^{-1}\hat{p}(\mathbf{x})\right\}\right\} = b$, we obtain $-\beta^{-1}\hat{p}(\mathbf{x}) \geq b = \hat{u}(\mathbf{x})$. In all cases (ii) follows. □

Another equivalent way to express (5.31) makes use of *Karush-Kuhn-Tucker multipliers*. Precisely, we have the following result.

Proposition 5.4. *Let $\beta \geq 0$. The variational inequality (5.31) is equivalent to the following Karush-Kuhn-Tucker conditions: there exist two functions λ_a and λ_b in $L^2(\Omega)$ such that, a.e. in Ω,*

$$\begin{cases} \beta u + p - \lambda_a + \lambda_b = 0 \\ a \leq u \leq b, \quad \lambda_a \geq 0, \ \lambda_b \geq 0 \\ \lambda_a (u - a) = \lambda_b (b - u) = 0 \qquad \textit{(complementarity conditions)}. \end{cases} \tag{5.44}$$

Proof. Define

$$\lambda_a = (\beta u + p)^+ = \max\{0, \beta u + p\}, \qquad \lambda_b = (\beta u + p)^- = \max\{0, -(\beta u + p)\};$$

then $\lambda_a \geq 0$, $\lambda_b \geq 0$ and the first two equations in (5.44) are satisfied. Assume that (5.31) is true. Recalling (5.41), if $\lambda_a(\mathbf{x}) > 0$ then $u(\mathbf{x}) = a$. Similarly, if $\lambda_b(\mathbf{x}) > 0$ then $u(\mathbf{x}) = b$.

Conversely, assume that (5.44) holds true. We distinguish the three cases $a < u(\mathbf{x}) < b$, $u(\mathbf{x}) = a$ and $u(\mathbf{x}) = b$. In the first case, the complementarity conditions give $\lambda_a(\mathbf{x}) = \lambda_b(\mathbf{x}) = 0$ and $\beta u(\mathbf{x}) + p(\mathbf{x}) = 0$. If $u(\mathbf{x}) = a$ and $v \in \mathcal{U}_{ad}$ we have $v(\mathbf{x}) \geq u(\mathbf{x})$, $\lambda_b(\mathbf{x}) = 0$ and $\beta u(\mathbf{x}) + p(\mathbf{x}) = \lambda_a \geq 0$. Similarly, if $u(\mathbf{x}) = b$ and $v \in \mathcal{U}_{ad}$ we have $v(\mathbf{x}) \leq u(\mathbf{x})$, $\lambda_a(\mathbf{x}) = 0$ and $\beta u(\mathbf{x}) + p(\mathbf{x}) = -\lambda_b \leq 0$. Then $(p + \beta u)(v - u) \geq 0$ a.e. in Ω, and (5.31) follows by integration over Ω. □

The following conditions are also equivalent to (5.44) (see the more general result reported in Lemma 9.1):

- there exists $\lambda \in L^2(\Omega)$ such that, a.e. in Ω,

$$\begin{cases} \beta u + p + \lambda = 0 \\ a \leq u \leq b \\ u = a \ \text{ where } \lambda < 0, \qquad u = b \ \text{ where } \lambda > 0; \end{cases} \tag{5.45}$$

- there exists $\lambda \in L^2(\Omega)$ such that, for any $c \in \mathbb{R}_+$, a.e. in Ω

$$\begin{cases} \beta u + p + \lambda = 0 \\ \lambda - \min\{0, \lambda + c(u - a)\} - \max\{0, \lambda + c(u - b)\} = 0. \end{cases} \tag{5.46}$$

Expressing the optimality condition as in (5.44) or (5.45) is instrumental to set up the primal-dual active set algorithm, which will be presented in Sect. 6.8. Formulation (5.46) is instead useful to interpret this algorithm as a semismooth Newton method.

5.3 A General Framework for Linear-quadratic OCPs

In this section we provide a general theoretical framework to analyze *linear-quadratic* OCPs, more precisely in presence of *control constraints*. We derive the system of optimality conditions, using either the adjoint method or the Lagrange multipliers method.

5.3.1 The Mathematical Setting

As we sketched in Sect. 5.1, the essential components of our class of OCPs are:

- a Hilbert space V for the state variables, a Hilbert space $\mathcal{U}$ for the control variables and a *convex, closed* subset $\mathcal{U}_{ad} \subseteq \mathcal{U}$ of admissible controls. We use the symbol $\langle \cdot, \cdot \rangle_{V^*,V}$ to denote the duality pairing between V^* and V (see Sect. A.2.2) and similarly for the duality between $\mathcal{U}^*$ and $\mathcal{U}$;
- a control u that influences the state variable y through the following state problem: find $y = y(u) \in V$ such that

$$a(y, \varphi) = \langle F + Bu, \varphi \rangle_{V^*,V} \qquad \forall \varphi \in V. \tag{5.47}$$

Here $a(\cdot, \cdot) : V \times V \to \mathbb{R}$ is a continuous and coercive bilinear form, that is, one for which there exist positive numbers M and α such that

$$|a(\varphi, \psi)| \leq M \|\varphi\|_V \|\psi\|_V \qquad \forall \varphi, \psi \in V$$

and

$$a(\varphi, \varphi) \geq \alpha \|\varphi\|_V^2 \qquad \forall \varphi \in V.$$

$F \in V^*$ and $B : \mathcal{U} \to V^*$ is a linear continuous operator, called *control operator*. In particular, we require that there exists $b \geq 0$ such that

$$\|Bu\|_{V^*} \leq b \|u\|_{\mathcal{U}} \qquad \forall u \in \mathcal{U}.$$

Thus, B maps a control u into an element Bu of V^*. Typically B is an "embedding" operator (see Sect. A.2.3) defined by

$$\langle Bu, \varphi \rangle_{V^*,V} = (u, \varphi)_{\mathcal{U}} \qquad \forall \varphi \in V.$$

Thanks to Lax-Milgram Theorem A.6, for each $u \in \mathcal{U}$, equation (5.47) has a unique solution $y = y(u) \in V$ and the following stability estimate holds

$$\|y(u)\|_V \leq \frac{1}{\alpha} \left(\|F\|_* + b \|u\|_{\mathcal{U}} \right). \tag{5.48}$$

Note that (5.48) states the *continuity* of the *control to state* map $u \longmapsto y(u)$;
- an *observation* operator $C : V \to Z$, where Z is a Hilbert space. Given a state y, Cy is the object we are interested to observe. A crucial hypothesis is that $C \in \mathcal{L}(V, Z)$, so that there exists $\gamma \geq 0$ such that

$$\|Cy\|_Z \leq \gamma \|y\|_V \qquad \forall y \in V; \tag{5.49}$$

- a *cost* (or *objective*) functional $\tilde{J} : V \times \mathcal{U} \to \mathbb{R}$, defined by

$$\tilde{J}(y, u) = \frac{1}{2} \|Cy - z_d\|_Z^2 + \frac{\beta}{2} \|u\|_{\mathcal{U}}^2, \tag{5.50}$$

where z_d is a prescribed (target) function in Z, $\beta \geq 0$, and the associated *reduced* cost functional

$$J(u) = \frac{1}{2} \|Cy(u) - z_d\|_Z^2 + \frac{\beta}{2} \|u\|_{\mathcal{U}}^2. \tag{5.51}$$

The optimal control problem can be formulated as an optimization problem following two different approaches[3]: *(i)* the full-space approach, where both y and u are optimization variable and the PDE constraint is explicitly specified, or *(ii)* the reduced-space approach, where only

[3] As already noted, the two formulations yield different approaches to set up numerical approximation procedures, which goes under the name of *Simultaneous Analysis and Design* or *Nested Analysis and Design*, respectively; see Sect. 6.1.

the control variable is an optimization variable, whereas the state variable is considered to be an implicit function of the control through the PDE constraint.

In the former case, y, u are two independent variables restricted by an equality constraint (the state equation) and by a control constraint ($u \in \mathcal{U}_{ad}$):

$$\begin{gathered} \tilde{J}(y,u) \to \min \\ \text{subject to} \\ y \in V, \quad u \in \mathcal{U}_{ad}, \\ a(y,\varphi) = \langle F + Bu, \varphi \rangle_{V^*,V} \qquad \forall \varphi \in V; \end{gathered} \tag{5.52}$$

a solution $(\hat{u}, \hat{y})$ is called an *optimal control-state pair.*

The latter formulation involves the minimization of the reduced cost functional subject only to a control constraint,

$$\begin{gathered} J(u) = \tilde{J}(y(u),u) \to \min \\ \text{subject to} \\ u \in \mathcal{U}_{ad}. \end{gathered} \tag{5.53}$$

In the latter case one looks for an optimal control $\hat{u}$; the corresponding optimal state is given by $\hat{y} = y(\hat{u})$.

5.3.2 A First Optimality Condition

To derive optimality conditions we use the second formulation (5.53), proceeding as we did in Sect. 5.1.2. Define $y_0(u) = y(u) - y(0)$ and recall that $u \mapsto y_0(u)$ is a linear map since $y_0(u) \in V$ satisfies the equation

$$a(y_0(u), \varphi) = \langle Bu, \varphi \rangle_{V^*,V} \qquad \forall \varphi \in V. \tag{5.54}$$

We write the (reduced) cost functional (5.51) as

$$\begin{aligned} J(u) &= \frac{1}{2}\|Cy_0(u) + Cy(0) - z_d\|_Z^2 + \frac{\beta}{2}\|u\|_{\mathcal{U}}^2 \\ &= \frac{1}{2}\|Cy_0(u)\|_Z^2 + \frac{\beta}{2}\|u\|_{\mathcal{U}}^2 + (Cy_0(u), Cy(0) - z_d)_Z + \frac{1}{2}\|Cy(0) - z_d\|_Z^2. \end{aligned}$$

Introducing the symmetric bilinear form

$$q(u,v) = (Cy_0(u), Cy_0(v))_Z + \beta(u,v)_{\mathcal{U}}, \tag{5.55}$$

the linear functional

$$Lu = (Cy_0(u), Cy(0) - z_d)_Z$$

and the constant

$$c = \frac{1}{2}\|Cy(0) - z_d\|_Z^2,$$

we can write

$$J(u) = \frac{1}{2}q(u,u) - Lu + c. \tag{5.56}$$

To proceed as in Sect. 5.1 (see in particular Remark 5.1) we have to check that q is continuous and nonnegative in $\mathcal{U}$ and that L is continuous in $\mathcal{U}$. In fact, using (5.48), (7.37) and Schwarz inequality we find

$$\begin{aligned}|q(u,v)| &\le \|Cy_0(u)\|_Z \|Cy_0(v)\|_Z + \beta \|u\|_{\mathcal{U}} \|v\|_{\mathcal{U}} \\ &\le \gamma^2 \|y_0(u)\|_V \|y_0(v)\|_V + \beta \|u\|_{\mathcal{U}} \|v\|_{\mathcal{U}} \le (\gamma^2 b^2 \alpha^{-2} + \beta) \|u\|_{\mathcal{U}} \|v\|_{\mathcal{U}}\end{aligned}$$

which shows the continuity of q. Moreover, q is nonnegative since

$$q(u,u) \ge \frac{\beta}{2} \|u\|_{\mathcal{U}}^2 \ge 0.$$

Concerning the linear form L, using the same inequalities, we have

$$\begin{aligned}|Lu| &= |(z_d - Cy(0), Cy_0(u))_Z| \le \|z_d - Cy(0)\|_Z \|Cy_0(u)\|_Z \\ &\le \gamma \|z_d - Cy(0)\|_Z \|y_0(u)\|_V \le b\gamma\alpha^{-1} \|z_d - Cy(0)\|_Z \|u\|_{\mathcal{U}},\end{aligned}$$

thus yielding the continuity of L (hence $L \in \mathcal{U}^*$).

We can now calculate

$$J'(u)v = \frac{d}{d\varepsilon} J(u + \varepsilon v)\Big|_{\varepsilon=0} = q(u,v) - Lv \quad \text{for all } v \in \mathcal{U} \tag{5.57}$$

and conclude that a necessary and sufficient condition for $\hat{u} \in \mathcal{U}_{ad}$ to be a minimizer is the following variational inequality

$$J'(\hat{u})(v - \hat{u}) = q(\hat{u}, v - \hat{u}) - L(v - \hat{u}) \ge 0 \quad \text{for all } v \in \mathcal{U}_{ad}, \tag{5.58}$$

formally identical to (5.29). The analogous of Proposition 5.3 reads as follows.

Proposition 5.5. *$\hat{u} \in \mathcal{U}_{ad}$ is an optimal control, that is a minimizer of (5.53), if and only if the following variational inequality holds*

$$J'(\hat{u})(v - \hat{u}) = (Cy(\hat{u}) - z_d, C(y(v) - y(\hat{u})))_Z + \beta(\hat{u}, v - \hat{u})_{\mathcal{U}} \ge 0 \qquad \forall v \in \mathcal{U}_{ad}. \tag{5.59}$$

5.3.3 Use of the Adjoint State

As already noticed, (5.59) is of least practical interest, due to the presence of the term $y(v) - y(\hat{u})$. To find a more convenient expression of $J'(u)$, we can write our reduced cost functional in the following augmented form using the state equation,

$$J(u) = \frac{1}{2} \|Cy(u) - z_d\|_Z^2 + \frac{\beta}{2} \|u\|_{\mathcal{U}}^2 - a(y(u), p) + \langle F + Bu, p \rangle_{V^*,V},$$

where the *multiplier (or co-state* or *adjoint state)* $p \in V$ is to be suitably chosen. Then, the linear part Lu in (5.56) becomes

$$\tilde{L}u = (Cy_0(u), Cy(0) - z_d)_Z - a(y_0(u), p) + \langle Bu, p \rangle_{V^*,V},$$

still a continuous functional in $\mathcal{U}$. Thus, replacing L with $\tilde{L}$, (5.57) yields the following formula for the derivative J' at an arbitrary point $\bar{u}$

$$J'(\bar{u})v = (Cy(\bar{u}) - z_d, Cy_0(v))_Z - a(y_0(v), p) + (\beta\bar{u}, v)_{\mathcal{U}} + \langle Bv, p \rangle_{V^*,V} \tag{5.60}$$

for any variation $v \in \mathcal{U}$. As done in Sect. 5.1.3, in order to get rid of the terms containing $y_0(v)$, we introduce the following *adjoint* problem:

Given $\bar{u}$, find $p \in V$ such that

$$a^*(p, \psi) = (Cy(\bar{u}) - z_d, C\psi)_Z \qquad \forall \psi \in V, \tag{5.61}$$

where

$$a^*(p, \psi) = a(\psi, p) \qquad \forall \psi, p \in V$$

is the adjoint bilinear form of $a(\cdot, \cdot)$.

By the Lax-Milgram Theorem, problem (5.61) admits a unique solution $\bar{p} = p(\bar{u})$. Indeed, the bilinear form $a^*(\cdot, \cdot)$ inherits the coercivity and continuity properties of $a(\cdot, \cdot)$. Moreover, the linear map $\psi \mapsto G\psi$ on the right-hand side of (5.61),

$$\psi \mapsto (Cy(\bar{u}) - z_d, C\psi)_Z = G\psi \tag{5.62}$$

is an element[4] of V^*. As a matter of fact, G is linear and also continuous, since from (7.37) we deduce

$$|G\psi| \leq \|Cy(\bar{u}) - z_d\|_Z \|C\psi\|_Z \leq \gamma \|Cy(\bar{u}) - z_d\|_Z \|\psi\|_V.$$

Inserting $p = \bar{p}$ into (5.60), we find the more convenient expression of $J'(\bar{u})$ we were looking for,

$$J'(\bar{u})v = (\beta\bar{u}, v)_{\mathcal{U}} + \langle Bv, \bar{p}\rangle_{V^*,V} \qquad \forall \bar{u}, v \in \mathcal{U}. \tag{5.63}$$

From (5.58), setting $\bar{u} = \hat{u}$ (and, consequently, $\bar{p} = p(\hat{u})$) in (5.63) we obtain the variational inequality

$$(\beta\hat{u}, v - \hat{u})_{\mathcal{U}} + \langle B(v - \hat{u}), \hat{p}\rangle_{V^*,V} \geq 0. \tag{5.64}$$

Since (5.64) is a necessary and sufficient optimality condition, we can summarize our conclusions in the following main theorem.

Theorem 5.3. *For the OCP* (5.47), (5.51) *and* (5.53), $\hat{u}$ *and* $\hat{y} = y(\hat{u})$ *are the optimal pair if and only if, together with the adjoint state* $\hat{p} = p(\hat{u}) \in V$, *they fulfill the following system of optimality conditions:*

$$\begin{cases} a(\hat{y}, \varphi) = \langle F + B\hat{u}, \varphi\rangle_{V^*,V} & \forall \varphi \in V \quad \textit{state equation} \\ a^*(\hat{p}, \psi) = (C\hat{y} - z_d, C\psi)_Z & \forall \psi \in V \quad \textit{adjoint equation} \\ (\beta\hat{u}, v - \hat{u})_{\mathcal{U}} + \langle B(v - \hat{u}), \hat{p}\rangle_{V^*,V} \geq 0 & \forall v \in \mathcal{U}_{ad} \quad \textit{minimum principle.} \end{cases} \tag{5.65}$$

Note that the observation operator C enters into the data of the adjoint problem; this is in fact a general property of the adjoint problem, no matter which observation operator is considered. Also remark that, from the expression (5.50) of the cost functional $\tilde{J}$, the right-

[4] More rigorously: to identify G, it is necessary to write this functional as a duality pairing between V^* and V. To do this, let us introduce the inverse of the Riesz map (see Theorem A.2) $\mathcal{R}_Z^{-1} : Z \to Z^*$ and the adjoint operator $C^* : Z^* \to V^*$ (see (A.12)). In this way, we can write

$$(Cy(u) - z_d, C\psi)_Z = \left(\mathcal{R}_Z^{-1}(Cy(u) - z_d), C\psi\right)_{Z^*,Z} = \left(C^*\mathcal{R}_Z^{-1}(Cy(u) - z_d), \psi\right)_{V^*,V},$$

so that $G = C^*\mathcal{R}_Z^{-1}(Cy(u) - z_d)$.

hand side of the adjoint equation in (5.65) is nothing but $\tilde{J}'_y(\hat{u},\hat{y})\,\psi$. Hence, the adjoint equation can be written as

$$a^*(\hat{p},\psi) = \tilde{J}'_y(\hat{u},\hat{y})\,\psi \qquad \forall\psi \in V.$$

Equation (5.63) provides also an important information for numerical approximation, since it allows to calculate the gradient[5] $\nabla J(u)$ of the cost functional, which provides the direction of maximum increase (or ascent) for J. To calculate $\nabla J(u)$, we have to write the expression

$$(\beta u, v)_{\mathcal{U}} + \langle Bv, p(u)\rangle_{V^*,V}$$

as a scalar product in $\mathcal{U}$. For that, we introduce the operator $B^* \in L(V,\mathcal{U}^*)$, the adjoint of B, and the Riesz map (see Sect. A.2.2) $\mathcal{R}_{\mathcal{U}} : \mathcal{U}^* \to \mathcal{U}$.

Proposition 5.6. *For any* $u \in U$,

$$\nabla J(u) = \mathcal{R}_{\mathcal{U}} B^* p(u) + \beta u, \tag{5.66}$$

where $p(u) \in V$ *is the solution of the adjoint problem (5.61).*

Proof. Since $J'(u) \in \mathcal{U}^*$, it is necessary to express the second term appearing in (5.63) as a scalar product in $\mathcal{U}$. Let $p(u) \in V$ be the solution of (5.61). Then we can write, by the definition of B^*,

$$\langle Bv, p(u)\rangle_{V^*,V} = \langle B^*p(u), v\rangle_{\mathcal{U}^*,\mathcal{U}} = (\mathcal{R}_{\mathcal{U}} B^* p(u), v)_{\mathcal{U}}.$$

From (5.63) we deduce

$$J'(u)\,v = (\beta u + \mathcal{R}_{\mathcal{U}} B^* p(u), v)_{\mathcal{U}} \qquad \forall v \in \mathcal{U}. \tag{5.67}$$

Thus, the Riesz element (belonging to $\mathcal{U}$) which represents $J'(u)$ is given by (5.66). □

Formula (5.66) has mainly a theoretical interest, due to the presence of the operator $\mathcal{R}_{\mathcal{U}} B^*$. However, in many cases, the gradient can be easily calculated by hands, as several examples in the following sections will highlight.

5.3.4 The Lagrange Multipliers Approach

To reformulate the optimality conditions (5.65) in a compact way, it is convenient to introduce the following Lagrangian functional

$$\mathcal{L}(y,p,u) = \frac{1}{2}\|Cy - z_d\|_Z^2 + \frac{\beta}{2}\|u\|_{\mathcal{U}}^2 - a(y,p) + \langle F + Bu, p\rangle_{V^*,V}$$

where now y, p, u are *independent* variables. By requiring that the derivatives of $\mathcal{L}$ with respect to the variables p, u, evaluated at $(\hat{y},\hat{p},\hat{u})$ vanish, we obtain:

$$\begin{aligned} &\mathcal{L}'_p(\hat{y},\hat{p},\hat{u})\,\varphi = -a(\hat{y},\varphi) + \langle F + B\hat{u}, \varphi\rangle_{V^*,V} = 0 \qquad \forall\varphi \in V,\\ &\mathcal{L}'_y(\hat{y},\hat{p},\hat{u})\,\psi = (C\hat{y} - z_d, C\psi)_Z - a(\psi,\hat{p}) = (C\hat{y} - z_d, C\psi)_Z - a^*(\hat{p},\psi) = 0 \qquad \forall\psi \in V \end{aligned}$$

which correspond to the state and the adjoint equations, respectively. Moreover

$$\mathcal{L}'_u(\hat{y},\hat{p},\hat{u})\,v = (\beta\hat{u}, v)_{\mathcal{U}} + \langle Bv, \hat{p}\rangle_{V^*,V},$$

[5] In this context, $\nabla J(u)$ is also called *sensitivity* of J. As we have seen in Sect. 4.1.4, the so-called sensitivity approach to compute this quantity is, however, substantially different and much more expensive than the adjoint approach we have addressed so far.

and we can conclude that the system (5.65) is equivalent to

$$\begin{cases} \mathcal{L}'_p(\hat{y},\hat{p},\hat{u})\,\varphi = 0 & \forall \varphi \in V \\ \mathcal{L}'_y(\hat{y},\hat{p},\hat{u})\,\psi = 0 & \forall \psi \in V \\ \mathcal{L}'_u(\hat{y},\hat{p},\hat{u})\,(v-\hat{u}) \geq 0 & \forall v \in \mathcal{U}_{ad}, \end{cases} \tag{5.68}$$

that is, $(\hat{y},\hat{p},\hat{u})$ is a *stationary point* for the Lagrangian $\mathcal{L}$.

5.3.5 Existence and Uniqueness of an Optimal Control (*)

The existence and uniqueness of the solution of the OCP (5.52) or (5.53) is stated in the following theorem, which is actually a particular case of the general Theorem 9.1. The proof that we offer here, however, is tailored on the quadratic case. The less interested reader can safely skip it and jump to Sect. 5.4.

Theorem 5.4. *If $\beta > 0$ there exists a unique optimal control $\hat{u} \in \mathcal{U}_{ad}$. If $\beta = 0$ and $\mathcal{U}_{ad}$ is bounded, there exists an optimal control $\hat{u} \in \mathcal{U}_{ad}$; if moreover the operator $C \circ y_0 \in \mathcal{L}(\mathcal{U}, Z)$ is one-to-one, the optimal control $\hat{u}$ is unique.*

On the basis of Proposition 5.5, to prove Theorem 5.4 it is enough to show that the variational inequality (5.58) has a unique solution $\hat{u} \in \mathcal{U}_{ad}$. When $\beta > 0$, this follows from the following theorem, which can be seen as a generalization of the Lax-Milgram theorem A.6; its proofs relies on the classical Contraction Mapping Theorem A.21.

Theorem 5.5 (Stampacchia). *Let $\mathcal{U}$ be a Hilbert space, $K \subseteq \mathcal{U}$ a closed, convex set and q a symmetric, continuous and coercive bilinear form on $\mathcal{U}$, that is, there exist positive numbers N, θ such that*

$$|q(u,v)| \leq N \|u\|_{\mathcal{U}} \|v\|_{\mathcal{U}} \qquad \forall u,v \in \mathcal{U}$$

and

$$q(u,u) \geq \theta \|v\|^2_{\mathcal{U}} \qquad \forall u \in K.$$

Then, if $L \in \mathcal{U}^*$, the variational inequality

$$q(u, v-u) \geq L(v-u) \qquad \forall v \in K \tag{5.69}$$

has a unique solution $u \in K$.

Proof. Since q is a continuous bilinear form, the linear functional $v \mapsto q(u,v)$ is an element of $\mathcal{U}^*$, obviously depending on u. Let $Au \in \mathcal{U}$ denote the Riesz element associated to this functional,

$$(Au, v)_{\mathcal{U}} = q(u,v) \qquad \forall v \in \mathcal{U}. \tag{5.70}$$

A is a linear and continuous operator, since from

$$\|Au\|^2_{\mathcal{U}} = (Au, Au)_{\mathcal{U}} = q(u, Au) \leq N \|u\|_{\mathcal{U}} \|Au\|_{\mathcal{U}}$$

it follows that

$$\|Au\|_{\mathcal{U}} \leq N \|u\|_{\mathcal{U}}. \tag{5.71}$$

Similarly, let z_L be the Riesz element associated to L. Then we can write (5.69) in the form

$$(Au - z_L, v-u)_{\mathcal{U}} \geq 0 \qquad \forall v \in K.$$

By adding and subtracting u we obtain the equivalent inequality

$$(u - [u - t(Au - z_L)], v - u)_{\mathcal{U}} \geq 0 \qquad \forall v \in K, \forall t > 0. \tag{5.72}$$

Introducing the projection operator onto K, Pr_K, we recognize that (5.72) is equivalent to the following fixed point equation in K

$$u = Pr_K[u - t(Au - z_L)]. \tag{5.73}$$

Since K is a closed subset of the Hilbert space $\mathcal{U}$, it is a complete metric space if endowed with the same metric of $\mathcal{U}$. To conclude our proof it is enough to show that for suitable values of t, the operator $u \to T(u) = Pr_K[u - t(Au - z_L)]$ is a strict contraction, thanks to the Contraction Mapping Theorem A.21.

Since a projection operator is *nonexpansive* (see Corollary 5.1),

$$\begin{aligned}\|Pr_K[u - t(Au - z_L)] - Pr_K[v - t(Av - z_L)]\|_{\mathcal{U}}^2 &\leq \|u - v - tA(u - v)\|_{\mathcal{U}}^2 \\ &= \|u - v\|_{\mathcal{U}}^2 - 2t\,(A(u - v), u - v)_{\mathcal{U}} + t^2\,\|A(u - v)\|_{\mathcal{U}}^2.\end{aligned}$$

Using (5.70) and (5.71), and thanks to the coercivity of q, we obtain

$$\begin{aligned}\|u - v\|_{\mathcal{U}}^2 - 2t\,(A(u - v), u - v)_{\mathcal{U}} + t^2\,\|A(u - v)\|_{\mathcal{U}}^2 &\leq \|u - v\|_{\mathcal{U}}^2 - 2tq\,(u - v, u - v) + N^2t^2\,\|u - v\|_{\mathcal{U}}^2 \\ &\leq \left(1 - 2\theta t + N^2t^2\right)\|u - v\|_{\mathcal{U}}^2.\end{aligned}$$

For $0 < t < 2\theta/N^2$ we have $1 - 2\theta t + N^2t^2 < 1$ and therefore T is a contraction. □

The case $\beta = 0$ is more delicate, since we only have

$$q\,(u, u) = \|Cy_0(u)\|_Z^2 \geq 0.$$

In general the bilinear form q is not coercive and in this case we cannot apply Theorem 5.5. However if K is also *bounded*, to achieve existence it suffices that q is a *nonnegative* bilinear form while, for uniqueness, it suffices that q is *positive*[6], that is

$$q\,(u, u) > 0 \;\text{ if } u \in K, u \neq 0.$$

Indeed, we have the following variant of Theorem 5.5.

Theorem 5.6. *Let $\mathcal{U}$ be a Hilbert space and $K \subseteq \mathcal{U}$ be a closed, bounded, convex set and q be a symmetric, continuous and nonnegative bilinear form on $\mathcal{U}$. Then, if $L \in \mathcal{U}^*$, the variational inequality (5.69) has a solution $u \in K$. If q is positive, this solution is unique.*

Proof. First we prove uniqueness. Let u and w be two solutions to (5.69), i.e.

$$q\,(u, v - u) \geq L\,(v - u) \qquad \forall v \in K,$$

$$q\,(w, v - w) \geq L\,(v - w) \qquad \forall v \in K.$$

Insert $v = \hat{w}$ and $v = \hat{u}$ into the first and the second inequality, respectively.
We get, by changing sign in the first inequality

$$q\,(u, u - w) \leq L\,(u - w) \;\text{ and }\; q\,(w, u - w) \geq L\,(u - w)\,.$$

Subtracting the two inequalities yields

$$q\,(u - w, u - w) \leq 0.$$

By the positivity of q, it follows that $u = w$.

To prove existence we use an approximation argument. For $j \geq 1$, consider the variational problems: find $u_j \in K$ such that

$$q_j\,(u_j, v - u_j) = q\,(u_j, v - u_j) + \frac{1}{j}\,(u_j, v - u_j)_{\mathcal{U}} \geq L\,(v - u_j) \qquad \forall v \in K. \tag{5.74}$$

Since q_j is coercive, by Theorem 5.5, for each j (5.74) has a unique solution.

---

[6] Indeed, a positive bilinear form is strictly convex.

Since K is bounded, the sequence of solutions $\{u_j\}$ has a subsequence $\{u_{j_m}\}$ weakly convergent to u. Being K closed and convex, it follows that $u \in K$ (see Exercise 1). We want to show that u is a solution of (5.69) by passing to the limit into the inequality

$$q(u_{j_m}, v - u_{j_m}) + \frac{1}{j}(u_{j_m}, v - u_{j_m})_{\mathcal{U}} \geq L(v - u_{j_m}), \tag{5.75}$$

that is

$$q(u_{j_m}, v) - q(u_{j_m}, u_{j_m}) + \frac{1}{j}(u_{j_m}, v - u_{j_m})_{\mathcal{U}} \geq Lv - Lu_{j_m}. \tag{5.76}$$

Since $\{u_{j_m}\}$ is bounded, $\frac{1}{j}(u_{j_m}, v - u_{j_m})_{\mathcal{U}} \to 0$ as $j \to \infty$. Moreover, $Lu_{j_m} \to Lu$ by definition of weak convergence, since $L \in \mathcal{U}^*$. Also $q(u_{j_m}, v) \to q(u, v)$ since, for fixed v, the map $u \mapsto q(u, v)$ is linear and continuous. Finally, since q is in particular nonnegative and symmetric, we have (see Exercise 2)

$$|q(u, v)| \leq \sqrt{q(u, u)\, q(v, v)}.$$

Then, we can write

$$q(u, u) = \lim_{m \to \infty} q(u_{j_m}, u) \leq \sqrt{q(u, u)} \lim_{m \to \infty} \inf \sqrt{q(u_{j_m}, u_{j_m})}$$

or

$$q(u, u) \leq \lim_{m \to \infty} \inf q(u_{j_m}, u_{j_m}),$$

which expresses the weak sequential lower continuity of q. Passing to the limit into (5.76), and rearranging terms, we get

$$q(u, v - u) \geq L(v - u)$$

which concludes the proof. □

Proof. (of Theorem 5.4) To apply Theorem 5.6 to the OCP (5.52) or (5.53), we need to check under which conditions $q(u, u) = \|Cy_0(u)\|_Z^2$ is a positive bilinear form. Now, $q(u, u) = \|Cy_0(u)\|_Z^2 = 0$ implies $Cy_0(u) = 0$ and this in turn implies $u = 0$ if the operator $C \circ y_0 : \mathcal{U} \to Z$ is one-to-one. □

Remark 5.3. If $B : \mathcal{U} \to V^*$ and C: $V \to Z$ are both one-to-one operators, then $C \circ y_0$ is also one-to-one. In fact, $Cy_0(u) = 0$ and the injectivity of C give $y_0(u) = 0$. From the state equation with $F = 0$, we infer $\langle Bu, v\rangle_{V^*,V} = 0$ for all $v \in V$, which means $Bu = 0$. Since B is one-to-one we deduce that $u = 0$. •

5.4 Variational Formulation and Well-posedness of Boundary Value Problems

Since in this chapter the state equations will be boundary value problems for second order elliptic equations in divergence form, we devote this section to recall their weak or variational formulation and to examine their well-posedness. Accordingly, a solution of the weak formulation is called a weak or variational solution of the original boundary value problem.

In a *bounded* and *Lipschitz* domain $\Omega \subset \mathbb{R}^d$ (an interval if $d = 1$) we consider the operator

$$\mathcal{E}\varphi = \underbrace{-\text{div}(\mathbf{A}\nabla\varphi)}_{diffusion} + \underbrace{\text{div}(\mathbf{b}\,\varphi) + \mathbf{c}\cdot\nabla\varphi}_{transport} + \underbrace{r\varphi}_{reaction}. \tag{5.77}$$

Usually, the first term in (5.77) models (thermal or molecular) diffusion in heterogeneous or anisotropic media featuring a constitutive law for a flux function $\mathbf{q} = -\mathbf{A}\nabla\varphi$ (Fourier or Fick law). The matrix $\mathbf{A} = \mathbf{A}(\mathbf{x}) = (a_{ij}(\mathbf{x}))$ is called *diffusion matrix*, and its dependence on $\mathbf{x}$ denotes anisotropic diffusion. The terms $\text{div}(\mathbf{b}\varphi)$ models drift or transport; if $\text{div}\mathbf{b} = 0$, then $\text{div}(\mathbf{b}\varphi)$ reduces to $\mathbf{b}\cdot\nabla\varphi$, which is of the same form of the third term $\mathbf{c}\cdot\nabla\varphi$. The vectors $\mathbf{b} = \mathbf{b}(\mathbf{x})$ and $\mathbf{c} = \mathbf{c}(\mathbf{x})$ have the dimensions of a *velocity*. Finally, $r\varphi$ is a reaction term: if, e.g., φ is the concentration of a substance, $r = r(\mathbf{x})$ represents its rate of decomposition ($r > 0$) or growth ($r < 0$), see [237].

Multiplying (5.77) by a smooth function ψ and integrating once by parts, we get the formula

$$\int_{\Omega} \mathcal{E}\varphi\,\psi\,d\mathbf{x} = -\int_{\Gamma} \frac{\partial \varphi}{\partial n_{\mathcal{E}}}\,\psi\,d\mathbf{x} + a(\varphi,\psi) \tag{5.78}$$

where

$$a(\varphi,\psi) = \int_{\Omega} \mathbf{A}\nabla\varphi\cdot\nabla\psi\,d\mathbf{x} - \int_{\Omega} \varphi\mathbf{b}\cdot\nabla\psi\,d\mathbf{x} + \int_{\Omega} \mathbf{c}\cdot\nabla\varphi\,\psi\,d\mathbf{x} + \int_{\Omega} r\varphi\psi\,d\mathbf{x} \tag{5.79}$$

is the bilinear form associated to $\mathcal{E}$, and

$$\frac{\partial \varphi}{\partial n_{\mathcal{E}}} = (\mathbf{A}\nabla\varphi - \mathbf{b}\varphi)\cdot\mathbf{n}$$

is the natural flux at the boundary associated to the terms $-\mathrm{div}(\mathbf{A}\nabla\varphi)$ and $\mathrm{div}(\mathbf{b}\varphi)$, and takes the name of *conormal derivative of* φ.

We will extensively use the adjoint bilinear form a^* of a, given by

$$a^*(\varphi,\psi) = \int_{\Omega} \mathbf{A}^{\top}\nabla\varphi\cdot\nabla\psi\,d\mathbf{x} - \int_{\Omega} \mathbf{b}\cdot\nabla\varphi\,\psi\,d\mathbf{x} + \int_{\Omega} \mathbf{c}\cdot\nabla\psi\,\varphi d\mathbf{x} + \int_{\Omega} r\varphi\psi\,d\mathbf{x}$$

obtained by exchanging the arguments φ and ψ in the analytic expression of a, that is $a^*(\varphi,\psi) = a(\psi,\varphi)$; a^* is the bilinear form associated to the *formal adjoint* operator[7] of $\mathcal{E}$, given by

$$\mathcal{E}^*\psi = -\mathrm{div}(\mathbf{A}^{\top}\nabla\psi) - \mathbf{b}\cdot\nabla\psi - \mathrm{div}(\mathbf{c}\psi) + r\psi. \tag{5.80}$$

Throughout this section, we assume the following hypotheses, under which all the integrals in the analytic expressions of a and a^* are well defined:

1. the differential operator $\mathcal{E}$ is *uniformly elliptic*, i.e. there exist positive numbers α and M such that
$$\alpha\,|\boldsymbol{\xi}|^2 \le \mathbf{A}(\mathbf{x})\,\xi\cdot\xi \le M\,|\boldsymbol{\xi}|^2 \qquad \forall\boldsymbol{\xi}\in\mathbb{R}^d \text{ a.e. in } \Omega; \tag{5.81}$$
2. the coefficients $\mathbf{b}$, $\mathbf{c}$ and r are all bounded, that is,
$$|\mathbf{b}(\mathbf{x})| \le \beta, \quad |\mathbf{c}(\mathbf{x})| \le \gamma, \quad |r(\mathbf{x})| \le \rho \qquad \text{a.e. in } \Omega. \tag{5.82}$$

The uniform ellipticity condition (5.81) states that $\mathbf{A}$ is *positive definite* in Ω. If $\mathbf{A}$ is symmetric its minimum eigenvalue is real and bounded from below by α, called *ellipticity constant*, and its maximum eigenvalue is bounded from above by M.

We give below the weak formulation of the most common boundary value problems, that can be achieved by following the steps 1-4 in Sect. 5.1.1. We also indicate some hypotheses for the well posedness of these problems.

Dirichlet problem

We start by considering the homogeneous Dirichlet problem:

$$\begin{cases} \mathcal{E}y = f + \mathrm{div}\mathbf{F} & \text{in } \Omega \\ y = 0 & \text{on } \Gamma = \partial\Omega. \end{cases} \tag{5.83}$$

[7] We refer to this operator as to the *formal* adjoint because no boundary condition is involved in its definition; this latter is given by the relation

$$(\mathcal{E}\varphi,\psi)_{L^2(\Omega)} = (\varphi,\mathcal{E}^*\psi)_{L^2(\Omega)} \qquad \forall\varphi,\psi\in C_0^\infty(\Omega).$$

If $f \in L^2(\Omega)$ and $\mathbf{F} \in L^2(\Omega)^d$, from Theorem A.9, the right hand side in the differential equation is an element of $H^{-1}(\Omega)$. Choosing $\varphi \in C_0^1(\Omega)$, from (5.78) we get

$$a(y,\varphi) = \int_\Omega (f + \mathrm{div}\mathbf{F})\varphi \, d\mathbf{x} = \int_\Omega f\varphi \, d\mathbf{x} - \int_\Omega \mathbf{F} \cdot \nabla\varphi \, d\mathbf{x} \tag{5.84}$$

since $\varphi = 0$ on Γ. Thus, the weak formulation of (5.83) is: given $f \in L^2(\Omega)$ and $\mathbf{F} \in L^2(\Omega)^d$, find $y \in H_0^1(\Omega)$ such that

$$a(y,\varphi) = F\varphi \qquad \forall \varphi \in H_0^1(\Omega), \tag{5.85}$$

where F is the linear functional in $H^{-1}(\Omega)$ defined by

$$\varphi \longmapsto (f,\varphi)_{L^2(\Omega)} - (\mathbf{F}, \nabla\varphi)_{L^2(\Omega)^d}.$$

If the Dirichlet condition is *nonhomogeneous*, we impose $y = g$ on Γ with $g \in H^{1/2}(\Gamma)$, $g \neq 0$ (see Sect. A.5.7). Then g is the trace on Γ of a function $\tilde{g} \in H^1(\Omega)$, called *lifting* of g. By setting $\tilde{y} = y - \tilde{g}$, we end up with a homogeneous Dirichlet problem for the equation $\mathcal{E}\tilde{y} = \tilde{f} + \mathrm{div}\tilde{\mathbf{F}}$, with

$$\tilde{f} = f + \mathbf{c} \cdot \nabla\tilde{g} + r\tilde{g} \quad \text{and} \quad \tilde{\mathbf{F}} = \mathbf{F} - \mathbf{A}\nabla\tilde{g} + \mathbf{b}\tilde{g}.$$

Neumann and Robin problems

In the case of a Neumann problem we prescribe the *conormal derivative (flux)* at the boundary

$$\begin{cases} \mathcal{E}y = f & \text{in } \Omega \\ \dfrac{\partial y}{\partial n_{\mathcal{E}}} = g & \text{on } \Gamma. \end{cases} \tag{5.86}$$

Choosing $\varphi \in C^1(\bar{\Omega})$, from (5.78) we get

$$a(y,\varphi) = \int_\Omega f\varphi \, d\mathbf{x} + \int_\Gamma \frac{\partial y}{\partial n_{\mathcal{E}}}\varphi \, d\mathbf{x} = \int_\Omega f\varphi \, d\mathbf{x} + \int_\Gamma g\varphi \, d\sigma. \tag{5.87}$$

Note that, if $g \in L^2(\Gamma)$, the last integral is well defined. Since, by Lemma A.3, $C^1(\overline{\Omega})$ is dense in $H^1(\Omega)$, the weak formulation of problem (5.86) is the following: given $f \in L^2(\Omega)$ and $g \in L^2(\Gamma)$, find $y \in H^1(\Omega)$ such that

$$a(y,\varphi) = F\varphi \qquad \forall \varphi \in H^1(\Omega), \tag{5.88}$$

where F is the linear functional defined by

$$\varphi \longmapsto (f,\varphi)_{L^2(\Omega)} + (g,\varphi)_{L^2(\Gamma)}. \tag{5.89}$$

Note that the Dirichlet condition is imposed to the solutions, while the Neumann condition is hidden into the weak formulation (5.88). For this reason, we say that the Neumann condition is *natural* while the Dirichlet one is *forced (or essential).*

In the case of a Robin problem the condition on Γ in (5.86) takes the form

$$\frac{\partial y}{\partial n_{\mathcal{E}}} + hy = g$$

where $h \in L^\infty(\Gamma)$, $h \geq 0$ a.e. on Γ.

The corresponding variational formulation is: given $f \in L^2(\Omega)$, $g \in L^2(\Gamma)$ and $h \in L^\infty(\Gamma)$, $h \geq 0$ a.e. on Γ, find $y \in H^1(\Omega)$ such that

$$\tilde{a}(y,\varphi) = a(y,\varphi) + (hy,\varphi)_{L^2(\Gamma)} = F\varphi \qquad \forall \varphi \in H^1(\Omega). \tag{5.90}$$

Mixed Dirichlet/Neumann problem

In this case we prescribe zero Dirichlet data on $\Gamma_D \subset \Gamma$ and the conormal derivative on $\Gamma_N = \bar{\Gamma} \setminus \Gamma_D$, that is,

$$\begin{cases} \mathcal{E}y = f & \text{in } \Omega \\ y = 0 & \text{on } \Gamma_D \\ \dfrac{\partial y}{\partial n_{\mathcal{E}}} = g & \text{on } \Gamma_N. \end{cases} \tag{5.91}$$

Here $V = H^1_{\Gamma_D}(\Omega)$. Choosing $\varphi \in C^1(\overline{\Omega})$ vanishing in a neighborhood of Γ_D, from (5.78) we get

$$a(y,\varphi) = \int_\Omega f\varphi\, d\mathbf{x} + \int_{\Gamma_N} \frac{\partial y}{\partial n_{\mathcal{E}}} \varphi\, d\mathbf{x} = \int_\Omega f\varphi\, d\mathbf{x} + \int_{\Gamma_N} g\varphi\, d\sigma.$$

Note that if $g \in L^2(\Gamma_N)$ the last integral is well defined.

Since the closure in $H^1(\Omega)$ of the space of test functions is dense in $H^1_{\Gamma_D}(\Omega)$, the weak formulation of problem (5.86) is the following: given $f \in L^2(\Omega)$ and $g \in L^2(\Gamma_N)$, find $y \in H^1_{\Gamma_D}(\Omega)$ such that

$$a(y,\varphi) = F\varphi \qquad \forall \varphi \in H^1_{\Gamma_D}(\Omega), \tag{5.92}$$

where L is the linear functional defined by

$$\varphi \longmapsto (f,\varphi)_{L^2(\Omega)} + (g,\varphi)_{L^2(\Gamma_N)}.$$

Existence, Uniqueness and Regularity

As all the above problems can be recast as abstract variational problems, their analysis can be carried out by invoking the Lax-Milgram Theorem A.6. Under the hypotheses (5.81), (5.82), a is a continuous bilinear form in $H^1(\Omega)$ and therefore also in $H^1_0(\Omega)$ and $H^1_{\Gamma_D}(\Omega)$. In fact, by a repeated use of Schwarz inequality, we can write ($L^2 = L^2(\Omega)$ or $L^2(\Omega)^d$):

$$\begin{aligned} |a(\varphi,\psi)| &\leq d^2 M \|\nabla\varphi\|_{L^2} \|\nabla\varphi\|_{L^2} + \beta \|\nabla\varphi\|_{L^2} \|\psi\|_{L^2} + \gamma \|\varphi\|_{L^2} \|\nabla\psi\|_{L^2} + \rho \|\varphi\|_{L^2} \|\psi\|_{L^2} \\ &\leq \left(d^2 M + \beta + \gamma + \rho\right) \|\varphi\|_{H^1(\Omega)} \|\psi\|_{H^1(\Omega)}. \end{aligned}$$

For the extra term in the expression of $\tilde{a}$ in problem (5.90), observe that, from the trace inequality (A.40), we get

$$\left|(h\varphi,\psi)_{L^2(\Gamma)}\right| \leq \|h\|_{L^\infty(\Gamma)} \|\varphi\|_{L^2(\Gamma)} \|\psi\|_{L^2(\Gamma)} \leq c_{tr}^2 \|h\|_{L^\infty(\Gamma)} \|\varphi\|_{H^1(\Omega)} \|\psi\|_{H^1(\Omega)}$$

so that also $\tilde{a}$ is continuous. The proof of the coercivity of a requires additional assumptions on the coefficients $\mathbf{b}, \mathbf{c}, r$. We shall indicate below some of these conditions.

Proposition 5.7. *Assume that* $f \in L^2(\Omega)$, $\mathbf{F} \in L^2(\Omega)^d$. *If* **b** *and* **c** *have Lipschitz components and*

$$\frac{1}{2}\mathrm{div}\,(\mathbf{b}-\mathbf{c}) + r \geq 0 \quad a.e.\ in\ \Omega, \tag{5.93}$$

problem (5.85) has a unique solution. Moreover,

$$\|\nabla y\|_{L^2(\Omega)^d} \leq \frac{1}{\alpha}\left\{C_P\,\|f\|_{L^2(\Omega)} + \|\mathbf{F}\|_{L^2(\Omega)^d}\right\}. \tag{5.94}$$

Proof. We already know that a is continuous and $F \in H^{-1}(\Omega)$. We have only to show that

$$a\,(y,y) = \int_\Omega \left\{\mathbf{A}\nabla y\cdot\nabla y - (\mathbf{b}-\mathbf{c})y\cdot\nabla y + ry^2\right\}d\mathbf{x} \geq \alpha\|y\|_V^2$$

for any $y \in V$. Since $y = 0$ on Γ, integrating by parts we obtain

$$\int_\Omega (\mathbf{b}-\mathbf{c})y\cdot\nabla y\;d\mathbf{x} = \frac{1}{2}\int_\Omega (\mathbf{b}-\mathbf{c})\cdot\nabla y^2 d\mathbf{x} = -\frac{1}{2}\int_\Omega \mathrm{div}(\mathbf{b}-\mathbf{c})\;y^2 d\mathbf{x}.$$

Therefore, from (5.81) and (5.93), it follows that a is coercive, since

$$a\,(y,y) \geq \alpha\int_\Omega |\nabla y|^2\,d\mathbf{x} + \int_\Omega\left[\frac{1}{2}\mathrm{div}(\mathbf{b}-\mathbf{c}) + r\right]y^2 d\mathbf{x} \geq \alpha\,\|\nabla y\|^2_{L^2(\Omega)^d}.$$

The Lax-Milgram Theorem A.6 gives existence, uniqueness and the stability estimate (5.94). □

Increasing the summability of the data we can obtain at least continuity up to the boundary of a Lipschitz domain. Precisely, the following slightly more general result holds for the Dirichlet problem (see, e.g., [104] for the proof) under the hypotheses of Proposition 5.7; note that no regularity is required on the coefficients.

Theorem 5.7. *Let* Ω *be a bounded, Lipschitz domain and* $y \in H_0^1(\Omega)$ *be the weak solution of* $\mathcal{E}y = f + \mathrm{div}\mathbf{F}$ *in* Ω. *If* $f \in L^p(\Omega)$ *and* $\mathbf{F} \in L^q(\Omega)^d$ *with* $p > d/2$ *and* $q > d$, $d \geq 2$, *then* $y \in C^{0,\sigma}(\overline{\Omega})$ *for some* $\sigma \in (0,1]$, *and*

$$\|y\|_{C^{0,\sigma}(\overline{\Omega})} \leq C\left(\|f\|_{L^p(\Omega)} + \|\mathbf{F}\|_{L^q(\Omega)^d}\right) \tag{5.95}$$

where σ *and* C *depend only on* $\alpha, \beta_0, \gamma_0, \alpha_0, M$ *and on both the diameter and the Lipschitz constant of* Ω.

Concerning the Robin or Neumann problems, we have instead the following result.

Proposition 5.8. *Assume that* $f \in L^2(\Omega)$, $g \in L^2(\Gamma)$, $0 \leq h \leq h_0$ *a.e. on* Γ. *If* **b** *and* **c** *have Lipschitz components and*

$$(\mathbf{b}-\mathbf{c})\cdot\mathbf{n} \leq 0 \quad a.e.\ on\ \Gamma \quad and \quad \frac{1}{2}\mathrm{div}\,(\mathbf{b}-\mathbf{c}) + r \geq c_0 > 0 \quad a.e\ in\ \Omega, \tag{5.96}$$

then, problem (5.88) has a unique solution. Moreover,

$$\|y\|_{H^1(\Omega)} \leq \frac{1}{\min\{\alpha, c_0\}}\left\{\|f\|_{L^2(\Omega)} + \overline{C}\,(d,\Omega)\,\|g\|_{L^2(\Gamma)}\right\}.$$

Proof. As in the proof of Proposition 5.7, we write

$$a(y,y) \geq \alpha \int_\Omega |\nabla y|^2 \, d\mathbf{x} - \frac{1}{2}\int_\Gamma (\mathbf{b}-\mathbf{c})\cdot \mathbf{n}\, y^2 d\sigma + \int_\Omega \left[\frac{1}{2}\mathrm{div}(\mathbf{b}-\mathbf{c}) + r\right] y^2 d\mathbf{x}$$

$$(\text{by } (5.96)) \geq \alpha \int_\Omega |\nabla y|^2 \, d\mathbf{x} + c_0 \int_\Omega y^2 d\mathbf{x} \geq \min\{\alpha, c_0\} \|y\|^2_{H^1(\Omega)}.$$

Finally, using the trace inequality, it is not difficult to check that

$$\left|(f,\varphi)_{L^2(\Omega)}\right| + \left|(g,\varphi)_{L^2(\Gamma)}\right| \leq \left(\|f\|_{L^2(\Omega)} + \overline{C}(d,\Omega)\|g\|_{L^2(\Gamma)}\right)\|\varphi\|_{H^1(\Omega)}$$

and therefore the functional (5.89) belongs to $H^1(\Omega)^*$. We can apply Lax-Milgram Theorem A.6 and conclude the proof. □

A result similar to the one reported in Theorem 5.7 holds for the case of Robin/Neumann conditions, see e.g. [115] for the proof.

Theorem 5.8. *Let Ω be a bounded, Lipschitz domain and assume that $a(\mathbf{x}) \geq c_0 > 0$ a.e. in Ω. Let $y \in H^1(\Omega)$ be the weak solution of*

$$\mathcal{E}y = f \text{ in } \Omega, \qquad \frac{\partial y}{\partial n_{\mathcal{E}}} + hy = g \text{ on } \Gamma.$$

If $f \in L^p(\Omega)$, $g \in L^q(\Gamma)$, with $p > d/2$, $q > d-1$, $d \geq 2$, then $y \in C(\overline{\Omega})$ and

$$\|y\|_{C(\overline{\Omega})} \leq C\left\{\|f\|_{L^p(\Omega)} + \|g\|_{L^q(\partial\Omega)}\right\}$$

where the constant C depends on $\alpha, \beta_0, \gamma_0, \alpha_0, h_0$, M and on both the diameter and the Lipschitz constant of Ω.

Finally, in the case of a mixed Dirichlet-Neumann problem we have the following result.

Proposition 5.9. *Assume that $f \in L^2(\Omega)$, $g \in L^2(\Gamma_N)$. If* **b** *and* **c** *have Lipschitz components in $\overline{\Omega}$ and*

$$(\mathbf{b}-\mathbf{c})\cdot\mathbf{n} \leq 0 \text{ a.e. on } \Gamma_N, \quad \frac{1}{2}\mathrm{div}(\mathbf{b}-\mathbf{c}) + r \geq 0, \text{ a.e. in } \Omega,$$

then problem (5.92) has a unique solution $y \in H^1_{\Gamma_D}(\Omega)$. Moreover,

$$\|\nabla y\|_{L^2(\Omega)^d} \leq \frac{1}{\alpha}\left\{C_P\|f\|_{L^2(\Omega)} + \overline{C}\|g\|_{L^2(\partial\Omega)}\right\}.$$

The proof is similar to the one of Proposition 5.8 (see Exercise 4).

Remark 5.4. The regularity for mixed problems is much more involved and requires subtle conditions along the border between Γ_D and Γ_N. We will not insist on this subject; see, e.g., [133] for further details. •

In the following sections we apply the theory just developed to various OCPs (whose analysis features an increasing mathematical complexity) governed by elliptic advection-diffusion-reaction problems, because of their relevance in applications. In particular, we deal with distributed or boundary controls, distributed or boundary observation operators, different boundary conditions and different quadratic cost functionals.

For each OCP, we will systematically carry out the analysis of the state equation, an existence and uniqueness analysis of the optimal control, and will derive the optimality conditions.

5.5 Distributed Observation and Control

We start by considering a simple case, where both the observation and the control are distributed. We first address the case of a state problem with Robin conditions; then, we turn to a different quadratic cost functional.

5.5.1 Robin Conditions

Let $\mathcal{U}_{ad} \subseteq \mathcal{U} = L^2(\Omega)$ be a closed, convex set and $f \in L^2(\Omega)$, $g \in L^2(\Gamma)$, $0 \le h \le h_o$ a.e. on Γ. Let moreover Ω_0 be an open subset of Ω and $z_d \in L^2(\Omega_0)$. Consider the minimization of the functional

$$\tilde{J}(y,u) = \frac{1}{2}\int_{\Omega_0} (y - z_d)^2 \, d\mathbf{x} + \frac{\beta}{2}\int_{\Omega} u^2 d\mathbf{x},$$

with $\beta \ge 0$, subject to $(y,u) \in H^1(\Omega) \times \mathcal{U}_{ad}$ and to the state problem

$$\begin{cases} \mathcal{E}y = -\mathrm{div}\,(\mathbf{A}\nabla y) + ry = f + u & \text{in } \Omega \\ \dfrac{\partial y}{\partial n_{\mathcal{E}}} + hy = g & \text{on } \Gamma. \end{cases} \tag{5.97}$$

The weak form of problem (5.97) reads

$$\begin{aligned} \tilde{a}(y,\varphi) &= (\mathbf{A}\nabla y, \nabla\varphi)_{L^2(\Omega)} + (ry,\varphi)_{L^2(\Omega)} + (hy,\varphi)_{L^2(\Gamma)} \\ &= (f+u,\varphi)_{L^2(\Omega)} + (g,\varphi)_{L^2(\Gamma)} \qquad \forall\varphi \in H^1(\Omega). \end{aligned}$$

With reference to the general framework of Sect. 5.3, here $Z = L^2(\Omega_0)$, $V = H^1(\Omega)$ and the observation operator $C : V \to Z$ is given by

$$Cy = \chi_{\Omega_0} y, \tag{5.98}$$

where χ_{Ω_0} is the characteristic function of Ω_0. We are thus interested in observing y only on the subset Ω_0 of Ω. Since $\|Cy\|_{L^2(\Omega)} \le \|y\|_{L^2(\Omega)} \le \|y\|_{H^1(\Omega)}$ we have $C \in \mathcal{L}(V,Z)$. However, if Ω_0 is strictly contained in Ω, C is not a one-to-one operator.

- *The state problem.* Assume that $r(\mathbf{x}) \ge c_0 > 0$. Then, by Proposition 5.8, there exists a unique weak solution $y = y(u) \in H^1(\Omega)$ for every $u \in L^2(\Omega)$ and we can consider the reduced cost functional $J(u) = \tilde{J}(y(u), u)$.
- *Existence and uniqueness of the optimal control.* If $\beta > 0$, by Theorem 5.4, there exists a unique optimal control $\hat{u} \in \mathcal{U}_{ad}$. If $\beta = 0$ and $\mathcal{U}_{ad}$ is bounded, there exists at least an optimal control, but uniqueness is not guaranteed. Indeed, the relation $Cy_0(u) = \chi_{\Omega_0} y_0(u) = 0$ gives $y_0(u) = 0$ a.e. only on the subset Ω_0 and we cannot infer that $u = 0$ all over Ω.
- *Optimality conditions.* We make use of the Lagrangian function

$$\mathcal{L}(y,p,u) = \frac{1}{2}\|y - z_d\|^2_{L^2(\Omega_0)} + \frac{\beta}{2}\|u\|^2_{L^2(\Omega)} - \tilde{a}(y,p) + (f+u,p)_{L^2(\Omega)} + (g,p)_{L^2(\Gamma)}.$$

By differentiating $\mathcal{L}(y,p,u)$, we find:

$$\begin{aligned}
\mathcal{L}'_p(\hat{y},\hat{p},\hat{u})\,\varphi &= -\tilde{a}(\hat{y},\varphi) + (f+u,\varphi)_{L^2(\Omega)} + (g,\varphi)_{L^2(\Gamma)} && \forall\varphi \in H^1(\Omega)\\
\mathcal{L}'_y(\hat{y},\hat{p},\hat{u})\,\psi &= -\tilde{a}^*(\hat{p},\psi) + (\hat{y}-z_d,\psi)_{L^2(\Omega_0)} && \forall\psi \in H^1(\Omega)\\
\mathcal{L}'_u(\hat{y},\hat{p},\hat{u})\,v &= (\beta\hat{u}+\hat{p},v-\hat{u})_{L^2(\Omega)} && \forall v \in \mathcal{U}_{ad}.
\end{aligned}$$

Thus, from Theorem 5.3 and relation (5.68), we obtain the following system of optimality conditions:

$$\begin{cases}
\tilde{a}(\hat{y},\varphi) = (f+u,\varphi)_{L^2(\Omega)} + (g,\varphi)_{L^2(\Gamma)} & \forall\varphi \in H^1(\Omega)\\
\tilde{a}^*(\hat{p},\psi) = (\hat{y}-z_d,\psi)_{L^2(\Omega_0)} & \forall\psi \in H^1(\Omega)\\
(\beta\hat{u}+\hat{p},v-\hat{u})_{L^2(\Omega)} \geq 0 & \forall v \in \mathcal{U}_{ad}.
\end{cases} \tag{5.99}$$

Moreover, for all $u \in \mathcal{U}$, the following formula holds for the gradient of the cost functional,

$$\nabla J(u) = \beta u + p(u) \in L^2(\Omega),$$

where $p = p(u)$ is the unique solution to the adjoint problem

$$\tilde{a}^*(p,\psi) = (y(u)-z_d,\psi)_{L^2(\Omega_0)} \quad \forall\psi \in H^1(\Omega).$$

Remark 5.5. The adjoint equation is the weak formulation of the Robin problem

$$\begin{cases}
\mathcal{E}^*\hat{p} = -\mathrm{div}\left(\mathbf{A}^\top\nabla\hat{p}\right) + r\hat{p} = (\hat{y}-z_d)\,\chi_{\Omega_0} & \text{in } \Omega\\
\dfrac{\partial\hat{p}}{\partial n_{\mathcal{E}^*}} + h\hat{p} = 0 & \text{on } \Gamma,
\end{cases}$$

which is clearly well posed in $H^1(\Omega)$. •

Recalling Sect. 5.2.2, if $\beta > 0$, from the variational inequality in system (5.104) we deduce that

$$\hat{u} = Pr_{\mathcal{U}_{ad}}\left(-\beta^{-1}\hat{p}\right),$$

i.e. $\hat{u}$ is the projection with respect to the $L^2(\Omega)$ scalar product of $-\beta^{-1}\hat{p}$ onto $\mathcal{U}_{ad}$. Let us examine some special cases.

(a) Let $\mathcal{U}_{ad} = L^2(\Omega)$, that is, no control constraint is imposed, and let $\beta > 0$, since $\mathcal{U}_{ad}$ is unbounded. The variational inequality reduces to the equation

$$(\hat{p}+\beta\hat{u},v)_{L^2(\Omega)} = 0 \qquad \forall v \in L^2(\Omega)$$

yielding

$$\hat{p}(\mathbf{x}) + \beta\hat{u}(\mathbf{x}) = 0 \quad \text{whence} \quad \hat{u}(\mathbf{x}) = -\beta^{-1}\hat{p}(\mathbf{x}) \quad \text{a.e. in } \Omega.$$

Inserting this equality into (5.68) we get a system in the variables $\hat{y},\hat{p}$ only:

$$\begin{cases}
-\mathrm{div}(\mathbf{A}\nabla\hat{y}) + r\hat{y} + \beta^{-1}\hat{p} = f & \text{in } \Omega, \quad \dfrac{\partial\hat{y}}{\partial n_{\mathcal{E}}} + h\hat{y} = g \ \text{ on } \Gamma\\
-\mathrm{div}\left(\mathbf{A}^\top\nabla\hat{p}\right) + r\hat{p} = (\hat{y}-z_d)\chi_{\Omega_0} & \text{in } \Omega, \quad \dfrac{\partial\hat{p}}{\partial n_{\mathcal{E}^*}} + h\hat{p} = 0 \ \text{ on } \Gamma.
\end{cases}$$

(b) $\mathcal{U}_{ad} = \left\{v \in L^2(\Omega) : v \geq 0 \text{ a.e. in } \Omega\right\}$, $\beta > 0$ (since $\mathcal{U}_{ad}$ is unbounded). From Lemma 5.1 with $a = 0$ and $b = +\infty$, it follows that

$$\hat{u}(\mathbf{x}) = Pr_{[0,\infty)}\left(-\beta^{-1}\hat{p}(\mathbf{x})\right) = \max\left\{0,-\beta^{-1}\hat{p}(\mathbf{x})\right\} = -\beta^{-1}\min\left\{0,\hat{p}(\mathbf{x})\right\}$$

a.e. in Ω. System (5.68) now reads

$$\begin{cases}
-\mathrm{div}(\mathbf{A}\nabla\hat{y}) + r\hat{y} + \beta^{-1}\min\{0,\hat{p}\} = f & \text{in } \Omega, \quad \hat{y} = 0 \ \text{ on } \Gamma\\
-\mathrm{div}\left(\mathbf{A}^\top\nabla\hat{p}\right) + r\hat{p} = (\hat{y}-z_d)\chi_{\Omega_0} & \text{in } \Omega, \quad \hat{p} = 0 \ \text{ on } \Gamma.
\end{cases}$$

(c) $\mathcal{U}_{ad} = \{v \in L^2(\Omega) : a \le v(\mathbf{x}) \le b \text{ a.e. in } \Omega\}$. Here $\mathcal{U}_{ad}$ is bounded in $L^2(\Omega)$ so that we may consider $\beta \ge 0$. From Lemma 5.1, if $\beta > 0$ we get, for a.e. $\mathbf{x} \in \Omega$,

$$\hat{u}(\mathbf{x}) = Pr_{[a,b]}\left(-\beta^{-1}\hat{p}(\mathbf{x})\right) = \min\left\{b, \max\left\{a, -\beta^{-1}\hat{p}(\mathbf{x})\right\}\right\} = \max\left\{a, \min\left\{b, -\beta^{-1}\hat{p}(\mathbf{x})\right\}\right\}.$$

In the case $\beta = 0$, a conclusion similar to the one of Remark 5.2 holds.

Remark 5.6. Note that in the three cases above, if $\beta > 0$ $\hat{u}$ enjoys better regularity than $L^2(\Omega)$, indeed $\hat{u} \in H^1(\Omega)$. •

5.5.2 Dirichlet Conditions, Energy Cost Functional

Given $z_d \in H_0^1(\Omega)$, we examine the minimization of the energy functional

$$\tilde{J}(y,u) = \frac{1}{2}\|y - z_d\|^2_{H^1(\Omega)} = \frac{1}{2}\int_\Omega \left\{(y - z_d)^2 + |\nabla(y - z_d)|^2\right\} d\mathbf{x}, \tag{5.100}$$

where $y = y(u)$ is the solution of the state problem:

$$\begin{cases} -\Delta y = f + u & \text{in } \Omega \\ y = 0 & \text{on } \Gamma \end{cases} \tag{5.101}$$

whose weak formulation is

$$a(y,\varphi) = (\nabla y, \nabla\varphi) = (f + u, \varphi)_{L^2(\Omega)}. \tag{5.102}$$

Here $Z = H_0^1(\Omega)$ and $Cy = y$, the identity operator in $H_0^1(\Omega)$. We choose for the time being $\mathcal{U} = L^2(\Omega)$ (even though $\beta = 0$) and let $\mathcal{U}_{ad} \subset L^2(\Omega)$ be a *bounded*, closed and convex set. Hence, for every $u \in L^2(\Omega)$, (5.102) has a unique solution $y = y(u) \in H_0^1(\Omega)$ and

$$\|\nabla y(u)\|_{L^2(\Omega)} \le C_P\left(\|f\|_{L^2(\Omega)} + \|u\|_{L^2(\Omega)}\right).$$

• *Existence and uniqueness of the optimal control.* Existence of an optimal control is guaranteed, since $\mathcal{U}_{ad}$ is bounded. To check uniqueness, according to Theorem 5.4, we examine the injectivity of the operator $C \circ y_0 \in \mathcal{L}(\mathcal{U}, Z)$, where $y_0(u) \in H_0^1(\Omega)$ is the solution to

$$a(y_0(u), \varphi) = (u, \varphi)_{L^2(\Omega)} \qquad \forall\varphi \in H_0^1(\Omega). \tag{5.103}$$

Now, from the state equation, $Cy_0(u) = y_0(u) = 0$ implies

$$0 = a(y_0(u), \varphi) = (u, \varphi)_{L^2(\Omega)} \qquad \forall\varphi \in H_0^1(\Omega).$$

Since $H_0^1(\Omega)$ is dense in $L^2(\Omega)$, we infer $u = 0$. We deduce that the optimal control is unique.

• *Optimality conditions.* The *Lagrangian* is

$$\mathcal{L}(y,p,u) = \frac{1}{2}\|y - z_d\|^2_{H^1(\Omega)} - a(y,p) + (p, f + u)_{L^2(\Omega)}.$$

By differentiating $\mathcal{L}(y,p,u)$ with respect to p, we find

$$\mathcal{L}'_p(\hat{y},\hat{p},\hat{u})\varphi = -a(\hat{y},\varphi) + (f + \hat{u}, \varphi)_{L^2(\Omega)}.$$

Similarly, by differentiating $\mathcal{L}(y,p,u)$ with respect to y, we obtain

$$\mathcal{L}'_y(\hat{y},\hat{p},\hat{u})\psi = (\hat{y} - z_d, \psi)_{H^1(\Omega)} - a(\psi,\hat{p}) = (\hat{y} - z_d, \psi)_{H^1(\Omega)} - a^*(\hat{p},\psi)$$

and, for each $v \in L^2(\Omega)$,
$$\mathcal{L}'_u(\hat{y}, \hat{p}, \hat{u})\, v = (\hat{p}, v)_{L^2(\Omega)}.$$

Thus, the system of optimality conditions for $\hat{u}$ and $\hat{y} = y(\hat{u})$, together with the adjoint state $\hat{p} \in H_0^1(\Omega)$, reads:

$$\begin{cases} a(\hat{y}, \varphi) = (f + u, \varphi)_{L^2(\Omega)} & \forall \varphi \in H_0^1(\Omega) \\ a^*(\hat{p}, \psi) = (\hat{y} - z_d, \psi)_{H^1(\Omega)} & \forall \psi \in H_0^1(\Omega) \\ (\hat{p}, v - \hat{u})_{L^2(\Omega)} \geq 0 & \forall v \in \mathcal{U}_{ad}. \end{cases} \tag{5.104}$$

Moreover, for all $u \in \mathcal{U}$, the following formula holds
$$\nabla J(u) = p(u), \tag{5.105}$$
where $p = p(u)$ is the unique solution to the adjoint problem
$$a^*(p, \psi) = (y(u) - z_d, \psi)_{H^1(\Omega)} \quad \forall \psi \in H_0^1(\Omega).$$

Remark 5.7. The adjoint equation for $\hat{p}$ is the weak formulation of the problem:
$$\begin{cases} \mathcal{E}^*\hat{p} = (I - \Delta)(\hat{y} - z_d) & \text{in } \Omega \\ \hat{p} = 0 & \text{on } \partial\Omega \end{cases}$$
which is clearly well posed since $(I - \Delta)(\hat{y} - z_d) \in H^{-1}(\Omega)$. •

Remark 5.8. An interesting case occurs when $\mathcal{U}_{ad}$ coincides with a ball, precisely
$$\mathcal{U}_{ad} = B_R = \left\{ u \in L^2(\Omega) : \|u\|_{L^2(\Omega)} \leq R \right\}.$$
In this case, $\hat{u} \in B_R$ satisfies the inequality
$$(\hat{p}, v - \hat{u})_{L^2(\Omega)} \geq 0 \qquad \forall v \in B_R. \tag{5.106}$$
If $\hat{p} \neq 0$ a.e. in Ω,
$$\hat{u} = -R \frac{\hat{p}}{\|\hat{p}\|_{L^2(\Omega)}}.$$
This follows from a geometrical argument: from (5.106) we see that the vectors $\hat{p}$ and $v - \hat{u}$ meet at an acute angle for any $v \in B_R$, and $\hat{u}$ given above is the only point that fulfills this property. •

An alternative choice for the control space*

If $\mathcal{U}_{ad}$ is unbounded, we can still prove existence and uniqueness of the optimal control, but in a slightly different setting. Let us choose $\mathcal{U} = H^{-1}(\Omega)$, $\mathcal{U}_{ad} \subseteq H^{-1}(\Omega)$, closed, convex, $f \in H^{-1}(\Omega)$, and write the state equation in the form
$$a(y, \varphi) = \langle f + u, \varphi \rangle_{H^{-1}(\Omega), H_0^1(\Omega)} \qquad \forall \varphi \in H_0^1(\Omega). \tag{5.107}$$
Referring to Theorem 5.4, the quadratic form (5.55) in the present case is
$$q(u, u) = \|y_0(u)\|_{H^1(\Omega)}^2.$$
Using the state equation we have
$$|\langle u, \varphi \rangle_{H^{-1}(\Omega), H_0^1(\Omega)}| = |a(y_0(u), \varphi)| \leq \|y_0(u)\|_{H^1(\Omega)} \|\varphi\|_{H^1(\Omega)} \qquad \forall \varphi \in H_0^1(\Omega),$$

from which

$$\|u\|_{H^{-1}(\Omega)} = \sup_{\|\varphi\|_{H^{-1}(\Omega)}=1} |\langle u,\varphi\rangle_{H^{-1}(\Omega),H_0^1(\Omega)}| \leq \|y_0(u)\|_{H^1(\Omega)}.$$

This means that

$$q(u,u) = \|y_0(u)\|_{H^1(\Omega)}^2 \geq \|u\|_{H^{-1}(\Omega)}^2 \tag{5.108}$$

and therefore q is continuous, symmetric and coercive in $\mathcal{U} = H^{-1}(\Omega)$.

Existence and uniqueness of the optimal control $\hat{u}$ then directly follow from Theorem 5.4. The Lagrangian now takes the form

$$\mathcal{L}(y,p,u) = \frac{1}{2}\|y - z_d\|_{H^1(\Omega)}^2 - a(y,p) + \langle f+u,p\rangle_{H^{-1}(\Omega),H_0^1(\Omega)}.$$

For every $\varphi,\psi \in H_0^1(\Omega)$,

$$\mathcal{L}_p(\hat{y},\hat{p},\hat{u})\varphi = -a(\hat{y},\varphi) + \langle f+\hat{u},p\rangle_{H^{-1}(\Omega),H_0^1(\Omega)},$$

while

$$\mathcal{L}_y(\hat{y},\hat{p},\hat{u})\psi = (\hat{y}-z_d,\psi)_{H^1(\Omega)} - a^*(\hat{p},\psi)$$

as in the previous case and for all $v \in H^{-1}(\Omega)$,

$$\mathcal{L}_v(\hat{y},\hat{p},\hat{u})v = \langle \hat{p},v\rangle_{H_0^1(\Omega),H^{-1}(\Omega)}. \tag{5.109}$$

The optimality system (5.68) reads, in this case:

$$\begin{cases} a(\hat{y},\varphi) = \langle f+\hat{u},\varphi\rangle_{H^{-1}(\Omega),H_0^1(\Omega)} & \forall\varphi \in H_0^1(\Omega) \\ a^*(\hat{p},\psi) = (\hat{y}-z_d,\psi)_{H^1(\Omega)} & \forall\psi \in H_0^1(\Omega) \\ \langle \hat{p},v-\hat{u}\rangle_{H_0^1(\Omega),H^{-1}(\Omega)} \geq 0 & \forall v \in \mathcal{U}_{ad}. \end{cases} \tag{5.110}$$

Regarding the gradient of the cost functional, from (5.109) we obtain

$$J'(\hat{u})v = \langle \hat{p},v\rangle_{H_0^1(\Omega),H^{-1}(\Omega)}.$$

Observe now that the operator $-\Delta : H_0^1(\Omega) \to H^{-1}(\Omega)$ coincides with the inverse of the Riesz operator in $H_0^1(\Omega)$, so that we can write

$$J'(\hat{u})v = \langle \hat{p},v\rangle_{H_0^1(\Omega),H^{-1}(\Omega)} = (-\Delta\hat{p},v)_{H^{-1}(\Omega)}.$$

Therefore,

$$\nabla J(u) = -\Delta p(u), \tag{5.111}$$

where $p(u) \in H_0^1(\Omega)$ solves

$$a^*(p,\psi) = (y(u)-z_d,\psi)_{H^1(\Omega)} \qquad \forall\psi \in H_0^1(\Omega).$$

By comparing the two formulas (5.105) and (5.111), we see that a more general functional setting produces a much more complex formula for the gradient of J.

5.6 Distributed Observation, Neumann Boundary Control

Let $\Omega_0 \subset \Omega$ be a subdomain of Ω of positive measure, $\Gamma_D \subset \Gamma$, $\Gamma_N = \Gamma \backslash \Gamma_D$, and $z_d \in L^2(\Omega_0)$. Consider the minimization of the following functional

$$\tilde{J}(y,u) = \frac{1}{2}\int_{\Omega_0} (y - z_d)^2 \, d\mathbf{x} + \frac{\beta}{2}\int_{\Gamma_N} u^2 d\sigma, \qquad \beta \geq 0,$$

subject to $(y,u) \in H^1_{\Gamma_D}(\Omega) \times \mathcal{U}_{ad}$, where $\mathcal{U}_{ad} \subseteq \mathcal{U} = L^2(\Gamma_N)$ is a closed, convex set, and to the state problem

$$\begin{cases} \mathcal{E}y = -\Delta y + \operatorname{div}(\mathbf{b}y) = f & \text{in } \Omega \\ y = 0 & \text{on } \Gamma_D \\ \dfrac{\partial y}{\partial n} - (\mathbf{b} \cdot \mathbf{n})\, y = u & \text{on } \Gamma_N, \end{cases} \tag{5.112}$$

where $f \in L^2(\Omega)$. With reference to the general framework of Sect. 5.3, here $Z = L^2(\Omega_0)$, $V = H^1_{\Gamma_D}(\Omega)$, endowed with the norm $\|\nabla y\|_{L^2(\Omega)^d}$, and the observation operator is the same as in (5.98). The control u is a boundary control, defined on Γ_N; the control operator $B : L^2(\Gamma_N) \to V^*$ is defined by

$$\langle Bu, \varphi \rangle_* = \int_{\Gamma_N} u\varphi d\sigma.$$

By the trace inequality (A.47) and the Poincaré inequality (A.37), we have

$$\left| \int_{\Gamma_N} u\varphi d\sigma \right| \leq \|u\|_{L^2(\Gamma_N)} \|y\|_{L^2(\Gamma_N)} \leq c_{tr} \|u\|_{L^2(\Gamma_N)} \|y\|_{H^1(\Omega)} \leq c_{tr} C_P \|u\|_{L^2(\Gamma_N)} \|\nabla y\|_{L^2(\Omega)^d}$$

so that $B \in \mathcal{L}(\mathcal{U}, V^*)$ with $\|Bu\|_* \leq c_{tr} C_P \|u\|_{L^2(\Gamma_N)}$.

- *The state equation.* According to (5.92), the variational formulation of problem (5.112) is

$$a(y,\varphi) = (\nabla y, \nabla \varphi)_{L^2(\Omega)} - (y, \mathbf{b} \cdot \nabla \varphi)_{L^2(\Omega)} = (f,\varphi)_{L^2(\Omega)} + (u,\varphi)_{L^2(\Gamma_N)} \qquad \forall \varphi \in H^1_{0,\Gamma_D}(\Omega)$$

for $y \in H^1_{\Gamma_D}(\Omega)$. Existence and uniqueness of the solution $y = y(u)$ of (5.112), for each $u \in \mathcal{U}$, follow from Proposition 5.9, provided $\mathbf{b}$ has Lipschitz components and

$$\mathbf{b} \cdot \mathbf{n} \leq 0 \ \text{ a.e. on } \Gamma_N, \qquad \operatorname{div}\mathbf{b} \geq 0 \quad \text{a.e. in } \Omega. \tag{5.113}$$

- *Existence/uniqueness of the optimal control.* By Theorem 5.4, if $\beta > 0$, there exists a unique optimal control $\hat{u}$. If $\beta = 0$ and $\mathcal{U}_{ad}$ is bounded, there exists an optimal control, however since the operator $u \longmapsto Cy_0(u)$ is not one-to-one, it may not be unique.
- *Optimality conditions.* The Lagrangian is

$$\mathcal{L}(y,p,u) = \frac{1}{2}\|(y - z_d)\|^2_{L^2(\Omega_0)} + \frac{\beta}{2}\|u\|^2_{L^2(\Gamma_N)} - a(y,p) + (f,p)_{L^2(\Omega)} + (u,p)_{L^2(\Gamma_N)}.$$

We have:

$$\begin{aligned} \mathcal{L}'_p(\hat{y},\hat{p},\hat{u})\varphi &= -a(\hat{y},\varphi) + (f,\varphi)_{L^2(\Omega)} + (\hat{u},\varphi)_{L^2(\Gamma_N)}, \\ \mathcal{L}'_y(\hat{y},\hat{p},\hat{u})\psi &= (\hat{y} - z_d, \psi)_{L^2(\Omega_0)} - a(\psi,\hat{p}), \\ \mathcal{L}'_u(\hat{y},\hat{p},\hat{u})(w - \hat{u}) &= (\beta\hat{u} + \hat{p}, v - \hat{u})_{L^2(\Gamma_N)}. \end{aligned}$$

Theorem 5.3 gives the following optimality system for $\hat{u} \in \mathcal{U}_{ad}$, $\hat{y} = y(\hat{u})$ and the adjoint state $\hat{p} \in H^1_{\Gamma_D}(\Omega)$:

$$\begin{cases} a(\hat{y},\varphi) = (f,\varphi)_{L^2(\Omega)} + (g+\hat{u},\varphi)_{L^2(\Gamma_N)} & \forall\varphi \in H^1_{\Gamma_D}(\Omega) \\ a^*(\hat{p},\psi) = (\hat{y}-z_d,\psi)_{L^2(\Omega_0)} & \forall\psi \in H^1_{0,\Gamma_D}(\Omega) \\ (\beta\hat{u}+\hat{p}, v-\hat{u})_{L^2(\Gamma_N)} \geq 0 & \forall v \in \mathcal{U}_{ad}. \end{cases} \tag{5.114}$$

Moreover, for every $u \in L^2(\Gamma_N)$,

$$\nabla J(u) = p(u) + \beta u \in L^2(\Gamma_N)$$

where $p = p(u)$ is the solution of the adjoint problem

$$a^*(p,\psi) = (y(u)-z_d,\psi)_{L^2(\Omega_0)} \qquad \forall\psi \in H^1_{\Gamma_D}(\Omega). \tag{5.115}$$

Remark 5.9. Observe that (5.115) is the variational formulation of the problem

$$\begin{cases} \mathcal{E}^*p = -\Delta p - \mathbf{b}\cdot\nabla p = \chi_{\Omega_0}(y-z_d) & \text{in } \Omega \\ p = 0 & \text{on } \Gamma_D \\ \dfrac{\partial p}{\partial n} = 0 & \text{on } \Gamma_N. \end{cases} \tag{5.116}$$

Under the conditions (5.113) a^* is coercive and therefore the adjoint problem has a unique weak solution, thanks to the Lax-Milgram Theorem. •

Remark 5.10. If $\beta > 0$, the variational inequality

$$(\beta\hat{u}+\hat{p}, v-\hat{u})_{L^2(\Gamma_N)} \geq 0$$

gives $\hat{u} = Pr_{\mathcal{U}_{ad}}\left(-\beta^{-1}\hat{p}\right) \in L^2(\Gamma_N)$. In particular, when

$$\mathcal{U}_{ad} = \left\{u \in L^2(\Gamma_N) : a \leq u \leq b \quad \text{a.e. on } \Gamma_N\right\}$$

we deduce that $\hat{u}(\sigma) = \max\left\{a, \min\left\{b, -\beta^{-1}\hat{p}(\sigma)\right\}\right\}$, a.e. on Γ_N. If $\beta = 0$, then $\hat{u}(\sigma) = a$ if $\hat{p}(\sigma) > 0$, $\hat{u}(\sigma) = b$ if $\hat{p}(\sigma) < 0$, on Γ_N. No information on $\hat{u}$ is available at points where $\hat{p} = 0$. •

5.7 Boundary Observation, Neumann Boundary Control

Let $\mathcal{U} = L^2(\Gamma)$ and $\mathcal{U}_{ad} \subseteq \mathcal{U}$ be a closed convex set. Given $f \in L^2(\Omega), z_d \in L^2(\Gamma)$, $h \in L^\infty(\Gamma)$, consider the minimization of the cost functional

$$\tilde{J}(y,u) = \frac{1}{2}\int_\Gamma (hy-z_d)^2\, d\sigma + \frac{\beta}{2}\int_\Gamma u^2 d\sigma, \qquad \beta \geq 0$$

subject to $(y,u) \in H^1(\Omega) \times \mathcal{U}_{ad}$ and to the state problem:

$$\begin{cases} \mathcal{E}y = -\Delta y + y = f & \text{in } \Omega \\ \dfrac{\partial y}{\partial n} = u & \text{on } \Gamma. \end{cases} \tag{5.117}$$

The (boundary) observation operator is $Cy = hy_{|\Gamma}$ and, from the trace inequality,

$$\|hy\|_{L^2(\Gamma)} \leq \|h\|_{L^\infty(\Gamma)}\|y\|_{L^2(\Gamma)} \leq c_{tr}\|h\|_{L^\infty(\Gamma)}\|y\|_{H^1(\Omega)}$$

whence $C \in \mathcal{L}(V,Z)$.

• *The state problem.* The weak formulation of problem (5.117) is

$$a(y,\varphi) = (\nabla y, \nabla\varphi)_{L^2(\Omega)} + (y,\varphi)_{L^2(\Omega)} = (f,\varphi)_{L^2(\Omega)} + (u,\varphi)_{L^2(\Gamma)} \qquad \forall\varphi \in H^1(\Omega), \quad (5.118)$$

with $y \in H^1(\Omega)$. Proposition 5.8 ensures the well-posedness of problem (5.118).

• *Existence and uniqueness of the optimal control.* Thanks to Theorem 5.4, if $\beta > 0$ there exists a unique optimal control $\hat{u}$. Let us now consider the case $\beta = 0$. If $\mathcal{U}_{ad}$ is bounded, there exists at least an optimal control. If $h \neq 0$ a.e. on Γ then the optimal control is also unique. Indeed, in this case, $Cy_0(u)_{|\Gamma} = hy_0(u)_{|\Gamma} = 0$ implies $y_0(u)_{|\Gamma} = 0$. Since y_0 satisfies $-\Delta y_0 + y_0 = 0$ in a weak sense, we infer $y_0(u) = 0$ in Ω. Hence, $\partial y_0/\partial n = 0$ on Γ.

• *Optimality conditions.* The Lagrangian is

$$\mathcal{L}(y,p,u) = \frac{1}{2}\|hy - z_d\|^2_{L^2(\Gamma)} + \frac{\beta}{2}\|u\|^2_{L^2(\Omega)} - a(y,p) + (f,p)_{L^2(\Omega)} + (u,p)_{L^2(\Gamma)}.$$

We have:

$$\begin{aligned}
&\mathcal{L}'_p(\hat{y},\hat{p},\hat{u})\varphi = -a(\hat{y},\varphi) + (f,\varphi)_{L^2(\Omega)} + (\hat{u},\varphi)_{L^2(\Gamma)},\\
&\mathcal{L}'_y(\hat{y},\hat{p},\hat{u})\psi = (h\hat{y} - z_d, h\psi)_{L^2(\Omega_0)} - a(\psi,\hat{p}),\\
&\mathcal{L}'_u(\hat{y},\hat{p},\hat{u})(w - \hat{u}) = (\beta\hat{u} + \hat{p}, v - \hat{u})_{L^2(\Gamma)}.
\end{aligned}$$

Now, from Theorem 5.3 we derive the following optimality system for $\hat{u} \in \mathcal{U}_{ad}$, $\hat{y} = y(\hat{u})$ and the adjoint state $\hat{p} \in H^1(\Omega)$:

$$\begin{cases}
a(\hat{y},\varphi) = (f,\varphi)_{L^2(\Omega)} + (\hat{u},\varphi)_{L^2(\Gamma)} & \forall\varphi \in H^1(\Omega)\\
a^*(\hat{p},\psi) = (h\hat{y} - z_d, h\psi)_{L^2(\Gamma)} & \forall\psi \in H^1(\Omega)\\
(\hat{p} + \beta\hat{u}, v - \hat{u})_{L^2(\Gamma)} \geq 0 & \forall v \in \mathcal{U}_{ad}.
\end{cases} \quad (5.119)$$

Moreover, for each $u \in \mathcal{U}_{ad}$,

$$\nabla J(u) = p(u)_{|\Gamma} + \beta u \in L^2(\Gamma), \quad (5.120)$$

where $p = p(u) \in H^1(\Omega)$ is the unique solution to the adjoint equation

$$a^*(p,\psi) = (hy(u) - z_d, h\psi)_{L^2(\Gamma)} \qquad \forall\psi \in H^1(\Omega).$$

Remark 5.11. The adjoint equation is the weak formulation of the Neumann problem

$$\begin{cases}
\mathcal{E}^*\hat{p} = -\Delta\hat{p} + \hat{p} = 0 & \text{in } \Omega\\
\dfrac{\partial\hat{p}}{\partial n} = h(h\hat{y} - z_d) & \text{on } \Gamma.
\end{cases}$$

Since $h \in L^\infty(\Gamma)$ then $h(h\hat{y} - z_d) \in L^2(\Gamma)$ and hence the adjoint equation has a unique solution. •

5.8 Boundary Observation, Distributed Control, Dirichlet Conditions

We examine here a case of flux observation at the boundary. Let $\mathcal{U} = L^2(\Omega)$, $\mathcal{U}_{ad} \subseteq \mathcal{U}$ be closed and convex. Given a smooth subset $\Gamma_0 \subseteq \Gamma$ and $z_d \in L^2(\Gamma_0)$, $f \in L^2(\Omega)$, consider the minimization of the functional

$$\tilde{J}(y,u) = \frac{1}{2}\int_{\Gamma_0}\left[\frac{\partial y}{\partial n} - z_d\right]^2 d\sigma + \frac{\beta}{2}\int_\Omega u^2 d\mathbf{x}$$

where $\beta > 0$, subject to the state problem

$$\begin{cases} \mathcal{E}y = -\Delta y = f + u & \text{in } \Omega \\ y = 0 & \text{on } \Gamma, \end{cases} \tag{5.121}$$

whose weak formulation is: find $y \in H_0^1(\Omega)$ such that

$$a(y, \varphi) = (f + u, \varphi)_{L^2(\Omega)} \qquad \forall \varphi \in H_0^1(\Omega). \tag{5.122}$$

As we shall see, some caution is needed when dealing with this optimal control problem. Indeed, in the use of the adjoint state to achieve the optimality conditions, some adjustement is required with respect to the general framework of Sect. 5.3.

• *The state problem.* Clearly, for each $u \in L^2(\Omega)$, (5.122) has a unique solution $y(u) \in H_0^1(\Omega)$. However, this minimal regularity of $y(u)$ is not enough to deduce that $\frac{\partial y}{\partial n}(u)$ has a trace on Γ_0 belonging to $L^2(\Gamma_0)$, a necessary condition for $J(u) = \tilde{J}(y(u), u)$ to make sense. To achieve a better (elliptic) regularity of $y(u)$, we assume that Ω is either a convex or a smooth (at least of class C^2) domain in $\mathbb{R}^d$. In this case the weak solution $y(u)$ enjoys a better regularity and indeed (see e.g. [237]) $y(u) \in Y = H^2(\Omega) \cap H_0^1(\Omega)$. Moreover

$$\|y(u)\|_{H^2(\Omega)} \leq c(d, \Omega)\left(\|f\|_{L^2(\Omega)} + \|u\|_{L^2(\Omega)}\right). \tag{5.123}$$

As a consequence, $-\Delta : Y \to L^2(\Omega)$ is a *continuous isomorphism*.

Since now $\nabla y(u) \in H^1(\Omega)$, each partial derivative has a trace in $L^2(\Gamma_0)$ and in particular

$$Cy(u) = \left.\frac{\partial y(u)}{\partial n}\right|_{\Gamma_0} = \nabla y(u)|_{\Gamma_0} \cdot \mathbf{n} \in L^2(\Gamma_0).$$

Thus, we can observe $\frac{\partial y(u)}{\partial n}$ on Γ_0 and the cost functional J is well defined. Moreover, $y(u)$ possesses second derivatives in $L^2(\Omega)$ and the equation $-\Delta y = f + u$ makes sense pointwise a.e. in Ω (we say that the equation holds in *strong form*).

• *Existence and uniqueness of the optimal control.* Since $\beta > 0$, Theorem 5.4 guarantees existence and uniqueness of an optimal control $\hat{u}$.

• *Optimality conditions.* The usual derivation of the optimal conditions would lead to an adjoint problem of the form: find $\hat{p} \in H_0^1(\Omega)$ such that

$$a(\hat{p}, \psi) = (\nabla \hat{p}, \nabla \psi) = \left(\frac{\partial \hat{y}(u)}{\partial n} - z_d, \frac{\partial \psi}{\partial n}\right)_{L^2(\Gamma_0)} \qquad \forall \psi \in H_0^1(\Omega) \tag{5.124}$$

which does not fit into the variational theory of Sect. 5.4, since the right hand side is not well defined for $\psi \in H_0^1(\Omega)$ ($\frac{\partial \psi}{\partial n} \notin L^2(\Gamma_0)$). On the other hand, if we choose Y instead of $H_0^1(\Omega)$ as state space, we see that the bilinear form is not coercive in Y. Therefore we abandon the variational formulation (5.124) and try to exploit the strong form of the state equation. Going back to the derivation of the optimality conditions in Sect. 5.3, Proposition 5.5 gives the following variational inequality,

$$J'(\hat{u})(v - \hat{u}) = \int_{\Gamma_0} \left(\frac{\partial y(\hat{u})}{\partial n} - z_d\right) \frac{\partial y_0(v - \hat{u})}{\partial n} d\sigma + \beta \int_\Omega \hat{u}(v - \hat{u}) d\mathbf{x} \geq 0 \tag{5.125}$$

for all $v \in \mathcal{U}_{ad}$.

Since $-\Delta y = f + u$ a.e. in Ω, we can write the augmented functional

$$J(u) = \frac{1}{2}\int_{\Gamma} \chi_{\Gamma_0}\left(\frac{\partial y(u)}{\partial n} - z_d\right)^2 d\sigma + \frac{\beta}{2}\int_{\Omega} u^2 d\mathbf{x} + \int_{\Omega} p(f + u + \Delta y(u)) d\mathbf{x}.$$

Setting $w(v) = y_0(v - \hat{u}) \in Y$, the optimality condition becomes

$$\int_{\Gamma} \chi_{\Gamma_0}\left(\frac{\partial y(\hat{u})}{\partial n} - z_d\right)\frac{\partial w(v)}{\partial n} d\sigma + \int_{\Omega} (\beta\hat{u} + p)(v - \hat{u}) d\mathbf{x} + \int_{\Omega} p\Delta w(v)\, d\mathbf{x} \geq 0. \quad (5.126)$$

Now we proceed formally, integrating twice by parts the last term. We find, recalling that $w(v) = 0$ on Γ,

$$\int_{\Omega} p\Delta w(v)\, d\mathbf{x} = \int_{\Gamma} p\frac{\partial w(v)}{\partial n} d\sigma - \int_{\Omega} \nabla w(v)\cdot\nabla p\, d\mathbf{x} = \int_{\Gamma} p\frac{\partial w(v)}{\partial n} d\sigma + \int_{\Omega} w(v)\,\Delta p\; d\mathbf{x}.$$

In order to get rid in (5.126) of the terms containing $w(v)$, we choose $p = \hat{p} \in Y$ as the solution of the following adjoint problem:

$$\begin{cases} \mathcal{E}^*\hat{p} = -\Delta\hat{p} = 0 & \text{in } \Omega \\ \hat{p} = -\chi_{\Gamma_0}\left(\dfrac{\partial\hat{y}(u)}{\partial n} - z_d\right) & \text{on } \Gamma. \end{cases} \quad (5.127)$$

Inserting $p = \hat{p}$ in (5.126) we find the variational inequality

$$\int_{\Omega} (\beta\hat{u} + \hat{p})(v - \hat{u})\; d\mathbf{x} \geq 0. \quad (5.128)$$

The optimality system is therefore given by the state equation, the adjoint equation (5.127) and the variational inequality (5.128).

Apparently everything works fine, but there is still a problem: the Dirichlet datum in the adjoint problem belongs only to $L^2(\Gamma)$ and in general is not the trace of a function of $H^1(\Omega)$. This means that this kind of problem cannot be solved via a variational formulation as in Sect. 5.4 and, in particular, there is no reason to expect that $\hat{p} \in Y$, which we implicitly assumed for the integration by parts. Hence, it is necessary to reconsider problem (5.127), as we do in the next subsection, introducing the notion of *very weak solution*, due to the low regularity of the Dirichlet data that induces a lower–than-H^1 regularity of the adjoint state.

According to Proposition 5.10 below, problem (5.127) has a unique *very weak solution*, which is defined as the function $\hat{p} \in L^2(\Omega)$ that satisfies

$$\int_{\Omega} \hat{p}\Delta\psi d\mathbf{x} = -\int_{\Gamma_0}\left(\frac{\partial\hat{y}}{\partial n} - z_d\right)\frac{\partial\psi}{\partial n} d\sigma \qquad \forall\psi \in Y \quad (5.129)$$

and

$$\|\hat{p}\|_{L^2(\Omega)} \leq c(d,\Omega)\left\|\frac{\partial\hat{y}}{\partial n} - z_d\right\|_{L^2(\Gamma_0)}.$$

Thus, (5.129) is the proper weak formulation of the adjoint problem (5.127).

Remark 5.12. We emphasize that the regularity of the multiplier in this section is lower than that of the multipliers considered so far: indeed, $y \in H^2(\Omega)$ whereas $p \in L^2(\Omega)$ only, so that there is a break in the symmetry between the regularity of the state and the adjoint solutions. This fact is not unexpected. Indeed, as we shall see in Chapter 9, the state equation can be written in the form $G(y,u) = 0$ with $G : V \times \mathcal{U} \to W$, and the multiplier p is chosen in the dual space W^*. For instance, in the variational setting, we can write the state equation under the form $G(y,u) = \mathcal{E}y - f - Bu = 0$ with $G : V \times \mathcal{U} \to V^*$ and p belongs to $(V^*)^* = V$. In the very weak formulation, we have instead $G(y,u) = \mathcal{E}y - f - Bu = 0$ with $G : Y \times \mathcal{U} \to L^2(\Omega)$ and p belongs exactly to $L^2(\Omega)^*$, usually identified with $L^2(\Omega)$ itself. •

5.9 Dirichlet Problems with L^2 Data. Transposition (or Duality) Method

The analysis of this case requires more tools of functional analysis. Consider the Dirichlet problem

$$\begin{cases} -\Delta w = f & \text{in } \Omega \\ w = g & \text{on } \Gamma \end{cases} \tag{5.130}$$

where $f \in L^2(\Omega)$, $g \in L^2(\Gamma)$. Assume that the domain Ω is *smooth* (say, of class C^2) and let $Y = H_0^1(\Omega) \cap H^2(\Omega)$. To derive a new weak formulation, multiply the differential equation by a test function ψ and integrate over Ω,

$$\int_\Omega -\Delta w \psi \, d\mathbf{x} = \int_\Omega f \psi \, d\mathbf{x}.$$

Integrating twice by parts we find

$$\int_\Omega -w \Delta \psi \, d\mathbf{x} - \int_\Gamma \psi \frac{\partial w}{\partial n} d\sigma + \int_\Gamma w \frac{\partial \psi}{\partial n} d\sigma = \int_\Omega f \psi \, d\mathbf{x}.$$

Choose $\psi \in Y$ in order to get rid of the terms containing $\frac{\partial w}{\partial n}$, on which there are no information. Taking into account the Dirichlet condition, we obtain the equation

$$\int_\Omega -w \Delta \psi \, d\mathbf{x} = \int_\Omega f \, \psi \, d\mathbf{x} - \int_\Gamma g \frac{\partial \psi}{\partial n} d\sigma. \tag{5.131}$$

Thus, if w is a regular solution of problem (5.130), then w satisfies (5.131) for all $\psi \in Y$. Viceversa, assume that f, g are smooth and w is a smooth function satisfying (5.131) for every $\psi \in Y$. Integrating back by parts we find

$$\int_\Omega \psi \Delta w \, d\mathbf{x} + \int_\Gamma w \frac{\partial \psi}{\partial n} d\sigma = \int_\Gamma g \frac{\partial \psi}{\partial n} d\sigma - \int_\Omega f \psi \, d\mathbf{x}. \tag{5.132}$$

In particular, (5.132) holds for each $\psi \in C_0^\infty(\Omega)$, whence $-\Delta w = f$ in Ω. Using now $\psi \in Y$, we infer

$$\int_\Gamma (w - g) \frac{\partial \psi}{\partial n} d\sigma = 0. \tag{5.133}$$

Since $\psi \in Y$ is arbitrary, (5.133) implies[8] $w = g$ on Γ.

---

[8] Let $h = w - g \in C^\infty(\Gamma)$. Let $z \in H^2(\Omega)$ be such that $\Delta z = 0$ in Ω, $z = h$ on Γ. Then

$$\int_\Omega \Delta \psi \cdot z \, d\mathbf{x} = \int_\Gamma h \frac{\partial \psi}{\partial n} d\sigma = 0.$$

Since $\Delta : Y \to L^2(\Omega)$ is a continuous isomorphism we deduce that $z = 0$ whence $h = 0$.

Therefore, for regular functions, the formulations (5.130) and (5.131) are equivalent. However, (5.131) requires only $w \in L^2(\Omega)$. We are led to the following notion of very weak solution:

Definition 5.1. The function $w \in L^2(\Omega)$ is a very weak solution to problem (5.130) if

$$-\int_\Omega w\Delta\psi \, d\mathbf{x} = \int_\Omega f\psi \, d\mathbf{x} - \int_\Gamma g\frac{\partial\psi}{\partial n}d\sigma \tag{5.134}$$

for all $\psi \in Y = H_0^1(\Omega) \cap H^2(\Omega)$.

We show that with the new formulation problem (5.130) is well posed. Indeed we have:

Proposition 5.10. *Problem (5.130) has a unique very weak solution $w \in L^2(\Omega)$. Moreover*

$$\|w\|_{L^2(\Omega)} \leq c(d,\Omega)\left\{\|f\|_{L^2(\Omega)} + \|g\|_{L^2(\Gamma)}\right\}.$$

The proof of Proposition 5.10 is a consequence of the following result.

Proposition 5.11. (Transposition (or Duality) method). *Let Y and H be two Hilbert spaces, $T : Y \to H$ a continuous isomorphism and $F \in Y^*$. The equation*

$$(w, T\psi)_H = F\psi \qquad \forall\psi \in Y \tag{5.135}$$

has a unique solution $\bar{w} \in H$. Moreover,

$$\|\bar{w}\|_H \leq C\|F\|_{Y^*}$$

where $C = \|T^{-1}\|_{\mathcal{L}(H,Y)}$.

Proof. Since T is a continuous isomorphism, for every $h \in H$ we have

$$\|T^{-1}h\|_Y \leq \|T^{-1}\|_{\mathcal{L}(H,Y)}\|h\|_H. \tag{5.136}$$

Now, setting $h = T\psi$, equation (5.135) is equivalent to

$$(w,h)_H = (F \circ T^{-1})h \qquad \forall h \in H. \tag{5.137}$$

From (5.136):

$$|(F \circ T^{-1})h| \leq \|F\|_{Y^*}\|T^{-1}\|_{\mathcal{L}(H,Y)}\|h\|_H$$

whence $F \circ T^{-1} \in H^*$ with $\|F \circ T^{-1}\|_{H^*} \leq \|T^{-1}\|_{\mathcal{L}(H,Y)}\|F\|_{Y^*}$. By Riesz Representation Theorem, (5.137) has a unique solution $\bar{w} \in H$ with

$$\|\bar{w}\|_H \leq \|T^{-1}\|_{\mathcal{L}(H,Y)}\|F\|_{Y^*}.$$

□

Remark 5.13. Proposition 5.11 is equivalent to the following statement: If $T : Y \to H$ is a continuous isomorphism then its adjoint (or dual) operator $T^* : H^* = H \to Y^*$ is also a continuous isomorphism. Indeed, equation (5.135) is equivalent to $T^*w = F$. This is the reason of the name transposition (or duality) method. •

Proof. (*Proposition 5.10*) The very weak formulation of problem (5.130) satisfies (5.134). It is now sufficient to apply Proposition 5.11 with $Y = H_0^1(\Omega) \cap H^2(\Omega)$, $H = L^2(\Omega)$, $T = -\Delta$ and F given by

$$F\psi = \int_\Omega f\psi \, d\mathbf{x} - \int_\Gamma g\frac{\partial\psi}{\partial n} d\sigma.$$

F belongs to Y^*, since

$$|F\psi| \le \|f\|_{L^2(\Omega)} \|\psi\|_{L^2(\Omega)} + \|g\|_{L^2(\Gamma)} \left\|\frac{\partial\psi}{\partial n}\right\|_{L^2(\Gamma)} \le c(d,\Omega)\left\{\|f\|_{L^2(\Omega)} + \|g\|_{L^2(\Gamma)}\right\} \|\psi\|_Y$$

and

$$\|F\|_{Y^*} \le c(d,\Omega)\left\{\|f\|_{L^2(\Omega)} + \|g\|_{L^2(\Gamma)}\right\}.$$

□

Remark 5.14. We can obtain a representation formula for the solution to problem (5.130). Let $\{u_k\}_{k\ge1}$ be an orthonormal basis in $L^2(\Omega)$ given by the Dirichlet eigenfunctions for the operator $(-\Delta)$, with corresponding eigenvalues $\{\lambda_k\}_{k\ge1}$, with $0 < \lambda_1 \le \lambda_2 \le \ldots$, $\lambda_k \to +\infty$. By elliptic regularity, all these eigenfunctions belong to Y. Then we can write

$$w = \sum_1^\infty \hat{w}_k u_k$$

where $\hat{w}_k = (w, u_k)_{L^2(\Omega)}$. Inserting $\psi = u_k$ into (5.131), we get

$$(w, -\Delta u_k)_{L^2(\Omega)} = \int_\Omega -w\Delta u_k \, d\mathbf{x} = \hat{f}_k - \int_\Gamma g\frac{\partial u_k}{\partial n} d\sigma = \hat{f}_k - g_k.$$

Since $(w, -\Delta u_k)_{L^2(\Omega)} = \lambda_k \hat{w}_k$, we infer that $\hat{w}_k = (\hat{f}_k - g_k)/\lambda_k$ and

$$w = \sum_1^\infty \frac{\hat{f}_k - g_k}{\lambda_k} u_k.$$

•

Remark 5.15. If $g \in H^{1/2}(\Gamma)$, the very weak solution y to problem (5.130) actually belongs to $H^1(\Omega)$ and coincides with the variational solution. In fact, let $z \in H^1(\Omega)$ be the variational solution. Then

$$\int_\Omega \nabla z \cdot \nabla\psi \, d\mathbf{x} = \int_\Omega f\psi \, d\mathbf{x} \qquad \forall\psi \in H_0^1(\Omega).$$

Choosing $\psi \in H_0^1(\Omega) \cap H^2(\Omega)$ we can integrate by parts to get

$$-\int_\Omega z\Delta\psi \, d\mathbf{x} + \int_\Gamma z\frac{\partial\psi}{\partial n} d\sigma = \int_\Omega f\psi \, d\mathbf{x} \qquad \forall\psi \in H_0^1(\Omega) \cap H^2(\Omega). \tag{5.138}$$

Using $z = g$ on Γ, we conclude that z is a very weak solution to (5.130) and therefore, by uniqueness, $z = y$. This remark will be useful in the sequel. •

Remark 5.16. The method works with more general hypotheses on f and g. Namely, since $\psi \in Y$, one could choose $f \in Y^*$ and substitute $\int_\Omega f\psi \, d\mathbf{x}$ in the Definition 5.1 by the duality $\langle f, \psi\rangle_{Y^*,Y}$. Similarly, since $\frac{\partial\psi}{\partial n} \in H^{1/2}(\Gamma)$ one could choose g in the dual space of $H^{1/2}(\Gamma)$, denoted by $H^{-1/2}(\Gamma)$, and substitute $\int_\Gamma g\frac{\partial\psi}{\partial n} d\sigma$ in the Definition 5.1 by the duality $\langle g, \frac{\partial\psi}{\partial n}\rangle_{H^{1/2},H^{-1/2}}$. Still, problem (5.130) has a unique very weak solution $w \in L^2(\Omega)$ and

$$\|w\|_{L^2(\Omega)} \le c(d,\Omega)\left\{\|f\|_{Y^*} + \|g\|_{H^{-1/2}(\Gamma)}\right\}.$$

•

5.10 Pointwise Observations

The transposition method introduced in Sect. 5.9 is appropriate to handle OCPs involving pointwise observations. Under suitable regularity hypotheses on the domain and the data of the state problem, the state variable is continuous and pointwise observations are well defined. Consider for instance the following problem:

$$\begin{cases} \mathcal{E}y = -\Delta y + ry = f & \text{in } \Omega \\ y = 0 & \text{on } \Gamma. \end{cases} \tag{5.139}$$

Let Ω be either a convex or a smooth domain (e.g. of class C^2) and r be bounded and non negative a.e. in Ω. Then the operator $\mathcal{E} : Y = H_0^1(\Omega) \cap H^2(\Omega) \to L^2(\Omega)$ is a *continuous isomorphism* between Y and $L^2(\Omega)$. Thus the variational solution y belongs to Y and

$$\|y\|_{H^2(\Omega)} \le C \|f\|_{L^2(\Omega)}, \tag{5.140}$$

where C depends on Ω and d only. As a consequence, the PDE in (5.139) makes sense pointwise a.e. in Ω. On the other hand, from the Sobolev embedding theorem A.16, we know that if $d \le 3$, Y is continuously embedded into $C(\overline{\Omega})$, that is

$$\|y\|_{C(\overline{\Omega})} \le C_1 \|y\|_{H^2(\Omega)}. \tag{5.141}$$

In particular, (5.140) implies that the *evaluation operator* $y \mapsto y(\boldsymbol{\xi})$ is well defined for all $\boldsymbol{\xi} \in \overline{\Omega}$ and belongs to Y^*. The evaluation operator at a point $\boldsymbol{\xi}$ is usually denoted by $\delta(\mathbf{x} - \boldsymbol{\xi})$ or $\delta_{\boldsymbol{\xi}}$ and is called the *Dirac delta at* $\boldsymbol{\xi}$. We can write $\langle \delta_{\boldsymbol{\xi}}, y \rangle = y(\boldsymbol{\xi})$ to denote the action of $\delta_{\boldsymbol{\xi}}$ on $y \in Y$. If $z \in C(\overline{\Omega})$, the operator $z\delta_{\boldsymbol{\xi}}$ is well defined as an element of Y^* by the relation

$$\langle z\delta_{\boldsymbol{\xi}}, y \rangle = \langle \delta_{\boldsymbol{\xi}}, zy \rangle = z(\boldsymbol{\xi})\, y(\boldsymbol{\xi}). \tag{5.142}$$

Thus, if $z_d \in C(\overline{\Omega})$ is a given target state and $\boldsymbol{\xi}_j \in \overline{\Omega}$, $j = 1, \ldots, N$, it makes sense to consider the problem

$$\tilde{J}(y, u) = \frac{1}{2} \sum_{j=1}^{N} \left(y(\boldsymbol{\xi}_j) - z_d(\boldsymbol{\xi}_j)\right)^2 + \frac{\beta}{2} \|u\|_{L^2(\Omega)}^2 \to \min \tag{5.143}$$

with $\beta > 0$, subject to $u \in \mathcal{U}_{ad}$, a closed convex subset of $L^2(\Omega)$, and to

$$\begin{cases} \mathcal{E}y = f + u & \text{in } \Omega, \\ y = 0 & \text{on } \Gamma. \end{cases} \tag{5.144}$$

Since $\beta > 0$, existence and uniqueness of an optimal control $\hat{u}$ follows from Theorem 5.4. To find the optimality conditions, we exploit the full regularity of the state variable and introduce the Lagrangian function

$$\mathcal{L}(y, p, u) = \frac{1}{2} \sum_{j=1}^{N} \left(y(\boldsymbol{\xi}_j) - z_d(\boldsymbol{\xi}_j)\right)^2 + \frac{\beta}{2} \|u\|_{L^2(\Omega)}^2 + (f + u - \mathcal{E}y, p)_{L^2(\Omega)}$$

with $p \in L^2(\Omega)$. The adjoint equation then reads

$$\mathcal{L}'_y(\hat{y}, \hat{p}, \hat{u})\, \psi = \sum_{j=1}^{N} \left(y(\boldsymbol{\xi}_j) - z_d(\boldsymbol{\xi}_j)\right) \psi(\boldsymbol{\xi}_j) - (\mathcal{E}\psi, p)_{L^2(\Omega)} = 0 \quad \forall \psi \in Y. \tag{5.145}$$

We recognize that $\hat{p}$ is the very weak solution of the problem

$$\begin{cases} \mathcal{E}\hat{p} = \sum_{j=1}^{N} \left(\hat{y}\left(\boldsymbol{\xi}_j\right) - z_d\left(\boldsymbol{\xi}_j\right)\right)\delta_{\boldsymbol{\xi}_j} & \text{in } \Omega \\ \hat{p} = 0 & \text{on } \Gamma, \end{cases} \tag{5.146}$$

which is well-posed according to Remark 5.16. The Euler equation takes the usual form

$$\mathcal{L}'_u\left(\hat{y}, \hat{p}, \hat{u}\right)v = \left(\beta\hat{u} + \hat{p}, v - \hat{u}\right)_{L^2(\Omega)} \geq 0 \qquad \forall v \in \mathcal{U}_{ad}, \tag{5.147}$$

which gives $\hat{u} = Pr_{\mathcal{U}_{ad}}\left(-\beta^{-1}\hat{p}\right)$. We summarize these results in the following proposition.

Proposition 5.12. *Let $\Omega \subset \mathbb{R}^d$, $d = 2, 3$, be either a convex or a domain of class C^2, and consider the OCP (5.143), (5.144). Then the control $\hat{u} \in \mathcal{U}_{ad}$ and the state $\hat{y} = y(\hat{u}) \in Y$ are optimal if and only if, together with the adjoint state $\hat{p} \in L^2(\Omega)$, they are solutions of the state equation (5.144), the adjoint equation (5.145) and the variational inequality (5.147). Moreover,*

$$\nabla J\left(\hat{u}\right) = \beta\hat{u} + \hat{p}.$$

5.11 Distributed Observation, Dirichlet Control

The case of OCPs with boundary Dirichlet control is more delicate to treat. For simplicity, we consider *unconstrained* OCPs for the Laplace operator with distributed observation. We show how to derive optimality conditions, depending on the choices of the cost functional and of the control space $\mathcal{U}$. We consider the minimization of the following functional

$$\tilde{J}(y, u) = \frac{1}{2}\|y - z_d\|_Z^2 + \frac{\beta}{2}\|u\|_{\mathcal{U}_0}^2 \tag{5.148}$$

where $\beta > 0$, $z_d \in Z$ is a given target state, $f \in L^2(\Omega)$ and $\mathcal{U}_0$ is a Hilbert space included into $\mathcal{U}$, with (y, u) subject to the state problem

$$\begin{cases} -\Delta y = f & \text{in } \Omega \\ y = u & \text{on } \Gamma. \end{cases} \tag{5.149}$$

Note that the norms of the penalization term and of the control space do not necessarily coincide.

5.11.1 *Case $\mathcal{U} = H^{1/2}(\Gamma)$*

Choosing $H^{1/2}(\Gamma)$ as control space looks rather natural, as this space is the proper space of traces for a well-posed variational formulation in $H^1(\Omega)$ of the Dirichlet problem (5.149). We examine the two options $\mathcal{U}_0 = H^{1/2}(\Gamma)$ and $\mathcal{U}_0 = L^2(\Gamma)$.

Option 1: $\mathcal{U}_0 = H^{1/2}(\Gamma)$

If we also choose $\mathcal{U}_0 = H^{1/2}(\Gamma)$ and $Z = L^2(\Omega)$, we can consider the OCP

$$\tilde{J}(y, u) = \frac{1}{2}\|y - z_d\|_{L^2(\Omega)}^2 + \frac{\beta}{2}\|u\|_{H^{1/2}(\Gamma)}^2 \to \min. \tag{5.150}$$

Since $\beta > 0$, existence and uniqueness of an optimal control $\hat{u}$ follow immediately.

• *Analysis of the state equation.* The standard variational formulation of the state problem requires an extension (lifting) $u_{ext} \in H^1(\Omega)$ of the Dirichlet data u. Then, setting $z = y - u_{ext} \in H_0^1(\Omega)$, the weak formulation of the state problem for z takes the form

$$(\nabla z, \nabla\varphi)_{L^2(\Omega)} = (f, \varphi)_{L^2(\Omega)} - (\nabla u_{ext}, \nabla\varphi)_{L^2(\Omega)} \qquad \forall\varphi \in H_0^1(\Omega). \tag{5.151}$$

By a straightforward application of the Lax-Milgram Theorem, this problem admits a unique solution $z = z(u) \in H_0^1(\Omega)$. The corresponding OCP problem in reduced form reads

$$J_0(u) = \frac{1}{2}\|z(u) + u_{ext} - z_d\|_{L^2(\Omega)}^2 + \frac{\beta}{2}\|u\|_{H^{1/2}(\Gamma)}^2 \to \min \tag{5.152}$$

with $u \in H^{1/2}(\Gamma)$.

• *Optimality conditions.* We introduce the Lagrangian functional

$$\mathcal{L}(z, u, p) = \frac{1}{2}\|z+u_{ext}-z_d\|_{L^2(\Omega)}^2+\frac{\beta}{2}\|u\|_{H^{1/2}(\Gamma)}^2+(f,p)_{L^2(\Omega)}-(\nabla u_{ext}, \nabla p)_{L^2(\Omega)}-(\nabla z, \nabla p)_{L^2(\Omega)}$$

with the multiplier $p \in H_0^1(\Omega)$ to be chosen.

Then, the *adjoint equation* is given by ($\hat{z} = z(\hat{u})$)

$$\mathcal{L}'_z(\hat{z}, \hat{u}, \hat{p})\,\psi = (\hat{z} + \hat{u}_{ext} - z_d, \psi)_{L^2(\Omega)} - (\nabla\hat{p}, \nabla\psi)_{L^2(\Omega)} = 0 \quad \forall\psi \in H_0^1(\Omega). \tag{5.153}$$

Thus $\hat{p}$ is the variational solution of the problem ($\hat{z} + \hat{u}_{ext} = \hat{y}$)

$$\begin{cases} -\Delta\hat{p} = \hat{y} - z_d & \text{in } \Omega \\ \hat{p} = 0 & \text{on } \Gamma. \end{cases} \tag{5.154}$$

The *Euler equation* is

$$\mathcal{L}'_u(\hat{z}, \hat{u}, \hat{p})\,v = (\hat{y} - z_d, v)_{L^2(\Omega)} + \beta(\hat{u}, v)_{H^{1/2}(\Gamma)} - (\nabla\hat{p}, \nabla v)_{L^2(\Omega)} = 0 \ \ \forall v \in H^1(\Omega). \tag{5.155}$$

To turn (5.155) into a somewhat more explicit form, we would like to integrate by parts the last term. Since $\hat{p}$ is in $H_0^1(\Omega)$ only, we need to give a meaning to the trace of its normal derivative on Γ. First observe that for functions belonging to the space

$$H_{div}(\Omega) = \{\mathbf{w} \in L^2(\Omega)^d \ : \ \operatorname{div}\mathbf{w} \in L^2(\Omega)\}$$

the trace $\mathbf{w}\cdot\mathbf{n}$ on Γ is well defined as an element of $H^{-1/2}(\Gamma)$ and the following integration by parts formula holds (see Theorem A.14):

Proposition 5.13. *Let Ω be a bounded Lipschitz domain. Then, for any $v \in H^1(\Omega)$, $\mathbf{w} \in H_{div}(\Omega)$,*

$$(\mathbf{w}, \nabla v)_{L^2(\Omega)} - \langle\mathbf{w}\cdot\mathbf{n}, v\rangle_{H^{-1/2},H^{1/2}} = (\operatorname{div}\mathbf{w}, v)_{L^2(\Omega)} \tag{5.156}$$

and moreover

$$\|\mathbf{w}\cdot\mathbf{n}\|_{H^{-1/2}(\Gamma)} \le C(d, \Omega)\left\{\|\mathbf{w}\|_{L^2(\Omega)} + \|\operatorname{div}\mathbf{w}\|_{L^2(\Omega)}\right\}.$$

As a consequence, by taking $\mathbf{w} = \nabla\hat{p}$, from (5.153) we have $\operatorname{div}\mathbf{w} = \hat{y} - z_d \in L^2(\Omega)$, whence $\frac{\partial\hat{p}}{\partial n} \in H^{-1/2}(\Gamma)$ and, in particular (letting hereon $\langle\cdot,\cdot\rangle_{H^{-1/2},H^{1/2}} = \langle\cdot,\cdot\rangle_{H^{-1/2}(\Gamma),H^{1/2}(\Gamma)}$)

$$(\nabla\hat{p}, \nabla v)_{L^2(\Omega)} - (\hat{y} - z_d, v)_{L^2(\Omega)} = \left\langle\frac{\partial\hat{p}}{\partial n}, v\right\rangle_{H^{-1/2},H^{1/2}} \qquad \forall v \in H^1(\Omega). \tag{5.157}$$

Using (5.157) we can write (5.155) in the form

$$\beta\left(\hat{u}, v\right)_{H^{1/2}(\Gamma)} - \left\langle \frac{\partial \hat{p}}{\partial n}, v \right\rangle_{H^{-1/2}, H^{1/2}} = 0 \qquad \forall v \in H^{1/2}\left(\Gamma\right). \tag{5.158}$$

Introducing the Riesz isometry (see Sect. A.2.2) $\Lambda = H^{-1/2} \to H^{1/2}$, we find

$$\left(\beta\hat{u} - \Lambda\frac{\partial \hat{p}}{\partial n}, v\right)_{H^{1/2}} = 0 \qquad \forall v \in H^{1/2}\left(\Gamma\right)$$

from which

$$\nabla J(\hat{u}) = \beta\hat{u} - \Lambda\frac{\partial \hat{p}}{\partial n} = 0 \quad \text{in } H^{1/2}\left(\Gamma\right). \tag{5.159}$$

In conclusion, we have proved the following result.

Proposition 5.14. *Consider the OCP (5.150), (5.149). Then the control $\hat{u} \in H^{1/2}(\Gamma)$ and the state $\hat{y} \in H^1(\Omega)$ are optimal if and only if, together with the adjoint state $\hat{p} \in H_0^1(\Omega)$, they satisfy the state equation (5.151), the adjoint equation (5.153), and the Euler equation (5.158) or (5.159).*

Option 2: $\mathcal{U}_0 = L^2(\Gamma)$

The presence of the fractional Sobolev norm in the expression of $\tilde{J}$ and the fractional inner product in the optimality conditions make the choice $\mathcal{U}_0 = H^{1/2}(\Gamma)$ hardly appealing, notably from the numerical point of view. Note instead that the duality term $\langle \frac{\partial \hat{p}}{\partial n}, v\rangle_{H^{-1/2},H^{1/2}}$ is not an issue. A possible way to overcome this inconvenience is to choose $\mathcal{U}_0 = L^2(\Gamma)$ and indeed, this option is successfully exploited in several applications (see e.g. [74] for further details).

In principle, weakening the norm in the control space reflects on the choice of a narrower observation space (with a stronger norm) in order to ensure existence and uniqueness of an optimal control $\hat{u} \in H^{1/2}(\Gamma)$. Indeed, in this *asymmetric* situation, the bilinear form appearing in (5.11) is given, for the case at hand, by

$$q\left(u, v\right) = \frac{1}{2}\left(y_0\left(u\right), y_0(v)\right)_Z + \frac{\beta}{2}\left(u, v\right)_{L^2(\Gamma)}.$$

This bilinear form is coercive if we make the choice $Z = H^1(\Omega)$; in fact, by the trace inequality (A.43), we have

$$\|u\|_{H^{1/2}(\Gamma)} \leq \|y_0\left(u\right)\|_{H^1(\Omega)}.$$

Hence, we consider the minimization problem

$$\tilde{J}(y, u) = \frac{1}{2}\|y - z_d\|^2_{H^1(\Omega)} + \frac{\beta}{2}\|u\|^2_{L^2(\Gamma)} \to \min \tag{5.160}$$

with $u \in L^2(\Gamma)$. Thanks to Theorem 5.4, even if $\beta = 0$, we can prove that there exists a unique optimal control $\hat{u} \in H^{1/2}(\Gamma)$

• *Optimality conditions.* Using the extension method for the state problem, we introduce the Lagrangian functional

$$\mathcal{L}\left(z, u, p\right) = \frac{1}{2}\|z + u_{ext} - z_d\|^2_{H^1(\Omega)} + \frac{\beta}{2}\|u\|^2_{L^2(\Gamma)} + \left(f, p\right)_{L^2(\Omega)} - \left(\nabla u_{ext}, \nabla p\right)_{L^2(\Omega)} - \left(\nabla z, \nabla p\right)_{L^2(\Omega)}$$

with the multiplier $p \in H_0^1(\Omega)$ to be chosen. Proceeding as in the previous case, the *adjoint equation* takes the form

$$(\nabla \hat{p}, \nabla \psi)_{L^2(\Omega)} = (\hat{y} - z_d, \psi)_{H^1(\Omega)} \quad \forall \psi \in H_0^1(\Omega) \tag{5.161}$$

which represents the variational formulation of the problem

$$\begin{cases} -\Delta \hat{p} = (I - \Delta)(\hat{y} - z_d) & \text{in } \Omega \\ \hat{p} = 0 & \text{on } \Gamma. \end{cases}$$

The associated *Euler equation* is given by

$$\beta (\hat{u}, v)_{L^2(\Gamma)} - \left\langle \frac{\partial \hat{p}}{\partial n}, v \right\rangle_{H^{-1/2}, H^{1/2}} = 0 \qquad \forall v \in H^{1/2}(\Gamma)$$

or, equivalently[9],

$$\left\langle \beta \hat{u} - \frac{\partial \hat{p}}{\partial n}, v \right\rangle_{H^{-1/2}, H^{1/2}} = 0 \qquad \forall v \in H^{1/2}(\Gamma). \tag{5.162}$$

Thus, from (5.162) we deduce

$$\beta \hat{u} - \frac{\partial \hat{p}}{\partial n} = 0. \tag{5.163}$$

Remark 5.17. Since $\hat{u} \in H^{1/2}(\Gamma)$, we see that actually $\frac{\partial \hat{p}}{\partial n} \in H^{1/2}(\Gamma)$ and, by elliptic regularity, $\hat{p} \in H^2(\Omega)$. See, e.g., [237]. •

We summarize everything in the following proposition.

Proposition 5.15. *Consider the OCP (5.149), (5.160). Then the control $\hat{u} \in H^{1/2}(\Gamma)$ and the state $\hat{y} \in H^1(\Omega)$ are optimal if and only if, together with the adjoint state $\hat{p} \in H_0^1(\Omega)$, $\hat{u}$ and $\hat{y}$ are solutions of the state problem (5.149), the adjoint problem (5.161), and the Euler equation (5.163).*

An alternative formulation*

An alternative approach makes use of a different formulation of the state problem, aimed to avoid the use of extensions of the Dirichlet datum. We illustrate the method in the case $\mathcal{U}_0 = H^{1/2}(\Gamma)$. The other case is similar. The idea (see [16]) is to exploit equation (5.156) introducing a *new unknown* $\xi \in H^{-1/2}(\Gamma)$, playing the role of $\frac{\partial y}{\partial n}$, and appending a multiplier $\eta \in H^{-1/2}(\Gamma)$ to the Dirichlet condition. We shall use this kind of strategy in Sect. 5.13.2. The new formulation, which is often referred to as *primal mixed formulation*, reads as follows.

Find $(y, \xi) \in W = H^1(\Omega) \times H^{-1/2}(\Gamma)$ such that

$$(\nabla y, \nabla \varphi)_{L^2(\Omega)} - \langle \xi, \varphi \rangle_{H^{-1/2}, H^{1/2}} = (f, \varphi)_{L^2(\Omega)} \tag{5.164}$$

$$\langle \eta, y - u \rangle_{H^{-1/2}, H^{1/2}} = 0 \tag{5.165}$$

for all $(\varphi, \eta) \in W$.

---

[9] We can consider the Hilbert triplet $H^{1/2}(\Gamma) \subset L^2(\Gamma) \subset H^{-1/2}(\Gamma)$, where $L^2(\Gamma)$ is identified with its dual. Then if $u \in L^2(\Gamma)$, $\langle u, v \rangle_{H^{-1/2}, H^{1/2}} = (u, v)_{L^2(\Gamma)}$ $\forall v \in H^{1/2}(\Gamma)$.

The variational and the primal mixed formulations are equivalent. Indeed, if y is the variational solution to (5.149), from (5.156) with $\mathbf{w} = \nabla y$, we infer that the pair $(y, \xi) = (y, \frac{\partial y}{\partial n})$ solves (5.164), (5.165) for all $(\varphi, \eta) \in W$. Conversely, let $(y, \xi) \in W$ solve (5.164), (5.165) for all $(\varphi, \eta) \in W$. Choosing first the test functions φ in $H_0^1(\Omega)$ we conclude that $-\Delta y = f$ in Ω. From (5.165) we infer $y = u$ on Γ, so that y is the unique variational solution to (5.149). Finally, from (5.156) and (5.164) with $\mathbf{w} = \nabla y$, we deduce that $\xi = \frac{\partial y}{\partial n}$.

We can now go back to our OCP problem, and state it as follows

$$\tilde{J}(y, \xi, u) = \frac{1}{2}\|y - z_d\|^2_{L^2(\Omega)} + \frac{\beta}{2}\|u\|^2_{H^{1/2}(\Gamma)} \to \min \tag{5.166}$$

subject to $(y, \xi, u) \in W \times H^{1/2}(\Gamma)$, and to the state problem (5.164), (5.165).

- *Optimality conditions.* To derive the optimality conditions we define the Lagrangian

$$\begin{aligned}\mathcal{L}(y, \xi, u, p, q) = &\frac{1}{2}\|y - z_d\|^2_{L^2(\Omega)} + \frac{\beta}{2}\|u\|^2_{H^{1/2}(\Gamma)} + (f, \varphi)_{L^2(\Omega)} \\ &- (\nabla y, \nabla p)_{L^2(\Omega)} + \langle \xi, p\rangle_{H^{-1/2}, H^{1/2}} + \langle q, y - u\rangle_{H^{-1/2}, H^{1/2}}\end{aligned}$$

where the multipliers $p \in H^1(\Omega)$ and $q \in H^{-1/2}(\Gamma)$ replace the test functions in (5.164), (5.165). We check that we reach the same conclusions in Proposition 5.14.

The *adjoint equation* is given by the system $(\hat{\mathcal{L}}'_{(\cdot)}\psi = \mathcal{L}'_{(\cdot)}(\hat{y}, \hat{\xi}, \hat{u}, \hat{p}, \hat{q})\,\psi)$

$$\hat{\mathcal{L}}'_y\psi = (\hat{y} - z_d, \psi)_{L^2(\Omega)} - (\nabla\hat{p}, \nabla\psi)_{L^2(\Omega)} + \langle \hat{q}, \psi\rangle_{H^{-1/2}, H^{1/2}} = 0 \qquad \forall\psi \in H^1(\Omega) \tag{5.167}$$

and

$$\hat{\mathcal{L}}'_\xi\eta = \langle \eta, \hat{p}\rangle_{H^{-1/2}, H^{1/2}} = 0 \qquad \forall\eta \in H^{1/2}(\Gamma). \tag{5.168}$$

From (5.167) and (5.168) we infer that $\hat{p}$ is the variational solution of the adjoint problem (5.154) while

$$\hat{q} = \frac{\partial\hat{p}}{\partial n} \quad \text{on } \Gamma. \tag{5.169}$$

The *Euler equation* takes the form

$$\hat{\mathcal{L}}'_u v = \beta\,(\hat{u}, v)_{H^{1/2}(\Gamma)} - \langle \hat{q}, v\rangle_{H^{-1/2}, H^{1/2}} = 0 \quad \forall v \in H^{1/2}(\Gamma)$$

which, by (5.169), coincides with (5.158). Finally, observe that from the vanishing of the derivatives $\mathcal{L}_p'$ and $\mathcal{L}_q'$, we recover the state problem under the formulation (5.164), (5.165).

5.11.2 Case $\mathcal{U} = \mathcal{U}_0 = L^2(\Gamma)$

We have seen in the previous subsection that by choosing $\mathcal{U} = H^{1/2}(\Gamma)$ and $\mathcal{U}_0 = L^2(\Gamma)$ instead of $H^{1/2}(\Gamma)$, the situation has improved since the use of fractional norms or inner products is avoided. The only little drawback is the appearance of the source term $(\nabla\hat{y} - \nabla z_d, \psi)_{L^2(\Omega)}$ in the adjoint equation. A further semplification can be achieved with the choice $\mathcal{U} = \mathcal{U}_0 = L^2(\Gamma)$, that calls into play very weak solutions of the state problem, as in Definition 5.1. Thus, we consider the minimization problem

$$\tilde{J}(y, u) = \frac{1}{2}\|y - z_d\|^2_{L^2(\Omega)} + \frac{\beta}{2}\|u\|^2_{L^2(\Gamma)} \to \min \tag{5.170}$$

subject to $(y, u) \in L^2(\Omega) \times L^2(\Gamma)$ and to the state equation

$$(y, \Delta\psi)_{L^2(\Omega)} = \left(u, \frac{\partial \psi}{\partial n}\right)_{L^2(\Gamma)} - (f, \psi)_{L^2(\Omega)} \qquad \forall \psi \in Y = H^2(\Omega) \cap H_0^1(\Omega). \tag{5.171}$$

Existence and uniqueness of an optimal control $\hat{u}$ follow by the standard theory, since $\beta > 0$.

• *Optimality conditions.* We introduce the Lagrangian functional

$$\mathcal{L}(y, u, p) = \frac{1}{2}\|y - z_d\|^2_{L^2(\Omega)} + \frac{\beta}{2}\|u\|^2_{L^2(\Gamma)} + \left(u, \frac{\partial p}{\partial n}\right)_{L^2(\Gamma)} - (f, \psi)_{L^2(\Omega)} - (y, \Delta p)_{L^2(\Omega)}$$

where the multiplier $p \in Y$ takes the place of ψ in (5.171). Then,

$$\mathcal{L}'_y(\hat{y}, \hat{u}, \hat{p})\,\varphi = -(\Delta\hat{p}, \varphi)_{L^2(\Omega)} + (\hat{y} - z_d, \varphi)_{L^2(\Omega)} = 0 \qquad \forall \varphi \in L^2(\Omega) \tag{5.172}$$

is the *adjoint equation*, that gives, since $\hat{p} \in Y$,

$$\begin{cases} -\Delta\hat{p} = \hat{y} - z_d & \text{in } \Omega \\ \hat{p} = 0 & \text{on } \Gamma. \end{cases}$$

The *Euler equation* takes the form

$$\mathcal{L}'_u(\hat{y}, \hat{u}, \hat{p})\, v = \left(\beta\hat{u} - \frac{\partial \hat{p}}{\partial n}, v\right)_{L^2(\Gamma)} = 0 \qquad \forall v \in L^2(\Gamma) \tag{5.173}$$

yielding

$$\beta\hat{u} - \frac{\partial \hat{p}}{\partial n} = 0 \qquad \text{on } \Gamma. \tag{5.174}$$

Although everything is cast in the standard theory, there is some price to pay. First of all, to use the transposition method, we need a stronger regularity of Ω, namely we require Ω to be either convex or of class C^2. Moreover, in principle, the numerical solution of equation (5.171) requires a special care, since the Dirichlet conditions do not fit the variational setting in a straightforward way, and the optimal order of convergence in the context of finite element discretization may be difficult to achieve. Nevertheless, since $\hat{p} \in Y$, the trace of $\nabla\hat{p}$ belongs to $H^{1/2}(\Gamma)$ and if Ω is a C^2 domain, the normal vector field $\mathbf{n}$ is in $C^1(\Gamma)$. This implies that $\frac{\partial \hat{p}}{\partial n} = \nabla\hat{p} \cdot \mathbf{n} \in H^{1/2}(\Gamma)$ which in turn, from (5.174), gives $\hat{u} \in H^{1/2}(\Gamma)$.

But then, owing to Remark 5.12, we infer that the optimal state $\hat{y}$ is actually a *variational solution* of the state equation and the usual numerical approximation methods can be used. We can summarize these conclusions as follows.

Proposition 5.16. *Let Ω be a C^2 domain. Consider the OCP (5.149), (5.170). Then, the control $\hat{u}$ and the state $\hat{y}$ are optimal if and only if $\hat{u} \in H^{1/2}(\Gamma)$, $\hat{y} \in H^1(\Omega)$ and together with the adjoint state $\hat{p} \in H^2(\Omega) \cap H_0^1(\Omega)$, they are solutions of the state problem (5.149), the adjoint problem (5.172), and the Euler equation (5.174).*

Remark 5.18. In this case we easily derive the formula

$$\nabla J(u) = \frac{\partial p(u)}{\partial n} + \beta u \quad \text{in } L^2(\Gamma)$$

where $p(u) \in Y$ solves the adjoint equation $-\Delta p = y(u) - z_d$ in Ω, $p(u) = 0$ on Γ. •

5.11.3 Case $\mathcal{U} = \mathcal{U}_0 = H^1(\Gamma)$

A way to avoid fractional norms and the use of very weak solution of the state problem (requiring extra regularity of the domain) is to choose $\mathcal{U} = \mathcal{U}_0 = H^1(\Gamma)$. $H^1(\Gamma)$ is made of functions belonging to $L^2(\Gamma)$ together with their *tangential gradient*, that is,

$$H^1(\Gamma) = \left\{ u \in L^2(\Gamma) : \nabla_\Gamma u \in L^2(\Gamma) \right\}.$$

The definition of tangential gradient can be found in Sect. 11.2.2. Endowed with the inner product

$$(u, v)_{H^1(\Gamma)} = (u, v)_{L^2(\Gamma)} + (\nabla_\Gamma u, \nabla_\Gamma v)_{L^2(\Gamma)},$$

$H^1(\Gamma)$ is a Hilbert space, continuously embedded into $H^{1/2}(\Gamma)$. We denote by $H^{-1}(\Gamma)$ the dual of $H^1(\Gamma)$. Then we may choose $Z = L^2(\Omega)$ so that our OCP becomes

$$\tilde{J}(y, u) = \frac{1}{2}\|y - z_d\|^2_{L^2(\Omega)} + \frac{\beta}{2}\|u\|^2_{H^1(\Gamma)} \to \min \tag{5.175}$$

with $u \in H^1(\Gamma)$. Clearly, since $\beta > 0$, we have existence and uniqueness of the optimal control $\hat{u} \in H^1(\Gamma)$.

• *Optimality conditions.* Using the extension method for the state problem and keeping the notations of the case $\mathcal{U} = \mathcal{U}_0 = H^{1/2}(\Gamma)$, we introduce the Lagrangian functional

$$\mathcal{L}(z, u, p) = \frac{1}{2}\|z + u_{ext} - z_d\|^2_{L^2(\Omega)} + \frac{\beta}{2}\|u\|^2_{H^1(\Gamma)} + (f, p)_{L^2(\Omega)} - (\nabla u_{ext}, \nabla p)_{L^2(\Omega)} - (\nabla z, \nabla p)_{L^2(\Omega)}.$$

The adjoint system still coincides with (5.172); however, the Euler equation becomes now a differential equation on Γ and using once more (5.157), it takes the form

$$\hat{\mathcal{L}}'_u v = \beta (\nabla_\Gamma \hat{u}, \nabla_\Gamma v)_{L^2(\Gamma)} + \beta (\hat{u}, v)_{L^2(\Gamma)} - \left\langle \frac{\partial \hat{p}}{\partial n}, v \right\rangle_{H^{-1/2}, H^{1/2}} = 0 \quad \forall v \in H^1(\Gamma). \tag{5.176}$$

We can summarize these conclusions as follows.

Proposition 5.17. *Consider the OCP (5.175), (5.149). Then the control $\hat{u} \in H^1(\Gamma)$ and the state $\hat{y} \in H^1(\Omega)$ are optimal if and only if, together with the adjoint state $\hat{p} \in H^1_0(\Omega)$, $\hat{u}$ and $\hat{y}$ are solutions of the state problem (5.149), the adjoint equation (5.172), and the Euler equation (5.176).*

Here the drawback is the appearance of the tangential gradients both in the cost functional and in the Euler equation. An application of this approach will be considered in Sect. 5.13.2.

5.12 A State-Constrained Control Problem

In this section we consider a state-constrained optimal control problem where both control and observation are distributed. Let Ω be a bounded Lipschitz domain and set $\mathcal{U}_{ad} = \mathcal{U} = L^2(\Omega)$; we want to solve the minimization problem

$$\tilde{J}(y, u) = \frac{1}{2}\|y - z_d\|^2_{L^2(\Omega)} + \frac{\beta}{2}\|u\|^2_{L^2(\Omega)} \to \min \tag{5.177}$$

with $\beta > 0$, subject to

$$\begin{cases} -\Delta y + y = u & \text{in } \Omega \\ \dfrac{\partial y}{\partial n} = 0 & \text{on } \Gamma \end{cases} \tag{5.178}$$

and to the *state constraints*

$$y_a(\mathbf{x}) \leq y(\mathbf{x}) \leq y_b(\mathbf{x}) \ \text{ a.e in } \Omega, \tag{5.179}$$

where $y_a, y_b \in L^\infty(\Omega)$. Hereon in this section we denote the optimal control problem (5.177)–(5.179) by (P). With reference to the general framework of Sect. 5.3, here $Z = L^2(\Omega)$, $V = H^1(\Omega)$ and the observation operator $C : V \to Z$ is given by $Cy = y$.

According to Proposition 5.8, for every $u \in L^2(\Omega)$ there exists a unique weak solution $y = y(u) \in H^1(\Omega)$ with

$$\|y(u)\|_{H^1(\Omega)} \leq C(d, \Omega) \|u\|_{L^2(\Omega)} \tag{5.180}$$

and we can consider the reduced cost functional $J(u) = \tilde{J}(y(u), u)$. For (5.179) to make sense, we restrict to dimensions $d = 2, 3$; indeed, in this case, from the elliptic regularity property (see Theorem 5.8) we have $y(u) \in C(\overline{\Omega})$.

The presence of the state constraints (5.179) is troublesome from both a theoretical and a numerical standpoint. Since $\beta > 0$, as long as there exists a *feasible* point u, that is one for which the corresponding $y(u)$ satisfies (5.179), there exists a unique optimal control $\hat{u}$; this follows from the general Theorem 5.4.

Hence, existence and uniqueness of the optimal control are not an issue. Things change significantly when we come to first order optimality conditions. Indeed, it is known from Karush-Kuhn-Tucker theory in Banach spaces that Lagrange multipliers associated to the state constraints exist under suitable assumptions that generalize those in the finite dimensional case (see Sect. 2.3). Even more crucially, these multipliers are *measures* [235] rather than, say, L^2–functions, a circumstance rather uncomfortable to be addressed numerically.

To overcome this difficulty a *Laurentiev-type* regularization technique can be invoked. This is a rather common approach in the treatment of inverse problems, consisting in three steps: first we introduce a regularization parameter $\lambda > 0$ and replace the state constraints (5.179) by

$$y_a(\mathbf{x}) \leq \lambda u(\mathbf{x}) + y(\mathbf{x}) \leq y_b(\mathbf{x}) \ \text{ a.e in } \Omega \tag{5.181}$$

obtaining what we call a *regularized* control problem, denoted by $(P)_\lambda$.

Then (step 2), we convert problem $(P)_\lambda$ into an equivalent one, with control box-constraints only, we check its well-posedness and derive the usual first-order optimality conditions under the Karush-Kuhn-Tucker (KKT) form. Finally (step 3), we let $\lambda \to 0$ to recover the solution of the original problem. The two following results address steps 2 and 3, respectively. Our presentation follows [197]; similar considerations can also be found in [198].

We first characterize the optimality conditions of problem $(P)_\lambda$, corresponding to the step 2 of the Laurentiev-type regularization method. The following result holds.

Theorem 5.9. *The regularized control problem $(P)_\lambda$ admits a unique optimal pair*

$$(\hat{u}_\lambda, \hat{y}_\lambda) \in L^2(\Omega) \times \left(H^1(\Omega) \cap C(\overline{\Omega})\right).$$

Moreover $(\hat{u}_\lambda, \hat{y}_\lambda)$ is optimal if and only if there exist nonnegative Lagrange multipliers $\eta_\lambda, \xi_\lambda \in L^2(\Omega)$ and an adjoint state $\hat{p}_\lambda \in H^1(\Omega) \cap C(\overline{\Omega})$ such that the following optimality system holds:

$$\begin{cases} -\Delta \hat{y}_\lambda + \hat{y}_\lambda = \hat{u}_\lambda & \text{in } \Omega, \quad \dfrac{\partial \hat{y}_\lambda}{\partial n} = 0 \text{ on } \Gamma & \text{state problem} \\ -\Delta \hat{p}_\lambda + \hat{p}_\lambda = \hat{y}_\lambda - z_d + \eta_\lambda - \xi_\lambda & \text{in } \Omega, \quad \dfrac{\partial \hat{p}_\lambda}{\partial n} = 0 \text{ on } \Gamma & \text{adjoint problem} \\ \hat{p}_\lambda + \beta \hat{u}_\lambda + \lambda(\eta_\lambda - \xi_\lambda) = 0 & & \\ y_a \leq \lambda \hat{u}_\lambda + \hat{y}_\lambda \leq y_b, \ \xi_\lambda \geq 0, \ \eta_\lambda \geq 0 \quad \text{a.e. in } \Omega & & \text{KKT conditions} \\ \xi_\lambda (\lambda \hat{u}_\lambda + \hat{y}_\lambda - y_a) = \eta_\lambda (y_b - \lambda \hat{u}_\lambda - \hat{y}_\lambda) = 0 & & \end{cases}$$

Proof. Let us introduce a new control variable by setting $v = v_\lambda = \lambda u + Su$, where $S : u \to y(u)$ is the (linear and symmetric) control-to-state map. Since $y(u) \in H^1(\Omega)$, a space that is compactly contained into $L^2(\Omega)$, when operating from $L^2(\Omega)$ into $L^2(\Omega)$, S is compact and its eigenvalues are all positive (see [237, Chapter 6]).

It follows that, for $\lambda > 0$, the operator $S_\lambda = \lambda I + S$ has a continuous inverse

$$S_\lambda^{-1} = (\lambda I + S)^{-1} : L^2(\Omega) \to L^2(\Omega).$$

In particular

$$\|v\|_{L^2(\Omega)} = \|S_\lambda u\|_{L^2(\Omega)} \leq C_0 \|u\|_{L^2(\Omega)} = C_0 \left\|S_\lambda^{-1} v\right\|_{L^2(\Omega)} \tag{5.182}$$

for a suitable constant C_0, depending only on d, λ and Ω.

Setting $B_\lambda = S S_\lambda^{-1}$ and inserting $S_\lambda^{-1} v = u$ and $y(u) = B_\lambda v$ into $J(u)$ we are led to the following OCP:

$$\begin{cases} J_\lambda(v) = \dfrac{1}{2} \|B_\lambda v - z_d\|^2_{L^2(\Omega)} + \dfrac{\beta}{2} \left\|S_\lambda^{-1} v\right\|^2_{L^2(\Omega)} \longrightarrow \min \\ \text{subject to} \\ v \in V_{ad} = \left\{v \in L^2(\Omega) : \ y_a \leq v \leq y_b \text{ a.e. in } \Omega\right\}. \end{cases} \tag{5.183}$$

Note that v is only subject to box-constraints. An optimal solution $\hat{v}_\lambda$ for (5.183) corresponds to an optimal pair $(\hat{u}_\lambda, \hat{y}_\lambda) = \left(S_\lambda^{-1} \hat{v}_\lambda, B_\lambda \hat{v}_\lambda\right)$ for problem $(P)_\lambda$ and, conversely, if $(\hat{u}_\lambda, \hat{y}_\lambda)$ is an optimal pair for $(P)_\lambda$, then $\hat{v}_\lambda = S_\lambda \hat{u}_\lambda$ is optimal for (5.183). The two problems are therefore equivalent.

Problem (5.183) fits into the framework of Sect. 5.2, as

$$J_\lambda(v) = \frac{1}{2} q(v, v) - Lv + c$$

provided we set

$$q(v, w) = (B_\lambda v, B_\lambda w)_{L^2(\Omega)} + \beta \left(S_\lambda^{-1} v, S_\lambda^{-1} w\right)_{L^2(\Omega)}, \qquad \text{and} \qquad Lv = (B_\lambda v, z_d)_{L^2(\Omega)}.$$

Now, q is symmetric and thanks to (5.180) and (5.182), it is also continuous and coercive (as $\beta > 0$) in $L^2(\Omega)$, while L is continuous in $L^2(\Omega)$. Then, Theorem 5.3 ensures existence and uniqueness of the optimal control $\hat{v}_\lambda$, characterized by the following variational inequality

$$(B_\lambda \hat{v}_\lambda - z_d, B_\lambda(v - \hat{v}_\lambda))_{L^2(\Omega)} + \left(\beta S_\lambda^{-1} \hat{v}_\lambda, S_\lambda^{-1}(v - \hat{v}_\lambda)\right)_{L^2(\Omega)} \geq 0 \quad \forall v \in V_{ad}. \tag{5.184}$$

Since both S and S_λ^{-1} are self-adjoint operators, we can write (5.184) under the form (since $B_\lambda = S S_\lambda^{-1}$)

$$\left(S_\lambda^{-1} \left[S(B_\lambda \hat{v}_\lambda - z_d)_{L^2(\Omega)} + \beta S_\lambda^{-1} \hat{v}_\lambda\right], v - \hat{v}_\lambda\right)_{L^2(\Omega)} \geq 0 \quad \forall v \in V_{ad}. \tag{5.185}$$

This latter inequality shows that

$$\nabla J_\lambda(\hat{v}_\lambda) = S_\lambda^{-1} \left[S(B_\lambda \hat{v}_\lambda - z_d) + \beta S_\lambda^{-1} \hat{v}_\lambda\right] = g_\lambda \in L^2(\Omega).$$

Set

$$\xi_\lambda = g_\lambda^+ = \max\{0, g_\lambda\} \text{ and } \eta_\lambda = \max\{0, -g_\lambda\}.$$

With the same argument used in the proof of Proposition 5.4, it turns out that (5.185) is equivalent to the following Karush-Kuhn-Tucker conditions, valid a.e. in Ω:

$$\begin{cases} S_\lambda^{-1}\left[S\left(B_\lambda \hat{v}_\lambda - z_d\right) + \beta S_\lambda^{-1}\hat{v}_\lambda\right] + \eta_\lambda - \xi_\lambda = 0 \\ y_a \le \hat{v}_\lambda \le y_b \\ \xi_\lambda \ge 0, \;\; \xi_\lambda\left(\hat{v}_\lambda - y_a\right) = 0 \\ \eta_\lambda \ge 0, \;\; \eta_\lambda\left(y_b - \hat{v}_\lambda\right) = 0. \end{cases} \tag{5.186}$$

Note that ξ_λ and η_λ play the role of Lagrange multipliers associated to the constraints $y_a \le v_\lambda$ and $v_\lambda \le y_b$, respectively. An important consequence of the regularization procedure is that both ξ_λ and η_λ are $L^2(\Omega)$ functions rather than measures.

Applying $S_\lambda = \lambda I + S$ to both sides of the first equation in (5.186) we get

$$S\left(B_\lambda \hat{v}_\lambda - z_d + \eta_\lambda - \xi_\lambda\right) + \beta S_\lambda^{-1}\hat{v}_\lambda + \lambda(\eta_\lambda - \xi_\lambda) = 0. \tag{5.187}$$

By expressing (5.187) in terms of the optimal pair $(\hat{u}_\lambda, \hat{y}_\lambda) = \left(S_\lambda^{-1}\hat{v}_\lambda, B_\lambda \hat{v}_\lambda\right)$ for problem $(P)_\lambda$, we get

$$S\left(\hat{y}_\lambda - z_d + \eta_\lambda - \xi_\lambda\right) + \beta \hat{u}_\lambda + \lambda(\eta_\lambda - \xi_\lambda) = 0.$$

Observe now that

$$\hat{p}_\lambda = S\left(\hat{y}_\lambda - z_d + \eta_\lambda - \xi_\lambda\right)$$

is the solution of the *adjoint problem*

$$\begin{cases} -\Delta \hat{p}_\lambda + \hat{p}_\lambda = \hat{y}_\lambda - z_d + \eta_\lambda - \xi_\lambda & \text{in } \Omega \\ \dfrac{\partial \hat{p}_\lambda}{\partial n} = 0 & \text{on } \Gamma. \end{cases} \tag{5.188}$$

Since $\eta_\lambda, \xi_\lambda \in L^2(\Omega)$, the adjoint problem (5.188) admits a unique solution $\hat{p}_\lambda \in H^1(\Omega) \cap C(\overline{\Omega})$. □

Regarding the recovery of the solution of the original problem (P) – which we have referred to as step 3 – the following result holds; see, e.g., [197] for its proof.

Theorem 5.10. *Let $\{\lambda_n\}, \lambda_n > 0$, be a sequence converging to 0 and $\hat{u}_n, \hat{u}$ be the optimal solution of problem $(P)_{\lambda_n}$ and of $(P)_\lambda$, respectively. Then $\hat{u}_n \to \hat{u}$ in $L^2(\Omega)$.*

Remark 5.19. A Laurentiev-type regularization can also be applied to the case of boundary control by looking for controls in the range of the adjoint operator S^*. In the case of Neumann control, (5.178) is replaced by

$$\begin{cases} -\Delta y + y = 0 & \text{in } \Omega \\ \dfrac{\partial y}{\partial n} = u & \text{on } \Gamma. \end{cases} \tag{5.189}$$

In this case, we can make the ansatz $u = S^* v$ with a new control v defined in Ω. This allows to express the pointwise state constraint (5.179) under the form $y_a \le SS^*v \le y_b$ a.e. in Ω, and to express the combined state/control constraint in the regularized problem as

$$y_a \le \lambda v + SS^* v \le y_b \;\; \text{a.e. in } \Omega.$$

See, e.g., [264] for further details. •

Remark 5.20. Alternative regularization techniques can be used. For instance, in the Moreau-Yosida regularization (see, e.g., [142]) one state constraint $\{y \le y_b\}$ is included in the penalized functional

$$\frac{1}{2}\int_\Omega (y - z_d)^2\, d\mathbf{x} + \frac{\beta}{2}\int_\Omega u^2 d\mathbf{x} + \frac{1}{2\gamma}\left\|\max\left\{0, \mu + \gamma\left(y - y_b\right)\right\}\right\|^2_{L^2(\Omega)}$$

where γ and μ are a penalization and a shift parameters, respectively. In this case, the optimality conditions lead to the adjoint problem

$$\begin{cases} -\Delta\hat{p} + \hat{p} = \hat{y} - z_d - \max\{0, \mu + \gamma(\hat{y} - y_b)\} & \text{in } \Omega \\ \dfrac{\partial \hat{p}}{\partial n} = 0 & \text{on } \Gamma \end{cases}$$

and to the Euler equation

$$\beta\hat{u} - \hat{p} = 0 \qquad \text{in } \Omega.$$

•

5.13 Control of Viscous Flows: the Stokes Case

In this section we consider two examples of OCPs for fluid flows modeled by the linear steady Stokes equations. The Stokes equations describe the flow of a Newtonian, incompressible viscous fluid confined in a (sufficiently smooth) domain $\Omega \subset \mathbb{R}^d$ when convective forces can be neglected. We start by considering an unconstrained problem, then we turn to a problem involving control constraints.

5.13.1 Distributed Velocity Control

In the first case we want to approach a given velocity distribution $\boldsymbol{z}_d$ over the bounded Lipschitz domain $\Omega \subset \mathbb{R}^d$, $d = 2, 3$ by controlling a distributed force $\boldsymbol{u}$. Thus we seek the solution of the following minimization problem

$$J(\boldsymbol{u}) = \frac{1}{2}\int_\Omega |\boldsymbol{v}(\boldsymbol{u}) - \boldsymbol{z}_d|^2\, d\mathbf{x} + \frac{\tau}{2}\int_\Omega |\boldsymbol{u}|^2\, d\mathbf{x} \to \min \tag{5.190}$$

where $\tau > 0$ and $(\boldsymbol{v}, \pi) = (\boldsymbol{v}(\boldsymbol{u}), \pi(\boldsymbol{u}))$ denotes the solution of the following steady Stokes system:

$$\begin{cases} -\nu\Delta\boldsymbol{v} + \nabla\pi = \boldsymbol{f} + \boldsymbol{u} & \text{in } \Omega \\ \operatorname{div}\boldsymbol{v} = 0 & \text{in } \Omega \\ \boldsymbol{v} = \boldsymbol{0} & \text{on } \Gamma_D \\ -\pi\mathbf{n} + \nu\dfrac{\partial\boldsymbol{v}}{\partial n} = \boldsymbol{h} & \text{on } \Gamma_N. \end{cases} \tag{5.191}$$

Here $\boldsymbol{v}$ and π denote the fluid velocity and pressure, respectively, while $\nu > 0$ denotes its kinematic viscosity; $f \in L^2(\Omega)^d$, $\mathbf{h} \in L^2(\Gamma_N)^d$ are given data. The first equation expresses the conservation of the linear momentum of the fluid, while the second equation, also denoted as incompressibility condition, enforces the mass conservation. Moreover, we impose the zero velocity profile over $\Gamma_D \subset \Gamma$ and the normal Cauchy stress $\boldsymbol{h}$ over $\Gamma_N = \Gamma \setminus \bar{\Gamma}_D$.

We define the velocity space $X = H^1_{\Gamma_D}(\Omega)^d$, the pressure space $Q = L^2(\Omega)$, and we set $V = X \times Q$; $y = (\boldsymbol{v}, \pi)$ denotes the state variable. Moreover, we take $\mathcal{U} = L^2(\Omega)^d$ as control space; here $\mathcal{U}_{ad} = \mathcal{U}$, that is, the OCP is unconstrained. Before addressing the well-posedness of problem (5.190)–(5.191), we provide further details about the weak formulation (and the well-posedness) of the state problem; indeed, this is the first time we are dealing with a state problem given by a system of PDEs, which moreover features a *saddle-point* form. This avoids the use of divergence free spaces, somewhat uncomfortable from a numerical point of view.

For $(\boldsymbol{v}, \boldsymbol{w}) \in X \times X$, $q \in Q$ we define the bilinear forms $e : X \times X \to \mathbb{R}$, $b : X \times Q \to \mathbb{R}$ as

$$e(\boldsymbol{v}, \boldsymbol{w}) = \int_\Omega \nu\nabla\boldsymbol{v} : \nabla\boldsymbol{w}\, d\mathbf{x}, \qquad b(\boldsymbol{v}, q) = -\int_\Omega q \operatorname{div}\boldsymbol{v}\, d\mathbf{x}, \tag{5.192}$$

where

$$\nabla \boldsymbol{v} : \nabla \boldsymbol{w} = \sum_{i,j=1}^{d} \frac{\partial v_i}{\partial x_j} \frac{\partial w_i}{\partial x_j} \qquad \forall \boldsymbol{v}, \boldsymbol{w} \in X.$$

and we set

$$(\nabla \boldsymbol{v}, \nabla \boldsymbol{w})_{L^2} = \int_\Omega \nabla \boldsymbol{v} : \nabla \boldsymbol{w} \, d\mathbf{x}, \qquad \|\nabla \boldsymbol{w}\|_{L^2}^2 = \int_\Omega \nabla \boldsymbol{v} : \nabla \boldsymbol{v} \, d\mathbf{x} = \|\boldsymbol{w}\|_X^2$$

since the Poincaré inequality holds in X.

Moreover, we define the linear functionals

$$F\boldsymbol{\varphi} = \int_\Omega (\boldsymbol{f} + \boldsymbol{u}) \cdot \boldsymbol{\varphi} \, d\mathbf{x} + \int_{\Gamma_N} \boldsymbol{h} \cdot \boldsymbol{\varphi} \, d\sigma.$$

The weak form of problem (5.191) thus reads: *find* $(\boldsymbol{v}, \pi) \in X \times Q$ *such that*

$$\begin{cases} e(\boldsymbol{v}, \boldsymbol{\varphi}) + b(\boldsymbol{\varphi}, \pi) = F\boldsymbol{\varphi} & \forall \boldsymbol{\varphi} \in X \\ b(\boldsymbol{v}, \phi) = 0 & \forall \phi \in Q. \end{cases} \tag{5.193}$$

System (5.193) is obtained by multiplying the first equation in (5.191) by a test function $\boldsymbol{\varphi} \in X$ and the second one by a test function $\phi \in Q$, exploiting the Green identity and integrating by parts.

• *Analysis of the state system (5.193).* The weak formulation (5.193) results into a mixed variational problem; its well-posedness can be proved by means of Theorem A.7, see Sect. A.4.2. Indeed the following conditions hold:

1. e is continuous over $X \times X$; in fact,

$$|e(\boldsymbol{v}, \boldsymbol{w})| \leq \nu \|\nabla \boldsymbol{v}\|_{L^2} \|\nabla \boldsymbol{w}\|_{L^2} = \nu \|\boldsymbol{v}\|_X \|\boldsymbol{w}\|_X;$$

2. e is coercive on X and, in particular, on $X_0 = \{\boldsymbol{v} \in X : b(\boldsymbol{v}, q) = 0 \ \forall q \in Q\}$, since

$$e(\boldsymbol{w}, \boldsymbol{w}) = \nu \|\nabla \boldsymbol{w}\|_{L^2}^2 = \nu \|\boldsymbol{w}\|_X^2;$$

3. b is continuous over $X \times Q$, since

$$|b(\boldsymbol{v}, q)| \leq \|q\|_Q \|\mathrm{div} \boldsymbol{v}\|_{L^2(\Omega)} \leq \sqrt{d} \|q\|_Q \|\nabla \boldsymbol{v}\|_{L^2} = \sqrt{d} \|q\|_Q \|\boldsymbol{v}\|_X;$$

4. b satisfies the inf-sup condition: there exists $\beta_0 > 0$ such that

$$\inf_{q \in Q} \sup_{\boldsymbol{w} \in X} \frac{b(\boldsymbol{w}, q)}{\|\boldsymbol{w}\|_X \|q\|_Q} \geq \beta_0. \tag{5.194}$$

The last condition is an immediate consequence of a general result stating that for any $q \in Q = L^2(\Omega)$ there exists $\boldsymbol{w} \in X$ such that $\mathrm{div} \boldsymbol{w} = q$ and $\|\boldsymbol{w}\|_{H^1(\Omega)} \leq C \|q\|_Q$, for a constant $C > 0$ independent of $\boldsymbol{w}$ and q [10]. Thanks to this result, we have that, given q, there exists $\boldsymbol{w}$ such that

$$\frac{b(\boldsymbol{w}, q)}{\|\boldsymbol{w}\|_X} = \frac{\|q\|_Q^2}{\|\boldsymbol{w}\|_X} \geq \frac{1}{C} \|q\|_Q;$$

[10] Let $q \in L^2(\Omega)$, fix $\boldsymbol{v}_0 \in X$ such that

$$\int_{\Gamma_N} \boldsymbol{v}_0 \cdot \mathbf{n} \, d\sigma = 1$$

and set $\boldsymbol{v} = \boldsymbol{v}_0 \int_\Omega q$. Then $\|\boldsymbol{v}\|_X \leq a \|q\|_{L^2(\Omega)}$, where $a = |\Omega|^{1/2} \|\boldsymbol{v}_0\|_X$, and

$$\int_\Omega \mathrm{div} \boldsymbol{v} = \int_{\Gamma_N} \boldsymbol{v} \cdot \mathbf{n} d\sigma = \int_\Omega q.$$

hence,

$$\sup_{v\in X}\frac{b(\boldsymbol{v},q)}{\|\boldsymbol{v}\|_X\|q\|_Q}\geq\frac{b(\boldsymbol{w},q)}{\|\boldsymbol{w}\|_X\|q\|_Q}\geq\frac{1}{C}\qquad\forall q\in Q.$$

By taking the infimum with respect to $q \in Q$ we get (5.194) with $\beta_0 = 1/C$. Thus there exists a unique solution to system (5.193). Moreover, the stability estimates (A.23) yield in this case

$$\|\boldsymbol{v}(\boldsymbol{u})\|_X\leq C_1\|F\|_{X^*},\qquad \|p(\boldsymbol{u})\|_Q\leq\tilde{C}_1\|F\|_{X^*},$$

where $C_1, \tilde{C}_1$ are positive constants depending only on ν, β_0. In particular, $\boldsymbol{v}_0(\boldsymbol{u}) = \boldsymbol{v}(\boldsymbol{u}) - \boldsymbol{v}(\mathbf{0})$ and $\pi_0(\boldsymbol{u}) = \pi(\boldsymbol{u}) - \pi(\mathbf{0})$ are such that

$$\|\boldsymbol{v}_0(\boldsymbol{u})\|_X\leq C_1\|\boldsymbol{u}\|_{\mathcal{U}},\qquad \|\pi_0(\boldsymbol{u})\|_Q\leq\tilde{C}_1\|\boldsymbol{u}\|_{\mathcal{U}}\qquad\forall\boldsymbol{u}\in\mathcal{U}.$$

Remark 5.21. Using divergence free spaces, we seek velocity functions in the closed subspace of X given by

$$X_{div}=\{\boldsymbol{v}\in H^1_{\Gamma_D}(\Omega)^d\ :\ \mathrm{div}\boldsymbol{v}=0\},$$

a Hilbert space endowed with the norm $\|\boldsymbol{v}\|_{X_{div}} = \|\nabla\boldsymbol{v}\|_{L^2}$. This eliminates the pressure on the first equation in (5.191), whose weak formulation is : find $\boldsymbol{v} \in X_{div}$ such that

$$e(\boldsymbol{v},\boldsymbol{\varphi})=(\boldsymbol{f}+\boldsymbol{u},\boldsymbol{\varphi})_{L^2(\Omega)^d}\qquad\forall\boldsymbol{\varphi}\in X_{div}. \tag{5.195}$$

By the Lax-Milgram theorem A.6, (5.195) has a unique solution that satisfies the condition $\mathrm{div}\boldsymbol{v} = 0$. To recover the pressure, we observe that the vector

$$\boldsymbol{G}=\Delta\boldsymbol{v}+\boldsymbol{f}+\boldsymbol{u}\in X^*$$

satisfies the equation

$$\langle\boldsymbol{G},\boldsymbol{\varphi}\rangle_{X^*,X}=0\qquad\forall\boldsymbol{\varphi}\in X_{div}.$$

Hence (see, e.g., [105, Lemma 2.1]) if Ω is a bounded, Lipschitz domain, there exists a unique $\pi \in L^2(\Omega)$ such that $\boldsymbol{G} = \nabla\pi$ and that $(\boldsymbol{v}, \pi)$, satisfies the Neumann boundary condition in a suitable sense [11]. •

• *Existence and uniqueness of an optimal control.* We can now analyze the well posedness of problem (5.190)–(5.191); a general result can be obtained as in the case of Theorem 5.4, once the OCP has been cast under the general form of a linear-quadratic OCP described in Sect. 5.3.

Thus $\pi = q - \mathrm{div}\boldsymbol{v}$ has zero mean value. By, e.g., [105, Theorem 5.1], there exists $\tilde{\boldsymbol{v}} \in H^1_0(\Omega)^d$ such that $\mathrm{div}\tilde{\boldsymbol{v}} = \pi$ with $\|\tilde{\boldsymbol{v}}\|_X \leq C\|\pi\|_{L^2(\Omega)}$, C independent of $\tilde{\boldsymbol{v}}$ and π. Then $\boldsymbol{w} = \tilde{\boldsymbol{v}} + \boldsymbol{v}$ satisfies $\mathrm{div}\boldsymbol{w} = q - \mathrm{div}\boldsymbol{v} + \mathrm{div}\boldsymbol{v} = q$. Moreover

$$\|\boldsymbol{w}\|_X\leq(C+bC+a)\|q\|_{L^2(\Omega)}$$

where $b = |\Omega|^{1/2}\|\mathrm{div}\boldsymbol{v}_0\|_{L^2(\Omega)}$.

[11] Let $\boldsymbol{\sigma} = -\nu\nabla\boldsymbol{v} + \pi I_d$ where I_d is the identity matrix in $\mathbb{R}^d$. The Stokes equation implies that

$$\boldsymbol{\sigma}\in H_{\mathrm{div}}(\Omega)=\{\mathbf{u}\in L^2(\Omega)^d\ :\ \mathrm{div}\mathbf{u}\in L^2(\Omega)\}$$

and therefore (see [176]) $\boldsymbol{\sigma}\cdot\mathbf{n}|_{\Gamma_N}$ is well defined as an element of $H_{00}^{-1/2}(\Gamma_N)^d$, the dual of the Hilbert space $H_{00}^{1/2}(\Gamma_N)^d$ where $H_{00}^{1/2}(\Gamma_N) = \{g \in L^2(\Gamma_N) : \exists u \in H^1_{\Gamma_D}(\Omega), u_{|\Gamma_N} = g\}$, see Sect. A.5.10. Using formula (A.51) one can show that the Neumann condition holds in the following sense

$$\langle\boldsymbol{\sigma}\cdot\mathbf{n}|_{\Gamma_N},\boldsymbol{\eta}\rangle_{\Gamma_N}=(\boldsymbol{h},\boldsymbol{\eta})_{L^2(\Gamma_N)^d}\qquad\forall\boldsymbol{\eta}\in H_{00}^{1/2}(\Gamma_N)^d$$

where $\langle\cdot,\cdot\rangle_{\Gamma_N}$ denotes the duality pairing between $H_{00}^{-1/2}(\Gamma_N)^d$ and $H_{00}^{1/2}(\Gamma_N)^d$.

In this case we denote by $B : \mathcal{U} \to X^* \times Q^*$ the control operator

$$B(\boldsymbol{u}) = (\boldsymbol{u}, 0),$$

and by $C : X \times Q \to \mathcal{Z} = X \times Q$ the observation operator

$$C(\boldsymbol{v}, \pi) = (\boldsymbol{v}, 0).$$

Remark 5.22. The image of the operator C is isometrically isomorphic (see Sect. A.1.3) to X, thus we can identify $\mathcal{Z}$ with X in this case. •

Moreover, we can write the cost functional (5.190) under the form (5.56) by setting

$$q(\boldsymbol{u}, \boldsymbol{\eta}) = (\boldsymbol{v}_0(\boldsymbol{u}), \boldsymbol{v}_0(\boldsymbol{\eta}))_{L^2(\Omega)^d} + \tau (\boldsymbol{u}, \boldsymbol{\eta})_{\mathcal{U}}, \tag{5.196}$$

$$L\boldsymbol{u} = (\boldsymbol{v}_0(\boldsymbol{u}), \boldsymbol{v}(\mathbf{0}) - \boldsymbol{z}_d)_{L^2(\Omega)^d}, \qquad c = \frac{1}{2} \|\boldsymbol{v}(\mathbf{0}) - \boldsymbol{z}_d\|^2_{L^2(\Omega)^d}.$$

q is continuous and coercive in $\mathcal{U}$ (as $\tau > 0$) and L is continuous in $\mathcal{U}$. Hence, there exists a unique solution of problem (5.190)–(5.191).

• *Optimality conditions.* As in the cases previously analyzed, we can recover the system of first-order (necessary and sufficient) optimality conditions in weak form, introducing the *Lagrangian* $\mathcal{L} = \mathcal{L}((\boldsymbol{v}, \pi), \boldsymbol{u}, (\boldsymbol{z}, q))$ given by

$$\begin{aligned} \mathcal{L}((\boldsymbol{v}, \pi), \boldsymbol{u}, (\boldsymbol{z}, q)) = & \frac{1}{2} \|\boldsymbol{v} - \boldsymbol{z}_d\|^2_{L^2(\Omega)^d} + \frac{\tau}{2} \|\boldsymbol{u}\|^2_{L^2(\Omega)^d} \\ & - e(\boldsymbol{v}, \boldsymbol{z}) - b(\boldsymbol{z}, \pi) - b(\boldsymbol{v}, q) + (\boldsymbol{f} + \boldsymbol{u}, \boldsymbol{z})_{L^2(\Omega)^d} + (\boldsymbol{h}, \boldsymbol{z})_{L^2(\Gamma_N)^d} \end{aligned} \tag{5.197}$$

where $(\boldsymbol{z}, q) \in X \times Q$ are the adjoint variables (or Lagrange multipliers related to the constraints represented by the equations in the state system).

By taking the derivatives of the Lagrangian with respect to the state variables and setting them equal to zero, we obtain

$$\begin{aligned} & \mathcal{L}'_{\boldsymbol{v}}((\hat{\boldsymbol{v}}, \hat{\pi}), \hat{\boldsymbol{u}}, (\hat{\boldsymbol{z}}, \hat{q}))\, \boldsymbol{\varphi} = (\hat{\boldsymbol{v}} - \boldsymbol{z}_d, \boldsymbol{\varphi})_{L^2(\Omega)^d} - e(\boldsymbol{\varphi}, \hat{\boldsymbol{z}}) - b(\boldsymbol{\varphi}, \hat{q}) = 0 \quad \forall \boldsymbol{\varphi} \in X, \\ & \mathcal{L}'_{\pi}((\hat{\boldsymbol{v}}, \hat{\pi}), \hat{\boldsymbol{u}}, (\hat{\boldsymbol{z}}, \hat{q}))\, \phi = -b(\hat{\boldsymbol{z}}, \phi) = 0 \qquad \forall \phi \in Q. \end{aligned} \tag{5.198}$$

Equations (5.198) yields the weak formulation of the adjoint problem: this latter is again a Stokes problem, with homogeneous Dirichlet/Neumann conditions, and source term (for the momentum equation) equal to $\hat{\boldsymbol{v}} - \boldsymbol{z}_d$:

$$\begin{cases} -\nu \Delta \hat{\boldsymbol{z}} + \nabla \hat{q} = \hat{\boldsymbol{v}} - \boldsymbol{z}_d & \text{in } \Omega \\ \operatorname{div} \hat{\boldsymbol{z}} = 0 & \text{in } \Omega \\ \hat{\boldsymbol{z}} = \mathbf{0} & \text{on } \Gamma_D \\ -\hat{q}\mathbf{n} + \nu \dfrac{\partial \hat{\boldsymbol{z}}}{\partial n} = \mathbf{0} & \text{on } \Gamma_N. \end{cases}$$

By taking the derivative of the Lagrangian with respect to the control variable, we obtain the Euler equation

$$\mathcal{L}'_{\boldsymbol{u}}((\hat{\boldsymbol{v}}, \hat{\pi}), \hat{\boldsymbol{u}}, (\hat{\boldsymbol{z}}, \hat{q}))\, \boldsymbol{\eta} = (\hat{\boldsymbol{z}} + \tau \hat{\boldsymbol{u}}, \boldsymbol{\eta})_{L^2(\Omega)^d} = 0 \qquad \forall \boldsymbol{\eta} \in L^2(\Omega)^d. \tag{5.199}$$

Finally, from (5.199) we obtain

$$J'(\hat{\boldsymbol{u}})\, \boldsymbol{\eta} = \int_{\Omega} (\hat{\boldsymbol{z}} + \tau \hat{\boldsymbol{u}}) \cdot \boldsymbol{\eta}\, d\mathbf{x} = 0 \qquad \forall \boldsymbol{\eta} \in L^2(\Omega)^d \tag{5.200}$$

thus yielding the expression of the gradient of the cost functional

$$\nabla J(\hat{\boldsymbol{u}}) = \hat{\boldsymbol{z}} + \tau \hat{\mathbf{u}}.$$

Remark 5.23. The case of a velocity profile $\boldsymbol{g} \neq \mathbf{0}$ on Γ_D requires an extension of the procedure described in this section. If $\boldsymbol{g} \in H_{00}^{1/2}(\Gamma_D)$ with $\int_{\Gamma_D} \boldsymbol{g} \cdot \mathbf{n} d\sigma = 0$, there exists a function $\boldsymbol{r}_g \in H^1_{\Gamma_N}(\Omega)^d$, called lifting of $\boldsymbol{g}$, such that $\boldsymbol{r}_g|_{\Gamma_D} = \boldsymbol{g}$ and $\text{div}\boldsymbol{r}_g = 0$ (see, e.g., [105, Chap. 1, Lemma 2.2]). The function $\tilde{\boldsymbol{v}} = \boldsymbol{v} - \boldsymbol{r}_g$ satisfies (5.191) with

$$\tilde{\boldsymbol{v}}|_{\Gamma_D} = \mathbf{0} \quad \text{and} -\pi\mathbf{n} + \nu\frac{\partial \tilde{\boldsymbol{v}}}{\partial n} = \boldsymbol{h} - \nu\frac{\partial \tilde{\boldsymbol{r}}_g}{\partial n} \in H_{00}^{-1/2}(\Gamma_N)^d.$$

The analysis of the state problem, as well as the derivation of the optimality conditions, can be easily adapted to the case where the velocity field is $\tilde{\boldsymbol{v}}$. •

Remark 5.24. An alternative treatment of optimal control problems for Stokes equations with respect to Lagrange multiplier techniques can be found in [126], where a *penalty/least-squares* method using least-squares variational principles to treat the PDE constraints is used. •

5.13.2 Boundary Velocity Control, Vorticity Minimization

As in the example introduced in Sect. 1.6, we consider the problem of reducing the vorticity in the proximity of a body immersed in a fluid through a velocity control operated on a portion of its boundary. Using the same notation of Sect. 1.6, we denote by $\Omega \subset \mathbb{R}^2$, the region occupied by the fluid, and by Γ_{in}, Γ_{out} and Γ_w the inflow, outflow and wall boundary portions, respectively, while $\Gamma = \partial\Omega$. The boundary of the body is denoted by Γ_B, whereas $\Gamma_c \subset \Gamma_B$ is the control boundary (see Fig. 5.1). The goal is to minimize the vorticity (and thus the drag, as a result) by controlling the velocity $\boldsymbol{u}$ on the boundary $\Gamma_c \subset \Gamma_B$. Thus, we seek the solution of the minimization problem[12]

$$J(\boldsymbol{u}) = \frac{1}{2}\int_\Omega |\nabla \times \boldsymbol{v}(\boldsymbol{u})|^2 \, d\mathbf{x} + \frac{\tau}{2}\int_{\Gamma_c} |\nabla \boldsymbol{u}|^2 \, d\sigma \to \min \tag{5.201}$$

where $\tau > 0$ and $(\boldsymbol{v}, \pi) = (\boldsymbol{v}(\boldsymbol{u}), \pi(\boldsymbol{u}))$ is the solution of the following steady Stokes system:

$$\begin{cases} -\nu\Delta\boldsymbol{v} + \nabla\pi = \boldsymbol{f} & \text{in } \Omega \\ \text{div}\boldsymbol{v} = 0 & \text{in } \Omega \\ \boldsymbol{v} = \boldsymbol{v}_\infty & \text{on } \Gamma_{in} \\ \boldsymbol{v} = \boldsymbol{u} & \text{on } \Gamma_c \\ \boldsymbol{v} = \mathbf{0} & \text{on } \Gamma_w \cup (\Gamma_B \setminus \Gamma_c) \\ -\pi\mathbf{n} + \nu\dfrac{\partial \boldsymbol{v}}{\partial n} = \mathbf{0} & \text{on } \Gamma_{out}. \end{cases} \tag{5.202}$$

We impose no-slip conditions[13] on both Γ_w and $\Gamma_B \setminus \Gamma_c$, no-stress conditions at the outflow Γ_{out} and Dirichlet conditions on the control boundary Γ_c. This is a Dirichlet boundary control problem, similarly to those addressed in Sect. 5.11. We consider suction/injection of fluid through the control boundary Γ_c only in the normal direction, imposing a no-slip condition in the tangential one.

---

12 Recall that in two dimensions, $\nabla \times \boldsymbol{v} = \frac{\partial v_2}{\partial x_1} - \frac{\partial v_1}{\partial x_2}$ is a scalar, while, for a scalar φ, its curl is the vector $\nabla \times \varphi = (\frac{\partial \varphi}{\partial x_2}, -\frac{\partial \varphi}{\partial x_1})$.

13 A better condition to impose on the non-physical boundary Γ_w would be a *symmetry condition*, prescribing the velocity to be *(i)* parallel to the boundary, $\boldsymbol{v} \cdot \mathbf{n} = 0$, and such that the tangential stress is vanishing, $(-\pi\mathbf{n} + \nu\frac{\partial \boldsymbol{v}}{\partial n}) \cdot \boldsymbol{t} = 0$. This introduces only slight differences in our weak formulation and functional setting.

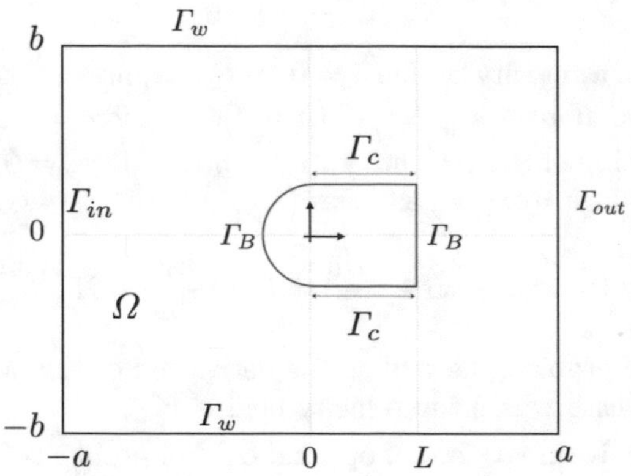

Fig. 5.1 Domain for the boundary velocity control problem

As in Sect. 5.11, the natural choice for the control space would be $\mathcal{U} = H^{1/2}(\Gamma_c)^2$. As a consequence we would need to penalize the vorticity by adding the $H^{1/2}(\Gamma_c)^2$-norm of the control variable in the cost functional. However, as seen in Sect. 5.11.1, a fractional inner product would arise from the optimality condition in the realization of the $H^{1/2}(\Gamma_c)$-inner product. See, e.g., [159] for further details. Also, the choice $\mathcal{U} = L^2(\Gamma_c)^2$ and the use of the transposition method is not appropriate given the low regularity of the solution of a mixed problem, even in presence of more regularity of the domain. Thus the best choice here seems to be $\mathcal{U} = H_0^1(\Gamma_c)^2$, requiring further regularity on the controls (see e.g. [128]).

Moreover we handle the Dirichlet boundary conditions (involving the control function) by a Lagrange multiplier approach, as illustrated in Sect. 5.11; see also [125]. Note that a similar approach to treat non-homogeneous Dirichlet boundary conditions could have been exploited also to formulate problem (5.191), although its boundary data do not involve the control function. We have however opted for a lifting-based approach in that case since this is the most widely used approach in the numerical implementation. For ease of notation, hereon we denote by $\mathcal{T}_1 = H_{00}^{1/2}(\Gamma_1)^2$ and by $\langle \cdot, \cdot \rangle_{\Gamma_1}$ the duality pairing between $\mathcal{T}_1$ and its dual $\mathcal{T}_1^*$.

• *Analysis of the state problem.* First of all, we look for the weak formulation of the state problem. Let e and b be the bilinear forms defined in (5.192). Define

$$\boldsymbol{g}(\boldsymbol{u}) = \begin{cases} \boldsymbol{u} & \text{on } \Gamma_c \\ \boldsymbol{0} & \text{on } \Gamma_w \cup (\Gamma_B \setminus \Gamma_c) \\ \boldsymbol{v}_\infty & \text{on } \Gamma_{in}. \end{cases}$$

Let $\Gamma_1 = \Gamma_c \cup \Gamma_{in}$, $\Gamma_0 = \Gamma_w \cup (\Gamma_B \setminus \Gamma_c)$. Assuming for simplicity that $\boldsymbol{v}_\infty \in H_0^1\left(\Gamma_{in}\right)^2$ we have

$$\boldsymbol{g}(\boldsymbol{u}) \in (H_0^1(\Gamma_1))^2 \subset \mathcal{T}_1$$

with continuous injection. Finally, we introduce the state space

$$X = H^1_{\Gamma_0}(\Omega)^2,$$

endowed with the norm $\|\boldsymbol{v}\|_X = \|\nabla \boldsymbol{v}\|_{L^2}$. Observe that if $\boldsymbol{v} \in X$, then $\boldsymbol{v}|_{\Gamma_1} \in \mathcal{T}_1$.

Our weak formulation of (5.202) reads as follows: *find* $\boldsymbol{v} \in X$, $\pi \in L^2(\Omega)$, $\boldsymbol{t} \in \mathcal{T}_1^*$ *such that*

$$\begin{cases} e(\boldsymbol{v}, \boldsymbol{\varphi}) + b(\boldsymbol{\varphi}, \pi) + \langle \boldsymbol{t}, \boldsymbol{\varphi} \rangle_{\Gamma_1} = (\boldsymbol{f}, \boldsymbol{\varphi})_{L^2(\Omega)^2} & \forall \boldsymbol{\varphi} \in X \\ b(\boldsymbol{v}, \phi) = 0 & \forall \phi \in L^2(\Omega) \\ \langle \boldsymbol{\lambda}, \boldsymbol{v} \rangle_{\Gamma_1} = \langle \boldsymbol{\lambda}, \boldsymbol{g}(\boldsymbol{u}) \rangle_{\Gamma_1} & \forall \boldsymbol{\lambda} \in \mathcal{T}_1^*. \end{cases} \tag{5.203}$$

Note that a formal integration by parts in the first equation gives

$$\pi \mathbf{n} - \nu \frac{\partial \boldsymbol{v}}{\partial n} = \boldsymbol{t} \qquad \text{on } \Gamma_1$$

and

$$\pi \mathbf{n} - \nu \frac{\partial \boldsymbol{v}}{\partial n} = \mathbf{0} \qquad \text{on } \Gamma_{out}.$$

For convenience, we group together the last two equations in (5.203) and obtain the following equivalent formulation: *find* $\boldsymbol{v} \in X$, $\pi \in L^2(\Omega)$, $\boldsymbol{t} \in \mathcal{T}_1^*$ *such that*

$$\begin{cases} e(\boldsymbol{v}, \boldsymbol{w}) + b(\boldsymbol{w}, \pi) + \langle \boldsymbol{t}, \boldsymbol{w} \rangle_{\Gamma_1} = (\boldsymbol{f}, \boldsymbol{w})_{L^2(\Omega)^2} & \forall \boldsymbol{w} \in X \\ b(\boldsymbol{v}, \phi) + \langle \boldsymbol{\lambda}, \boldsymbol{v} \rangle_{\Gamma_1} = \langle \boldsymbol{\lambda}, \boldsymbol{g}(\boldsymbol{u}) \rangle_{\Gamma_1} & \forall (\phi, \boldsymbol{\lambda}) \in Q \end{cases} \tag{5.204}$$

where $Q = L^2(\Omega) \times \mathcal{T}_1^*$.

Upon defining the bilinear form $\tilde{b} : X \times Q \to \mathbb{R}$ as

$$\tilde{b}(\boldsymbol{w}, (\phi, \boldsymbol{\lambda})) = b(\boldsymbol{w}, \phi) + \langle \boldsymbol{\lambda}, \boldsymbol{w} \rangle_{\Gamma_1} \tag{5.205}$$

and the norm

$$\|(\phi, \boldsymbol{\lambda})\|_Q = \sqrt{\|\phi\|^2_{L^2(\Omega)} + \|\boldsymbol{\lambda}\|^2_{\mathcal{T}_1^*}},$$

we rewrite problem (5.204) in the saddle-point form : *find* $(\boldsymbol{v}, (\pi, \boldsymbol{t})) \in X \times Q$ *such that*

$$\begin{cases} e(\boldsymbol{v}, \boldsymbol{\varphi}) + \tilde{b}(\boldsymbol{\varphi}, (\pi, \boldsymbol{t})) = F_1 \boldsymbol{\varphi} & \forall \boldsymbol{\varphi} \in X \\ \tilde{b}(\boldsymbol{v}, (\phi, \boldsymbol{\lambda})) = F_2(\phi, \boldsymbol{\lambda}) & \forall (\phi, \boldsymbol{\lambda}) \in Q, \end{cases} \tag{5.206}$$

with $F_1 \in X^*$ *and* $F_2 \in Q^*$ *given by*

$$F_1 \boldsymbol{w} = (\boldsymbol{f}, \boldsymbol{w})_{L^2(\Omega)^2}, \qquad F_2(\phi, \boldsymbol{\lambda}) = \langle \boldsymbol{\lambda}, \boldsymbol{g}(\boldsymbol{u}) \rangle_{\Gamma_1}.$$

The well-posedness of problem (5.206) follows from Theorem A.7. As in the case discussed in the previous section, the inf-sup condition on $\tilde{b}(\cdot, \cdot)$ is the hardest assumption to verify; we prove that

Lemma 5.2. *There exists a constant* $\tilde{\beta} > 0$ *such that*

$$\inf_{0 \neq (\phi, \boldsymbol{\lambda}) \in Q} \sup_{0 \neq \boldsymbol{v} \in X} \frac{\tilde{b}(\boldsymbol{v}, (\phi, \boldsymbol{\lambda}))}{\|\boldsymbol{v}\|_X \|(\phi, \boldsymbol{\lambda})\|_Q} \geq \tilde{\beta}. \tag{5.207}$$

Proof. Given $(\tilde{\phi}, \tilde{\boldsymbol{\lambda}}) \in Q$, according to Riesz Theorem there exists a unique $\boldsymbol{\lambda}^* \in \mathcal{T}_1$ such that

$$\langle \tilde{\boldsymbol{\lambda}}, \boldsymbol{\theta} \rangle_{\Gamma_1} = (\boldsymbol{\lambda}^*, \boldsymbol{\theta})_{H^{1/2}_{00}(\Gamma_1)} \quad \forall \boldsymbol{\theta} \in \mathcal{T}_1,$$

with $\|\boldsymbol{\lambda}^*\|_{\mathcal{T}_1} = \|\tilde{\boldsymbol{\lambda}}\|_{\mathcal{T}_1^*}$. Then we choose $\tilde{\boldsymbol{v}} \in X$ such that

$$b(\tilde{\boldsymbol{v}}, \phi) = (\tilde{\phi}, \phi)_{L^2(\Omega)} \qquad \forall \phi \in L^2(\Omega)$$
$$\tilde{\boldsymbol{v}} = -\boldsymbol{\lambda}^* \qquad \text{on } \Gamma_1$$
$$\|\tilde{\boldsymbol{v}}\|_X \leq \tilde{C}(\|\tilde{\phi}\|_{L^2(\Omega)} + \|\boldsymbol{\lambda}^*\|_{\mathcal{T}_1}. \tag{5.208}$$

Such $\tilde{\boldsymbol{v}}$ can be obtained as follows. Let $\boldsymbol{r} \in X$ be such that its trace on $\partial\Omega$ is equal to

$$\boldsymbol{h} = \begin{cases} -\boldsymbol{\lambda}^* & \text{on } \Gamma_1 \\ \mathbf{0} & \text{on } \Gamma_0 \\ \boldsymbol{g} & \text{on } \Gamma_{out} \end{cases}$$

where

$$\boldsymbol{g} = \left(-\int_\Omega \tilde{\phi}\, d\mathbf{x} + \int_{\Gamma_1} \boldsymbol{\lambda}^* \cdot \mathbf{n}\, d\sigma \right) \boldsymbol{g}_0 \tag{5.209}$$

and $\int_{\Gamma_{out}} \boldsymbol{g}_0 \cdot \mathbf{n} = 1$. Let us now set $\tilde{\boldsymbol{v}} = \boldsymbol{w} + \boldsymbol{r}$, with $\boldsymbol{w} \in H_0^1(\Omega)$ to be determined in order to have

$$\mathrm{div}\tilde{\boldsymbol{v}} = \mathrm{div}\boldsymbol{w} + \mathrm{div}\boldsymbol{r} = \tilde{\phi},$$

that is, $\mathrm{div}\boldsymbol{w} = \tilde{\phi} - \mathrm{div}\boldsymbol{r}$. We have, from (5.209),

$$\int_\Omega \tilde{\phi}\, d\mathbf{x} - \int_\Omega \mathrm{div}\boldsymbol{r}\, d\mathbf{x} = \int_\Omega \tilde{\phi}\, d\mathbf{x} + \int_{\Gamma_1} \boldsymbol{\lambda}^* \cdot \mathbf{n} -\, d\sigma \int_{\Gamma_{out}} \boldsymbol{g} \cdot \mathbf{n}\, d\sigma = 0.$$

Hence, there exists $\boldsymbol{w} \in H_0^1(\Omega)$ with $\mathrm{div}\boldsymbol{w} = \tilde{\phi} - \mathrm{div}\boldsymbol{r}$ so that $\mathrm{div}\tilde{\boldsymbol{v}} = \tilde{\phi}$ as required.
Hence, using (5.208)

$$\begin{aligned}\tilde{b}(\tilde{\boldsymbol{v}}, (\tilde{\phi}, \tilde{\boldsymbol{\lambda}}))) &= b(\tilde{\boldsymbol{v}}, \tilde{\phi}) + \langle \tilde{\boldsymbol{\lambda}}, \tilde{\boldsymbol{v}} \rangle_{H^{-1/2}, H^{1/2}} = \|\tilde{\phi}\|^2_{L^2(\Omega)} + \|\boldsymbol{\lambda}^*\|^2_{\mathcal{T}_1} \\ &\geq \frac{1}{2}(\|\tilde{\phi}\|_{L^2(\Omega)} + \|\boldsymbol{\lambda}^*\|_{\mathcal{T}_1})^2 \geq \frac{1}{2\tilde{C}}(\|\tilde{\phi}\|_{L^2(\Omega)} + \|\boldsymbol{\lambda}^*\|_{\mathcal{T}_1})\|\tilde{\boldsymbol{v}}\|_X \\ &\geq \frac{1}{2\tilde{C}}(\|\tilde{\phi}\|^2_{L^2(\Omega)} + \|\tilde{\boldsymbol{\lambda}}\|^2_{\mathcal{T}_1^*})^{1/2}\|\tilde{\boldsymbol{v}}\|_X = \frac{1}{2\tilde{C}}\|(\tilde{\phi}, \tilde{\boldsymbol{\lambda}})\|_Q \|\tilde{\boldsymbol{v}}\|_X.\end{aligned}$$

Thus (5.207) follows with $\tilde{\beta} = \frac{1}{2\tilde{C}}$, by taking the supremum over $\tilde{\boldsymbol{v}}$ and the infimum over $(\tilde{\phi}, \tilde{\boldsymbol{\lambda}})$. □

Hence, by applying Theorem A.7 (see Exercise 11 for the proof of the other assumptions), we have that for any $\boldsymbol{u} \in \mathcal{U}$ the state problem (5.206) has a unique weak solution. Moreover,

$$\|\boldsymbol{v}\|_X + \|\pi\|_{L^2(\Omega)} + \|\boldsymbol{t}\|_{\mathcal{T}_1^*} \leq C\left(\|\boldsymbol{f}\|_{L^2(\Omega)} + \|\boldsymbol{g}(\boldsymbol{u})\|_{\mathcal{T}_1}\right) \tag{5.210}$$

where C depends only on ν, $\tilde{\beta}$.

• *Existence and uniqueness of an optimal control.* First note that, if $\boldsymbol{v}_0(\boldsymbol{u}) = \boldsymbol{v}(\boldsymbol{u}) - \boldsymbol{v}(\mathbf{0})$, $\pi_0(\boldsymbol{u}) = \pi(\boldsymbol{u}) - \pi(\mathbf{0})$ and $\boldsymbol{t}_0(\boldsymbol{u}) = \boldsymbol{t}(\boldsymbol{u}) - \boldsymbol{t}(\mathbf{0})$, we have from (5.210), for any $\boldsymbol{u} \in \mathcal{U}$,

$$\|\boldsymbol{v}_0(\boldsymbol{u})\|_X + \|\pi_0(\boldsymbol{u})\|_{L^2(\Omega)} + \|\boldsymbol{t}_0\|_{\mathcal{T}_1^*} \leq C\|\boldsymbol{u}\|_{\mathcal{U}}$$

since $\|\boldsymbol{g}(\boldsymbol{u})\|_{\mathcal{T}_1} \leq C\|\mathbf{u}\|_{H_0^1(\Gamma_c)^2}$. Thus the solution of the OCP (5.201)–(5.202) exists and is unique since $\tau > 0$.

• *Optimality conditions.* To characterize the system of optimality conditions, we introduce the Lagrangian functional

$$\begin{aligned}\mathcal{L}\left((\boldsymbol{v}, \pi, \boldsymbol{t}), \boldsymbol{u}, (\boldsymbol{z}, q, \boldsymbol{r})\right) = \tilde{J}\left(\boldsymbol{v}, \boldsymbol{u}\right) + (\boldsymbol{f}, \boldsymbol{z})_{L^2(\Omega)} - e\left(\boldsymbol{v}, \boldsymbol{z}\right) \\ -\tilde{b}(\boldsymbol{z}, (\boldsymbol{t}, \pi)) + \langle \boldsymbol{r}, \boldsymbol{g}(\boldsymbol{u}) \rangle_{\Gamma_1} - \tilde{b}(\boldsymbol{v}, (q, \boldsymbol{r}))\end{aligned} \tag{5.211}$$

where $(\boldsymbol{z}, q, \boldsymbol{r}) \in X \times L^2(\Omega) \times \mathcal{T}_1^*$ is the adjoint variable (or Lagrange multiplier related to the constraint represented by the state equations). To highlight the dependence on $\boldsymbol{u}$, we rewrite the Lagrangian as

$$\mathcal{L}\left((\boldsymbol{v},\pi,\boldsymbol{t}),\boldsymbol{u},(\boldsymbol{z},q,\boldsymbol{r})\right)=\frac{1}{2}\int_{\Omega}|\nabla\times\boldsymbol{v}|^2\,d\mathbf{x}+\frac{\tau}{2}\|\boldsymbol{u}\|^2_{H^1_0(\Gamma_c)^2}$$
$$+(\boldsymbol{f},\boldsymbol{z})_{L^2(\Omega)^2}-e\left(\boldsymbol{v},\boldsymbol{z}\right)-b(\boldsymbol{z},\pi)-\langle\boldsymbol{t},\boldsymbol{z}\rangle_{\Gamma_1}$$
$$+\langle\boldsymbol{r},\boldsymbol{u}\rangle_{\Gamma_c}+\langle\boldsymbol{r},\boldsymbol{v}_\infty\rangle_{\Gamma_{in}}-\langle\boldsymbol{r},\boldsymbol{v}\rangle_{\Gamma_1}-b(\boldsymbol{v},q).$$

By differentiating the Lagrangian with respect to the adjoint variables and setting its derivatives to zero, we obtain (with $\hat{\mathcal{L}}'_{(\cdot)}=\mathcal{L}'_{(\cdot)}((\hat{\boldsymbol{v}},\hat{\pi},\hat{\boldsymbol{t}}),\hat{\boldsymbol{u}},(\hat{\boldsymbol{z}},\hat{q},\hat{\boldsymbol{r}}))$)

$$\hat{\mathcal{L}}'_{\boldsymbol{z}}\boldsymbol{\psi}=(\boldsymbol{f},\boldsymbol{\psi})_{L^2(\Omega)}-e\left(\hat{\boldsymbol{v}},\boldsymbol{\psi}\right)-b(\boldsymbol{\psi},\hat{\pi})-\langle\hat{\boldsymbol{t}},\boldsymbol{\psi}\rangle_{\Gamma_1}=0 \qquad \forall\boldsymbol{\psi}\in X$$
$$\hat{\mathcal{L}}'_{(q,\boldsymbol{r})}(\xi,\boldsymbol{\kappa})=\langle\boldsymbol{\kappa},\hat{\boldsymbol{u}}\rangle_{\Gamma_c}+\langle\boldsymbol{\kappa},\boldsymbol{v}_\infty\rangle_{\Gamma_{in}}-\langle\boldsymbol{\kappa},\hat{\boldsymbol{v}}\rangle_{\Gamma_1}-b(\hat{\boldsymbol{v}},\xi)=0 \qquad \forall(\xi,\boldsymbol{\kappa})\in Q$$

which is nothing but the weak form (5.204) of the state problem. Differentiating now the Lagrangian with respect to the state variables and setting its derivatives to zero, we obtain

$$\begin{aligned}&\hat{\mathcal{L}}'_{\boldsymbol{v}}\boldsymbol{\varphi}=\int_{\Omega}(\nabla\times\hat{\boldsymbol{v}})\cdot(\nabla\times\boldsymbol{\varphi})d\mathbf{x}-e(\boldsymbol{\varphi},\hat{\boldsymbol{z}})-\langle\boldsymbol{r},\boldsymbol{\varphi}\rangle_{\Gamma_1}-b(\boldsymbol{\varphi},\hat{q})=0 \quad \forall\boldsymbol{\varphi}\in X\\&\hat{\mathcal{L}}'_{(\pi,\boldsymbol{t})}(\phi,\boldsymbol{\lambda})=\langle\boldsymbol{\lambda},\hat{\boldsymbol{z}}\rangle_{\Gamma_1}+b(\hat{\boldsymbol{z}},\phi)=0 \qquad \forall(\phi,\boldsymbol{\lambda})\in Q.\end{aligned} \tag{5.212}$$

Finally, differentiating the Lagrangian with respect to the control variable, we obtain the following Euler equation

$$\hat{\mathcal{L}}'_{\boldsymbol{u}}\boldsymbol{\eta}=-\langle\hat{\boldsymbol{r}},\boldsymbol{\eta}\rangle_{\Gamma_c}+\tau(\hat{\boldsymbol{u}},\boldsymbol{\eta})_{H^1_0(\Gamma_c)^2}=0 \qquad \forall\boldsymbol{\eta}\in\mathcal{U}. \tag{5.213}$$

Equations (5.212) represent the weak formulation of the following adjoint problem, which reads (see Exercise 12):

$$\begin{cases}-\nu\Delta\hat{\boldsymbol{z}}+\nabla\hat{q}=\nabla\times(\nabla\times\hat{\boldsymbol{v}}) & \text{in }\Omega\\ \text{div}\hat{\boldsymbol{z}}=0 & \text{in }\Omega\\ \hat{\boldsymbol{z}}=\mathbf{0} & \text{on }\Gamma_1\cup\Gamma_0\\ -\hat{q}\mathbf{n}+\nu\dfrac{\partial\hat{\boldsymbol{z}}}{\partial n}-(\nabla\times\hat{\boldsymbol{v}})\mathbf{n}^\perp=\mathbf{0} & \text{on }\Gamma_{out},\end{cases} \tag{5.214}$$

with

$$\hat{\boldsymbol{r}}=\hat{q}\mathbf{n}-\nu\frac{\partial\hat{\boldsymbol{z}}}{\partial n}+(\nabla\times\hat{\boldsymbol{v}})\mathbf{n}^\perp \qquad \text{on }\Gamma_1,$$

where $\mathbf{n}^\perp=(-n_2,n_1)$ is a tangential vector to the boundary. Recall that $\Gamma_1=\Gamma_c\cup\Gamma_{in}$ and $\Gamma_0=\Gamma_w\cup(\Gamma_B\setminus\Gamma_c)$.

The Euler equation (5.213) constitutes a differential equation for $\hat{\boldsymbol{u}}$. From (5.213) we can deduce that the derivative of the reduced functional $J(\boldsymbol{u})=\tilde{J}(\boldsymbol{v}(\boldsymbol{u}),\boldsymbol{u})$ at $\boldsymbol{u}=\hat{\boldsymbol{u}}\in\mathcal{U}_{ad}$ is given by

$$J'(\hat{\boldsymbol{u}})\boldsymbol{\eta}=(\tau\hat{\boldsymbol{u}},\boldsymbol{\eta})_{H^1_0(\Gamma_c)^d}+\langle\hat{\boldsymbol{r}},\boldsymbol{\eta}\rangle_{\Gamma_c} \qquad \forall\boldsymbol{\eta}\in\mathcal{U}_{ad}.$$

Then, denoting by $\mathcal{T}_c=H^{1/2}_{00}(\Gamma_c)^d$, and by $\langle\cdot,\cdot\rangle_{\Gamma_c}$ the duality pairing between $\mathcal{T}_c$ and its dual $\mathcal{T}_c^*$, we can introduce the embedding operator $B:\mathcal{T}_c^*\to H^{-1}(\Gamma_c)^d$ defined by

$$\langle B\hat{\boldsymbol{r}},\boldsymbol{\eta}\rangle_{H^{-1}(\Gamma_c)^d,H^1_0(\Gamma_c)^d}=\langle\hat{\boldsymbol{r}},\boldsymbol{\eta}\rangle_{\Gamma_c} \quad \forall\boldsymbol{\eta}\in H^1_0(\Gamma_c)^d.$$

In this way, we find that

$$\nabla J(\hat{\boldsymbol{u}}) = \tau\hat{\boldsymbol{u}} + \hat{\boldsymbol{w}}$$

where $\hat{\boldsymbol{w}} \in H_0^1(\Gamma_c)^d$ is the unique weak solution to

$$-\Delta\hat{\boldsymbol{w}} = B\hat{\boldsymbol{r}} \quad \text{in } \Gamma_c. \tag{5.215}$$

Example 5.1. Assuming that $\Gamma_c = \{(x_1, k) \, : \, 0 < x_1 < L\}$ – see Fig. 5.1 – and that the control is under the form $\boldsymbol{u} = (0, u)$, $u \in H_0^1(0, L)$, (5.213) becomes: find $\hat{u} \in H_0^1(0, L)$ such that

$$\tau \int_0^L \hat{u}'\eta' dx_1 = \langle \hat{r}_2, \eta \rangle_{\Gamma_c} \qquad \forall \eta \in H_0^1(0, L)$$

which is the weak formulation of

$$\begin{cases} -\tau\hat{u}'' = \hat{r}_2 & \text{in } (0, L) \\ \hat{u}(0) = \hat{u}(L) = 0. \end{cases}$$

provided r_2 is sufficiently smooth. •

Remark 5.25. Note that $-\hat{q}\mathbf{n} + \nu\frac{\partial\hat{\boldsymbol{z}}}{\partial n} - (\nabla \times \hat{\boldsymbol{v}})\mathbf{n}^{\perp}$ is well defined in $\mathcal{T}_1^*$. Indeed, set $\hat{\boldsymbol{w}} = \nabla\hat{\boldsymbol{v}} - (\nabla\hat{\boldsymbol{v}})^{\top}$ and observe that, since $\operatorname{div}\hat{\boldsymbol{v}} = 0$, $\nabla \times \nabla \times \hat{\boldsymbol{v}} = -\Delta\hat{\boldsymbol{v}} = -\operatorname{div}\hat{\boldsymbol{w}}$. Then the first equation in (5.214) can be written in the form

$$\operatorname{div}(-\nu\nabla\hat{\boldsymbol{z}} + \hat{q}I + \hat{\boldsymbol{w}}) = \mathbf{0}$$

and therefore

$$\nu\frac{\partial\hat{\boldsymbol{z}}}{\partial n} - \hat{q}\mathbf{n} - \hat{\boldsymbol{w}}\mathbf{n} = \nu\frac{\partial\hat{\boldsymbol{z}}}{\partial n} - \hat{q}\mathbf{n} - (\nabla \times \hat{\boldsymbol{v}})\mathbf{n}^{\perp} \in \mathcal{T}_1^*.$$ •

Remark 5.26. (*The Stokes solver*). We have seen that the weak solution $(\boldsymbol{v}, \pi, \boldsymbol{t}) \in V = X \times L^2(\Omega) \times \mathcal{T}_1^*$ of the Stokes problem

$$\begin{cases} e(\boldsymbol{v}, \boldsymbol{\varphi}) + b(\boldsymbol{\varphi}, \pi) + \langle \mathbf{t}, \boldsymbol{\varphi} \rangle_{\Gamma_1} = \langle \mathbf{F}, \boldsymbol{\varphi} \rangle_{X^*,X} & \forall \boldsymbol{\varphi} \in X \\ b(\boldsymbol{v}, \phi) = (h, \phi)_{L^2(\Omega)} & \forall \phi \in L^2(\Omega) \\ \langle \boldsymbol{\lambda}, \boldsymbol{v} \rangle_{\Gamma_1} = \langle \boldsymbol{\lambda}, \mathbf{G} \rangle_{\Gamma_1} & \forall \boldsymbol{\lambda} \in \mathcal{T}_1^*, \end{cases} \tag{5.216}$$

for *any* $(\mathbf{F}, h, \mathbf{G}) \in V^* = X^* \times L^2(\Omega) \times \mathcal{T}_1$ exists and is unique. Moreover,

$$\|\boldsymbol{v}\|_X + \|\pi\|_{L^2(\Omega)} + \|\mathbf{t}\|_{\mathcal{T}_1^*} \le C\left\{\|\mathbf{F}\|_{X^*} + \|h\|_{L^2(\Omega)} + \|\mathbf{G}\|_{\mathcal{T}_1}\right\} \tag{5.217}$$

where C only depends on the constants $\nu, \tilde{\beta}$. Then, we can introduce the operator $R : V^* \to V$ such that, for every $(\mathbf{F}, h, \mathbf{G}) \in V^*$,

$$R(\mathbf{F}, h, \mathbf{G}) = (\boldsymbol{v}, \pi, \mathbf{t}) \in V \tag{5.218}$$

is the unique solution of (5.216). R is called the Stokes solver. By (5.217), R is a (linear) continuous isomorphism between V^* and V. Moreover, it is possible to show that R is self-adjoint (see Exercise 13). •

5.14 Exercises

1. Let $\mathcal{U}$ be a Hilbert space and $K \subseteq \mathcal{U}$ be a closed, convex subset. Let $\{u_m\}_{m\ge1} \subset K$ be such that $u_m \rightharpoonup u$ in $\mathcal{U}$, that is $(u_m - u, v)_{\mathcal{U}} \to 0$ as $m \to \infty$, for all $v \in \mathcal{U}$. Show that $v \in K$ (this means that K is *weakly sequentially closed*).

[*Hint.* Let $z = \Pr_K u$. Then $(z - u, u_m - z) \geq 0$, $m \geq 1$. Pass to the limit and conclude that $u = z$.]

2. Let q be a nonnegative and symmetric bilinear form on the Hilbert space $\mathcal{U}$. Show the following Schwarz-type inequality

$$|q(u, v)| \leq \sqrt{q(u, u)\, q(v, v)}$$

with equality if and only if u and v are colinear. Deduce that $u \longmapsto q(u, u)$ is a convex function, strictly convex if q is positive.
[*Hint:* observe that, for any real number t, $0 \leq q(u + tv, u + tv) = \ldots$].

3. Consider the optimality condition

$$\int_\Omega (\hat{p} + \beta\hat{u})(v - \hat{u})\, d\mathbf{x} \geq 0 \qquad \forall v \in \mathcal{U}_{ad} \tag{5.219}$$

found in the constrained problem of Sect. 5.2.3. Show that if the space of admissible controls is given by $\mathcal{U}_{ad} = \left\{v \in L^2(\Omega : v \geq 0 \text{ a.e. in } \Omega)\right\}$, then (5.219) is equivalent to the following system of complementarity conditions:

$$\begin{cases} \hat{u} \geq 0 \\ \hat{p} + \beta\hat{u} \geq 0 \qquad \text{a.e. in } \Omega. \\ \hat{u}(\hat{p} + \beta\hat{u}) = 0 \end{cases}$$

4. Show that the Dirichlet/Neumann problem (5.91) is well posed if

$$(\mathbf{b} - \mathbf{c}) \cdot \mathbf{n} \leq 0 \text{ a.e. on } \Gamma_N, \quad \frac{1}{2}\text{div}\,(\mathbf{b} - \mathbf{c}) + r \geq 0 \text{ a.e. in } \Omega.$$

5. Let $z_d \in L^2(\Omega)$ be given and $\Omega_0 \subset \Omega$. Analyze the problem

$$\tilde{J}(y, u) = \frac{1}{2}\int_{\Omega_0} |\nabla y - \nabla z_d|^2\, d\mathbf{x} + \frac{\beta}{2}\int_\Gamma u^2 d\sigma \to \min,$$

$\beta \geq 0$, subject to $u \in \mathcal{U}_{ad} \subset L^2(\Gamma)$, convex, closed, and to

$$\begin{cases} \mathcal{E}y = -\text{div}(\mathbf{A}\nabla y) = 0 & \text{in } \Omega \\ \dfrac{\partial y}{\partial n_{\mathcal{E}}} + y = u & \text{on } \Gamma. \end{cases} \tag{5.220}$$

In particular, examine the special cases

$$\mathcal{U}_{ad} = \left\{u \in L^2(\Gamma),\ u \geq 1 \text{ a.e. on } \Gamma\right\} \quad \text{and} \quad \mathcal{U}_{ad} = \left\{u \in L^2(\Gamma),\ \|u\|_{L^2(\Gamma)} \leq 1\right\}.$$

6. Let $\mathcal{U}_{ad} \subseteq L^2(\Omega)$, $W_{ad} \subseteq L^2(\Gamma)$ both closed and convex sets. Analyze the problem

$$\tilde{J}(y, u) = \frac{1}{2}\int_\Omega (y - z_d)^2\, d\mathbf{x} + \frac{\beta_1}{2}\int_\Omega u^2 d\sigma + \frac{\beta_2}{2}\int_\Gamma w^2 d\sigma \to \min,$$

$\beta_1, \beta_2 > 0$, subject to $(u, w) \in \mathcal{U}_{ad} \times W_{ad}$ and to

$$\begin{cases} -\Delta y + y = f + u & \text{in } \Omega \\ \dfrac{\partial y}{\partial n} = w & \text{on } \Gamma. \end{cases}$$

7. Let $\Gamma_D \subset \Gamma$, $\Gamma_N = \Gamma \backslash \Gamma_D$, $\mathcal{U}_{ad} \subseteq L^2(\Gamma_N)$ be a closed and convex set. Analyze the problem

$$\tilde{J}(y,u) = \frac{1}{2}\int_\Omega |\nabla y - \nabla z_d|^2 \, d\mathbf{x} + \frac{1}{2}\int_{\Gamma_N} u^2 d\mathbf{x} \to \min$$

subject to

$$\begin{cases} -\Delta y + \mathbf{b}\cdot\nabla y = f & \text{in } \Omega \\ y = 0 & \text{on } \Gamma_D \\ \dfrac{\partial y}{\partial n} + hy = u & \text{on } \Gamma_N \end{cases}$$

by choosing an appropriate functional setting and formulating reasonable hypotheses on the data.

8. Let $\Omega = (0,1)\times(0,1)$. Analyze the problem

$$\tilde{J}(y,u) = \frac{1}{2}\int_\Omega (y - z_d)^2 \, d\mathbf{x} + \frac{1}{2}\int_\Gamma (u^2 + |\nabla u|^2) d\mathbf{x} \to \min,$$

subject to

$$\begin{cases} -\Delta y = f & \text{in } \Omega \\ \dfrac{\partial y}{\partial n} + y = u & \text{on } \Gamma. \end{cases}$$

9. *Approximate controllability.* Given $z_d \in L^2(\Omega)$, consider the OCP

$$\tilde{J}(y,u) = \int_\Omega (y - z_d)^2 \, d\mathbf{x} \to \min,$$

subject to $(y,u) \in H_0^1(\Omega) \times L^2(\Omega)$ and to

$$\begin{cases} -\Delta y = u & \text{in } \Omega \\ y = 0 & \text{on } \Gamma. \end{cases}$$

Show that, given $\varepsilon > 0$, it is possible to find u_ε such that $\tilde{J}(y(u_\varepsilon), u_\varepsilon) < \varepsilon$, through the following steps:

- Show that if $w \in L^2(\Omega)$ and

$$\int_\Omega y(u)\, w \, d\mathbf{x} = 0 \quad \forall u \in L^2(\Omega) \tag{5.221}$$

then $w = 0$.
[*Hint:* y satisfies the equation

$$(\nabla y, \nabla \varphi)_{L^2(\Omega)^2} = (u, \varphi)_{L^2(\Omega)} \qquad \forall \varphi \in H_0^1(\Omega). \tag{5.222}$$

Let z be the weak solution to $-\Delta z = w$ in Ω and $z = 0$ on Γ. Then

$$(\nabla z, \nabla \varphi)_{L^2(\Omega)^2} = (w, \varphi)_{L^2(\Omega)} \qquad \forall \varphi \in H_0^1(\Omega). \tag{5.223}$$

Inserting $\varphi = z$ and $\varphi = y$ into (5.222) and (5.223), respectively, and taking into account (5.221), deduce that $(u,z)_{L^2(\Omega)} = 0$ for each $u \in L^2(\Omega)$.]
- Use the result above to deduce that the set S of solutions $y(u)$, as u varies over $L^2(\Omega)$, is dense in $L^2(\Omega)$.
[*Hint:* By contradiction assume that the closure in $L^2(\Omega)$ of S is a closed linear proper subspace of $L^2(\Omega)$. Deduce that there exists an element $z \neq 0$, orthogonal to each element of S.]

10. Verify that the Neumann problem (5.86) admits a unique solution also in the case where $g \in H^{-1/2}(\Gamma)$.
11. Verify the other assumptions of Theorem A.7 on the bilinear forms appearing in (5.206).
12. Show that the strong form of the adjoint problem (5.212) is provided by (5.214).
[*Hint:* in two dimensions we have, for smooth vector fields $\boldsymbol{v}$ and $\boldsymbol{\varphi}$,

$$\int_\Omega (\nabla \times \boldsymbol{\varphi}) \cdot \boldsymbol{w} \, d\mathbf{x} - \int_\Omega \boldsymbol{\varphi} \cdot (\nabla \times \boldsymbol{w}) \, d\mathbf{x} = \int_\Gamma (\boldsymbol{\varphi} \cdot \mathbf{n}^\perp) \boldsymbol{w} d\sigma$$

and, as a consequence, by taking $\boldsymbol{w} = \nabla \times \boldsymbol{v}$,

$$\int_\Omega (\nabla \times \boldsymbol{v}) \cdot (\nabla \times \boldsymbol{\varphi}) \, d\mathbf{x} = \int_\Omega (\nabla \times \nabla \times \boldsymbol{v}) \cdot \boldsymbol{\varphi} \, d\mathbf{x} + \int_\Gamma (\nabla \times \boldsymbol{v}) \mathbf{n}^\perp \cdot \boldsymbol{\varphi} d\sigma \tag{5.224}$$

where $\mathbf{n}^\perp = (-n_2, n_1)$ is a tangential vector to the boundary. Use (5.224) and integrate by parts the terms $-e(\hat{\boldsymbol{v}}, \boldsymbol{\varphi}) - b(\boldsymbol{\varphi}, \hat{p})$ in (5.212).]
13. Prove that the Stokes solver R defined in (5.218) is self-adjoint.
[*Hint:* let $(\tilde{\mathbf{F}}, \tilde{h}, \tilde{\mathbf{G}}) \in V^*$ and $R(\tilde{\mathbf{F}}, \tilde{h}, \tilde{\mathbf{G}}) = (\tilde{\boldsymbol{v}}, \tilde{\pi}, \tilde{\boldsymbol{t}})$. We have to show that

$$\left\langle (\boldsymbol{v}, \pi, \boldsymbol{t}), (\tilde{\mathbf{F}}, \tilde{h}, \tilde{\mathbf{G}}) \right\rangle_{V,V^*} = \langle (\mathbf{F}, h, \mathbf{G}), (\tilde{\boldsymbol{v}}, \tilde{\pi}, \tilde{\boldsymbol{t}}) \rangle_{V^*,V} .$$

Indeed, we have: $(\tilde{\boldsymbol{v}}, \tilde{\pi}, \tilde{\boldsymbol{t}}) \in X \times L^2(\Omega) \times \mathcal{T}_1^*$ is such that

$$\begin{cases} e(\tilde{\boldsymbol{v}}, \boldsymbol{\varphi}) + b(\boldsymbol{\varphi}, \tilde{\pi}) + \langle \tilde{\boldsymbol{t}}, \boldsymbol{\varphi} \rangle_{\Gamma_1} = \langle \tilde{\mathbf{F}}, \boldsymbol{\varphi} \rangle_{X^*,X} & \forall \boldsymbol{\varphi} \in X \\ b(\tilde{\boldsymbol{v}}, \phi) = (\tilde{h}, \phi)_{L^2(\Omega)} & \forall \phi \in L^2(\Omega) \\ \langle \boldsymbol{\lambda}, \tilde{\boldsymbol{v}} \rangle_{\Gamma_1} = \langle \boldsymbol{\lambda}, \tilde{\mathbf{G}} \rangle_{\Gamma_1} & \forall \boldsymbol{\lambda} \in \mathcal{T}_1^* . \end{cases} \tag{5.225}$$

Substitute $(\boldsymbol{\varphi}, \phi, \boldsymbol{\lambda}) = (\tilde{\boldsymbol{v}}, \tilde{\pi}, \tilde{\boldsymbol{t}})$ into (5.216) and $(\boldsymbol{\varphi}, \phi, \boldsymbol{\lambda}) = (\boldsymbol{v}, \pi, \boldsymbol{t})$ into (5.225) and compare the expressions so obtained.]

Chapter 6
Numerical Approximation of Linear-Quadratic OCPs

In this chapter we address the numerical solution of OCPs with quadratic functional and linear state equation. For the sake of exposition we consider two different examples of state problems: a scalar advection-diffusion (-reaction) problem, and the Stokes system for viscous incompressible flows. After a quick overview in Sect. 6.1, in Sect. 6.2 we describe two general approaches which go under the name of *optimize-then-discretize* and *discretize-then-optimize*. In Sect. 6.3 we solve the OCPs by employing the numerical *iterative methods* for unconstrained optimization introduced in Sect. 3.1; the generalization of this approach to the constrained case – the so-called *projected methods* – is treated in Sect. 6.4. Some numerical results are shown in Sect. 6.5. Then, in Sect. 6.6 we consider a different strategy, based on the so-called *all-at-once (or one-shot) methods*, yielding the solution of a saddle point system in the unconstrained case, and the *primal-dual active set method* in the constrained case, see Sect. 6.8. Some numerical examples are shown in Sects. 6.7–6.9. Finally, we derive both a priori and a posteriori error estimates for linear-quadratic OCPs in Sects. 6.10–6.11, respectively.

6.1 A Classification of Possible Approaches

Optimal control problems addressed in Chapter 5 can be written as PDE-constrained optimization problems under the following general form

$$\tilde{J}(y,u) \to \min \quad \text{subject to} \quad G(y,u) = 0, \quad y \in V, \quad u \in \mathcal{U}_{ad}$$

where $\mathcal{U}_{ad} \subseteq \mathcal{U}$ denotes the space of admissible controls, expressed by a set of (inequality) control constraints, and the state problem $G(y,u) = 0$ plays the role of an equality *constraint*. We recall that two strategies can be pursued: we can either

1. either keep the constraint $G(y,u) = 0$ and treat both y and u as a single optimization variable, $x = (y,u)$,
2. or eliminate the PDE constraint by introducing the *control-to-state* map $u \mapsto y(u)$, and replace y by $y(u)$, the solution of $G(y(u),u) = 0$, thus keeping only u as optimization variable.

The choice is often driven by external reasons: for instance, if we aim at building an optimization loop by using an existing PDE solver providing $y = y(u)$, keeping only u as optimization variable is recommended.

A. Manzoni et al., *Optimal Control of Partial Differential Equations*,
Applied Mathematical Sciences 207, https://doi.org/10.1007/978-3-030-77226-0_6

Depending on the way the PDE constraint is treated, we can use:

- *iterative methods*, also referred to as *black-box methods*, where the PDE constraint is eliminated and the only optimization variable is the control u;
- *all-at-once methods*, also referred to as *one-shot methods*, where the PDE constraint is explicitly kept in the problem formulation and both y and u are treated as optimization variables.

In this chapter we present both iterative and all-at-once methods for linear-quadratic OCPs; extensions to nonlinear problems will be considered in Chapter 9.6. A chapter preview is reported in the table below.

Iterative (or black-box) methods	**All-at-once (or one-shot) methods**
- Treat u as optimization variable	- Treat both y and u as optimization variables
Unconstrained OCPs	
Descent methods (Sect. 6.3)	*Quadratic Programming (Sect. 6.6)*
- Iterative update of control	- Solution of a large linear system
- Evaluation of the gradient of the cost functional based on the adjont solution	- Efficient preconditioning for saddle-point problems
Constrained OCPs	
Projected Descent methods (Sect. 6.4)	*Primal-Dual Active Set methods (Sect. 6.8)*
- Projection of the updated control on the set of admissible controls	- Solution of a quadratic subproblem at each step
	- Update of active/inactive constraint sets

Even though for pedagogical reasons our presentation is mainly focused on the discrete (algebraic) problem, the methods above can also be formulated on the infinite dimensional problem (that is, on function spaces); we will first formulate the primal-dual active set method on the infinite-dimensional constrained OCP to highlight relevant mathematical properties of its construction.

6.2 Optimize & Discretize, or the Other Way Around?

The interplay between optimization and discretization, which this section is devoted to, is an essential component of numerical approximation of OCPs. For the sake of exposition, we consider the OCP (5.1)–(5.2), here recalled for the reader's convenience:

$$\tilde{J}(y,u) = \frac{1}{2}\int_\Omega (y - z_d)^2 d\mathbf{x} + \frac{\beta}{2}\int_\Omega u^2 d\mathbf{x} \;\rightarrow\; \min \tag{6.1a}$$

$$\text{subject to } u \in \mathcal{U}_{ad},\; y \in V:$$

$$\begin{cases} -\Delta y + \mathbf{b}\cdot\nabla y = f + u & \text{in } \Omega \\ y = 0 & \text{on } \Gamma, \end{cases} \tag{6.1b}$$

where $\beta > 0$ and $\mathcal{U}_{ad} \subset \mathcal{U} = L^2(\Omega)$ denotes the set of admissible controls.

The weak formulation of the state problem (6.1a) reads: find $y = y(u) \in V = H_0^1(\Omega)$ such that

$$a(y, \varphi) = (f + u, \varphi)_{L^2(\Omega)} \qquad \forall \varphi \in V$$

with

$$a(y, \varphi) = \int_\Omega (\nabla y \cdot \nabla \varphi + \mathbf{b} \cdot \nabla y \varphi)\, d\mathbf{x}; \tag{6.2}$$

$f \in L^2(\Omega)$ and $\mathbf{b} \in L^\infty(\Omega)^d$ are given data, with $\mathrm{div}\mathbf{b} \leq 0$.

If we use the *optimize-then-discretize* (OtD) strategy, we carry out optimization first (at the continuous level, see, e.g., (5.25)) and then discretize the resulting optimality system. With the *discretize-then-optimize* (DtO) strategy, it is the other way around, that is we first discretize the state system, and then minimize the cost functional subject to the corresponding algebraic system, following the procedure described in Sect. 4.1. The two approaches do not necessary yield identical solutions in general.

Before moving on, we introduce some notation regarding PDE discretization used throughout the book. Readers less-acquainted with the numerical approximation of PDEs can refer to Sect. B.2.

To compute the solution of a problem like the state equation in (6.1) we use the finite element (FE) method, that we briefly recall in Sect. B.4 for the reader's convenience. We therefore need to introduce suitable, finite-dimensional subspaces of both the state and the control space. From now on, we denote by $V_h \subset V$ a family of finite-dimensional subspaces of V, such that dim $V_h = N_y$; the subscript $h > 0$ is a characteristic discretization parameter (most typically, the mesh size). In the same way, we denote by $\mathcal{U}_h \subset \mathcal{U}$ a family of finite-dimensional subspaces of $\mathcal{U}$, such that dim $\mathcal{U}_h = N_u$. The dependence of both N_y and N_u on h is understood. For instance, we can set $V_h = X_h^r \cap V$, where

$$X_h^r = \left\{ v_h \in C(\overline{\Omega}) : v_h|_K \in \mathbb{P}_r \ \ \forall K \in \mathcal{T}_h \right\}, \qquad r = 1, 2, \ldots \tag{6.3}$$

is the space of globally continuous functions that are polynomials of degree at most r on the single elements of the triangulation $\mathcal{T}_h$. Here we restrict to Lipschitz domains $\Omega \subset \mathbb{R}^d$ $(d = 2, 3)$ with polygonal shape, for which $\mathcal{T}_h$ represents a partition with non-overlapping elements K, triangles ($d = 2$) or tetrahedra ($d = 3$); h will denote the maximum diameter of the elements K. Note that $N_y <$ dim X_h^r when Dirichlet boundary conditions are imposed; in the case of a Neumann (or Robin) problem, $N_y =$ dim X_h^r.

For the case at hand, we choose $\mathcal{U}_h = X_h^r$, thus yielding $N_y = N_u$, although different options make sense too. Note that, in general, the dimension N_u of the control variable may differ from N_y, that of the state variable: this can happen, for instance, when dealing with boundary control problems.

We also point out that we do not explicitly consider as further *discretization* stage the evaluation of quadrature formulas needed to compute the arrays (namely, matrices and vectors involving integrals) arising from the discretization of a PDE problem. In other words, we refer to a discrete PDE problem as to the one obtained by replacing the infinite dimensional spaces with the corresponding finite dimensional ones, but still keeping the exact integral operators. This is called *Galerkin method*, as opposed to *generalized Galerkin methods* (or Galerkin methods with numerical integration) where integrals are replaced by numerical quadratures (see, e.g., [218]).

For the sake of exposition we focus on the case of distributed control in the unconstrained case, hence $\mathcal{U}_{ad} = \mathcal{U}$; the treatment of control constraints makes the specific object of Sects. 6.4 and 6.8. The numerical approximation of Dirichlet boundary control problems, as well as of state constrained OCPs (that is, OCPs with constraints on the solution y of the state problem), requires more advanced tools and we only provide references to technical papers addressing these aspects.

6.2.1 Optimize Then Discretize

In the *optimize then discretize* approach (OtD, in short), we first recover a system of optimality conditions at the continuous level. For the case at hand, Proposition 5.2 (in weak formulation) yields: find $y \in V$, $u \in \mathcal{U}$ and $p \in V$ such that:

$$\begin{cases} a(y,\varphi) = (f+u,\varphi)_{L^2(\Omega)} & \forall \varphi \in V \\ a^*(p,\psi) = -(y - z_d,\psi)_{L^2(\Omega)} & \forall \psi \in V \\ (\beta u - p, v)_{L^2(\Omega)} = 0 & \forall v \in \mathcal{U}. \end{cases} \tag{6.4}$$

Remark 6.1. Note that the adjoint variable has now an opposite sign than that appearing in the optimality conditions considered throughout Chapter 5. Although the two formulations are equivalent, the choice operated here is more convenient for the numerical approximation. •

Then, we approximate the spaces V by V_h and $\mathcal{U}$ by $\mathcal{U}_h$ (two suitable finite-dimensional spaces) and replace (6.4) with the following problem: find $y_h \in V_h$, $u_h \in \mathcal{U}_h$ and $p_h \in V_h$ such that:

$$\begin{cases} a(y_h,\varphi_h) = (f+u_h,\varphi_h)_{L^2(\Omega)} & \forall \varphi_h \in V_h \\ a^*(p_h,\psi_h) = -(y_h - z_d,\psi_h)_{L^2(\Omega)} & \forall \psi_h \in V_h \\ (\beta u_h - p_h, v_h)_{L^2(\Omega)} = 0 & \forall v_h \in \mathcal{U}_h. \end{cases} \tag{6.5}$$

Here the same space V_h is used to approximate both state and control variables; choosing two different spaces V_h, $\tilde{V}_h$ is also possible, see Sect. 6.2.3. Moreover, V_h is used to approximate the trial space (where we seek the solution) as well as the test space, thus yielding a Galerkin problem[1] for both the state and the adjoint system in (6.5). We denote by $\{\varphi_j,\ j=1,\dots,N_y\}$ and $\{\psi_j,\ j=1,\dots,N_u\}$ a basis for V_h and $\mathcal{U}_h$, respectively. Set

$$y_h(\mathbf{x}) = \sum_{j=1}^{N_y} y_{h,j}\varphi_j(\mathbf{x}) \in V_h, \qquad p_h(\mathbf{x}) = \sum_{j=1}^{N_y} p_{h,j}\varphi_j(\mathbf{x}) \in V_h \tag{6.6}$$

as discrete state and adjoint variables, respectively, and

$$u_h(\mathbf{x}) = \sum_{j=1}^{N_u} u_{h,j}\psi_j(\mathbf{x}) \in \mathcal{U}_h \tag{6.7}$$

as discrete control variable, and introduce the *stiffness* matrix $\mathbb{A} \in \mathbb{R}^{N_y \times N_y}$

$$(\mathbb{A})_{ij} = a(\varphi_j,\varphi_i), \qquad i,j = 1,\dots,N_y, \tag{6.8}$$

the *control* matrix $\mathbb{B} \in \mathbb{R}^{N_y \times N_u}$

$$(\mathbb{B})_{ij} = (\psi_j,\varphi_i)_{L^2(\Omega)}, \qquad i = 1,\dots,N_y,\ j = 1,\dots,N_u \tag{6.9}$$

and the vector $\mathbf{f} \in \mathbb{R}^{N_y}$

$$(\mathbf{f})_i = (f,\varphi_i)_{L^2(\Omega)}, \qquad i = 1,\dots,N_y.$$

Then $(6.5)_1$ can be equivalently written as a linear system,

$$\mathbb{A}\mathbf{y} = \mathbf{f} + \mathbb{B}\mathbf{u}, \tag{6.10}$$

with $\mathbf{y} = (y_{h,1},\dots,y_{h,N_y})^\top \in \mathbb{R}^{N_y}$ and $\mathbf{u} = (u_{h,1},\dots,u_{h,N_u})^\top \in \mathbb{R}^{N_u}$.

[1] A more general Petrov-Galerkin problem is obtained instead when the trial and the test spaces do not coincide; further details can be found in Sect. B.2.

The discrete adjoint problem $(6.5)_2$ can be written in matrix form (see Exercise 1) as

$$\mathbb{A}^\top \mathbf{p} = -\mathbb{M}(\mathbf{y} - \mathbf{z}_d), \tag{6.11}$$

where $\mathbf{p} = (p_{h,1}, \dots, p_{h,N_y})^T \in \mathbb{R}^{N_y}$ while $\mathbf{z}_d \in \mathbb{R}^{N_y}$ is the vector of components[2]

$$(\mathbf{z}_d)_i = (z_d, \varphi_i)_{L^2(\Omega)}, \qquad i = 1, \dots, N_y.$$

We have exploited the property $a^*(\varphi_j, \varphi_i) = a(\varphi_i, \varphi_j)$ and defined the finite element (state) mass matrix $\mathbb{M} \in \mathbb{R}^{N_y \times N_y}$,

$$(\mathbb{M})_{ij} = (\varphi_j, \varphi_i)_{L^2(\Omega)}, \qquad i, j = 1, \dots, N_y. \tag{6.12}$$

Finally, equation $(6.5)_3$, expressing the minimization property, can be rewritten as

$$\beta \mathbb{N} \mathbf{u} - \mathbb{B}^\top \mathbf{p} = \mathbf{0} \tag{6.13}$$

where $\mathbb{N} \in \mathbb{R}^{N_u \times N_u}$ is the finite element (control) mass matrix,

$$(\mathbb{N})_{ij} = (\psi_j, \psi_i)_{L^2(\Omega)}, \qquad i, j = 1, \dots, N_u. \tag{6.14}$$

Summarizing, the vectors $(\mathbf{y}, \mathbf{u}, \mathbf{p}) \in \mathbb{R}^{N_y} \times \mathbb{R}^{N_u} \times \mathbb{R}^{N_y}$ corresponding to the optimal state, control, and adjoint variables solve the following *discrete KKT system*:

$$\begin{cases} \mathbb{A}\mathbf{y} = \mathbf{f} + \mathbb{B}\mathbf{u} \\ \mathbb{A}^\top \mathbf{p} = -\mathbb{M}(\mathbf{y} - \mathbf{z}_d) \\ \beta \mathbb{N}\mathbf{u} - \mathbb{B}^T \mathbf{p} = \mathbf{0}, \end{cases} \tag{6.15}$$

that is

$$\begin{bmatrix} \mathbb{M} & 0 & \mathbb{A}^\top \\ 0 & \beta\mathbb{N} & -\mathbb{B}^\top \\ \mathbb{A} & -\mathbb{B} & 0 \end{bmatrix} \begin{bmatrix} \mathbf{y} \\ \mathbf{u} \\ \mathbf{p} \end{bmatrix} = \begin{bmatrix} \mathbb{M}\,\mathbf{z}_d \\ \mathbf{0} \\ \mathbf{f} \end{bmatrix}. \tag{6.16}$$

Note that the reordering[3] of the equations yields a block-symmetric matrix in (6.16); moreover, as we will see in Sect. 6.6, by considering $\mathbf{y}$ and $\mathbf{u}$ as components of a unique optimization variable will unveal an extremely convenient structure. By block Gaussian elimination of the vectors $\mathbf{y}$ and $\mathbf{p}$, (6.13) yields the reduced system

$$\mathbb{H}\mathbf{u} = \mathbf{b} \tag{6.17}$$

where $(\mathbb{A}^{-\top} = (\mathbb{A}^{-1})^\top)$

$$\mathbb{H} = \beta\mathbb{N} + \mathbb{B}^\top \mathbb{A}^\top \mathbb{M} \mathbb{A}^{-1} \mathbb{B} \in \mathbb{R}^{N_u \times N_u} \tag{6.18}$$

is the Hessian of the reduced cost functional (or *reduced Hessian* matrix), and

$$\mathbf{b} = \mathbb{B}^\top \mathbb{A}^\top \mathbb{M}(\mathbf{z}_d - \mathbb{A}^{-1}\mathbf{f}) \in \mathbb{R}^{N_u}.$$

Remark 6.2. In the case of a distributed control over the whole domain Ω, provided the same discretization is used for the state and the control variable, we have that $\mathbb{B} = \mathbb{M} = \mathbb{N}$ and all the blocks in the system (6.16) are square matrices. •

---

[2] The target function $z_d \in L^2(\Omega)$ could also be replaced by an approximation $z_{dh} \in V_h$. For instance, we can consider z_{dh} as the piecewise finite element approximation of degree r of z_d, the continuous function $\Pi_h^r z_d$ obtained as the L^2 projection or the *Clément interpolant* or *Scott-Zhang interpolant* of z_d over $\mathcal{T}_h$ [66, 242].

[3] Should we have operated with the opposite sign for the adjoint variable, see Remark 6.1, we would have ended up with a similar matrix with however different opposite signs for the extra-diagonal matrices $\mathbb{A}$ and $\mathbb{B}$.

A very similar system of equations is obtained when a cost functional like (5.65), involving an observation operator $C : V \to Z$, is minimized, with $z_d \in Z$. For instance, instead of (6.1a), we could consider the following problem (we refer to the general framework of Sect. 5.3.1):

$$\tilde{J}(y,u) = \frac{1}{2}\int_\Omega (Cy - z_d)^2 d\mathbf{x} + \frac{\beta}{2}\int_\Omega u^2 d\mathbf{x} \;\to\; \min \tag{6.19a}$$

$$\text{subject to } u \in \mathcal{U}_{ad},\ y \in V:$$

$$\begin{cases} -\Delta y + \mathbf{b}\cdot\nabla y = f + u & \text{in } \Omega \\ y = 0 & \text{on } \partial\Omega, \end{cases} \tag{6.19b}$$

The associated system of optimality conditions related to (6.19) (in presence of control constraints) has been derived in (5.65). To get the corresponding discrete version, we can proceed similarly to what done for problem (6.1), by introducing a further finite dimensional space Z_h (a priori different from V_h, depending on the observation operator at hand) to approximate the observation space Z. Let dim $Z_h = N_z$ and $\{\xi_j\,,\, j = 1,\dots,N_z\}$ denote a basis of Z_h. We obtain the following discrete system of optimality conditions: find $y_h \in V_h$, $p_h \in V_h$ and $u_h \in \mathcal{U}_h$ such that:

$$\begin{cases} a\left(y_h, \varphi_h\right) = (f, \varphi_h)_{L^2(\Omega)} + (u_h, \varphi_h)_{L^2(\Omega)} & \forall \varphi_h \in V_h \\ a^*\left(p_h, \psi_h\right) = -\left(C y_h - z_d, C\psi_h\right)_{L^2(\Omega)} & \forall \psi_h \in V_h \\ \left(\beta u_h, v_h\right)_{L^2(\Omega)} - \left(v_h, p_h\right)_{L^2(\Omega)} = 0 & \forall v_h \in \mathcal{U}_h. \end{cases} \tag{6.20}$$

Problem (6.20) can be equivalently rewritten from an algebraic standpoint as follows (see Exercise 2): denoting as usual by $(\mathbf{y},\mathbf{u},\mathbf{p}) \in \mathbb{R}^{N_y} \times \mathbb{R}^{N_u} \times \mathbb{R}^{N_y}$ the vectors of degrees of freedom corresponding to the optimal state, control, and adjoint variables, they satisfy

$$\begin{bmatrix} \mathbb{C}^\top \mathbb{M}\mathbb{C} & 0 & \mathbb{A}^\top \\ 0 & \beta\mathbb{N} & -\mathbb{B}^\top \\ \mathbb{A} & -\mathbb{B} & 0 \end{bmatrix} \begin{bmatrix} \mathbf{y} \\ \mathbf{u} \\ \mathbf{p} \end{bmatrix} = \begin{bmatrix} \mathbb{C}^\top \mathbb{M}\tilde{\mathbf{z}}_d \\ \mathbf{0} \\ \mathbf{f} \end{bmatrix}. \tag{6.21}$$

The *observation* matrix $\mathbb{C} \in \mathbb{R}^{N_y \times N_z}$ usually represents a (discrete) restriction operator. The mass matrix $\mathbb{M} \in \mathbb{R}^{N_z \times N_z}$ is now given by

$$(\mathbb{M})_{ij} = (\xi_j, \xi_i)_Z, \qquad i,j = 1,\dots,N_z.$$

All the conclusions and the results presented in the following sections are valid for any linear-quadratic OCP under the form (6.19); nevertheless, for the sake of simplicity we restrict ourselves to the model problem (6.1). The extension to the case (6.19) is straightforward.

Example 6.1. For the problem addressed in Sect. 5.5.1, $Z = L^2(\Omega_0)$,where Ω_0 is a given subset of Ω, whereas $V = H^1(\Omega)$. The observation operator $C : V \to Z$ is the restriction operator $Cy = \chi_{\Omega_0} y$, hence $\mathbb{C} \in \mathbb{R}^{N_y \times N_z}$ is a suitable rectangular matrix, and $\mathbb{M} \in \mathbb{R}^{N_z \times N_z}$ is given by

$$(\mathbb{M})_{ij} = (\xi_j, \xi_i)_{L^2(\Omega_0)}, \qquad i,j = 1,\dots,N_z.$$

•

Example 6.2. For the problem of Sect. 5.5.2, $Z = V = H^1_0(\Omega)$ and $C : V \to Z$ is the identity operator. Hence, $\mathbb{C} = \mathbb{I} \in \mathbb{R}^{N_y \times N_y}$ is the identity matrix, and

$$(\mathbb{M})_{ij} = (\varphi_j, \varphi_i)_{H^1_0(\Omega)} = a(\varphi_j, \varphi_i) + (\varphi_j, \varphi_i)_{L^2(\Omega)}, \qquad i,j = 1,\dots,N_y,$$

that is, $\mathbb{M}$ is the sum of the stiffness matrix $\mathbb{A}$ and the $L^2(\Omega)$ mass matrix. •

Remark 6.3. In the case of a boundary control (say, of Neumann type) some of the blocks in system (6.16) are no longer squared. Consider for instance the problem addressed in Sect. 5.6 and assume that the observation is distributed on the whole domain Ω. In this case N_u is the number of Neumann boundary nodes, which is typically much smaller than N_y, the number of nodes of the state variable. •

6.2.2 Discretize then Optimize

Alternatively to the previous strategy we may first discretize problem (6.1a)–(6.1b) by substituting all functional spaces by finite dimensional ones. By so doing we end up with the following discrete OCP:

$$\begin{gathered}\tilde{J}_h(y_h,u_h)=\frac{1}{2}\int_\Omega (y_h-z_d)^2 d\mathbf{x}+\frac{\beta}{2}\int_\Omega u_h^2\, d\mathbf{x}\ \rightarrow\ \min\\ \text{subject to } u_h\in\mathcal{U}_h,\\ y_h\in V_h\ :\quad a\,(y_h,\varphi_h)=(f+u_h,\varphi_h)_{L^2(\Omega)}\qquad \forall\varphi_h\in V_h.\end{gathered}\tag{6.22}$$

This is the so-called *discretize then optimize* approach (DtO, in short). With the same notation of the previous section, problem (6.22) can be rewritten in algebraic form as follows:

$$\begin{gathered}\tilde{J}_h(\mathbf{y},\mathbf{u})=\frac{1}{2}(\mathbf{y}-\mathbf{z}_d)^\top\mathbb{M}(\mathbf{y}-\mathbf{z}_d)+\frac{\beta}{2}\mathbf{u}^\top\mathbb{N}\mathbf{u}\ \rightarrow\ \min\\ \text{subject to } \mathbf{u}\in\mathbb{R}^{N_u},\\ \mathbf{y}\in\mathbb{R}^{N_y}\ :\quad \mathbb{A}\mathbf{y}=\mathbf{f}+\mathbb{B}\mathbf{u}.\end{gathered}\tag{6.23}$$

The algebraic OCP (6.23) is indeed very similar to the problems already analyzed in Sect. 4.1. It is characterized by a quadratic cost function and a set of linear equality constraints; since no further constraints are imposed on the optimization variables, the optimization problem is unconstrained. Since $\mathbb{A}$ is non-singular, we can formally invert the state system to get $\mathbf{y}=\mathbb{A}^{-1}(\mathbf{f}+\mathbb{B}\mathbf{u})$, then replace it in $\tilde{J}_h(\mathbf{y},\mathbf{u})$ to obtain the reduced cost function

$$J_h(\mathbf{u})=\tilde{J}_h(\mathbf{y}(\mathbf{u}),\mathbf{u})=\tilde{J}_h(\mathbb{A}^{-1}\left(\mathbf{f}+\mathbb{B}\mathbf{u}\right),\mathbf{u}).\tag{6.24}$$

According to Theorem 4.1, the solution of the optimization problem (6.23) is unique and is characterized by the following optimality conditions, as in (6.15):

$$\begin{cases}\mathbb{A}\mathbf{y}=\mathbf{f}+\mathbb{B}\mathbf{u}\\ \mathbb{A}^\top\mathbf{p}=-\mathbb{M}(\mathbf{y}-\mathbf{z}_d)\\ \beta\mathbb{N}\mathbf{u}-\mathbb{B}^\top\mathbf{p}=\mathbf{0}.\end{cases}\tag{6.25}$$

Indeed, by deriving (6.24), we obtain (see Exercise 3)

$$\nabla J_h(\mathbf{u})=\beta\mathbb{N}\mathbf{u}+\mathbb{B}^\top(\mathbb{A}^{-\top}\mathbb{M}(\mathbb{A}^{-1}(\mathbf{f}+\mathbb{B}\mathbf{u})-\mathbf{z}_d))=\beta\mathbb{N}\mathbf{u}-\mathbb{B}^\top\mathbf{p}\tag{6.26}$$

provided that $\mathbf{p}$, the discrete adjoint vector, fulfills

$$\mathbb{A}^{-\top}\mathbb{M}(\mathbb{A}^{-1}(\mathbf{f}+\mathbb{B})-\mathbf{z}_d)=\mathbf{p}\quad\Leftrightarrow\quad\mathbb{A}^T\mathbf{p}=-\mathbb{M}(\mathbf{y}-\mathbf{z}_d).$$

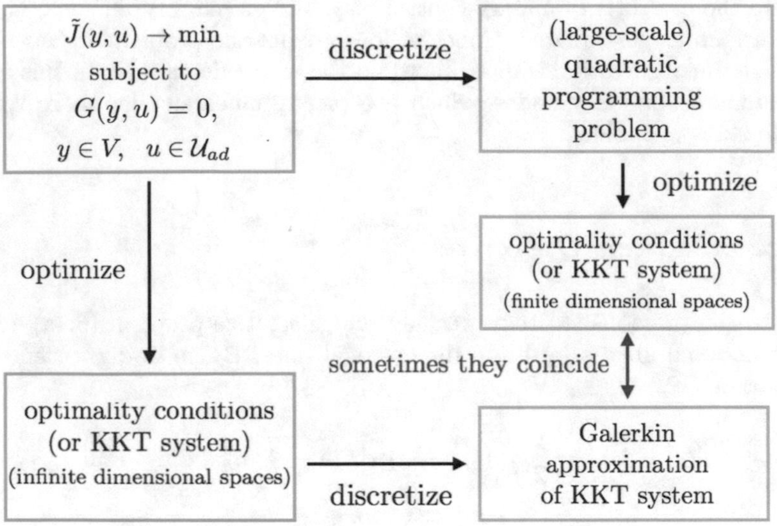

Fig. 6.1 OtD and DtO approaches in the case of linear-quadratic OCPs; the final discrete problems concide if a Galerkin approximation of both state and adjoint problems is adopted

We point out that, by following the DtO approach, the discretization p_h of the adjoint variable p is automatically determined by the discrete test space used in the variational formulation of the state problem appearing in (6.23). Similarly, the discrete test space of the adjoint problem is equal to the discrete trial space of the state problem. Thus, if we consider a Galerkin approximation of the state equation, we find that $p_h \in V_h$; this is also clear if we recall the role of the adjoint variable in the Lagrangian functional.

6.2.3 Pro's and Con's

In the case of linear-quadratic OCPs, the OtD and the DtO approaches generate the same problem, if a Galerkin approximation of both state and adjoint problems is adopted. More specifically, if we choose the same discrete space to approximate the state and the adjoint space in the OtD approach, the system of optimality conditions (6.15) is equal to (6.25). A graphical sketch of the two approaches is reported in Fig. 6.1. A case where the two approaches yield different algebraic systems is discussed in the next subsection.

If the state and the adjoint variables are taken in two different finite-dimensional spaces, the discrete optimality system is not the same. Indeed, by choosing as adjoint space $\tilde{V}_h \neq V_h$, $\dim\tilde{V}_h = N_p$, with N_p possibly different than N_y, and corresponding basis $\{\tilde{\varphi}_j \,,\, j = 1, \dots, N_p\}$, the OtD approach yields the following discrete adjoint problem

$$\tilde{\mathbb{A}}\mathbf{p} = -\tilde{\mathbb{T}}(\mathbf{y} - \mathbf{z}_d), \tag{6.27}$$

instead of (6.11), where $\mathbf{p} = (p_{h,1}, \dots, p_{h,N_p})^T \in \mathbb{R}^{N_p}$ is the discrete adjoint vector, while $\tilde{\mathbb{A}} \in \mathbb{R}^{N_p \times N_p}$ and $\tilde{\mathbb{T}} \in \mathbb{R}^{N_p \times N_y}$ are the matrices

$$(\tilde{\mathbb{A}})_{ij} = a(\tilde{\varphi}_j, \tilde{\varphi}_i), \qquad i,j = 1, \dots, N_p,$$

$$(\tilde{\mathbb{T}})_{ij} = (\varphi^j, \tilde{\varphi}_i)_{L^2(\Omega)}, \qquad i = 1, \dots, N_p,\; j = 1, \dots, N_y,$$

that replace $\mathbb{A}^\top$ and $\mathbb{M}$, respectively. Similarly, instead of (6.13) we get

$$\beta \mathbb{N}\mathbf{u} - \tilde{\mathbb{B}}^\top \mathbf{p} = \mathbf{0}$$

with $\tilde{\mathbb{B}} \in \mathbb{R}^{N_p \times N_u}$,

$$(\tilde{\mathbb{B}})_{ij} = (\psi_j, \tilde{\varphi}_i)_{L^2(\Omega)}, \qquad i = 1, \dots, N_p,\ j = 1, \dots, N_u.$$

In compact form,

$$\begin{bmatrix} \tilde{\mathbb{T}} & 0 & \tilde{\mathbb{A}} \\ 0 & \beta\mathbb{N} & -\tilde{\mathbb{B}}^\top \\ \mathbb{A} & -\mathbb{B} & 0 \end{bmatrix} \begin{bmatrix} \mathbf{y} \\ \mathbf{u} \\ \mathbf{p} \end{bmatrix} = \begin{bmatrix} \tilde{\mathbb{T}}\mathbf{z}_d \\ \mathbf{0} \\ \mathbf{f} \end{bmatrix}. \tag{6.28}$$

The corresponding reduced Hessian matrix $\mathbb{H} = \beta\mathbb{N} + \tilde{\mathbb{B}}^\top \tilde{\mathbb{A}}^{-1} \tilde{\mathbb{T}} \mathbb{A}^{-1} \mathbb{B} \in \mathbb{R}^{N_u \times N_u}$ is no longer symmetric, even if $\mathbb{A}$ is symmetric[4]. Moreover, there is no optimization problem featuring (6.28) as optimality system. In conclusion, optimization and discretization steps do not commute in this case.

Using different discretizations for the adjoint and the state variables could be motivated by the property that the adjoint variable is often more regular than the state variable (the latter tight to the regularity of the control function u; see e.g. [148]).

Last, but not least, the numerical strategy pursued to compute the solution of an OCP should be *consistent* with the infinite dimensional system of optimality conditions. For instance, the identity

$$\beta u - p = 0 \qquad \text{a.e. in } \Omega$$

that links the adjoint and the control variables in the case of problem (6.1) should ideally be preserved at the discrete level: this implies that the discrete control space and the discrete adjoint space coincide. Moreover, there are situations where approximating the adjoint variable in the same trial space of the state variable can be troublesome, see, e.g., [205, 204].

Concerning the numerical approximation of OCPs with Dirichlet boundary control, a special care must be paid when the control space is $\mathcal{U} = L^2(\Gamma)$, and the very-weak solution of the state problem is involved in the system of optimality conditions (see Sect. 5.9). In general, a Dirichlet problem with L^2 boundary data cannot be approximated by standard Galerkin techniques – see e.g. [33]. However, things are better in the case of OCPs with Dirichlet controls, since the solution of the OCP enjoys more regularity than the one required by the problem formulation. Indeed, the optimal state $\hat{y} \in H^1(\Omega)$ is a variational solution of the state problem, so that we can exploit the standard finite element approximation as in the case of distributed or Neumann controls (see Sect. 5.11).

A final comment is in order. So far in this Chapter we have mainly considered unconstrained OCPs since they also represent a first step towards the resolution of more general and involved problems. In fact, when tackling constrained OCPs, various techniques like active set strategies, penalty methods and Sequential Quadratic Programming (SQP) for more general nonlinear OCPs (see Chapter 9.6) involve the solution of a sequence of simpler problems with the same structure of (6.4) (or (6.15)). Developing suitable methods for these latter problems is crucial to design efficient algorithms for more challenging problems. The discrete formulation of an OCP with (pointwise) control constraints will be considered later in Sect. 6.8, directly at an algebraic level.

[4] In the case of the OtD approach, if $\mathbb{A}$ is symmetric then $\mathbb{H}$ is symmetric as well.

6.2.4 The case of Advection Diffusion Equations with Dominating Advection

An example in which DtO and OtD lead to different solutions is that of OCPs with state problem like the one in (6.1) and suitable numerical stabilization techniques are employed on both state and adjoint equations.

More specifically, let us consider as state problem the following advection-diffusion equation (with distributed control):

$$\begin{cases} \mathcal{E}y = -\mathrm{div}(\nu\nabla y) + \mathbf{b}\cdot\nabla y = u & \text{in } \Omega \\ y = 0 & \text{on } \Gamma_D \\ \nu\dfrac{\partial y}{\partial n} = 0 & \text{on } \Gamma_N, \end{cases} \tag{6.29}$$

where $\nu(\mathbf{x}) \geq \nu_0$ for all $\mathbf{x} \in \Omega$, $\mathrm{div}\mathbf{b} \leq 0$, $\Gamma = \bar{\Gamma}_D \cup \bar{\Gamma}_N$, $\Gamma_D \cap \Gamma_N = \emptyset$ and $\Gamma_N = \{\mathbf{x} \in \partial\Omega : \mathbf{b}(\mathbf{x})\cdot\mathbf{n}(\mathbf{x}) > 0\}$ is the outflow boundary. This represents a simplified version of the problem described in Sect. 1.3 dealing with the control of pollutant emissions from chimneys. We assume that transport dominates over diffusion; more precisely, that the Péclet number

$$Pe(\mathbf{x}) = \frac{\|\mathbf{b}(\mathbf{x})\|_{L^\infty(\Omega)} L}{2\nu_0}, \qquad \mathbf{x} \in \Omega$$

is very large; here L denotes the linear dimension of the domain. We consider the following cost functional to be minimized,

$$\tilde{J}(y,u) = \frac{1}{2}\int_{\Omega_0} (ky - z_d)^2\, d\mathbf{x} + \frac{\beta}{2}\int_\Omega u^2\, d\mathbf{x}, \tag{6.30}$$

for a suitable function $k \in L^\infty(\Omega)$. Similarly to the case of problem (6.1), the system of optimality conditions reads: find $y \in V$, $u \in \mathcal{U}$ and $p \in V$ such that

$$\begin{cases} a(y,\varphi) = (u,\varphi)_{L^2(\Omega)} & \forall \varphi \in V \\ a^*(p,\psi) = (ky - z_d, k\psi)_{L^2(\Omega_0)} & \forall \psi \in V \\ (p + \beta u, v)_{L^2(\Omega)} = 0 & \forall v \in \mathcal{U}, \end{cases} \tag{6.31}$$

with $V = H^1_{\Gamma_D}(\Omega)$, $\mathcal{U} = L^2(\Omega)$, and

$$a(y,\varphi) = \int_\Omega (\nu\nabla y\cdot\nabla\varphi + \mathbf{b}\cdot\nabla y\varphi)\, d\mathbf{x}, \qquad a^*(p,\psi) = a(\psi,p).$$

In strong form the adjoint problem reads:

$$\begin{cases} \mathcal{E}^* p = -\mathrm{div}(\nu\nabla p + \mathbf{b}p) = \chi_{\Omega_0} k(ky - z_d) & \text{in } \Omega \\ p = 0 & \text{on } \Gamma_D \\ \nu\dfrac{\partial p}{\partial n} + \mathbf{b}\cdot\mathbf{n}p = 0 & \text{on } \Gamma_N. \end{cases} \tag{6.32}$$

In both OtD and DtO approaches, to solve the state and the adjoint problems we use the Galerkin–Least–Squares (GLS) finite element formulation, a stabilized and strongly consistent method necessary to prevent numerical oscillations that would otherwise affect the Galerkin method for larger Péclet numbers, see e.g. [218, Chapter 12].

Optimize then Discretize

We discretize the optimality system (6.31) as follows: find $y_h \in V_h$, $u_h \in \mathcal{U}_h$, $p_h \in V_h$ such that:

$$\begin{cases} a(y_h, \varphi_h) + s_h(y_h, \varphi_h) = (u_h, \varphi_h)_{L^2(\Omega)} & \forall \varphi_h \in V_h \\ a^*(p_h, \psi_h) + s_h^*(p_h, \psi_h) = (ky_h - z_d, k\psi_h)_{L^2(\Omega_0)} & \forall \psi_h \in V_h \\ (p_h + \beta u_h, v_h)_{L^2(\Omega)} = 0 & \forall v_h \in \mathcal{U}_h, \end{cases} \tag{6.33}$$

with

$$s_h(y_h, \varphi_h) = \sum_{K \in T_h} \int_K \delta_K R(y_h; u_h)\, \mathcal{E}\varphi_h \, d\mathbf{x}, \tag{6.34}$$

$$s_h^*(p_h, \psi_h) = \sum_{K \in T_h} \int_K \delta_K R^*(p_h; y_h)\, \mathcal{E}^*\psi_h \, d\mathbf{x}. \tag{6.35}$$

Here $\delta_K > 0$ is a suitable stabilization parameter, to be chosen according to the *local Péclet number* $\mathbb{P}e_h = \|\mathbf{b}\|_{L^\infty(K)} h_K / 2\|\nu\|_{L^\infty(K)}$, $h_K = diam(K)$, whereas

$$R(y; u) = \mathcal{E}y - u, \qquad R^*(p; y) = \mathcal{E}^*p - C_0 y, \qquad C_0 y = \chi_{\Omega_0} k\, (k\, y - z_d).$$

The terms s_h and s_h^* aim at controlling the numerical oscillations that would affect the pure Galerkin method (see [218]). Other stabilization strategies are of course available. The discretized system (6.33) is *strongly consistent*, meaning that if (y, u, p) solve (6.31) and are such that $y, p \in H^2(K)$ for any $K \in \mathcal{T}_h$, then (y, u, p) also solve (6.33). Note however that, due to the presence of the GLS terms $s_h(y_h, \varphi_h)$ and $s_h^*(p_h, \psi_h)$, and to the right-hand side of $(6.33)_2$, (6.33) cannot represent the system of optimality conditions for a discrete optimization problem. As a matter of fact, let us rewrite $(6.33)_1$ as

$$a_h(y_h, \varphi_h) = (u_h, \varphi_h)_{L^2(\Omega)} + \sum_{K \in T_h} \int_K \delta_K u_h \, \mathcal{E}\varphi_h \, d\mathbf{x} \quad \forall \varphi_h \in V_h$$

where

$$a_h(y_h, \varphi_h) = a(y_h, \varphi_h) + \sum_{K \in T_h} \int_K \delta_K \mathcal{E}y_h \, \mathcal{E}\varphi_h \, d\mathbf{x}$$

and, similarly, $(6.33)_2$ as

$$a_h^*(p_h, \psi_h) = (ky_h - z_d, k\psi_h)_{L^2(\Omega)} + \sum_{K \in T_h} \int_K \delta_K C_0 y_h \, \mathcal{E}^*\psi_h \, d\mathbf{x} \quad \forall \varphi_h \in V_h$$

where

$$a_h^*(p_h, \psi_h) = a^*(p_h, \psi_h) + \sum_{K \in T_h} \int_K \delta_K \mathcal{E}^* p_h \, \mathcal{E}^*\psi_h \, d\mathbf{x}.$$

Since the advection term has different forms in the state $\mathcal{E}y$ and adjoint $\mathcal{E}^*p$ operators, $a_h(\varphi_h, y_h) \neq a_h^*(y_h, \varphi_h)$.

Discretize then Optimize

If we follow the DtO approach, we first discretize (by the same stabilized GLS formulation as in $(6.33)_1$–(6.34)), then we recover the optimality conditions by introducing the (now, discrete) Lagrangian functional

$$\mathcal{L}_h(y_h, u_h, p_h) = J(y_h, u_h) + (u_h, p_h)_{L^2(\Omega)} - a(y_h, p_h) - s_h(y_h, p_h). \tag{6.36}$$

By differentiating $\mathcal{L}_h$ with respect to p_h, y_h, and u_h, respectively, we obtain the following optimality system: find $y_h \in V_h$, $p_h \in V_h$ and $u_h \in \mathcal{U}_h$ such that

$$\begin{cases} a(y_h, \varphi_h) + s_h(y_h, \varphi_h) = (u_h, \varphi_h)_{L^2(\Omega)} & \forall \varphi_h \in V_h \\ a_h(\psi_h, p_h) = (ky_h - z_d, k\psi_h)_{L^2(\Omega)} & \forall \varphi_h \in V_h \\ (p_h + \beta u_h, v_h)_{L^2(\Omega)} + \sum_{K \in \mathcal{T}_h} \int_K \delta_K v_h \mathcal{E} p_h \, d\mathbf{x} = 0 & \forall v_h \in \mathcal{U}_h. \end{cases} \tag{6.37}$$

Note that the discrete state equation $(6.37)_1$ is equal to the one, $(6.33)_1$, obtained with the OtD approach. Instead, the discrete adjoint equation $(6.37)_2$ involves a stabilization term

$$\tilde{s}_h^*(p_h, \psi_h) = \sum_{K \in \mathcal{T}_h} \int_K \delta_K \mathcal{E}\psi_h \, \mathcal{E} p_h \, d\mathbf{x} \tag{6.38}$$

different than the one given by (6.35), and appearing in $(6.33)_2$.

There are therefore substantial differences between the discrete adjoint equation $(6.37)_2$ and the discretized adjoint equation $(6.33)_2$, as well as between the discrete optimality condition $(6.37)_3$ and the discretized optimality condition $(6.33)_3$:

- while the discrete state equation is strongly consistent (as it happened for $(6.33)_1$), $(6.37)_2$ is not, that is, it is not satisfied if y_h and p_h are replaced by y and p, respectively. In fact, the discrete adjoint equation does not result from a weighted residual method applied to the continuous adjoint problem;
- equation $(6.37)_2$ is not the adjoint of equation $(6.33)_1$;
- the optimality condition $(6.37)_3$ is not satisfied if u_h and p_h are replaced by u and p, respectively.

To overcome this lack of consistency, we can follow the method proposed in [78] and introduce a different, stabilized Lagrangian functional

$$\mathcal{L}_h^s(y_h, p_h, u_h) = \mathcal{L}(y_h, p_h, u_h) + S_h(y_h, p_h, u_h), \tag{6.39}$$

where

$$S_h(y, p, u) = \sum_{K \in \mathcal{T}_h} \int_K \delta_K R(y; u) \; R^*(p; y) \, d\mathbf{x}. \tag{6.40}$$

By differentiating $\mathcal{L}_h^s$ we obtain the new discretized state and adjoint equations

$$a(y_h, \varphi_h) + \tilde{\tilde{s}}_h(y_h, \varphi_h) = (u_h, \varphi_h)_{L^2(\Omega)} \qquad \forall \varphi_h \in \mathcal{V}_h \tag{6.41}$$

$$a^*(p_h, \psi_h) + \tilde{\tilde{s}}_h^*(p_h, \psi_h) = (ky_h - z_d, k\psi_h)_{L^2(\Omega_0)} \qquad \forall \psi_h \in \mathcal{V}_h, \tag{6.42}$$

which are still written under the form of $(6.33)_1$, $(6.33)_2$, respectively, yet involving the new stabilization terms

$$\tilde{\tilde{s}}_h(y_h, \varphi_h; u_h) = -\sum_{K \in \mathcal{T}_h} \int_K \delta_K R(y_h; u_h) \; \mathcal{E}^* \varphi_h \, d\mathbf{x}, \tag{6.43}$$

$$\tilde{\tilde{s}}_h^*(p_h, \psi_h; y_h) = -\sum_{K \in \mathcal{T}_h} \int_K \delta_K \left(R^*(p_h; y_h) \; \mathcal{E}\psi_h - R(y_h; u_h) \; G\psi_h \right) d\mathbf{x}, \tag{6.44}$$

instead than s_h, s_h^*, respectively, with $G\psi = \chi_{\Omega_0} k^2 \psi$.

The optimality condition now becomes

$$(p_h + \beta u_h, v_h)_{L^2(\Omega)} - \sum_{K\in\mathcal{T}_h} \int_K \delta_K \; R^*(p_h; y_h) v_h \, d\mathbf{x} = 0 \qquad \forall v_h \in \mathcal{U}_h. \tag{6.45}$$

In this way, we still have a strongly consistent method for both the discrete state (6.41) and adjoint (6.42) problems. More importantly, equation (6.42) is indeed the adjoint of equation (6.41). Furthermore, the strong consistency property is also reflected in the optimality condition (6.45), which is indeed fulfilled if u_h and p_h are replaced by the optimal control u and the corresponding adjoint p, since $R^*(p; y) = 0$.

6.2.5 The case of Stokes Equations

Let us consider the following OCP for the Stokes equations (indeed very similar to the one addressed in Sect. 5.13.1) with distributed control and observation:

$$\begin{aligned} \tilde{J}(\boldsymbol{v}, \pi, \boldsymbol{u}) = \frac{1}{2} \int_\Omega |\boldsymbol{v} - \boldsymbol{v}_d|^2 \, d\mathbf{x} + \frac{\delta}{2} \int_\Omega (\pi - \pi_d)^2 \, d\mathbf{x} + \frac{\alpha}{2} \int_\Omega |\boldsymbol{u}|^2 \, d\mathbf{x} \;\rightarrow\; \min \\ \text{subject to } (\boldsymbol{v}, \pi) \in V, \; \boldsymbol{u} \in \mathcal{U}, \\ \begin{cases} -\nu \Delta \boldsymbol{v} + \nabla \pi = \boldsymbol{f} + \boldsymbol{u} & \text{in } \Omega \\ \operatorname{div} \boldsymbol{v} = 0 & \text{in } \Omega \\ \boldsymbol{v} = \boldsymbol{g} & \text{on } \Gamma; \end{cases} \end{aligned} \tag{6.46}$$

$\boldsymbol{v}_d \in L^2(\Omega)^d$ and $\pi_d \in L^2(\Omega)$ are given target velocity and pressure fields, while $\delta > 0$, $\alpha > 0$ are given parameters. The second term of $\mathcal{J}$ penalizes the pressure, driving it to be as close as possible to π_d; its role will become clear in the sequel.

For the numerical approximation of the OCP (6.46), we can follow either the OtD or the DtO approach; here we adopt the latter choice, the former yielding indeed the same result, provided the discrete spaces for both the state and the adjoint problems be the same (see Exercise 4).

Let us introduce a stable pair of finite element spaces (X_h, Q_h) for the velocity and the pressure (such as the Taylor-Hood $\mathbb{P}_2$–$\mathbb{P}_1$ finite elements; see Sect. B.2.3), of dimension N_v, N_π, respectively; hence, $V_h = X_h \times Q_h$ has dimension $N_y = N_v + N_\pi$. Since we deal with distributed controls, here we can choose $\mathcal{U}_h = X_h$. The discretized state equation in algebraic form reads

$$\underbrace{\begin{bmatrix} \mathbb{E} & \mathbb{B}^\top \\ \mathbb{B} & 0 \end{bmatrix}}_{\mathbb{A}_S} \begin{bmatrix} \mathbf{v} \\ \boldsymbol{\pi} \end{bmatrix} = \begin{bmatrix} \mathbb{M}_v \\ 0 \end{bmatrix} \begin{bmatrix} \mathbf{u} \\ 0 \end{bmatrix} + \underbrace{\begin{bmatrix} \mathbf{f} \\ \mathbf{g} \end{bmatrix}}_{\mathbf{f}_s}. \tag{6.47}$$

Here $(\mathbf{v}, \boldsymbol{\pi}, \mathbf{u})$ are the vectors of degrees of freedom corresponding to the optimal state (velocity and pressure) and control variables, respectively, $\mathbb{M}_v \in \mathbb{R}^{N_v \times N_v}$ is the velocity mass matrix, $\mathbb{E} \in \mathbb{R}^{N_v \times N_v}$ and $\mathbb{B} \in \mathbb{R}^{N_v \times N_\pi}$ is the matrix representations of the (vector) Laplace and divergence operators, $\mathbf{f} \in \mathbb{R}^{N_v}$ and $\mathbf{g} \in \mathbb{R}^{N_\pi}$ account for the source terms and non-homogeneous Dirichlet boundary conditions.

The discrete counterpart of (6.46) reads

$$\tilde{J}_h(\mathbf{v},\boldsymbol{\pi},\mathbf{u}) = \frac{1}{2}\mathbf{v}^\top \mathbb{M}_v\mathbf{v} - \mathbf{v}^\top\mathbb{M}_v\mathbf{v}_d + \frac{\delta}{2}\boldsymbol{\pi}^\top\mathbb{M}_\pi\boldsymbol{\pi} + \frac{\alpha}{2}\mathbf{u}^\top\mathbb{M}_v\mathbf{u} + \frac{1}{2}\mathbf{v}_d^\top\mathbb{M}_v\mathbf{v}_d \to \min$$

$$\text{subject to } \boldsymbol{u}\in\mathbb{R}^{N_v}, \qquad (6.48)$$

$$\mathbf{v}\in\mathbb{R}^{N_v},\ \boldsymbol{\pi}\in\mathbb{R}^{N_\pi}: \quad \begin{cases}\mathbb{E}\mathbf{v} + \mathbb{B}^\top\boldsymbol{\pi} = \mathbb{M}_v\mathbf{u} + \mathbf{f}\\ \mathbb{B}\mathbf{v} = \mathbf{g},\end{cases}$$

where $\mathbb{M}_\pi$ is the pressure mass matrix. Introducing the discrete adjoint velocity $\mathbf{z}$ and pressure $\mathbf{q}$, similarly to Sect. 6.2.2, we obtain the following optimality system:

- *state problem*

$$\begin{cases}\mathbb{E}\mathbf{v} + \mathbb{B}^\top\boldsymbol{\pi} = \mathbb{M}_v\mathbf{u} + \mathbf{f}\\ \mathbb{B}\mathbf{v} \qquad\quad = \mathbf{g};\end{cases} \tag{6.49}$$

- *adjoint problem*

$$\begin{cases}\mathbb{E}\mathbf{z} + \mathbb{B}^\top\mathbf{q} = -\mathbb{M}_v(\mathbf{v} - \mathbf{v}_d)\\ \mathbb{B}\mathbf{z} \qquad\quad = -\delta\mathbb{M}_\pi(\boldsymbol{\pi} - \boldsymbol{\pi}_d);\end{cases} \tag{6.50}$$

- *optimality condition*

$$-\mathbb{M}_v\mathbf{z} + \alpha\mathbb{M}_v\mathbf{u} = \mathbf{0}. \tag{6.51}$$

Equivalently, (6.49)–(6.51) can be rewritten in matrix form as follows:

$$\begin{bmatrix}\mathbb{M}_v & 0 & 0 & \mathbb{E} & \mathbb{B}^\top\\ 0 & \delta\mathbb{M}_\pi & 0 & \mathbb{B} & 0\\ 0 & 0 & \alpha\mathbb{M}_v & -\mathbb{M}_v & 0\\ \mathbb{E} & \mathbb{B}^\top & -\mathbb{M}_v & 0 & 0\\ \mathbb{B} & 0 & 0 & 0 & 0\end{bmatrix}\begin{bmatrix}\mathbf{v}\\ \boldsymbol{\pi}\\ \mathbf{u}\\ \mathbf{z}\\ \mathbf{q}\end{bmatrix} = \begin{bmatrix}\mathbb{M}_v\mathbf{v}_d\\ \delta\mathbb{M}_\pi\boldsymbol{\pi}_d\\ \mathbf{0}\\ \mathbf{f}\\ \mathbf{g}\end{bmatrix}. \tag{6.52}$$

If we introduce the *aggregated* variables for the state $\mathbf{Y}$ (velocity and pressure) and the adjoint $\mathbf{P}$ (adjoint velocity and pressure),

$$\mathbf{Y} = \begin{bmatrix}\mathbf{v}\\ \boldsymbol{\pi}\end{bmatrix}, \qquad \mathbf{P} = \begin{bmatrix}\mathbf{z}\\ \mathbf{q}\end{bmatrix},$$

we can highlight the same block structure of the system of optimality conditions (6.16). In fact, system (6.52) can be written as

$$\begin{bmatrix}\mathbb{M} & 0 & \mathbb{A}_S\\ 0 & \alpha\mathbb{M}_v & -\mathbb{V}^\top\\ \mathbb{A}_S & -\mathbb{V} & 0\end{bmatrix}\begin{bmatrix}\mathbf{Y}\\ \mathbf{u}\\ \mathbf{P}\end{bmatrix} = \begin{bmatrix}\mathbf{f}_a\\ \mathbf{0}\\ \mathbf{f}_s\end{bmatrix}, \tag{6.53}$$

provided we define

$$\mathbb{M} = \begin{bmatrix}\mathbb{M}_v & 0\\ 0 & \delta\mathbb{M}_\pi\end{bmatrix}, \qquad \mathbb{V} = \begin{bmatrix}\mathbb{M}_v\\ 0\end{bmatrix}, \qquad \mathbf{f}_a = \begin{bmatrix}\mathbb{M}_v\mathbf{v}_d\\ \delta\mathbb{M}_\pi\boldsymbol{\pi}_d\end{bmatrix}, \qquad \mathbf{f}_s = \begin{bmatrix}\mathbf{f}\\ \mathbf{g}\end{bmatrix}. \tag{6.54}$$

6.3 Iterative Methods (I): Unconstrained OCPs

So far, we have derived the discrete optimality system for an unconstrained linear-quadratic OCP; both OtD and DtO approaches yield large linear systems to solve, with a saddle-point structure. Employing an efficient preconditioned iterative solver to compute its solution is the goal of *all-at-once* methods. We will describe them in Sect. 6.6.

In this section we address a simpler class of methods, the so-called *iterative (or black-box, or reduced space) methods*, for the numerical solution of unconstrained OCPs, which rely on classical algorithms for numerical optimization, such as descent or trust-region methods (see Sects. 3.1–3.2). Iterative methods consider the state variable y as a function of u, the control, and then recast the OCP into an unconstrained optimization problem for u,

$$J(u) = \tilde{J}(y(u), u) \rightarrow \min \quad \text{subject to } u \in \mathcal{U}. \tag{6.55}$$

Once u is available, $y(u)$ is computable by solving the state equation. Although an iterative method can also be set on the infinite-dimensional problem (6.55) in a Banach space setting (see, e.g., [148, Chapter 2]), here we introduce it on the discrete counterpart of (6.55),

$$J_h(\mathbf{u}) = \tilde{J}_h(\mathbf{y}(\mathbf{u}), \mathbf{u}) \rightarrow \min \quad \text{subject to } \mathbf{u} \in \mathbb{R}^{N_u}. \tag{6.56}$$

The optimization variable $\mathbf{u}$ is the vector of nodal values of the discrete control u_h and N_u is the dimension of the discrete control space. For the sake of simplicity we focus on the model problem (6.1); the extension to advection-diffusion-reaction equations or Stokes equations is straightforward. When a descent method is employed, given an initial guess $\mathbf{u}_0 \in \mathbb{R}^{N_u}$, we need to find iteratively a sequence $\{\mathbf{u}_k\}$ such that

$$\mathbf{u}_{k+1} = \mathbf{u}_k + \tau_k \, \mathbf{d}_k$$

where $\mathbf{d}_k$ is a descent direction (that can depend on the gradient ∇J_h and/or the Hessian[5] $\mathbb{H}$ of the objective function J_h) and $\tau_k > 0$ is a suitable step size which should ensure that

$$J_h(\mathbf{u}_k + \tau_k \mathbf{d}_k) < J_h(\mathbf{u}_k).$$

Below we apply the algorithms of Sect. 3.1 to problem (6.56).

Steepest descent method

As seen in Sect. 3.1.1, the *steepest descent method* (also called *gradient method*) uses the direction

$$\mathbf{d}_k = -\nabla J_h(\mathbf{u}_k)$$

at each step. It is globally convergent (that is for every choice of $\mathbf{u}_0$) but only linearly: it might therefore require a large number of steps to converge to the solution within a given tolerance. Moreover, it requires to evaluate the gradient ∇J_h at each step. The evaluation of $J_h(\bar{\mathbf{u}})$ at an arbitrary point $\bar{\mathbf{u}}$ (the generic iterate $\mathbf{u}_k$) requires to solve the state problem in order to compute $\bar{\mathbf{y}} = \mathbf{y}(\bar{\mathbf{u}})$. In the same way, evaluating the *reduced gradient* $\nabla J_h(\bar{\mathbf{u}})$ requires to solve both the state and the adjoint problem, that is, two potentially large-scale linear systems (whose size depends on the dimension of the discrete space V_h). Note that $\nabla J_h(\bar{\mathbf{u}})$ depends on the inner product we have chosen in $\mathbb{R}^{N_u}$.

---

[5] If the functional is quadratic, i.e. of the form

$$J_h(\mathbf{u}) = \frac{1}{2}\mathbf{u}^\top \mathbb{H}\mathbf{u} - \mathbf{b}^\top \mathbf{u},$$

the (reduced) Hessian matrix $\mathbb{H} = \beta\mathbb{N} + \mathbb{B}^\top \mathbb{A}^{-\top}\mathbb{M}\mathbb{A}^{-1}\mathbb{B} \in \mathbb{R}^{N_u \times N_u}$ defined in (6.18) does not depend on $\mathbf{u}$. Evaluating its action $\mathbb{H}\mathbf{w}$ on a generic vector $\mathbf{w}$ involves, however, the solution of a state and an adjoint problem, see Sect. 6.3.1.

Conjugate gradient method

The *conjugate gradient (CG) method* improves the gradient method by using descent directions that are conjugate with respect to $\mathbb{H}$, i.e. such that

$$\mathbf{d}_k^\top \, \mathbb{H} \, \mathbf{d}_{k-1} = 0. \tag{6.57}$$

In the case of a more general (non quadratic) cost functional, $\mathbb{H} = \mathbb{H}(\mathbf{u}_k)$, the k-th descent direction is determined as

$$\mathbf{d}_k = -\nabla J_h(\mathbf{u}_k) + \beta_k \mathbf{u}_{k-1}$$

where $\beta_k \in \mathbb{R}$ is chosen so that (6.57) is fulfilled. For a quadratic functional,

$$\beta_k = \frac{\nabla J_h(\mathbf{u}_k)^\top \mathbb{H} \mathbf{d}_{k-1}}{\mathbf{d}_{k-1}^\top \mathbb{H} \mathbf{d}_{k-1}}.$$

In the case of a general nonlinear cost functionals, different nonlinear CG methods, corresponding to different choices of β_k, can be employed, such as the Fletcher-Reeves and the Polak-Ribière formulae, see equation (3.6) and Sect. 3.1.1.

Newton and quasi-Newton methods

Among descent methods, Newton and quasi-Newton methods are those most employed. In the Newton method we minimize the quadratic functional

$$m(\mathbf{d}_k) = J(\mathbf{u}_k) + \mathbf{d}_k^\top \nabla J_h(\mathbf{u}_k) + \frac{1}{2} \mathbf{d}_k^\top (\mathbb{H}(\mathbf{u}_k)) \mathbf{d}_k$$

which represents an approximation of $J(\mathbf{u}_k)$ in a neighborhood of $\mathbf{u}_k$. Provided the Hessian matrix is positive definite, the descent direction $\mathbf{d}_k$ that minimizes $m(\mathbf{d}_k)$ is

$$\mathbf{d}_k = -(\mathbb{H}(\mathbf{u}_k))^{-1} \nabla J_h(\mathbf{u}_k),$$

that is, the descent direction is obtained by solving the linear system

$$(\mathbb{H}(\mathbf{u}_k)) \mathbf{d}_k = -\nabla J_h(\mathbf{u}_k). \tag{6.58}$$

Newton methods converge quadratically, but only locally (that is for appropriate choice of $\mathbf{u}_0$ sufficiently close to the minimum). Note that for a general nonlinear cost functional the Hessian matrix $\mathbb{H} = \mathbb{H}(\mathbf{u}_k)$ must be reassembled at each step. In any case, $\mathbb{H}$ is usually not formed explicitly due to the large computational effort that would be required. For instance, in the case of a linear-quadratic OCP, (6.58) can be more conveniently written as (see Exercise 5)

$$\begin{bmatrix} \mathbb{M} & 0 & -\mathbb{A}^\top \\ 0 & \beta \mathbb{N} & \mathbb{B}^\top \\ -\mathbb{A} & \mathbb{B} & 0 \end{bmatrix} \begin{bmatrix} \delta \mathbf{y} \\ \mathbf{d}_k \\ \delta \mathbf{p} \end{bmatrix} = \begin{bmatrix} 0 \\ \mathbb{B}^\top \mathbf{p}_k - \beta \mathbb{N} \mathbf{u}_k \\ 0 \end{bmatrix}.$$

Quasi-Newton methods provide instead an alternative to the exact calculation of the Hessian matrix $\mathbb{H}(\mathbf{u}_k)$. An approximate inverse $\mathbb{B}_k^{-1}$ is used to replace $(\mathbb{H}(\mathbf{u}_k))^{-1}$ in equation (6.58); one of the most popular formulas for updating $\mathbb{B}_k$ is the BFGS (from Broyden-Fletcher-Goldfarb-Shanno) update, see equation (3.14), whereas a possible way to update the inverse of $\mathbb{B}_k$ instead of $\mathbb{B}_k$ itself was provided in (3.14). Quasi-Newton methods need to store gradient information at each step to approximate second derivatives; due to the large dimension of the discretized OCP, the limited-memory BFGS (LM-BFGS) should be employed for the sake of computational efficiency (see [208, Section 7.2]).

The different choices are summarized in Tab. 6.1; see also Algorithm 6.1.

steepest descent	$\mathbf{d}_k = -\nabla J_h(\mathbf{u}_k)$
conjugate gradient	$\mathbf{d}_k = -\nabla J_h(\mathbf{u}_k) + \beta_k \mathbf{u}_{k-1}$
Newton	$\mathbf{d}_k = -\mathbb{H}^{-1}\nabla J_h(\mathbf{u}_k)$
quasi-Newton	$\mathbf{d}_k = -\mathbb{B}_k^{-1}\nabla J_h(\mathbf{u}_k)$

Table 6.1 Choice of descent directions for the solution of an unconstrained OCP

Note that, when using an iterative method, the PDE is nested in the optimization routine; this method is therefore recommended if a PDE solver is already available. This solver must provide updated state variables that satisfy the PDE constraint, and corresponding sensitivity information (e.g., the gradient ∇J_h). The three equations forming the optimality system are then uncoupled, and the state and the adjoint problems are iteratively solved, to provide those quantities required to evaluate ∇J_h and, possibly, the Hessian matrix $\mathbb{H}$. We remark that, in an iterative method, the optimization solver needs only control variables to be stored, and PDE constraints are always (implicitly) satisfied.

Algorithm 6.1 Iterative algorithm for unconstrained OCPs

Input: initial guess $\mathbf{u}_0$

compute $\mathbf{y}_0$ by solving the state equation with $\mathbf{u} = \mathbf{u}_0$
compute $\mathbf{p}_0$ by solving the adjoint equation with $\mathbf{y} = \mathbf{y}_0$
compute $J_h(\mathbf{u}_0)$ and $\nabla J_h(\mathbf{u}_0)$
set $k = 0$
while $|\nabla J_h(\mathbf{u}_k)| / |\nabla J_h(\mathbf{u}_0)| > \text{tol}$ or $|J_h(\mathbf{u}_k) - J_h(\mathbf{u}_{k-1})| > \text{tol}$
  compute the search direction $\mathbf{d}_k$ according to a descent method
  compute the step length τ_k with a line-search routine
  set $\mathbf{u}_{k+1} = \mathbf{u}_k + \tau_k \mathbf{d}_k$
  compute $\mathbf{y}_{k+1}$ by solving the state equation with $\mathbf{u} = \mathbf{u}_{k+1}$
  compute $\mathbf{p}_{k+1}$ by solving the adjoint equation with $\mathbf{y} = \mathbf{y}_{k+1}$
  compute $J_h(\mathbf{u}_{k+1})$ and $\nabla J_h(\mathbf{u}_{k+1})$
  $k = k + 1$
end while

Remark 6.4. Employing a finite difference approximation of the gradient may seem an attractive alternative, because of its ease of implementation. However, this often induces limited accuracy and large computational costs when dealing with, e.g., distributed control variables, whose dimension N_u depends on the discretization of the control u. Finite difference approximations of the gradient can be more reliable when dealing with parametrized control variables that are expressed in terms of few *control parameters* (or *design variables*), with respect to which the cost function is indeed minimized. •

For an OCP related with a Stokes problem (see Sect. 6.2.5), at each step of an iterative algorithm two Stokes problems have to be solved; numerical results are reported in Sect. 6.5.2.

Choice of of the step length

For linear-quadratic problems the calculation of the *optimal step length* τ_k acoording to the (line-search) criterium (3.16) is straightforward. In fact, recalling that $\mathbf{y}$ and $\mathbf{u}$ represent the nodal values of state and control variables, we have

$$\phi(\tau) = J_h(\mathbf{u}_k + \tau \mathbf{d}_k) = \frac{1}{2}\|\mathbf{y}_k + \tau(\mathbf{y}(\mathbf{d}_k) - \mathbf{y}(\mathbf{0})) - \mathbf{z}_d\|_{\mathbb{M}}^2 + \frac{\beta}{2}\|\mathbf{u}_k + \tau \mathbf{d}_k\|_{\mathbb{N}}^2$$

where $\mathbf{y}(\mathbf{d}_k)$ and $\mathbf{y}(\mathbf{0})$ denote the numerical solutions of the state equation when $\mathbf{u} = \mathbf{d}_k$ or $\mathbf{u} = \mathbf{0}$, respectively. Then a simple calculation (see Exercise 6) yields

$$\tau_k = \arg\min_{\tau > 0} \phi(\tau) = -\frac{\beta(\mathbf{u}_k, \mathbf{d}_k)_{\mathbb{N}} + (\mathbf{y}_k - \mathbf{z}_d, \mathbf{y}(\mathbf{d}_k) - \mathbf{y}(\mathbf{0}))_{\mathbb{M}}}{\|\mathbf{y}(\mathbf{d}_k) - \mathbf{y}(\mathbf{0})\|_{\mathbb{M}}^2 + \beta\|\mathbf{d}_k\|_{\mathbb{N}}^2}. \tag{6.59}$$

In this case, in step 7 of Algorithm 6.1 we can compute τ_k without invoking an inexact line-search routine. For more general problems, alternative step length selection strategies in step 4 are based, e.g., on Wolfe conditions, see Sect. 3.1.2, or the backtracking Armijo line-search: for given parameters $\sigma \in (0,1)$ – usually $\sigma \in (10^{-5}, 10^{-1})$ – and $\rho \in [1/10, 1/2)$, we accept the first step length $\tau = \rho^j, j = 0, 1, \ldots$ which satisfies the descent criterion

$$\varphi(\tau) \leq \varphi(0) + \sigma\tau\varphi'(0)$$

where

$$\varphi(\tau) = J(\mathbf{u}_k + \tau \mathbf{d}_k), \qquad \varphi'(0) = (\nabla J(\mathbf{u}_k), \mathbf{d}_k).$$

Further details about the evaluation of the descent directions in the case of control constraints can be found, e.g., in [253].

For the sake of completeness, we also report the algorithm of the steepest descent method with the backtracking Armijio line-search in Algorithm 6.2. In particular, we denote by τ_0 and τ_{min} the initial and the minimum step size, respectively, and by K_{max} the maximum number of allowed iterations.

Algorithm 6.2 Steepest descent method with backtracking Armijio line-search for unconstrained OCPs

---

Input: initial guess $\mathbf{u}_0$, σ, ρ, τ_0, τ_{min}

1: compute $\mathbf{y}_0$ by solving the state equation with $\mathbf{u} = \mathbf{u}_0$
2: compute $\mathbf{p}_0$ by solving the adjoint equation with $\mathbf{y} = \mathbf{y}_0$
3: compute $J_h(\mathbf{u}_0)$ and $\nabla J_h(\mathbf{u}_0)$
4: set $k = 0$
5: **while** $|\nabla J_h(\mathbf{u}_k)| / |\nabla J_h(\mathbf{u}_0)| >$ tol and $k < K_{max}$
6:     set $\tau = \tau_0$
7:     **while** $\tau > \tau_{min}$
8:         $\mathbf{u}_{k+1} = \mathbf{u}_k - \tau \nabla J_h(\mathbf{u}_k)$
9:         compute $\mathbf{y}_{k+1}$ by solving the state equation with $\mathbf{u} = \mathbf{u}_{k+1}$
10:         compute $\mathbf{p}_{k+1}$ by solving the adjoint equation with $\mathbf{y} = \mathbf{y}_{k+1}$
11:         compute $J_h(\mathbf{u}_{k+1})$ and $\nabla J_h(\mathbf{u}_{k+1})$
12:         **if** $J_h(\mathbf{u}_{k+1}) \leq J(\mathbf{u}_k) - \sigma\tau(\nabla J_h(\mathbf{u}_k), \nabla J_h(\mathbf{u}_k))$
13:             set $k = k + 1$
14:             break
15:         **else**
16:             $\tau = \rho\tau$
17:         **end if**
18:     **end while**
19: **end while**

6.3.1 Relation with Solving the Reduced Hessian Problem

Any iterative method for the minimization of $J_h(\mathbf{u})$ (see (6.56)) can be regarded as an iterative method to solve the linear system (6.17) iteratively. For instance, an iteration of the steepest descent algorithm described above is equivalent to an iteration of the non-stationary Richardson method for the solution of (6.17):

$$\mathbf{u}_{k+1} = \mathbf{u}_k + \tau_k \mathbf{r}_k, \qquad \mathbf{r}_k = \mathbf{b} - \mathbb{H}\mathbf{u}_k. \tag{6.60}$$

As a matter of fact, the steepest descent update reads

$$\mathbf{u}_{k+1} = \mathbf{u}_k + \tau_k \mathbf{d}_k, \qquad \mathbf{d}_k = -\nabla J_h(\mathbf{u}_k) = \mathbb{B}^\top \mathbf{p}_k - \beta \mathbb{N}\mathbf{u}_k.$$

Using the adjoint equation (6.11), $\mathbf{p}_k = -\mathbb{A}^{-\top}(\mathbb{M}\mathbf{y}_k - \mathbb{M}\mathbf{z}_d)$, thus

$$\mathbf{u}_{k+1} = \mathbf{u}_k + \tau_k \big(\mathbb{B}^\top \mathbb{A}^{-\top}\mathbb{M}\mathbf{z}_d - \beta \mathbb{N}_h \mathbf{u}_k - \mathbb{B}^\top \mathbb{A}^{-\top}\mathbb{M}\mathbf{y}_k\big).$$

Then, using the state equation (6.10), $\mathbf{y}_k = \mathbb{A}^{-1}(\mathbb{B}\mathbf{u}_k + \mathbf{f})$, substituting and rearranging we obtain

$$\begin{aligned}\mathbf{u}_{k+1} &= \mathbf{u}_k + \tau_k(\mathbb{B}^\top \mathbb{A}^{-\top}\mathbb{M}\mathbf{z}_d - \mathbb{B}^\top \mathbb{A}^{-\top}\mathbb{M}\mathbb{A}^{-1}\mathbf{f} - \beta\mathbb{N}\mathbf{u}_k - \mathbb{B}^\top \mathbb{A}^{-\top}\mathbb{M}\mathbb{A}^{-1}\mathbb{B}\mathbf{u}_k)\\ &= \mathbf{u}_k + \tau_k(\mathbf{b} - \mathbb{H}\mathbf{u}_k),\end{aligned}$$

that is (6.60). This result is not surprising if we write the algebraic counterpart of the reduced cost functional: in fact, by substituting $\mathbf{y}(\mathbf{u}) = \mathbb{A}^{-1}(\mathbb{M}\mathbf{u} + \mathbf{f})$ in $\tilde{J}_h(\mathbf{y}, \mathbf{u})$, we obtain

$$\begin{aligned}J_h(\mathbf{u}) &= \frac{1}{2}\mathbf{u}^\top \mathbb{M}\mathbb{A}^{-\top}\mathbb{M}\mathbb{A}^{-1}\mathbb{M}\mathbf{u} - \mathbf{u}^\top \mathbb{M}\mathbb{A}^{-\top}\mathbb{M}\mathbf{z}_d + \frac{\beta}{2}\mathbf{u}^\top \mathbb{N}\mathbf{u}\\ &\quad + \mathbf{u}^\top \mathbb{M}\mathbb{A}^{-\top}\mathbb{M}\mathbb{A}^{-1}\mathbf{f} + \frac{1}{2}\mathbf{z}_d^\top \mathbb{M}\mathbf{z}_d + \frac{1}{2}\mathbf{f}\mathbb{A}^{-\top}\mathbb{M}\mathbb{A}^{-1}\mathbf{f} - \mathbf{f}^\top \mathbb{A}^{-\top}\mathbb{M}\mathbf{z}_d\\ &= \frac{1}{2}\mathbf{u}^\top \mathbb{H}\mathbf{u} - \mathbf{u}^\top \mathbf{b} + c.\end{aligned}$$

The constant c accounts for all the $\mathbf{u}$-independent terms since the minimizer of J_h is independent of c. If $\mathbb{H}$ is a symmetric positive definite matrix, minimizing the quadratic functional $J_h(\mathbf{u})$ is equivalent to solving the linear system (see e.g. [222])

$$\nabla J_h(\mathbf{u}) = \mathbb{H}\mathbf{u} - \mathbf{b} = \mathbf{0}.$$

Because of the difficulty in assembling $\mathbb{H}$, the linear system (6.17) is typically solved by iterative methods, in particular CG, that can take advantage of the symmetry of $\mathbb{H}$; see, e.g., [236]. No matter which iterative method is used, at each iteration a matrix-vector multiplication $\mathbb{H}\mathbf{r}$ has to be calculated for the current value $\mathbf{r}$ of the system residual. This vector can be calculated without explicitly constructing the entries of $\mathbb{H}$ by proceeding as follows. We first compute the solution of the state equation

$$\mathbb{A}\mathbf{y}(\mathbf{r}) = \mathbb{B}\mathbf{r},$$

then that of the solution of an adjoint equation

$$\mathbb{A}^T \mathbf{p}(\mathbf{r}) = \mathbb{M}\mathbf{y}(\mathbf{r}).$$

The desired vector is then recovered using the identity

$$\mathbb{H}\mathbf{r} = \beta\mathbb{N}\mathbf{r} + \mathbb{B}^\top \mathbf{p}(\mathbf{r}).$$

Finally, upon convergence to the optimal control variable, the last step consists in solving the state and the adjoint equation to retrieve the state and adjoint solutions, respectively.

6.4 Iterative Methods (II): Control Constraints

Iterative methods can also be used for constrained OCPs with control constraints on the control function under the form of box constraints, that is, with

$$\mathcal{U}_{ad} = \{\mathbf{u} \in \mathbb{R}^{N_u} : u_{a,i} \le u_i \le u_{b,i},\ i = 1, \dots, N_u\} \subset \mathcal{U}. \tag{6.61}$$

By using a more compact notation, the numerical problem (6.56) becomes

$$J_h(\mathbf{u}) = \tilde{J}_h(\mathbf{y}(\mathbf{u}), \mathbf{u}) \ \to \ \min \qquad \text{subject to } \mathbf{a} \le \mathbf{u} \le \mathbf{b} \tag{6.62}$$

where $\mathbf{u} \in \mathbb{R}^{N_u}$, $\mathbf{a}, \mathbf{b} \in \mathbb{R}^{N_u}$ are two vectors representing the bounds at the discrete level.

Following Sect. 3.3, the basic idea of a *projection method* is to use, at each iteration, a descent direction $\mathbf{d}_k$ computed for the unconstrained problem, and then projecting the new iterate onto the set of admissible controls defined by the inequality constraints.

Then step 8 in Algorithm 6.1 is now

$$\mathbf{u}_{k+1} = Pr_{\mathcal{U}_{ad}}(\mathbf{u}_k + \tau_k \mathbf{d}_k); \tag{6.63}$$

$\tau_k > 0$ is a suitable step length and $Pr_{\mathcal{U}_{ad}}$ denotes the projection onto $\mathcal{U}_{ad}$, see Sect. 4.2.2. Although it is often unfeasible in practice, for box constraints the projection can be easily calculated as follows,

$$Pr_{\mathcal{U}_{ad}}\mathbf{w}(\mathbf{x}) = Pr_{[\mathbf{a},\mathbf{b}]}\mathbf{w}(\mathbf{x}) = \max\{\mathbf{a}, \min\{\mathbf{w}(\mathbf{x}), \mathbf{b}\}\}.$$

This equality has to be understood componentwise, that is,

$$(Pr_{\mathcal{U}_{ad}}\mathbf{w})_i(\mathbf{x}) = (Pr_{[\mathbf{a},\mathbf{b}]}\mathbf{w})_i(\mathbf{x}) = \max\{a_i, \min\{w_i(\mathbf{x}), b_i\}\}, \quad i = 1, \dots, N_u.$$

For instance, when using the projected gradient method, (6.63) becomes

$$\mathbf{u}_{k+1} = Pr_{\mathcal{U}_{ad}}(\mathbf{u}_k - \tau_k \nabla J(\mathbf{u}_k)).$$

If we use the backtracking Armijo line-search, in this case we would proceed as follows: for given parameters $\sigma \in (0,1)$, $\rho \in [1/10, 1/2)$, we accept the first step length $\tau = \rho^j, j = 0, 1, \dots$ which satisfies the descent criterion

$$\varphi(\tau) \le \varphi(0) + \sigma\tau\varphi'(0)$$

where

$$\varphi(\tau) = J(Pr_{[\mathbf{a},\mathbf{b}]}(\mathbf{u}_k + \tau\mathbf{d}_k)), \qquad \varphi'(0) = (\nabla J(\mathbf{u}_k), \tilde{\mathbf{d}}_k), \qquad \mathbf{d}_k = -\nabla J(\mathbf{u}_k).$$

Note that

$$\tilde{\mathbf{d}}_k = \begin{cases} \mathbf{0} & \text{if either } \mathbf{u}_k \notin [\mathbf{a}, \mathbf{b}] \text{ or } (\mathbf{u}_k = \mathbf{b} \text{ and } \mathbf{d}_k \ge \mathbf{0}) \text{ or } (\mathbf{u}_k = \mathbf{a} \text{ and } \mathbf{d}_k \le \mathbf{0}) \\ \mathbf{d}_k & \text{otherwise} \end{cases}$$

is a modified descent direction. All iterative methods described in Sect. 6.3 can therefore be employed for the numerical approximation of OCPs in presence of control constraints, too.

6.5 Numerical Examples

In this section we consider some numerical test cases dealing with the application of the iterative methods discussed so far to OCPs governed by either (advection-)diffusion equations or Stokes equations. We limit oursevles to the case of unconstrained OCPs; further applications of iterative methods to constrained OCPs will be considered in Chapter 10.

6.5.1 OCPs governed by Advection-Diffusion Equations

Test case 1: Laplace Equation

To investigate the role of the choice of the descent direction and the regularization parameter $\beta > 0$, we consider the distributed OCP (6.1) in the case $\mathbf{b} = \mathbf{0}$:

$$\tilde{J}(y,u) = \frac{1}{2}\int_{\Omega}(y - z_d)^2 d\mathbf{x} + \frac{\beta}{2}\int_{\Omega} u^2 d\mathbf{x} \ \rightarrow \ \min \tag{6.64a}$$

$$\text{subject to } u \in \mathcal{U},\ y \in V:$$

$$\begin{cases} -\Delta y = f + u & \text{in } \Omega \\ y = 0 & \text{on } \Gamma, \end{cases} \tag{6.64b}$$

where $\beta > 0$, $\mathcal{U} = L^2(\Omega)$ denotes the set of admissible controls, and $V = H_0^1(\Omega)$. For the case at hand, we take $\Omega = (0,1)^2$ and the target function

$$z_d(\mathbf{x}) = 10x_1(1 - x_1)x_2(1 - x_2).$$

To discretize state, control and adjoint variables, we use linear ($\mathbb{P}_1$) FE spaces, built on a mesh made by 17361 vertices and 34320 triangular elements, resulting in $N_y = N_u = N_p = 16961$ degrees of freedom for the three variables. To solve problem (6.64) we employ either the steepest descent method or the BFGS method, including the backtracking Armijio line search method, considering different values of the regularization parameter.

The optimal state, adjoint and control numerical solutions, computed with the BFGS method in the cases $\beta = 10^{-3}$ and $\beta = 10^{-6}$, are displayed in Fig. 6.2. We highlight that, if the regularization parameter is not sufficiently small, the difference between the target z_d and the computed optimal state might be relevant (see Fig. 6.2, top). Smaller values of the cost functional at optimum are reached for smaller values of β, however requiring a larger number of iterations when using the steepest descent method, see Fig. 6.3. We also highlight that the behavior of this latter method reported a huge decrease of the cost functional in the first, few iterations, followed by a mild decrease in the remaining iterations, often occurs when relying on inexact (e.g., backtracking) line search procedures.

On the contrary, the BFGS method requires very few iterations to achieve convergence – measured in terms of the decrease $|\nabla J_h(\mathbf{u}_k)| \,/\, |\nabla J_h(\mathbf{u}_0)|$ of the norm of the gradient of the cost functional, if compared to the case without control (that is, setting $\mathbf{u}_0 = \mathbf{0}$). For the sake of comparison, here we have stopped the BFGS iterations when the cost functional reaches the same value of the optimal cost obtained with the steepest descent method. For the case at hand, the tolerance on the stopping criterion of the descent method is tol $= 10^{-3}$, whereas for the backtracking Armijio line search method we select $\rho = 0.7$, $\sigma = 10^{-5}$, $\tau_{min} = 0.01$ in the case of the BFGS method, $\tau_{min} = 0.2$ in the case of the steepest descent method. To keep computational costs limited, a maximum number $K_{max} = 5$ of step size rescaling is allowed at each iteration.

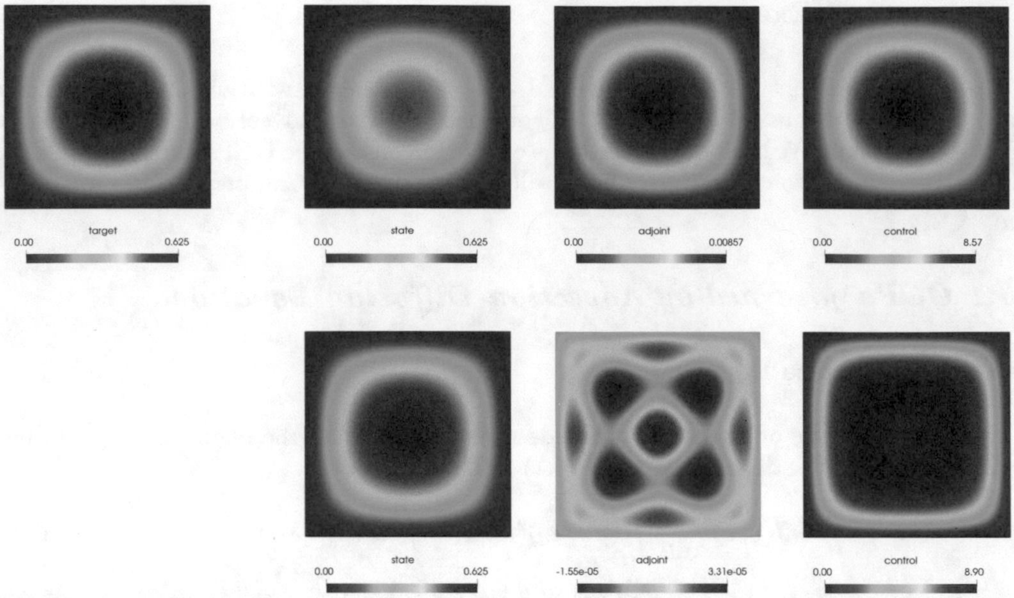

Fig. 6.2 Unconstrained OCP for the Laplace equation, numerical approximation obtained by the BFGS method. Top: target function z_d; optimal state $\hat{y}$, adjoint $\hat{p}$ and control $\hat{u}$ for $\beta = 10^{-3}$. Bottom: optimal state $\hat{y}$, adjoint $\hat{p}$ and control $\hat{u}$ for $\beta = 10^{-6}$.

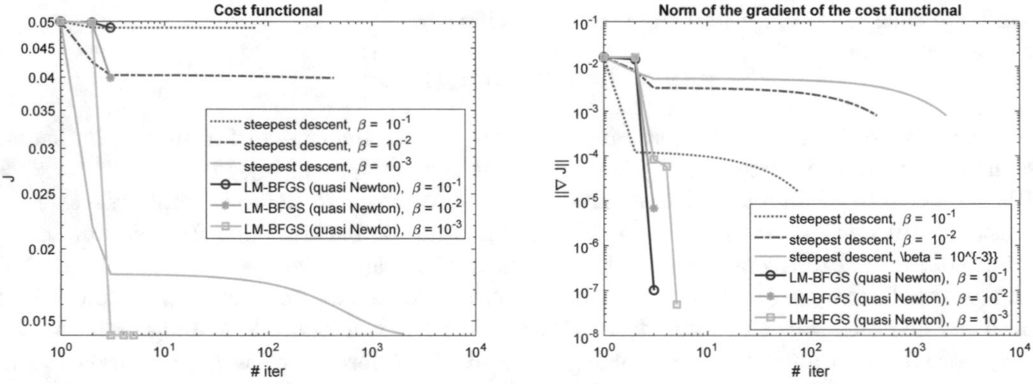

Fig. 6.3 Unconstrained OCP for the Laplace equation, numerical approximation obtained by iterative methods. Left: values of the cost functional $J_h(\mathbf{u}_k)$ along the iterations; right: values of the norm of the gradient $|\nabla J_h(\mathbf{u}_k)|$ of the cost functional during the iterations. The problem has been solved using the steepest descent method and the LM-BFGS method, for different values of the regularization parameter $\beta = 10^{-1}, 10^{-2}, 10^{-3}$

Test case 2: Advection-Diffusion Equation

To investigate the role of the observation region $\Omega_0 \subset \Omega$ and of boundary controls, we consider a boundary OCP for a mixed Dirichlet-Neumannn advection-diffusion equation, namely the Graetz flow problem (see, e.g., [205, 187] and references therein). This latter is a benchmark problem in heat transfer, dealing with forced steady heat convection and heat conduction in a rectangular channel, whose walls are kept at different temperature. The flow has an imposed temperature on Γ_D, and a known convective field (in our case, a given parabolic velocity profile $\mathbf{b}(\mathbf{x}) = (1.5x_2(1 - x_2^2), 0)$, such that $\text{div}\mathbf{b} = 0$).

The OCP is then similar to the one addressed in Sect. 5.6:

$$\tilde{J}(y,u) = \frac{1}{2}\int_{\Omega_0}(y - z_d)^2 d\mathbf{x} + \frac{\beta}{2}\int_{\Gamma_c} u^2 d\sigma \;\rightarrow\; \min \tag{6.65a}$$

$$\text{subject to } u \in \mathcal{U},\; y \in V:$$

$$\begin{cases} -\mathrm{div}(\mu\nabla y) + \mathrm{div}(\mathbf{b}y) = 0 & \text{in } \Omega \\ y = 1 & \text{on } \Gamma_D \\ \dfrac{\partial y}{\partial n} - (\mathbf{b}\cdot\mathbf{n})\, y = u & \text{on } \Gamma_c \\ \dfrac{\partial y}{\partial n} - (\mathbf{b}\cdot\mathbf{n})\, y = 0 & \text{on } \Gamma_{out}, \end{cases} \tag{6.65b}$$

where $\beta = 10^{-3}$, $V = H^1_{\Gamma_D}(\Omega)$, $\mathcal{U} = L^2(\Gamma_c)$ denotes the set of admissible controls. Here $\mu = 1/10$, $\Omega = (0, 1+l) \times (0,1)$, where $l = 3$, $\Gamma_D \subset \Gamma$, $\Gamma_N = \Gamma_c \cup \Gamma_{out} = \Gamma \backslash \Gamma_D$, and $\Gamma_c \subset \Gamma_N$ is the portion of the Neumann boundary where we apply the boundary control (see Fig. 6.4).

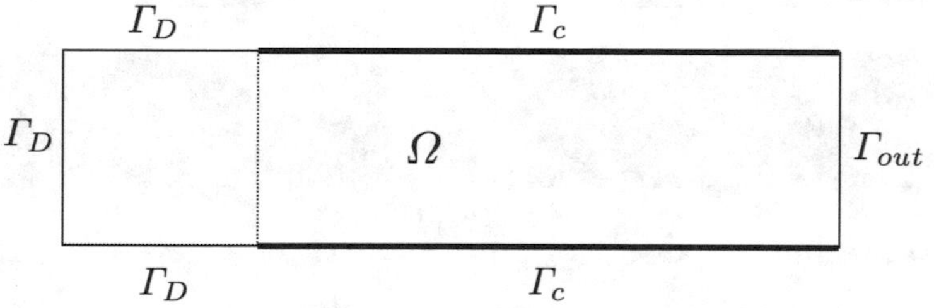

Fig. 6.4 Test case 2: domain and boundaries

Here the state variable plays the role of adimensional temperature, which is imposed on Γ_D; the control acts as heat flux on the boundary Γ_c, while an insulation condition (no flux) is imposed on Γ_{out}. We consider two different options regarding the observation region Ω_0,

$$\Omega_0^1 = \{(x_1, x_2)\;: x_1 > 1, x_2 < 0.3 \text{ or } x_2 > 0.7\} \qquad \text{or} \qquad \Omega_0^2 = \{(x_1, x_2)\;: x_1 > 1\}$$

and the target function,

$$z_d^1 = 2 \qquad \text{or} \qquad z_d^2 = 4x_2(1 - x_2).$$

To discretize state, control and adjoint variables, we use linear ($\mathbb{P}_1$) FE spaces, built on a mesh made by 11028 vertices and 21653 triangular elements, resulting in $N_y = N_p = 10907$ degrees of freedom for the state and the control variables, whereas the dimension of the control space is $N_u = 243$. In all the resulting scenarios, the BFGS method is employed to solve the unconstrained OCP. Numerical results are reported in Table 6.2 and Fig. 6.5.

case	Ω_0	z_d	# iter	$J_h(\mathbf{u}_0)$	$J_h(\hat{\mathbf{u}})$	$1 - J_h(\hat{\mathbf{u}})/J_h(\mathbf{u}_0)$	$\lvert\nabla J(\hat{\mathbf{u}})\rvert/\lvert\nabla J_h(\mathbf{u}_0)\rvert$
(a)	Ω_0^1	z_d^1	59	316.6661	3.5682	98.87%	0.0250
(b)	Ω_0^2	z_d^1	54	192.8385	2.6368	98.63%	0.0569
(c)	Ω_0^1	z_d^2	48	29.2188	0.7826	97.32%	0.0753
(d)	Ω_0^2	z_d^2	87	45.9996	0.8464	98.16%	0.0387

Table 6.2 Unconstrained OCP for an advection-diffusion equation, numerical approximation obtained by the BFGS method. Number of iterations required by the BFGS method to converge, values of the cost functional in absence of control ($\mathbf{u}_0 = \mathbf{0}$) and for the optimal control, cost reduction $1 - J_h(\hat{\mathbf{u}})/J_h(\mathbf{u}_0)$, and reduction $|\nabla J(\hat{\mathbf{u}})|/|\nabla J_h(\mathbf{u}_0)|$ of the norm of the gradient ∇J_h compared to the case in absence of control

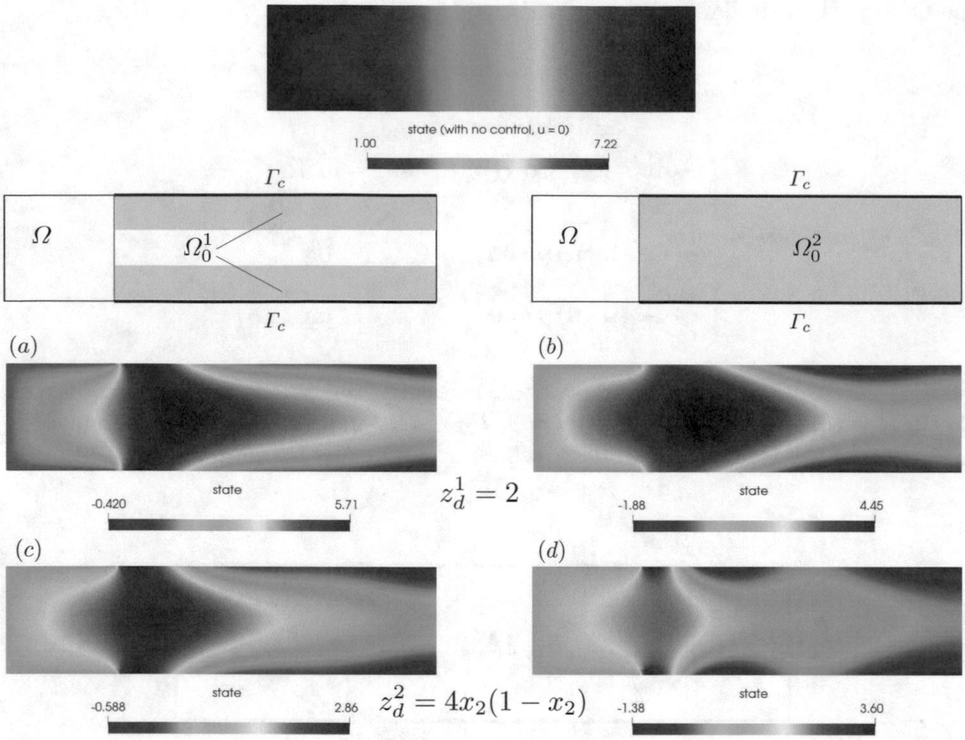

Fig. 6.5 Unconstrained OCP for an advection-diffusion equation, numerical approximation using the BFGS method. Uncontrolled state (top) and optimal states in different scenarios. Case (a): $\Omega_0 = \Omega_0^1$, $z_d = z_d^1$. Case (b): $\Omega_0 = \Omega_0^2$, $z_d = z_d^1$. Case (c): $\Omega_0 = \Omega_0^1$, $z_d = z_d^2$. Case (d): $\Omega_0 = \Omega_0^2$, $z_d = z_d^2$

Compared to the previous distributed OCP, getting close to the target – and then small values of the cost functional – appears to be harder in the case of boundary controls; on the other hand, the regularization parameter has a minor impact on the convergence of the optimization algorithm. The higher number of iterations required to achieve convergence is motivated by the fact that, for the case at hand, a constant step size τ is chosen.

6.5.2 OCPs governed by the Stokes Equations

Let us consider the following unconstrained OCP governed by Stokes equations:

$$\tilde{J}(\boldsymbol{v}, \pi, \boldsymbol{u}) = \frac{1}{2}\int_\Omega |\boldsymbol{v} - \boldsymbol{v}_d|^2 \, d\mathbf{x} + \frac{\alpha}{2}\int_\Omega |\boldsymbol{u}|^2 \, d\mathbf{x} \; \rightarrow \; \min$$

$$\text{subject to}$$

$$\begin{cases} -\nu \Delta \boldsymbol{v} + \nabla \pi = \boldsymbol{f} + \boldsymbol{u} & \text{in } \Omega \\ \operatorname{div} \boldsymbol{v} = 0 & \text{in } \Omega \\ \boldsymbol{v} = \mathbf{0} & \text{on } \partial\Omega \setminus \Gamma_{right} \\ \boldsymbol{v} \cdot \mathbf{t} = 0, \qquad (\mathbf{T}(\boldsymbol{v}, \pi)\mathbf{n}) \cdot \mathbf{n} = 0 & \text{on } \Gamma_{right}, \end{cases} \tag{6.66}$$

where $\mathbf{T}(\boldsymbol{v}, \pi)\mathbf{n} = -\pi\mathbf{n} + \nu\frac{\partial \boldsymbol{v}}{\partial n}$, whereas $\mathbf{n}$ and $\mathbf{t}$ are the (outward directed) normal and the tangent unit vectors on the boundary $\partial\Omega$.

For the case at hand, $\Omega = (0,1) \times (0,2)$, $\Gamma_{right} = \{(x_1, x_2) \in \partial\Omega : x_1 = 1\}$ and $\boldsymbol{f} = 0\mathbf{e}_1 - 1\mathbf{e}_2$, with the target $\boldsymbol{v}_d = x_2\mathbf{e}_1 + 0\mathbf{e}_2$, denoting by $(\mathbf{e}_1, \mathbf{e}_2)$ the standard basis in $\mathbb{R}^2$.

Numerical results have been obtained considering the steepest descent method with fixed step size $\tau = 5$, for different values of the regularization coefficient α, using the inf-sup compatible pair $\mathbb{P}_2 - \mathbb{P}_1$ of FE spaces, built on a mesh made by 5140 vertices and 10278 triangular elements. The effect of the control function is to induce the flow to align along the horizontal direction (see Fig. 6.6).

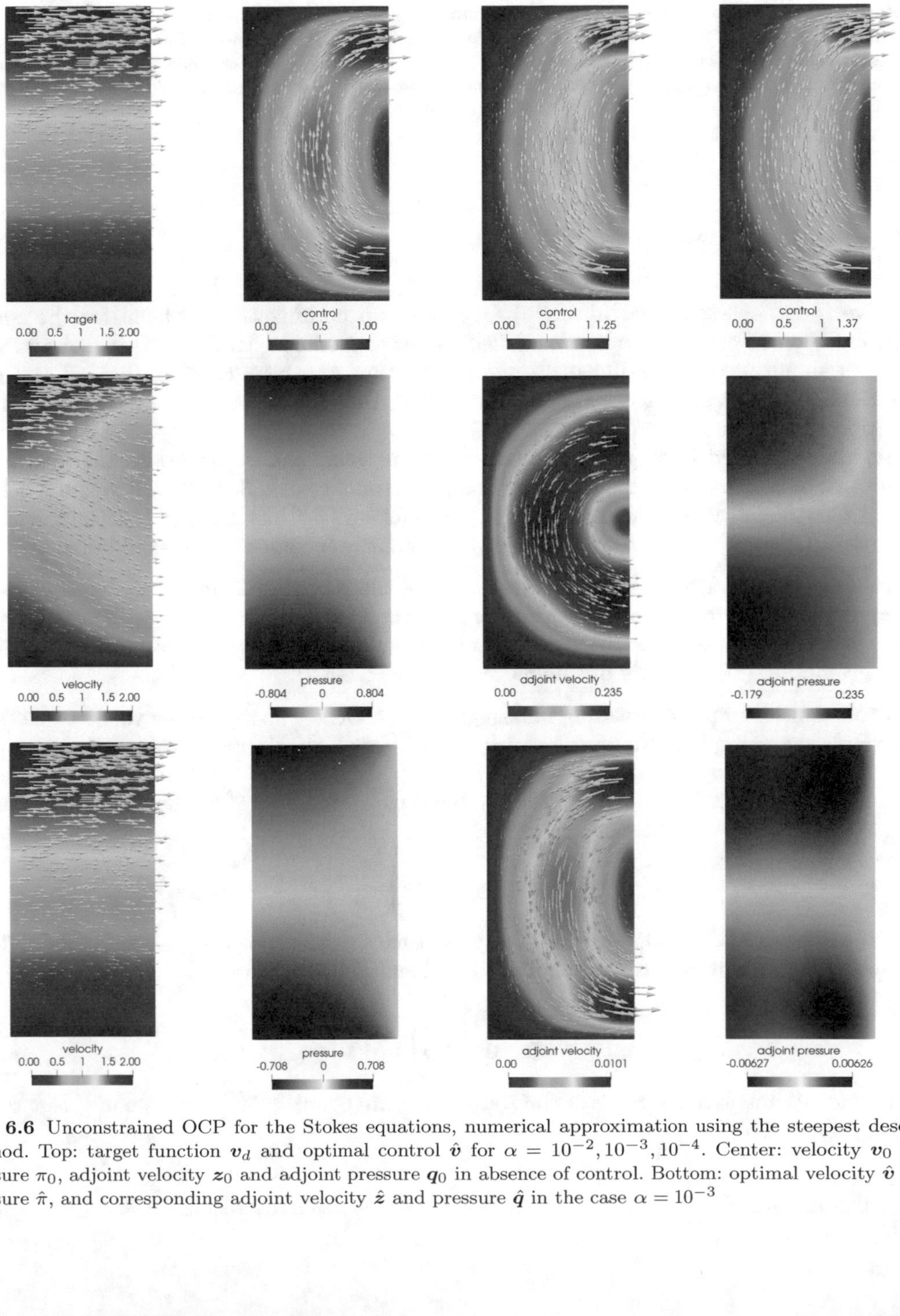

Fig. 6.6 Unconstrained OCP for the Stokes equations, numerical approximation using the steepest descent method. Top: target function $\boldsymbol{v}_d$ and optimal control $\hat{\boldsymbol{v}}$ for $\alpha = 10^{-2}, 10^{-3}, 10^{-4}$. Center: velocity $\boldsymbol{v}_0$ and pressure π_0, adjoint velocity $\boldsymbol{z}_0$ and adjoint pressure $\boldsymbol{q}_0$ in absence of control. Bottom: optimal velocity $\hat{\boldsymbol{v}}$ and pressure $\hat{\pi}$, and corresponding adjoint velocity $\hat{\boldsymbol{z}}$ and pressure $\hat{\boldsymbol{q}}$ in the case $\alpha = 10^{-3}$

The number of iterations to reach convergence, the distance $\|\hat{\boldsymbol{v}} - \boldsymbol{v}_d\|_{L^2(\Omega)^2}$ between the optimal state velocity and the target velocity, and the norm of the optimal control $\hat{\boldsymbol{u}}$ are reported in Table 6.3. The smaller the regularization parameter, the stronger the impact of the control and the higher the decrease of the cost functional.

β	$J_h(\hat{u})$	# iter	$\|\hat{\boldsymbol{u}}\|_{L^2(\Omega)}$
10^{-2}	0.00317078	29	0.752571
10^{-3}	0.000383566	70	0.827743
10^{-4}	$5.9189 \cdot 10^{-5}$	90	0.847255

Table 6.3 Unconstrained OCP for the Stokes equations, numerical approximation through the steepest descent method. Values of the cost functional for the optimal control $J_h(\hat{\boldsymbol{u}})$, number of iterations required by the steepest descent method to converge, norm of the optimal control $\|\hat{\boldsymbol{u}}\|_{L^2(\Omega)}$, for different values of the regularization parameter α. The value of the cost functional in the absence of control ($\boldsymbol{u}_0 = \mathbf{0}$) is $J_h(\boldsymbol{u}_0) = 0.076222$

6.6 All-at-once Methods (I)

Iterative methods recast the (discrete) OCP into an optimization problem for the control $\mathbf{u}$. Instead, *all-at-once* methods, also called *one-shot* methods, treat both control and state variables simultaneously as independent optimization variables, coupled through the PDE constraint, and therefore solve the state problem, the adjoint problem and the optimality condition as a unique system.

For unconstrained linear-quadratic OCPs, the problem can be naturally analyzed in the framework of equality constrained quadratic programming (see Sect. 3.4.1), yielding a KKT system of optimality conditions in saddle point form. This has already been addressed for algebraic OCPs in Sect. 4.1.2. Saddle point problems are encountered in several kind of OCPs:

- problems with control constraints treated using the primal-dual active set strategy yield a sequence of subproblems under this form (see Sect. 6.8);
- the same also happens for more general nonlinear problems when the sequential quadratic programming (SQP) method is employed (see Sect. 9.6.2).

Let us consider an unconstrained linear-quadratic OCP whose discrete version (6.23) can be formulated as a quadratic programming problem with equality constraint

$$\frac{1}{2}\mathbf{x}^\top \mathbb{J}\mathbf{x} - \mathbf{h}^\top \mathbf{x} \to \min \qquad \text{subject to} \qquad \mathbf{x} \in \mathbb{R}^{N_y+N_u}, \;\; \mathbb{D}\mathbf{x} = \mathbf{f}, \tag{6.67}$$

where

$$\mathbf{x} = \begin{bmatrix} \mathbf{y} \\ \mathbf{u} \end{bmatrix}, \quad \mathbb{J} = \begin{bmatrix} \mathbb{M} & 0 \\ 0 & \beta\mathbb{N} \end{bmatrix}, \quad \mathbf{h} = \begin{bmatrix} \mathbb{M}\mathbf{z}_d \\ \mathbf{0} \end{bmatrix}, \quad \mathbb{D} = [\mathbb{A} \;\; -\mathbb{M}].$$

Note that both OtD and DtO approaches yield this kind of problems. As seen in Sect. 3.4.1, the optimality conditions[6] corresponding to problem (6.67) are given by

$$\begin{bmatrix} \mathbb{J} & \mathbb{D}^\top \\ \mathbb{D} & 0 \end{bmatrix} \begin{bmatrix} \mathbf{x} \\ \mathbf{p} \end{bmatrix} = \begin{bmatrix} \mathbf{h} \\ \mathbf{f} \end{bmatrix} \tag{6.68}$$

where the adjoint variable $\mathbf{p}$ plays the role of Lagrange multiplier. This is equivalent to the KKT system (6.16).

[6] Note that the analogy with a saddle-point problem can also be seen at the continuous level, see Sect. 6.10.

In the case of unconstrained linear-quadratic OCPs, all-at-once methods provide the solution of the OCP by solving the (coupled) linear system (6.68) simultaneously. The matrix in (6.68) has a saddle-point structure; it is block symmetric, indefinite, usually ill-conditioned and often very large and sparse. Efficient preconditioned iterative solvers (such as Krylov subspace methods, like MINRES or GMRES) are the matter of choice for the solution of (6.68). Below we describe some preconditioning strategies which are best suited in the case of saddle-point problems arising from the discretization of OCPs.

Let us rewrite (6.68) with shorthand notation $\mathcal{K}\chi = \mathfrak{b}$, with obvious meaning of symbols. A left preconditioner for $\mathcal{K}$ is a matrix $\mathcal{P}$ such that:

(i) $\mathcal{P}^{-1}\mathcal{K}$ has better spectral properties than $\mathcal{K}$ (i.e. lowest condition number and/or more clustered eigenvalues),
(ii) $\mathcal{P}^{-1}\chi$ is cheap to evaluate for any given vector χ.

The preconditioned system reads

$$\mathcal{P}^{-1}\mathcal{K}\chi = \mathcal{P}^{-1}\mathfrak{b}; \tag{6.69}$$

right or symmetric preconditioners could be devised as well (see, e.g., [218]). Constructing efficient preconditioners is a highly problem-dependent task: "... the construction of high-quality preconditioners necessitates exploiting the block structure of the problem, together with detailed knowledge about the origin and structure of the various blocks" [30, Sec. 10]. For the sake of simplification, we can say that the two main preconditioning strategies are *domain decomposition* (DD, see e.g., [4] and references therein) and *multigrid* techniques (MG, see [42, 43] and the monographs [50, 132] for a general introduction). Both have proven to be among the fastest methods for solving large-scale PDE problems. Regarding their application to PDE-constrained optimization, here we focus on *multigrid preconditioners*; see, e.g., [241, 228] for a general overview on other possible approaches and, e.g., [138, 20] for DD preconditioners.

MG preconditioners can be applied to the whole KKT system or separately to every block. Preconditioning each single block enables to exploit existing MG (either algebraic or geometric) preconditioners for the matrices $\mathbb{D}$ and $\mathbb{E}$ as building blocks of the preconditioner, in particular optimal multigrid preconditioners for the matrix $\mathbb{E}$ encoding the discretized PDE operator. See for instance [42, 43, 240] for more on the former approach.

In Sect. 6.6.1 we will focus on elliptic problems and present two different block diagonal preconditioners following the work in [226, 279]. Similarly, in Sect. 6.6.2 we discuss the one-shot approach for the optimal control of the Stokes equations principally referring to [229]. Numerical results are shown in Sect. 6.9.2.

Remark 6.5. All-at-once (or one-shot) methods fit the Simultaneous Analysis and Design (SAND) framework, introduced in Sect. 4.2. Depending on the way the full system (6.68) is actually solved, a further distinction between *reduced SAND* and *full SAND* approaches can be made. Reduced SAND stems from a suitable factorization (or block elimination) of the KKT system; the most known strategies are the null space and the range space methods, see e.g. [208]. For instance, the range space method solves for the multiplier estimates first, and then compute optimization variables. Full SAND solves (6.68) simultaneously. Our description actually conforms to the full SAND approach. •

6.6.1 OCPs Governed by Scalar Elliptic Equations

A distributed control problem for the Laplace equation like the one in (6.16) yields a system of KKT optimality conditions under the form (6.68). Here we assume that $\mathbb{B} = \mathbb{N} = \mathbb{M}$ (distributed control and observation), so that all the blocks appearing in the matrices $\mathbb{D}$ and

$\mathbb{J}$ are square. System (6.16) thus reads

$$\begin{bmatrix} \mathbb{M} & 0 & \mathbb{A}^\top \\ 0 & \beta\mathbb{M} & -\mathbb{M} \\ \mathbb{A} & -\mathbb{M} & 0 \end{bmatrix} \begin{bmatrix} \mathbf{y} \\ \mathbf{u} \\ \mathbf{p} \end{bmatrix} = \begin{bmatrix} \mathbb{M}\mathbf{z}_d \\ \mathbf{0} \\ \mathbf{f} \end{bmatrix}. \tag{6.70}$$

Under the assumption that $\mathbb{D} = (\mathbb{A} \ \ -\mathbb{M})$ has full rank, the system above is nonsingular, symmetric and indefinite. the MINRES iterative algorithm can be effectively used. To preserve symmetry in the preconditioned system, a preconditioner has to be symmetric and positive definite; in particular, effective *block-diagonal* preconditioners are often employed.

An efficient iterative method for this kind of system is MINRES [210]. An alternative is represented by the *Bramble-Pasciak* CG method [49], well-suited for saddle point problems like the ones arising in PDE-constrained optimization. It uses a *block-triangular* preconditioner and a suitable inner product in which the preconditioned matrix is symmetric and, under certain conditions, positive definite. Finally, we mention the possibility to define *constraint* preconditioners which allow to use the *preconditioned projected conjugate gradient* (PPCG) method [111].

Below we describe block-diagonal, block-triangular and constraint preconditioners, *well-designed* for linear systems arising from linear-quadratic OCPs. Earlier approaches, where available preconditioners are only exploited for the state operator, can be found in [22, 131, 39].

Block-diagonal Preconditioners

A basic block diagonal preconditioner for saddle point systems is given by (see, e.g., [30])

$$\mathcal{P}_d = \begin{bmatrix} \mathbb{J} & 0 \\ 0 & \mathbb{S} \end{bmatrix} \tag{6.71}$$

where $\mathbb{S} = \mathbb{D}\mathbb{J}^{-1}\mathbb{D}^\top$ is the so-called Schur complement. Note that $\mathbb{S}$ is nonsingular as both $\mathbb{J}$ and $\mathbb{A}$ are nonsingular. In our case

$$\mathcal{P}_d = \begin{bmatrix} \mathbb{J} & 0 \\ 0 & \mathbb{D}\mathbb{J}^{-1}\mathbb{D}^\top \end{bmatrix} = \begin{bmatrix} \mathbb{M} & 0 & 0 \\ 0 & \beta\mathbb{M} & 0 \\ 0 & 0 & \frac{1}{\beta}\mathbb{M} + \mathbb{A}\mathbb{M}^{-1}\mathbb{A}^\top \end{bmatrix}. \tag{6.72}$$

The matrix $\mathcal{T} = \mathcal{P}_d^{-1}\mathcal{K}$ is nonsingular and satisfies (see [202])

$$(\mathcal{T} - \mathbb{I})\Big(\mathcal{T} - \frac{1}{2}(1+\sqrt{5})\mathbb{I}\Big)\Big(\mathcal{T} - \frac{1}{2}(1-\sqrt{5})\mathbb{I}\Big) = 0. \tag{6.73}$$

$\mathcal{T}$ is diagonalizable and has at most three distinct eigenvalues, $\{1, \frac{1}{2}(1+\sqrt{5}), \frac{1}{2}(1-\sqrt{5})\}$; both GMRES and MINRES algorithms applied to the preconditioned system with matrix $\mathcal{T}$ will terminate after at most 3 iterations. This brilliantly fulfills the first request *(i)*. Unfortunately, forming the preconditioned system is as expensive as computing a factorization of $\mathcal{K}$ – that is, the second request *(ii)* is not met at all.

Relation (6.73) implies that a good preconditioner is found when the (1,1) block $\mathbb{J}$ and the Schur complement of $\mathcal{K}$ are well approximated. With this aim, $\mathcal{P}_d$ can be replaced by

$$\hat{\mathcal{P}}_d = \begin{bmatrix} \hat{\mathbb{J}} & 0 \\ 0 & \hat{\mathbb{S}} \end{bmatrix} \tag{6.74}$$

where $\widehat{\mathbb{J}}$ and $\widehat{\mathbb{S}}$ represent suitable approximations of $\mathbb{J}$ and $\mathbb{S}$.

Note that the Schur block $\mathbb{S} = \frac{1}{\beta}\mathbb{M} + \mathbb{A}\mathbb{M}^{-1}\mathbb{A}^\top$ is the only block of $\mathcal{P}_d$ that contains the PDE matrix. Indeed, blocks (1,1) and (2,2) are simple mass matrices that can be inverted without difficulties. Following [226], for small enough β (say, $10^{-1} \lesssim \beta \lesssim 10^{-7}$) the term $\frac{1}{\beta}\mathbb{M}$ is smaller than $\mathbb{A}\mathbb{M}^{-1}\mathbb{A}^\top$ (see Exercise 7). This suggests us to use the following preconditioner

$$\widehat{\mathcal{P}}_d = \begin{bmatrix} \mathbb{M} & 0 & 0 \\ 0 & \beta\mathbb{M} & 0 \\ 0 & 0 & \mathbb{A}\mathbb{M}^{-1}\mathbb{A}^\top \end{bmatrix}. \tag{6.75}$$

Solving the (left) preconditioned system (6.69) with $\mathcal{P} = \widehat{\mathcal{P}}_d$ at each iteration requires to compute the preconditioned residual $\mathbf{z}$ by solving $\mathcal{P}_d\mathbf{z} = \mathbf{r}$, where $\mathbf{z}$ and the residual $\mathbf{r}$ are given by

$$\mathbf{z} = \begin{bmatrix} \mathbf{z}_y \\ \mathbf{z}_u \\ \mathbf{z}_p \end{bmatrix}, \qquad \mathbf{r} = \begin{bmatrix} \mathbf{r}_y \\ \mathbf{r}_u \\ \mathbf{r}_p \end{bmatrix},$$

respectively. Formally, we obtain

$$\mathbf{z} = \begin{bmatrix} \mathbb{M}^{-1} & 0 & 0 \\ 0 & \frac{1}{\beta}\mathbb{M}^{-1} & 0 \\ 0 & 0 & \mathbb{A}^{-\top}\mathbb{M}\mathbb{A}^{-1} \end{bmatrix} \mathbf{r}.$$

This expression highlights that we still need to solve for $\mathbb{A}$ and $\mathbb{A}^\top$, that is the state and the adjoint problem. A cheaper version of this preconditioner can be obtained by replacing the exact matrices $\mathbb{M}$ and $\mathbb{A}$ with suitable approximations $\hat{\mathbb{A}}$ and $\hat{\mathbb{M}}$. The new preconditioner reads

$$\ddot{\mathcal{P}}_d = \begin{bmatrix} \hat{\mathbb{M}} & 0 & 0 \\ 0 & \beta\hat{\mathbb{M}} & 0 \\ 0 & 0 & \hat{\mathbb{A}}\mathbb{M}^{-1}\hat{\mathbb{A}}^\top \end{bmatrix}. \tag{6.76}$$

Thanks to its mesh independent condition number, a good approximation $\hat{\mathbb{M}}$ for the mass matrix $\mathbb{M}$ is for instance provided by the lumped mass matrix or even by the diagonal of the mass matrix. Other options exist, for instance based on performing a few iterations of the symmetric Gauss-Seidel method, possibly accelerated by Chebyshev semi-iterations (see [226, 241] for further details).

Concerning $\hat{\mathbb{A}}$, a possibility is to regard it as an inexact solver of $\mathbb{A}$ obtained by carrying out a fixed (low) number of multigrid iterations, say k. Such $\hat{\mathbb{A}}$ provides an optimal preconditioner for $\mathbb{A}$ in the case the latter is a finite element discretization of the Laplace equation (e.g. [88, 50, 132]). To summarize, and following [226], we can choose $\hat{\mathbb{A}}$ and $\hat{\mathbb{M}}$ such that, for instance:

- $\hat{\mathbb{A}}$ involves k Algebraic Multigrid (AMG) V-cycles,
- $\hat{\mathbb{M}}$ involves carrying out m iterations of the symmetric Gauss-Seidel method (in short SGS(m)).

For the algebraic multigrid method we can use, e.g., the `HSL_MI20` package [47], while the symmetric Gauss-Seidel method and the Chebyshev semi-iterative method are implemented in Algorithms 6.3 and 6.4, respectively.

Remark 6.6. Using the lumped mass matrix would result in a preconditioner that is less costly and less time-consuming than SGS preconditioning. However, its spectral properties are not as good, yielding a higher number of iterations of the outer Krylov solver and therefore a higher computational time. •

Algorithm 6.3 Symmetric Gauss-Seidel for the system $\mathbb{A}\mathbf{x} = \mathbf{b}$: $\mathbf{x} = \texttt{SGS}(\mathbb{A}, \mathbf{b}, m)$

Input: $\mathbb{A}$, $\mathbf{b}$, $\mathbf{x}_0$, m

set $\mathbb{D}_A = \texttt{diag}(\mathbb{A})$, $\mathbb{E}_A = -\texttt{tril}(\mathbb{A})$, $\mathbb{F}_A = -\texttt{triu}(\mathbb{A})$, $k = 0$
while $k < m$
    $\mathbf{x}^{(k+1/2)} = (\mathbb{D}_A - \mathbb{E}_A)\backslash(\mathbb{F}_A\mathbf{x}^{(k)}) + (\mathbb{D}_A - \mathbb{E}_A)\backslash\mathbf{b}$
    $\mathbf{x}^{(k+1)} = (\mathbb{D}_A - \mathbb{F}_A)\backslash(\mathbb{E}_A\mathbf{x}^{(k+1/2)}) + (\mathbb{D}_A - \mathbb{F}_A)\backslash\mathbf{b}$
    $k = k + 1$
end while

Algorithm 6.4 Chebyshev semi-iterative method (l steps) for $\mathbb{M}\mathbf{w} = \mathbf{b}$

Input: $\mathbb{M}$, $\mathbf{b}$, l, ω

set $\mathbb{D} = \texttt{diag}(\mathbb{M})$, $\mathbf{g} = \omega\mathbb{D}^{-1}\mathbf{b}$, and the relaxation parameter ω
set $\mathbb{S} = (\mathbb{I} - \omega\mathbb{D}^{-1}\mathbb{M})$
set $\mathbf{w}_0 = \mathbf{0}$ and $\mathbf{w}_1 = \mathbb{S}\mathbf{w}_{k-1} + \mathbf{g}$
set $c_0 = 2$ and $c_1 = \omega$
for $k = 2, \dots, l$
    $c_{k+1} = \omega c_k - \frac{1}{4}c_{k-1}$
    $\eta_{k+1} = \omega\frac{c_k}{c_{k+1}}$
    $\mathbf{w}_{k+1} = \eta_{k+1}(\mathbb{S}\mathbf{w}_k + \mathbf{g} - \mathbf{w}_{k-1}) + \mathbf{w}_{k-1}$
end for

Note that both $\hat{\mathbb{A}}$ and $\hat{\mathbb{M}}$ are not formed explicitly but are defined through the action of their inverses; in fact, when we have to solve the system $\hat{\mathbb{A}}\mathbf{w} = \mathbf{z}$, instead of forming the matrix $\hat{\mathbb{A}}$, we apply a routine, say $\mathbf{w} = \texttt{amg-precon}(\mathbb{A}, \mathbf{z})$, that implements the multigrid method.

In Algorithm 6.5 we summarize the main steps involved for the computation of $\mathbf{z}$ such that $\hat{\mathcal{P}}_d\mathbf{z} = \mathbf{r}$, where

$$\mathbf{z} = \begin{bmatrix} \mathbf{z}_y \\ \mathbf{z}_u \\ \mathbf{z}_p \end{bmatrix}, \qquad \mathbf{r} = \begin{bmatrix} \mathbf{r}_y \\ \mathbf{r}_u \\ \mathbf{r}_p \end{bmatrix}.$$

The use of the block diagonal preconditioner $\hat{\mathcal{P}}_d$ allows to achieve a convergence rate independent of the mesh size h [226], a property that is referred to as *optimality.* Unfortunately, the preconditioner $\hat{\mathcal{P}}_d$ is not "robust" with respect to the regularization parameter β, meaning that its spectral preconditioning properties are not independent of β. In particular for smaller β the number of iterations of the outer solver is higher. See the numerical tests in Sect. 6.7.

In [279] a preconditioner is designed for solving the reduced system

$$\begin{bmatrix} \mathbb{M} & \mathbb{A}^\top \\ \mathbb{A} & -\beta^{-1}\mathbb{M} \end{bmatrix} \begin{bmatrix} \mathbf{y} \\ \mathbf{p} \end{bmatrix} = \begin{bmatrix} \mathbb{M}\mathbf{z}_d \\ \mathbf{f} \end{bmatrix} \tag{6.77}$$

Algorithm 6.5 Application of the block preconditioner $\hat{\mathcal{P}}_d$: $\hat{\mathcal{P}}_d\mathbf{z} = \mathbf{r}$

Input: $\mathbb{A}$, $\mathbb{M}$, β, $\mathbf{r}_y$, $\mathbf{r}_u$, $\mathbf{r}_p$

$\mathbf{z}_y = \texttt{SGS}(\mathbb{M}, \mathbf{r}_y, m)$
$\mathbf{z}_u = \beta^{-1}\texttt{SGS}(\mathbb{M}, \mathbf{r}_u, m)$
$\mathbf{w}_1 = \texttt{amg-precon}(\mathbb{A}^\top, \mathbf{r}_p)$
$\mathbf{w}_2 = \mathbb{M}\mathbf{w}_1$
$\mathbf{z}_p = \texttt{amg-precon}(\mathbb{A}, \mathbf{w}_2)$

obtained by substituting $\mathbf{u} = \beta^{-1}\mathbf{p}$ into the original optimality system (6.70). The proposed block diagonal preconditioner is

$$\mathcal{P}_{dr} = \begin{bmatrix} \mathbb{M} + \beta^{1/2}\mathbb{A} & 0 \\ 0 & \beta^{-1}\mathbb{M} + \beta^{-1/2}\mathbb{A}^\top \end{bmatrix}.$$

The application of $\mathcal{P}_{dr}$ still requires the exact inversion of the two blocks; in order to ensure an efficient evaluation of $\mathcal{P}_{dr}^{-1}\mathbf{r}$, the inexact "inversion" of the two diagonal blocks is approximately accomplished by a suitable number of multigrid iterations. The approximation of $\mathcal{P}_{dr}$ that is obtained in this way is denoted $\hat{\mathcal{P}}_{dr}$. To compare the expression of $\hat{\mathcal{P}}_{dr}$ with the one of the preconditioner $\hat{\mathcal{P}}_d$ previously introduced, let us point out that the analogue of $\mathcal{P}_d$ when dealing with system (6.77) would be given by

$$\mathcal{P}_{d2} = \begin{bmatrix} \mathbb{M} & 0 \\ 0 & \beta^{-1}\mathbb{M} + \mathbb{A}\mathbb{M}^{-1}\mathbb{A}^\top \end{bmatrix},$$

which can be approximated by

$$\hat{\mathcal{P}}_{d2} = \begin{bmatrix} \hat{\mathbb{M}} & 0 \\ 0 & \hat{\mathbb{A}}\mathbb{M}^{-1}\hat{\mathbb{A}}^\top \end{bmatrix}.$$

Remark 6.7. In the case of boundary control and distributed observation (see, e.g., the problem discussed in Sect. 5.6), in system (6.16) $\mathbb{M} \neq \mathbb{N}$ and the block $\mathbb{B}$ is rectangular. Similarly to (6.72), an *ideal* preconditioner would be in this case

$$\mathcal{P}_d^b = \begin{bmatrix} \mathbb{M} & 0 & 0 \\ 0 & \beta\mathbb{N} & 0 \\ 0 & 0 & \frac{1}{\beta}\mathbb{B}\mathbb{N}^{-1}\mathbb{B}^\top + \mathbb{A}\mathbb{M}^{-1}\mathbb{A}^\top \end{bmatrix}.$$

As done with (6.72), a cheaper version of $\mathcal{P}_d^b$ can be obtained by the same procedure. The preconditioner that we find in this way has a similar structure to (6.76), precisely (see, e.g., [226])

$$\hat{\mathcal{P}}_d^b = \begin{bmatrix} \hat{\mathbb{M}} & 0 & 0 \\ 0 & \beta\hat{\mathbb{N}} & 0 \\ 0 & 0 & \hat{\mathbb{A}}\mathbb{M}^{-1}\hat{\mathbb{A}}^\top \end{bmatrix}.$$ •

Block-triangular Preconditioners

For the saddle-point problem (6.68), a block triangular preconditioner of the form

$$\hat{\mathcal{P}}_{BT} = \begin{bmatrix} \hat{\mathbb{J}} & 0 \\ \mathbb{D} & -\hat{\mathbb{S}} \end{bmatrix}$$

where $\hat{\mathbb{J}} \approx \mathbb{J}$, $\hat{\mathbb{S}} \approx \mathbb{S}$, and $\mathbb{S} = \mathbb{D}\mathbb{J}^{-1}\mathbb{D}^\top$ is the Schur complement of (6.68), was proposed by Bramble and Pasciak in [49] to be used for conjugate gradient (CG) iterations.

The matrix $\hat{\mathcal{K}} = \mathcal{P}_{BT}^{-1}\mathcal{K}$ is in fact symmetric and positive definite with respect to the (nonstandard) inner product $(\mathbf{v}_1, \mathbf{v}_2)_{\mathcal{H}} = \sqrt{\mathbf{v}_1^\top \mathcal{H}\mathbf{v}_2}$, where $\mathcal{H}$ is the matrix

$$\mathcal{H} = \begin{bmatrix} \mathbb{J} - \hat{\mathbb{J}} & 0 \\ 0 & \hat{\mathbb{S}} \end{bmatrix}$$

provided $\mathbb{J} - \hat{\mathbb{J}}$ and $\hat{\mathbb{S}}$ are symmetric and positive definite blocks.

Algorithm 6.6 Bramble Pasciak CG method for the system $\mathbb{A}\mathbf{x} = \mathbf{b}$

Input: $\mathbb{A}$, $\mathbf{b}$, $\mathbf{x}^{(0)} = \mathbf{0}$

1: set $\mathbf{r}^{(0)} = \mathcal{P}\backslash(\mathbf{b} - \mathcal{K}\mathbf{x}^{(0)})$, $\mathbf{p}^{(0)} = \mathbf{r}^{(0)}$, $k = 0$
2: **for** $k = 0, 1, \dots$
3: $\quad \alpha = \dfrac{(\mathbf{r}^{(k)}, \mathbf{r}^{(k)})_{\mathcal{H}}}{(\mathcal{P}\backslash\mathcal{K}\mathbf{p}^{(k)}, \mathbf{r}^{(k)})_{\mathcal{H}}}$
4: $\quad \mathbf{x}^{(k+1)} = \mathbf{x}_k + \alpha\mathbf{p}^{(k)}$
5: $\quad \mathbf{r}^{(k+1)} = \mathbf{r}^{(k)} - \alpha\mathcal{P}\backslash\mathcal{K}\mathbf{p}^{(k)}$
6: $\quad \beta = \dfrac{(\mathbf{r}^{(k+1)}, \mathbf{r}^{(k+1)})_{\mathcal{H}}}{(\mathbf{r}^{(k)}, \mathbf{r}^{(k)})_{\mathcal{H}}}$
7: $\quad \mathbf{p}^{(k+1)} = \mathbf{r}^{(k+1)} + \beta\mathbf{p}^{(k)}$
8: **end for**

When applied to the OCP system (6.70), the block triangular preconditioner reads (see [227, 228])

$$\hat{\mathcal{P}}_{BT} = \begin{bmatrix} \gamma_0\hat{\mathbb{M}} & 0 & 0 \\ 0 & \beta\gamma_0\hat{\mathbb{M}} & 0 \\ \mathbb{A} & -\mathbb{M} & -\hat{\mathbb{S}} \end{bmatrix}$$

whereas the nonstandard inner product $(\mathbf{v}_1, \mathbf{v}_2)_{\mathcal{H}}$ is realized by choosing

$$\mathcal{H} = \begin{bmatrix} \mathbb{M} - \gamma_0\hat{\mathbb{M}} & 0 & 0 \\ 0 & \beta(\mathbb{M} - \gamma_0\hat{\mathbb{M}}) & 0 \\ \mathbb{A} & -\mathbb{M} & \hat{\mathbb{S}} \end{bmatrix},$$

where $\gamma_0 > 0$ is a suitable scaling factor ensuring that $\mathbb{M} - \gamma_0\hat{\mathbb{M}}$ is positive definite. The Bramble Pasciak CG method is reported in Algorithm 6.6.

With the Bramble-Pasciak CG method, multigrid-based preconditioners can be used for each block, as only the action of the inverse of $\hat{\mathbb{D}}$ is needed when evaluating the inner product with $\mathcal{H}$. Moreover, the Schur complement matrix $\mathbb{S} = \frac{1}{\beta}\mathbb{M} + \hat{\mathbb{A}}\mathbb{M}^{-1}\hat{\mathbb{A}}^\top$ can be approximated by $\hat{\mathbb{S}} = \hat{\mathbb{A}}\mathbb{M}^{-1}\hat{\mathbb{A}}^\top$, similarly to what done in (6.76). See, e.g., [227, 228] for more details and [86] for a discussion related to different solvers derived from the Bramble-Pasciak CG method.

Constraint Preconditioners

Another class of symmetric indefinite preconditioners for the projected conjugate gradient (PPCG) method [111, 84] are the so-called *constraint preconditioners* [157, 85]:

$$\mathcal{P}_{CP} = \begin{bmatrix} \mathbb{G} & \mathbb{D}^\top \\ \mathbb{D} & 0 \end{bmatrix}$$

where $\mathbb{G} \in \mathbb{R}^{(N_y+N_u)\times(N_y+N_u)}$ is a symmetric matrix. The inclusion of the exact representation of the $(1,2)$ and $(2,1)$ matrix blocks in the preconditioner, which are associated with the constraints of the OCP, yields a more favorable distribution of the eigenvalues of the preconditioned system (6.69). Their name comes from the observation that these blocks are unchanged from the original system. The PPCG method for the saddle-point system (6.68) is reported in Algorithm 6.7.

Algorithm 6.7 Projected Preconditioned Conjugate Gradient (PPCG) method

Input: $\mathbb{J}$, $\mathbb{D}$, $\mathbf{g}$, $\mathbf{h}$, $\mathbf{x}^{(0)}$ such that $\mathbb{D}\mathbf{x}^{(0)} = \mathbf{g}$

1: set $\tilde{\mathbf{r}}^{(0)} = \mathbb{J}\mathbf{x}^{(0)} - \mathbf{h}$

2: solve $\mathcal{P}_{CP} \begin{bmatrix} \mathbf{g}^{(0)} \\ \mathbf{v}^{(0)} \end{bmatrix} = \begin{bmatrix} \mathbf{r}^{(0)} \\ \mathbf{0} \end{bmatrix}$

3: $\mathbf{d}^{(0)} = -\mathbf{g}^{(0)}$, $\mathbf{r}^{(0)} = \tilde{\mathbf{r}}^{(0)} - \mathbb{D}^\top \mathbf{v}^{(0)}$

4: **for** $k = 0, 1, \ldots$

5: $\quad \alpha = \dfrac{(\mathbf{r}^{(k)}, \mathbf{g}^{(k)})}{(\mathbf{d}^{(k)}, \mathbb{J}\mathbf{d}^{(k)})}$

6: $\quad \mathbf{x}^{(k+1)} = \mathbf{x}^{(k)} + \alpha \mathbf{d}^{(k)}$

7: $\quad \tilde{\mathbf{r}}^{(k+1)} = \mathbf{r}^{(k)} + \alpha \mathbb{J}\mathbf{d}^{(k)}$

8: $\quad$ solve $\mathcal{P}_{CP} \begin{bmatrix} \mathbf{g}^{(k+1)} \\ \mathbf{v}^{(k+1)} \end{bmatrix} = \begin{bmatrix} \tilde{\mathbf{r}}^{(k+1)} \\ \mathbf{0} \end{bmatrix}$

9: $\quad \delta = \dfrac{(\tilde{\mathbf{r}}^{(k+1)}, \mathbf{g}^{(k+1)})}{(\mathbf{r}^{(k)}, \mathbf{g}^{(k)})}$

10: $\quad \mathbf{d}^{(k+1)} = -\mathbf{g}^{(k+1)} + \beta \mathbf{d}^{(k)}$

11: $\quad \mathbf{r}^{(k+1)} = \tilde{\mathbf{r}}^{(k+1)} - \mathbb{D}^\top \mathbf{v}^{(k+1)}$

12: **end for**

A possible approximation of the matrix $\mathcal{P}_{CP}$ is

$$\hat{\mathcal{P}}_{CP} = \begin{bmatrix} \beta \hat{\mathbb{A}}^\top \mathbb{M}^{-1} \hat{\mathbb{A}} & 0 & \mathbb{A}^\top \\ 0 & 0 & -\hat{\mathbb{M}} \\ \mathbb{A} & -\hat{\mathbb{M}} & 0 \end{bmatrix}$$

where $\hat{\mathbb{A}}$ denotes the approximation of $\mathbb{A}$ based e.g. on k AMG V-cycles and $\hat{\mathbb{M}}$ stands for m steps of the symmetric Gauss-Seidel method; see, e.g., [226, 157].

Remark 6.8. Symmetric indefinite preconditioners for the projected conjugate gradient method can also be designed in the case where $\mathbb{J}$ is positive definite only on $ker(\mathbb{D})$. This is a typical situation for certain classes of PDE-constrained optimization problems, see [241]. •

6.6.2 OCPs governed by Stokes Equations

As done in (6.67), the optimality system related to the Stokes OCP (6.46) – no matter whether an OtD or a DtO approach is used – could have also been obtained by rewriting as

$$\frac{1}{2}\mathbf{X}^\top \mathbb{J}_S \mathbf{X} - \mathbf{h}_S^\top \mathbf{X} \rightarrow \min \qquad \text{subject to} \qquad \mathbf{X} \in \mathbb{R}^{N_y + N_u}, \; \mathbb{D}_S \mathbf{X} = \mathbf{f}_S, \tag{6.78}$$

the discrete OCP, where

$$\mathbb{J}_S = \begin{bmatrix} \mathbb{M} & 0 \\ 0 & \alpha \mathbb{M}_v \end{bmatrix}, \qquad \mathbb{D}_S = [\mathbb{A}_S \quad -\mathbb{V}], \qquad \mathbf{X} = \begin{bmatrix} \mathbf{Y} \\ \mathbf{U} \end{bmatrix}, \qquad \mathbf{h}_S = \begin{bmatrix} \mathbf{f}_a \\ \mathbf{0} \end{bmatrix}. \tag{6.79}$$

Matrices $\mathbb{A}_S$, $\mathbb{M}$ and $\mathbb{V}$ are defined in (6.47) and (6.54), respectively. Here $\mathbf{X}$ denotes the vector related to state and control variables.

Problem (6.78) is equivalent to the saddle-point system

$$\begin{bmatrix} \mathbb{J}_S & \mathbb{D}_S^\top \\ \mathbb{D}_S & 0 \end{bmatrix} \begin{bmatrix} \mathbf{X} \\ \mathbf{P} \end{bmatrix} = \begin{bmatrix} \mathbf{h}_S \\ \mathbf{f}_S \end{bmatrix}, \tag{6.80}$$

where $\mathbf{P}$ denotes the vector related to the adjoint variable. $\mathbb{D}_S$ is non-singular, and so is the matrix $\mathcal{K}_S$ in the left hand side of (6.80) provided $\mathbb{D}_S$ has full rank. This is true provided the observation is made on the whole domain Ω and $\delta \neq 0$, that is, the pressure term is present in the functional. If the observation is restricted to a portion of the domain, or else the pressure term in the functional is missing, the diagonal preconditioner we are going to discuss can not be used as is, since it requires the Schur complement to be nonsingular.

By exploiting the analogy of (6.53) with the optimality system (6.68) obtained in the case of scalar elliptic equations, a block diagonal preconditioner can be set up as follows

$$\mathcal{P}_d = \begin{bmatrix} \mathbb{J}_S & 0 \\ 0 & \mathbb{S} \end{bmatrix} \tag{6.81}$$

where $\mathbb{S} = \mathbb{D}_S \mathbb{J}_S^{-1} \mathbb{D}_S^\top$ is the (non-singular) Schur complement in this case.

According to (6.79), we obtain

$$\mathcal{P}_d = \begin{bmatrix} \mathbb{J}_S & 0 \\ 0 & \mathbb{D}_S \mathbb{J}_S^{-1} \mathbb{D}_S^\top \end{bmatrix} = \begin{bmatrix} \mathbb{M} & 0 & 0 \\ 0 & \alpha \mathbb{M}_v & 0 \\ 0 & 0 & \alpha^{-1} \mathbb{V} \mathbb{M}_v^{-1} \mathbb{V}^\top + \mathbb{A}_S \mathbb{M}^{-1} \mathbb{A}_S^\top \end{bmatrix}.$$

As in the elliptic case, $\mathcal{P}_d$ needs to be replaced by a convenient approximation

$$\hat{\mathcal{P}}_d = \begin{bmatrix} \hat{\mathbb{J}}_S & 0 \\ 0 & \hat{\mathbb{S}} \end{bmatrix}, \tag{6.82}$$

where $\hat{\mathbb{J}}_S$ and $\hat{\mathbb{S}}$ represent suitable approximations of $\mathbb{J}_S$ and $\mathbb{S}$, respectively. Except that for small values of α, the term $\mathbb{A}_S \mathbb{M}^{-1} \mathbb{A}_S^\top$ dominates $\alpha^{-1} \mathbb{V} \mathbb{M}_v \mathbb{V}^\top$ (see [229]). However, solving a system with $\mathbb{A}_S \mathbb{M}^{-1} \mathbb{A}_S^\top$ requires two solves with the discrete Stokes matrix, which is not cheap. We can therefore consider the following preconditioner

$$\hat{\mathcal{P}}_d = \begin{bmatrix} \hat{\mathbb{M}} & 0 & 0 \\ 0 & \alpha \hat{\mathbb{M}}_v & 0 \\ 0 & 0 & \hat{\mathbb{A}}_S \hat{\mathbb{M}}^{-1} \hat{\mathbb{A}}_S^\top \end{bmatrix}, \tag{6.83}$$

where $\mathbb{M}$, $\mathbb{M}_v$ and $\mathbb{A}_S$ have been replaced by suitable approximations $\hat{\mathbb{M}}$, $\hat{\mathbb{M}}_v$ and $\hat{\mathbb{A}}_S$.

For the approximation of $\mathbb{M}$ (and $\mathbb{M}_v$) we could simply use the lumped mass matrix (or even the diagonal of the mass matrix), otherwise a few iterations of the symmetric Gauss-Seidel method. Regarding $\hat{\mathbb{A}}_S$, recall that $\mathbb{A}_S$ represents the Stokes operator, a saddle-point matrix for which the task of preconditioning is not straightforward. Moreover, we need to ensure that not only $\hat{\mathbb{A}}_S$ is an effective preconditioner for $\mathbb{A}_S$, but also that $\hat{\mathbb{A}}_S \hat{\mathbb{M}}^{-1} \hat{\mathbb{A}}_S^\top$ is a good approximation of $\mathbb{A}_S \mathbb{M}^{-1} \mathbb{A}_S^\top$. A good choice is to take $\hat{\mathbb{A}}_S$ implicitly defined by the application of a preconditioned iterative method [229, 48]. For instance, a preconditioned Krylov subspace method can be used, for which several effective preconditioners are available. However, using a Krylov subspace method as inner solver for the Stokes saddle-point matrix leads to a nonstationary preconditioner for the whole system, which is allowed only if we use a flexible outer method, for example the flexible GMRES[7] method. A possible alternative is to use as inner solver a stationary iterative method, like the Uzawa method (see e.g. [30, 255]).

[7] By flexible iterative method we mean an iterative procedure that allows the preconditioner to vary from step to step. See e.g. [236] for the description of the flexible GMRES method.

In this case, to solve the system $\mathbb{A}_S \mathbf{w} = \mathbf{d}$ the action of the approximation $\hat{\mathbb{A}}_S$ is implicitly defined by the following iterative scheme

$$\mathbf{w}^{(k+1)} = \mathbf{w}^{(k)} + \mathcal{M}^{-1}\mathbf{r}^{(k)}, \quad k \geq 0$$

where the matrix $\mathcal{M}$ is given by

$$\mathcal{M} = \begin{bmatrix} \hat{\mathbb{E}} & 0 \\ \mathbb{B} & -\hat{\mathbb{M}}_\pi \end{bmatrix}$$

and the residual is defined as $\mathbf{r}^{(k)} = \mathbf{d} - \mathbb{A}_S\mathbf{w}^{(k)}$. $\hat{\mathbb{E}}$ and $\hat{M}_\pi$ represent suitable approximations of $\mathbb{E}$ and $\mathbb{M}_\pi$ (being $\mathbb{M}_\pi$ itself an approximation of the Schur complement for the Stokes system (6.47), $\mathbb{S}_S = \mathbb{B}\mathbb{E}^{-1}\mathbb{B}^\top$). In particular for $\hat{\mathbb{M}}_\pi$ we can use one of the several alternatives already mentioned, while an effective choice for $\hat{\mathbb{E}}$ is to use a fixed number of multigrid iterations. The preconditioner $\hat{\mathcal{P}}_d$ is robust with respect to the mesh size h but not to the parameters α and δ, see e.g. [229]. Other alternatives might include inexact Uzawa methods, see, e.g., [89].

In Algorithm 6.8 we summarize the main steps required by the application of the preconditioner $\hat{\mathcal{P}}_d$; for the sake of simplicity, we use the exact mass matrices instead of their approximation. The Uzawa method, whose call takes the form $\mathbf{x}_{zq} = \texttt{uzawa}(\mathbb{A}_S, \mathbf{b}_{zq}, m_{max})$, is reported in Algorithm 6.9. Note that in this case

$$\mathbf{z} = \begin{bmatrix} \mathbf{z}_v \\ \mathbf{z}_\pi \\ \mathbf{z}_u \\ \mathbf{z}_{zq} \end{bmatrix}, \quad \mathbf{r} = \begin{bmatrix} \mathbf{r}_v \\ \mathbf{r}_\pi \\ \mathbf{r}_u \\ \mathbf{r}_z \\ \mathbf{r}_q \end{bmatrix}, \quad \mathbf{b}_{zq} = \begin{bmatrix} \mathbf{b}_z \\ \mathbf{b}_q \end{bmatrix}, \quad \mathbf{z}_{zq} = \begin{bmatrix} \mathbf{z}_z \\ \mathbf{z}_q \end{bmatrix}, \quad \mathbf{r}_{zq} = \begin{bmatrix} \mathbf{r}_z \\ \mathbf{r}_q \end{bmatrix}.$$

Algorithm 6.8 Application of the block preconditioner $\hat{\mathcal{P}}_d$: $\hat{\mathcal{P}}_d\mathbf{z} = \mathbf{r}$

Input: $\mathbb{A}_S$, $\mathbb{M}_v$, $\mathbb{M}_\pi$, α, $\mathbf{r}_v$, $\mathbf{r}_\pi$, $\mathbf{r}_u$, $\mathbf{r}_z$, $\mathbf{r}_q$

1: $\mathbf{z}_v = \mathbb{M}_v \backslash \mathbf{r}_v$
2: $\mathbf{z}_\pi = \delta^{-1}\mathbb{M}_\pi \backslash \mathbf{r}_\pi$
3: $\mathbf{z}_u = \alpha^{-1}\mathbb{M}_v \backslash \mathbf{r}_u$
4: $\mathbf{w}_{zq} = \texttt{uzawa}(\mathbb{A}_S, \mathbf{r}_{zq}, k_{max})$
5: $\mathbf{w}_{zq} = \mathbb{M} \backslash \mathbf{w}_{zq}$
6: $\mathbf{z}_{zq} = \texttt{uzawa}(\mathbb{A}_S, \mathbf{w}_{zq}, k_{max})$

Algorithm 6.9 Uzawa method: $\mathbf{x}_{zq} = \texttt{uzawa}(\mathbb{A}_S, \mathbf{b}_{zq}, k_{max})$

Input: $\mathbb{A}_S$, $\mathbb{M}_{pi}$, $\mathbf{b}_{zq}$, k_{max}

1: set $\mathbf{x}^{(0)} = \mathbf{0}$, $\mathbf{r}^{(0)} = \mathbf{b}_{zq} - \mathbb{A}_S\mathbf{x}^{(0)}$, $k = 1$
2: **while** $k < k_{max}$
3: solve the linear system $\mathcal{M}\mathbf{z}_{zq}^{(m)} = \mathbf{r}_{zq}^{(m)}$, i.e.

$$\mathbf{z}_w^{(k)} = \texttt{amg-precon}(\mathbb{E}, \mathbf{r}_w^{(k)})$$
$$\mathbf{z}_q^{(k)} = \mathbb{M}_\pi \backslash (\mathbb{B}\mathbf{z}_w^{(k)} - \mathbf{r}_q^{(k)})$$

4: update the solution $\mathbf{x}_{zq}^{(k+1)} = \mathbf{x}_{zq}^{(k)} + \mathbf{z}_{zq}^{(k)}$
5: update the residual $\mathbf{r}_{zq}^{(k+1)} = \mathbf{r}_{zq}^{(k)} - \mathbb{A}_S\mathbf{z}_{zq}^{(k)}$
6: set $k = k + 1$
7: **end while**

Remark 6.9. As done for elliptic OCPs, see equation (6.77), one can also consider preconditioners for the reduced system obtained after eliminating the optimality equation in the full KKT system. With this reduced formulation, there is no need to add the pressure term in the functional; moreover, observations of the velocity only on a portion of the domain (instead than on the whole domain) are admissible, see for instance [279]. •

6.7 Numerical Examples

We now report some results obtained with the all-at-once method on OCPs governed by elliptic equations. We compare the performances of different preconditioners on two test cases, governed by either a Laplace problem or a diffusion-reaction problem, adapted from the examples proposed in [226], and implemented starting from the code provided by the authors[8].

Test case 1: Laplace Equation

Let us denote by $\Omega = (0,1)^2$, by

$$z_d = \begin{cases} (2x_1 - 1)^2(2x_2 - 1)^2 & (x_1, x_2) \in (0, 1/2]^2 \\ 0 & \text{otherwise} \end{cases}$$

the target function, and consider the following OCP:

$$\tilde{J}(y,u) = \frac{1}{2}\int_\Omega (y - z_d)^2 d\mathbf{x} + \beta \int_\Omega u^2 d\mathbf{x} \;\rightarrow\; \min \tag{6.84a}$$

where

$$\begin{cases} -\Delta y = u & \text{in } \Omega \\ y = z_d & \text{on } \Gamma, \end{cases} \tag{6.84b}$$

where $\beta > 0$ is a suitable regularization parameter. We rely on a DtO approach and $\mathbb{Q}_1$ conforming piecewise bilinear finite elements in Ω, obtaining the following discrete OCP:

$$\begin{aligned} \tilde{J}_h(\mathbf{y},\mathbf{u}) = \frac{1}{2}\|y_h - z_d\|^2_{L^2(\Omega)} + \beta\|u_h\|^2_{L^2(\Omega)} = \frac{1}{2}\mathbf{y}^\top \mathbb{M}\mathbf{y} - \mathbf{y}^\top\mathbf{b} + c + \beta\mathbf{u}^\top\mathbb{M}\mathbf{u} \;\rightarrow\; \min \\ \text{subject to } \mathbf{u} \in \mathbb{R}^{N_u}, \\ \mathbf{y} \in \mathbb{R}^{N_y} \;:\; \mathbb{K}\mathbf{y} = \mathbb{M}\mathbf{u} + \mathbf{d}. \end{aligned} \tag{6.85}$$

Here we denote by $\mathbb{K} \in \mathbb{R}^{N_y \times N_y}$ and $\mathbb{M} \in \mathbb{R}^{N_y \times N_y}$ the stiffness and the mass matrix, respectively, with

$$(\mathbb{K})_{ij} = \int_\Omega \nabla\varphi_j \cdot \nabla\varphi_i, \quad i,j = 1,\ldots,N_y, \tag{6.86}$$

by $\mathbf{b} = \mathbb{M}\mathbf{z}_d$, $c = \|z_d\|^2_{L^2(\Omega)}$ and $\mathbf{d} \in \mathbb{R}^{N_y}$ a vector containing the terms coming from the boundary values of y_h. Introducing the Lagrangian

$$\mathcal{L}(\mathbf{y},\mathbf{u},\mathbf{p}) = \tilde{J}_h(\mathbf{y},\mathbf{u}) + \mathbf{p}^\top(\mathbb{K}\mathbf{y} - \mathbb{M}\mathbf{u} - \mathbf{d})$$

and using the KKT optimality conditions (that is, the stationarity conditions of $\mathcal{L}$) we find that $\mathbf{y}$, $\mathbf{u}$, $\mathbf{p}$ are the solution of the following saddle point system:

[8] For further details, see `https://github.com/tyronerees/poisson-control`.

$$\underbrace{\begin{bmatrix} 2\beta\mathbb{M} & 0 & \mathbb{M}^\top \\ 0 & \mathbb{M} & \mathbb{K}^\top \\ \mathbb{M} & \mathbb{K} & 0 \end{bmatrix}}_{\mathcal{K}} \begin{bmatrix} \mathbf{u} \\ \mathbf{y} \\ \mathbf{p} \end{bmatrix} = \begin{bmatrix} \mathbf{0} \\ \mathbf{b} \\ \mathbf{d} \end{bmatrix}. \tag{6.87}$$

To solve this system, we employ the MINRES method with one of the following (optimal) block-diagonal preconditioners,

$$\widehat{\mathcal{P}}_{dg} = \begin{bmatrix} 2\beta\widetilde{\mathbb{M}} & 0 & 0 \\ 0 & \widetilde{\mathbb{M}} & 0 \\ 0 & 0 & \widetilde{\mathbb{K}}\mathbb{M}^{-1}\widetilde{\mathbb{K}}^\top \end{bmatrix}, \qquad \widehat{\mathcal{P}}_{da} = \begin{bmatrix} 2\beta\widetilde{\mathbb{M}} & 0 & 0 \\ 0 & \widetilde{\mathbb{M}} & 0 \\ 0 & 0 & \widehat{\mathbb{K}}\mathbb{M}^{-1}\widehat{\mathbb{K}}^\top \end{bmatrix} \tag{6.88}$$

where $\widetilde{\mathbb{K}}$ and $\widehat{\mathbb{K}}$ are two approximations of the stiffness matrix obtained with the geometric or the algebraic multigrid method, respectively. In particular, $\widetilde{\mathbb{K}}$ denontes two geometric V-cycles and $\widehat{\mathbb{K}}$ denotes two algebraic multigrid (AMG) V-cycles of the `HSL` package `HSL_MI20` package [47] applied through a `Matlab` interface. For both multigrid methods, the smoother is given by the relaxed Jacobi method, that is, if we have to solve a linear system $\mathbb{A}\mathbf{x} = \mathbf{b}$ for some matrix $\mathbb{A}$ and vectors $\mathbf{x}, \mathbf{b}$, we iterate as follows,

$$\mathbf{x}^{(k+1)} = (\mathbb{I} - \omega\mathbb{D}^{-1}\mathbb{A})\mathbf{x}^{(k)} + \omega\mathbb{D}^{-1}\mathbf{b}$$

where $\mathbb{D} = diag(\mathbb{A})$, $\mathbb{I}$ is the identity matrix, $\mathbf{x}^{(0)} = \mathbf{0}$ and ω is a relaxation parameter. As suggested in [226], we consider two pre- and post- smoothing steps of relaxed Jacobi to ensure a reasonable overall efficiency, with the optimal relaxation parameter $\omega = 8/9$. The approximation $\widetilde{\mathbb{M}}$ of the mass matrix is instead obtained through the Chebyshev semi-iteration method – that is, a solve with $\mathbb{M}$ is approximated by few steps of relaxed Jacobi, accelerated by the Chebyshev semi-iteration, see [226] for further details. We highlight that the approximate mass matrix solve is the the cheapest part of the application of the preconditioner, and also a lumped mass matrix, or even the diagonal of $\mathbb{M}$ would lead to an optimal preconditioner.

As an alternative to the MINRES method, we also consider the PPCG method, employing the following constraint preconditioner,

$$\widehat{\mathcal{P}}_{ca} = \begin{bmatrix} 0 & 0 & -\widetilde{\mathbb{M}} \\ 0 & 2\beta\widehat{\mathbb{K}}^\top\mathbb{M}^{-1}\widehat{\mathbb{K}} & \mathbb{K}^\top \\ -\widetilde{\mathbb{M}} & \mathbb{K} & 0 \end{bmatrix}$$

where, as for the diagonal preconditioner, $\widehat{\mathbb{K}}$ stands for the approximation of the solves with $\mathbb{K}$ by two AMG V-cycles of the `HSL` package `HSL_MI20`, and $\widetilde{\mathbb{M}}$ denotes few iterations of the Chebyshev semi-iterative method.

Aiming at showing the mesh-independent properties of these preconditioners, we compare the time to solve the system (6.90), for different values of β, and different spatial discretizations (obtained by taking $h = 2^{-2}, 2^{-3}, \dots, 2^{-8}$, using a direct method (implemented in the \ (backslash) function in `Matlab`, the MINRES method with preconditioner $\widehat{\mathcal{P}}_{dg}$ and $\widehat{\mathcal{P}}_{da}$, and the PPCG method with preconditioner $\widehat{\mathcal{P}}_{ca}$, possibly including $\mathbb{G} = diag(\mathcal{K})$ – even if this latter preconditioner might reveal prohibitive to use for small values of h. In particular, we take $\beta = 10^{-2}, 5\cdot 10^{-5}, 10^{-5}, 10^{-8}$. The MINRES method is stopped when the relative residual (in the 2-norm) has reached the desired tolerance, whereas the PPCG method is terminated when $\mathbf{r}^\top\mathbf{g}$ has reached the desired tolerance relative to its initial value (see Algorithm 6.7). In both cases, we choose 10^{-8} as tolerance.

Results are reported in Fig. 6.7 and Table 6.4. The number of iterations required by the iterative methods does not increase as the mesh is refined with the MINRES method or the PPCG method with using the $\widehat{\mathcal{P}}_{ca}$ preconditioner. Moreover, the smaller the value of β, the larger the number of iterations required to converge (keeping fixed the tolerance). The geometric MG preconditioner ($\widehat{\mathcal{P}}_{dg}$) converges in about the same number of iterations as the corresponding AMG preconditioner ($\widehat{\mathcal{P}}_{da}$), but the latter is faster[9] than the former (at least for coarser meshes). The employed preconditioners work effectively with regularization parameter $\beta = 10^{-2}, 5 \cdot 10^{-5}, 10^{-5}$, offering mesh size-independent convergence until $\beta = 10^{-5}$. For $\beta = 10^{-8}$ the number of iterations grows remarkably for both methods, however tending to a constant value for smaller values of h. On the other hand, for any fixed value of h, more and more iterations are required to achieve convergence as β becomes smaller.

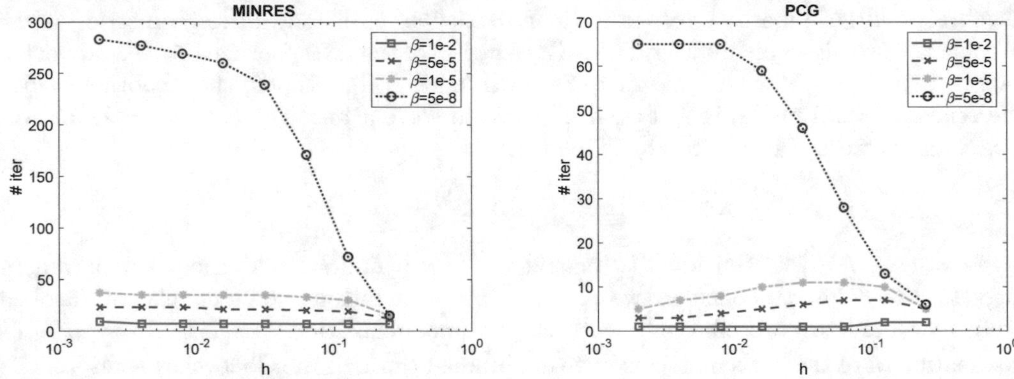

Fig. 6.7 Case 1. Number of iterations for the MINRES method with $\widehat{\mathcal{P}}_{da}$ and PPCG with $\widehat{\mathcal{P}}_{ca}$ to solve problem (6.84) for $\beta = 10^{-2}, 5 \cdot 10^{-5}, 10^{-5}, 10^{-8}$, as function of the mesh size h.

Test case 2: Diffusion-Reaction Equation

We now consider an OCP for a diffusion-reaction equation, which takes the same form of problem (6.84), except that now the state problem (6.84b) is replaced by

$$\begin{cases} -\Delta y + y = u & \text{in } \Omega \\ y = z_d & \text{on } \Gamma, \end{cases} \tag{6.89}$$

In this case, the KKT optimality conditions yield the following saddle point system:

$$\underbrace{\begin{bmatrix} 2\beta \mathbb{M} & 0 & \mathbb{M}^\top \\ 0 & \mathbb{M} & (\mathbb{K}+\mathbb{M})^\top \\ \mathbb{M} & \mathbb{K}+\mathbb{M} & 0 \end{bmatrix}}_{\mathcal{K}} \begin{bmatrix} \mathbf{u} \\ \mathbf{y} \\ \mathbf{p} \end{bmatrix} = \begin{bmatrix} \mathbf{0} \\ \mathbf{b} \\ \mathbf{d} \end{bmatrix}. \tag{6.90}$$

To solve this system, we employ the MINRES method with one of the following (optimal) block-diagonal preconditioners,

[9] Results have been obtained on a 2.2 GHz Intel Core i7 processor, 16GB RAM.

$\beta = 10^{-2}$						
h	$3N_y$	backslash	MINRES ($\widehat{\mathcal{P}}_{dg}$)	MINRES ($\widehat{\mathcal{P}}_{da}$)	PPCG ($\widehat{\mathcal{P}}_{ca}$)	PPCG ($\widehat{\mathcal{P}}_{ca}$ with $\mathbb{G} = diag(\mathcal{K})$)
2^{-2}	27	0.0001	0.0930 (12)	0.0021 (12)	0.0025 (3)	0.0011 (6)
2^{-3}	147	0.0009	0.1038 (12)	0.0032 (12)	0.0015 (3)	0.0054 (11)
2^{-4}	675	0.0027	0.1366 (12)	0.0071 (12)	0.0032 (3)	0.0221 (11)
2^{-5}	2883	0.0085	0.1102 (12)	0.0247 (12)	0.0107 (3)	0.1150 (11)
2^{-6}	11907	0.0488	0.1824 (12)	0.1009 (12)	0.0441 (3)	0.9520 (10)
2^{-7}	48387	0.2681	0.5454 (12)	0.4469 (13)	0.2089 (3)	6.2209 (10)
2^{-8}	195075	1.5194	2.0213 (12)	2.0513 (14)	1.0019 (3)	42.5664 (10)
2^{-9}	783363	11.4919	8.9402 (12)	10.5010 (16)	4.6273 (3)	344.9240 (9)

$\beta = 5 \cdot 10^{-5}$						
h	$3N_y$	backslash	MINRES ($\widehat{\mathcal{P}}_{dg}$)	MINRES ($\widehat{\mathcal{P}}_{da}$)	PPCG ($\widehat{\mathcal{P}}_{ca}$)	PPCG ($\widehat{\mathcal{P}}_{ca}$ with $\mathbb{G} = diag(\mathcal{K})$)
2^{-2}	27	0.0001	0.0762 (14)	0.0034 (20)	0.0015 (6)	0.0009 (6)
2^{-3}	147	0.0008	0.0830 (28)	0.0075 (32)	0.0038 (10)	0.0053 (11)
2^{-4}	675	0.0027	0.0895 (30)	0.0192 (34)	0.0102 (11)	0.0230 (11)
2^{-5}	2883	0.0091	0.1346 (30)	0.0680 (36)	0.0288 (11)	0.1123 (11)
2^{-6}	11907	0.0520	0.2989 (30)	0.2916 (36)	0.1049 (10)	0.8468 (10)
2^{-7}	48387	0.3436	1.2287 (32)	1.1829 (36)	0.4998 (10)	5.6151 (10)
2^{-8}	195075	1.6133	5.0477 (32)	4.9689 (36)	1.8249 (8)	30.3726 (10)
2^{-9}	783363	10.7542	23.1803 (32)	23.3629 (38)	8.3284 (8)	179.6498 (9)

$\beta = 10^{-5}$						
h	$3N_y$	backslash	MINRES ($\widehat{\mathcal{P}}_{dg}$)	MINRES ($\widehat{\mathcal{P}}_{da}$)	PPCG ($\widehat{\mathcal{P}}_{ca}$)	PPCG ($\widehat{\mathcal{P}}_{ca}$ with $\mathbb{G} = diag(\mathcal{K})$)
2^{-2}	27	0.0001	0.0765 (18)	0.0029 (22)	0.0018 (8)	0.0012 (6)
2^{-3}	147	0.0011	0.0891 (44)	0.0110 (50)	0.0052 (15)	0.0056 (10)
2^{-4}	675	0.0030	0.1023 (48)	0.0318 (58)	0.0143 (19)	0.1049 (11)
2^{-5}	2883	0.0097	0.1754 (51)	0.1117 (60)	0.0471 (19)	0.8305 (10)
2^{-6}	11907	0.0545	0.4482 (51)	0.4876 (60)	0.1765 (18)	5.2758 (10)
2^{-7}	48387	0.3113	1.9730 (52)	1.9447 (60)	0.7429 (16)	6.2209 (10)
2^{-8}	195075	1.5165	8.4082 (54)	8.3950 (60)	2.9093 (14)	29.1385 (10)
2^{-9}	783363	8.6826	38.5304 (54)	39.3321 (62)	13.2998 (14)	172.6486 (9)

Table 6.4 Case 1. Comparison of times (and iterations) for solving problem (6.84) with $\beta = 10^{-2}, 5{\cdot}10^{-5}, 10^{-5}$ for different mesh sizes h (with $3N_y$ unknowns) to a tolerance of 10^{-8} for `Matlab`'s backslash method, MINRES method with preconditioners $\widehat{\mathcal{P}}_{dg}$ or $\widehat{\mathcal{P}}_{da}$, and PPCG with preconditioners $\widehat{\mathcal{P}}_{ca}$, or constraint preconditioner including $\mathbb{G} = diag(\mathcal{K})$.

$$\widehat{\mathcal{P}}_{dg} = \begin{bmatrix} 2\beta\widetilde{\mathbb{M}} & 0 & 0 \\ 0 & \widetilde{\mathbb{M}} & 0 \\ 0 & 0 & (\widetilde{\mathbb{K}+\mathbb{M}})\mathbb{M}^{-1}(\widetilde{\mathbb{K}+\mathbb{M}})^\top \end{bmatrix}, \qquad \widehat{\mathcal{P}}_{da} = \begin{bmatrix} 2\beta\widetilde{\mathbb{M}} & 0 & 0 \\ 0 & \widetilde{\mathbb{M}} & 0 \\ 0 & 0 & (\widehat{\mathbb{K}+\mathbb{M}})\mathbb{M}^{-1}(\widehat{\mathbb{K}+\mathbb{M}})^\top \end{bmatrix} \tag{6.91}$$

or the PPCG method with the constraint preconditioner

$$\widehat{\mathcal{P}}_{ca} = \begin{bmatrix} 0 & 0 & -\widetilde{\mathbb{M}} \\ 0 & 2\beta(\widehat{\mathbb{K}+\mathbb{M}})^\top\mathbb{M}^{-1}(\widehat{\mathbb{K}+\mathbb{M}}) & (\mathbb{K}+\mathbb{M})^\top \\ -\widetilde{\mathbb{M}} & \mathbb{K}+\mathbb{M} & 0 \end{bmatrix}.$$

Compared to case 1, the only difference is represented by the approximation of the solves with the stiffness matrix, now given by $\mathbb{K}+\mathbb{M}$ instead of $\mathbb{K}$ because of the presence of the mass matrix, which exploit the same multigrid solvers as before. Results are reported in Fig. 6.8 and look similar to the ones obtained for the previous case of the OCP governed by the Laplace equation. Computational times depend on both h and β in the same way as the ones reported in Table 6.4 and are not reported for the sake of space.

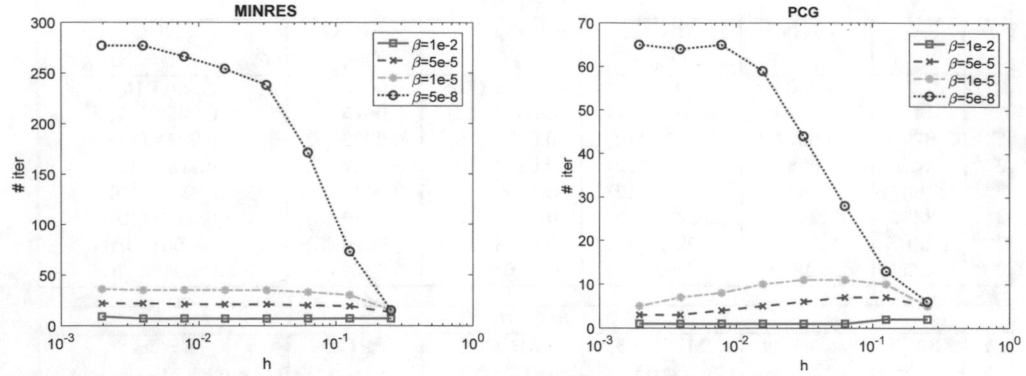

Fig. 6.8 Case 2. Number of iterations for the MINRES method with $\widehat{\mathcal{P}}_{da}$ and PPCG with $\widehat{\mathcal{P}}_{ca}$ to solve problem (6.84a)–(6.89) for $\beta = 10^{-2}, 5 \cdot 10^{-5}, 10^{-5}, 10^{-8}$, as function of the mesh size h.

6.8 All-at-once Methods (II): Control Constraints

A more general strategy to approximate the solution of constrained OCPs is the so-called *primal-dual active set* (PDAS) method, that we will present here in the case of box constraints. First contributions on the PDAS method for optimal control problems are given in [34, 143, 35]; for a rigorous formulation of the PDAS algorithm in Hilbert spaces see [34, 162].

For the sake of brevity, we confine to the Optimize-then-Discretize approach, first providing a formulation of the method at the continuous level, then introducing a suitable discretization. Due to the presence of the control constraints, the optimality system features a nonlinear nature, so that a sequence of saddle-point problems – similar to the ones arising in the unconstrained case – have to be solved, each one characterized by a set of active and inactive control constraints, which can entirely change from one iteration to the next. At each step, a problem has to be solved without inequality constraints, and we can take advantage of the preconditioners introduced in Sect. 6.6. For this reason, we consider the PDAS method as an all-at-once method. Later on, we will also highlight some analogies between this method and the more general sequential quadratic programming (SQP) method, described in Sect. 9.6.2, and employed for the numerical solution of nonlinear OCPs.

We consider problem (6.1) with

$$\mathcal{U}_{ad} = \{u \in \mathcal{U} \ : \ a(\mathbf{x}) \leq u(\mathbf{x}) \leq b(\mathbf{x}) \text{ a.e. in } \Omega\}.$$

Here $\mathcal{U} = L^2(\Omega)$, and $a, b \in L^\infty(\Omega)$. Different (e.g., unilateral) constraints on the control can also be considered. The optimality system for this problem reads: find $y \in V$, $u \in \mathcal{U}$ and $p \in V$ such that:

$$\begin{cases} a\left(y, \varphi\right) = \left(f + u, \varphi\right)_{L^2(\Omega)} & \forall \varphi \in V \\ a^*\left(p, \psi\right) = -\left(y - z_d, \psi\right)_{L^2(\Omega)} & \forall \psi \in V \\ \left(\beta u - p, v - u\right)_{L^2(\Omega)} \geq 0 & \forall v \in \mathcal{U}. \end{cases} \tag{6.92}$$

Recall (see Proposition 5.4) that the variational inequality appearing in the system above is equivalent to the following KKT conditions: there exist two functions λ_a and λ_b in $L^2(\Omega)$ such that a.e. in Ω

$$\begin{cases} \beta u - p - \lambda_a + \lambda_b = 0 \\ a \leq u \leq b, \ \ \lambda_a \geq 0, \ \ \lambda_b \geq 0 \\ \lambda_a \left(u - a\right) = \lambda_b \left(b - u\right) = 0 \end{cases} \tag{6.93}$$

or, alternatively (see equation (5.45)), there exists $\lambda \in L^2(\Omega)$ such that:

$$\begin{cases} \beta u - p + \lambda = 0 \\ a \leq u \leq b \\ u = a \ \text{ where } \lambda < 0, \qquad u = b \ \text{ where } \lambda > 0. \end{cases} \tag{6.94}$$

The last two equations in (6.94) can be equivalently rewritten as

$$\lambda - \min\{0, \lambda + c(u-a)\} - \max\{0, \lambda + c(u-b)\} = 0 \qquad \forall c \in \mathbb{R}_+. \tag{6.95}$$

Let us define the active sets

$$\mathcal{A}^+ = \{x \in \Omega \ : \ \lambda + c(u-b) > 0\}, \qquad \mathcal{A}^- = \{x \in \Omega \ : \ \lambda + c(u-a) < 0\}.$$

If they were known, the optimality system would reduce to

$$\begin{cases} a\,(y, \varphi) = (f + u, \varphi)_{L^2(\Omega)} & \forall \varphi \in V \\ a^*\,(p, \psi) = -\,(y - z_d, \psi)_{L^2(\Omega)} & \forall \psi \in V \\ \beta u - p + \lambda = 0 & \text{a.e. in } \ \Omega \\ u = b \ \text{ in } \mathcal{A}^+, \qquad \lambda = 0 \ \text{ in } \mathcal{I}, \qquad u = a \ \text{ in } \mathcal{A}^-, \end{cases} \tag{6.96}$$

where $\mathcal{I} = \Omega \setminus \{\mathcal{A}^+ \cup \mathcal{A}^-\}$ denotes the *inactive* set, that is, the set of points in Ω where the control constraints are not active. Problem (6.96) can then be solved. More in general, however, the PDAS method starts by making a guess of the optimal active sets, then move from one step to the following one by solving a quadratic subproblem in which some of the inequality constraints are treated as equality constraints on the basis of a prediction made according to the complementarity conditions. More in detail:

1. set (y_0, u_0, λ_0), $k = 0$, and initialize active and inactive sets as follows:

$$\mathcal{A}_0^+ = \{x \in \Omega \ : \ \lambda_0 + c(u_0 - b) > 0\}, \quad \mathcal{A}_0^- = \{x \in \Omega \ : \ \lambda_0 + c(u_0 - a) < 0\},$$

$$\mathcal{I}_0 = \{x \in \Omega \ : \ \lambda_0 + c(u_0 - b) \leq 0 \leq \lambda_0 + c(u_0 - a)\};$$

2. until $k < k_{max}$, a maximum number of iterations, do:

   a. solve for $(y_{k+1}, u_{k+1}, p_{k+1}, \lambda_{k+1})$ the following KKT system:

$$\begin{cases} a\,(y_{k+1}, \varphi) = (f + u_{k+1}, \varphi)_{L^2(\Omega)} & \forall \varphi \in V \\ a^*\,(p_{k+1}, \psi) = -\,(y_{k+1} - z_d, \psi)_{L^2(\Omega)} & \forall \psi \in V \\ \beta u_{k+1} - p_{k+1} + \lambda_{k+1} = 0 & \text{a.e. in } \ \Omega \\ u_{k+1} = b \ \text{ in } \mathcal{A}_k^+, \qquad \lambda_{k+1} = 0 \ \text{ in } \mathcal{I}_k, \qquad u_{k+1} = a \ \text{ in } \mathcal{A}_k^-, \end{cases} \tag{6.97}$$

   b. given $(y_{k+1}, u_{k+1}\lambda_{k+1})$, update active and inactive sets as follows:

$$\mathcal{A}_{k+1}^+ = \{x \in \Omega \ : \ \lambda_{k+1} + c(u_{k+1} - b) > 0\},$$

$$\mathcal{A}_{k+1}^- = \{x \in \Omega \ : \ \lambda_{k+1} + c(u_{k+1} - a) < 0\},$$

$$\mathcal{I}_{k+1} = \{x \in \Omega \ : \ \lambda_{k+1} + c(u_{k+1} - b) \leq 0 \leq \lambda_{k+1} + c(u_{k+1} - a)\};$$

   c. check convergence: if $\mathcal{A}_{k+1}^+ = \mathcal{A}_k^+$, $\mathcal{A}_{k+1}^- = \mathcal{A}_k^-$ and $\mathcal{I}_{k+1} = \mathcal{I}_k$ then end the loop, otherwise set $k \leftarrow k + 1$ and go back to point $2a$.

For the approximation of the PDAS method, introduce suitable discrete spaces V_h and $\mathcal{U}_h$ of dimension $N_y = \dim V_h$ and $N_u = \dim \mathcal{U}_h$, to approximate the state space V and the control space $\mathcal{U}$, respectively. $\mathcal{U}_h$ is also the space for approximating the multiplier λ, and we assume that the adjoint space is equal to the state space, that is, $\tilde{V}_h = V_h$. The discretized version of problem (6.97) reads (in algebraic form):

$$\begin{cases} \mathbb{A}\mathbf{y}_{k+1} = \mathbf{f} + \mathbb{M}\mathbf{u}_{k+1} \\ \mathbb{A}^\top \mathbf{p}_{k+1} = -\mathbb{M}(\mathbf{y}_{k+1} - \mathbf{z}_d) \\ \beta\mathbf{u}_{k+1} - \mathbb{M}\mathbf{p}_{k+1} + \boldsymbol{\lambda}_{k+1} = \mathbf{0} \\ \mathbf{u}_{k+1} = \mathbf{b} \text{ in } \mathcal{A}_k^+, \qquad \boldsymbol{\lambda}_{k+1} = \mathbf{0} \text{ in } \mathcal{I}_k, \qquad \mathbf{u}_{k+1} = \mathbf{a} \text{ in } \mathcal{A}_k^-, \end{cases} \tag{6.98}$$

with obvious meaning of notation. In particular, the active and inactive sets are subsets of the mesh nodes, that is,

$$\mathcal{A}_k^+ = \{i \in \{1, \ldots, N_u\} \ : \ (\boldsymbol{\lambda}_k)_i + c(\mathbf{u}_k - \mathbf{b})_i > 0\},$$

$$\mathcal{A}_k^- = \{i \in \{1, \ldots, N_u\} \ : \ (\boldsymbol{\lambda}_k)_i + c(\mathbf{u}_k - \mathbf{a})_i < 0\},$$

$$\mathcal{I}_k = \{i \in \{1, \ldots, N_u\} \ : \ (\boldsymbol{\lambda}_k)_i + c(\mathbf{u}_k - \mathbf{b})_i \leq 0 \leq (\boldsymbol{\lambda}_k)_i + c(\mathbf{u}_k - \mathbf{a})_i\}.$$

Equivalently, (6.98) can be rewritten as

$$\begin{bmatrix} \mathbb{M} & 0 & \mathbb{A}^\top & 0 \\ 0 & \beta\mathbb{M} & -\mathbb{M} & \mathbb{I} \\ \mathbb{A} & -\mathbb{M} & 0 & 0 \\ 0 & c\mathbb{I}_{\mathcal{A}_k} & 0 & \mathbb{I}_{\mathcal{I}_k} \end{bmatrix} \begin{bmatrix} \mathbf{y}_{k+1} \\ \mathbf{u}_{k+1} \\ \mathbf{p}_{k+1} \\ \boldsymbol{\lambda}_{k+1} \end{bmatrix} = \begin{bmatrix} \mathbb{M}\mathbf{z}_d \\ 0 \\ \mathbf{f} \\ c\boldsymbol{\psi}_k \end{bmatrix} \tag{6.99}$$

where:

- the matrix $\mathbb{I}_{\mathcal{A}_k} \in \mathbb{R}^{N_u \times N_u}$ is diagonal,

$$(\mathbb{I}_{\mathcal{A}_k})_{jj} = \begin{cases} 1 & \text{if } j \in \mathcal{A}_k^+ \cup \mathcal{A}_k^- \\ 0 & \text{otherwise;} \end{cases}$$

- the matrix $\mathbb{I}_{\mathcal{I}_k} \in \mathbb{R}^{N_u \times N_u}$ is diagonal,

$$(\mathbb{I}_{\mathcal{I}_k})_{jj} = \begin{cases} 1 & \text{if } j \in \mathcal{I}_k \\ 0 & \text{otherwise;} \end{cases}$$

- the vector $\boldsymbol{\psi}_k \in \mathbb{R}^{N_u}$ is given by $\boldsymbol{\psi}_k = \chi_{\mathcal{A}_k^-}\mathbf{a} + \chi_{\mathcal{A}_k^+}\mathbf{b}$, that is,

$$(\boldsymbol{\psi}_k)_j = \begin{cases} (\mathbf{a})_j & \text{if } j \in \mathcal{A}_k^- \\ 0 & \text{if } j \in \mathcal{I}_k \\ (\mathbf{b})_j & \text{if } j \in \mathcal{A}_k^+ \end{cases}$$

This follows from rewriting $(6.98)_4$ as

$$\mathbb{I}_{\mathcal{A}_k}\mathbf{u}_{k+1} + \mathbb{I}_{\mathcal{I}_k}\boldsymbol{\lambda}_{k+1} = \boldsymbol{\psi}_k. \tag{6.100}$$

By using the similar shorthand notation of the unconstrained case, (6.99) can be rewritten as $\mathcal{K}^k\boldsymbol{\chi} = \mathfrak{b}^k$. The matrices $\mathbb{I}_{\mathcal{A}_k}$ and $\mathbb{I}_{\mathcal{I}_k}$ change at every iteration, as they depend on the active and inactive sets. The whole matrix $\mathcal{K}^k$ changes therefore at every iteration, and so does the right-hand side $\mathfrak{b}^k$, because of the modifications occurring in the components $\boldsymbol{\psi}_k$. The PDAS method is implemented in Algorithm 6.10.

Algorithm 6.10 PDAS method for constrained OCPs

Input: initial guess $\mathbf{y}_0$, $\mathbf{u}_0$, $\mathbf{p}_0$, $\boldsymbol{\lambda}_0$, $c \in \mathbb{R}_+$, k_{max}

Initialize active and inactive sets:

$$\begin{aligned} \mathcal{A}_0^+ &= \{i \in \{1,\dots,N_u\} \ : \ (\boldsymbol{\lambda}_0)_i + c(\mathbf{u}_0 - \mathbf{b})_i > 0\}, \\ \mathcal{A}_0^- &= \{i \in \{1,\dots,N_u\} \ : \ (\boldsymbol{\lambda}_0)_i + c(\mathbf{u}_0 - \mathbf{a})_i < 0\}, \\ \mathcal{I}_0 &= \{i \in \{1,\dots,N_u\} \ : \ (\boldsymbol{\lambda}_0)_i + c(\mathbf{u}_0 - \mathbf{b})_i \le 0 \le (\boldsymbol{\lambda}_0)_i + c(\mathbf{u}_0 - \mathbf{a})_i\} \end{aligned}$$

set $k = 0$ and `converged = false`

while `converged = false` and $k < k_{max}$

  solve (6.99) for $(\mathbf{y}_{k+1}, \mathbf{u}_{k+1}, \mathbf{p}_{k+1}, \boldsymbol{\lambda}_{k+1})$

  given $(\mathbf{y}_{k+1}, \mathbf{u}_{k+1}, \boldsymbol{\lambda}_{k+1})$, update active and inactive sets:

$$\begin{aligned} \mathcal{A}_{k+1}^+ &= \{i \in \{1,\dots,N_u\} \ : \ (\boldsymbol{\lambda}_{k+1})_i + c(\mathbf{u}_{k+1} - \mathbf{b})_i > 0\}, \\ \mathcal{A}_{k+1}^- &= \{i \in \{1,\dots,N_u\} \ : \ (\boldsymbol{\lambda}_{k+1})_i + c(\mathbf{u}_{k+1} - \mathbf{a})_i < 0\}, \\ \mathcal{I}_{k+1} &= \{i \in \{1,\dots,N_u\} \ : \ (\boldsymbol{\lambda}_{k+1})_i + c(\mathbf{u}_{k+1} - \mathbf{b})_i \le 0 \le (\boldsymbol{\lambda}_{k+1})_i + c(\mathbf{u}_{k+1} - \mathbf{a})_i\} \end{aligned}$$

  check convergence:

  if $\mathcal{A}_{k+1}^+ = \mathcal{A}_k^+$, $\mathcal{A}_{k+1}^- = \mathcal{A}_k^-$ and $\mathcal{I}_{k+1} = \mathcal{I}_k$

    `converged = true`

  else

    k = k+1;

  end if

end while

Note that the matrix $\mathcal{K}^k$ is no longer symmetric, as in the unconstrained case (see equation (6.68)). To restore its symmetry, we note that from (6.100)

$$(\boldsymbol{\lambda}_{k+1})_j = 0 \qquad \forall j \in \mathcal{I}_k.$$

By eliminating these components from (6.99) we obtain

$$\begin{bmatrix} \mathbb{M} & 0 & \mathbb{A}^\top & 0 \\ 0 & \beta\mathbb{M} & -\mathbb{M} & \tilde{\mathbb{P}}_{\mathcal{A}_k} \\ \mathbb{A} & -\mathbb{M} & 0 & 0 \\ 0 & \tilde{\mathbb{P}}_{\mathcal{A}_k}^\top & 0 & 0 \end{bmatrix} \begin{bmatrix} \mathbf{y}_{k+1} \\ \mathbf{u}_{k+1} \\ \mathbf{p}_{k+1} \\ \boldsymbol{\lambda}_{k+1}^{\mathcal{A}_k} \end{bmatrix} = \begin{bmatrix} \mathbb{M}\mathbf{z}_d \\ 0 \\ \mathbf{f} \\ \tilde{\boldsymbol{\psi}}_k^{\mathcal{A}} \end{bmatrix}. \tag{6.101}$$

Here $\boldsymbol{\lambda}_{k+1}^{\mathcal{A}_k}$ denotes the restriction of $\boldsymbol{\lambda}_{k+1}$ to the active set $\mathcal{A}_k$, $\tilde{\mathbb{P}}_{\mathcal{A}_k} \in \mathbb{R}^{(N_y+N_u)\times N_{\mathcal{A}_k}}$ is a rectangular matrix, $\tilde{\mathbb{P}}_{\mathcal{A}_k} = \begin{bmatrix} 0 & \mathbb{P}_{\mathcal{A}_k} \end{bmatrix}$; $\mathbb{P}_{\mathcal{A}_k} \in \mathbb{R}^{N_u \times N_{\mathcal{A}_k}}$ is also rectangular, consisting of those rows of the matrix $\mathbb{I}_{\mathcal{A}_k}$ which correspond to the active indexes. In the same way, $\tilde{\boldsymbol{\psi}}_k^{\mathcal{A}} \in \mathbb{R}^{N_{\mathcal{A}_k}}$ is given by

$$\tilde{\boldsymbol{\psi}}_k^{\mathcal{A}} = \mathbb{P}_{\mathcal{A}_k^-}\mathbf{a} + \mathbb{P}_{\mathcal{A}_k^+}\mathbf{b}$$

where $\mathbb{P}_{\mathcal{A}_k^-} \in \mathbb{R}^{N_u \times N_{\mathcal{A}_k^-}}$ and $\mathbb{P}_{\mathcal{A}_k^+} \in \mathbb{R}^{N_u \times N_{\mathcal{A}_k^+}}$ are rectangular matrices consisting of those rows of $\mathbb{I}_{\mathcal{A}_k}$ which correspond to the indexes of $\mathcal{A}_k^-$ and $\mathcal{A}_k^+$, respectively.

The linear system (6.101) now exhibits the same saddle-point structure of the optimality system arising in the unconstrained case; indeed, we are just adding a further Lagrange multiplier. Note that the variable $\boldsymbol{\lambda}_{k+1}$ is the Lagrange multiplier associated with the discrete constraint $\mathbf{a} \le \mathbf{u} \le \mathbf{b}$, and the relationships

$$\boldsymbol{\lambda}_{k+1} = \tilde{\mathbb{P}}_{\mathcal{A}_k^+}\boldsymbol{\lambda}_{k+1}^{\mathcal{A}_k}, \qquad \boldsymbol{\lambda}_{k+1}^{\mathcal{A}_k} = \tilde{\mathbb{P}}_{\mathcal{A}_k^+}^\top\boldsymbol{\lambda}_{k+1}$$

hold. Finally, by solving the system $\mathbb{M}\boldsymbol{\xi}_{k+1} = \boldsymbol{\lambda}_{k+1}$, we find the nodal values $\boldsymbol{\xi}_{k+1}$ of a finite element function which approximates the multiplier $\boldsymbol{\lambda}_{k+1}$ in the continuous system (6.97).

Both block-diagonal and block-triangular preconditioners can be efficiently used to solve the system (6.101) arising as part of the primal-dual active set method, see, e.g., [141, 228, 252].

Remark 6.10. The primal-dual active set method can also be interpreted as a (regularized) semismooth Newton method. This follows by solving approximately the (non differentiable) nonlinear equation

$$\lambda - \min\{0, \lambda + c(u-a)\} - \max\{0, \lambda + c(u-b)\} = 0 \qquad \forall c \in \mathbb{R}_+$$

by a semismooth Newton algorithm (in the sense made precise in [143]). •

Regularized State-Constrained Problems: Mixed Constraints

A primal-dual active set strategy can also be employed to approximate mixed control-state constraints[10]. For the sake of exposition, we consider the regularized control problem (5.177)–(5.179), where the state constraints are replaced by the Laurentiev-type regularization (5.181). Such a problem is recalled below for reader's convenience:

$$\begin{aligned} &\min_{y\in V, u\in \mathcal{U}_{ad}\subseteq\mathcal{U}} \tilde{J}(y,u) = \frac{1}{2}\int_\Omega (y-z_d)^2 d\mathbf{x} + \frac{\beta}{2}\int_\Omega u^2 d\mathbf{x} \\ &\text{subject to} \begin{cases} -\Delta y + = f + u & \text{in } \Omega \\ \dfrac{\partial y}{\partial n} = 0 & \text{on } \partial\Omega \end{cases} \\ &\text{and} \quad y_a \le \varepsilon u + y(u) \le y_b \qquad \text{a.e in } \Omega. \end{aligned} \tag{6.102}$$

Note that the regularization parameter has now been denoted by $\varepsilon > 0$ since λ already denotes the Lagrange multiplier related to the mixed state-control constraint.

Using Theorem 5.9 and condition 5.46, we have that the optimality system for problem (6.102) reads: find $y \in V$, $u \in \mathcal{U}$, $p \in V$, $\lambda \in \mathcal{U}$ such that, for any $c \in \mathbb{R}_+$:

$$\begin{cases} a(y,\varphi) = (f+u,\varphi)_{L^2(\Omega)} & \forall \varphi \in V \\ a^*(p,\psi) = -(y-z_d,\psi)_{L^2(\Omega)} - (\lambda,\psi)_{L^2(\Omega)} & \forall \psi \in V \\ \beta u - p + \varepsilon\lambda = 0 & \text{a.e. in } \Omega \\ \lambda - \min\{0, \lambda - c(y_a - \varepsilon u - y)\} - \max\{0, \lambda + c(\varepsilon u + y - y_b)\} = 0 & \text{a.e. in } \Omega. \end{cases} \tag{6.103}$$

The generic step of the PDAS method reads:

$$\begin{cases} a(y_{k+1},\varphi) = (f+u_{k+1},\varphi)_{L^2(\Omega)} & \forall \varphi \in V \\ a^*(p_{k+1},\psi) = -(y_{k+1}-z_d,\psi)_{L^2(\Omega)} & \forall \psi \in V \\ \beta u_{k+1} - p_{k+1} + \varepsilon\lambda_{k+1} = 0 & \text{a.e. in } \Omega \\ \varepsilon u_{k+1} + y_{k+1} = y_b \ \text{ in } \mathcal{A}_k^+, \quad \lambda_{k+1} = 0 \ \text{ in } \mathcal{I}_k, \quad \varepsilon u_{k+1} + y_{k+1} = y_a \ \text{ in } \mathcal{A}_k^-, \end{cases}$$

where the active and inactive sets now take the form

$$\mathcal{A}_-^k = \{\mathbf{x} \in \Omega : \lambda - c(y_a - \varepsilon u - y) < 0\},$$

[10] Mixed control-state constraints can be viewed as a possible way to regularize problems with pure state constraints, see Sect. 5.12.

$$\mathcal{A}_+^k = \{\mathbf{x} \in \Omega \, : \, \lambda + c(\varepsilon u + y - y_b) > 0\},$$
$$\mathcal{I} = \Omega \setminus \{\mathcal{A}_a \cup \mathcal{A}_b\}.$$

From an algebraic standpoint, the discretization of the problem above can be carried out as in the case of control constraints, yielding the following saddle-point system:

$$\begin{bmatrix} \mathbb{M} & 0 & \mathbb{A}^\top & \tilde{\mathbb{P}}_{\mathcal{A}_k}^\top \\ 0 & \beta\mathbb{M} & -\mathbb{M} & \varepsilon\tilde{\mathbb{P}}_{\mathcal{A}_k} \\ \mathbb{A} & -\mathbb{M} & 0 & 0 \\ \tilde{\mathbb{P}}_{\mathcal{A}_k}^\top & \varepsilon\tilde{\mathbb{P}}_{\mathcal{A}_k}^\top & 0 & 0 \end{bmatrix} \begin{bmatrix} \mathbf{y}_{k+1} \\ \mathbf{u}_{k+1} \\ \mathbf{p}_{k+1} \\ \boldsymbol{\lambda}_{k+1}^{\mathcal{A}_k} \end{bmatrix} = \begin{bmatrix} \mathbb{M}\mathbf{z}_d \\ 0 \\ \mathbf{f} \\ \tilde{\psi}_k^{\mathcal{A}} \end{bmatrix}, \tag{6.104}$$

with the matrix $\tilde{\mathbb{P}}_{\mathcal{A}_k}^\top$ and the vector $\tilde{\psi}_k^{\mathcal{A}}$ defined similarly than in (6.101). Preconditioned iterative solvers (with either block-diagonal or block-triangular preconditioners) can therefore be employed within a PDAS algorithm also in the case of state constraints provided a Laurentiev regularization is operated, as done in the more general case of mixed state-control constraints. See, e.g., [141, 197] for further details.

6.9 Numerical Examples

In this section we report some numerical results obtained with the PDAS method, regarding constrained OCPs governed by either the Laplace equation or the Stokes equations.

6.9.1 OCPs Governed by the Laplace Equation

Test case 1

Let us denote by $\Omega = (0,1)^2$, by $a > 0$ a bound on the control function, by $z_d(\mathbf{x}) = \sin(2\pi x_1)\sin(2\pi x_2)$ the target function, and consider the following OCP:

$$\tilde{J}(y,u) = \frac{1}{2}\int_\Omega (y - z_d)^2 d\mathbf{x} + \frac{\beta}{2}\int_\Omega u^2 d\mathbf{x} \;\rightarrow\; \min \tag{6.105a}$$
$$\text{subject to } u \in \mathcal{U}_{ad} = \{u \in L^2(\Omega) \, : -a \le u \le a\}, \; y \in V :$$
$$\begin{cases} -\Delta y = u & \text{in } \Omega \\ y = z_d & \text{on } \Gamma, \end{cases} \tag{6.105b}$$

where $\beta > 0$ is a suitable regularization parameter. With the choice of z_d made above, the *best possible* control (i.e., the one giving a state solution equal to z_d, the desired value, disregarding the discretization error) is given by $u_{ex} = 8\pi^2 \sin(2\pi x_1)\sin(2\pi x_2)$.

As initial guesses for the PDAS algorithm, we consider $u = 0$ and $\lambda = 0$. This means that at the first iteration the inactive set $\mathcal{I}_0$ coincides with the whole domain, i.e. the bounds are nowhere active. Tests have been performed using $\mathbb{Q}_1$ conforming piecewise bilinear finite elements in Ω, built on a quadrilateral mesh, and considering different values for the regularization parameter $\beta = 10^{-2}, 10^{-4}, 10^{-6}, 10^{-8}$, and different bounds on the control $a = 100, 70, 50, 30$, in order to analyze their effect on the OCP solution and the PDAS algorithm.

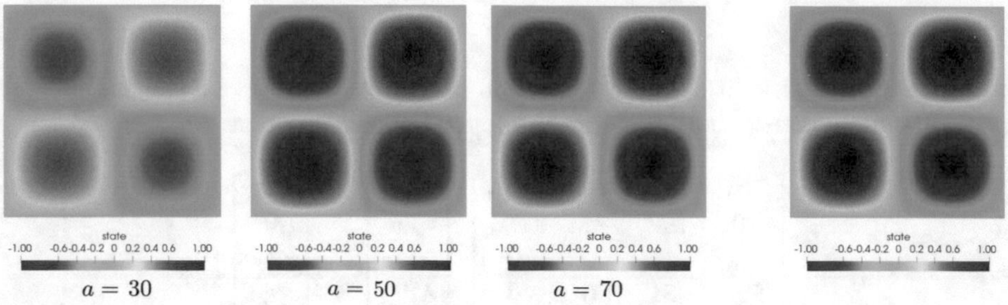

Fig. 6.9 Case 1. Optimal state $\hat{y}$ for $\beta = 10^{-8}$ and different control bounds: from left to right, $a = 30, 50, 70$ and the unconstrained case

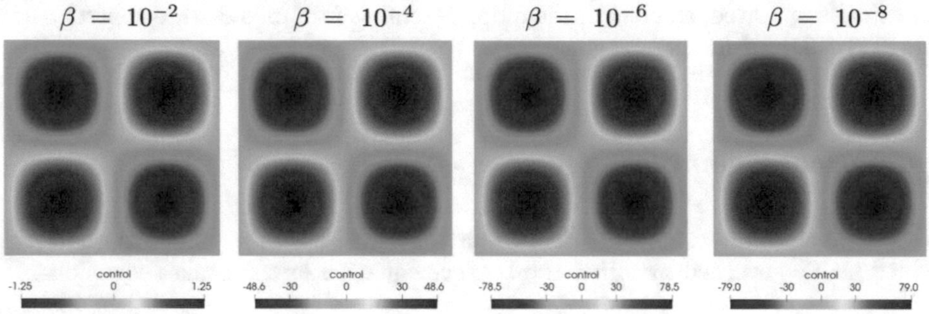

Fig. 6.10 Case 1. Optimal control $\hat{u}$ in the unconstrained case for $\beta = 10^{-2}, 10^{-4}, 10^{-6}, 10^{-8}$ (from left to right)

Numerical results are shown in Figures 6.9–6.13. In the case $a = 100$, the PDAS method converges in a single iteration, since the best possible control u_{ex} is such that $\|u_{ex}\|_{L^\infty(\Omega)} \leq 80$; however, the norm of the control highly depends on the regularization parameter β (see Fig. 6.10).

The stricter the bound a, the higher the number of iterations required to the PDAS method to converge. In particular, the optimal control is determined by forcing the control to fulfill the bound constraints in some regions (i.e., the ones corresponding to the active sets $\mathcal{A}_+$ and $\mathcal{A}_-$), and, at the same time, by widening these regions as far as β allows: indeed, smaller values of β allow a larger control norm, and the possibility, for the control, to act on wider regions of the domain (Fig. 6.11).

The larger the regularization parameter α, the lower the number of iterations required to the PDAS method to converge and the norm of the control, and the larger the value of the distance $\|\hat{y} - z_d\|_{L^2(\Omega)}$ between the optimal state and the target (Fig. 6.12).

Finally, we can notice that, by keeping fixed the control bound a and β, the number of iterations required to achieve convergence increases for decreasing mesh sizes h (Fig. 6.13).

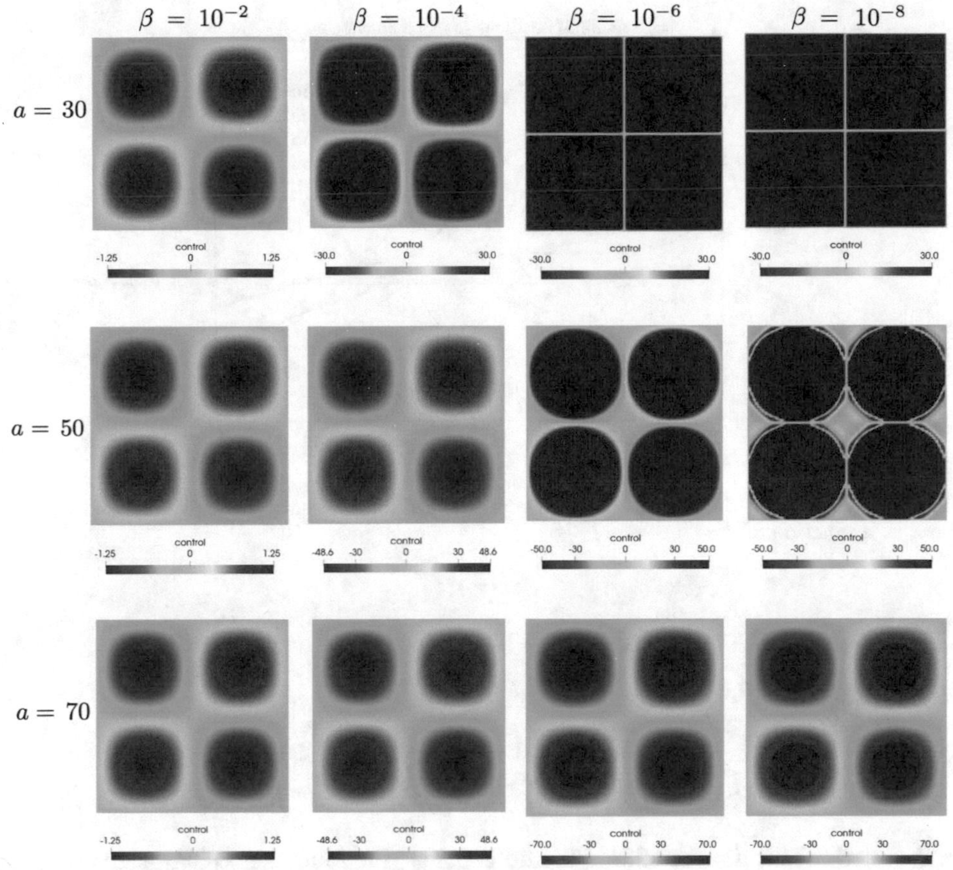

Fig. 6.11 Case 1. Optimal control $\hat{u}$ in the constrained case for different control bounds: from top to bottom, $a = 30, 50, 70$, and different regularization parameters, $\beta = 10^{-2}, 10^{-4}, 10^{-6}, 10^{-8}$ (from left to right)

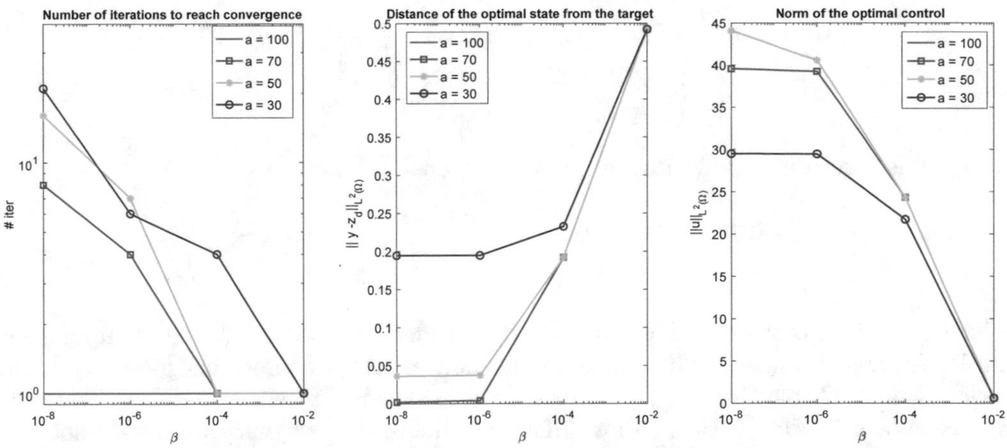

Fig. 6.12 Case 1. From left to right: number of iterations to reach convergence, distance $\|\hat{y} - z_d\|_{L^2(\Omega)}$ between the optimal state and the target, norm of the optimal control $\hat{u}$, for different control bounds and different regularization parameters. In the center and right plots, the values obtained in the cases $a = 100$ and $a = 70$ are almost superimposed

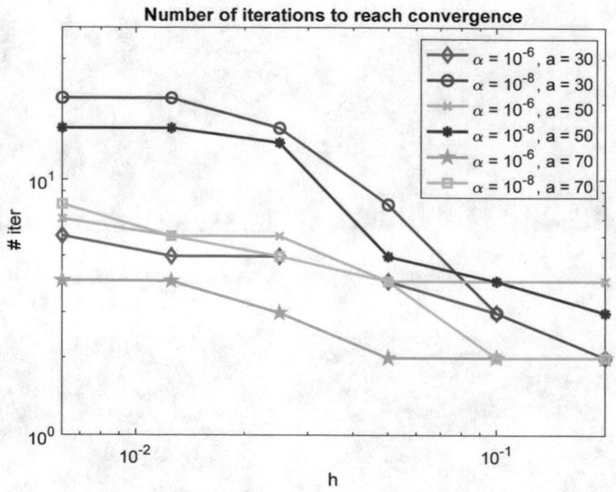

Fig. 6.13 Case 1. Number of iterations to reach convergence, for different control bounds and regularization parameters, as a function of the mesh size h

Test case 2

Let us denote by $\Omega = (0,1)^2$, by

$$z_d(\mathbf{x}) = -x_1 \exp\left(-\sum_{i=1}^{d}(x_i - 0.5)^2\right)$$

the target function, for $d = 2$ and $d = 3$, and consider the following OCP, originally proposed in [228]:

$$\tilde{J}(y,u) = \frac{1}{2}\int_\Omega (y - z_d)^2 d\mathbf{x} + \frac{\beta}{2}\int_\Omega u^2 d\mathbf{x} \rightarrow \min \tag{6.106a}$$

$$\text{subject to } u \in \mathcal{U}_{ad} = \{u \in L^2(\Omega) : a \le u \le b\},\ y \in V:$$

$$\begin{cases} -\Delta y = u & \text{in } \Omega = (0,1)^d \\ y = z_d & \text{on } \Gamma, \end{cases} \tag{6.106b}$$

where $\beta > 0$ is a suitable regularization parameter, and

$$a = \begin{cases} -0.15 & x_1 < 0.5 \\ -0.2 & x_1 \ge 0.5, \end{cases} \qquad b = -0.01 \exp\left(-\sum_{i=1}^{d} x_i^2\right).$$

We solve the OCP in both cases $d = 2$ and $d = 3$, considering $u = 0$ and $\lambda = 0$ as initial guesses for the PDAS algorithm. Tests have been performed using $\mathbb{Q}_1$ conforming piecewise bilinear finite elements in Ω, built on a quadrilateral or a hexahedral mesh, respectively. Numerical results are shown in Figs. 6.14–6.15 for different values of the regularization parameter $\beta = 10^{-2}, 10^{-4}, 10^{-6}$; the number of iterations to reach convergence, the distance $\|\hat{y} - z_d\|_{L^2(\Omega)}$ between the optimal state and the target, and the norm of the optimal control $\hat{u}$, behave as functions of β similarly to the ones obtained in Test case 1. Note that, in presence of a discontinuous bound function a, large values of the regularization coefficients might limit the norm of the control, as in the case reported in Fig. 6.14 (top).

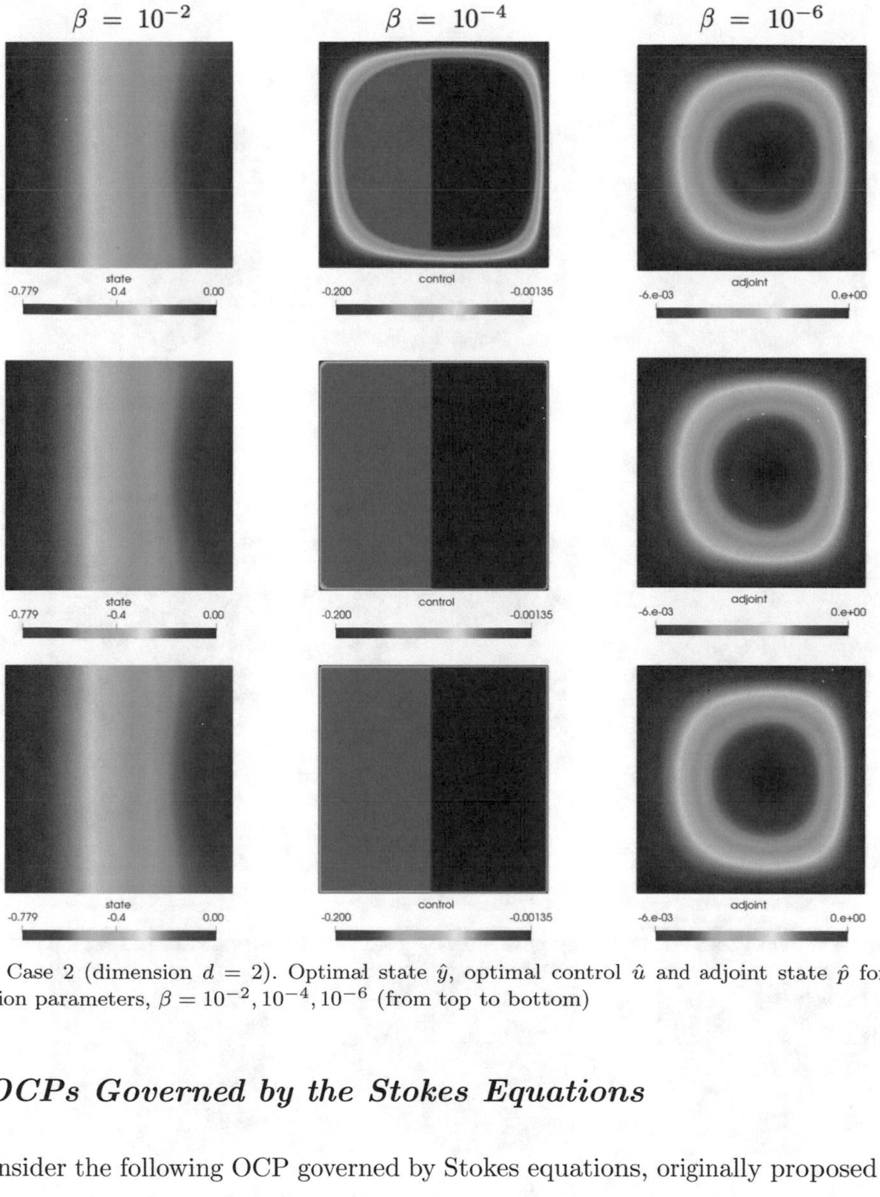

Fig. 6.14 Case 2 (dimension $d = 2$). Optimal state $\hat{y}$, optimal control $\hat{u}$ and adjoint state $\hat{p}$ for different regularization parameters, $\beta = 10^{-2}, 10^{-4}, 10^{-6}$ (from top to bottom)

6.9.2 OCPs Governed by the Stokes Equations

Let us consider the following OCP governed by Stokes equations, originally proposed in [229]:

$$
\begin{aligned}
&\tilde{J}(\boldsymbol{v}, \pi, \boldsymbol{u}) = \frac{1}{2}\int_\Omega |\boldsymbol{v} - \boldsymbol{v}_d|^2\, d\mathbf{x} + \frac{\alpha}{2}\int_\Omega |\boldsymbol{u}|^2\, d\mathbf{x} \;\rightarrow\; \min \\
&\qquad\text{subject to } (\boldsymbol{v}, \pi) \in V,\; \boldsymbol{u} \in \mathcal{U}_{ad}, \\
&\qquad\begin{cases} -\nu\Delta \boldsymbol{v} + \nabla \pi = \boldsymbol{f} + \boldsymbol{u} & \text{in } \Omega \\ \operatorname{div} \boldsymbol{v} = 0 & \text{in } \Omega \\ \boldsymbol{v} = \boldsymbol{g} & \text{on } \Gamma, \end{cases}
\end{aligned}
\tag{6.107}
$$

where $\Omega = (0,1)^2$, $\Gamma_{right} = \{(x_1, x_2) \in \partial\Omega \ : \ x_1 = 1\}$ and

$$
\boldsymbol{g} = \begin{cases} \mathbf{0} & \text{on } \Gamma_{right} \\ -\mathbf{e}_2 & \text{on } \Gamma \setminus \Gamma_{right}, \end{cases}
$$

denoting by $(\mathbf{e}_1, \mathbf{e}_2)$ the standard basis in $\mathbb{R}^2$.

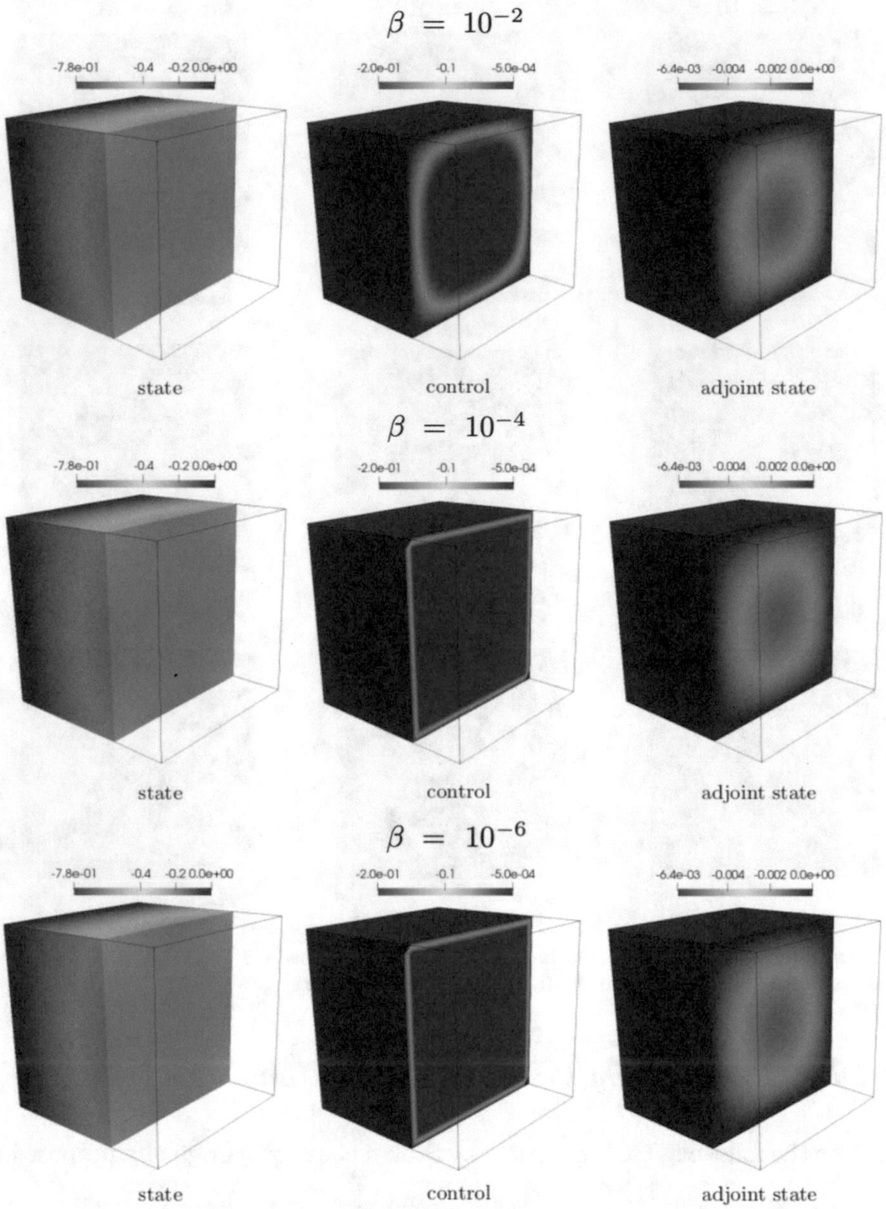

Fig. 6.15 Case 2 (dimension $d = 3$). Optimal state $\hat{y}$, optimal control $\hat{u}$ and adjoint state $\hat{p}$ for different regularization parameters, $\beta = 10^{-2}, 10^{-4}, 10^{-6}$ (from top to bottom)

As target function we consider

$$\boldsymbol{v}_d = (x_2 - 0.5)\mathbf{e}_1 - (x_1 - 0.5)\mathbf{e}_2,$$

we set $V = H_0^1(\Gamma_{right})^2 \times L^2(\Omega)$, and consider

$$\mathcal{U}_{ad} = \{\boldsymbol{u} \in L^2(\Omega)^2 \ : -550 \leq u_i \leq 550 \text{ a.e. in } \Omega, \ i = 1, 2\}.$$

Numerical results have been obtained for different values of the regularization coefficient α, using the inf-sup compatible pair $\mathbb{Q}_2 - \mathbb{Q}_1$ of FE spaces, whose dimensions for the case at hand are $N_v = 51842$ and $N_\pi = 6561$, respectively.

Note that, since the target velocity is not compatible with the boundary conditions prescribed on the state velocity in problem (6.107), we cannot expect to reduce the value of the cost functional below a certain value. The effect of the control function is to move the vortex from the right side of the domain towards the center; in particular, smaller values of α yield a more remarkable shift (see Fig. 6.16).

The number of iterations to reach convergence, the distance $\|\hat{\boldsymbol{v}} - \boldsymbol{v}_d\|_{L^2(\Omega)^2}$ between the optimal state velocity and the target velocity, and the norm of the optimal control $\hat{\boldsymbol{u}}$ plotted versus α, behave similarly to the ones obtained in Test case 1, see Table 6.5.

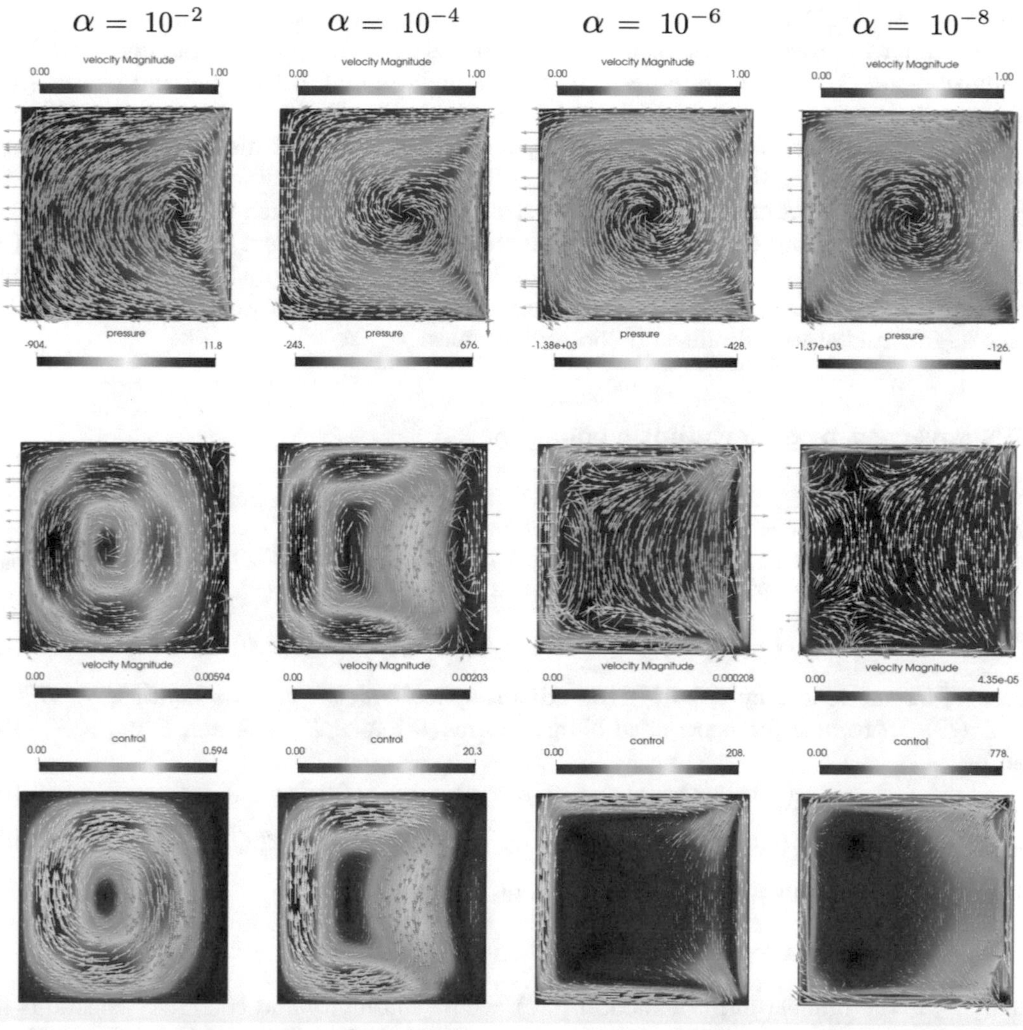

Fig. 6.16 Case 3 (Stokes). Optimal velocity and pressure $(\hat{\boldsymbol{v}}, \hat{\pi})$, adjoint velocity $\hat{\boldsymbol{z}}$ and optimal control $\hat{\boldsymbol{u}}$ and for different regularization parameters, $\beta = 10^{-2}, 10^{-4}, 10^{-6}, 10^{-8}$ (from top to bottom, from left to right). In the case of state variables, the vector fields in the top row are oriented as the velocity field, and colored according to the pressure values. Note that pressure peaks are localized along the boundary of the domain and for this reason are not clearly visible in this figure

β	$J_h(\hat{u})$	# iter	$\|\hat{u}\|_{L^2(\Omega)}$	$\|\hat{y}-z_d\|_{L^2(\Omega)}$
10^{-2}	0.0579601	1	0.326406	0.338902
10^{-4}	0.0430436	1	10.3743	0.274453
10^{-6}	0.0244265	1	72.2504	0.208885
10^{-8}	0.0176069	6	308.912	0.185093

Table 6.5 Case 3. Values of the cost functional for the optimal control $J_h(\hat{u})$, number of iterations required by the steepest descent method to converge, norm of the optimal control $\|\hat{u}\|_{L^2(\Omega)}$, value of the cost $\|\hat{y}-z_d\|_{L^2(\Omega)}$, for different values of the regularization parameter β. The value of the cost functional in the absence of control ($u = 0$) is $J_h(u_0) = 0.0585097$

6.10 A priori Error Estimates

In this section we provide some *a priori* error estimates for the Galerkin-FE solution of a linear-quadratic OCP. A quite general (and intuitive) fact is that error estimates for optimal controls cannot improve those known for the solutions of the associated state equation. Early contributions on the subject date back to the 70's and are due to Falk [94] and Geveci [102]. L^2-error estimates are obtained which exploit the H^1-regularity of the optimal control[11] and the optimal regularity of the state variable, both for the case of distributed controls and Neumann boundary controls.

Here we mainly focus on the case of unconstrained OCPs, for which this kind of results can be derived by exploiting the saddle-point nature of the KKT system (6.4), as well as of its discrete version (6.5), and Theorem B.2. A similar approach is adopted by Becker, Kapp and Rannacher in [23], where the KKT system is indeed rewritten as a weakly coercive problem for the aggregated state, adjoint and control variables.

OCPs governed by scalar elliptic equations

We start by considering the case of an OCP for a scalar elliptic equation with distributed control, like (6.1). First of all, we want to recast (6.4) in the form of a saddle-point problem[12]. With that in mind, let us denote by $\boldsymbol{x} = (y, u) \in X$, where $X = V \times \mathcal{U}$ is the product space between the state space V and the control space $\mathcal{U}$, equipped with the scalar product

$$(\boldsymbol{x}, \boldsymbol{w})_X = (y, z)_V + (u, v)_{\mathcal{U}} \qquad \forall \boldsymbol{x} = (y, u), \boldsymbol{w} = (z, v) \in X;$$

moreover, let us denote by $Q(= V)$ the adjoint space. For the case at hand, $V = H_0^1(\Omega)$, $\mathcal{U} = L^2(\Omega)$. Introduce the *aggregated* bilinear forms $\mathcal{D} : X \times X \to \mathbb{R}$ and $\mathcal{E} : Q \times X \to \mathbb{R}$, given by

$$\mathcal{D}(\boldsymbol{x}, \boldsymbol{w}) = (y, z)_{L^2(\Omega)} + \beta(u, v)_{L^2(\Omega)} \qquad \forall \boldsymbol{x}, \boldsymbol{w} \in X, \tag{6.108}$$

$$\mathcal{E}(q, \boldsymbol{x}) = a(y, q) - (u, q)_{L^2(\Omega)} \qquad \forall \boldsymbol{x} \in X, \forall q \in Q, \tag{6.109}$$

with $\beta \geq 0$, and the functionals $\mathcal{H} \in X^*$, $\mathcal{G} \in Q^*$

$$\langle \mathcal{H}, \boldsymbol{w} \rangle_{X^*,X} = (z_d, z)_{L^2(\Omega)} + (0, v)_{L^2(\Omega)}, \qquad \langle \mathcal{G}, q \rangle_{Q^*,Q} = (f, q)_{L^2(\Omega)}.$$

Here $a(\cdot, \cdot)$ is the bilinear form defined in (6.2), whose continuity and coercivity constants are denoted by M and α, respectively. Then, we can rewrite (6.4) as follows: find $(\boldsymbol{x}, p) \in X \times Q$ such that

$$\begin{cases} \mathcal{D}(\boldsymbol{x}, \boldsymbol{w}) + \mathcal{E}(p, \boldsymbol{w}) = \langle \mathcal{H}, \boldsymbol{w} \rangle_{X^*,X} & \forall \boldsymbol{w} \in X \\ \mathcal{E}(q, \boldsymbol{x}) \qquad\qquad\quad\;\; = \langle \mathcal{G}, q \rangle_{Q^*,Q} & \forall q \in Q. \end{cases} \tag{6.110}$$

[11] Assuming $f \in L^2(\Omega)$, $a, b \in L^\infty(\Omega) \cap H^1(\Omega)$, so that the optimal control belongs to $H^1(\Omega)$.

[12] This is indeed the continuous counterpart of what we did from an algebraic standpoint in Sect. 6.6.

Proposition 6.1. *The bilinear forms $\mathcal{D}(\cdot,\cdot)$ and $\mathcal{E}(\cdot,\cdot)$ satisfy the hypotheses of Theorem A.7; hence, problem (6.110) is well-posed.*

Proof. The symmetry and the non-negativity of $\mathcal{D}(\cdot,\cdot)$ are evident. Thanks to the continuity of the bilinear form $a(\cdot,\cdot)$, $\mathcal{D}(\cdot,\cdot)$ is continuous on $X \times X$, as

$$\begin{aligned}|\mathcal{D}(\boldsymbol{x},\boldsymbol{w})| \le \|y\|_{L^2(\Omega)}\|z\|_{L^2(\Omega)} + \beta\|u\|_{\mathcal{U}}\|v\|_{\mathcal{U}} &\le C_P^2\|y\|_V\|z\|_V + \beta\|u\|_{\mathcal{U}}\|v\|_{\mathcal{U}} \\ &\le (C_P^2+\beta)\|\boldsymbol{x}\|_X\|\boldsymbol{w}\|_X,\end{aligned}$$

where C_P is the Poincaré constant. $\mathcal{D}(\cdot,\cdot)$ is strongly coercive on X_0: in fact, if $\boldsymbol{w}=(z,v)\in X_0$, then

$$\mathcal{E}(q,\boldsymbol{w}) = 0 \quad \forall q \in Q \quad \Leftrightarrow \quad a(z,q) = (v,q)_{L^2(\Omega)} \quad \forall q \in Q.$$

Using the Lax-Milgram Lemma we have $\|z\|_V \le \frac{1}{\alpha}\|v\|_{\mathcal{U}}$, from which we deduce

$$\begin{aligned}\mathcal{D}(\boldsymbol{w},\boldsymbol{w}) &= \|z\|^2_{L^2(\Omega)} + \beta\|v\|^2_{\mathcal{U}} \ge \|z\|^2_{L^2(\Omega)} + \frac{\beta}{2}\|v\|^2_{\mathcal{U}} + \frac{\beta}{2}\|v\|^2_{\mathcal{U}} \\ &\ge \|z\|^2_{L^2(\Omega)} + \frac{\beta}{2}\|v\|^2_{\mathcal{U}} + \frac{\beta}{2}\alpha^2\|z\|^2_V \ge \frac{\beta}{2}\max\left(1,\alpha^2\right)\left(\|z\|^2_V + \|v\|^2_{\mathcal{U}}\right).\end{aligned}$$

Moreover, $\mathcal{E}(\cdot,\cdot)$ is continuous on $X \times X$ since

$$|\mathcal{E}(\boldsymbol{x},q)| \le |a(y,q)| + |(u,q)_{L^2(\Omega)}| \le M\|y\|_V\|q\|_Q + \|u\|_{\mathcal{U}}\|q\|_Q \le (M+1)\|\boldsymbol{x}\|_X\|q\|_Q.$$

Furthermore, $\mathcal{E}(\cdot,\cdot)$ is inf-sup stable on $X \times Q$, since

$$\sup_{0\ne\boldsymbol{x}\in X}\frac{\mathcal{E}(q,\boldsymbol{x})}{\|\boldsymbol{x}\|_X} = \sup_{0\ne(z,v)\in V\times\mathcal{U}}\frac{a(y,q)-(u,q)_{L^2(\Omega)}}{(\|y\|^2_V+\|u\|^2_{\mathcal{U}})^{1/2}} \underset{(y,u)=(q,0)}{\ge} \frac{a(q,q)}{\|q\|_Y} \ge \alpha\|q\|_V = \alpha\|q\|_Q.$$

□

Remark 6.11. Thanks to the equivalence between saddle-point problems and constrained minimization problems stated in Proposition A.9, the result above allows the original OCP to be rewritten under the form of a constrained minimization problem, which is nothing but the counterpart of problem (6.67) at the continuous level. •

Remark 6.12. The analogy between the optimality system and a saddle-point problem holds at the continuous level for a constrained OCP, too, provided a primal-dual active set algorithm is used. In fact, by defining $\mathbf{q} = (q, \lambda_{\mathcal{A}})$ and the *aggregated* bilinear forms $\mathcal{D}(\boldsymbol{x},\boldsymbol{w})$ as in (6.108), and

$$\mathcal{E}_{\mathcal{A}}(\mathbf{q},\boldsymbol{x}) = a(y,q) - (v,q)_{L^2(\Omega)} + (u,\lambda_{\mathcal{A}})_{L^2(\mathcal{A})} \qquad \forall\boldsymbol{x}\in X, \forall\mathbf{q}\in Q\times L^2(\mathcal{A})$$

instead of (6.109), where $\mathcal{A}$ denotes the active set, it is possible to show that $\mathcal{E}_{\mathcal{A}}$ also fulfills the assumptions of Theorem A.7, see [141]. Because of the nature of the PDAS algorithm, the main difference with respect to the unconstrained case is that this saddle-point problem is changing at each iteration as long as the active set changes. •

Let us now consider the Galerkin approximation of the saddle-point problem (6.110); we define $X_h = V_h \times \mathcal{U}_h \subset X$ and $Q_h \subset Q$ for suitable finite element spaces V_h, $\mathcal{U}_h$ and Q_h; moreover, we assume that $Q_h = V_h$. The Galerkin-FE approximation (6.5) of (6.4) can be rewritten as: find $(\boldsymbol{x}_h, p_h) \in X_h \times Q_h$ such that

$$\begin{cases}\mathcal{D}(\boldsymbol{x}_h,\boldsymbol{w}_h) + \mathcal{E}(p_h,\boldsymbol{w}_h) = \langle\mathcal{H},\boldsymbol{w}_h\rangle_{X^*,X} & \forall\boldsymbol{w}_h\in X_h \\ \mathcal{E}(q_h,\boldsymbol{x}_h) = \langle\mathcal{G},q_h\rangle_{Q^*,Q} & \forall q_h\in Q_h\end{cases} \tag{6.111}$$

Applying Theorem B.2, we obtain the following *convergence* result.

Theorem 6.1. *If $V_h = Q_h$, the bilinear forms $\mathcal{D}(\cdot,\cdot)$ and $\mathcal{E}(\cdot,\cdot)$ satisfy the hypotheses of Theorem B.2. Then, for every $h > 0$, problem (6.111) has a unique solution $(\boldsymbol{x}_h, p_h) \in X_h \times Q_h$. Moreover, if $(\boldsymbol{x}, p) \in X \times Q$ denotes the unique solution of (6.110), the following optimal error inequality holds*

$$\begin{aligned}\|y - y_h\|_V + &\|p - p_h\|_V + \|u - u_h\|_{\mathcal{U}} \\ &\leq C\left(\inf_{w_h \in V_h} \|y - w_h\|_V + \inf_{q_h \in V_h} \|p - q_h\|_V + \inf_{v_h \in \mathcal{U}_h} \|u - v_h\|_{\mathcal{U}}\right)\end{aligned} \quad (6.112)$$

where $C = C(M, \alpha, \beta)$ is independent of h. Furthermore, if we use finite elements of local degree r for approximating both the state and adjoint variables, and of local degree s for approximating the control variable, the following a priori error estimate holds

$$\begin{aligned}\|y - y_h\|_V + &\|p - p_h\|_V + \|u - u_h\|_{\mathcal{U}} \\ &\leq c_1 h^r (|y|_{H^{r+1}(\Omega)} + |p|_{H^{r+1}(\Omega)}) + c_2 h^s |u|_{H^s(\Omega)}\end{aligned} \quad (6.113)$$

provided that $y, p \in H^{r+1}(\Omega)$, $u \in H^s(\Omega)$, $r \geq 1$, $s \geq 0$.

Proof. We only sketch the main points, see Exercise 8 for more details. $\mathcal{D}(\cdot,\cdot)$ and $\mathcal{E}(\cdot,\cdot)$ fulfill the assumptions of Theorem B.2 thanks to the property that $a(\cdot,\cdot)$ is continuous on the discrete subspaces $V_h \times Q_h$ and strongly coercive in V_h, provided that $V_h = Q_h$, and then follow the main lines of the proof of Proposition 6.1. Then, the well-posedness of problem (6.111) as well as the error inequality (6.120) directly follow from Theorem B.2, whereas (6.113) is a direct consequence of the a priori error estimates for the FE approximation of elliptic problems, see Theorem B.4. □

OCPs governed by the Stokes equations

Similar results can be obtained in the case of an unconstrained linear-quadratic OCP for the Stokes equations. Here we only sketch the main points, see, e.g., [233] for more. In the case of the problem of Sect. 5.13.1, the system of optimality conditions can be reformulated as a saddle point system. To see that, we denote by $\boldsymbol{x} = ((\boldsymbol{v}, \pi), \boldsymbol{u}) \in \mathcal{X} = V \times \mathcal{U}$ the variable grouping the state $(\boldsymbol{v}, \pi) \in V$ and the control $\boldsymbol{u} \in \mathcal{U}$, by $\mathcal{Q} = V$ the adjoint space, being in this case $V = X \times Q$ the product space between the velocity and the pressure space. Then, let us define:

- the bilinear form $a : V \times V \to \mathbb{R}$

$$a((\boldsymbol{v}, \pi), (\boldsymbol{\varphi}, \phi)) = e(\boldsymbol{v}, \boldsymbol{\varphi}) + b(\boldsymbol{\varphi}, \pi) + b(\boldsymbol{v}, \phi) \quad \forall (\boldsymbol{v}, \pi), (\boldsymbol{\varphi}, \phi) \in V \quad (6.114)$$

  which is continuous and weakly coercive on $V \times V$ (see Exercise 9);
- for any $\boldsymbol{x} = ((\boldsymbol{v}, \pi), \boldsymbol{u}) \in \mathcal{X}$ and corresponding test functions $\boldsymbol{\zeta} = ((\boldsymbol{\varphi}, \phi), \boldsymbol{\eta}) \in \mathcal{X}$, the bilinear forms

$$\mathcal{D}(\boldsymbol{x}, \boldsymbol{\zeta}) = (\boldsymbol{v}, \boldsymbol{\varphi})_{L^2(\Omega)} + \tau(\boldsymbol{u}, \boldsymbol{\eta})_{L^2(\Omega)} \qquad \forall \boldsymbol{x}, \boldsymbol{\zeta} \in \mathcal{X}, \quad (6.115)$$

$$\mathcal{E}(\boldsymbol{x}, (\boldsymbol{\varphi}, \phi)) = a((\boldsymbol{v}, \pi), (\boldsymbol{\varphi}, \phi)) - (\boldsymbol{u}, \boldsymbol{\varphi})_{L^2(\Omega)} \qquad \forall \boldsymbol{x} \in \mathcal{X},\ \forall (\boldsymbol{\varphi}, \phi) \in \mathcal{Q}, \quad (6.116)$$

  which satisfy the assumptions of Theorem A.7 (see Exercise 9).

Then, $(\boldsymbol{x}, (\boldsymbol{z}, q)) \in \mathcal{X} \times \mathcal{Q}$ is an optimal solution of problem (5.190)–(5.191) if and only if

$$\begin{cases} \mathcal{D}(\boldsymbol{x}, \boldsymbol{\zeta}) + \mathcal{E}(\boldsymbol{\zeta}, (\boldsymbol{z}, q)) = \langle \mathcal{H}, \boldsymbol{\zeta} \rangle_{\mathcal{X}^*, \mathcal{X}} & \forall \boldsymbol{\zeta} \in \mathcal{X} \\ \mathcal{E}(\boldsymbol{x}, (\boldsymbol{\varphi}, \phi)) = 0 & \forall (\boldsymbol{\varphi}, \phi) \in \mathcal{Q}, \end{cases} \quad (6.117)$$

with $\mathcal{H} = ((z_d, 0), \mathbf{0}) \in \mathcal{X} = V \times \mathcal{U}$. Similarly to what done above for the elliptic problem, we can introduce the Galerkin approximation of problem (6.117) by considering two finite-dimensional spaces $\mathcal{X}_h \subset \mathcal{X}$, $\mathcal{Q}_h \subset \mathcal{Q}$, such that $\mathcal{X}_h = \mathcal{V}_h \times \mathcal{U}_h$, $\mathcal{Q}_h = V_h$: find $(\boldsymbol{x}_h, (\boldsymbol{z}_h, q_h)) \in \mathcal{X}_h \times \mathcal{Q}_h$ such that

$$\begin{cases} \mathcal{D}(\boldsymbol{x}_h, \zeta_h) + \mathcal{E}(\zeta_h, (\boldsymbol{z}_h, q_h)) = \langle \mathcal{H}, \zeta_h \rangle_{\mathcal{X}^*, \mathcal{X}} & \forall \zeta_h \in \mathcal{X}_h \\ \mathcal{E}(\boldsymbol{x}_h, (\boldsymbol{\psi}_h, \xi_h)) \qquad\qquad = 0 & \forall (\boldsymbol{\psi}_h, \xi_h) \in \mathcal{Q}_h. \end{cases} \tag{6.118}$$

A necessary condition to ensure that $\mathcal{D}(\cdot,\cdot)$ and $\mathcal{E}(\cdot,\cdot)$ fulfill the assumptions of Theorem B.2 is that V_h is an inf-sup stable subspace of V, that is, the couple of discrete spaces (X_h, Q_h) satisfy the discrete inf-sup condition (B.16), here reported for convenience,

$$\inf_{q_h \in Q_h} \sup_{w_h \in X_h} \frac{\mathcal{E}(\zeta_h, (\boldsymbol{z}_h, q_h))}{\|\zeta_h\|_{\mathcal{X}} \|(\boldsymbol{z}_h, q_h)\|_{Q}} \geq \beta^s, \tag{6.119}$$

for a suitable constant $\beta^s > 0$ independent of h. Then, a result similar to that reported in Proposition 6.1 holds for the Stokes case as well.

Theorem 6.2. *If the couple of discrete spaces (X_h, Q_h) satisfy the discrete inf-sup condition* (B.16) *and if $\mathcal{Q}_h = V_h$, the bilinear forms $\mathcal{D}(\cdot,\cdot)$ and $\mathcal{E}(\cdot,\cdot)$ satisfy the hypotheses of Theorem B.2. Then, for every $h > 0$, problem* (6.118) *has a unique solution $(\boldsymbol{x}_h, (\boldsymbol{z}_h, q_h)) \in \mathcal{X}_h \times \mathcal{Q}_h$. Moreover, if $(\boldsymbol{x}, (\boldsymbol{z}, q)) \in \mathcal{X} \times \mathcal{Q}$ denotes the unique solution of* (6.117)*, the following error inequality holds*

$$\begin{aligned} &\|\boldsymbol{v} - \boldsymbol{v}_h\|_X + \|\pi - \pi_h\|_Q + \|\boldsymbol{u} - \boldsymbol{u}_h\|_{\mathcal{U}} \\ &\quad + \|\boldsymbol{z} - \boldsymbol{z}_h\|_X + \|q - q_h\|_Q \leq C \Big(\inf_{\varphi_h \in V_h} \|\boldsymbol{v} - \varphi_h\|_V + \inf_{\phi_h \in Q_h} \|\pi - \phi_h\|_Q \\ &\qquad + \inf_{\eta_h \in \mathcal{U}_h} \|\boldsymbol{u} - \eta_h\|_{\mathcal{U}} \\ &\qquad + \inf_{\psi_h \in V_h} \|\boldsymbol{z} - \psi_h\|_V + \inf_{\xi_h \in Q_h} \|q - \xi_h\|_Q \Big) \end{aligned}$$

where $C = C(\nu, \tau, \beta^s)$ is independent of h.

OCPs with Control or State Constraints

In the case of constrained OCPs, deriving a priori error estimates becomes much more involved. Other approaches which do not rely on the saddle-point nature of the KKT system are also available in literature, aiming at the derivation of error bounds for state, adjoint and control variables separately. In this case, usually, the control space is first discretized, then the discrete state space is considered. The effect of an independent discretization of the control and state spaces has also been investigated:

1. in [145] the discretization of the state space alone is considered. Denoting by S_h a suitable discretization of the control-to-state map $y = Su$ by means of continuous, piecewise linear finite elements, a bound
$$\|u - u_h\|_{\mathcal{U}} \leq Ch^2,$$
with $C > 0$, independent of h, is proven, where u is the optimal control for the continuous problem and $u_h = \arg\min_{u \in \mathcal{U}_{ad}} \tilde{J}(S_h u, u)$;

2. in [232] no discretization of the state space is considered, whereas the effects of the discretization of boundary controls is analyzed. A space $\mathcal{U}_h$ of continuous, piecewise linear finite elements is considered to approximate $\mathcal{U} = L^2(\Gamma)$, leading to the corresponding set

$$\mathcal{U}_{h,ad} = \{u \in \mathcal{U}_h \ : \ a \leq u \leq b \text{ a.e. on } \Gamma\};$$

then the estimate

$$\|u - u_h\|_{\mathcal{U}} \leq Ch^{3/2}$$

is recovered, where u is the optimal control for the continuous problem and $u_h = \arg\min_{\mathcal{U}_{h,ad}} \tilde{J}(Su_h, u_h)$. For piecewise constant control approximations the estimate is instead of the form

$$\|u - u_h\|_{\mathcal{U}} \leq Ch.$$

The literature about error estimates for Dirichlet boundary control problems is not ample. Deckelnick, Günther and Hinze consider in [75] the Dirichlet boundary control of an elliptic equation with L^2-boundary controls subject to pointwise bounds on the controls – that is, $\mathcal{U}_{ad} = \{u \in L^2(\Gamma) : a \leq u(\mathbf{x}) \leq b \text{ a.e. on } \Gamma\}$ – and the cost functional (5.170). In this case, denoting by $u \in L^2(\Gamma)$ the optimal control, by $u_h \in \mathcal{U}_{ad,h}$ the solution of the discrete problem, where $\mathcal{U}_{ad,h}$ is the discrete counterpart of $\mathcal{U}_{ad}$ and by y and y_h the corresponding states (provided the state solution is approximated by piecewise linear, continuous finite elements) the following error estimate holds,

$$\|u - u_h\|_{L^2(\Gamma)} + \|y - y_h\|_{L^2(\Omega)} \leq Ch\sqrt{|\log h|}$$

for a suitable constant $C > 0$ independent of h. In the case $\Omega \subset \mathbb{R}^2$, and under additional conditions on the mesh regularity, an improved error bound under the form

$$\|u - u_h\|_{L^2(\Gamma)} + \|y - y_h\|_{L^2(\Omega)} \leq Ch^{3/2}$$

is also derived, reflecting a superconvergence effect; see also [58]. May, Rannacher, and Vexler also consider in [191] the case of Dirichlet boundary control without control constraints on two-dimensional convex polygonal domains, showing optimal error estimates for both the state and the adjoint state by using duality techniques and an optimal error estimate in $H^{-1/2}(\Gamma)$ for the control.

Regarding instead the case of both control and state constrained elliptic OCPs under the form (5.177)–(5.179), only few attempts have been made to develop a rigorous finite element analysis. Deckelnick and Hinze prove in [76] the following error bounds,

$$\|u - u_h\|_{L^2(\Omega)}, \quad \|y - y_h\|_{H^1(\Omega)} \leq \begin{cases} Ch|\log h|, & \text{if } d = 2, \\ C\sqrt{h}, & \text{if } d = 3 \end{cases}$$

in the case of a control discretization made by piecewise constant finite elements, and state discretization by piecewise linear, continuous finite elements; similar results can also be found in [196]. Error estimates for the Lavrentiev-regularized problem are instead provided in [147].

6.11 A Posteriori Error Estimates

A posteriori error bounds allow to quantify the discretization error through computable quantities (for instance, the *residual* of the approximate solution) once the discrete problem has been solved. This information is usually exploited within a mesh refinement procedure – this is called *a posteriori adaptivity*. Error control and adaptive mesh design is a central topic in numerical analysis, which is beyond the scope of the present introduction; see, e.g. [218, Chapter 4] for a general introduction to *a posteriori adaptivity*, and [3, 25, 112, 266] for more on this subject.

We now describe a possible strategy, known as *dual weighted residual* (DWR) method, to derive a posteriori error estimates for OCPs where the state problem is given by a scalar elliptic PDE like (6.1). DWR methods have been introduced by Rannacher and coworkers (see [25] and references therein) for the more general sake of error control and adaptive mesh design in PDE problems, and provide a way to derive estimates based on suitable functionals of the solution of the problem, by taking advantage of several ideas arising in the field of optimal control.

Here we exploit the DWR method to estimate the discretization error on the solution of an unconstrained OCP. For the sake of simplicity, we consider a pure diffusion problem, setting $\mathbf{b} = \mathbf{0}$; see e.g. [23] for the extension to the case of advection-diffusion-reaction problems. The case of control-constrained OCPs, which would require further care when formulating the discrete system of KKT conditions, is discussed in [267].

DWR methods are based on the assumption that the discretization error on the OCP problem can be identified with the error on the cost functional $|\tilde{J}(\hat{y}, \hat{u}) - \tilde{J}(\hat{y}_h, \hat{u}_h)|$, taking however into account the whole set of KKT conditions for the state, the adjoint and the control variable. Indeed, the cost functional depends directly on y and indirectly on u and p so that it is possible to estimate the distance between the discrete optimum $(\hat{y}_h, \hat{u}_h, \hat{p}_h)$ and the continuous one $(\hat{y}, \hat{u}, \hat{p})$, by taking into account these relationships.

Let us use the notation $x = (y, u, p)$ and $x_h = (y_h, u_h, p_h)$ and express the error on the cost functional as

$$\tilde{J}(y, u) - \tilde{J}(y_h, u_h) = \mathcal{L}(y, p, u) - \mathcal{L}(y_h, u_h, p_h) = \mathcal{L}(x) - \mathcal{L}(x_h).$$

Note that, provided a Galerkin approximation is used, the error on the OCP problem can be seen as the error on the Lagrangian.

According to the Fundamental Theorem of Calculus, we have

$$\begin{aligned}\mathcal{L}(x) - \mathcal{L}(x_h) &= \int_0^1 \frac{d}{ds}\mathcal{L}(sx + (1-s)x_h)ds = \int_0^1 \nabla\mathcal{L}(sx + (1-s)x_h)(x - x_h)ds \\ &= \nabla\mathcal{L}(x_h)(x - x_h) + \int_0^1 \left(\nabla\mathcal{L}(sx + (1-s)x_h) - \nabla\mathcal{L}(x_h)\right)(x - x_h)ds.\end{aligned}$$

In the case of a linear state equation and a quadratic cost functional (with respect to y), $\nabla\mathcal{L}(x)$ is linear with respect to x, so that the previous equation directly yields the following useful relationship,

$$\tilde{J}(y, u) - \tilde{J}(y_h, u_h) = \frac{1}{2}\nabla\mathcal{L}(x)(x - x_h) + \frac{1}{2}\nabla\mathcal{L}(x_h)(x - x_h). \tag{6.120}$$

Since the exact solution $\hat{x}$ is such that $\nabla\mathcal{L}(\hat{x}) = 0$, we end up with the following

Proposition 6.2. *The optimal solution $\hat{x} = (\hat{y}, \hat{u}, \hat{p})$ of the linear-quadratic OCP (6.1) and the corresponding approximation $\hat{x}_h = (\hat{y}_h, \hat{u}_h, \hat{p}_h)$ fulfill system (6.4) and (6.5), respectively, and are such that*

$$\tilde{J}(\hat{y}, \hat{u}) - \tilde{J}(\hat{y}_h, \hat{u}_h) = \frac{1}{2}\nabla\mathcal{L}(\hat{x}_h)(\hat{x} - \hat{x}_h). \tag{6.121}$$

To turn (6.121) into a computable expression , let us assume that, as usual, the triangulation $\mathcal{T}_h$ of the domain Ω underlying the finite element spaces is made by elements denoted by K, and denote by $h_K = diam(K)$, see Sect. B.4. Moreover, we denote by e the edge of a generic triangle K and define the *jump* of the normal derivative of y_h through the internal side e the quantity

$$\left[\frac{\partial y_h}{\partial n}\right] = \nabla y_h|_{K_1} \cdot \mathbf{n}_1 + \nabla y_h|_{K_2} \cdot \mathbf{n}_2 = (\nabla y_h|_{K_1} - \nabla y_h|_{K_2}) \cdot \mathbf{n}_1$$

where K_1 and K_2 are the two triangles sharing the side e, whose normal outgoing unit vectors are given by $\mathbf{n}_1$ and $\mathbf{n}_2$, respectively, with $\mathbf{n}_1 = -\mathbf{n}_2$. By convention, such a definition can be extended to the boundary edges $e \in \Gamma$ by setting the jump equal to zero, yielding the so-called *generalized jump*. By introducing the residuals

$$\begin{aligned}
R_y(\hat{y}_h) &= f + \hat{u}_h + \Delta \hat{y}_h, \qquad & r_y(\hat{y}_h) &= -\frac{1}{2}\left[\frac{\partial \hat{y}_h}{\partial n}\right] \quad \text{on } \partial K \setminus \Gamma,\\
R_p(\hat{p}_h) &= \hat{y}_h - z_d + \Delta \hat{p}_h, \qquad & r_p(\hat{p}_h) &= -\frac{1}{2}\left[\frac{\partial \hat{p}_h}{\partial n}\right] \quad \text{on } \partial K \setminus \Gamma,\\
R_u(\hat{p}_h, \hat{u}_h) &= \hat{p}_h,
\end{aligned}$$

and defining the weights

$$\begin{aligned}
\omega_K^p &= \|\hat{p} - \hat{p}_h\|_K + h_K^{1/2}\|\hat{p} - \hat{p}_h\|_{\partial K},\\
\omega_K^y &= \|\hat{y} - \hat{y}_h\|_K + h_K^{1/2}\|\hat{y} - \hat{y}_h\|_{\partial K},\\
\omega_K^u &= \|\hat{u} - \hat{u}_h\|_K,
\end{aligned}$$

following [23] we obtain the result below.

Theorem 6.3. *The following weighted a posteriori error estimate holds on the cost functional $\tilde{J}$,*

$$|\tilde{J}(\hat{y}, \hat{u}) - \tilde{J}(\hat{y}_h, \hat{u}_h)| \leq \eta_{\omega p}(\hat{y}_h, \hat{u}_h, \hat{p}_h) = \sum_{K \in \mathcal{T}_h} \eta_K(\hat{y}_h, \hat{u}_h, \hat{p}_h) \tag{6.122}$$

where

$$\eta_K(\hat{y}_h, \hat{u}_h, \hat{p}_h) = \omega_K^p \rho_K^y + \omega_K^y \rho_K^p + \omega_K^u \rho_K^u$$

is the local error indicator, with

$$\begin{aligned}
\rho_K^y &= \|R_y(\hat{y}_h)\|_K + h_K^{-1/2}\|r_y(\hat{y}_h)\|_{\partial K},\\
\rho_K^p &= \|R_p(\hat{p}_h)\|_K + h_K^{-1/2}\|r_p(\hat{p}_h)\|_{\partial K},\\
\rho_K^u &= \|R_u(\hat{y}_h)\|_K.
\end{aligned}$$

Proof. For the sake of notation we omit the symbol ^ throughout the proof. First of all we can write

$$\begin{aligned}
\tilde{J}(y,u) - \tilde{J}(y_h,u_h) &= \frac{1}{2}\Big[(f, p - p_h)_{L^2(\Omega)} + (u_h, p - p_h)_{L^2(\Omega)} - a(y_h, p - p_h)\Big]\\
&+ \frac{1}{2}\Big[(y_h - z_d, y - y_h)_{L^2(\Omega)} - a(y - y_h, p_h)\Big]\\
&+ \frac{1}{2}\Big[(\beta u_h, u - u_h)_{L^2(\Omega)} + (u - u_h, p_h)_{L^2(\Omega)}\Big] = \frac{1}{2}[I_y + I_p + I_u].
\end{aligned}$$

Using integration by parts elementwise, we have

$$
\begin{aligned}
I_y &= (f, p - p_h)_{L^2(\Omega)} + (u_h, p - p_h)_{L^2(\Omega)} - a(y_h, p - p_h) \\
&= \int_\Omega f(p - p_h)d\mathbf{x} + \int_\Omega u_h(p - p_h)d\mathbf{x} - \int_\Omega \nabla y_h \cdot \nabla(p - p_h)d\mathbf{x} \\
&= \int_\Omega (f + u_h)(p - p_h)d\mathbf{x} - \sum_{K\in\mathcal{T}_h} \int_K \nabla y_h \cdot \nabla(p - p_h)d\mathbf{x} \\
&= \sum_{K\in\mathcal{T}_h} \int_K \Delta y_h(p - p_h)d\mathbf{x} - \sum_{K\in\mathcal{T}_h} \int_{\partial K} \frac{\partial y_h}{\partial n}(p - p_h)d\sigma + \sum_{K\in\mathcal{T}_h} \int_K (f + u_h)(p - p_h)d\mathbf{x};
\end{aligned}
$$

in particular, regarding the boundary integrals appearing in the last expression, we can express the boundary ∂K of each element as the union of its edges, and find that

$$
\begin{aligned}
\sum_{K\in\mathcal{T}_h} \int_{\partial K} \frac{\partial y_h}{\partial n}(p - p_h)d\sigma &= \sum_{K\in\mathcal{T}_h} \sum_{e\in\partial K} \int_e \frac{\partial y_h}{\partial n}(p - p_h)d\sigma = \frac{1}{2} \sum_{K\in\mathcal{T}_h} \sum_{e\subset\partial K} \int_e \left[\frac{\partial y_h}{\partial n}\right](p - p_h)d\sigma \\
&= \frac{1}{2} \sum_{K\in\mathcal{T}_h} \int_{\partial K} \left[\frac{\partial y_h}{\partial n}\right](p - p_h)d\sigma,
\end{aligned}
$$

where the factor $1/2$ takes into account the fact that each internal side of the grid is shared by two elements. Hence,

$$
\begin{aligned}
I_y &= \sum_{K\in\mathcal{T}_h} \left(\int_K (f + u_h + \Delta y_h)(p - p_h)d\mathbf{x} - \frac{1}{2} \int_{\partial K} \left[\frac{\partial y_h}{\partial n}\right](p - p_h)d\sigma \right) \\
&= \sum_{K\in\mathcal{T}_h} \left((R_y(y_h), p - p_h)_{L^2(K)} + (r_y(y_h), p - p_h)_{L^2(\Gamma)} \right).
\end{aligned}
$$

Similarly, it is possible to show that

$$
\begin{aligned}
I_p &= (y_h - z_d, y - y_h)_{L^2(\Omega)} - a(y - y_h, p_h) \\
&= \sum_{K\in\mathcal{T}_h} \left(\int_K (y_h - z_d + \Delta p_h)(y - y_h)d\mathbf{x} - \frac{1}{2} \int_{\partial K} \left[\frac{\partial p_h}{\partial n}\right](y - y_h)d\sigma \right) \\
&= \sum_{K\in\mathcal{T}_h} \left((R_p(p_h), y - y_h)_{L^2(K)} + (r_p(p_h), y - y_h)_{L^2(\Gamma)} \right)
\end{aligned}
$$

and

$$
\begin{aligned}
I_u &= (\beta u_h, u - u_h)_{L^2(\Omega)} + (u - u_h, p_h)_{L^2(\Omega)} \\
&= \sum_{K\in\mathcal{T}_h} \int_K (\beta u_h + \beta u_h, u - u_h)_{L^2(\Omega)} = \sum_{K\in\mathcal{T}_h} (R_u(u_h), u - u_h)_{L^2(K)}.
\end{aligned}
$$

□

To evaluate the terms involving the exact solution (y, u, p) we can replace this latter by higher-order reconstructions of the approximated solution (y_h, u_h, p_h). For instance, in [24, 25] a Galerkin approximation of the OCP based on linear finite elements is considered, hence the errors $y - y_h$, $u - u_h$ and $p - p_h$, are estimated through the interpolation errors operating a quadratic reconstruction of (y, u, p).

Remark 6.13. A slightly different derivation of the a posteriori error estimates can be found, e.g., in [23], still relying on the method we have considered, which aims at estimating the error on the cost functional. •

Remark 6.14. Other approaches aiming instead at estimating the sum of errors on state, adjoint and control variables can be found in [174, 180, 181]. The result is indeed quite similar, since the error on the cost functional can be bounded by suitable combinations of the errors on state, adjoint and control variables. •

A posteriori error estimates like those derived above can be exploited when performing mesh adaptivity in the numerical solution of OCP problems. The error contribution arising from the numerical discretization can be treated separately from the one arising from the iterative optimization process, in view of mesh adaptation. By splitting the error on the cost functional into the sum of an iteration error plus a discretization error, the adaptive strategy is performed on the latter once the former is reduced below a given threshold. Further details can be found, e.g., in [23, 78] and references therein. For the derivation of a posteriori error estimates in the case of OCPs for the Stokes equation, see, e.g., [182].

6.12 Exercises

1. Observing that the algebraic counterpart of the adjoint bilinear form $a^*(\cdot,\cdot)$ of $a(\cdot,\cdot)$ is given by the transpose matrix $A^\top$, show that the discrete adjoint problem $(6.5)_2$ can be rewritten under the form (6.11).
2. Following the same procedure used to derive the expression of the linear system (6.15) from the discrete system of optimality conditions (6.5), show that (6.20) can be equivalently rewritten in the form of system (6.21).
3. Compute the derivative of (6.24) with respect to $\mathbf{u}$ and show that $\nabla J_h(\mathbf{u}) = \beta\mathbb{N}\mathbf{u} - \mathbb{B}^T\mathbf{p}$ provided that the discrete adjoint vector fulfills $\mathbb{A}^T\mathbf{p} = -\mathbb{M}(\mathbf{y} - \mathbf{z}_d)$.
4. Write the system of first-order optimality conditions for the OCP (6.46), then discretize it using a stable pair of finite element spaces (X_h, Q_h). Under which assumptions on the discretization of the state problem, the adjoint problem and the optimality condition is it possible to recover a linear system equivalent to (6.52)?
5. Show that, in the case of a linear-quadratic OCP, problem (6.58) can be rewritten under the form
$$\begin{bmatrix} \mathbb{M} & 0 & -\mathbb{A}^\top \\ 0 & \beta\mathbb{N} & \mathbb{B}^\top \\ -\mathbb{A} & \mathbb{B} & 0 \end{bmatrix} \begin{bmatrix} \delta\mathbf{y} \\ \mathbf{d}_k \\ \delta\mathbf{p} \end{bmatrix} = \begin{bmatrix} 0 \\ \mathbb{B}^\top\mathbf{p}_k - \beta\mathbb{N}\mathbf{u}_k \\ 0 \end{bmatrix}.$$
To do that, perform block elimination and obtain $\delta\mathbf{p}$ as a function of $\delta\mathbf{y}$, then express this latter as a function of $\mathbf{d}_k$.
6. Show that for a quadratic cost functional under the form
$$J(\mathbf{u}) = \frac{1}{2}\mathbf{u}^\top\mathbb{H}\mathbf{u} - \mathbf{b}^\top\mathbf{u}$$
an exact line search method yields the expression (6.59) for the step length. With this goal, show that by minimizing $J(\mathbf{u}_k + \tau_k\mathbf{d}_k)$ with respect to τ_k one obtains
$$\tau_k = -\frac{\nabla J_h(\mathbf{u}_k)^\top\mathbf{d}_k}{\mathbf{d}_k^\top\mathbb{H}\mathbf{d}_k}.$$
Then, obtain (6.59) by substituting into the equation above the expressions (6.26) and (6.18) of the gradient $\nabla J_h(\mathbf{u}_k)$ and the reduced Hessian $\mathbb{H}$, respectively.
7. Verify that
$$\frac{1}{\beta}\mathbb{M} + \mathbb{A}\mathbb{M}^{-1}\mathbb{A}^\top \approx \mathbb{A}\mathbb{M}^{-1}\mathbb{A}^\top$$
by comparing the order of magnitude of the entries of $\mathbb{M}$ and $\mathbb{A}$ in terms of the mesh size h. Show in particular that if $h = 0.1$ or $h = 0.01$, the approximation is valid for $\beta \gg 10^{-4}$ or $\beta \gg 10^{-8}$, respectively.

[*Hint.* Recall that, by denoting with h the mesh size, $M_{ij} = \int_K \varphi_i \varphi_j \, d\mathbf{x} \approx h^2$, $|\nabla \varphi_i| \approx \frac{1}{h}$ so that $K_{ij} = \int_K \nabla \varphi_i \cdot \nabla \varphi_j \, d\mathbf{x} = O(1)$.]

8. Check that the aggregated bilinear form $\mathcal{D}(\cdot,\cdot)$ defined in (6.108) is continuous on $X_h \times X_h$ and strongly coercive over $X_0^h = \{\boldsymbol{w}_h \in X_h \, : \, \mathcal{E}(\boldsymbol{w}_h, q_h) = 0 \quad \forall q_h \in Q_h\} \subset X_h$. Moreover, check that the aggregated bilinear form $\mathcal{E}(\cdot,\cdot)$ defined in (6.109) is continuous and inf-sup stable over $X_h \times Q_h$.
9. Show that the bilinear form $a : V \times V \to \mathbb{R}$ defined in (6.114) is continuous and weakly coercive on $V \times V$, and that the bilinear forms defined in (6.115)–(6.116) satisfy the assumptions of Theorem A.7.

Chapter 7
Quadratic Control Problems Governed by Linear Evolution PDEs

In this chapter we address time-dependent linear-quadratic OCPs, that is OCPs involving a quadratic cost functional and a state system described by a linear initial-boundary value problem (IBVP). Both unconstrained and constrained problems will be considered. We follow the same road map of Chapter 5: we first analyze a simple case, then we formulate a class of OCPs by introducing a more general mathematical setting, and finally we present some meaningful instances. In Sect. 7.1 we analyze an unconstrained linear-quadratic OCP governed by the evolutionary heat equation, involving a distributed heat source as control function. We prove its well-posedness and derive a system of optimality conditions, following an *adjoint-based* approach and, subsequently, recovering them by the Lagrange multiplier method. The extension to the constrained case is presented in Sect. 7.2, whereas an example related to initial control – that is, the identification of the initial condition – is presented in Sect. 7.3. A general framework for linear-quadratic OCPs governed by parabolic PDEs under divergence form is presented in Sect. 7.4; both the mathematical setting and the optimality conditions are then generalized to problems featuring e.g. side controls and side observations, as well as Cauchy-Neumann and Cauchy-Robin boundary conditions, like the one presented in Sect. 7.5. Finally, we discuss a simple instance of OCP related with the time-dependent Stokes equations in Sect. 7.6, and briefly mention the case of optimal control for the wave equation in Sect. 7.7.

7.1 Optimal Heat Source (1): an Unconstrained Case

Let $\Omega \subset \mathbb{R}^d$ be a bounded Lipschitz domain (an interval if $d = 1$), $T > 0$ and consider the space-time cylinder (see Fig. 7.1) $Q_T = \Omega \times (0, T)$, whose lateral surface is denoted by $S_T = \Gamma \times (0, T)$, $\Gamma = \partial\Omega$. Consider the minimization problem

$$\tilde{J}(y,u) = \frac{1}{2}\int_{\Omega} \left(y(\mathbf{x},T) - z_d(\mathbf{x})\right)^2 d\mathbf{x} + \frac{\beta}{2}\int_{Q_T} u^2 d\mathbf{x}dt \to \min \tag{7.1}$$

with $\beta > 0$, where $z_d \in L^2(Q_T)$ is a given target, $u \in L^2(Q_T)$ and $y = y(\mathbf{x},t;u)$ solves the *state problem:*

$$\begin{cases} y_t - \Delta y + y = f + u & \text{in } Q_T \\ \dfrac{\partial y}{\partial n} = 0 & \text{on } S_T \\ y(\mathbf{x},0) = g(\mathbf{x}) & \text{in } \Omega. \end{cases} \tag{7.2}$$

We will use the symbols y_t or $\partial_t y$ to denote the partial derivative of y with respect to t.

A. Manzoni et al., *Optimal Control of Partial Differential Equations*,
Applied Mathematical Sciences 207, https://doi.org/10.1007/978-3-030-77226-0_7

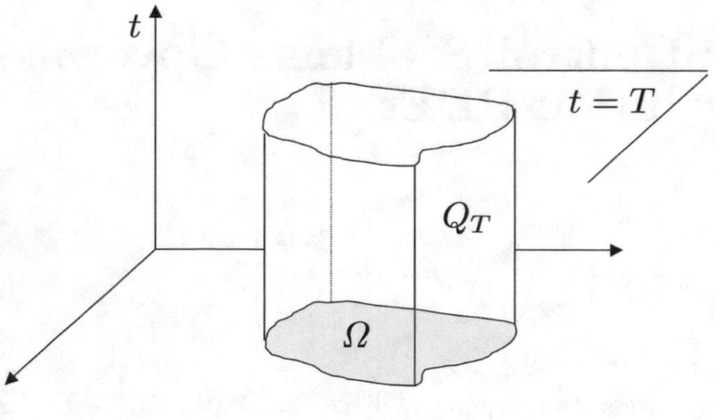

Fig. 7.1 Space-time cylinder Q_T

Thus, we want to control the distributed heat source u in Q_T, in order to minimize the distance from a given target z_d of the final observation of y, given by $y(\mathbf{x},T)$. We could as well interpret the problem above as an inverse problem, where an unknown source u has to be determined from the measurements $y(\mathbf{x},T)$ of the temperature at a final time T.

This linear-quadratic evolution control problem is the analogue of the quadratic cost problem considered in Sect.4.6, when only the final payoff $\psi(\mathbf{y}(T))$ is present. We will recognize other analogies between these two problems concerning the optimality conditions.

7.1.1 Analysis of the State System

For the well-posedness analysis of the state problem (7.2) we introduce a weak formulation. Let us proceed formally, assuming that every function we deal with is smooth enough to undergo the requested operations. As we did several times in Chapter 5, we multiply the diffusion equation by a smooth function $\varphi = \varphi(\mathbf{x}, t)$ and integrate over Q_T to find

$$\int_{Q_T} \{y_t\varphi - \Delta y\ \varphi + y\varphi\}\ d\mathbf{x}dt = \int_{Q_T} (f+u)\varphi\ d\mathbf{x}dt.$$

Integrating by parts the second term with respect to $\mathbf{x}$ and taking into account the homogeneous Neumann condition $\frac{\partial y}{\partial n} = 0$ on S_T, we get

$$\int_{Q_T} \{y_t\varphi + \nabla y \cdot \nabla\varphi\ + y\varphi\}\ d\mathbf{x}dt = \int_{Q_T} (f+u)\varphi\ d\mathbf{x}dt. \tag{7.3}$$

To introduce a correct functional setting, note that the integrals at the left hand side of (7.3) make sense if $y, y_t \in L^2(Q_T)$ and $\nabla y \in L^2(Q_T)^d$, that is $y \in H^1(Q_T)$, with the test function φ such that $\varphi \in L^2(Q_T), \nabla\varphi \in L^2(Q_T)^d$. The conditions on φ are equivalent to require that $\varphi(\cdot, t) \in H^1(\Omega)$ for all $t \in [0, T]$ and that $\|\varphi(t,\cdot)\|_{H^1(\Omega)} \in L^2(0,T)$: with shorthand notation we write $\varphi \in L^2(0,T; H^1(\Omega))$.

The initial condition holds a.e. in Ω and it is attained in the *mean square* sense,

$$\int_{\Omega} (y(\mathbf{x}, t) - g(\mathbf{x}))^2 \, d\mathbf{x} \to 0 \quad \text{as } t \to 0. \tag{7.4}$$

The above arguments motivate our *first weak formulation.*

Definition 7.1. Let $u \in L^2(Q_T)$ and $g \in H^1(\Omega)$; then $y \in H^1(Q_T)$ is a weak solution of problem (7.2) if:

1. for every $\varphi \in L^2(0, T; H^1(\Omega))$ (7.3) holds;
2. $y(\mathbf{x},0) = g$ in the sense of (7.4).

We can easily check that, if y, g, f are smooth functions and y is a weak solution, then y solves problem (7.2) in a classical sense (see Exercise 1). Thus, weak and classical formulations are equivalent.

For the numerical treatment of problem 7.2, it is often more convenient to adopt an alternative formulation in which space and time are separated. To find it, choose $\varphi(\mathbf{x},t) = \phi(\mathbf{x})\,\eta(t)$ with $\phi \in H^1(\Omega)$ and $\eta \in L^2(0,T)$. Clearly $\varphi \in L^2(0, T; H^1(\Omega))$ and can be inserted into (7.3) to obtain

$$\int_0^T \left(\int_{\Omega} \{y_t\phi + \nabla y \cdot \nabla\phi + y\phi\} \, d\mathbf{x} \right) \eta(t)\, dt = \int_0^T \left(\int_{\Omega} (f+u)\phi \, d\mathbf{x} \right) \eta(t)\, dt.$$

Since η is an arbitrary function of $L^2(0,T)$, we deduce the following equation

$$\int_{\Omega} \{y_t(\mathbf{x},t)\,\phi(\mathbf{x}) + \nabla y(\mathbf{x},t) \cdot \nabla\phi(\mathbf{x}) + y(\mathbf{x},t)\,\phi(\mathbf{x})\} \, d\mathbf{x} = \int_{\Omega} (f(\mathbf{x},t) + u(\mathbf{x},t))\phi(\mathbf{x})\, d\mathbf{x} \tag{7.5}$$

a.e. in $(0,T)$. On the other hand, it can be proved that, if (7.5) holds for every $\phi \in H^1(\Omega)$ and a.e. in $(0,T)$ then (7.3) holds for every $v \in L^2(0,T; H^1(\Omega))$. Thus, we have the following definition, equivalent to Definition 7.1:

Definition 7.2. A function $y \in H^1(Q_T)$ is a weak solution of problem (7.2) if:

1. for every $\phi \in H^1(\Omega)$ and a.e. $t \in (0,T)$, (7.5) holds;
2. $u(0) = g$ in the sense of (7.4).

The well-posedness of IBVPs, will be adressed in Sect. 7.4.1. From the results in that section we deduce the following proposition.

Proposition 7.1. *Let $g \in H^1(\Omega)$, $f, u \in L^2(Q_T)$. Then problem (7.2) has a unique weak solution $y \in H^1(Q_T)$. Moreover, the function $t \mapsto \|y(\cdot,t)\|^2_{L^2(\Omega)}$ is continuous in $[0,T]$, and the following estimates hold:*

$$\max_{t\in[0,T]} \|y(\cdot,t)\|^2_{L^2(\Omega)} + \|\nabla y\|^2_{L^2(Q_T)^d} \le \|g\|^2_{L^2(\Omega)} + \|f\|^2_{L^2(Q_T)} + \|u\|^2_{L^2(Q_T)} \tag{7.6}$$

and

$$\|y_t\|^2_{L^2(Q_T)} \le 2\|g\|^2_{H^1(\Omega)} + 4\|f\|^2_{L^2(Q_T)} + 4\|u\|^2_{L^2(Q_T)}. \tag{7.7}$$

7.1.2 Existence and Uniqueness of the Optimal Control. Optimality Conditions

For fixed f, g and any control $u \in L^2(Q_T)$, let $y = y(\mathbf{x},t;u) = y(u)$ be the weak solution to the state problem (7.2). By substituting $y(u)$ into $\tilde{J}(y,u)$, we obtain the reduced cost functional

$$J(u) = \tilde{J}(y(u), u) = \frac{1}{2}\int_\Omega (y(\mathbf{x},T;u) - z_d(\mathbf{x}))^2 \, d\mathbf{x} + \frac{\beta}{2}\int_{Q_T} u^2 d\mathbf{x}dt. \tag{7.8}$$

We set $y_0(u) = y(u) - y(0)$, which is the solution of (7.2) with $f = 0$, $g = 0$; $u \longmapsto y_0(u)$ is a linear map. Then

$$\begin{aligned} J(u) &= \frac{1}{2}\int_\Omega (y_0(\mathbf{x},T;u) + y(\mathbf{x},T;0) - z_d(\mathbf{x}))^2 \, d\mathbf{x} + \frac{\beta}{2}\int_{Q_T} u^2 d\mathbf{x}dt \\ &= \frac{1}{2}q(u,u) - Lu + c, \end{aligned} \tag{7.9}$$

where q is the symmetric bilinear form

$$q(u,w) = \int_\Omega y_0(\mathbf{x},T;u)\, y_0(\mathbf{x},T;w)\, d\mathbf{x} + \beta\int_{Q_T} uw \, d\mathbf{x}dt,$$

L is the linear functional

$$Lu = \int_\Omega (z_d(\mathbf{x}) - y(\mathbf{x},T;0))\, y_0(\mathbf{x},T;u)\, d\mathbf{x}$$

and $c = \frac{1}{2}\int_\Omega (y(\mathbf{x},T;0) - z_d(\mathbf{x}))^2 \, d\mathbf{x}$.

The bilinear form $q(\cdot,\cdot)$ is coercive and continuous in $L^2(Q_T)$. Indeed, since $\beta \geq 0$,

$$q(u,u) = \int_\Omega y_0(\mathbf{x},T;u)^2 \, d\mathbf{x} + \beta\int_{Q_T} u^2 \, d\mathbf{x}dt \geq \beta \|u\|^2_{L^2(Q_T)},$$

while, from (7.7) and (7.6), it holds

$$\begin{aligned} |q(u,w)| &\leq \|y_0(\cdot,T;u)\|_{L^2(\Omega)} \|y_0(\cdot,T;w)\|_{L^2(\Omega)} + \beta\|u\|_{L^2(Q_T)} \|w\|_{L^2(Q_T)} \\ &\leq (1+\beta)\left(\|u\|^2_{L^2(Q_T)} + \|w\|^2_{L^2(Q_T)}\right). \end{aligned}$$

Moreover, the linear form L is continuous in $L^2(Q_T)$, since

$$|Lu| \leq \|z_d - y(\cdot,T;0)\|_{L^2(\Omega)} \|y_0(\cdot,T;u)\|_{L^2(\Omega)} \leq \|z_d - y(\cdot,T;0)\|_{L^2(\Omega)} \|u\|_{L^2(Q_T)}.$$

The reduced cost functional J in (7.9) has therefore the same structure as the one in Sect. 5.1.2 for the elliptic case. From the results in that section, we deduce that there exists a unique optimal control $\hat{u}$, with corresponding optimal state $\hat{y} = y(\hat{u})$, characterized by the following *Euler equation*

$$J'(\hat{u})\, w = q(\hat{u}, w) - Lw = 0 \qquad \forall w \in L^2(Q_T)$$

which, after some adjustments, becomes

$$J'(\hat{u})\, w = \int_\Omega (\hat{y}(\mathbf{x},T) - z_d(\mathbf{x}))\, y_0(\mathbf{x},T;w)\, d\mathbf{x} + \beta\int_{Q_T} \hat{u}w \, d\mathbf{x}dt = 0 \qquad \forall w \in L^2(Q_T).$$

7.1.3 Use of the Adjoint State

As in the elliptic case, we look for an expression of $J'(\hat{u})\,w$ which does not contain y_0. With this aim, we write the reduced cost functional (7.8) in the following augmented form, using the formulation (7.3) of the state equation for $y(u)$ with the test function v replaced by a multiplier $p = p(\mathbf{x},t)$ to be chosen later,

$$\begin{aligned} J(u) &= \int_\Omega (y(\mathbf{x},T;u) - z_d(\mathbf{x}))^2\, d\mathbf{x} + \frac{\beta}{2}\int_{Q_T} u^2 d\mathbf{x}dt \\ &+ \int_{Q_T} \{(f+u)p - y_t(u)p - \nabla y(u)\cdot\nabla p - y(u)\,p\}\, d\mathbf{x}dt. \end{aligned}$$

Note that in this way we have just added zero to the reduced functional J, which can thus be written under the form

$$J(u) = \frac{1}{2}q(u,u) - \tilde{L}u + \tilde{c},$$

where $\tilde{L}$ is still a linear and continuous functional, given by

$$\tilde{L}u = \int_\Omega (y(\mathbf{x},T;0) - z_d(\mathbf{x}))y_0(\mathbf{x},T;u)\, d\mathbf{x} + \int_{Q_T} \{up - \partial_t y_0(u)p - \nabla y_0(u)\cdot\nabla p - y_0(u)\,p\}\, d\mathbf{x}dt$$

and

$$\tilde{c} = \frac{1}{2}\int_\Omega (y(\mathbf{x},T;0) - z_d(\mathbf{x}))^2\, d\mathbf{x} + \int_{Q_T} \{fp - y_t(0)p - \nabla y(0)\cdot\nabla p - y(0)\,p\}\, d\mathbf{x}dt.$$

The Euler equation for the augmented functional is then

$$J'(\hat{u})\,w = q(\hat{u},w) - \tilde{L}w = 0 \qquad \forall w \in L^2(Q_T),$$

that is,

$$\begin{aligned} J'(\hat{u})\,w &= \int_\Omega (\hat{y}(\mathbf{x},T) - z_d(\mathbf{x}))\, y_0(\mathbf{x},T;w)\, d\mathbf{x} + \int_{Q_T} (\beta\hat{u} + p)w\, d\mathbf{x}dt \\ &- \int_{Q_T} \{\partial_t y_0(w)p + \nabla y_0(w)\cdot\nabla p + y_0(w)\,p\}\, d\mathbf{x}dt = 0 \qquad \forall w \in L^2(Q_T). \end{aligned}$$

Integrating by parts, and noticing that $y_0(\mathbf{x},0;w) = 0$, we find

$$\int_{Q_T} \partial_t y_0(w)p\, d\mathbf{x}dt = \int_\Omega y_0(\mathbf{x},T;w)p(\mathbf{x},T)\, d\mathbf{x} - \int_{Q_T} p_t y_0(w)\, d\mathbf{x}dt.$$

Thus we get

$$\begin{aligned} J'(\hat{u})\,w &= \int_\Omega (\hat{y}(\mathbf{x},T) - z_d(\mathbf{x}) - p(\mathbf{x},T))\, y_0(\mathbf{x},T;w)\, d\mathbf{x} + \int_{Q_T} (\beta\hat{u} + p)w\, d\mathbf{x}dt \\ &+ \int_{Q_T} \{p_t y_0(w) - \nabla p\cdot\nabla y_0(w) - p y_0(w)\}\, d\mathbf{x}dt = 0. \end{aligned} \tag{7.10}$$

To get rid of the terms containing y_0, we choose $p = \hat{p} \in H^1(Q_T)$ as the weak solution of the following *adjoint problem*:

$$\begin{cases} -\hat{p}_t - \Delta\hat{p} + \hat{p} = 0 & \text{in } Q_T \\ \dfrac{\partial \hat{p}}{\partial n} = 0 & \text{on } S_T \\ \hat{p}(\mathbf{x},T) = \hat{y}(\mathbf{x},T) - z_d(\mathbf{x}) & \text{in } \Omega. \end{cases} \tag{7.11}$$

Thus $\hat{p}$ satisfies the equation

$$\int_{Q_T} \{\hat{p}_t\varphi - \nabla\hat{p}\cdot\nabla\varphi - \hat{p}\varphi\}\, d\mathbf{x}dt = 0 \qquad \forall\varphi \in L^2(0,T;H^1(\Omega)) \tag{7.12}$$

or, equivalently,

$$\int_\Omega \{\hat{p}_t(\mathbf{x},t)\,\phi(\mathbf{x}) - \nabla\hat{p}(\mathbf{x},t)\cdot\nabla\phi(\mathbf{x}) - \hat{p}(\mathbf{x},t)\,\phi(\mathbf{x})\}\, d\mathbf{x} = 0 \ \ \forall\phi\in H^1(\Omega) \text{ and a.e. } t\in(0,T). \tag{7.13}$$

Both (7.12) and (7.13) account for the Neumann condition on S_T. The *final* condition holds a.e. in Ω and it is attained in the mean square sense, as in (7.4)

$$\int_\Omega (\hat{p}(\mathbf{x},t) - \hat{y}(\mathbf{x},T) + z_d(\mathbf{x}))^2\, d\mathbf{x} \to 0 \ \text{ as } t\to T. \tag{7.14}$$

Note that (7.11) is a *final value problem* for the *backward in time* heat equation and therefore (7.11) is well posed. With this choice of p, the first and the last integral in (7.10) vanish and the Euler equation reduces to

$$J'(\hat{u})\,w = \int_{Q_T} (\beta\hat{u} + \hat{p})w\; d\mathbf{x}dt = 0 \qquad \forall w\in L^2(Q_T)$$

whence

$$\beta\hat{u} + \hat{p} = 0 \ \text{ a.e. in } Q_T. \tag{7.15}$$

Moreover,

$$\nabla J(\hat{u}) = \hat{p} + \beta\hat{u}.$$

Summarizing, we have proven the following result.

Proposition 7.2. *For the OCP* (7.1)–(7.2), *the control* $\hat{u}$ *and the state* $\hat{y} = y(\hat{u})$ *are optimal if and only if, together with the adjoint state (or multiplier)* $\hat{p}$, *they are weak solutions of the optimality conditions given by the state system (7.2), the adjoint problem (7.11) and the Euler equation (7.15).*

7.1.4 The Lagrange Multipliers Approach

Also in the parabolic case, with a slightly stronger effort, we can recover the optimality conditions in terms of a Lagrangian function given by

$$\begin{aligned} \mathcal{L}(y,u,p) &= \frac{1}{2}\int_\Omega (y(\mathbf{x},T) - z_d(\mathbf{x}))^2\, d\mathbf{x} + \frac{\beta}{2}\int_{Q_T} u^2 d\mathbf{x}dt \\ &+ \int_{Q_T} \{fp + up - y_t p - \nabla y\cdot\nabla p\ - yp\}\ d\mathbf{x}dt \end{aligned} \tag{7.16}$$

where y, u, p have to be considered as *independent variables* belonging to the spaces $V_{ad} = \{H^1(Q_T) : y(0) = g\}$, $L^2(Q_T)$ and $H^1(Q_T)$, respectively. Note that $\mathcal{L}$ is linear with respect to p. By requiring that the derivative of $\mathcal{L}$ with respect to p vanishes on any variation $\varphi \in H^1(Q_T)$ we get

$$\begin{aligned}\mathcal{L}'_p(\hat{y}, \hat{u}, \hat{p})\,\varphi &= \frac{d}{d\varepsilon}\mathcal{L}(\hat{y}, \hat{u}, \hat{p} + \varepsilon\varphi)\Big|_{\varepsilon=0} \\ &= \int_{Q_T} \{(f + \hat{u})\varphi - \hat{y}_t\varphi - \nabla\hat{y}\cdot\nabla\varphi - \hat{y}\varphi\}\; d\mathbf{x}dt = 0 \qquad \forall\varphi \in H^1(Q_T).\end{aligned}$$

Since $H^1(Q_T)$ is dense[1] into $L^2(0, T; H^1(\Omega))$, this corresponds to the state equation.

Let us now compute

$$\mathcal{L}'_y(\hat{y}, \hat{u}, \hat{p})\,\psi = \frac{d}{d\varepsilon}\mathcal{L}(\hat{y} + \varepsilon\psi, \hat{u}, \hat{p})\Big|_{\varepsilon=0}.$$

The variations ψ must belong to $H^1(Q_T)$ with $\psi(\mathbf{x},0) = 0$ to ensure that $\hat{y} + \varepsilon\psi \in V_{ad}$. We have

$$\begin{aligned}\mathcal{L}'_y(\hat{y}, \hat{u}, \hat{p})\,\psi &= \int_\Omega (\hat{y}(\mathbf{x},T) - z_d(\mathbf{x}))\,\psi(\mathbf{x},T)\,d\mathbf{x} + \int_{Q_T} \{-\psi_t\hat{p} - \nabla\psi\cdot\nabla\hat{p} - \psi\hat{p}\}\; d\mathbf{x}dt \\ &= \int_\Omega (\hat{y}(\mathbf{x},T) - z_d(\mathbf{x}) - \hat{p}(\mathbf{x},T))\,\psi(\mathbf{x},T)\,d\mathbf{x} + \int_{Q_T} \{\hat{p}_t\psi - \nabla\hat{p}\cdot\nabla\psi - \hat{p}\psi\}\; d\mathbf{x}dt.\end{aligned}$$

Choosing variations ψ of the form $\psi(\mathbf{x}, t) = \phi(\mathbf{x})\,\eta(t)$ with $\phi \in H^1(\Omega), \eta \in C^1([0, T]), \eta(0) = 0 = \eta(T)$, from $\mathcal{L}'_y(\hat{y}, \hat{u}, \hat{p})\,\psi = 0$, we get

$$\int_0^T \left(\int_\Omega \{\hat{p}_t(\mathbf{x}, t)\,\phi(\mathbf{x}) - \nabla\hat{p}(\mathbf{x}, t)\cdot\nabla\phi(\mathbf{x}) - \hat{p}(\mathbf{x}, t)\,\phi(\mathbf{x})\}\,d\mathbf{x}\right)\eta(t)\;\; dt = 0.$$

Since η is arbitrary,

$$\int_\Omega \{\hat{p}_t(\mathbf{x}, t)\,\phi(\mathbf{x}) - \nabla\hat{p}(\mathbf{x}, t)\cdot\nabla\phi(\mathbf{x}) - \hat{p}(\mathbf{x}, t)\,\phi(\mathbf{x})\}\,d\mathbf{x} = 0 \qquad \forall\phi \in H^1(\Omega),\ \text{a.e. in } (0, T),$$

which is indeed the weak form (7.13) of the adjoint problem (7.11). If we now choose variations ψ of the form $\psi(\mathbf{x}, t) = \phi(\mathbf{x})\,\eta(t)$ with $\phi \in H^1(\Omega),\ \eta \in C^1([0, T]), \eta(0) = 0,\ \eta(T) \neq 0$, we are left with

$$\int_\Omega (\hat{y}(\mathbf{x},T) - z_d(\mathbf{x}) - \hat{p}(\mathbf{x},T))\,\phi(\mathbf{x})\,d\mathbf{x} = 0 \qquad \forall\phi \in H^1(\Omega).$$

Since $H^1(\Omega)$ is dense in $L^2(\Omega)$ we get the final condition

$$\hat{p}(\mathbf{x},T) = \hat{y}(\mathbf{x},T) - z_d(\mathbf{x}) \qquad \text{a.e. in } \Omega.$$

From Theorem A.19 *(b)*, we recover condition (7.14). Finally,

$$\mathcal{L}'_u(\hat{y}, \hat{u}, \hat{p})\,v = \int_{Q_T} (\beta\hat{u} + \hat{p})v\,d\mathbf{x}dt = 0 \qquad \forall v \in L^2(Q_T)$$

is equivalent to the equation $\hat{p} + \beta\hat{u} = 0$, a.e. in Q_T.

---

[1] See Lemma A.3 in Sect. A.5.1.

Summarizing, the optimality conditions (7.2), (7.11) and (7.15) can be written in terms of the Lagrangian (7.16) as follows:

$$\begin{cases} \mathcal{L}'_p(\hat{y},\hat{u},\hat{p})\,\varphi = 0 & \forall \varphi \in H^1(Q_T) \\ \mathcal{L}'_y(\hat{y},\hat{u},\hat{p})\,\psi = 0 & \forall \psi \in H^1(Q_T), \quad \psi(\mathbf{x},0)=0 \\ \mathcal{L}'_u(\hat{y},\hat{u},\hat{p})\,v = 0 & \forall v \in L^2(Q_T). \end{cases} \tag{7.17}$$

7.2 Optimal Heat Source (2): a Constrained Case

Suppose now that the control u represents a heat source subject to the constraint $u \in \mathcal{U}_{ad}$, a closed convex subset of $L^2(Q_T)$. Reasoning as in the elliptic case (see Sect. 5.2) we infer that the necessary and sufficient optimality conditions for $\hat{u} \in \mathcal{U}_{ad}$ to be an optimal control are given by system (7.17) where however the last equation is replaced by the variational inequality

$$\mathcal{L}'_u(\hat{y},\hat{u},\hat{p})(w-\hat{u}) = \int_{Q_T} (\beta\hat{u}+\hat{p})(w-\hat{u})\; d\mathbf{x}dt \ge 0 \qquad \forall w \in \mathcal{U}_{ad}. \tag{7.18}$$

This condition characterizes $\hat{u} \in \mathcal{U}_{ad}$ as the projection of $-\beta^{-1}\hat{p}$ onto $\mathcal{U}_{ad}$, that is

$$\hat{u} = Pr_{\mathcal{U}_{ad}}\left(-\beta^{-1}\hat{p}\right).$$

In the relevant case of *box constraints*, we can repeat the same considerations done for the elliptic case in Sect. 5.2. For instance, if

$$\mathcal{U}_{ad} = \left\{u \in L^2(Q_T) : a \le u(\mathbf{x},t) \le b, \text{ a.e. in } Q_T\right\} \tag{7.19}$$

a, b constants[2], then, a.e. in Q_T

$$\hat{u}(\mathbf{x},t) = \beta^{-1}\max\{a, \min\{b, -\hat{p}(\mathbf{x},t)\}\} = \beta^{-1}\min\{b, \max\{a, -\hat{p}(\mathbf{x},t)\}\}.$$

Existence and uniqueness of an optimal control follows from the general results of Sect. 7.4.

7.3 Initial control

If we want to control the *initial state* we are led to the following minimization problem,

$$\tilde{J}(y,u) = \frac{1}{2}\int_\Omega (y(\mathbf{x},T) - z_d(\mathbf{x}))^2\, d\mathbf{x} + \frac{\beta}{2}\int_\Omega u^2 d\mathbf{x} \to \min \tag{7.20}$$

with $\beta > 0$, where $u \in \mathcal{U}_{ad}$, a closed convex subset of $L^2(\Omega)$, and $y = y(\mathbf{x},t;u)$ solves the *state problem*:

$$\begin{cases} y_t - \Delta y + y = f & \text{in } Q_T \\ \dfrac{\partial y}{\partial n} = 0 & \text{on } S_T \\ y(\mathbf{x},0) = u(\mathbf{x}) & \text{in } \Omega. \end{cases} \tag{7.21}$$

---

[2] Or $a, b \in L^\infty(Q_T)$.

If regarded as an inverse problem, this amounts to reconstruct the initial data from a final temperature measurement. Performing the same calculations as in Sect. 7.1.3 (see Exercise 2), we obtain the adjoint equation (7.12) while, instead of (7.18), we now have the variational inequality

$$\int_{\Omega} (\beta \hat{u}(\mathbf{x}) + \hat{p}(\mathbf{x}, 0))(w(\mathbf{x}) - \hat{u}(\mathbf{x}))d\mathbf{x} \geq 0 \qquad \forall w \in \mathcal{U}_{ad} \tag{7.22}$$

which gives $\hat{u}(\mathbf{x}) = Pr_{\mathcal{U}_{ad}}\left(-\beta^{-1}\hat{p}(\mathbf{x}, 0)\right)$ a.e. in Ω.

7.4 General Framework for Linear-Quadratic OCPs Governed by Parabolic PDEs

In this section we provide a general theoretical framework to analyze a linear-quadratic OCP governed by a parabolic IBVP, in presence of control constraints. As done in Sect. 5.3, we will derive the system of optimality conditions, using either the adjoint method or the method of Lagrange multipliers. Several examples related to advection, diffusion and reaction problems will be extensively discussed in the next sections.

7.4.1 Initial-boundary Value Problems for Uniformly Parabolic Linear Equations

Let $\mathbf{A} = \mathbf{A}(\mathbf{x},t)$ be a square matrix of order $d \times d$, $\mathbf{c} = \mathbf{c}(\mathbf{x}, t)$ a vector field in $\mathbb{R}^d$, $r = r(\mathbf{x},t)$ and $f = f(\mathbf{x},t)$ two real functions. We consider equations *in divergence form* of the type

$$\mathcal{P}y = y_t + \mathcal{E}y = y_t - \operatorname{div}(\mathbf{A}(\mathbf{x}, t)\nabla y) + \mathbf{c}(\mathbf{x},t) \cdot \nabla y + r(\mathbf{x},t)\, y \tag{7.23}$$

and make the following assumptions:

(*a*) $\mathcal{P}$ is *uniformly parabolic*: there exists $\alpha > 0$ such that

$$\alpha |\boldsymbol{\xi}|^2 \leq \sum_{i,j=1}^{d} a_{ij}(\mathbf{x}, t)\xi_i\xi_j \quad \text{and} \quad |a_{ij}(\mathbf{x},t)| \leq M, \quad \forall \boldsymbol{\xi} \in \mathbb{R}^d, \text{ a.e. in } Q_T; \tag{7.24}$$

(*b*) the coefficients $\mathbf{c}$ and r are bounded (i.e. they all belong to $L^\infty(Q_T)$), with

$$|\mathbf{c}| \leq \gamma_0, \ |r| \leq r_0, \text{ a.e. in } Q_T. \tag{7.25}$$

We consider initial-boundary value problems of the form:

$$\begin{cases} y_t + \mathcal{E}y = F & \text{in } Q_T \\ y(\mathbf{x},0) = g(\mathbf{x}) & \text{in } \Omega \\ \mathcal{B}y = 0 & \text{on } S_T \end{cases} \tag{7.26}$$

where $\mathcal{B}y$ stands for $\mathcal{B}y = \mathbf{A}\nabla y \cdot \nu + hy$ for the Neumann/Robin condition, with $h \in L^\infty(S_T)$, $h \geq 0$, $\mathcal{B}y = y$ for the Dirichlet condition.

In Sect. 7.1.1 we introduced a *weak formulation* of the Neumann problem for the heat equation with a reaction term. To achieve a greater generality, necessary to address more significant applications, it is convenient to separate space and time and look at functions $v = v(\mathbf{x}, t)$ as a function of t with values into a suitable Hilbert space V, determined by the type of boundary conditions, that is $v : [0, T] \to V$.

When we adopt this convention, we write $v(t)$ instead of $v(\mathbf{x},t)$ and $y_t(t)$ or $\partial_t y(t)$ instead of $y_t(\mathbf{x},t)$ and $\partial_t y(\mathbf{x},t)$, respectively. This approach requires the use of integrals for functions with values in Hilbert spaces, and of Sobolev spaces involving time; see Sect. A.6.

To give a weak formulation of problem (7.26), we follow the usual procedure: choose a set of test functions suitably related to the boundary conditions, multiply the differential equation by a test function φ, integrate over the space-time cylinder Q_T and integrate by parts the divergence form term.

Introducing the bilinear form

$$a(y,\varphi;t) = \int_\Omega \{\mathbf{A}(t)\nabla y \cdot \nabla\varphi + (\mathbf{c}(t)\cdot\nabla y)\,\varphi + r(t)y\varphi\}\,d\mathbf{x}$$

or, in the case of Robin condition,

$$a(y,\varphi;t) = \int_\Omega \{\mathbf{A}(t)\nabla y \cdot \nabla\varphi + (\mathbf{c}(t)\cdot\nabla y)\,\varphi + r(t)y\varphi\}\,d\mathbf{x} + \int_\Gamma h(t)y\varphi\,d\sigma$$

we end up, formally, with the following equation

$$\int_0^T (y_t(t),\varphi(t))_{L^2}\,dt + \int_0^T a(y,\varphi;t)\,dt = \int_0^T (f(t),\varphi(t))_{L^2}\,dt. \tag{7.27}$$

Here we adopt more natural and flexible hypotheses than those stated in Sect. 7.1.1.

We are given a Hilbert triplet[3] $V \hookrightarrow H \hookrightarrow V^*$, where $H = L^2(\Omega)$ and V is a Sobolev space, $H_0^1(\Omega) \subseteq V \subseteq H^1(\Omega)$. We denote by $\langle\cdot,\cdot\rangle_*$ the duality between V and V^* and by $\|\cdot\|_*$ the norm in V^*. The choice of V depends on the boundary condition we are dealing with. Classical choices are $V = H_0^1(\Omega)$ for the homogeneous Dirichlet condition, $V = H^1(\Omega)$ for the Neumann or Robin condition, $V = H^1_{\Gamma_D}(\Omega)$ in the case of mixed Dirichlet/Robin or Neumann conditions (where Γ_D denotes the portion of the boundary holding a Dirichlet condition).

If the initial data g and the source term f belong to H and $L^2(0,T;V^*)$, respectively, our solution belongs to the Hilbert space

$$H^1(0,T;V,V^*) = \{y : y \in L^2(0,T;V),\; y_t \in L^2(0,T;V^*)\} \tag{7.28}$$

endowed with the scalar product

$$(y,\varphi)_{H^1(0,T;V,V^*)} = \int_0^T \{(y(t),\varphi(t))_V + (y_t(t),\varphi_t(t))_{V^*}\}\,dt$$

and the norm

$$\|y\|_{H^1(0,T;V,V^*)} = \left(\|y\|^2_{L^2(0,T;V)} + \|y_t\|^2_{L^2(0,T;V^*)}\right)^{1/2}.$$

Thanks to Theorem A.19 (b), $H^1(0,T;V,V^*) \hookrightarrow C([0,T];H)$, so that

$$\|u\|_{C([0,T];H)} \le C(T)\,\|u\|_{H^1(0,T;V,V^*)}\,.$$

A first consequence is that in equation (7.27) the L^2–inner products $(y_t(t),\varphi(t))_{L^2(\Omega)}$ and $(f(t),\varphi(t))_{L^2(\Omega)}$ must be replaced by the duality pairings $\langle y_t(t),\varphi(t)\rangle_*$ and $\langle f(t),\varphi(t)\rangle_*$. Moreover, the initial condition $y(0) = g$ means $\|y(t) - g\|_H \to 0$ as $t \to 0$.

Thus, a weak formulation of problem (7.26) reads as follows.

[3] See Sect. A.2.3 for the definition of Hilbert triplet.

Definition 7.3. *Let $f \in L^2(0,T;V^*)$ and $g \in H$; $y \in H^1(0,T;V,V^*)$ is called a weak solution to problem (7.26) if*

$$y(0) = g \tag{7.29}$$

and, for all $\varphi \in L^2(0,T;V)$,

$$\int_0^T \langle y_t(t), \varphi(t)\rangle_* \, dt + \int_0^T a(y(t), \varphi(t); t)\, dt = \int_0^T \langle f(t), \varphi(t)\rangle_* \, dt \tag{7.30}$$

or, equivalently,

$$\langle y_t(t), \phi\rangle_* + a(y(t), \phi; t) = \langle f(t), \phi\rangle_* \qquad \forall\, \phi \in V, \text{ a.e. } t \in (0,T). \tag{7.31}$$

Existence and uniqueness of a weak solution to problem (7.26) are guaranteed by the following hypotheses on the bilinear form a:

(i) *continuity*: there exists $\mathcal{M} = \mathcal{M}(T) > 0$, such that

$$|a(\phi, \eta; t)| \le \mathcal{M}\, \|\phi\|_V \|\eta\|_V \qquad \forall \phi, \eta \in V, \text{ a.e. } t \in (0,T);$$

(ii) $V - H$ *weak coercivity*: there exist two positive numbers λ and α_0 such that

$$a(\phi, \phi; t) + \lambda \|\phi\|_H^2 \ge \alpha_0 \|\phi\|_V^2 \qquad \forall \phi \in V, \text{ a.e. } t \in (0,T);$$

this is sometimes called *Gärding inequality.*

(iii) $t-$*measurability*: the map $t \mapsto a(\phi, \eta; t)$ is measurable for all $\phi, \eta \in V$ (see Sect. A.5).

Remark 7.1. Note that, under the assumption (i), for every $\phi, \eta \in L^2(0,T;V)$, the function $t \mapsto B(\phi(t), \eta(t); t)$ belongs to $L^1(0,T)$. •

Under the hypotheses (7.24) and (7.25), by following what done in the elliptic case in Sect. 5.4, we can show that the bilinear form a is continuous, with $\mathcal{M}$ depending only on d, T and on M, γ_0, r_0 that is, on the dimension d and the size of the coefficients a_{ij}, c_j, r (and of $\|h\|_{L^\infty(S_T)}$ in the case of Robin condition).

Concerning the weak coercivity, from (7.25) we have, for every $\varepsilon > 0$ (L^2 denotes either $L^2(\Omega)$ or $L^2(\Omega)^d$)

$$\int_\Omega (\mathbf{c} \cdot \nabla \phi)\, \phi d\mathbf{x} \ge -\gamma_0 \|\nabla \phi\|_{L^2} \|\phi\|_{L^2} \ge -\frac{\gamma_0}{2}\left[\varepsilon \|\nabla \phi\|_{L^2}^2 + \frac{1}{\varepsilon}\|\phi\|_{L^2}^2\right]$$

and

$$\int_\Omega r\phi^2 d\mathbf{x} \ge -r_0 \|\phi\|_{L^2}^2,$$

whence, as $h \ge 0$ a.e. on $\partial\Omega$,

$$a(\phi, \phi; t) \ge \left(\alpha - \frac{\gamma_0 \varepsilon}{2}\right) \|\nabla \phi\|_{L^2}^2 - \left(\frac{\gamma_0}{2\varepsilon} + r_0\right) \|\phi\|_{L^2}^2. \tag{7.32}$$

If $\gamma_0 = 0$, i.e. $\mathbf{c} = \mathbf{0}$, assumption (ii) holds provided $\lambda > r_0$. If $\gamma_0 > 0$, choose in (7.32)

$$\varepsilon = \frac{\alpha}{\gamma_0} \quad \text{and} \quad \lambda = 2\left(\frac{\gamma_0}{2\varepsilon} + r_0\right).$$

Then

$$a(\phi, \phi; t) + \lambda \|\phi\|_{L^2}^2 \ge \frac{\alpha}{2} \|\nabla \phi\|_{L^2}^2 + \frac{\lambda}{2} \|\phi\|_{L^2}^2 \ge \min\left\{\frac{\alpha}{2}, \frac{\lambda}{2}\right\} \|\phi\|_{H^1(\Omega)}^2,$$

that is, a is *weakly coercive* independently of the choice of V, at least[4] with $\alpha_0 = \min\left\{\frac{\alpha}{2}, \frac{\lambda}{2}\right\}$. Finally, for fixed ϕ, η in V, the function $t \longmapsto a(\phi, \eta; t)$ is measurable thanks to Fubini Theorem.

The main theorem is the following[5].

Theorem 7.1. *Let $f \in L^2(0,T;V^*)$, $g \in H$. Then problem (7.29), 7.30) or (7.31) has a unique weak solution $y \in H^1(0,T;V,V^*)$. Moreover the following estimates hold,*

$$\max_{t\in[0,T]} \|y(t)\|_H^2, \ \alpha_0 \int_0^T \|y(s)\|_V^2 \, ds \le e^{2\lambda T} \left\{ \|g\|_H^2 + \frac{1}{\alpha_0} \int_0^T \|f(s)\|_*^2 \, ds \right\} \tag{7.33}$$

and

$$\int_0^T \|y_t(s)\|_*^2 \, ds \le C_0 \|g\|_H^2 + (\alpha_0^{-1} C_0 + 2) \int_0^T \|f(s)\|_*^2 \, ds \tag{7.34}$$

where $C_0 = 2\alpha_0^{-1} \mathcal{M}^2 e^{2\lambda T}$.

The estimates (7.33) and (7.34) deteriorate as T becomes large, unless the bilinear form is coercive. Indeed, in this case $\lambda = 0$ and no time dependence pops up in those constants. This is very useful for studying asymptotic properties of the solution as $t \to +\infty$.

Remark 7.2 (Regularity). If $\mathbf{A}$ is symmetric, time-independent and $f \in L^2(0,T;H)$, $g \in V$, then $y_t \in L^2(0,T;H)$, $y \in L^\infty(0,T;V)$ and

$$\alpha_0 \operatorname*{ess\ sup}_{t\in[0,T]} \|y(t)\|_V^2 + \int_0^T \|y_t(s)\|_H^2 \, ds \le C \left\{ \|g\|_V^2 + \int_0^T \|f(s)\|_H^2 \, ds \right\}, \tag{7.35}$$

where $C = C(\mathcal{M}, \lambda, T, \alpha_0)$; see Sect. A.5.1 for the definition of essential supremum. If in addition Ω is a C^2 domain and the coefficients a_{ij} (also h for the Robin condition) are Lipschitz continuous, then $y \in L^2\left(0,T;H^2(\Omega)\right)$ and

$$\int_0^T \|y_t(s)\|_H^2 \, ds \le C \left\{ \|g\|_V^2 + C \int_0^T \|f(s)\|_H^2 \, ds \right\},$$

where C depends also on the curvature of Γ and on the Lipschitz constant of the coefficients. These results apply in particular to problem (7.2) where $\mathbf{A}$ is the identity matrix, $\lambda = 0$, $V = H^1(\Omega)$. Indeed, note that $y_t \in L^2(0,T;H)$ and $y \in L^\infty(0,T;V)$ imply $y \in H^1(Q_T)$. •

7.4.2 The Mathematical Setting for OCPs

We point out the main differences with respect to the elliptic case. The essential components of time-dependent linear-quadratic OCPs are:

- a separable Hilbert space V, $H_0^1(\Omega) \subseteq V \subseteq H^1(\Omega)$, $H = L^2(\Omega)$ and a Hilbert space for the state variable y, $Y = H^1(0,T;V,V^*)$, defined in (7.28);
- a Hilbert space $\mathcal{U}$ for the control variable and a *convex, closed* subset $\mathcal{U}_{ad} \subseteq \mathcal{U}$ of admissible controls. We may distinguish *distributed, side or initial controls* defined in a subset of Q_T, S_T or $\Omega \times \{0\}$, respectively;

[4] This is a rather "crude" estimate. Better estimates could be available in each particular case.

[5] See e.g. [237, Thm. 10.6], Sect.10.3 for its proof and further details.

- a *control operator* $B : \mathcal{U} \to L^2(0,T;V^*)$, linear and continuous. In particular, we require that there exists $b \geq 0$ such that

$$\|Bu\|_{L^2(0,T;V^*)} \leq b \|u\|_{\mathcal{U}} \qquad \forall u \in \mathcal{U}.$$

Thus, B maps a control function u into an element Bu of $L^2(0,T;V^*)$. Usually B is an embedding operator, for instance $B : L^2(0,T;H) = L^2(Q_T) \to L^2(0,T;V^*)$ is defined by

$$\int_0^T \langle Bu(t), \varphi(t) \rangle_* \, dt = \int_0^T (u(t), \varphi(t))_H \, dt = \int_{Q_T} u\varphi \; d\mathbf{x}dt \;\; \forall \varphi \in L^2(0,T;V).$$

If the control function coincides with the initial value only, we set $B = 0$;

- the state equation: a uniformly parabolic equation

$$y_t + \mathcal{E}y = F + Bu \qquad \text{in } Q_T = \Omega \times (0,T) \tag{7.36}$$

supplied with boundary conditions on S_T and an initial condition $y(0) = g$. In general, $F \in L^2(0,T;V^*)$, $g \in H$ and we write the state problem in the weak form (7.30) or (7.31) with $f = F + Bu$. Thanks to Theorem 7.1, for each $u \in \mathcal{U}$, the state problem has a unique weak solution $y = y(u) \in Y$ and the stability estimates (7.33), (7.34) hold. We point out that these stability estimates ensure the continuity of the *control-to-state* map $u \longmapsto y(u)$ from $\mathcal{U}$ into Y;

- a linear continuous *observation* operator $C : Y \to Z$, where Z is a Hilbert space. Thus there exists $\gamma \geq 0$ such that

$$\|Cy\|_Z \leq \gamma \|y\|_Y \qquad \forall y \in Y. \tag{7.37}$$

Some remarkable examples follow:

(**a**) *distributed observation* in $Q_T : Z = L^2(0,T;H) = L^2(Q_T)$ and $Cy = y$ (the *embedding of* $L^2(0,T;V)$ into Z);

(**b**) *final observation:* $Z = H$ and $Cy = \kappa y(T)$, where $\kappa \in L^\infty(\Omega)$. Note that

$$\begin{aligned}\|\kappa y(T)\|_H &\leq \|\kappa\|_{L^\infty(\Omega)} \|y(T)\|_H \leq \|\kappa\|_{L^\infty(\Omega)} \max_{t\in[0,T]} \|y(t)\|_H \\ &\leq C_0 \|\kappa\|_{L^\infty(\Omega)} \|y\|_Y,\end{aligned}$$

for some constant C_0, since $Y \hookrightarrow C([0,T];H)$;

(**c**) *pointwise observations*: $Z = L^2(0,T)^N$ and C defined by

$$(Cy)(t) = (b_1 y(x_1,t), \ldots, b_N y(x_N;t))$$

with $b_j \in \mathbb{R}$, $x_j \in \overline{\Omega}$, $j = 1, \ldots, N$. For instance, in dimension $d = 1$, $V = H^1(a,b) \hookrightarrow C([a,b])$ so that $y \in L^2(0,T;C([a,b]))$ and, for each x_j, the function $t \longmapsto y(x_j,t;)$ is well defined as an element in $L^2(0,T)$. Similar considerations hold in dimension $d = 2, 3$ if $f \in L^2(0,T;H)$, Ω and the coefficients of the equation are sufficiently smooth (see Remark 7.2). Indeed in this case $y \in L^2(0,T;H^2(\Omega))$ and $H^2(\Omega) \hookrightarrow C(\overline{\Omega})$;

- a cost (or objective) functional $\tilde{J} : Z \times \mathcal{U} \to \mathbb{R}$, defined by

$$\tilde{J}(y,u) = \frac{1}{2} \|Cy - z_d\|_Z^2 + \frac{\beta}{2} \|u\|_{\mathcal{U}}^2, \tag{7.38}$$

where z_d is a prescribed (target) function in Z, $\beta \geq 0$, and the associated reduced cost functional is

$$J(u) = \frac{1}{2} \|Cy(u) - z_d\|_Z^2 + \frac{\beta}{2} \|u\|_{\mathcal{U}}^2. \tag{7.39}$$

The optimal control problem can be formulated in two ways, exactly as in the elliptic case: in the first one, y, u are two independent variables restricted by an equality constraint (the state equation) and by a control constraint ($u \in \mathcal{U}_{ad}$):

$$\tilde{J}(y,u) \to \min$$
$$\text{subject to } y \in Y, u \in \mathcal{U}_{ad}$$

and to

$$\int_0^T \langle y_t(t), \varphi(t)\rangle_* \, dt + \int_0^T a(y(t), \varphi(t); t)\, dt = \int_0^T \langle F(t) + Bu(t), \varphi(t)\rangle_* \, dt \tag{7.40}$$

$\forall \varphi \in L^2(0,T;V)$ *or, equivalently,*

$$\langle y_t(t), \phi\rangle_* + a(y(t), \phi; t) = \langle F(t) + Bu(t), \phi\rangle_* \quad \forall \phi \in V, \text{ a.e. } t \in (0,T), \tag{7.41}$$

with

$$y(0) = g. \tag{7.42}$$

A minimizer $(\hat{u}, \hat{y})$ is called an *optimal control-state pair*.

The second formulation involves the minimization of the reduced cost functional subject only to a control constraint and is formally identical to the reduced cost formulation (5.53) in the elliptic case:

$$J(u) \to \min$$
$$\text{subject to } u \in \mathcal{U}_{ad}.$$

In the latter case, once an optimal control $\hat{u}$ is found, the corresponding optimal state is retrieved by setting $\hat{y} = y(\hat{u})$.

7.4.3 Optimality Conditions

As in the elliptic case, the reduced cost functional (7.39) can be written in the form

$$J(u) = \frac{1}{2} q(u,u) - Lu + c$$

where q is the symmetric bilinear form

$$q(u,v) = (Cy_0(u), Cy_0(v))_Z + \beta(u,v)_{\mathcal{U}},$$

L is the linear and continuous functional given by

$$Lu = (z_d - Cy(0), y_0(u))_Z$$

and

$$c = \frac{1}{2} \|y(T;0) - z_d\|_Z^2,$$

where $y_0(u) = y(u) - y(0)$ is the weak solution of the state problem with $F = 0$, $g = 0$. If $\beta > 0$, existence and uniqueness of the optimal control then follow from Theorem 5.4. If $\beta = 0$ and $\mathcal{U}_{ad}$ is bounded there exists at least an optimal control which is also unique if the operator $u \to C \circ y_0(u)$ is injective.

Moreover, for any $w \in \mathcal{U}$, after small rearrangements,

$$J'(u)w = q(u,w) - Lw + \beta(u,w)_{\mathcal{U}} = (Cy(u) - z_d, Cy_0(w))_Z + \beta(u,w)_{\mathcal{U}}. \tag{7.43}$$

From Proposition 5.5 a necessary and sufficient condition of optimality reads

$$J'(\hat{u})(v - \hat{u}) = (Cy(\hat{u}) - z_d, Cy_0(v - \hat{u}))_Z + \beta(\hat{u}, v - \hat{u})_{\mathcal{U}} \geq 0 \qquad \forall v \in \mathcal{U}_{ad}. \tag{7.44}$$

As already done in the elliptic case, to get a better condition that does not contain $y_0(v - \hat{u})$, we write the functional J in the augmented form, using the state equation (7.40) with the test function replaced by a multiplier $p \in Y$,

$$\begin{aligned} J(u) = &\frac{1}{2}\|Cy(u) - z_d\|_Z^2 + \frac{\beta}{2}\|u\|_{\mathcal{U}}^2 + \\ &+ \int_0^T \{\langle F(t) + Bu(t), p(t)\rangle_* - \langle y_t(t;u), p(t)\rangle_* - a(y(t;u), p(t); t)\}\, dt. \end{aligned}$$

Note that the last integral vanishes, due to the state equation. Using the integration by parts formula (see Theorem A.19)

$$\int_0^T \langle y_t(t;u), p(t)\rangle_* dt = (y(T;u), p(T))_H - (y(0;u), p(0))_H - \int_0^T \langle y(t;u), p_t(t)\rangle_* \, dt,$$

we have

$$\begin{aligned} J(u) = &\frac{1}{2}\|Cy(u) - z_d\|_Z^2 + \frac{\beta}{2}\|u\|_{\mathcal{U}}^2 + \int_0^T \langle F(t) + Bu(t), p(t)\rangle_* \, dt \\ &- (y(T;u), p(T))_H + (y(0;u), p(0))_H + \int_0^T \{\langle y(t;u), p_t(t)\rangle_* - a(y(t;u), p(t); t)\}\, dt. \end{aligned}$$

Computing again $J'(\hat{u})$ along any direction $w \in \mathcal{U}$ and recalling (7.43), we find, setting $\hat{y} = y(\hat{u})$,

$$\begin{aligned} J'(\hat{u})w = &(C\hat{y} - z_d, Cy_0(w))_Z + (\beta\hat{u}, w)_{\mathcal{U}} + \int_0^T \langle Bw(t), p(t)\rangle_* \, dt \\ &- (p(T), y_0(T;w))_H + \int_0^T \{\langle p_t(t), y_0(t;w)\rangle_* - a(y_0(t;w), p(t); t)\}\, dt \end{aligned}$$

since $y_0(0;w) = 0$ for all w.

We now choose the multiplier $\hat{p} = p(t;\hat{u})$ as the solution of a suitable adjoint problem (that depends on the form of the observation operators), in order to get rid ot the terms containing y_0. We refer to the three cases (**a**), (**b**), (**c**) above. Further relevant cases will be considered in the next sections.

(**a**) $Z = L^2(0,T;H) = L^2(Q_T)$ and $Cy = y$, the *embedding of* $L^2(0,T;V)$ into $L^2(Q_T)$. We have

$$(C\hat{y} - z_d, Cy_0(w))_Z = \int_0^T (\hat{y}(t) - z_d(t), y_0(w))_{L^2(\Omega)}\, dt.$$

Then, if $\hat{p}$ is the solution of the following adjoint equation: $\forall \varphi \in L^2(0,T;V)$,

$$\int_0^T \{\langle \hat{p}_t(t), \varphi(t)\rangle_* - a(\varphi(t), \hat{p}(t); t)\}\, dt = \int_0^T (\hat{y}(t) - z_d(t), \varphi(t))_{L^2(\Omega)}\, dt \qquad (7.45)$$

with $\hat{p}(T) = 0$, all the terms containing y_0 disappear in the expression of $J'(\hat{u})\, w$; equivalently, (7.45) can be written as:

$$\begin{cases} -\langle \hat{p}_t(t), \psi\rangle_* + a^*(\hat{p}(t), \psi; t) = (\hat{y}(t) - z_d(t), \psi)_{L^2(\Omega)} & \forall \psi \in V \text{ and a.e. in } (0,T) \\ \hat{p}(T) = 0. \end{cases} \qquad (7.46)$$

(**b**) $Z = H$ and $Cy = \kappa y(T)$, where $\kappa \in L^\infty(\Omega)$. We have

$$(C\hat{y} - z_d, Cy_0(w))_Z = (\kappa \hat{y}(T) - z_d, \kappa y_0(T; w))_{L^2(\Omega)}.$$

Then, by the same considerations made in (**a**), we choose $\hat{p}$ as the solution of the adjoint problem:

$$\begin{cases} -\langle \hat{p}_t(t), \psi\rangle_* + a^*(\hat{p}(t), \psi; t) = 0 & \forall \psi \in V \text{ and a.e. in } (0,T) \\ \hat{p}(T) = \kappa(\kappa \hat{y}(T) - z_d). \end{cases} \qquad (7.47)$$

(**c**) *Pointwise observations:* $Z = (L^2(0,T))^N$ and

$$Cy(t) = (b_1 y(x_1, t), \ldots, b_N y(x_N; t)).$$

Then, with z_d fixed in $L^2(0,T)^N$, we have

$$(C\hat{y} - z_d, Cy_0(w))_Z = \sum_{j=1}^N \int_0^T b_j(b_j \hat{y}(x_j, t) - z_{dj}(t)) y_0(x_j, t; w))\, dt$$

and $\hat{p}$ is the solution of the adjoint problem:

$$\begin{cases} -\langle \hat{p}_t, \psi\rangle_* + a^*(\hat{p}(t), \psi; t) \\ \qquad = \sum_{j=1}^N b_j \psi(x_j) \int_0^T (b_j \hat{y}(x_j, t) - z_j(t)) dt \quad \forall \psi \in V, \text{ a.e. in } (0,T) \\ \hat{p}(T) = 0. \end{cases} \qquad (7.48)$$

In all cases, by a straightforward application of Theorem 7.1, all the above adjoint problems admit a unique weak solution $\hat{p} = p(\hat{u}) \in Y$. With this choice of $\hat{p}$, we are led to the following theorem.

Theorem 7.2. *The control $\hat{u}$ and the state $\hat{y} = y(\hat{u})$ are optimal if and only if, together with the adjoint state $\hat{p}$, they solve the optimality system given by the state problem, the adjoint problem and the variational inequality*

$$J'(\hat{u})(v - \hat{u}) = (\beta \hat{u}, v - \hat{u})_{\mathcal{U}} + \int_0^T \langle B(v(t) - \hat{u}(t)), \hat{p}(t)\rangle_*\, dt \geq 0 \qquad \forall v \in \mathcal{U}_{ad}. \quad (7.49)$$

By introducing the *Lagrangian function*

$$\mathcal{L}(y,u,p) = \frac{1}{2}\|Cy - z_d\|_Z^2 + \frac{\beta}{2}\|u\|_{\mathcal{U}}^2 + \int_0^T \{\langle F(t) + Bu(t), p(t)\rangle_* - \langle y_t(t), p(t)\rangle_* - a(y(t), p(t); t)\}\, dt, \tag{7.50}$$

the adjoint problem can be obtained by requiring that

$$\mathcal{L}'_y(\hat{y},\hat{u},\hat{p})\,\psi = \tilde{J}'_y(\hat{y},\hat{u})\,\psi - \int_0^T \{\langle \psi_t(t), \hat{p}(t)\rangle_* + a(\psi(t), \hat{p}(t); t)\}\, dt = 0$$

$\forall \psi \in Y, \psi(0) = 0$, and performing an integration by parts (see next sections).

The variational inequality corresponds to

$$\mathcal{L}'_u(\hat{y},\hat{u},\hat{p})(v - \hat{u}) = (\beta\hat{u}, v - \hat{u})_{\mathcal{U}} + \int_0^T \langle B(v(t) - \hat{u}(t)), \hat{p}(t)\rangle_*\, dt \geq 0 \quad \forall v \in \mathcal{U}_{ad}. \tag{7.51}$$

Remark 7.3. Introducing the adjoint operator

$$B^* : L^2(0,T;V) \to \mathcal{U}^*$$

and the canonical isometry $R_{\mathcal{U}} : \mathcal{U}^* \to \mathcal{U}$ (i.e. the Riesz map, see Sect. A.2.2), we can write

$$\int_0^T \langle B(v(t) - \hat{u}(t)), \hat{p}(t)\rangle_*\, dt = (R_{\mathcal{U}} B^* \hat{p}, v - \hat{u})_{\mathcal{U}}$$

so that

$$J'(\hat{u})(v - \hat{u}) = (\beta\hat{u} + R_{\mathcal{U}} B^* \hat{p}, v - \hat{u})_{\mathcal{U}} \geq 0 \qquad \forall v \in \mathcal{U}_{ad}.$$

Thus,

$$\nabla J(\hat{u}) = \beta\hat{u} + R_{\mathcal{U}} B^* \hat{p}.$$

•

7.5 Further Applications to Equations in Divergence Form

In this section we present some additional examples of time-dependent linear-quadratic OCPs for equations in divergence form, with control functions, observations and boundary conditions different than those considered so far, without however carrying out a comprehensive analysis as we did instead in Chapter 5 for OCPs governed by elliptic PDEs. Some advanced tools introduced in Chapter 5 can also be adapted to time-dependent OCPs to face more involved cases.

All these examples share similar mathematical notations and properties. However, each of them has specific features. For this reason, we prefer to provide the reader with a comprehensive list, even though this might look a little repetitive.

7.5.1 Side Control, Final and Distributed Observation, Cauchy-Robin Conditions

We consider the minimization of the following cost functional

$$\tilde{J}(y,u) = \frac{1}{2}\int_{Q_T} |y - y_d|^2 \, d\mathbf{x}dt + \frac{1}{2}\int_{\Omega} |y(T) - z_d|^2 \, d\mathbf{x} + \frac{\beta}{2}\int_{S_T} u^2 d\sigma dt \tag{7.52}$$

with $\beta \geq 0$; y_d and z_d are two target states, and y satisfies the state problem

$$\begin{cases} y_t - \operatorname{div}(\mathbf{A}(\mathbf{x})\nabla y) = f & \text{in } Q_T \\ \dfrac{\partial y}{\partial n} + hy = u & \text{on } S_T \\ y(\mathbf{x},0) = g(\mathbf{x}) & \text{in } \Omega. \end{cases} \tag{7.53}$$

The interpretation is that we look for the optimal radiation in order to minimize the distance of both y and $y(T)$ from the two target states y_d and z_d, respectively.

We assume $f \in L^2(Q_T)$, $g \in L^2(\Omega)$ and $h \in L^\infty(S_T)$, $h \geq 0$ a.e. with respect to the surface measure on S_T. The control variable is constrained to $u \in \mathcal{U}_{ad} \subseteq \mathcal{U} = L^2(S_T)$, with $\mathcal{U}_{ad}$ being closed and convex.

• *The state equation.* Let $V = H^1(\Omega)$. Then, a weak formulation of the Neumann problem (7.53) is: find $u \in Y = H^1(0,T;V,V^*)$ such that for all $\varphi \in V$, a.e. in $(0,T)$,

$$\langle y_t(t), \varphi\rangle_* + (\mathbf{A}\nabla y(t), \nabla\varphi)_{L^2(\Omega)} + (h(t)\,y(t), \varphi)_{L^2(\Gamma)} = (u(t), \varphi)_{L^2(\Gamma)}$$

and $y(0) = g$. The control operator $B : L^2(S_T) \to L^2(0,T;V^*)$ is defined for a.e. $t \in (0,T)$ by

$$\langle Bu(t), \varphi\rangle_{V^*} = (u(t), \varphi)_{L^2(\Gamma)} \qquad \forall\varphi \in V.$$

Thanks to the trace inequality (A.40), one can prove that B is continuous. Then, thanks to Theorem (7.1), for all $u \in L^2(S_T)$, problem (7.53) has a unique weak solution $y(u) \in Y$.

• *Existence and uniqueness of the optimal control.* The observation operator is defined as

$$Cy = (y, y(T)) \in Z$$

where $Z = L^2(Q_T) \times L^2(\Omega)$, with norm $\|(y,y(T))\|_Z^2 = \|y\|_{L^2(Q_T)}^2 + \|y(T)\|_{L^2(\Omega)}^2$. Since

$$\|(y,y(T))\|_Z^2 \leq (1 + c_{tr})\,\|y\|_Y^2\,,$$

C is continuous from Y into Z.

If $\beta > 0$ there exists a unique optimal control. When $\beta = 0$ and $\mathcal{U}_{ad}$ is bounded, we consider the operator $u \to C(y_0(u))$. If $C(y_0(u)) = 0$, then $y_0(u) = 0$ in Q_T and therefore also $u = 0$. Thus also in this case there exists a unique optimal control.

• *Optimality conditions.* To write the adjoint problem, we can combine the cases (**a**) and (**b**) of Sect. 7.4.3. We conclude that the adjoint equation in weak form is:

$$\begin{cases} -\langle \hat{p}_t(t), \varphi\rangle_* + a^*(\hat{p}(t), \varphi) = (\hat{y}(t) - y_d(t), \varphi)_{L^2(\Omega)} & \forall\varphi \in V \text{ and a.e. in } (0,T) \\ \hat{p}(T) = \hat{y}(t) - z_d \end{cases}$$

where

$$a^*(\hat{p}(t), \varphi) = \left(\mathbf{A}^\top \nabla\hat{p}, \nabla\varphi\right)_{L^2(\Omega)} + (h\hat{p}, \varphi)_{L^2(\Gamma)}\,.$$

The strong form of this adjoint problem is:

$$\begin{cases} -\hat{p}_t - \operatorname{div}\left(\mathbf{A}^\top \nabla \hat{p}\right) = \hat{y} - y_d & \text{in } Q_T \\ \dfrac{\partial \hat{p}}{\partial n} + h\hat{p} = 0 & \text{on } S_T \\ \hat{p}\left(\mathbf{x},T\right) = \hat{y}\left(T\right) - z_d. & \text{in } \Omega. \end{cases} \tag{7.54}$$

The variational inequality (7.49) reads

$$J'\left(\hat{u}\right)\left(v - \hat{u}\right) = \left(\beta\hat{u} + \hat{p}, v - \hat{u}\right)_{L^2(S_T)} \geq 0 \qquad \forall v \in \mathcal{U}_{ad}. \tag{7.55}$$

We also deduce that

$$\nabla J\left(\hat{u}\right) = \beta\hat{u} + \hat{p} \qquad \text{on } S_T.$$

We summarize everything in the following proposition.

Proposition 7.3. *The control $\hat{u}$ and the state function $\hat{y} = y(\hat{u})$ are optimal for the OCP (7.52)–(7.53) if and only if, together with the adjoint state $\hat{p}$, they solve the state problem (7.53), the adjoint problem (7.54) and the variational inequality (7.55).*

Remark 7.4. If

$$\mathcal{U}_{ad} = \left\{u \in L^2\left(S_T\right) \,:\, a \leq u \leq b \text{ a.e. on } S_T\right\}$$

and $\beta > 0$, then

$$\hat{u} = Pr_{\mathcal{U}_{ad}}\left(-\beta^{-1}\hat{p}\right) \qquad \text{a.e. on } S_T.$$

If $\beta = 0$, then (see Sect. 5.5.1), a.e. on S_T,

$$\hat{u}\left(\sigma,t\right)\begin{cases} = a & \text{if } \hat{p}\left(\sigma,t\right) > 0 \\ \in [a,b] & \text{if } \hat{p}\left(\sigma,t\right) = 0 \\ = b & \text{if } \hat{p}\left(\sigma,t\right) < 0. \end{cases}$$

•

If we introduce the Lagrangian function

$$\begin{aligned} \mathcal{L}\left(y,u,p\right) = {} & \frac{1}{2}\int_{Q_T}\left(y - y_d\right)^2 d\mathbf{x}dt + \frac{1}{2}\int_\Omega \left(y\left(T\right) - z_d\right)^2 d\mathbf{x} + \frac{\beta}{2}\int_{S_T} u^2 d\sigma dt \\ & + \int_0^T \left\{fp - \langle y_t, p\rangle_* - \left(\mathbf{A}\nabla y, \nabla p\right)_{L^2(\Omega)} - \left(hy - u, p\right)_{L^2(\Gamma)}\right\} dt, \end{aligned} \tag{7.56}$$

the adjoint equation is obtained by requiring that $\mathcal{L}'_y\left(\hat{y},\hat{u},\hat{p}\right)\psi = 0$, for all $\psi \in Y$, $\psi\left(0\right) = 0$. We have

$$\begin{aligned} \mathcal{L}'_y\left(\hat{y},\hat{u},\hat{p}\right)\psi = {} & \int_0^T\int_\Omega \left(\hat{y} - y_d\right)\psi d\mathbf{x}dt + \int_\Omega \left(\hat{y}\left(T\right) - z_d\right)\psi\left(T\right) d\mathbf{x} \\ & - \int_0^T \left\{\langle \psi_t, \hat{p}\rangle_* + \left(\mathbf{A}\nabla\psi, \nabla\hat{p}\right)_{L^2(\Omega)} + \left(h\psi, \hat{p}\right)_{L^2(\Gamma)}\right\} dt \\ = {} & \int_0^T \left\{\left(\hat{y} - y_d, \psi\right)_{L^2(\Omega)} + \langle \hat{p}_t, \psi\rangle_* - \left(\mathbf{A}^\top\nabla\hat{p}, \nabla\psi\right)_{L^2(\Omega)} - \left(h\hat{p}, \psi\right)_{L^2(\Gamma)}\right\} dt \\ & + \left(\hat{y}\left(T\right) - z_d - \hat{p}\left(T\right), \psi\left(T\right)\right)_{L^2(\Omega)}. \end{aligned}$$

For any ψ of the form $\psi(t) = \varphi\eta(t)$ with $\varphi \in V$, $\eta \in C^1([0,T])$, $\eta(0) = 0$, we get

$$\begin{aligned}\int_0^T \Big\{ (\hat{y} - y_d, \varphi)_{L^2(\Gamma)} + \langle \hat{p}_t, \varphi\rangle_* - \left(\mathbf{A}^\top \nabla\hat{p}, \nabla\varphi\right)_{L^2(\Omega)} - (h\hat{p}, \varphi)_{L^2(\Gamma)} \Big\} \eta(t)\, dt \\ + (y(T) - z_d - \hat{p}(T), \varphi)_{L^2(\Omega)}\, \eta(T) = 0.\end{aligned} \tag{7.57}$$

If we assume that $\eta(T) = 0$, thanks to the arbitrariness of η, we deduce the adjoint equation in the weak form

$$-\langle \hat{p}_t, \varphi\rangle_* + \left(\mathbf{A}^\top \nabla\hat{p}, \nabla\varphi\right)_{L^2(\Omega)} + (h\hat{p}, \varphi)_{L^2(\Gamma)} = (\hat{y} - y_d, \varphi)_{L^2(\Gamma)} \quad \forall\varphi \in V \text{ and a.e. in } (0,T).$$

If instead $\eta(T) \neq 0$, (7.57) reduces to

$$(\hat{y}(T) - z_d - \hat{p}(T), \varphi)_{L^2(\Omega)} = 0 \qquad \forall\varphi \in V.$$

Since V is dense in $L^2(\Omega)$ we get the final condition

$$\hat{p}(T) = \hat{y}(T) - z_d.$$

Finally, the optimality condition is the variational inequality

$$\mathcal{L}'_u(\hat{y}, \hat{u}, \hat{p})(v - \hat{u}) = (\beta\hat{u} + \hat{p}, v - \hat{u})_{L^2(S_T)} \geq 0 \qquad \forall v \in \mathcal{U}_{ad}$$

from which we can also deduce the following expression for the gradient of the cost functional,

$$\nabla J(\hat{u}) = \beta\hat{u} + \hat{p} \qquad \text{on} S_T.$$

7.5.2 Time-distributed Control, Side Observation, Cauchy-Neumann Conditions

, In some applications we may be faced to minimize the cost functional

$$\tilde{J}(y, u_1, \ldots, u_N) = \frac{1}{2}\int_{S_T} |y - y_d|^2 \, d\sigma dt + \sum_{j=1}^N \frac{\beta_j}{2}\int_0^T u_j^2(t)\, dt$$

with a given $y_d \in L^2(0,T;S_T)$ and $\beta_j > 0$, $j = 1, \ldots, N$, subject to a state equation like

$$\begin{cases} y_t - \operatorname{div}(\mu(\mathbf{x})\nabla y) = \sum_{j=1}^N s_j(\mathbf{x})\, u_j(t) & \text{in } Q_T \\ \dfrac{\partial y}{\partial n} = 0 & \text{on } S_T \\ y(\mathbf{x},0) = g(\mathbf{x}) & \text{in } \Omega. \end{cases} \tag{7.58}$$

Here $\mu \in L^\infty(\Omega)$, $\mu(\mathbf{x}) \geq \mu_0 > 0$ a.e. in Ω, $g \in L^2(\Omega)$ and the functions s_j, called *control-shape functions,* all belong to $L^2(\Omega)$. In the special case $s_j(\mathbf{x}) = \delta(\mathbf{x} - \mathbf{x}_j)$, the *delta measure concentrated at* $\mathbf{x}_j$, we are dealing with a *space-pointwise* control, distributed in time.

The control function $\mathbf{u} = (u_1, \ldots, u_N)$ is subject to the box constraint

$$\mathcal{U}_{ad} = \left\{ \mathbf{u} \in L^2(0,T)^N \, : \, a_j \leq u_j(t) \leq b_j \text{ a.e. in } (0,T),\ j = 1, \ldots, N \right\}.$$

- *The state equation.* Let $V = H^1(\Omega)$; a weak formulation of problem (7.58) reads

$$\langle y_t(t), \varphi\rangle_* + (\mu\nabla y, \nabla\varphi)_{L^2(\Omega)} = \sum_{j=1}^N (s_j, \varphi)_{L^2(\Omega)}\, u_j(t) \quad \forall\varphi \in V \text{ and a.e. in } (0,T)$$

with $y(0) = g$. The control operator $B : L^2(0,T)^N \to L^2(0,T;V^*)$ is an embedding operator. Thanks to Theorem 7.1, for any $\mathbf{u} \in L^2(0,T)^N$, there exists a unique weak solution $y(\mathbf{u}) \in Y = H^1(0,T;V,V^*)$.

• *Esistence and uniqueness of the optimal control.* Here $Z = L^2(S_T)$ and the observation operator is $Cy = y_{|S_T}$. This is a boundary observation. Note that, by the trace inequality (A.40), we have

$$\|y\|_{L^2(S_T)} \leq c_{tr} \|y\|_{L^2(0,T;V)} \leq c_{tr} \|y\|_Y .$$

Since the hypotheses of Theorem 5.4 are fulfilled, there exists a unique optimal control $\hat{u} \in \mathcal{U}_{ad}$.

• *Optimality conditions.* To write the optimality conditions, we introduce the Lagrangian function

$$\mathcal{L}(y,\mathbf{u},p) = \frac{1}{2}\int_{S_T} (y - y_d)^2\, d\sigma dt + \sum_{j=1}^{N} \frac{\beta_j}{2}\int_0^T u_j^2(t)\, dt \tag{7.59}$$

$$+ \int_0^T \left\{ \sum_{j=1}^{N} (s_j, p(t))_{L^2(\Omega)}\, u_j(t) - \langle y_t(t), p(t)\rangle_* - (\mu\nabla y(t), \nabla p(t))_{L^2(\Omega)} \right\} dt.$$

The adjoint equation corresponds to $\mathcal{L}'_y(\hat{y},\hat{\mathbf{u}},\hat{p})\,\psi = 0$, for all $\psi \in Y$, $\psi(0) = 0$. We have:

$$\mathcal{L}'_y(\hat{y},\hat{\mathbf{u}},\hat{p})\,\psi = \int_0^T \int_\Gamma (\hat{y} - y_d)\,\psi d\sigma dt + \int_0^T \left\{ -\langle \psi_t(t), \hat{p}(t)\rangle_* - (\mu\nabla\psi(t), \nabla\hat{p}(t))_{L^2(\Omega)} \right\} dt$$

$$= \int_0^T \left\{ (\hat{y} - y_d, \psi)_{L^2(\Gamma)} + \langle \hat{p}_t(t), \psi(t)\rangle_* - (\mu\nabla\hat{p}(t), \nabla\psi(t))_{L^2(\Omega)} \right\} dt - (\psi(T), \hat{p}(T))_{L^2(\Omega)} .$$

By choosing ψ of the form $\psi(t) = \varphi\eta(t)$ with $\varphi \in V$, $\eta \in C^1([0,T]), \eta(0) = 0 = \eta(T)$, the equation $\mathcal{L}'_y(\hat{y},\hat{\mathbf{u}},\hat{p})\,\psi = 0$, yields

$$\int_0^T \left\{ (\hat{y} - y_d, \varphi)_{L^2(\Gamma)} + \langle \hat{p}_t(t), \varphi\rangle_* - (\mu\nabla\hat{p}(t), \nabla\varphi)_{L^2(\Omega)} \right\} \eta(t)\, dt = 0.$$

Hence, thanks to the arbitrariness of η,

$$-\langle \hat{p}_t(t), \varphi\rangle_* + (\mu\nabla\hat{p}(t), \nabla\varphi)_{L^2(\Omega)} = (\hat{y} - y_d, \varphi)_{L^2(\Gamma)} \quad \forall\varphi \in V \text{ and a.e. in } (0,T). \tag{7.60}$$

Choosing now ψ of the form $\psi(t) = \varphi\eta(t)$ with $\varphi \in V$, $\eta \in C^1([0,T]), \eta(0) = 0$, $\eta(T) = 1$, we deduce that $(\varphi, \hat{p}(T))_{L^2(\Omega)} = 0$ for all $\varphi \in V$. Since V is dense in $L^2(\Omega)$ we get the final condition

$$\hat{p}(T) = 0. \tag{7.61}$$

The strong form of the adjoint problem is:

$$\begin{cases} -\hat{p}_t - \operatorname{div}(\mu(\mathbf{x})\nabla\hat{p}) = 0 & \text{in } Q_T \\ \dfrac{\partial\hat{p}}{\partial n} = \hat{y} - y_d & \text{on } S_T \\ \hat{p}(T) = 0 & \text{in } \Omega. \end{cases} \tag{7.62}$$

The variational inequality comes from

$$\mathcal{L}'_{\mathbf{u}}(\hat{y},\hat{\mathbf{u}},\hat{p})(\mathbf{v} - \hat{\mathbf{u}}) \geq 0 \qquad \mathbf{v} \in \mathcal{U}_{ad},$$

which in this case turns into

$$\sum_{j=1}^{N}\int_0^T\left(\beta_j\hat{u}_j(t)+(s_j,\hat{p}(t))_{L^2(\Omega)}\right)(v_j(t)-\hat{u}_j(t))\geq 0\qquad\forall\mathbf{v}\in\mathcal{U}_{ad}.\tag{7.63}$$

Hence, each component u_j of the control vector function can be characterized as the projection of $-\beta^{-1}(s_j,\hat{p}(t))_{L^2(\Omega)}$ onto $[a_j,b_j]$ (see Sect. 5.2.2), i.e.

$$u_j(t)=Pr_{[a_j,b_j]}\left(-\beta_j^{-1}(s_j,\hat{p}(t))_{L^2(\Omega)}\right)=Pr_{[a_j,b_j]}\left(-\frac{1}{\beta_j}\int_\Omega s_j(\mathbf{x})\hat{p}(\mathbf{x},t)\,d\mathbf{x}\right).\tag{7.64}$$

Remark 7.5. Note that if $s_j(\mathbf{x})=\delta(\mathbf{x}-\mathbf{x}_j)$, (7.64) gives

$$u_j(t)=Pr_{[a_j,b_j]}\left(-\beta_j^{-1}\hat{p}(\mathbf{x}_j,t)\right).$$

•

7.6 Optimal Control of Time-Dependent Stokes Equations

We consider now an optimal control problem for the time-dependent Stokes equations. We consider an unconstrained distributed control problem, similar to the one of Sect. 5.13.1, where we want to reach a given velocity profile $\boldsymbol{z}_d$ over the bounded domain $\Omega\subset\mathbb{R}^d$, $d=2,3$, for (almost every) $t\in(0,T)$, by controlling a distributed force $\boldsymbol{u}=\boldsymbol{u}(\mathbf{x},t)\in L^2(Q_T)^d$. We assume that Ω is a polyhedron or a sufficiently smooth domain. The goal is to minimize over $L^2(Q_T)^d$ the cost functional

$$J(\boldsymbol{u})=\frac{1}{2}\int_{Q_T}|\boldsymbol{v}(\boldsymbol{u})-\boldsymbol{z}_d|^2\,d\mathbf{x}dt+\frac{\tau}{2}\int_{Q_T}|\boldsymbol{u}|^2\,d\mathbf{x}dt\tag{7.65}$$

where $\tau>0$ and $(\boldsymbol{v},\pi)=(\boldsymbol{v}(\boldsymbol{u}),\pi(\boldsymbol{u}))$ is the solution of the following unsteady Stokes system:

$$\begin{cases}\boldsymbol{v}_t-\nu\Delta\boldsymbol{v}+\nabla\pi=\boldsymbol{f}+\boldsymbol{u} & \text{in } Q_T\\ \operatorname{div}\boldsymbol{v}=0 & \text{in } Q_T\\ \boldsymbol{v}=\boldsymbol{h} & \text{on } S_D=\Gamma_D\times(0,T)\\ -\pi\mathbf{n}+\nu\dfrac{\partial\boldsymbol{v}}{\partial n}=\boldsymbol{c} & \text{on } S_N=\Gamma_N\times(0,T)\\ \boldsymbol{v}(\mathbf{x},0)=\boldsymbol{g} & \text{in } \Omega.\end{cases}\tag{7.66}$$

Here Γ_D and Γ_N are two subsets such that $\Gamma=\Gamma_D\cup\Gamma_N$, $\Gamma_D\cap\Gamma_N=\emptyset$. We impose the velocity profile $\boldsymbol{h}$ over S_D and the normal Cauchy stress $\boldsymbol{c}$ over S_N.

• *Saddle point formulation and analysis of the state equation.* To handle the inhomogeneous Dirichlet condition we rewrite the state problem into a parabolic saddle point formulation, introducing as a multiplier the restriction $\boldsymbol{\lambda}$ of the Cauchy stress $-\pi\mathbf{n}+\nu\frac{\partial\boldsymbol{v}}{\partial n}$ on S_D. In this way we can avoid to use a lifting for the Dirichlet data and divergence-free spaces for the velocity. As we explain in Remark 7.6 below, this leads to choose[6]

$$X=\left\{\boldsymbol{\varphi}\in(H^1(\Omega))^d:\boldsymbol{\varphi}_{|\Gamma_D}\in H_{00}^{1/2}(\Gamma_D)^d\right\}\tag{7.67}$$

as velocity space, $H=L^2(\Omega)^d$ and $Q=L^2(\Omega)$ as pressure space.

[6] See Sect. A.5.10 for the definition of the space $H_{00}^{1/2}(\Gamma_D)^d$, $\Gamma_D\subset\Gamma$, and their duals.

We keep the same notation of Sect. 5.13 and denote by

$$(\nabla \boldsymbol{v}, \nabla \boldsymbol{w})_{L^2} = \int_\Omega \nabla \boldsymbol{v} : \nabla \boldsymbol{w} d\mathbf{x}, \qquad \|\nabla \boldsymbol{v}\|_{L^2}^2 = \int_\Omega \nabla \boldsymbol{v} : \nabla \boldsymbol{v} d\mathbf{x}.$$

X is a closed subset of $H^1(\Omega)^d$ and therefore is a Hilbert space with respect to the same norm. Moreover, X is dense in H with compact embedding and $X \subset H \subset X^*$ form a Hilbert triplet.

For simplicity, we assume that the problem data are regular enough, namely $\boldsymbol{f} \in L^2(0,T;H) = L^2(Q_T)^d$, $\boldsymbol{h} \in H^1(0,T;H_{00}^{1/2}(\Gamma_D)^d)$, $\boldsymbol{c} \in L^2(0,T;L^2(\Gamma_N)^d)$ and $\boldsymbol{g} \in X$. Moreover, we assume that $\boldsymbol{h}$ and $\boldsymbol{g}$ satisfy the following compatibility conditions:

$$\mathrm{div}\boldsymbol{g} = 0 \text{ in } \Omega \tag{7.68}$$

and

$$\langle \boldsymbol{s}, \boldsymbol{h}(0) - \boldsymbol{g} \rangle_{\Gamma_D} = 0 \text{ for all } \boldsymbol{s} \in H_{00}^{-1/2}(\Gamma_D)^d. \tag{7.69}$$

By the symbol $\langle \cdot , \cdot \rangle_{\Gamma_D}$ we denote the duality pairing between $H_{00}^{1/2}(\Gamma_D)^d$ and its dual $H_{00}^{-1/2}(\Gamma_D)^d$.

Remark 7.6. Under the above assumptions, we shall find a velocity field $\boldsymbol{v}$ in $H^1(0,T;H) \cap L^2(0,T;X)$. Since $\boldsymbol{f}(t)$, $\boldsymbol{u}(t)$, and $\boldsymbol{v}_t(t)$ all belong to H for a.e. $t \in (0,T)$, the first equation in (7.66) implies that

$$\mathrm{div}\left(-\nu \nabla \boldsymbol{v}(t) + \pi(t) I\right) \in L^2(\Omega)^d$$

for a.e. $t \in (0,T)$, where I is the identity matrix. Hence the trace $\boldsymbol{\lambda}(t) = \left(-\pi \mathbf{n} + \nu \frac{\partial \boldsymbol{v}}{\partial n}\right)_{|\Gamma_D}$ belongs to $H_{00}^{-1/2}(\Gamma_D)^d$ (see Sect. A.5.10). Thus, in the integration by parts leading to the saddle point formulation below, one is forced to assume that the trace of a test function belongs to $H_{00}^{1/2}(\Gamma_D)^d$. Coherently, X becomes the correct choice for both test and velocity spaces. Also note that $H^1(0,T;H_{00}^{1/2}(\Gamma_D)) \subset C([0,T];H_{00}^{1/2}(\Gamma_D))$, hence $\boldsymbol{h}(0)$ makes sense. •

For $t \in (0,T)$ and $\boldsymbol{\varphi} \in X$, define now the linear functional $F(t) \in X^*$ by

$$\langle F(t), \boldsymbol{\varphi} \rangle_{X^*,X} = \int_\Omega (\boldsymbol{f}(t) + \boldsymbol{u}(t)) \cdot \boldsymbol{\varphi}\, d\mathbf{x} + \int_{\Gamma_N} \boldsymbol{c}(t) \cdot \boldsymbol{\varphi}\, d\sigma.$$

The saddle point formulation of problem (7.66) is obtained: first, by multiplying the first equation in (7.66) by a test function $\boldsymbol{\varphi} \in X$, integrating over Ω, integrating by parts and taking into account the boundary conditions; then, by multiplying the other equations by test functions $q, \boldsymbol{s}, \boldsymbol{\varphi}$ in $L^2(\Omega)$, $H_{00}^{-1/2}(\Gamma_D)^d$ and X, respectively, and integrating over Ω. Thus, the saddle point formulation of the problem results as follows: *find* $\boldsymbol{v} \in L^2(0,T;X) \cap H^1(0,T;H)$, $\pi \in L^2(0,T;Q)$ *and a multiplier* $\boldsymbol{\lambda} \in L^2(0,T;H_{00}^{-1/2}(\Gamma_D)^d)$ *such that, for a.e.* $t \in (0,T)$ *and every* $\boldsymbol{\varphi} \in X, (q,\boldsymbol{s}) \in M = L^2(\Omega) \times H_{00}^{-1/2}(\Gamma_D)^d$,

$$\begin{cases} (\boldsymbol{v}_t(t), \boldsymbol{\varphi})_H + \nu (\nabla \boldsymbol{v}(t), \nabla \boldsymbol{\varphi})_H - (\pi(t), \mathrm{div}\boldsymbol{\varphi})_{L^2(\Omega)} - \langle \boldsymbol{\lambda}(t), \boldsymbol{\varphi} \rangle_{\Gamma_D} = \langle F(t), \boldsymbol{\varphi} \rangle_{X^*,X} \\ (\mathrm{div}\boldsymbol{v}(t), q)_{L^2(\Omega)} = 0 \\ \langle \boldsymbol{s}, \boldsymbol{v}(t) \rangle_{\Gamma_D} = \langle \boldsymbol{s}, \boldsymbol{h}(t) \rangle_{\Gamma_D} \\ (\boldsymbol{v}(0), \boldsymbol{\varphi})_H = (\boldsymbol{g}, \boldsymbol{\varphi})_H. \end{cases} \tag{7.70}$$

A triplet $(\boldsymbol{v}, \pi, \boldsymbol{\lambda})$ that fulfills system (7.70) is named a *saddle point solution* to (7.66). Indeed, for smooth data and solution, it can be shown that the saddle point formulation is equivalent to (7.66).

Introducing the bilinear forms $e : X \times X \to \mathbb{R}$ and $b : X \times M \to \mathbb{R}$ given by

$$e(\boldsymbol{v}, \boldsymbol{\varphi}) = \nu (\nabla \boldsymbol{v}, \nabla \boldsymbol{\varphi})_H, \qquad b(\boldsymbol{\varphi}, (q, \boldsymbol{s})) = -(q, \mathrm{div}\boldsymbol{\varphi})_{L^2(\Omega)} - \langle \boldsymbol{s}, \boldsymbol{\varphi} \rangle_{\Gamma_D} \tag{7.71}$$

we can write (7.70) in the more compact form

$$\begin{cases} (\boldsymbol{v}_t(t), \boldsymbol{\varphi})_H + e(\boldsymbol{v}(t), \boldsymbol{\varphi}) + b(\boldsymbol{\varphi}, (\pi(t), \boldsymbol{\lambda}(t))) = \langle F(t), \boldsymbol{\varphi} \rangle_{X^*, X} \\ b(\boldsymbol{v}(t), (q, \boldsymbol{s})) = \langle \boldsymbol{s}, \boldsymbol{h}(t) \rangle_{\Gamma_D} \\ (\boldsymbol{v}(0), \boldsymbol{\varphi})_H = (\boldsymbol{g}, \boldsymbol{\varphi})_H . \end{cases} \tag{7.72}$$

Note that the set

$$Z = \{\boldsymbol{z} \in X : b(\boldsymbol{z}, (q, \boldsymbol{s})) = 0 \text{ for all } (q, \boldsymbol{s}) \in M\}$$

is contained in $H^1_{\Gamma_D}(\Omega)^d$ and therefore the bilinear form e is strictly coercive on Z, i.e.

$$e(\mathbf{z}, \mathbf{z}) \geq \nu \|\boldsymbol{z}\|_X^2 \qquad \text{for all } \boldsymbol{z} \in Z.$$

Moreover, from Lemma 5.207, it follows that

$$\inf_{(q,\mathbf{s}) \in M} \sup_{\boldsymbol{w} \in X} \frac{b(\mathbf{w}, (q, \mathbf{s}))}{\|\boldsymbol{w}\|_X \|(q, \mathbf{s})\|_Q} \geq \tau > 0.$$

The following result holds.

Theorem 7.3. *There exists a unique saddle point solution $(\boldsymbol{v}, \pi, \boldsymbol{\lambda})$ to (7.70). Moreover*

$$\begin{aligned} &\|\boldsymbol{v}\|_{L^\infty(0,T;X)}, \|\boldsymbol{v}\|_{H^1(0,T;H)}, \|\pi\|_{L^2(0,T;H)}, \|\boldsymbol{\lambda}\|_{L^2(0,T;H_{00}^{-1/2}(\Gamma_D)^d)} \\ &\leq C \left\{ \|\boldsymbol{g}\|_X + \|\boldsymbol{f}\|_{L^2(0,T;H)} + \|\boldsymbol{u}\|_{L^2(0,T;H)} + \|\boldsymbol{h}\|_{H^1(0,T;H_{00}^{1/2}(\Gamma_D)^d)} \right\} \end{aligned}$$

where C depends only on Ω, ν and β.

This result is a direct consequence of the abstract parabolic saddle point Theorem below (see [63] for the proof).

Theorem 7.4. *Let $X \subset H \subset X^*$ be a Hilbert triplet, with X compactly embedded into H. Let M be a Hilbert space and $e : X \times X \to \mathbb{R}$, $b : X \times M \to \mathbb{R}$ be two continuous bilinear forms satisfying the following conditions:*

(a) $\forall z \in Z = \{z \in X \ : \ b(z, \mu) = 0 \ \forall \mu \in M\}$,

$$e(z, z) \geq \nu \|z\|_X^2 ;$$

(b) $\forall v, w \in X$

$$e(v, w) \leq \frac{1}{2} e(v, v) + \frac{1}{2} e(w, w);$$

(c) there exists $\beta > 0$ such that the following inf-sup condition holds,

$$\inf_{\mu \in M} \sup_{w \in X} \frac{b(w, \mu)}{\|w\|_X \|\mu\|_M} \geq \beta.$$

If $F \in L^2(0, T; H)$, $G \in H^1(0, T; M^*)$, $g \in X$ and

$$b(g, \mu) = \langle \mu, G(0) \rangle_{M,M^*} \qquad \forall \mu \in M \tag{7.73}$$

then the problem

$$\begin{cases} \langle v_t(t), \varphi\rangle_{X^*,X} + e(v(t), \varphi) + b(\varphi, \zeta(t)) = \langle F(t), \varphi\rangle_{X^*,X} & \forall \varphi \in X \\ b(v(t), \mu) = \langle \mu, G(t)\rangle_{M,M^*} & \forall \mu \in M \\ (v(0), \varphi)_H = (g, \varphi)_H & \forall \varphi \in X \end{cases} \tag{7.74}$$

has a unique solution $(v, \zeta) \in \left[L^\infty(0,T;X) \cap H^1(0,T;H)\right] \times L^2(0,T;M)$ and

$$\|v\|_{L^\infty(0,T;X)}, \|v\|_{H^1(0,T;H)}, \|\zeta\|_{L^2(0,T;M)}$$
$$\leq C(\nu, \beta)\left\{\|g\|_X + \|F\|_{L^2(0,T;H)} + \|G\|_{H^1(0,T;M))}\right\}.$$

Remark 7.7. In the application to our saddle point problem, the bilinear forms $e : X \times X \to \mathbb{R}$ and $b : X \times M \to \mathbb{R}$ are defined in (7.71), with X as in (7.67) and $M = L^2(\Omega) \times H_{00}^{-1/2}(\Gamma_D)^d$.

Moreover, $G(t) = (\text{div}\boldsymbol{v}(t), \boldsymbol{v}(t))_\Gamma = (0, \boldsymbol{h}(t))$. Note that the condition (7.73) corresponds to the compatibility conditions (7.68) and (7.69): the initial data must be divergence free and the boundary data must be compatible with initial data on Γ for $t = 0$. •

• *Existence and uniqueness of an optimal control.* We can apply the general result of Theorem 5.4, since the cost functional (7.65) can be easily expressed under the form (5.56) by setting (using the notation of that section)

$$q(\boldsymbol{u}, \boldsymbol{\eta}) = (\boldsymbol{v}_0(\boldsymbol{u}), \boldsymbol{v}_0(\boldsymbol{\eta}))_{L^2(Q_T)^d} + \tau(\boldsymbol{u}, \boldsymbol{\eta})_{L^2(Q_T)^d}, \tag{7.75}$$

$$L\boldsymbol{u} = (\boldsymbol{v}_0(\boldsymbol{u}), \boldsymbol{v}(\mathbf{0}) - \boldsymbol{z}_d)_{L^2(Q_T)^d}, \qquad c = \frac{1}{2}\|\boldsymbol{v}(\mathbf{0}) - \boldsymbol{z}_d\|^2_{L^2(Q_T)^d},$$

where $\boldsymbol{v}_0$ is the saddle point solution to the state problem corresponding to $\boldsymbol{f} = \boldsymbol{g} = \boldsymbol{h} = \mathbf{0}$. Indeed, it is simple to prove that q is continuous and positive in $L^2(Q_T)^d$ (as $\tau > 0$) and that L is continuous in $L^2(Q_T)^d$ (see Exercise 4). Hence, there exists a unique optimal control $\hat{\boldsymbol{u}}$ of problem (7.65)–(7.66) with corresponding optimal state $(\hat{\boldsymbol{v}}, \hat{\pi}) = (\boldsymbol{v}(\hat{\boldsymbol{u}}), \pi(\hat{\boldsymbol{u}}))$ and multiplier $\hat{\boldsymbol{\lambda}} = \boldsymbol{\lambda}(\hat{\boldsymbol{u}})$.

• *Optimality conditions.* As done in Sect. 5.13.2, we can recover the system of first-order (necessary and sufficient) optimality conditions in weak form, introducing the Lagrangian $\mathcal{L} = \mathcal{L}((\boldsymbol{v}, \pi, \boldsymbol{\lambda}), \boldsymbol{u}, (\boldsymbol{z}, q, \boldsymbol{\sigma}))$ given by

$$\begin{aligned} \mathcal{L}((\boldsymbol{v}, \pi, \boldsymbol{\lambda}), \boldsymbol{u}, (\boldsymbol{z}, q, \boldsymbol{\sigma})) &= \frac{1}{2}\int_{Q_T} |\boldsymbol{v} - \boldsymbol{z}_d|^2 \, d\mathbf{x}dt + \frac{\tau}{2}\int_{Q_T} |\boldsymbol{u}|^2 \, d\mathbf{x}dt \\ &- \int_0^T \left\{(\boldsymbol{v}_t(t), \boldsymbol{z}(t)) + e(\boldsymbol{v}(t), \boldsymbol{z}(t)) - (\pi(t), \text{div}\boldsymbol{z}(t))_{L^2(\Omega)} - \langle\boldsymbol{\lambda}(t), \boldsymbol{z}(t)\rangle_{\Gamma_D}\right\} dt \\ &+ \int_0^T \left\{(q(t), \text{div}\boldsymbol{v}(t))_{L^2(\Omega)} + \langle\boldsymbol{\sigma}(t), \boldsymbol{h}(t) - \boldsymbol{v}(t)\rangle_{\Gamma_D}\right\} dt \\ &+ \int_0^T (\boldsymbol{f}(t) + \boldsymbol{u}(t), \boldsymbol{z}(t))_H + (\boldsymbol{c}(t), \boldsymbol{z}(t))_{L^2(\Gamma_N)^d} \, dt, \end{aligned}$$

where $\boldsymbol{z} \in L^2(0,T;X) \cap H^1(0,T;H)$ and $(\pi, \boldsymbol{\sigma}) \in L^2(0,T;M) \times L^2(0,T;H_{00}^{-1/2}(\Gamma_D)^d)$ are the adjoint variables (or Lagrange multipliers) related to the constraints represented by the equations in the state problem (7.70).

We now differentiate the Lagrangian with respect to the state variables $\boldsymbol{v}$, π, $\boldsymbol{\lambda}$ and set each derivative, evaluated at the optimal state and adjoint variables, equal to zero. We put $\hat{\mathcal{L}}'_{(\cdot)} = \mathcal{L}'_{(\cdot)}((\hat{\boldsymbol{v}}, \hat{\pi}, \hat{\boldsymbol{\lambda}}), \hat{\boldsymbol{u}}, (\hat{\boldsymbol{z}}, \hat{q}, \hat{\boldsymbol{\sigma}}))$.

- $\hat{\mathcal{L}}'_{\boldsymbol{v}}\boldsymbol{w} = 0$ for all $\boldsymbol{w} \in L^2(0,T;X) \cap H^1(0,T;H)$, $\boldsymbol{w}(0) = \mathbf{0}$, gives

$$\hat{\mathcal{L}}'_{\boldsymbol{v}}\boldsymbol{w} = \int_0^T (\hat{\boldsymbol{v}}(t) - \boldsymbol{z}_d, \boldsymbol{w}(t))_H dt - \int_0^T \{(\boldsymbol{w}_t(t), \hat{\boldsymbol{z}}(t))_H + e(\boldsymbol{w}(t), \hat{\boldsymbol{z}}(t)\} dt +$$
$$+ \int_0^T \left\{ (\hat{q}(t), \mathrm{div}\boldsymbol{w}(t))_{L^2(\Omega)} - \langle \hat{\boldsymbol{\sigma}}(t), \boldsymbol{w}(t) \rangle_{\Gamma_D} \right\} dt = 0. \tag{7.76}$$

Integrating by parts we get

$$\int_0^T (\boldsymbol{w}_t(t), \hat{\boldsymbol{z}}(t))_H \, dt = - \int_0^T (\boldsymbol{w}(t), \hat{\boldsymbol{z}}_t(t))_H \, dt + (\boldsymbol{w}(T), \hat{\boldsymbol{z}}(T))_H .$$

Thus, if we choose $\boldsymbol{w}(t) = a(t)\boldsymbol{\varphi}$ with $a \in C^1([0,T])$, $a(0) = 0$, $\boldsymbol{\varphi} \in X$, as a test function in (7.76), we obtain the following equation, for a.e. $t \in (0,T)$:

$$-(\hat{\boldsymbol{z}}_t(t), \boldsymbol{\varphi})_H + \nu(\nabla\hat{\boldsymbol{z}}(t), \nabla\boldsymbol{\varphi})_{L^2} - (\hat{q}(t), \mathrm{div}\boldsymbol{\varphi})_{L^2(\Omega)} - \langle \hat{\boldsymbol{\sigma}}(t), \boldsymbol{\varphi} \rangle_{\Gamma_D} = (\hat{\boldsymbol{v}}(t) - \boldsymbol{z}_d, \boldsymbol{\varphi})_H \tag{7.77}$$

and the final condition

$$\hat{\boldsymbol{z}}(T) = 0. \tag{7.78}$$

- $\hat{\mathcal{L}}'_{\pi} r = 0$ for all $r \in L^2(Q_T)$ yields

$$\hat{\mathcal{L}}'_{\pi} r = \int_0^T (r(t), \mathrm{div}\hat{\boldsymbol{z}}(t))_{L^2(\Omega)} = 0$$

which implies

$$\mathrm{div}\hat{\boldsymbol{z}} = 0 \quad \text{in } Q_T. \tag{7.79}$$

- $\hat{\mathcal{L}}'_{\boldsymbol{\lambda}}\boldsymbol{\mu} = 0$ for all $\boldsymbol{\mu} \in L^2(0,T; H_{00}^{-1/2}(\Gamma_D)^d)$ gives

$$\langle \boldsymbol{\mu}(t), \hat{\boldsymbol{z}}(t) \rangle_{\Gamma_D} = 0$$

which implies

$$\hat{\boldsymbol{z}} = \mathbf{0} \text{ on } S_D. \tag{7.80}$$

We recognize that system (7.77), (7.78), (7.79) and (7.80) represents the saddle point formulation of the following adjoint problem:

$$\begin{cases} \hat{\boldsymbol{z}}_t + \nu\Delta\hat{\boldsymbol{z}} - \nabla q = -\hat{\boldsymbol{v}}(t) + \boldsymbol{z}_d & \text{in } Q_T \\ \mathrm{div}\hat{\boldsymbol{z}} = 0 & \text{in } Q_T \\ \hat{\boldsymbol{z}} = \mathbf{0} & \text{on } S_D \\ -q\mathbf{n} + \nu\dfrac{\partial\hat{\boldsymbol{z}}}{\partial n} = \mathbf{0} & \text{on } S_N \\ \hat{\boldsymbol{z}}(\mathbf{x}, T) = \mathbf{0} & \text{in } \Omega. \end{cases} \tag{7.81}$$

This latter is a backward-in-time Stokes problem with final condition (that is, at the final time T), homogeneous Dirichlet/Neumann conditions on the boundary of Ω for all times, and source term (for the momentum equation) equal to $\boldsymbol{z}_d - \hat{\boldsymbol{v}}(t)$. After a time reversing transformation $t \to T - t$, Theorem 7.4 ensures existence and uniqueness of a saddle point solution $\hat{\mathbf{z}} \in L^\infty(0,T; H^1_{\Gamma_D}(\Omega)^d) \cap H^1(0,T;H), \hat{q} \in L^2(\Omega)$, with $\hat{\boldsymbol{\sigma}} = -\hat{q}\mathbf{n} + \nu\frac{\partial \hat{\mathbf{z}}}{\partial n}$ on Γ_D.

- Finally, $\hat{\mathcal{L}}'_{\mathbf{u}}\boldsymbol{\eta} = 0$ for all $\boldsymbol{\eta} \in H$ yields the following Euler equation

$$\hat{\mathcal{L}}'_{\mathbf{u}}\boldsymbol{\eta} = \int_0^T (\hat{\mathbf{z}}(t) + \tau\hat{\mathbf{u}}(t), \boldsymbol{\eta})_H = 0 \tag{7.82}$$

from which we obtain the expression of the gradient of the cost functional

$$\nabla J(\hat{\mathbf{u}}) = \hat{\mathbf{z}} + \tau\hat{\mathbf{u}} \qquad \text{in } L^2(Q_T).$$

Finally, by requiring that the derivatives of the Lagrangian with respect to the adjoint variables vanish, we obtain the state system (7.70).

Other examples of optimal control problems for the unsteady Stokes equations dealing with the case of Neumann control and initial control, can be found in Exercises 5 and 6, respectively.

7.7 Optimal Control of the Wave Equation

In this section we adapt the foregoing theory to an example of linear-quadratic OCP governed by the wave equation. Consider the minimization problem

$$\tilde{J}(y,u) = \frac{1}{2}\int_\Omega (y(T) - z_d)^2\, d\mathbf{x} + \frac{\beta}{2}\int_{Q_T} u^2 d\mathbf{x}dt \to \min \tag{7.83}$$

with $\beta \geq 0$, where $z_d \in L^2(Q_T)$ is a given target, u belongs to a closed, convex subset $\mathcal{U}_{ad}$ of $L^2(Q_T)$, and $y = y(\mathbf{x},t;u)$ solves the *state problem*:

$$\begin{cases} y_{tt} - \Delta y = f + u & \text{in } Q_T \\ y = 0 & \text{on } S_T \\ y(\mathbf{x},0) = g(\mathbf{x}),\ y_t(\mathbf{x},0) = h(\mathbf{x}) & \text{in } \Omega. \end{cases} \tag{7.84}$$

- *Analysis of the state problem.* For the well-posedness analysis of the state problem we introduce a weak formulation. We multiply the wave equation by a smooth function $v = v(\mathbf{x},t)$ vanishing on Γ, and integrate over Q_T. After an integration by parts in space, we find

$$\int_0^T (y_{tt}(t), v(t))_H\, dt + \int_0^T (\nabla y(t), \nabla v(t))_H\, dt = \int_0^T (f(t), v(t))_H\, dt, \tag{7.85}$$

where $H = L^2(\Omega)$. Again, the natural space for y is $L^2(0,T;H_0^1(\Omega))$. Thus, for a.e. $t > 0$, $y(t) \in V = H_0^1(\Omega)$, and $\Delta u(t) \in V^* = H^{-1}(\Omega)$. On the other hand, since

$$y_{tt} = \Delta y + f + u,$$

it is natural to require $y_{tt} \in L^2(0,T;V^*)$ and consequently the first integrand in (7.85) has to be interpreted as a duality, $\langle y_{tt}(t), v(t)\rangle_*$.

Accordingly, a reasonable assumption is $y_t \in L^2(0,T;H)$, an intermediate space between $L^2(0,T;V)$ and $L^2(0,T;V^*)$. Thus, we look for solutions y such that

$$y \in H^1(0,T;V,H),\ \ y_{tt} \in L^2(0,T;V^*). \tag{7.86}$$

For simplicity[7], we assume $f \in L^2(0,T;H)$, $g \in V$, and $h \in H$. Since y satisfies (7.86), it follows that

$$y \in C([0,T];H) \quad \text{and} \quad y_t \in C([0,T];V^*)$$

and therefore the initial conditions actually mean

$$\|y(t) - g\|_H \to 0, \ \|y_t(t) - h\|_{V^*} \to 0 \ \text{ as } t \to 0. \tag{7.87}$$

The above considerations lead to the following definition, analogous to Definition 7.1 for the heat equation:

Definition 7.4. Let $f \in L^2(0,T;H)$, $g \in V$, $h \in H$; $y \in H^1(0,T;V,H)$ is called a weak solution to problem (7.84) if $y_{tt} \in L^2(0,T;V^*)$ and:

1. for all $v \in L^2(0,T;V)$,
$$\int_0^T \langle y_{tt}(t), v(t)\rangle_* \, dt + \int_0^T (\nabla y(t), \nabla v(t))_{L^2(\Omega)^d} \, dt = \int_0^T (f(t) + u(t), v(t))_H \, dt;$$
2. $y(0) = g$, $y_t(0) = h$ (in the sense of (7.87)).

As in the case of the heat equation, condition 1 of Definition 7.4 can can be stated in the following equivalent form:

1′. for all $\varphi \in V$ and a.e. $t \in (0,T)$,

$$\langle y_{tt}(t), \varphi\rangle_* + (\nabla y(t), \nabla \varphi)_{L^2(\Omega)^d} = (f(t) + u(t), \varphi)_H \, . \tag{7.88}$$

Remark 7.8. If f, g, h are smooth functions and y is a smooth weak solution of problem (7.84) then it is actually a classical solution. •

The following theorem holds[8]:

Theorem 7.5. *Let $f \in L^2(0,T;H)$, $g \in V$, $h \in H$. There is a unique weak solution of problem (7.84). Moreover $y \in C([0,T];V)$, $y_t \in C([0,T];H)$ and*

$$\|y\|_{C([0,T];H)}, \ \|y_t\|_{C([0,T];H)}, \ \|y_{tt}\|_{L^2(0,T;V^*)} \le c\left\{\|f\|_{L^2(0,T;H)} + \|\nabla g\|_{L^2(\Omega)^d} + \|h\|_H\right\}$$

with $c = c(T, \Omega)$.

• *Existence and uniqueness of the optimal control.* The foregoing theory applies without problem and we deduce that if $\beta > 0$ there exists a unique optimal control pair $\hat{y}, \hat{u}$; when $\beta = 0$ and $\mathcal{U}_{ad}$ is bounded, we consider the operator $u \to y_0(u)$. If $y_0(u) = 0$ then, clearly $y_0(u) = 0$ in Q_T and therefore also $u = 0$. Thus also in this case there exists a unique optimal control pair.

• *Optimality conditions.* We introduce the Lagrangian

$$\mathcal{L}(y,u,p) = \frac{1}{2}\int_\Omega (y(T) - z_d)^2 \, d\mathbf{x} + \frac{\beta}{2}\int_{Q_T} u^2 d\mathbf{x}dt + \int_0^T \left\{(f+u)p - \langle y_{tt}, p\rangle_* - (\nabla y, \nabla p)_{L^2(\Omega)^d}\right\} dt$$

[7] We could also assume $f \in L^2(0,T;V^*)$, however in this case we would need to rely on the transposition method and look for *very weak solutions*, as done in Sect. 5.9.

[8] See e.g. Thm. 10.6 in [237, Sect.10.6] and [176, Vol.1, Chap. 3] for its proof and further details.

with $p \in H^1(0,T;V,H)$, $p_{tt} \in L^2(0,T;V^*)$. The adjoint problem is obtained by requiring that $\mathcal{L}'_y(\hat{y},\hat{u},\hat{p})\psi = 0$, for all $\psi \in H^1(0,T;V,H)$, $\psi(0) = \psi_t(0) = 0$. We find

$$\mathcal{L}'_y(\hat{y},\hat{u},\hat{p})\psi = (\hat{y}(T) - z_d, \psi(T))_H - \int_0^T \left\{ \langle \psi_{tt}, \hat{p} \rangle_* + (\nabla\psi, \nabla\hat{p})_{L^2(\Omega)^d} \right\} dt = 0. \tag{7.89}$$

Integrating by parts we get, since $\psi(0) = \psi_t(0) = 0$,

$$\begin{aligned}\int_0^T \langle \psi_{tt}, \hat{p} \rangle_* \, dt &= -\int_0^T (\hat{p}_t, \psi_t)_H \, dt + (\hat{p}(T), \psi_t(T))_H \\ &= \int_0^T (\hat{p}_{tt}, \psi)_* \, dt - (\hat{p}_t(T), \psi(T))_H + (\hat{p}(T), \psi_t(T))_H .\end{aligned}$$

Substituting into (7.89), we have

$$\int_0^T \{\langle (\hat{p}_{tt}, \psi)_H \rangle_* + (\nabla\hat{p}, \nabla\psi)_H\} \, dt + (\hat{y}(T) - z_d + \hat{p}_t(T), \psi(T))_H - (\psi_t(T), \hat{p}(T))_H = 0$$

for every $\psi \in H^1(0,T;V,H)$, $\psi(0) = \psi_t(0) = 0$. Choosing $\psi = a(t)\phi$ with $\phi \in V$ and $a \in C^1([0,T])$, $a(0) = a'(0) = 0$, we conclude that the adjoint problem is

$$\begin{cases} \langle \hat{p}_{tt}(t), \phi \rangle_* + (\nabla\hat{p}(t), \nabla\phi)_H = 0 & \forall\phi \in V \text{ and a.e. in } (0,T) \\ \hat{p}(T) = 0, \; \hat{p}_t(T) = z_d - \hat{y}(T) & \text{in } \Omega \end{cases} \tag{7.90}$$

in weak form, and

$$\begin{cases} \hat{p}_{tt} - \Delta\hat{p} = 0 & \text{in } Q_T \\ \hat{p} = 0 & \text{on } S_T \\ \hat{p}(T) = 0, \; \hat{p}_t(T) = z_d - \hat{y}(T) & \text{in } \Omega \end{cases} \tag{7.91}$$

in strong form. Since $z_d - \hat{y}(T) \in H$, problem (7.90) has a unique weak solution $\hat{p} \in H^1(0,T;V,H)$, $\hat{p}_{tt} \in L^2(0,T;V^*)$. Finally, the optimality condition is given by the variational inequality

$$\mathcal{L}'_u(\hat{y},\hat{u},\hat{p})(v - \hat{u}) = (\beta\hat{u} + \hat{p}, v - \hat{u})_{L^2(Q_T)} \geq 0 \qquad \forall v \in \mathcal{U}_{ad} \tag{7.92}$$

from which we deduce that the gradient of the cost functional is

$$\nabla J(\hat{u}) = \beta\hat{u} + \hat{p}.$$

Theorem 7.6. *The control $\hat{u}$ and the state function $\hat{y} = y(\hat{u})$ are optimal for the OCP (7.52)–(7.53) if and only if, together with the adjoint state $\hat{p}$, they solve the state system (7.88), the adjoint problem (7.90) and the variational inequality (7.92).*

Cost functionals involving the value of y_t at the final time are more delicate and involve adjoint problems with initial states in V^*.

7.8 Exercises

1. Show that if y, g, f are smooth functions and y is a weak solution to problem (7.2) in the sense of Definition 7.1, then y solves problem (7.2) in a classical sense, too, that is,

$$\begin{cases} y_t - \Delta y + y = f + u & \text{in } Q_T \\ \dfrac{\partial y}{\partial n} = 0 & \text{on } S_T \\ y(\mathbf{x},0) = g(\mathbf{x}) & \text{in } \Omega. \end{cases}$$

2. Derive the system of optimality conditions for the initial control problem (7.20)–(7.21), that is,

$$\tilde{J}(y,u) = \frac{1}{2}\int_\Omega (y(T) - z_d)^2 \, d\mathbf{x} + \frac{\beta}{2}\int_\Omega u^2 d\mathbf{x} \to \min$$

with $\beta > 0$, where $u \in \mathcal{U}_{ad} \subset L^2(\Omega)$ and $y = y(t;u)$ solves

$$\begin{cases} y_t - \Delta y + y = f & \text{in } Q_T \\ \dfrac{\partial y}{\partial n} = 0 & \text{on } S_T \\ y(\mathbf{x},0) = u(\mathbf{x}) & \text{in } \Omega. \end{cases}$$

[*Hint.* Introduce a suitable Lagrangian functional $\mathcal{L}$ and require that the derivatives of $\mathcal{L}$ with respect to p, y and u vanish on any admissible variation. In particular, show that the adjoint problem in this case is given by (7.12) while

$$\int_\Omega (\beta \hat{u}(\mathbf{x}) + \hat{p}(\mathbf{x},0))(w(\mathbf{x}) - \hat{u}(\mathbf{x})) \, d\mathbf{x} \geq 0 \qquad \forall w \in \mathcal{U}_{ad}. \tag{7.93}$$

gives the minimum principle.]

3. *Transposition method for the heat equation.* Consider the problem

$$\begin{cases} y_t - \Delta y = F & \text{in } Q_T \\ y = g & \text{on } S_T \\ y(\mathbf{x},0) = 0 & \text{in } \Omega \end{cases} \tag{7.94}$$

with $F \in L^2(0,T;V^*)$, $V = H^2(\Omega) \cap H_0^1(\Omega)$, $g \in L^2(S_T)$; here Ω is a smooth domain. A unique (very weak) solution can be constructed via the transposition method as follows (see Sect. 5.9). We ask the reader to fill in the details in the steps below (also recalling Remark 7.2).

   a. After a time transformation $t \mapsto T - t$, for every $f \in L^2(Q_T)$ there exists a unique weak solution $\psi = S(f) \in H^1(0,T;V,H)$, $H = L^2(\Omega)$, to the backward-in-time heat equation

$$-\psi_t - \Delta\psi = f,$$

   with $\psi(T) = 0$. Let X be the image of $S(f)$, that is

$$X = \left\{\psi : \psi = S(f),\ f \in L^2(Q_T)\right\}.$$

   Show that X is a Hilbert space when endowed with the inner product $(\psi_1, \psi_2)_X = (f_1, f_2)_{L^2(Q_T)}$ and that the backward-in-time heat operator

$$\psi \to \mathcal{H}^*(\psi) = -\psi_t - \Delta\psi$$

is a continuous isomorphism between X and $L^2(Q_T)$.

b. Multiply by $\psi \in X$ the heat equation in problem (7.94), integrate over Q_T, integrate by parts in space and time and use the boundary and initial conditions to find

$$\int_0^T (y, -\psi_t - \Delta\psi)_H \, dt = \int_0^T \langle F, \psi\rangle_{V^*,V} \, dt - \int_0^T \left(g, \frac{\partial\psi}{\partial n}\right)_{L^2(S_T)} dt. \tag{7.95}$$

c. Define $y \in L^2(Q_T)$ as a very weak solution to (7.94) if (7.95) holds for every $\psi \in X$. Show that there exists a unique very weak solution y to (7.94) and that

$$\|y\|_{L^2(Q_T)} \leq C(T, \Omega)\left\{\|g\|_{L^2(S_T)} + \|f\|_{L^2(0,T;V^*)}\right\}.$$

[*Hint.* The operator $L(\psi) = \int_0^T \langle F, \psi\rangle_{V^*,V} \, dt - \int_0^T \left(g, \frac{\partial\psi}{\partial n}\right)_{L^2(S_T)} dt$ belongs to X^*. Equation (7.95) reads

$$(y, \mathcal{H}^*(\psi))_{L^2(Q_T)} = L(\psi) \qquad \forall\psi \in X.$$

Apply now Proposition 5.11 to conclude.]

4. Denote by $\boldsymbol{v}_0$ the velocity component of the saddle point solution to the Stokes problem (7.72) corresponding to $\boldsymbol{f} = \boldsymbol{g} = \boldsymbol{h} = \boldsymbol{0}$. Show that the bilinear form

$$q(\boldsymbol{u}, \boldsymbol{\eta}) = (\boldsymbol{v}_0(\boldsymbol{u}), \boldsymbol{v}_0(\boldsymbol{\eta}))_{L^2(Q_T)^d} + \tau(\boldsymbol{u}, \boldsymbol{\eta})_{L^2(Q_T)^d}$$

is continuous and nonnegative in $L^2(Q_T)^d$ (as $\tau > 0$) and that the linear form

$$L\boldsymbol{u} = (\boldsymbol{v}_0(\boldsymbol{u}), \boldsymbol{v}(\boldsymbol{0}) - \boldsymbol{z}_d)_{L^2(Q_T)^d}$$

is continuous in $L^2(Q_T)^d$.

5. Analyze the following control problem for the time-dependent Stokes equations ($S_T = \Gamma \times (0, T)$):

$$J(\mathbf{c}) = \frac{1}{2}\int_{Q_T} |\boldsymbol{v}(\mathbf{c}) - \boldsymbol{z}_d|^2 \, d\mathbf{x}dt + \frac{\tau}{2}\int_{S_T} |\mathbf{c}|^2 \, d\sigma dt \to \min$$

where $(\boldsymbol{v}, \pi) = (\boldsymbol{v}(\mathbf{c}), \pi(\mathbf{c}))$ is the solution of the following unsteady Stokes system:

$$\begin{cases} \boldsymbol{v}_t - \nu\Delta\boldsymbol{v} + \nabla\pi = \boldsymbol{f} & \text{in } Q_T \\ \text{div}\boldsymbol{v} = 0 & \text{in } Q_T \\ -\pi\mathbf{n} + \nu\dfrac{\partial\boldsymbol{v}}{\partial n} = \mathbf{c} & \text{on } S_T \\ \boldsymbol{v}(\mathbf{x}, 0) = \boldsymbol{0} & \text{in } \Omega. \end{cases}$$

6. Consider the Stokes system

$$\begin{cases} \boldsymbol{v}_t - \nu \Delta \boldsymbol{v} + \nabla \pi = \boldsymbol{f} & \text{in } Q_T \\ \text{div}\boldsymbol{v} = 0 & \text{in } Q_T \\ \boldsymbol{v} = \mathbf{0} & \text{on } \Gamma_D \times (0,T) \\ -\pi \mathbf{n} + \nu \dfrac{\partial \boldsymbol{v}}{\partial n} = \mathbf{0} & \text{on } \Gamma_N \times (0,T) \\ \boldsymbol{v}(\mathbf{x},0) = \boldsymbol{\xi} & \text{in } \Omega \end{cases}$$

where the initial datum $\boldsymbol{\xi} \in X$ (see (7.67)) plays the role of control function. Write the optimality conditions for the minimization of the cost functional

$$J(\boldsymbol{\xi}) = \frac{1}{2}\int_{\Omega} |\boldsymbol{v}(T;\boldsymbol{\xi}) - \boldsymbol{z}_d|^2 \, d\mathbf{x} + \frac{\tau}{2}\int_{\Omega} \left(|\nabla \boldsymbol{\xi}|^2 + |\boldsymbol{\xi}|^2\right) d\mathbf{x}.$$

7. Let $f \in L^2(Q_T)$. Analyze the problem

$$\tilde{J}(y,u) = \frac{1}{2}\int_{Q_T} (y - z_d)^2 \, d\mathbf{x}dt + \frac{\beta}{2}\int_{S_T} u^2 d\sigma dt \to \min \tag{7.96}$$

with $\beta > 0$, where $z_d \in L^2(Q_T)$ is a given target, u belongs to a closed, convex subset $\mathcal{U}_{ad}$ of $L^2(S_T)$, and $y = y(u)$ solves the state problem:

$$\begin{cases} y_{tt} - \Delta y = f & \text{in } Q_T \\ \dfrac{\partial y}{\partial n} = u & \text{on } S_T \\ y(\mathbf{x},0) = 0, \; y_t(\mathbf{x},0) = 0 & \text{in } \Omega. \end{cases}$$

Chapter 8
Numerical Approximation of Quadratic OCPs Governed by Linear Evolution PDEs

In this chapter we address the numerical solution of time-dependent linear-quadratic OCPs, that is OCPs involving a quadratic cost functional and a state system described by a linear initial-boundary value problem (IBVP). We will closely follow the same road map of Chapter 6. In Sect. 8.1 we revisit the *optimize-then-discretize* and the *discretize-then-optimize* approach; we will show how time discretization might affect the numerical approximation of the system of optimality conditions. Solution approaches based on *iterative methods* will be presented in Sect. 8.2, showing some numerical results in Sect. 8.3, whereas those using *all-at-once (or one-shot) methods* in Sect. 8.4. .

8.1 Optimize & Discretize, or Discretize & Optimize, Revisited

In this chapter we focus on the numerical approximation of time-dependent linear-quadratic OCPs, confining to the case where the state system is given by a IBVP for a parabolic PDE; similar considerations can also be drawn when dealing with other linear evolution PDEs, such as time-dependent Stokes equations. The numerical approximation of OCPs for the wave (and, more generally speaking, linear hyperbolic) equations will not be addressed; the interested reader is referred, e.g., to [277, 92, 93].

A similar classification to the one introduced in Sect. 6.1 also holds in the case of time-dependent linear-quadratic OCPs. Indeed, also in this case, depending on the way the PDE constraint is treated, we can use:

- *iterative methods*, where the PDE constraint is eliminated and the only optimization variable is the control u;
- *all-at-once methods*, where the PDE constraint is explicitly kept in the problem formulation and both y and u are treated as optimization variables.

We highlight, however, that algebraic systems arising from the use of all-at-once methods become, in the case of time-dependent problems, more involved to treat, because of their (potentially very large) size: indeed, these methods compute the solution to OCPs in a single iteration by treating all time-steps at the same time.

As in the case of linear-quadratic OCPs governed by elliptic PDEs, the interplay between optimization and discretization yields slightly different numerical problems that must be solved, yet sharing a very similar structure. The need of solving two coupled dynamical systems resulting from the state and the adjoint problem, represent the main difference if compared to the case of OCPs for elliptic PDEs, and introduce a further level of difficulty.

A. Manzoni et al., *Optimal Control of Partial Differential Equations*,
Applied Mathematical Sciences 207, https://doi.org/10.1007/978-3-030-77226-0_8

For the sake of effort, we will just emphasize the distinguishing features related with the numerical solution of time-dependent linear-quadratic OCPs, exploiting as much what we already introduced in Chapter 6. Our OCP reads as follows:

$$\tilde{J}(y,u) = \frac{1}{2}\int_{Q_T} |y - y_d|^2\, d\mathbf{x}dt + \frac{\beta}{2}\int_{Q_T} u^2 d\mathbf{x}dt \to \min \tag{8.1}$$

with $\beta > 0$, where $y_d \in L^2(Q_T)$ is a given target, $u \in L^2(Q_T)$ and $y = y(\mathbf{x},t;u)$ solves the *state problem:*

$$\begin{cases} y_t - \Delta y = f + u & \text{in } Q_T \\ y = 0 & \text{on } S_T \\ y(\mathbf{x},0) = g(\mathbf{x}) & \text{in } \Omega. \end{cases} \tag{8.2}$$

According to (7.5), the weak formulation of the state problem (8.2) reads: for a.e. $t \in (0,T)$, find $y(t) \in V$ such that

$$\begin{cases} (\dot{y}(t),\varphi)_{L^2(\Omega)} + a(y(t),\varphi;t) = (f(t),\varphi)_{L^2(\Omega)} & \forall \varphi \in V \\ y(0) = g; \end{cases} \tag{8.3}$$

in this case $V = H_0^1(\Omega)$, and

$$a(y,\varphi) = \int_\Omega \nabla y \cdot \nabla \varphi d\mathbf{x}$$

is the bilinear form associated with the Laplace operator. The extension to more general elliptic operators is straightforward. For ease of notation, we might denote the time-derivative of $y(t)$ by $\dot{y}(t)$ instead of $\partial_t y(t)$.

Remark 8.1. Often, to simplify their definition, control functions $u = u(\mathbf{x},t)$ are expressed as the product $u(\mathbf{x},t) = \eta(\mathbf{x})\tilde{u}(t)$ of two functions: a time-dependent function $\tilde{u}(t)$, and a time-independent function $\eta(\mathbf{x})$ supported over those portions of the domain where the control acts. If the action of the control in space is determined a priori, only its dynamics in time expressed by $\tilde{u}(t)$ has to be adjusted to minimize the cost functional, thus simplifying the OCP substantially. •

8.1.1 Optimize Then Discretize

If we use the *optimize-then-discretize* (OtD) strategy, we carry out optimization first (at the continuous level) and then discretize the resulting optimality system. Hence, we first recover a system of optimality conditions at the continuous level; for the case at hand, Proposition 7.2 yields the following system of optimality conditions. For a.e. $t \in (0,T)$, find $y(t) \in V$, $u(t) \in \mathcal{U}$ and $p(t) \in V$ such that:

$$\begin{cases} (\dot{y}(t),\varphi)_{L^2(\Omega)} + a(y(t),\varphi;t) = (f(t) + u(t),\varphi)_{L^2(\Omega)} & \forall \varphi \in V \\ y(0) = g \\ -(\dot{p}(t),\psi)_{L^2(\Omega)} + a^*(p(t),\psi;t) = -(y(t) - y_d(t),\psi)_{L^2(\Omega)} & \forall \psi \in V \\ p(T) = 0 \\ (\beta u(t) - p(t),v)_{L^2(\Omega)} = 0 & \forall v \in \mathcal{U}. \end{cases} \tag{8.4}$$

Then, we approximate the spaces V by V_h and $\mathcal{U}$ by $\mathcal{U}_h$ (two suitable finite-dimensional spaces) and replace (8.4) with the following problem: for a.e. $t \in (0,T)$, find $y_h(t) \in V_h$, $u_h(t) \in \mathcal{U}_h$ and $p_h(t) \in V_h$ such that:

$$
\begin{cases}
(\dot{y}_h(t), \varphi)_{L^2(\Omega)} + a(y_h(t), \varphi; t) = (f(t), \varphi_h)_{L^2(\Omega)} + (u_h(t), \varphi_h)_{L^2(\Omega)} & \forall \varphi_h \in V_h \\
y_h(0) = g_h \\
-(\dot{p}_h(t), \psi_h)_{L^2(\Omega)} + a^*(p_h(t), \psi_h; t) = -(y_h(t) - y_d(t), \psi_h)_{L^2(\Omega)} & \forall \psi_h \in V_h \\
p_h(T) = 0 \\
(\beta u_h(t) - p_h(t), v_h)_{L^2(\Omega)} = 0 & \forall v_h \in \mathcal{U}_h.
\end{cases}
\tag{8.5}
$$

Here g_h denotes a suitable approximation of g in V_h. Note that the same space V_h is used to approximate both state and control variables; moreover, V_h is used to approximate the trial space (where we seek the solution) as well as the test space, thus yielding a Galerkin spatial approximation.

The spatial discretization of the optimality conditions (8.4) thus yields two systems of ODEs, coupled through the right-hand side of the adjoint problem, and a further condition involving both the adjoint state and the control, to be fulfilled a.e. $t \in (0, T)$.

Introducing a basis for V_h, $\{\varphi_j,\ j = 1, \dots, N_y\}$, and a basis for $\mathcal{U}_h$, $\{\psi_j,\ j = 1, \dots, N_u\}$, respectively, and proceeding as in Sect. 6.2.1, it is possible to rewrite system (8.5) expressing the ODE systems in matrix form (see Sect. B.3 in the Appendix B, and Exercise 1 for a detailed derivation):

$$
\begin{cases}
\mathbb{M}\dot{\mathbf{y}}(t) + \mathbb{A}\mathbf{y}(t) = \mathbf{f}(t) + \mathbb{B}\mathbf{u}(t) & \forall t \in (0, T) \\
\mathbf{y}(0) = \mathbf{g} \\
-\mathbb{M}\dot{\mathbf{p}}(t) + \mathbb{A}^\top \mathbf{p}(t) = -\mathbb{M}(\mathbf{y}(t) - \mathbf{y}_d(t)) & \forall t \in (0, T) \\
\mathbf{p}(T) = \mathbf{0} \\
\beta\mathbb{N}\mathbf{u}(t) - \mathbb{B}^\top \mathbf{p}(t) = \mathbf{0},
\end{cases}
\tag{8.6}
$$

where:

- $\mathbf{y}(t) = (y_{h,1}(t), \dots, y_{h,N_y}(t))^\top \in \mathbb{R}^{N_y}$, $\mathbf{u}(t) = (u_{h,1}(t), \dots, u_{h,N_u}(t))^\top \in \mathbb{R}^{N_u}$, $\mathbf{p}(t) = (p_{h,1}(t), \dots, p_{h,N_y}(t))^\top \in \mathbb{R}^{N_y}$ are the vectors corresponding to the optimal state, control, and adjoint variables, at time t;
- $\mathbb{A} \in \mathbb{R}^{N_y \times N_y}$ is the *stiffness* matrix defined in (6.8);
- $\mathbb{B} \in \mathbb{R}^{N_y \times N_u}$ is the control matrix defined in (6.9);
- the vector $\mathbf{f}(t) \in \mathbb{R}^{N_y}$ encodes the source term of the state problem, at time t;
- $\mathbb{M} \in \mathbb{R}^{N_y \times N_y}$ is the FE (state) mass matrix defined in (6.12);
- $\mathbb{N} \in \mathbb{R}^{N_u \times N_u}$ is the FE (control) mass matrix defined in (6.14);
- $\mathbf{y}_d(t) \in \mathbb{R}^{N_y}$ encodes the target function, at time t.

Let us now introduce a uniform partition of the time interval $(0, T)$ into subintervals $[t^n, t^{n+1}]$, $n = 0, \dots, N_t - 1$, with $t^0 = 0$ and $t^n = t^0 + n\Delta t$, of length Δt – hence, $t^{N_t} = T$. Then, performing a time discretization of the ODE systems appearing in (8.6) using a backward Euler scheme (see Sect. B.3 in the Appendix B), we end up with the following sequence of algebraic systems to be solved at each time step (the superscript n indicates that the considered quantity is evaluated at time t^n).

Given $\mathbf{g}$, find $\mathbf{y}^{n+1}$ such that, for $n = 0, 1, \dots, N_t - 1$,

$$
\begin{cases}
\mathbb{M}\dfrac{\mathbf{y}^{n+1} - \mathbf{y}^n}{\Delta t} + \mathbb{A}\mathbf{y}^{n+1} = \mathbf{f}^{n+1} + \mathbb{B}\mathbf{u}^{n+1}, & n = 0, 1, \dots, N_t - 1 \\
\mathbf{y}^0 = \mathbf{g}.
\end{cases}
\tag{8.7}
$$

Putting all of equation (8.7) together, we obtain the *one-shot* discretization for N_t time-steps under the following form:

$$\underbrace{\begin{bmatrix} \frac{\mathbb{M}}{\Delta t}+\mathbb{A} & & & \\ -\frac{\mathbb{M}}{\Delta t} & \frac{\mathbb{M}}{\Delta t}+\mathbb{A} & & \\ & \ddots & \ddots & \\ & & -\frac{\mathbb{M}}{\Delta t} & \frac{\mathbb{M}}{\Delta t}+\mathbb{A} \end{bmatrix}}_{\mathcal{K}} \begin{bmatrix} \mathbf{y}^1 \\ \mathbf{y}^2 \\ \vdots \\ \mathbf{y}^{N_t} \end{bmatrix} - \underbrace{\begin{bmatrix} \mathbb{B} & & & \\ & \mathbb{B} & & \\ & & \ddots & \\ & & & \mathbb{B} \end{bmatrix}}_{\mathcal{B}} \begin{bmatrix} \mathbf{u}^1 \\ \mathbf{u}^2 \\ \vdots \\ \mathbf{u}^{N_t} \end{bmatrix} = \begin{bmatrix} \mathbf{f}^1 + \frac{\mathbb{M}}{\Delta t}\mathbf{g} \\ \mathbf{f}^2 \\ \vdots \\ \mathbf{f}^{N_t} \end{bmatrix}. \tag{8.8}$$

Remark 8.2. Note that in the case control functions are expressed as the product $u(\mathbf{x},t) = \eta(\mathbf{x})\tilde{u}(t)$ of a time-dependent function $\tilde{u}(t)$, and a time-independent function $\eta(\mathbf{x})$ supported over those portions of the domain where the control acts, space discretization only acts on $\eta(\mathbf{x})$. In this case, denoting by $\eta_h \in V_h$ the approximation of η in V_h, and by $\boldsymbol{\eta} = (\eta_{h,1}, \dots, \eta_{h,N_y})^\top$ the vector corresponding to η_h, the approximated control takes the form $\mathbf{u}(t) = \boldsymbol{\eta}\tilde{u}(t)$, so that only its dynamics in time – expressed by $\tilde{u}(t)$ – has to be adjusted to minimize the cost functional. If the control is given by a scalar time-dependent function $\tilde{u}(t)$, we have that $N_u = 1$, since the product $\mathbb{B}\boldsymbol{\eta}$ is equal at each time step, and can be computed once for all. •

Proceeding in a similar way on the adjoint system, we need to solve the following algebraic system at each time step. Find $\mathbf{p}^n$ such that, for $n = N_t - 1, \dots, 1, 0$:

$$\begin{cases} \mathbb{M}\dfrac{\mathbf{p}^n - \mathbf{p}^{n+1}}{\Delta t} + \mathbb{A}^\top \mathbf{p}^n = -\mathbb{M}(\mathbf{y}^n - \mathbf{y}_d^n), & n = N_t - 1, \dots, 1, 0 \\ \mathbf{z}^{N_t} = \mathbf{0}. \end{cases} \tag{8.9}$$

Note that (8.9) is implicit. By considering all the time steps simulteaneously, we obtain the *one-shot* discretization for N_t time-steps under the following form:

$$\underbrace{\begin{bmatrix} \frac{\mathbb{M}}{\Delta t}+\mathbb{A}^\top & -\frac{\mathbb{M}}{\Delta t} & & \\ & \ddots & \ddots & \\ & & \frac{\mathbb{M}}{\Delta t}+\mathbb{A}^\top & -\frac{\mathbb{M}}{\Delta t} \\ & & & \frac{\mathbb{M}}{\Delta t}+\mathbb{A}^\top \end{bmatrix}}_{\mathcal{K}^\top} \begin{bmatrix} \mathbf{p}^0 \\ \mathbf{p}^1 \\ \vdots \\ \mathbf{p}^{N_t-1} \end{bmatrix} + \underbrace{\begin{bmatrix} \mathbb{M} & & & \\ & \mathbb{M} & & \\ & & \ddots & \\ & & & \mathbb{M} \end{bmatrix}}_{\mathcal{M}} \begin{bmatrix} \mathbf{y}^0 \\ \mathbf{y}^1 \\ \vdots \\ \mathbf{y}^{N_t-1} \end{bmatrix} = \begin{bmatrix} \mathbb{M}\mathbf{y}_d^0 \\ \mathbb{M}\mathbf{y}_d^1 \\ \vdots \\ \mathbb{M}\mathbf{y}_d^{N_t-1} \end{bmatrix}. \tag{8.10}$$

Finally, from the optimality condition, we have that

$$\beta\mathbb{N}\mathbf{u}^n - \mathbb{B}^\top\mathbf{p}^n = \mathbf{0}, \qquad n = 1, \dots, N_t,$$

that is,

$$\beta \underbrace{\begin{bmatrix} \mathbb{N} & & & \\ & \mathbb{N} & & \\ & & \ddots & \\ & & & \mathbb{N} \end{bmatrix}}_{\mathcal{N}} \begin{bmatrix} \mathbf{u}^1 \\ \mathbf{u}^2 \\ \vdots \\ \mathbf{u}^{N_t} \end{bmatrix} - \underbrace{\begin{bmatrix} \mathbb{B}^\top & & & \\ & \mathbb{B}^\top & & \\ & & \ddots & \\ & & & \mathbb{B}^\top \end{bmatrix}}_{\mathcal{B}^\top} \begin{bmatrix} \mathbf{p}^1 \\ \mathbf{p}^2 \\ \vdots \\ \mathbf{p}^{N_t} \end{bmatrix} = \begin{bmatrix} \mathbf{0} \\ \mathbf{0} \\ \vdots \\ \mathbf{0} \end{bmatrix}.$$

Using other implicit methods (e.g., the Crank-Nicolson scheme) for time discretization would lead to a similar form (see Exercise 2).

If we look at the matrices appearing in the equations (8.8)–(8.10), we note a clear analogy with the block structure of the KKT system (6.16) obtained in the case of an elliptic OCP; however, due to the different time steps involved in the state and the adjoint systems, things are more delicate here. A straightforward manner to recover the block structure of the KKT system would be to evaluate the state $\mathbf{y}^n$ and the target $\mathbf{y}_d^n$ in the adjoint problem (8.10) at time t^{n+1} instead than at time t^n, and require that

$$\beta \mathbb{N}\mathbf{u}^n - \mathbb{B}^\top \mathbf{p}^{n-1} = \mathbf{0}, \qquad n = 1, \ldots, N_t,$$

so that the unknown vectors appearing in the three systems would indeed be the same. However, this way of proceeding might yield some inconsistency. For this reason, we will come back to the OtD strategy later on, after introducing the alternative DtO approach. Note also that the way the cost functional – involving time integrals – is approximated through numerical quadrature formulas does not impact on the OtD approach, however it is crucial in the DtO approach, as different KKT systems will be obtained.

8.1.2 Discretize Then Optimize

With the *discretize-then-optimize* (DtO) strategy, we first discretize the state system, and then minimize the cost functional subject to the corresponding algebraic system, following the procedure described in Sect. 4.1. Hence, let us go back to the discretized state system (8.7), assume to define $\mathbf{y}^0$ as the projection of the initial condition onto the FE space, and to express the all-at-once form for the state system involving all the time steps $\mathbf{y}^0, \ldots, \mathbf{y}^{N_t}$ as unknowns, including the initial time step. According to [255], it is possible to express the state system as:

$$\underbrace{\begin{bmatrix} \frac{\mathbb{M}}{\Delta t} + \mathbb{A} & & & \\ -\frac{\mathbb{M}}{\Delta t} & \frac{\mathbb{M}}{\Delta t} + \mathbb{A} & & \\ & \ddots & \ddots & \\ & & -\frac{\mathbb{M}}{\Delta t} & \frac{\mathbb{M}}{\Delta t} + \mathbb{A} \end{bmatrix}}_{\mathcal{K}} \begin{bmatrix} \mathbf{y}^0 \\ \mathbf{y}^1 \\ \vdots \\ \mathbf{y}^{N_t} \end{bmatrix}$$
$$- \underbrace{\begin{bmatrix} \mathbb{B} & & & \\ & \mathbb{B} & & \\ & & \ddots & \\ & & & \mathbb{B} \end{bmatrix}}_{\mathcal{B}} \begin{bmatrix} \mathbf{u}^0 \\ \mathbf{u}^1 \\ \vdots \\ \mathbf{u}^{N_t} \end{bmatrix} = \underbrace{\begin{bmatrix} \left(\frac{\mathbb{M}}{\Delta t} + \mathbb{A}\right) \mathbf{g} \\ \mathbf{f}^1 \\ \vdots \\ \mathbf{f}^{N_t} \end{bmatrix}}_{\mathbf{F}}, \tag{8.11}$$

that is,

$$\mathcal{K}\mathbf{Y} - \mathcal{B}\mathbf{U} = \mathbf{F}.$$

The matrix $\mathcal{K}$ defines the block-lower triangular matrix in (8.11) and $\mathcal{B}$ the rectangular matrix acting on the discretized control; here we denote by

$$\mathbf{Y} = \begin{bmatrix} \mathbf{y}^0 \\ \mathbf{y}^1 \\ \vdots \\ \mathbf{y}^{N_t} \end{bmatrix}, \qquad \mathbf{U} = \begin{bmatrix} \mathbf{u}^0 \\ \mathbf{u}^1 \\ \vdots \\ \mathbf{u}^{N_t} \end{bmatrix}, \qquad \mathbf{P} = \begin{bmatrix} \mathbf{p}^0 \\ \mathbf{p}^1 \\ \vdots \\ \mathbf{p}^{N_t} \end{bmatrix}$$

the vectors containing the state, control and adjoint variables, respectively. Equation (8.11) represents the discretization of the state system; the adjoint of (8.11) will represent the time-evolution described by (8.9), however the final condition for the adjoint system will be different when using a DtO approach compared to the OtD strategy.

Then, we discretize the functional (8.1) in space using a standard FE approach in space; we highlight that the way it is discretized in time will impact on the resulting adjoint problem in the DtO approach.

Assuming that the same FE space is used to discretize both the state and the control space, and employing a trapezoidal rule for the time discretization leads to

$$\tilde{J}_h(\mathbf{Y}, \mathbf{U}) = \frac{\Delta t}{2}(\mathbf{Y} - \mathbf{Y}_d)^\top \mathcal{M}_{1/2}(\mathbf{Y} - \mathbf{Y}_d) + \frac{\beta \Delta t}{2} \mathbf{U}^\top \mathcal{M}_{1/2} \mathbf{U} \tag{8.12}$$

where

$$\mathcal{M}_{1/2} = \begin{bmatrix} \frac{1}{2}\mathbb{M} & & & & \\ & \mathbb{M} & & & \\ & & \ddots & & \\ & & & \mathbb{M} & \\ & & & & \frac{1}{2}\mathbb{M} \end{bmatrix}, \qquad \mathbf{Y}_d = \begin{bmatrix} \mathbf{y}_d^0 \\ \mathbf{y}_d^1 \\ \vdots \\ \mathbf{y}_d^{N_t} \end{bmatrix}.$$

The Lagragian for the discrete problem using the trapezoidal rule can now be written as

$$\mathcal{L}_h(\mathbf{Y}, \mathbf{U}, \mathbf{P}) = \tilde{J}_h(\mathbf{Y}, \mathbf{U}) - \mathbf{P}^\top (\mathbf{F} + \mathcal{B}\mathbf{U} - \mathcal{K}\mathbf{Y}) \tag{8.13}$$

and the corresponding KKT system results as follows (see Exercise 3):

$$\begin{bmatrix} \Delta t \mathcal{M}_{1/2} & 0 & \mathcal{K}^\top \\ 0 & \beta \Delta t \mathcal{M}_{1/2} & -\mathcal{B}^\top \\ \mathcal{K} & -\mathcal{B} & 0 \end{bmatrix} \begin{bmatrix} \mathbf{Y} \\ \mathbf{U} \\ \mathbf{P} \end{bmatrix} = \begin{bmatrix} \Delta t \mathcal{M}_{1/2} \mathbf{Y}_d \\ \mathbf{0} \\ \mathbf{F} \end{bmatrix}. \tag{8.14}$$

In the case of a control function acting over the whole domain, $\mathcal{B} = \mathcal{M}$, where

$$\mathcal{M} = \begin{bmatrix} \mathbb{M} & & & & \\ & \mathbb{M} & & & \\ & & \ddots & & \\ & & & \mathbb{M} & \\ & & & & \mathbb{M} \end{bmatrix}.$$

Note that the adjoint equation obtained through the OtD approach will in general differ from the discretization of the adjoint PDE obtained in the previous section, as the mass matrix in the $(1,1)$-block has the factor $1/2$ in the first and last block (this is due to the trapezoidal rule), while this would in general not be obtained from the discretization of the right hand side appearing in (8.9) when using an OtD approach.

Moreover, also the final condition for the adjoint problem (8.9) has to be represented by the first equation in (8.20). Note that the last line of the first equation in (8.20) gives

$$(\mathbb{M} + \Delta t \mathbb{A}^\top)\mathbf{p}^{N_t} = \frac{(\Delta t)^2}{2}\mathbb{M}(\mathbf{y}^{N_t} - \mathbf{y}_d^{N_t})$$

which does not necessarily coincide with the final condition of the adjoint system; the final condition is indeed fulfilled only in the limit $\Delta t \to 0$ – in other words, the final condition of the adjoint equation is satisfied to first order accuracy in Δt when using the trapezoidal rule.

Using instead a (left) rectangular rule to approximate the first integral appearing in the cost functional would yield

$$\tilde{J}_h(\mathbf{Y}, \mathbf{U}) = \frac{\Delta t}{2}(\mathbf{Y} - \mathbf{Y}_d)^\top \mathcal{M}_0(\mathbf{Y} - \mathbf{Y}_d) + \frac{\beta \Delta t}{2}\mathbf{U}^\top \mathcal{M}_{1/2}\mathbf{U} \tag{8.15}$$

where

$$\mathcal{M}_0 = \begin{bmatrix} \mathbb{M} & & & & \\ & \mathbb{M} & & & \\ & & \ddots & & \\ & & & \mathbb{M} & \\ & & & & 0 \end{bmatrix}.$$

By proceeding in the same way as before, we would end up with the following KKT system (see Exercise 4):

$$\begin{bmatrix} \Delta t \mathcal{M}_0 & 0 & \mathcal{K}^\top \\ 0 & \beta \Delta t \mathcal{M}_{1/2} & -\mathcal{B}^\top \\ \mathcal{K} & -\mathcal{B} & 0 \end{bmatrix} \begin{bmatrix} \mathbf{Y} \\ \mathbf{U} \\ \mathbf{P} \end{bmatrix} = \begin{bmatrix} \Delta t \mathcal{M}_0 \mathbf{Y}_d \\ \mathbf{0} \\ \mathbf{F} \end{bmatrix}. \tag{8.16}$$

This would allow for a discretization that preserves the final condition of the adjoint PDE (8.9) as the last line of the first block in (8.16) gives

$$(\mathbb{M} + \Delta t \mathbb{A}^\top)\mathbf{p}^{N_t} = \mathbf{0}$$

and hence the correct final condition for the adjoint PDE.

The Case of a Final Observation

In the case of a cost functional involving an observation at the final time, like

$$\tilde{J}(y, u) = \frac{1}{2}\int_\Omega |y\,(T) - z_d|^2\, d\mathbf{x} + \frac{\beta}{2}\int_{Q_T} u^2 d\mathbf{x}dt \tag{8.17}$$

the system of first-order KKT conditions can be derived in the same way as done in this section, and reads as follows (see Exercise 5):

$$\begin{bmatrix} \Delta t \mathcal{M}_T & 0 & \mathcal{K}^\top \\ 0 & \beta \Delta t \mathcal{M}_{1/2} & -\mathcal{B}^\top \\ \mathcal{K} & -\mathcal{B} & 0 \end{bmatrix} \begin{bmatrix} \mathbf{Y} \\ \mathbf{U} \\ \mathbf{P} \end{bmatrix} = \begin{bmatrix} \Delta t \mathcal{M}_T \mathbf{Z}_d \\ \mathbf{0} \\ \mathbf{F} \end{bmatrix}, \tag{8.18}$$

where

$$\mathcal{M}_T = \begin{bmatrix} 0 & & & & \\ & 0 & & & \\ & & \ddots & & \\ & & & 0 & \\ & & & & \mathbb{M} \end{bmatrix}.$$

The Case of Time-Dependent Stokes Equations

Optimal control problems involving the time-dependent Stokes equations, as those described in Sect. 7.6, can in principle be treated in the same way. For instance, minimizing the cost functional

$$J(\boldsymbol{u}) = \frac{1}{2}\int_{Q_T} |\boldsymbol{v}(\boldsymbol{u}) - \boldsymbol{v}_d|^2\, d\mathbf{x}dt + \frac{\tau}{2}\int_{Q_T} |\boldsymbol{u}|^2\, d\mathbf{x}dt \tag{8.19}$$

subject to the time-dependent Stokes system (7.66), would lead to the following system of first-order KKT conditions when dealing with the trapezoidal rule for the numerical approximation of the first integral appearing in (8.19):

$$\begin{bmatrix} \Delta t\mathcal{M}_{1/2} & 0 & \mathcal{K}^\top \\ 0 & \tau\Delta t\mathcal{M}_{1/2} & -\mathcal{B}^\top \\ \mathcal{K} & -\mathcal{B} & 0 \end{bmatrix} \begin{bmatrix} \mathbf{Y} \\ \mathbf{U} \\ \mathbf{P} \end{bmatrix} = \begin{bmatrix} \Delta t\mathcal{M}_{1/2}\mathbf{V}_d \\ \mathbf{0} \\ \mathbf{F} \end{bmatrix}. \tag{8.20}$$

where

$$\mathbf{Y} = \begin{bmatrix} \mathbf{v}^0 \\ \boldsymbol{\pi}^0 \\ \mathbf{y}^1 \\ \boldsymbol{\pi}^1 \\ \vdots \\ \mathbf{y}^{N_t} \\ \boldsymbol{\pi}^{N_t} \end{bmatrix}, \qquad \mathbf{U} = \begin{bmatrix} \mathbf{u}^0 \\ \mathbf{u}^1 \\ \vdots \\ \mathbf{u}^{N_t} \end{bmatrix}, \qquad \mathbf{P} = \begin{bmatrix} \mathbf{z}^0 \\ \mathbf{q}^0 \\ \mathbf{p}^1 \\ \mathbf{q}^1 \\ \vdots \\ \mathbf{p}^{N_t} \\ \mathbf{q}^{N_t} \end{bmatrix}, \qquad \mathbf{V}_d = \begin{bmatrix} \mathbf{v}_d^0 \\ \mathbf{0} \\ \mathbf{v}_d^1 \\ \mathbf{0} \\ \vdots \\ \mathbf{v}_d^{N_t} \\ \mathbf{0} \end{bmatrix}$$

are the vectors for the state $\mathbf{Y}$ (velocity and pressure), the control $\mathbf{U}$, the adjoint $\mathbf{P}$ (adjoint velocity and pressure) and the target function $\mathbf{Z}_d$, whereas

$$\mathcal{K} = \begin{bmatrix} \frac{\mathbb{M}}{\Delta t} + \mathbb{A}_S & & & \\ -\frac{\mathbb{M}}{\Delta t} & \frac{\mathbb{M}}{\Delta t} + \mathbb{A}_S & & \\ & \ddots & \ddots & \\ & & -\frac{\mathbb{M}}{\Delta t} & \frac{\mathbb{M}}{\Delta t} + \mathbb{A}_S \end{bmatrix}, \qquad \mathcal{B} = \begin{bmatrix} \mathbb{B}_S & & & & \\ & \mathbb{B}_S & & & \\ & & \ddots & & \\ & & & \mathbb{B}_S & \\ & & & & \mathbb{B}_S \end{bmatrix},$$

with

$$\mathbb{A}_S = \begin{bmatrix} \mathbb{E} & \mathbb{B}^\top \\ \mathbb{B} & 0 \end{bmatrix}, \qquad \mathbb{M} = \begin{bmatrix} \mathbb{M}_v & 0 \\ 0 & 0 \end{bmatrix}, \qquad \mathbb{B}_S = \begin{bmatrix} \mathbb{M}_v \\ 0 \end{bmatrix}.$$

We recall that $\mathbb{M}_v$ is the velocity mass matrix, $\mathbb{E}$ and $\mathbb{B}$ the matrix representations of the (vector) Laplace and divergence operators, whereas $\mathbf{F}$ accounts for the non-homogeneous Dirichlet boundary conditions; see Sect. 6.2.5. The interested reader can found a detailed derivation of the algebraic system of equations in [255].

8.2 Iterative Methods

So far, we have derived the discrete optimality system for an unconstrained time-dependent linear-quadratic OCP; both OtD and DtO approaches yield potentially large linear systems to solve – indeed, their dimension is equal to $3N_yN_t$ in the case the same discrete space is used for the state and the (distributed) control – presenting a saddle-point structure. As in the case of elliptic OCPs, employing an efficient preconditioned iterative solver to compute its solution is the goal of *all-at-once* methods – we will address them in Sect. 8.4.

In this section, instead, we revisit *iterative (or black-box, or reduced space) methods*, for the numerical solution of unconstrained time-dependent OCPs, which rely on classical algorithms for numerical optimization. Also in this case, we can formulate iterative methods starting from the discrete counterpart of the OCP, once the PDE constraint has been eliminated (by expressing $\mathbf{Y} = \mathbf{Y}(\mathbf{U})$ and the control $\mathbf{U}$ is considered as the only optimization variable,

$$J_h(\mathbf{U}) = \tilde{J}_h(\mathbf{Y}(\mathbf{U}), \mathbf{U}) \ \rightarrow \ \min \quad \text{subject to } \mathbf{U} \in \mathbb{R}^{N_t N_u}. \tag{8.21}$$

Here $\mathbf{U}$ collects the vectors of nodal values of the discrete control u_h at each time step t^n, $n = 0, \dots, N_t$, where N_u is the dimension of the discrete control space.

Using the compact notation makes the extension of the iterative methods of Sect. 6.3 to the time-dependent case straightforward. In this case, when a descent method is employed, given an initial guess $\mathbf{U}_0 \in \mathbb{R}^{N_u}$, we need to find iteratively a sequence $\{\mathbf{U}_k\}$ such that

$$\mathbf{U}_{k+1} = \mathbf{U}_k + \tau_k \, \mathbf{D}_k$$

where $\mathbf{D}_k$ represents a descent direction (that can depend on the gradient ∇J_h and/or the Hessian $\mathbb{H}$ of the objective function J_h) and $\tau_k > 0$ is a suitable step size which should ensure that

$$J_h(\mathbf{U}_k + \tau_k \mathbf{D}_k) < J_h(\mathbf{U}_k).$$

This means that, for any optimization step, we must evaluate the descent direction at each time step, so that the control $\mathbf{u}_k^n$, $n = 0, 1, \dots, N_t$ can be updated accordingly.

By replicating what done in Sect. 3.1 and Sect. 6.3, we obtain the different options summarized in Tab. 8.1. The corresponding iterative algorithm is reported in Algorithm 8.1.

steepest descent	$\mathbf{D}_k = -\nabla J_h(\mathbf{U}_k)$
conjugate gradient	$\mathbf{D}_k = -\nabla J_h(\mathbf{U}_k) + \beta_k \mathbf{U}_{k-1}$
Newton	$\mathbf{D}_k = -\mathbb{H}^{-1} \nabla J_h(\mathbf{U}_k)$
quasi-Newton	$\mathbf{D}_k = -\mathbb{B}_k^{-1} \nabla J_h(\mathbf{U}_k)$

Table 8.1 Choice of descent directions for the solution of an unconstrained time-dependent OCP

Also in this case, when using an iterative method, the PDE is nested in the optimization routine; the three equations forming the optimality system are then uncoupled, and the state and the adjoint problems are iteratively solved:

- control vectors $\mathbf{u}_k^0, \dots, \mathbf{u}_k^{N_t}$ computed at the previous optimization step, at all time steps, enter as data in the state system, that has to be solved (forward) in time
- once the state vectors $\mathbf{y}_k^0, \dots, \mathbf{y}_k^{N_t}$ have been computed, they are used as data for the adjoint system, that has to be solved backward in time;
- once the adjoint vectors $\mathbf{p}_k^0, \dots, \mathbf{p}_k^{N_t}$ have been computed, it is possible to evaluate the gradient and, possibly, the Hessian matrix $\mathbb{H}$.

Algorithm 8.1 Iterative algorithm for unconstrained OCPs, case of linear evolution PDEs

Input: initial guess $\mathbf{U}_0$

compute $\mathbf{Y}_0$ by solving the state equation with $\mathbf{U} = \mathbf{U}_0$ forward in time
compute $\mathbf{p}_0$ by solving the adjoint equation with $\mathbf{Y} = \mathbf{Y}_0$ backward in time
compute $J_h(\mathbf{U}_0)$ and $\nabla J_h(\mathbf{U}_0)$
set $k = 0$
while $|\nabla J_h(\mathbf{U}_k)| / |\nabla J_h(\mathbf{U}_0)| > \text{tol}$ or $|J_h(\mathbf{U}_k) - J_h(\mathbf{U}_{k-1})| > \text{tol}$
    compute the search direction $\mathbf{D}_k$ according to a descent method
    compute the step length τ_k with a line-search routine
    set $\mathbf{U}_{k+1} = \mathbf{U}_k + \tau_k \mathbf{D}_k$
    compute $\mathbf{Y}_{k+1}$ by solving the state equation with $\mathbf{U} = \mathbf{U}_{k+1}$ forward in time
    compute $\mathbf{P}_{k+1}$ by solving the adjoint equation with $\mathbf{Y} = \mathbf{Y}_{k+1}$ backward in time
    compute $J_h(\mathbf{U}_{k+1})$ and $\nabla J_h(\mathbf{U}_{k+1})$
    $k = k + 1$
end while

Note that different numerical approximations of the cost functional $J(u) = \tilde{J}(y(u), u)$ yield different expressions of $J_h(\mathbf{U})$, and therefore of its gradient. For instance, if the trapezoidal rule is used, as in (8.12), we obtain

$$\nabla J_h(\mathbf{U}) = \beta \Delta t \mathcal{M}_{1/2} \mathbf{U} - \mathcal{B}^\top \mathbf{P}.$$

Also when dealing with linear evolution PDEs, iterative methods can be adapted to the case of constrained OCPs , with control constraints on the control function expressed under the form of box constraints, similarly to what we have done in Sect. 6.4.

8.3 Numerical Examples

In this section we show the numerical results for two optimal control problems dealing with the heat equation, obtained using a quasi-Newton method exploiting the BFGS update for the update of the matrix $\mathbb{B}_k^{-1}$ and a cubic line search procedure at each step.

In particular, we aim at determining an optimal distributed control for a heat conduction problem in $\Omega = (0, 1)^2$, where we set the initial temperature distribution $g = 0$, the distributed source $f = 0$ and homogeneous Dirichlet boundary conditions on S_T; the heat conduction coefficient is taken equal to 1 for the sake of simplicity.

We deal with two different settings, defining target functions as the ones used in [254], to which we refer for some numerical results obtained with all-at-once methods. In particular, we consider:

- *case 1*: a distributed observation (that is, over the whole time interval $t \in (0, T)$) and a discontinuous cross-shaped target of the form

$$y_d = \begin{cases} -64t\, e^{-(x-0.5)^2-(y-0.5)^2} & x_1 \in (0.4, 0.6) \text{ or } x_2 \in (0.4, 0.6) \\ 0 & \text{otherwise} \end{cases}$$

  aiming at minimizing the cost functional (8.1), that is,

$$\tilde{J}(y, u) = \frac{1}{2} \int_{Q_T} |y - y_d|^2 \, d\mathbf{x}dt + \frac{\beta}{2} \int_{Q_T} u^2 d\mathbf{x}dt;$$

- *case 2*: a final observation (that is, at the final time $t = T$) and a discontinuous cross-shaped target of the form

$$z_d = \begin{cases} -x_1 e^{-(x-0.5)^2-(y-0.5)^2} & x_1 \in (0.4, 0.6) \text{ or } x_2 \in (0.4, 0.6) \\ 0 & \text{otherwise} \end{cases}$$

aiming at minimizing the cost functional (8.28), that is,

$$\tilde{J}(y,u) = \frac{1}{2}\int_\Omega |y(T) - z_d|^2 \, d\mathbf{x} + \frac{\beta}{2}\int_{Q_T} u^2 d\mathbf{x}dt$$

We adopt a OtD approach by first deriving the system of first-order optimality conditions. These latter are given by system (8.4) for case 1; for case 2, they read as follows:

$$\begin{cases} (\dot{y}(t), \varphi)_{L^2(\Omega)} + a(y(t), \varphi; t) = (f(t) + u(t), \varphi)_{L^2(\Omega)} & \forall \varphi \in V \\ y(0) = g \\ -(\dot{p}(t), \psi)_{L^2(\Omega)} + a^*(p(t), \psi; t) = 0 & \forall \psi \in V \\ p(T) = -(y(T) - z_d) \\ (\beta u(t) - p(t), v)_{L^2(\Omega)} = 0 & \forall v \in \mathcal{U}; \end{cases} \tag{8.22}$$

the only difference compared to case 1 stands in the form of the adjoint problem (see Sect. 7.1.3).

We then discretize the system of optimality conditions. For the spatial discretization of both the state and the adjoint problem, we apply the Galerkin-FE method on a finite-dimensional space $V_h \subset V = H_0^1(\Omega)$ of dimension $\dim(V_h) = \dim(W_h) = N_h$, where h denotes a parameter related to the mesh size of the computational grid, using linear finite elements. Regarding time discretization, we rely on the Cranck-Nicholson method, after introducing a partition of $N_t = T/\Delta t$ time intervals of length Δt. In both cases, we consider different settings, precisely:

(i) a mesh with 1342 triangular elements and 712 vertices, resulting in a state dimension $N_y = 632$ (corresponding to the internal vertices) and a dimension $N_u = 712$ of the control $\mathbf{u}^n$ at each time step, and a time discretization with $\Delta t = 0.05$; this results in a control vector $\mathbf{U}$ of dimension $N_t N_u = 20 \cdot 712 = 14240$;

(ii) a mesh with 3704 triangular elements and 1921 vertices, resulting in a state dimension $N_y = 1785$ (corresponding to the internal vertices) and a dimension $N_u = 1921$ of the control $\mathbf{u}^n$ at each time step, and a time discretization with $\Delta t = 0.01$; this results in a control vector $\mathbf{U}$ of dimension $N_t N_u = 100 \cdot 1921 = 192100$.

For each of these two settings, and both cases 1 and 2, we consider different values of the regularization parameter $\beta = 10^{-2}, 10^{-3}, 10^{-4}$, to assess its impact on the numerical optimization procedure. The iterative optimization method is stopped when a desired accuracy on the maximum norm $|\nabla J_h(\mathbf{U})|_\infty < \varepsilon$ of the gradient of the cost functional is reached; we fix $\varepsilon = 10^{-12}$.

For the case 1 with distributed observation, numerical results obtained with setting *(ii)* are reported in Figs. 8.1–8.2, showing different snapshots of the optimal state, adjoint and control variables obtained at $t = 0.6$ and $t = 0.9$, for the two values $\beta = 10^{-3}$ and $\beta = 10^{-4}$.

For the case 2 with final observation, numerical results obtained with setting *(ii)* are reported in Fig. 8.3, showing the optimal state, adjoint and control variables obtained at the final time $t = 1$, for the two values $\beta = 10^{-2}$ and $\beta = 10^{-3}$.

In both cases, it can be appreciated that the optimal temperature better converges to the target temperature distribution as the regularization parameter β decreases.

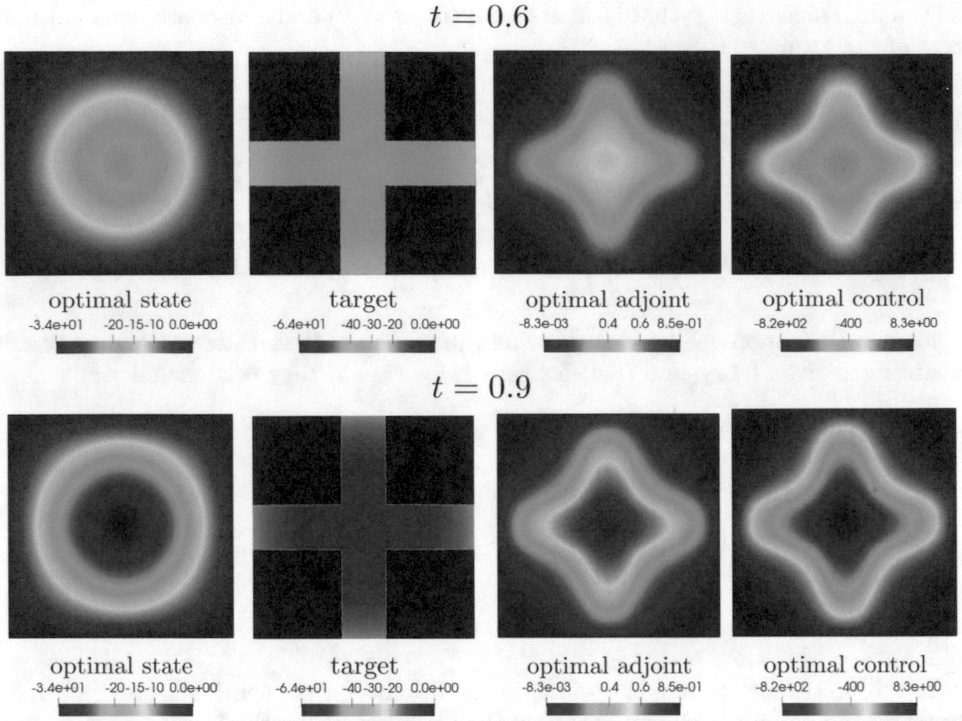

Fig. 8.1 Case 1. Target y_d, optimal state $\hat{y}$, adjoint $\hat{p}$ and control $\hat{u}$ at $t = 0.6$ and $t = 0.9$, for $\beta = 10^{-3}$.

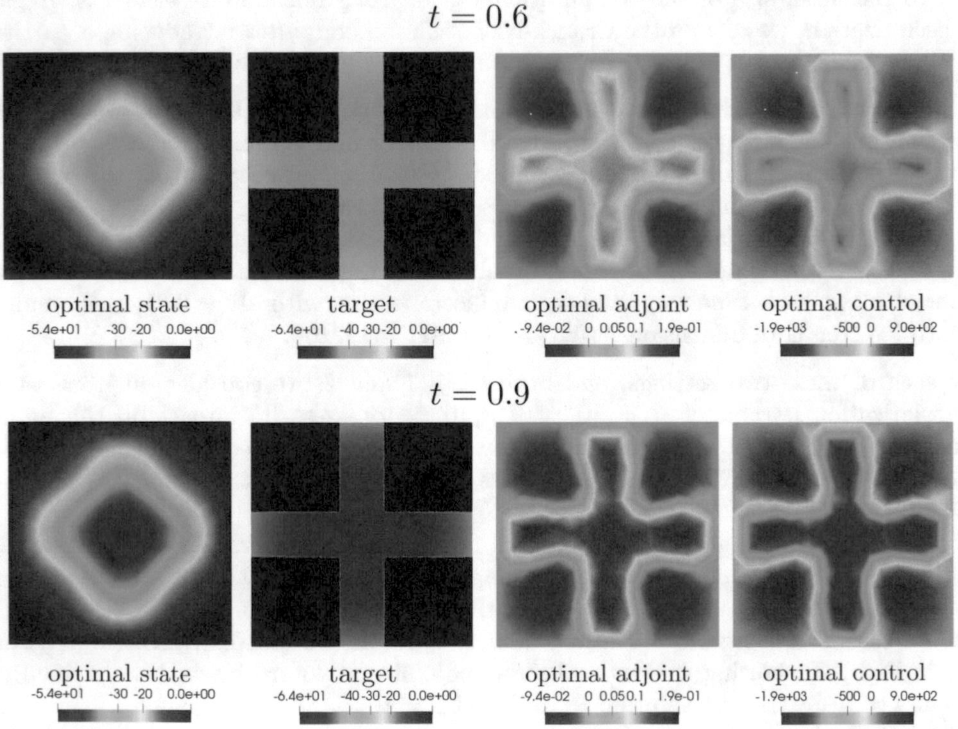

Fig. 8.2 Case 1. Target y_d, optimal state $\hat{y}$, adjoint $\hat{p}$ and control $\hat{u}$ at $t = 0.6$ and $t = 0.9$, for $\beta = 10^{-4}$.

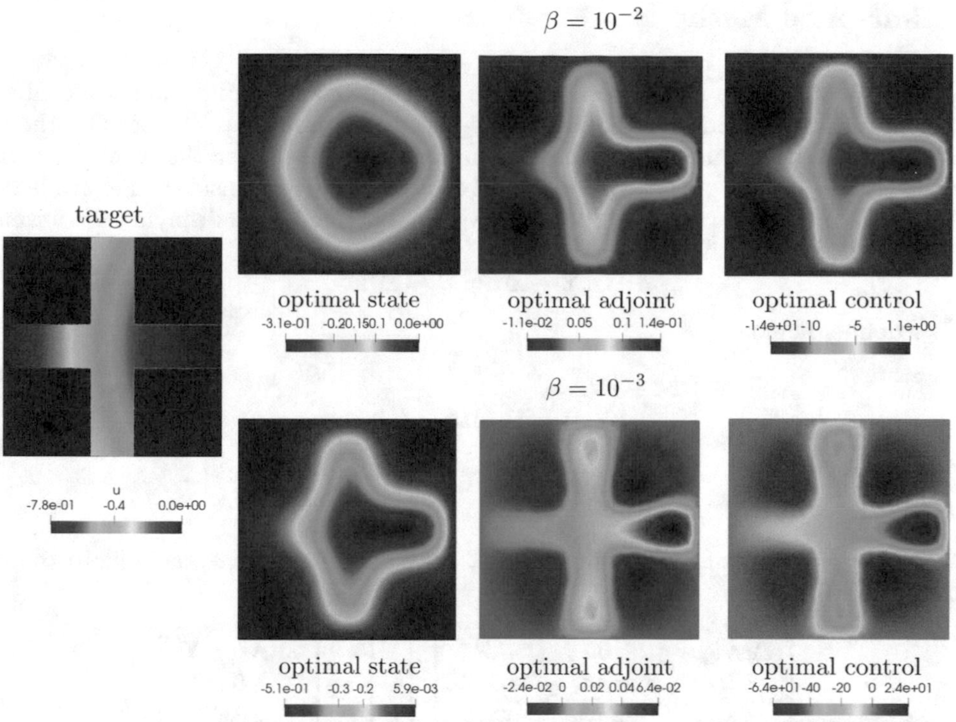

Fig. 8.3 Case 2. Target z_d, Optimal state $\hat{y}(T)$, adjoint $\hat{p}(T)$ and control $\hat{u}(T)$ at $t = T$, with $T = 1$, for two different values of the regularization parameter: $\beta = 10^{-2}$ (top), $\beta = 10^{-3}$ (bottom).

The values of the cost functional at the optimum, and the number of iterations required by the BFGS method to converge, for several values of β, are reported in Tab. 8.2 and Tab. 8.3 regarding cases 1 and 2, respectively. We can note that the regularization parameter has a strong impact on the convergence velocity of the iterative method: indeed, more than four times iterations are required to reach the same accuracy degree of when β decreases of two orders of magnitude.

	$\beta = 10^{-2}$	$\beta = 10^{-3}$	$\beta = 10^{-4}$
$h = 0.1$, $\Delta t = 0.05$	147.6619 (36)	91.9129 (63)	55.4118 (176)
$h = 0.05$, $\Delta t = 0.01$	159.5637 (46)	104.9392 (82)	67.2146 (206)

Table 8.2 Case 1. Values of the cost functional for the optimal control $J_h(\hat{\mathbf{U}})$ and number of iterations required by the BFGS method to converge, for different values of the regularization parameter β.

	$\beta = 10^{-2}$	$\beta = 10^{-3}$	$\beta = 10^{-4}$
$h = 0.1$, $\Delta t = 0.05$	0.018837 (64)	0.0074174 (153)	0.0019042 (318)
$h = 0.05$, $\Delta t = 0.01$	0.022603 (66)	0.012369 (151)	0.0050182 (308)

Table 8.3 Case 2. Values of the cost functional for the optimal control $J_h(\hat{\mathbf{U}})$ and number of iterations required by the BFGS method to converge, for different values of the regularization parameter β.

8.4 All-at-once Methods

In this section we discuss a possible extension of the preconditioning framework introduced in Sect. 6.6 for linear quadratic OCPs to time-dependent problems. We focus on the case of problem (8.1)–(8.2), assuming that the distributed control act over the whole domain, and follow the DtO approach. As shown in [213, 254], it is worthy to consider an equivalent – and slightly more convenient – form for the state system, obtained by multiplying the formulation in (8.11) by Δt, thus yielding

$$\mathcal{K}\mathbf{Y} - \Delta t\mathcal{M}\mathbf{U} = \Delta t\mathbf{F},$$

where we define, hereon,

$$\mathcal{K} = \begin{bmatrix} \mathbb{M} + \Delta t\mathbb{A} & & & \\ -\mathbb{M} & \mathbb{M} + \Delta t\mathbb{A} & & \\ & \ddots & \ddots & \\ & & -\mathbb{M} & \mathbb{M} + \Delta t\mathbb{A} \end{bmatrix}.$$

Hence, we obtain a system of first-order KKT optimality conditions, as done in the case of system (8.20), which now reads as:

$$\underbrace{\begin{bmatrix} \Delta t\mathcal{M}_{1/2} & 0 & \mathcal{K}^\top \\ 0 & \beta\Delta t\mathcal{M}_{1/2} & -\Delta t\mathcal{M} \\ \mathcal{K} & -\Delta t\mathcal{M} & 0 \end{bmatrix}}_{\mathcal{A}} \begin{bmatrix} \mathbf{Y} \\ \mathbf{U} \\ \mathbf{P} \end{bmatrix} = \begin{bmatrix} \Delta t\mathcal{M}_{1/2}\mathbf{Y}_d \\ \mathbf{0} \\ \Delta t\mathbf{F} \end{bmatrix}. \tag{8.23}$$

All-at-once methods provide the solution of the OCP by solving the (coupled) linear system (8.23) simultaneously. More than in the case of elliptic PDEs, a direct solver might run out of memory fairly quickly – for a reasonable sized spatial discretization, even in two spatial dimensions – as the dimension of the matrix $\mathcal{A}$ appearing in (8.23) crucially depends on the temporal discretization.

Since $\mathcal{A}$ has a saddle-point structure, is block symmetric, indefinite, usually ill-conditioned and extremely large, efficient preconditioned iterative solvers (such as Krylov subspace methods, like MINRES or GMRES) empowered by a suitable block preconditioner are the matter of choice for the solution of (8.23). In order for the preconditioned system to maintain the symmetric and indefinite nature of the problem, we need it to be symmetric and positive definite. A reasonable choice is to opt for a symmetric block-diagonal preconditioner, under the form

$$\mathcal{P}_d = \begin{bmatrix} \Delta t\mathcal{M}_{1/2} & 0 & 0 \\ 0 & \beta\Delta t\mathcal{M}_{1/2} & 0 \\ 0 & 0 & \mathcal{S} \end{bmatrix},$$

where $\mathcal{S}$ denotes the (negative) Schur complement of system (8.23), which is given by

$$\mathcal{S} = \frac{1}{\Delta t}\mathcal{K}\mathcal{M}_{1/2}^{-1}\mathcal{K}^\top + \frac{\Delta t}{\beta}\mathcal{M}\mathcal{M}_{1/2}^{-1}\mathcal{M}.$$

Note that this is the only block of $\mathcal{P}_d$ that contains the PDE matrix – blocks (1,1) and (2,2) are, as in the case of preconditioner (6.75), simple mass matrices that can be inverted without difficulties. Two possible options have been proposed in [213, 254] to approximate the diagonal blocks of $\mathcal{P}_d$, that is, the mass matrices – blocks (1,1) and (2,2) – and the Schur complement.

Option 1. We can replace $\mathcal{M}_{1/2}$ by $\mathcal{M}$ and approximate the Schur complement $\mathcal{S}$ by neglecting the term $\Delta t\beta^{-1}\mathcal{M}$ as already proposed for non time-dependent problems, thus getting

$$\hat{\mathcal{P}}_d = \begin{bmatrix} \Delta t\mathcal{M} & 0 & 0 \\ 0 & \beta\Delta t\mathcal{M} & 0 \\ 0 & 0 & \hat{\mathcal{S}} \end{bmatrix},$$

having defined

$$\hat{\mathcal{S}} = \frac{1}{\Delta t}\mathcal{K}\mathcal{M}^{-1}\mathcal{K}^{\top}. \tag{8.24}$$

Note that in this case the evaluation of the preconditioned residual involves formally the inverse of $\hat{\mathcal{S}}$, that is, $\hat{\mathcal{S}}^{-1} = \Delta t\,\mathcal{K}^{-\top}\mathcal{M}\mathcal{K}^{-1}$, where $\mathcal{K}$ is lower triangular. The inverse of $\mathcal{K}$ is therefore relatively cheap to implement. Since $\mathcal{K}$ is a lower block triangular matrix, both the application of $\mathcal{K}^{-1}$ and $\mathcal{K}^{-\top}$ are simple forward and backward substitutions, which only involve the solution of a linear system with the block $\mathbb{L} = \mathbb{M} + \Delta t\mathbb{K}$. Formally speaking, we can replace the action of $\mathcal{K}^{-1}$ by that of $\hat{\mathcal{K}}^{-1}$, that consists in replacing $\mathbb{L}$ by a multigrid approximation[1] $\hat{\mathbb{L}}$. Hence, we can set

$$\hat{\mathcal{S}}^{-1} = \hat{\mathcal{K}}^{-\top}\mathcal{M}\hat{\mathcal{K}}^{-1}$$

with the preconditioner in its final form

$$\hat{\mathcal{P}}_d = \begin{bmatrix} \Delta t\hat{\mathcal{M}} & 0 & 0 \\ 0 & \beta\Delta t\hat{\mathcal{M}} & 0 \\ 0 & 0 & \hat{\mathcal{S}} \end{bmatrix}. \tag{8.25}$$

Thanks to its mesh independent condition number, a good approximation $\hat{\mathcal{M}}$ for the mass matrix $\mathcal{M}$ is for instance provided by the lumped mass matrix or even by the diagonal of the mass matrix, even if other options are also available, such as the so-called Chebyshev semi-iterations (see Sect. 6.6.1 and references therein).

As in the elliptic case, also for time-dependent OCPs the use of the block diagonal preconditioner $\hat{\mathcal{P}}_d$ allows to achieve a convergence rate independent of the mesh size h, a property that is referred to as *optimality*. Indeed, it is possible to show (see [254] for the detailed proof) that the eigenvalues of $\mathcal{P}_d^{-1}$ are given by 1 and eigenvalues depending on the term $\Delta t\mathcal{K}^{-\top}\mathcal{M}\mathcal{K}^{-1}(\frac{1}{\Delta t}\mathcal{K}\mathcal{M}^{-1}\mathcal{K}^{\top} + \Delta t\beta^{-1}\mathcal{M})$; these latter only have a benign dependence on Δt and are independent of the discretization parameter h.

Option 2. We can consider the same approximation for the mass matrices as in Option 1, and approximate the Schur complement $\mathcal{S}$ by

$$\hat{\mathcal{S}} = \frac{1}{\Delta t}\left(\mathcal{K} + \frac{\Delta t}{\sqrt{\beta}}\mathcal{M}\right)\mathcal{M}_{1/2}^{-1}\left(\mathcal{K} + \frac{\Delta t}{\sqrt{\beta}}\mathcal{M}\right)^{\top}, \tag{8.26}$$

see [214]. In this case the eigenvalues of $\hat{\mathcal{S}}^{-1}\mathcal{S}$ belong to the interval $[1/2, 1]$. The resulting preconditioner can then be expressed under the form (8.25), replacing (8.24) by (8.26).

[1] Note that $\mathbb{L}$ is symmetric and positive definite. A typical approximation for $\mathbb{L} = \mathbb{M} + \Delta t\mathbb{A}$ would be the use of one algebraic multigrid (AMG) V-cycle involving, e.g. few steps of the Chebyshev smoother or the Gauss-Seidel smoother. The former option is usually preferred, because of the importance of the mass matrix in $\mathbb{L}$, at least for small Δt, given that Chebyshev methods are known to be very efficient for mass matrices.

Since

$$\mathcal{K} + \frac{\Delta t}{\sqrt{\beta}}\mathcal{M} = \begin{bmatrix} \tilde{\mathbb{L}} & & & \\ -\mathbb{M} & \tilde{\mathbb{L}} & & \\ & \ddots & \ddots & \\ & & -\mathbb{M} & \tilde{\mathbb{L}} \end{bmatrix}$$

with $\tilde{\mathbb{L}} = (1 + \frac{\Delta t}{\sqrt{\beta}})\mathbb{M} + \Delta t\mathbb{A}$, the inversion of this matrix and also its transpose involve forward and backward substitutions. As done in Option 1 for the blocks $\mathbb{L}$ appearing in the matrix $\mathcal{K}$, the diagonal blocks $\tilde{\mathbb{L}}$ will in fact be approximated by employing, e.g., algebraic multigrid techniques. Compared to Option 1, Option 2 yields a preconditioned system for which the number of MINRES iterations is almost independent of the mesh size, the regularization parameter β and, in the case of a distributed control, the time-step, too.

The strategy for constructing a block-diagonal preconditioner described so far can be easily extended to different situations. In particular:

- slightly different preconditioners are obtained in the case an OtD approach is considered or, equivalently, a (left) rectangular rule is used to approximate the first integral appearing in the cost functional. In this case, following the same procedure that led us to system (8.23), yields

$$\begin{bmatrix} \Delta t\mathcal{M}_0 & 0 & \mathcal{K}^\top \\ 0 & \beta\Delta t\mathcal{M}_{1/2} & -\mathcal{M} \\ \mathcal{K} & -\mathcal{M} & 0 \end{bmatrix} \begin{bmatrix} \mathbf{Y} \\ \mathbf{U} \\ \mathbf{P} \end{bmatrix} = \begin{bmatrix} \Delta t\mathcal{M}_0\mathbf{Y}_d \\ \mathbf{0} \\ \Delta t\mathbf{F} \end{bmatrix},$$

which is the analogue of (8.16). The main issue in this case is due to the rank-deficiency of the (1,1) block – indeed, $\mathcal{M}_0$ is not invertible. Option 2 described above yields the following preconditioner,

$$\hat{\mathcal{P}}_d = \begin{bmatrix} \Delta t\hat{\mathcal{M}}_0 & 0 & 0 \\ 0 & \beta\Delta t\hat{\mathcal{M}}_{1/2} & 0 \\ 0 & 0 & \hat{\mathcal{S}} \end{bmatrix}. \tag{8.27}$$

where now $\hat{\mathcal{M}}_0$ and $\hat{\mathcal{M}}_{1/2}$ denote approximation to $\mathcal{M}_0^\gamma$ and $\mathcal{M}_{1/2}$, respectively, being

$$\mathcal{M}_0^\gamma = \begin{bmatrix} \mathbb{M} & & & \\ & \mathbb{M} & & \\ & & \ddots & \\ & & & \mathbb{M} \\ & & & & \gamma\mathbb{M} \end{bmatrix},$$

for some constant γ such that $0 < \gamma \ll 1$, and

$$\hat{\mathcal{S}} = \frac{1}{\Delta t}\left(\mathcal{K} + \frac{\Delta t}{\sqrt{\beta}}\mathcal{M}_1\right)\mathcal{M}_0^{-1}\left(\mathcal{K} + \frac{\Delta t}{\sqrt{\beta}}\mathcal{M}_1\right)^\top.$$

being

$$\mathcal{M}_1 = \begin{bmatrix} \sqrt{2}\mathbb{M} & & & \\ & \mathbb{M} & & \\ & & \ddots & \\ & & & \mathbb{M} \\ & & & & \sqrt{2\gamma}\mathbb{M} \end{bmatrix}.$$

See, e.g., [254] for further details.

- In the case that the desired state is observed on a subset $\Omega_{obs} \subset \Omega$, the matrix $\mathcal{M}_{1/2}$ is a block-diagonal matrix consisting of matrices $\mathbb{M}_{obs}$, where this matrix only contains contributions from the domain Ω_{obs}. This means that the matrix $\mathbb{M}_{obs}$ is only semi-definite. For the preconditioning of this matrix, it is possible to introduce a small parameter $\nu > 0$, to obtain the preconditioning matrix $\widetilde{\mathbb{M}}_{obs} = \mathbb{M}_{obs} + \nu\mathbb{I}$, where $\mathbb{I}$ is an identity matrix associated with the degrees of freedom in the part $\Omega \setminus \Omega_{obs}$. In general, ν will depend on the mesh parameter h, the time-step Δt, and the regularization parameter β. See, e.g., [255] for further details.

Remark 8.3. The construction of a block-diagonal preconditioner in the case of a cost functional involving an observation at the final time, like (8.28), would require a different approximation of the (1,1) block, however yielding a similar preconditioner to the one discussed above. In the case of OCPs involving the time-dependent Stokes equations, the construction carried out so far can be easily adapted, taking advantage of several considerations made in Sect. 6.6.2 for the approximation of the Schur complement – that, in this case, would entail two solves with the discrete Stokes matrix, at each time step. See, e.g., [255] for further details. •

It should now be clear that the numerical approximation of OCPs governed by linear evolution PDEs borrows several toolboxes from the approaches already used in the case of OCPs for elliptic PDEs. On the other hand, a wealth of possible approaches can be obtained depending on the interplay between optimization and discretization tasks, as well as because of different options that can be followed when dealing with space and time discretization. In this chapter we have limited ourselves to show how the strategies proposed in Chapter 6 can be extended to the time-dependent case, provided state, control and adjoint discrete variables at all time steps are grouped together. Several other aspects, including, e.g., adaptivity [192], the treatment of different type of boundary conditions and/or constraints [109, 107], a priori [193, 194, 195] and a posteriori [183, 179, 108] error estimates, as well as the numerical approximation of feedback control problems and their connection with the Riccati Equations [18, 19, 28] of Chapter 4, would require a more advanced theoretical and numerical background. The interested reader can find further details in the referred papers.

8.5 Exercises

1. Proceeding as in Sect. 6.2.1, show that the semi-discrete problem (8.5) can be equivalently rewritten as the system (8.6).
2. Recover the *one-shot* discretization for N_t time steps of both the state and the adjoint system using the Crank-Nicolson scheme (recalled in Sect. B.3 of Appendix B).
3. Show that, similarly to what we did in Sect. 4.2.1, deriving the Lagrangian functional (8.13) with respect to $\mathbf{Y}$, $\mathbf{U}$ and $\mathbf{P}$ yields the first-order KKT system under the form

$$\begin{cases} \mathcal{K}\mathbf{Y} = \mathbf{F} + \mathcal{B}\mathbf{U} \\ \mathcal{K}^\top\mathbf{P} = -\Delta t\mathcal{M}_{1/2}(\mathbf{Y} - \mathbf{Y}_d) \\ \beta\Delta t\mathcal{M}_{1/2}\mathbf{U} - \mathcal{B}^\top\mathbf{P} = \mathbf{0} \end{cases}$$

   that is, system (8.20).
4. Proceeding as in Exercise 3 and introducing a suitable Lagrangian functional, show that the minimization of the cost functional 8.15 under the constraint expressed by the state system (8.11) yields the first-order KKT system (8.16).

5. Show that the minimization problem

$$\tilde{J}\left(y,u\right)=\frac{1}{2}\int_{\Omega}\left|y(T)-z_{d}\right|^{2}d\mathbf{x}dt+\frac{\beta}{2}\int_{Q_{T}}u^{2}d\mathbf{x}dt\rightarrow\min \tag{8.28}$$

with $\beta>0$, where $y_d\in L^2\left(Q_T\right)$ is a given target, $u\in L^2\left(Q_T\right)$ and $y=y\left(\mathbf{x},t;u\right)$ solves the *state problem:*

$$\begin{cases} y_t-\Delta y=f+u & \text{in } Q_T\\ y=0 & \text{on } S_T\\ y\left(\mathbf{x},0\right)=g\left(\mathbf{x}\right) & \text{in } \Omega. \end{cases} \tag{8.29}$$

yields the system of first-order KKT conditions (8.18). With this goal, proceed by first discretizing with respect to space and time the state system (8.29), use the trapezoidal rule to discretize the cost functional (8.28), introduce a suitable Lagrangian functional and derive it with respect to state, control and adjoint variables.

Part III
More General PDE-constrained Optimization Problems

Chapter 9
A Mathematical Framework for Nonlinear OCPs

In this chapter we provide a reference theoretical setting for investigating the well-posedness of a rather general class of OCPs and the corresponding set of optimality conditions. After an introductory section, we present some general results on *optimization in Banach and Hilbert spaces* in Sects. 9.2–9.3. In Sect. 9.2 we address the question of existence/uniqueness of a minimizer. A natural idea is to extend the result from Theorem 2.1 using *weak* rather than strong *convergence* in infinitely many dimensions. The search of optimality conditions requires differential calculus in these spaces, whose basic notions are presented in Sect. A.7.

In particular, the new tools are applied to derive first-order optimality conditions under the form of *variational inequalities* in Sect. 9.2.3. In Sect. 9.3 we deal with control constrained OCPs. After carrying out a well posedness analysis (Sect. 9.3.1), we extend both the adjoint method and the Lagrange multipliers method in Sect. 9.3.2 and Sect. 9.3.3, stating second order optimality conditions, too, in Sect. 9.3.4. We then analyze two model problems: a distributed control of a semilinear state equation (Sect. 9.4) and a least squares approximation of a reaction coefficient in a time-dependent diffusion-reaction equation (Sect. 9.5). Finally, some numerical approximation strategies to deal with nonlinear OCPs are described in Sect. 9.6 and Sect. 9.8, whereas some numerical results are reported in Sect. 9.7.

9.1 Motivation

So far we have considered linear-quadratic OCPs (for elliptic and evolution PDEs) in a Hilbert space setting. Both well-posedness analysis and the derivation of the system of first-order necessary optimality conditions have been obtained after expressing the reduced cost functional as a quadratic functional and, consequently, by exploiting basic properties such as the continuity and the coercivity of some bilinear forms.

To cover a broader range of applications we need to develop a more general framework for the analysis of OCPs where the state system may consist of a nonlinear PDE, the control function may belong to a Banach space, and/or the cost functional is no longer quadratic. The results that we will derive extend those of Chapters 5 and 7 for the linear-quadratic case.

For instance, consider the case where $\Omega \subset \mathbb{R}^d$, $d \leq 3$, is a bounded, smooth domain, $\beta \geq 0$ and $z_d \in L^6(\Omega)$ a given target and we want to solve the following OCP

$$F(y, u) = \frac{1}{6} \|y - z_d\|_{L^6(\Omega)}^6 + \frac{\beta}{2} \|u\|_{L^2(\Omega)}^2 \to \min$$

subject to

$$u \in \mathcal{U}_{ad} = \left\{ u \in L^2(\Omega) : a \leq u(\mathbf{x}) \leq b, \text{ a.e. in } \Omega \right\},$$

A. Manzoni et al., *Optimal Control of Partial Differential Equations*,
Applied Mathematical Sciences 207, https://doi.org/10.1007/978-3-030-77226-0_9

and to the state system

$$\begin{cases} -\Delta y + y + y^3 = u & \text{in } \Omega \\ \dfrac{\partial y}{\partial n} = 0 & \text{on } \Gamma = \partial\Omega. \end{cases} \tag{9.1}$$

In this case the state system is given by a nonlinear (actually, semilinear) PDE, and the cost functional is no longer quadratic. Typical examples of nonlinear PDEs arise, e.g., in fluid dynamics (Navier-Stokes equations) and in structural mechanics (non-elastic materials). Non quadratic cost functionals may be required to ensure the well posedness of some OCPs (e.g. in fluid dynamics). This problem will be analyzed in Sect. 9.4.

On the other hand, the role of control function can be played by some coefficients of (a linear) state system which have to be identified or estimated from a set of measurements. For instance, consider the case of a simple diffusion-reaction equation

$$\begin{cases} y_t - \Delta y + u(\mathbf{x})\, y = f(\mathbf{x}) & \text{in } Q_T = \Omega \times (0,T) \\ y = 0 & \text{on } \Gamma \times (0,T) \\ y(\mathbf{x},0) = g(\mathbf{x}) & \text{in } \Omega \end{cases}$$

where $y = y(u)$ is the state variable, while the reaction coefficient

$$u \in \mathcal{U}_{ad} = \{u \in L^\infty(\Omega) : a \le u \le b \text{ a.e. in } \Omega\}$$

is the unknown control function, that has to be recovered by minimizing (on $\mathcal{U}_{ad}$)

$$F(y) = \frac{1}{2}\int_\Omega (y(\mathbf{x},T) - z_d(\mathbf{x}))^2 d\mathbf{x}.$$

This identification problem will be analyzed in Sect. 9.5.

In both examples considered above, the cost functional is convex. However, the reduced functionals obtained from F by replacing y with $y(u)$, are in general nonconvex, due to the nonlinearity of the control-to-state map $u \mapsto y(u)$. As a consequence, first order necessary conditions are no longer sufficient for optimality and, in principle, one has to resort to higher order conditions to ensure optimality. Moreover, the search of necessary/sufficient conditions of optimality requires a notion of differentability for both the control-to-state mapping and the cost functional.

9.2 Optimization in Banach and Hilbert Spaces

9.2.1 Existence and Uniqueness of Minimizers

Let X be a *reflexive* Banach space and $F : X \to \mathbb{R}$ be a functional (linear or nonlinear) on X. Given $K \subseteq X$, we consider the problem

$$F(x) \to \min \qquad \text{subject to } x \in K. \tag{9.2}$$

The first question we address is to find conditions on K and X under which an optimal $\hat{x}$ exists. Let us re-consider the case of optimization in finite dimensions.

The main assumptions of Theorem 2.1 of Sect. 2.1.1 are:

(*a*) K is a *closed* set;
(*b*) F is *lower semicontinuous*, that is

$$F(x) \le \liminf_{k\to+\infty} F(x_k) \tag{9.3}$$

whenever $x_k \to x$ or, equivalently, the lower level sets

$$E_c = \{x : F(x) \le c\} \tag{9.4}$$

are closed for every $c \in \mathbb{R}$;

(*c*) if K is unbounded, then

$$F(x) \to +\infty \quad \text{as} \quad \|x\| \to +\infty. \tag{9.5}$$

Let us now define

$$\mu = \inf_{x \in K} F(x) \tag{9.6}$$

and let $\{x_k\}_{k \ge 1} \subset K$ be a *minimizing sequence*, that is

$$F(x_k) \to \mu \quad \text{as} \quad k \to +\infty. \tag{9.7}$$

From (c) $\{x_k\}$ is bounded, then there exists a subsequence $\{x_{k_j}\}_{j \ge 1}$ which converges to some element $\hat{x} \in K$ when $k_j \to +\infty$. From this fact and assumption (b), we deduce that

$$\mu \le F(\hat{x}) \le \liminf_{j \to +\infty} F(x_{k_j}) = \lim_{j \to +\infty} F(x_{k_j}) = \mu, \tag{9.8}$$

whence $\hat{x}$ is a minimizer.

Let us now examine more closely the main steps in the proof of Theorem 2.1, in order to extend them to the minimization of a function over an infinite dimensional space. The proof of *existence* essentially consists of three steps:

1. $\{x_{k_j}\}$ *is bounded.* This is trivial if K itself is bounded, whereas it follows from the *coercivity* assumption (c) if K is unbounded. No problem arises in the infinite dimensional context.
2. *Extraction of a convergent subsequence.* If the dimension of X is finite, any bounded sequence contains a subsequence converging to some limit $\hat{x}$. Since K is closed, $\hat{x} \in K$. Everything indeed works because we are actually minimizing F over a *compact* set: K or $K \cap E_c$, where E_c is a non empty lower level set of F.
   If $\dim X = \infty$, the matter is more involved since in this case closed and bounded sets are not necessarily compact, and the possibility to extract a convergent subsequence from a bounded sequence is very seldom possible (see Sect. A.3.1 in Appendix A). Indeed, compactness turns out to be a very strong requirement.
   The situation improves significantly if we content ourselves of extracting *weakly* convergent sequences, rather than *norm* (or *strongly*) convergent ones. In fact, if X is a *reflexive Banach space* (see Proposition A.4), from a bounded sequence it is possible to extract a *weakly* convergent sequence. By weakening the notion of convergence, we gain in compactness; however, another problem arises. Assuming that a subsequence x_{k_j} weakly converges to some $\hat{x} \in X$, we cannot always deduce that $\hat{x} \in K$: indeed, a closed set is in general not *weakly sequentially closed* (see Example A.6). Thus, instead of (a), we require that
   (a') *K is weakly sequentially closed.*
   By Proposition A.6, if we add *convexity*, then a closed set K is also *weakly sequentially closed.* Thus, alternatively we may assume that
   (a'') *K is closed and convex.*
   Recall that if K is a convex set, we say that F is *convex* in K if for all $x, y \in K$ and all $t \in [0, 1]$,
   $$F(x + t(y - x)) \le F(x) + t(F(y) - F(x)). \tag{9.9}$$
   F is *strictly convex* if the inequality (9.9) is strict for all $t \in (0, 1)$ and all $x \ne y$.

3. *Lower semicontinuity.* The lower semicontinuity of F enters in the final step (9.8), to conclude that $\hat{x}$ is a minimizer. If dim $X = \infty$, we deal with *weakly* convergent sequences $\{x_{k_j}\}$, and therefore we require that (9.8) still holds when strong convergence is replaced by weak convergence. Instead of (b), we require that

   (b') F is *weakly sequentially lower semicontinuous.*

The meaning of (b') is provided in the following definition.

Definition 9.1. F is weakly sequentially lower semicontinuous if

$$F(x) \leq \lim \inf_{k \to +\infty} F(x_k) \tag{9.10}$$

whenever $x_k \rightharpoonup x$; equivalently, the lower level sets (9.4) are weakly sequentially closed for every $c \in \mathbb{R}$.

Thanks to the new assumptions (a') (or (a'')) and (b'), existence can be proved (see Theorem 9.1 below).

We know from Theorem 2.1 that the uniqueness of a minimizer follows from the *strict* convexity of the functional F and the dimension of X does not matter, we do not require to weaken any assumption. Hence, a first existence and uniqueness result for the minimization problem (9.2) is the following one.

Theorem 9.1. *Let X be a reflexive Banach space, $K \subseteq X$ be a weakly sequentially closed set and $F : X \to \mathbb{R}$. Assume that:*

1. $\inf_{x \in K} F(x) = \mu > -\infty$;
2. *if K is unbounded, the following condition holds*

$$F(x) \to +\infty \quad \text{as} \quad \|x\| \to +\infty;$$

3. *F is sequentially lower semicontinuous with respect to the weak convergence, that is*

$$\text{if } x_j \rightharpoonup x \text{ for } j \to \infty \text{ then } F(x) \leq \lim \inf_{j \to \infty} F(x_j).$$

Then there exists $\hat{x} \in K$ such that

$$F(\hat{x}) = \min_{x \in K} F(x). \tag{9.11}$$

Moreover, if F is strictly convex, $\hat{x}$ is the unique minimizer of F over K.

Proof. We can proceed as done in the previous section. Consider a minimizing sequence $\{x_k\} \subset K$, for which (9.7) holds. By hypothesis 2, the sequence $\{x_k\}$ is bounded. Since X is reflexive, by Theorem A.4, there exists a subsequence $\{x_{k_j}\} \subset K$ such that

$$x_{k_j} \rightharpoonup \hat{x} \quad \text{as} \quad k_j \to \infty$$

and $\hat{x} \in K$ because K is weakly sequentially closed. We use now the weak lower semicontinuity of F to conclude that

$$F(\hat{x}) \leq \lim \inf_{k_j \to \infty} F(x_{k_j}) = \mu.$$

Hence $F(\hat{x}) = \mu$ and $\hat{x}$ is a minimizer.

Finally, if in addition F is strictly convex over K, $\hat{x}$ is the unique minimizer. This follows by using the same argument of the finite dimensional case. □

Remark 9.1. We considered only *reflexive* Banach spaces because of the weak compactness of their bounded subsets. If X is not reflexive but it is the *dual of a separable space*[1], then bounded subsets of X are weakly* sequentially compact (according to the Banach-Alaoglu Theorem A.5). Thus Theorem 9.1 still holds if K is a weakly* closed subset of K and F is weakly* sequentially lower semicontinuous. •

9.2.2 Convexity and Lower Semicontinuity

If F is convex, the lower level sets introduced in (9.4) are convex for every $c \in \mathbb{R}$; indeed, for every $x, y \in E_c$

$$F\left(x + t(y - x)\right) \leq F\left(x\right) + t\left(F\left(y\right) - F\left(x\right)\right) \leq c\left(1 - t\right) + tc = c,$$

then $x + t(y - x) \in E_c$ for every $t \in [0, 1]$. The fact that a closed convex set of a Banach space X is weakly sequentially closed (see Proposition A.6) implies that lower semicontinuous and convex functionals are also weakly sequentially lower semicontinuous. More precisely,

Proposition 9.1. *Let $F : U \to \mathbb{R}$ be convex and lower semicontinuous. Then F is weakly sequentially lower semicontinuous.*

Proof. Every E_c is convex (thanks to the convexity of F) and closed (because of the lower semicontinuity of F). Thanks to Proposition A.6, E_c is also weakly sequentially closed and therefore F is weakly sequentially lower semicontinuous. □

Example 9.1. (Lower semicontinuity of a norm). Consider the map $F : X \to \mathbb{R}$,

$$F : x \to \|x\| .$$

F is called the *norm functional*; it is continuous since if $x_k \to x$, then $\|x_k\| \to \|x\|$. It is also convex (although not strictly convex) since, by the triangle inequality,

$$F\left(x + t(y - x)\right) = \|x + t(y - x)\| \leq t\left\|x\right\| + (1 - t)\left\|y\right\| = tF\left(x\right) + (1 - t)\,F\left(y\right).$$

for every $t \in [0, 1]$ and $x, y \in X$. Hence, F is weakly sequentially lower semicontinuous, that is, if $x_k \rightharpoonup x$,

$$\|x\| \leq \lim\inf_{k \to +\infty} \|x_k\| .$$

•

Example 9.2. Quadratic cost functionals like the ones considered in Chapter 5 are sequentially lower semicontinuous with respect to the weak convergence. •

Example 9.3. Let $X = L^p\left(\Omega\right), 1 \leq p \leq \infty$, and consider

$$K = \{u \in L^p\left(\Omega\right) : a\left(\mathbf{x}\right) \leq u\left(\mathbf{x}\right) \leq b\left(\mathbf{x}\right) \text{ a.e. in } \Omega\}, \text{ for } a, b \in L^p\left(\Omega\right).$$

K is obviously convex. To check that K is closed, let $\{u_k\} \subset K$ and $u_k \to u$ in $L^p\left(\Omega\right)$ as $k \to +\infty$. Then:

1. if $p = \infty$, $u_k \to u$ a.e. in Ω, hence $a\left(\mathbf{x}\right) \leq u\left(\mathbf{x}\right) \leq b\left(\mathbf{x}\right)$ a.e. in Ω;
2. if $1 \leq p < \infty$, up to a subsequence $u_k \to u$ a.e. in Ω, and again $a\left(\mathbf{x}\right) \leq u\left(\mathbf{x}\right) \leq b\left(\mathbf{x}\right)$ a.e. in Ω.

---

[1] For instance $L^\infty\left(\Omega\right)$ is not reflexive but it is the dual of $L^1\left(\Omega\right)$, which is separable.

In both cases we conclude that K is closed in $L^p(\Omega)$. Since $L^p(\Omega)$, for $1 < p < \infty$, is reflexive, then K is weakly sequentially closed. As $L^1(\Omega)$ and $L^\infty(\Omega)$ are not reflexive, in principle we cannot draw any conclusion. However, if $p = \infty$ and Ω is bounded, we can deduce that K is *weakly* sequentially closed.* In fact, let $\{u_k\} \subset K$ and $u_k \overset{*}{\rightharpoonup} u$ that is

$$\int_\Omega u_k v \to \int_\Omega u_k v \quad \text{for all } v \in L^1(\Omega). \tag{9.12}$$

Since $L^2(\Omega) \subset L^1(\Omega)$, (9.12) holds in particular for all $v \in L^2(\Omega)$. Thus $u_k \rightharpoonup u$ in $L^2(\Omega)$, whence $u \in K$ since K is weakly sequentially closed in $L^2(\Omega)$. •

9.2.3 First Order Optimality Conditions

We now extend Theorem 2.2 (valid for functions of n variables) to the case of Gâteaux or Fréchet differentiable functionals. See Sect. A.7 in Appendix A for the definition of Fréchet and Gâteaux differentials of a functional F, for which we use the symbols F' and $F'_{\mathcal{G}}$, respectively.

Theorem 9.2. *Let $\mathcal{U}$ be a Banach space, $\mathcal{U}_{ad} \subseteq \mathcal{U}$ be a closed and convex subset and F be a* Gâteaux differentiable *functional over $\mathcal{U}_{ad}$.*

1. *If $F(\hat{u}) = \min_{v \in \mathcal{U}_{ad}} F(v)$, then the following variational inequality holds,*
$$\langle F'_{\mathcal{G}}(\hat{u}), v - \hat{u} \rangle_* \geq 0 \qquad \forall v \in \mathcal{U}_{ad}. \tag{9.13}$$
2. *If F is convex and (9.13) holds, then*
$$F(\hat{u}) = \min_{v \in \mathcal{U}_{ad}} F(v).$$

Proof. 1. Let us consider $t \in (0,1)$ and $v \in \mathcal{U}_{ad}$. Since $\mathcal{U}_{ad}$ is a convex set, $\hat{u} + t(v - \hat{u}) \in \mathcal{U}_{ad}$, so that $F(\hat{u}) \leq F[\hat{u} + t(v - \hat{u})]$. Then,

$$\lim_{t \to 0+} \frac{F[\hat{u} + t(v - \hat{u})] - F(\hat{u})}{t} \geq 0$$

or, equivalently,

$$\langle F'_{\mathcal{G}}(\hat{u}), v - \hat{u} \rangle_* \geq 0 \qquad \forall v \in \mathcal{U}_{ad}.$$

2. As F is convex, thanks to Theorem A.20, for each $v \in \mathcal{U}_{ad}$ we have

$$F(v) - F(\hat{u}) \geq \langle F'_{\mathcal{G}}(\hat{u}), v - \hat{u} \rangle_* .$$

Thus, if (9.13) holds, we have that $F(v) \geq F(\hat{u})$ for any $v \in \mathcal{U}_{ad}$.

□

Remark 9.2. If F is a convex functional, we can provide further equivalent characterizations of a minimizer (see Exercise 1). •

9.2.4 Second Order Optimality Conditions*

For a quadratic functional, first-order conditions are not only necessary, but also sufficient for optimality. This is no longer true for a general nonlinear functional. For this reason, we state an additional result which, provided a suitable second order optimality condition is verified, ensures that a stationary point of F is also a (strict local) minimizer of F, according to the following definition.

Definition 9.2. Let $\mathcal{U}$ be a Banach space. We say that $\hat{u} \in \mathcal{U}$ is a local minimizer of F if there exists a neighborhood $B(\hat{u})$ such that

$$J(\hat{u}) \le J(v) \qquad \forall v \in B(\hat{u}).$$

The local minimizer is strict if $J(\hat{u}) < J(v)\ \forall v \in B(\hat{u}),\ v \neq \hat{u}$.

In the unconstrained case, the following sufficient condition is the analogue of Theorem 2.3.

Theorem 9.3. *Let $\mathcal{U}$ be a Banach space and F be a Fréchet differentiable functional, such that $F'(\hat{u}) = 0$. If F is twice Fréchet differentiable at $\hat{u}$, and if there exists $\alpha > 0$ such that*

$$\langle F''(\hat{u})w, w\rangle_{\mathcal{U}^*,\mathcal{U}} \ge \alpha \|w\|^2_{\mathcal{U}} \qquad \forall w \in \mathcal{U}, \tag{9.14}$$

then $\hat{u}$ is a strict local minimizer of F.

Proof. From Taylor formula (see A.74), we can write

$$F(\hat{u}+w) - F(\hat{u}) = \frac{1}{2}\langle F''(\hat{u})w, w\rangle_{\mathcal{U}^*,\mathcal{U}} + \varepsilon(\hat{u}, w)\|w\|^2_{\mathcal{U}}$$

with $\varepsilon(\hat{u}, w) \to 0$ as $\|w\|_{\mathcal{U}} \to 0$. Let r be small enough such that $|\varepsilon(\hat{u}, w)| < \alpha/2$ if $\|w\|_{\mathcal{U}} \le r$. Then, from (9.13) we get

$$F(\hat{u}+w) - F(\hat{u}) \ge \left(\frac{\alpha}{2} - \varepsilon(\hat{u}, w)\right)\|w\|^2_{\mathcal{U}} > 0$$

if $\|w\|_{\mathcal{U}} \le r$. □

Although these conditions are hardly useful by themselves in practice, they play a significant role in numerical approximation (see Sect. 9.6.2).

Remark 9.3. For constrained optimization problems, the coercivity requirement (9.14) on the whole space $\mathcal{U}$ is too strong. However, this requirement can be relaxed as follows. As done in the finite dimensional case, we can introduce a cone of critical directions. If $\mathcal{U} = L^2(\Omega)$ with box constraints, $\mathcal{U}_{ad} = \{u \in L^2(\Omega) : a \le u \le b, \text{ a.e. on } \Omega\}$, the critical cone associated to $\hat{u}$ is defined by

$$C(\hat{u}) = \{v \in L^2(\Omega) : J'(\hat{u})v = 0 \text{ and } v \text{ satisfies the sign conditions } v \le 0 \text{ if } \hat{u} = b,\ v \ge 0 \text{ if } \hat{u} = a\}.$$

Then, if in Theorem 9.3 the coercivity condition (9.14) is restricted to $w \in C(\hat{u})$, then $\hat{u}$ is a local minimum on $\mathcal{U}_{ad}$. For a comprehensive discussion on the role of second order conditions see, e.g., [59]. •

9.3 Control Constrained OCPs

Let $\mathcal{U}$, V be two *reflexive* Banach spaces and W be a Banach space. Let $F : V \times \mathcal{U} \to \mathbb{R}$ and $G : V \times \mathcal{U} \to W$. We consider the following constrained optimization problem:

$$F(y,u) \to \min \quad \text{subject to} \quad (y,u) \in V \times \mathcal{U}_{ad}, \quad G(y,u) = 0 \text{ in } W, \tag{9.15}$$

where $\mathcal{U}_{ad} \subseteq \mathcal{U}$ is a (nonempty) *closed, convex* set. If $\mathcal{U}_{ad} \subset \mathcal{U}$ then $u \in \mathcal{U}_{ad}$ represents a *control constraint.* $G(y,u) = 0$ plays the role of a (generally nonlinear) state equation. We say that a point (y,u) is *feasible* if it belongs to $V \times \mathcal{U}_{ad}$ and $G(y,u) = 0$.

9.3.1 Existence

Existence of an optimal pair $(\hat{y}, \hat{u})$ follows from the general Theorem 9.1. One has to check that:

(1) $\inf_{(y,u) \in V \times \mathcal{U}} F(y,u) = \mu > -\infty$ and the set of feasible points in nonempty;
(2) a minimizing sequence $\{(y_k, u_k)\}$ is bounded in $V \times \mathcal{U}$;
(3) the set of feasible points is weakly sequentially closed in $V \times \mathcal{U}$;
(4) F is sequentially weakly lower semicontinuous.

Theorem 9.4. *Under assumptions* (1) – (4) *there exists a (optimal) solution* $(\hat{y}, \hat{u})$ *to problem (9.15).*

Remark 9.4. According to Remark 9.1, if $V \times \mathcal{U}$ is not a reflexive space, but it is the dual of a separable space, theorem 9.4 still holds by replacing weakly by weakly* in conditions (3) and (4) above. •

Remark 9.5. Theorem 9.4 also holds when both control and state constraints are present, this latter being expressed under the form $y \in V_{ad}$ with $V_{ad} \subset V$. In this case the set of feasible points is given by

$$F_{ad} = \{(y,u) \in V \times \mathcal{U} \ : \ (y,u) \in V_{ad} \times \mathcal{U}_{ad}, \ G(y,u) = 0\}. \quad \bullet$$

Example 9.4. *The linear-quadratic case.* The linear-quadratic case (5.52) of Sect. 5.3 fits well in the above framework: here V, $\mathcal{U}$ are Hilbert spaces and the state equation corresponds to the abstract variational problem

$$a(y,\varphi) = \langle Bu + f, \varphi \rangle_{V^*,V} \qquad \forall \varphi \in V \tag{9.16}$$

where a is a continuous and coercive bilinear form, $B : \mathcal{U} \to V^*$, $f \in V^*$. Indeed, since a is continuous, for fixed $y \in V$, the linear form $\varphi \longmapsto a(y,\varphi)$ defines an element $Ay \in V^*$ such that $a(y,\varphi) = \langle Ay, \varphi \rangle_{V^*,V}$. Hence equation (9.16) can be written as an equation in $W = V^*$ in the equivalent form

$$G(y,u) = Ay - Bu - f = 0.$$

Moreover, assuming that Z is a Hilbert space and $C \in \mathcal{L}(V,Z)$, the cost functional

$$F(y,u) = \frac{1}{2} \|Cy - z_d\|_Z^2 + \frac{\beta}{2} \|u\|_{\mathcal{U}}^2,$$

is convex and continuous. Thus Theorem 9.4 applies. •

9.3.2 First Order Conditions. Adjoint Equation. Multipliers

To derive first order optimality conditions, we recast the OCP (9.15) under a *reduced form*. Clearly this is not always possible; for instance, the state system must be uniquely solvable. Then, we can apply the results introduced in Sect. 9.2. Although in principle we deal with global minimizers, necessary first order conditions holds for *local minimizers* too, i.e. for control functions that minimize the cost functional in a neighborhood of $\hat{u}$. We make the following assumptions:

(*i*) $F : V \times \mathcal{U} \to \mathbb{R}$ and $G : V \times \mathcal{U} \to W$ are $\mathcal{F}$-differentiable;

(*ii*) for each u in a neighborhood $B_{\mathcal{U}}$ of $\mathcal{U}_{ad}$, the state equation $G(y,u) = 0$ in W defines a unique control-to-state map $u \mapsto y(u)$, that is $G(y(u),u) = 0$ for any $u \in B_{\mathcal{U}}$; moreover, the control-to-state map is $\mathcal{G}-$differentiable;

(*iii*) the partial derivative $G'_y(y(u),u) \in \mathcal{L}(V,W)$ is a continuous isomorphism[2] between V and W, for every $u \in B_{\mathcal{U}}$. In particular (see, e.g., [274]) this is equivalent to require that the adjoint operator $G'_y(y(u),u)^* \in \mathcal{L}(W^*,V^*)$ is a continuous isomorphism between W^* and V^*, for every $u \in B_{\mathcal{U}}$.

Remark 9.6. If moreover G is continuously (respectively, twice continuously) $\mathcal{F}$-differentiable, by the Implicit Function Theorem A.23 the map $u \in B_{\mathcal{U}} \mapsto y(u) \in V$ is continuously (respectively, twice continuously) $\mathcal{F}$-differentiable. •

Under hypothesis (*ii*), we can introduce the *reduced formulation* of problem (9.15),

$$J(u) = F(y(u),u) \to \min \qquad \text{subject to } u \in \mathcal{U}_{ad}. \tag{9.17}$$

Note that by (i), (ii) and the chain rule (A.67), J is $\mathcal{G}-$differentiable in $B_{\mathcal{U}}$ and

$$J'_{\mathcal{G}}(u) = F'_y(y(u),u) \circ y'_{\mathcal{G}}(u) + F'_u(y(u),u) \in \mathcal{U}^*.$$

A direct application of Theorem 9.2 to problem (9.15) yields the following result.

Theorem 9.5. *Under the assumptions* (*i*), (*ii*), (*iii*), *a solution* $\hat{u}$ *of problem* (9.15) *fulfills the following optimality condition*

$$\left\langle J'_{\mathcal{G}}(\hat{u}), v - \hat{u} \right\rangle_* \geq 0 \qquad \forall v \in \mathcal{U}_{ad}. \tag{9.18}$$

Remark 9.7. In the linear-quadratic case, the reduced functional is strictly convex if $\beta > 0$, as it is the composition of a strictly convex P with the affine linear control-to state-map. Thus, we can deduce uniqueness of the minimizer. Furthermore, (9.18) is also a sufficient optimality condition. However, in the nonlinear case, even if the cost functional is strictly convex, the reduced functional is not in general convex, because of the nonlinearity of the control-to-state map. Thus, the variational inequality (9.18) is only a necessary optimality condition. •

As in the linear-quadratic case, we need to make the optimality condition (9.18) more effective, to exploit it numerically. Also in this case, we proceed by introducing a suitable adjoint problem to write the derivative $J'_{\mathcal{G}}(u)$ in a more convenient form.

[2] A linear operator $T : X \to Y$ is a continuous *isomorphism* if it is bijective and both T and T^{-1} are bounded. A *homeomorphism* (or topological isomorphism) is a continuous map between topological spaces that has a continuous inverse.

Proposition 9.2. *The following formula holds,*

$$J'_{\mathcal{G}}(u) = F'_u(y(u), u) + G'_u(y(u), u)^* p(u) \qquad \forall u \in B_{\mathcal{U}} \tag{9.19}$$

where $p(u) \in W^*$ *is the solution of the adjoint equation*

$$G'_y(y(u), u)^* p(u) = -F'_y(y(u), u). \tag{9.20}$$

Proof. Using the identity $G(y(u), u) = 0$ for all $u \in B_{\mathcal{U}}$, we can write the reduced cost functional in the augmented form

$$J(u) = F(y(u), u) + \langle p, G(y(u), u)\rangle_{W^*,W}$$

where $p \in W^*$ is a multiplier to be chosen later on. We differentiate (à la Gâteaux) the above expression along the direction $v \in \mathcal{U}$; using the chain rule (A.67) we get: ,

$$\begin{aligned}\langle J'_{\mathcal{G}}(u), v\rangle_{\mathcal{U}^*,\mathcal{U}} &= \langle F'_y(y(u), u), y'_{\mathcal{G}}(u)v\rangle_{V^*,V} + \langle p, G'_y(y(u), u)\, y'_{\mathcal{G}}(u)v\rangle_{W^*,W} \\ &\quad + \langle F'_u(y(u), u), v\rangle_{\mathcal{U}^*,\mathcal{U}} + \langle p, G'_u(y(u), u)v\rangle_{W^*,W}.\end{aligned}$$

Let $G'_u(y(u), u)^* \in \mathcal{L}(W^*, \mathcal{U}^*)$ be the adjoint of the operator $G_u(y(u), u) \in \mathcal{L}(\mathcal{U}, W)$; then we can write

$$\begin{aligned}\langle J'_{\mathcal{G}}(u), v\rangle_{\mathcal{U}^*,\mathcal{U}} &= \langle F'_y(y(u), u) + G'_y(y(u), u)^* p, y'_{\mathcal{G}}(u)v\rangle_{V^*,V} \\ &\quad + \langle F'_u(y(u), u) + G'_u(y(u), u)^* p, v\rangle_{\mathcal{U}^*,\mathcal{U}}\end{aligned} \tag{9.21}$$

As in the linear case, in this expression, the *expensive* terms are those containing $y'_{\mathcal{G}}(u)v$. Indeed, their evaluation would entail the solution of a (linear) PDE problem for any v, the so-called *sensitivity* equations. To avoid this, and to get rid of those terms, choose $p = p(u)$ as the solution of the *adjoint problem*

$$F'_y(y(u), u) + G'_y(y(u), u)^* p(u) = 0. \tag{9.22}$$

Since $G'_y(y(u), u)^*$ is a continuous isomorphism between W^* and V^*, equation (9.22) has a unique solution $p = p(u) \in W^*$. Inserting $p = p(u)$ into (9.21) we obtain

$$\langle J'_{\mathcal{G}}(u), v\rangle_{\mathcal{U}^*,\mathcal{U}} = \langle F'_u(y(u), u) + G'_u(y(u), u)^* p(u), v\rangle_{\mathcal{U}^*,\mathcal{U}} \qquad \forall v \in \mathcal{U}$$

from which (9.19) follows. □

Remark 9.8. Equation (9.20) may be often written in the *weak form*

$$\langle p, G_y(y(u), u)v\rangle_{W^*,W} = -\langle F_y(y(u), u), v\rangle_{\mathcal{U}^*,\mathcal{U}} \qquad \forall v \in \mathcal{U} \tag{9.23}$$

which is much easier to handle. •

The following theorem is an immediate consequence of Proposition 9.2 and Theorem 9.5.

Theorem 9.6. *Under the assumptions (i), (ii), (iii), if* $\hat{u}$ *is a solution of problem (9.15), then there exists a unique multiplier* $\hat{p} \in W^*$ *such that* $\hat{u}$, $\hat{p}$ *and* $\hat{y} = y(\hat{u})$ *fulfill the following optimality system:*

$$\begin{cases} G(\hat{y}, \hat{u}) = 0 & \textit{(state equation)} \\ G'_y(\hat{y}, \hat{u})^* \hat{p} = -F'_y(\hat{y}, \hat{u}) & \textit{(adjoint equation)} \\ \langle F'_u(\hat{y}, \hat{u}) + G'_u(\hat{y}, \hat{u})^* \hat{p}, v - \hat{u}\rangle_{\mathcal{U}^*,\mathcal{U}} \geq 0 \quad \forall v \in \mathcal{U}_{ad}. & \textit{(variational inequality)} \end{cases} \tag{9.24}$$

Introducing the Lagrangian functional

$$\mathcal{L} : V \times \mathcal{U} \times W^* \to \mathbb{R}, \quad \mathcal{L}(y, u, p) = F(y, u) + \langle p, G(y, u)\rangle_{W^*,W}, \tag{9.25}$$

where y, u, p are now independent variables, system (9.24) can be equivalently written in the following way:

$$\begin{cases} \mathcal{L}'_p(\hat{y},\hat{u},\hat{p}) = 0 & \text{(state equation)} \\ \mathcal{L}'_y(\hat{y},\hat{u},\hat{p}) = 0 & \text{(adjoint equation)} \\ \langle \mathcal{L}'_u(\hat{y},\hat{u},\hat{p}), v - \hat{u}\rangle_{\mathcal{U}^*,\mathcal{U}} \geq 0 \quad \forall v \in \mathcal{U}_{ad} & \text{(variational inequality).} \end{cases} \tag{9.26}$$

Remark 9.9. In the unconstrained case $\mathcal{U}_{ad} \equiv \mathcal{U}$, the variational inequality reduces to the equation

$$\langle F'_u(y,u) + G'_u(y,u)^* p(u), v\rangle_{\mathcal{U}^*,\mathcal{U}} = 0 \qquad \forall v \in \mathcal{U}$$

or

$$F'_u(y,u) + G'_u(y,u)^* p(u) = 0$$

so that (9.24) can be viewed as the Euler-Lagrange system for the Lagrangian functional, that is, the solutions of (9.24) are the stationary points of the Lagrangian function. •

Remark 9.10. We consider once again the linear-quadratic case. We have

$$G(y,u) = Ay - Bu - f, \qquad F(y,u) = \frac{1}{2}\|Cy - z_d\|_Z^2 + \frac{\beta}{2}\|u\|_{\mathcal{U}}^2$$

so that

$$G'_y(y,u) = A, \quad G'_y(y,u)^* = A^* \quad \text{and } G_u(y,u) = -B, \quad G'_u(y,u)^* = -B^*$$

while, denoting by $\mathcal{R}_Z$ the canonical Riesz isomorphism between Z^* and Z,

$$F'_y(y,u)\psi = (Cy - z_d, C\psi)_Z = \langle C^*\mathcal{R}_Z^{-1}(Cy - z_d), \psi\rangle_{V^*,V}.$$

Thus the adjoint equation reads

$$G'_y(y,u)^* p + F'_y(y,u) = A^*p + C^*\mathcal{R}_Z^{-1}(Cy - z_d) = 0$$

or, more explicitly,

$$a^*(p,\psi) = (Cy - z_d, C\psi)_Z,$$

while the variational inequality becomes

$$\langle F'_u(y,u) + G'_u(y,u)^* p, v - u\rangle_{\mathcal{U}^*,\mathcal{U}} = (\beta u - \mathcal{R}_{\mathcal{U}}B^*p, v - u)_{\mathcal{U}} \geq 0 \qquad \forall v \in \mathcal{U}_{ad}$$

from which we can obtain the expression of the gradient of the cost functional,

$$\nabla J(u) = \beta u - \mathcal{R}_{\mathcal{U}}B^*p.$$

Thus we recover the results of Sect. 5.3. •

9.3.3 Karush-Kuhn-Tucker Conditions

In the case of box constraints on the control, $\mathcal{U}_{ad} = \{u \in \mathcal{U} : a \leq u(\mathbf{x}) \leq b \text{ a.e. in } \Omega\}$, with $\mathcal{U} = L^2(\Omega)$, $a, b \in L^\infty(\Omega)$, we can express the optimality system (9.26) by means of the *Karush-Kuhn-Tucker multipliers*, as in the linear-quadratic case (see Proposition 5.4).

Since in this case $\mathcal{U} = \mathcal{U}^*$, we have from the third equation in (9.24),

$$J'_{\mathcal{G}}(u) = \nabla J(u) = \mathcal{L}'_u(y, u, p) = F_u(y, u) + \mathcal{G}_u(y, u)^* p.$$

The following result (a slight variation of Lemma 1.12 in [148]) provides an alternative characterization of the optimality conditions also in view of numerical approximation.

Lemma 9.1. *Let $\hat{u} \in \mathcal{U}_{ad}$. The following conditions are equivalent:*

(i) for every $v \in \mathcal{U}_{ad}$

$$(\nabla J(\hat{u}), v - \hat{u})_{L^2(\Omega)} \geq 0; \tag{9.27}$$

(ii) for all $v \in \mathcal{U}_{ad}$

$$\nabla J(\hat{u})(v - \hat{u}) \geq 0 \quad a.e \text{ in } \Omega;$$

(iii) there exist $\hat{\lambda}_a, \hat{\lambda}_b \in L^2(\Omega)$ such that

$$\begin{aligned} &\nabla J(\hat{u}) + \hat{\lambda}_b - \hat{\lambda}_a = 0, \\ &\hat{\lambda}_a \geq 0, \quad \hat{\lambda}_a(\hat{u} - a) = 0, \\ &\hat{\lambda}_b \geq 0, \quad \hat{\lambda}_b(b - \hat{u}) = 0; \end{aligned} \tag{9.28}$$

(iv) there exists $\hat{\lambda} \in L^2(\Omega)$ such that

$$\begin{aligned} &\nabla J(\hat{u}) + \hat{\lambda} = 0, \\ &\hat{u} = a \quad \text{where } \hat{\lambda} < 0, \qquad \hat{u} = b \quad \text{where } \hat{\lambda} > 0; \end{aligned} \tag{9.29}$$

(v) there exists $\hat{\lambda} \in L^2(\Omega)$ such that, for any real number $c \geq 0$,

$$\begin{aligned} &\nabla J(\hat{u}) + \hat{\lambda} = 0, \\ &\hat{\lambda} - \min\{0, \hat{\lambda} + c(\hat{u} - a)\} - \max\{0, \hat{\lambda} + c(\hat{u} - b)\} = 0; \end{aligned} \tag{9.30}$$

(vi) for all $\gamma > 0$,

$$\hat{u} = Pr_{\mathcal{U}_{ad}}(\hat{u} - \gamma \nabla J(\hat{u}))$$

where $Pr_{\mathcal{U}_{ad}} : L^2(\Omega) \to \mathcal{U}_{ad}$ is the (pointwise) projector operator onto $\mathcal{U}_{ad}$, that is,

$$Pr_{\mathcal{U}_{ad}} w(\mathbf{x}) = \max\{a(\mathbf{x}), \min\{w(\mathbf{x}), b(\mathbf{x})\}\} = \min\{b(\mathbf{x}), \max\{w(\mathbf{x}), a(\mathbf{x})\}\}.$$

Proof. *(i)* $\Longleftrightarrow$ *(ii)* follows from Lemma 5.41.

(ii) $\Longleftrightarrow$ *(iii)*. Assume *(ii)* holds. Set

$$\hat{\lambda}_b = \max\{-\nabla J(\hat{u}), 0\}, \hat{\lambda}_a = \max\{\nabla J(\hat{u}), 0\}.$$

Then $\nabla J(\hat{u}) + \hat{\lambda}_b - \hat{\lambda}_a = 0$, $a \leq \hat{u} \leq b$ and $\hat{\lambda}_a \geq 0$, $\hat{\lambda}_b \geq 0$. If $\hat{u}(\mathbf{x}) > a(\mathbf{x})$ then *(ii)* gives $\nabla J(\hat{u})(\mathbf{x}) = 0$ and therefore $\hat{\lambda}_a(\mathbf{x}) = 0$. Similarly, if $\hat{u}(\mathbf{x}) < b(\mathbf{x})$ then *(ii)* gives $\nabla J(\hat{u})(\mathbf{x}) = 0$ and therefore $\hat{\lambda}_b(\mathbf{x}) = 0$. Thus *(ii)* $\Longrightarrow$ *(iii)*.

On the other hand, let *(iii)* hold. If $a(\mathbf{x}) < \hat{u}(\mathbf{x}) < b(\mathbf{x})$, then $\hat{\lambda}_b(\mathbf{x}) = \hat{\lambda}_a(\mathbf{x}) = 0$ and $\nabla J(\hat{u})(\mathbf{x}) = 0$. If $a(\mathbf{x}) = \hat{u}(\mathbf{x}) < b(\mathbf{x})$, then $\hat{\lambda}_b(\mathbf{x}) = 0$, $\nabla J(\hat{u})(\mathbf{x}) \geq 0$ and $\nabla J(\hat{u})(\mathbf{x})(v(\mathbf{x}) - a(\mathbf{x})) \geq 0$. If $a(\mathbf{x}) < \hat{u}(\mathbf{x}) = b(\mathbf{x})$, then $\hat{\lambda}_a(\mathbf{x}) = 0$, $\nabla J(\hat{u})(\mathbf{x}) \leq 0$ and $\nabla J(\hat{u})(\mathbf{x})(v(\mathbf{x}) - b(\mathbf{x})) \geq 0$. Thus *(iii)* $\Longrightarrow$ *(ii)*.

(iii) $\Longleftrightarrow$ *(iv)*. The implication *(iii)* $\Longrightarrow$ *(iv)* follows just by setting $\hat{\lambda} = \hat{\lambda}_b - \hat{\lambda}_a$. Vice versa, choose $\hat{\lambda}_b = \max\{-\nabla J(\hat{u}), 0\}, \hat{\lambda}_a = \max\{\nabla J(\hat{u}), 0\}$ and use $\nabla J(\hat{u}) + \hat{\lambda} = 0$. If $\hat{\lambda}_a(\mathbf{x}) > 0$, then $\hat{\lambda}_b(\mathbf{x}) = 0$ so that $\hat{\lambda} < 0$ which implies $\hat{u}(\mathbf{x}) = a$. Similarly $\hat{\lambda}_b(\mathbf{x}) > 0$ implies $\hat{u}(\mathbf{x}) = b$. Thus *(iv)* $\Longrightarrow$ *(iii)*.

(iv) $\Longleftrightarrow$ *(v)*. Assume *(iv)* holds. If $\hat{u}(\mathbf{x}) < b$ then $\hat{\lambda}(\mathbf{x}) \leq 0$. If $\hat{\lambda}(\mathbf{x}) = 0$,

$$\hat{\lambda}(\mathbf{x}) - \min\{0, \hat{\lambda}(\mathbf{x}) + c(\hat{u}(\mathbf{x}) - a)\} - \max\{0, \hat{\lambda}(\mathbf{x}) + c(\hat{u}(\mathbf{x}) - b)\} = \\ - \min\{0, c(\hat{u}(\mathbf{x}) - a)\} - \max\{0, c(\hat{u}(\mathbf{x}) - b)\} = 0;$$

if $\hat{\lambda}(\mathbf{x}) < 0$ then $\hat{u}(\mathbf{x}) = a$ and

$$\hat{\lambda}(\mathbf{x}) - \min\{0, \hat{\lambda}(\mathbf{x}) + c(\hat{u}(\mathbf{x}) - a)\} - \max\{0, \hat{\lambda}(\mathbf{x}) + c(\hat{u}(\mathbf{x}) - b)\} = \hat{\lambda}(\mathbf{x}) - \min\{0, \hat{\lambda}(\mathbf{x})\} = 0.$$

By a similar argument, if $\hat{u}(\mathbf{x}) > a$ then $\hat{\lambda}(\mathbf{x}) \geq 0$, and

$$\hat{\lambda}(\mathbf{x}) - \min\{0, \hat{\lambda}(\mathbf{x}) + c(\hat{u}(\mathbf{x}) - a)\} - \max\{0, \hat{\lambda}(\mathbf{x}) + c(\hat{u}(\mathbf{x}) - b)\} = 0;$$

if $\hat{u}(\mathbf{x}) = b$ then $\hat{\lambda}(\mathbf{x}) \geq 0$, so that

$$\hat{\lambda}(\mathbf{x}) - \min\{0, \hat{\lambda}(\mathbf{x}) + c(b - a)\} - \max\{0, \hat{\lambda}(\mathbf{x})\} = \hat{\lambda}(\mathbf{x}) - \hat{\lambda}(\mathbf{x}) = 0;$$

if $\hat{u}(\mathbf{x}) = a$ then $\hat{\lambda}(\mathbf{x}) \leq 0$, so that

$$\hat{\lambda}(\mathbf{x}) - \min\{0, \hat{\lambda}(\mathbf{x})\} - \max\{0, \hat{\lambda}(\mathbf{x}) + c(a - b)\} = \hat{\lambda}(\mathbf{x}) - \hat{\lambda}(\mathbf{x}) = 0.$$

This proves that *(iv)* $\Longrightarrow$ *(v)*. Vice versa, assume *(v)* holds. Then, if $\hat{\lambda}(\mathbf{x}) < 0$, from

$$0 = \hat{\lambda}(\mathbf{x}) - \min\{0, \hat{\lambda}(\mathbf{x}) + c(\hat{u}(\mathbf{x}) - a)\} - \max\{0, \hat{\lambda}(\mathbf{x}) + c(\hat{u}(\mathbf{x}) - b)\} = c(\hat{u}(\mathbf{x}) - a)$$

we deduce $\hat{u}(\mathbf{x}) = a$. Similarly we infer that if $\hat{\lambda}(\mathbf{x}) > 0$, then $\hat{u}(\mathbf{x}) = b$. Thus *(v)* $\Longrightarrow$ *(iv)*.

(i) $\Longleftrightarrow$ *(vi)*. It is enough to write the condition

$$(\nabla J(\hat{u}), v - \hat{u})_{L^2(\Omega)} \geq 0 \qquad \forall v \in \mathcal{U}_{ad}$$

in the equivalent form

$$(\hat{u} - (\hat{u} - \gamma \nabla J(\hat{u})), v - \hat{u})_{L^2(\Omega)} \geq 0 \qquad \forall v \in \mathcal{U}_{ad},$$

showing that $\hat{u} - \gamma \nabla J(\hat{u})$ is the projection of $\hat{u}$ onto $\mathcal{U}_{ad}$. □

9.3.4 Second Order Conditions

Let us assume that, in addition to conditions *(i)*, *(ii)* and *(iii)* displayed in Sect. 9.3.2, F and G are twice continuously $\mathcal{F}$−differentiable. Then also the control-to-state map $u \mapsto y(u)$ is twice continuously $\mathcal{F}$−differentiable. In the case of an OCP like (9.15), the positivity condition (9.14) can be better specified by requiring it to be fulfilled by the solutions of the linearized state equation (see (9.34) below).

We first show a formula for the second derivative of the reduced functional $J(u) = F(y(u), u)$, in terms of the Lagrangian function

$$\mathcal{L} : V \times \mathcal{U} \times W^* \to \mathbb{R}, \quad \mathcal{L}(y, u, p) = F(y, u) + \langle p, G(y, u) \rangle_{W^*, W}, \tag{9.31}$$

which involves the adjoint state $p(u)$, solution to (9.20), and avoids an explicit dependence on $y''(u)$.

Proposition 9.3. *Let $J : V \times \mathcal{U} \to \mathbb{R}$ and $G : V \times \mathcal{U} \to W$ be twice continuously $\mathcal{F}$-differentiable. The following formula holds for $J''(u) : \mathcal{U} \to \mathcal{L}(\mathcal{U}, \mathcal{U}^*)$:*

$$\langle J''(u)v_1, v_2 \rangle_{\mathcal{U}^*, \mathcal{U}} = \langle \mathcal{L}''_{yy}(y(u), u, p(u))z_2, z_1 \rangle_{V^*, V} + \langle \mathcal{L}''_{yu}(y(u), u, p(u))v_2, z_1 \rangle_{V^*, V} \\ + \langle \mathcal{L}''_{uy}(y(u), u, p(u))z_2, v_1 \rangle_{\mathcal{U}^*, \mathcal{U}} + \langle \mathcal{L}''_{uu}(y(u), u, p(u))v_2, v_1 \rangle_{\mathcal{U}^*, \mathcal{U}} \tag{9.32}$$

where $z_1 = y'(u)v_1$, $z_2 = y'(u)v_2$, $\forall v_1, v_2 \in \mathcal{U}$.

Proof. We have the identity

$$J(u) = \mathcal{L}(y(u), u, p) \qquad \text{for every } p \in W^*.$$

By differentiating along the direction $v_1 \in \mathcal{U}$, we obtain

$$\langle J'(u), v_1\rangle_{\mathcal{U}^*,\mathcal{U}} = \langle \mathcal{L}'_y(y(u), u, p), y'(u)v_1\rangle_{V^*,V} + \langle \mathcal{L}'_u(y(u), u, p), v_1\rangle_{\mathcal{U}^*,\mathcal{U}}.$$

Differentiating the expression above along the direction $v_2 \in \mathcal{U}$ we get

$$\begin{aligned}\langle J''(u)v_1, v_2\rangle_{\mathcal{U}^*,\mathcal{U}} = {} & \langle \mathcal{L}'_y(y(u), u, p), y''(u)[v_1, v_2]\rangle_{V^*,V} + \langle \mathcal{L}''_{yy}(y(u), u, p)y'(u)v_2, y'(u)v_1\rangle_{V^*,V} \\ & + \langle \mathcal{L}''_{yu}(y(u), u, p)v_2, y'(u)v_1\rangle_{V^*,V} + \langle \mathcal{L}''_{uy}(y(u), u, p)y'(u)v_2, v_1\rangle_{\mathcal{U}^*,\mathcal{U}} \\ & + \langle \mathcal{L}''_{uu}(y(u), u, p)v_2, v_1\rangle_{\mathcal{U}^*,\mathcal{U}}.\end{aligned}$$

Choosing $p = p(u)$ so that $\mathcal{L}'_y(y(u), u, p(u)) = 0$, the term containing $y''(u)$ vanishes and we end up with

$$\begin{aligned}\langle J''(u)v_1, v_2\rangle_{\mathcal{U}^*,\mathcal{U}} = {} & \langle \mathcal{L}''_{yy}(y(u), u, p(u))y'(u)v_2, y'(u)v_1\rangle_{V^*,V} + \langle \mathcal{L}''_{yu}(y(u), u, p(u))v_2, y'(u)v_1\rangle_{V^*,V} \\ & + \langle \mathcal{L}''_{uy}(y(u), u, p(u))y'(u)v_2, v_1\rangle_{\mathcal{U}^*,\mathcal{U}} + \langle \mathcal{L}''_{uu}(y(u), u, p(u))v_2, v_1\rangle_{\mathcal{U}^*,\mathcal{U}}.\end{aligned}$$

□

We can equivalently express formula (9.32) as

$$\begin{aligned}J''(u) = {} & y'(u)^*\mathcal{L}''_{yy}(y(u), u, p(u))y'(u) + y'(u)^*\mathcal{L}''_{yu}(y(u), u, p(u)) \\ & + \mathcal{L}''_{uy}(y(u), u, p(u))y'(u) + \mathcal{L}''_{uu}(y(u), u, p(u))\end{aligned}$$

or, in a more compact way,

$$J''(u) = T(u)^*\mathcal{L}''_{(y,u)}(y(u), u, p(u))T(u)$$

by introducing the abridged notation

$$\mathcal{L}''_{(y,u)} = \begin{pmatrix} \mathcal{L}''_{yy} & \mathcal{L}''_{yu} \\ \mathcal{L}''_{uy} & \mathcal{L}''_{uu} \end{pmatrix}, \qquad T(u) = \begin{pmatrix} y'(u) \\ I_{\mathcal{U}} \end{pmatrix}$$

where $I_{\mathcal{U}}$ is the identity operator in $\mathcal{U}$. Note that $\mathcal{L}''_{(y,u)}$ is a bilinear form on $V \times \mathcal{U}$ into $\mathbb{R}$.

Remark 9.11. An efficient calculation of $J''(u)v$ (the action of $J''(u)$ on a generic $v \in \mathcal{U}$) can be carried out as follows:

1. given $y(u) \in V$, evaluate the sensitivity $\delta_v y = y'(u)v$ by solving the linearized state problem
$$G'_y(y(u), u)\delta_v y + G'_u(y(u), u)v = 0;$$
2. compute
$$w_1 = \mathcal{L}''_{yy}(y(u), u, p(u))\delta_v y + \mathcal{L}''_{yu}(y(u), u, p(u))v;$$
3. solve the adjoint problem
$$G'_y(y(u), u)p_{w_1} = -w_1;$$
4. finally, set
$$J''(u)v = \mathcal{L}''_{uy}(y(u), u, p(u))\delta_v y + \mathcal{L}''_{uu}(y(u), u, p(u))v + G'_u(y(u), u)^* p_{w_1}.$$

•

This calculation represents indeed a basic step at each iteration of the Newton method, when solving a nonlinear OCP numerically.

We can now state a result providing a second order sufficient condition to be fulfilled by a triple (y, u, p) in order to be a local minimum. Indeed, directly from Proposition 9.3 and exploiting the result of Proposition 9.3, it is possible to prove the following result.

Theorem 9.7. *Let $F : V \times \mathcal{U} \to \mathbb{R}$ and $G : V \times \mathcal{U} \to W$ be twice continuously Fréchet differentiable, and let $(\hat{y}, \hat{u}, \hat{p})$ be a solution to the optimality system (9.24). If there exists a constant $\alpha > 0$ such that*

$$\left\langle \mathcal{L}''_{(y,u)}(\hat{y}, \hat{u}, \hat{p}) \begin{pmatrix} \hat{z} \\ v \end{pmatrix}, \begin{pmatrix} \hat{z} \\ v \end{pmatrix} \right\rangle_{V^* \times \mathcal{U}^*, V \times \mathcal{U}} \geq \alpha \|v\|^2_{\mathcal{U}} \tag{9.33}$$

for all $v \in \mathcal{U}$ and $\hat{z} = \hat{z}(v) \in V$ that satisfies the linearized equation

$$G'_y(\hat{y}, \hat{u})z + G'_u(\hat{y}, \hat{u})v = 0 \qquad \text{in } W, \tag{9.34}$$

then there exist two constants $\varepsilon, \sigma > 0$ such that

$$F(y, u) \geq F(\hat{y}, \hat{u}) + \sigma \|u - \hat{u}\|^2_{\mathcal{U}} \qquad \forall u \in \mathcal{U} \cap B_\varepsilon(\hat{u}).$$

In particular, $(\hat{y}, \hat{u})$ is a local minimizer of F.

Remark 9.12. Conditions (9.33)–(9.34) express the coercivity of $\mathcal{L}''_{(y,u)}(\hat{y}, \hat{u}, \hat{p})$ along the directions tangential to the constraints given by the state equation. This sufficient condition is the infinite dimensional counterpart of the similar condition in Proposition 2.2. In spite of its theoretical relevance, the verification of the coercivity condition might be rather involved. However, the computation of the second derivative play a significant role in numerical approximations. •

9.4 Distributed Control of a Semilinear State Equation

Let $\Omega \subset \mathbb{R}^d$, $d \leq 3$, be a bounded, smooth (e.g. C^2) domain, $\beta \geq 0$ and $z_d \in L^6(\Omega)$ a given target. We consider the following control problem:

$$F(y, u) = \frac{1}{6} \|y - z_d\|^6_{L^6(\Omega)} + \frac{\beta}{2} \|u\|^2_{L^2(\Omega)} \to \min \tag{9.35}$$

subject to

$$u \in \mathcal{U}_{ad} = \left\{ u \in L^2(\Omega) : a \leq u(\mathbf{x}) \leq b, \text{ a.e. in } \Omega \right\}, \tag{9.36}$$

and to the semilinear state system (9.1), here reported for the reader's convenience:

$$\begin{cases} -\Delta y + y + y^3 = u & \text{in } \Omega \\ \dfrac{\partial y}{\partial n} = 0 & \text{on } \Gamma. \end{cases} \tag{9.37}$$

Analysis of the state system. Problem (9.37) is a Neumann problem for a diffusion-reaction equation, with reaction given by $c(y) = y + y^3$. We look for weak solutions, that is for functions $y \in V = H^1(\Omega)$ satisfying the relation

$$\int_\Omega \left[\nabla y \cdot \nabla \varphi + \left(y + y^3 \right) \varphi \right] = \int_\Omega u\varphi \qquad \forall \varphi \in V. \tag{9.38}$$

Note that, since $d \leq 3$, the Sobolev embedding Theorem A.15 gives $\|y\|_{H^1(\Omega)} \leq C_S \|y\|_V$, and

$$\left| \int_\Omega y^3 \varphi \right| \leq \left(\int_\Omega y^6 \right)^{1/2} \left(\int_\Omega \varphi^2 \right)^{1/2} = \|y\|^3_{L^6(\Omega)} \|\varphi\|_{L^2(\Omega)}$$

so that all integrals in (9.38) are well defined.

The following result holds[3].

Proposition 9.4. *For every $u \in L^2(\Omega)$, problem (9.37) has a unique weak solution $y = y(u)$. Moreover $y(u) \in H^2(\Omega) \cap C(\overline{\Omega})$ and*

$$\|y\|_{H^2(\Omega)} + \|y\|_{C(\overline{\Omega})} \leq C(\Omega)\|u\|_{L^2(\Omega)}. \tag{9.39}$$

Proof. To prove existence we use Leray-Schauder Theorem A.8. Let $w \in V$ be fixed. By Lax-Milgram Theorem, the linear problem

$$-\Delta z + z = -w^3 + u$$

has a unique weak solution $z = T(w) \in V$. Since Ω is smooth and the right hand side is in $L^2(\Omega)$, from elliptic regularity it follows that $z \in H^2(\Omega)$ and

$$\|z\|_{H^2(\Omega)} \leq c(\Omega)\left\{\left\|w^3\right\|_{L^2(\Omega)} + \|u\|_{L^2(\Omega)}\right\} = c(\Omega)\left\{\|w\|^3_{L^6(\Omega)} + \|u\|_{L^2(\Omega)}\right\}. \tag{9.40}$$

Since $H^2(\Omega)$ is compactly embedded into V, the operator $T : V \to V$ is compact. To check the continuity of T, observe that, if $w_n \to w$ in V, then $\|w_n\|_{L^6(\Omega)}, \|w\|_{L^6(\Omega)} \leq M$ and from (9.40),

$$\|T(w_n) - T(w)\|_{H^2(\Omega)} \leq c(\Omega)\left\|w_n^3 - w^3\right\|_{L^2(\Omega)}.$$

Now, $\left|w_n^3 - w^3\right| = \left|(w_n - w)\left(w_n^2 + w_n w + w^2\right)\right| \leq |w_n - w|\left(2w_n^2 + 2w^2\right)$, and using Hölder inequality with exponents $p = 3, q = 3/2$ we get

$$\begin{aligned}
\left\|w_n^3 - w^3\right\|^2_{L^2(\Omega)} &\leq \|w_n - w\|^2_{L^6(\Omega)}\left\|2w_n^2 + 2w^2\right\|^2_{L^3(\Omega)} \\
&\leq 8\left(\|w_n\|^4_{L^6(\Omega)} + \|w\|^4_{L^6(\Omega)}\right)\|w_n - w\|^2_{L^6(\Omega)} \\
&\leq 8M^4\|w_n - w\|^2_{L^6(\Omega)} \leq 8C_S M^4\|w_n - w\|^2_V.
\end{aligned}$$

Therefore, T is (Lipschitz) continuous in V.

Finally, for every $s \in (0, 1]$ and every z solution of $z = sT(z)$ we have, inserting $\varphi = z$ and $y = z/s$ into (9.38),

$$\int_\Omega \left(|\nabla z|^2 + z^2 + s^{-2}z^4\right) \leq s\int_\Omega |uz| \leq \frac{1}{2}\|z\|^2_{L^2(\Omega)} + \frac{1}{2}\|u\|^2_{L^2(\Omega)}$$

from which

$$\|z\|^2_V \leq \|u\|_{L^2(\Omega)}.$$

The Leray-Schauder Theorem implies that T has a fixed point $y(u)$, which therefore is a weak solution to (9.37). Moreover, since $T(V) \subset H^2(\Omega)$ it follows that $y(u) \in H^2(\Omega)$. The Sobolev embedding Theorem A.15 gives the continuity of $y(u)$ in $\overline{\Omega}$ and the estimate (9.39).

To show uniqueness, suppose that y_1 and y_2 are weak solutions. Then $w = y_1 - y_2$ satisfies the equation

$$\int_\Omega [\nabla w \cdot \nabla\varphi + w\varphi] + \int_\Omega \left[y_1^2 + y_1 y_2 + y_2^2\right] w\varphi = 0 \qquad \forall\varphi \in V.$$

Choosing $\varphi = w$, we get

$$\int_\Omega \left[|\nabla w|^2 + w^2\right] = 0$$

from which $w = 0$. □

Remark 9.13. Note that the control-to-state map $u \mapsto y(u)$ is Lipschitz continuous. Indeed

$$h(u, v) = y(u + v) - y(u)$$

[3] When Ω is a bounded Lipschitz domain, the well posedness of problem (9.37) follows from the Browder-Minty theory of monotone operators, since $c : \mathbb{R} \to \mathbb{R}$ is continuous and monotonically increasing. The Browder-Minty theorem actually extends the result of the Lax-Milgram Theorem to the case of nonlinear monotone operators. See, e.g., [231, Theorem 10.49] for a statement of the Browder-Minty theorem and Chapter 10 of the same reference for a comprehensive introduction to the subject.

satisfies the equation

$$-\Delta h + (1 + c(u, v))h = v$$

with $\frac{\partial h}{\partial n} = 0$ on $\partial\Omega$, where $c(u,v) = y^2(u+v) + y(u+v)\,y(u) + y^2(u)$. Since c is bounded and nonnegative, we get, after multiplying by h and integrating by parts,

$$\int_\Omega (|\nabla h|^2 + h^2) \le \int_\Omega vh.$$

Using Schwarz inequality we find

$$\|h\|_V \le \|v\|_{L^2(\Omega)} \tag{9.41}$$

which is the Lipschitz continuity of $u \mapsto y(u)$. •

$\mathcal{F}$–differentiability of the control-to-state map. We must show that, for every $u \in L^2(\Omega)$, there exists a bounded, linear operator $y'(u) : L^2(\Omega) \to V$, such that, for every $v \in L^2(\Omega)$,

$$\frac{\|y(u+v) - y(u) - y'(u)\,v\|_V}{\|v\|_{L^2(\Omega)}} \to 0 \quad \text{as } \|v\|_{L^2(\Omega)} \to 0. \tag{9.42}$$

The expression of $y'(u)$ can be guessed, by formally taking the derivative with respect to u, along the direction v, of both equations in (9.37). We find that $y'(u)$ is the linear map that associates to $v \in L^2(\Omega)$ the weak solution $z = y'(u)\,v$ of the problem

$$\begin{cases} -\Delta z + z + 3y^2 z = v & \text{in } \Omega \\ \dfrac{\partial z}{\partial n} = 0 & \text{on } \Gamma. \end{cases} \tag{9.43}$$

For a rigorous proof, see Exercise 4.

Existence of an optimal control. We recast our OCP into the framework (9.15). First introduce the following operators:

a) the continuous isomorphism $A : V \to V^*$, associated to the bilinear form

$$a(y,\varphi) = \int_\Omega (\nabla y \cdot \nabla\varphi + y\varphi)$$

through the relation

$$a(y,\varphi) = \langle Ay, \varphi\rangle_{V^*,V} \qquad \forall y, \varphi \in V.$$

Since a is symmetric, we have $A = A^*$;

b) the operator $\Phi : V \to L^2(\Omega)$ given by

$$y \mapsto y^3.$$

From Example A.14, $\Phi \in C^1(V)$ and, for all $\varphi, y_0 \in V$, there holds

$$\Phi'(y_0)\varphi = 3y_0^2\varphi;$$

c) the embedding operator $B : L^2(\Omega) \to V^*$ defined by

$$\langle Bv, \varphi\rangle_{V^*,V} = \int_\Omega v\varphi.$$

Then problem (9.37) is equivalent to the equation

$$G(y,u) = Ay + B[\Phi(y) - u] = 0 \text{ in } V^*,$$

so that our OCP fits into the formulation (9.15). We thus have:

Proposition 9.5. *There exists an optimal pair* $(\hat{y}, \hat{u}) \in H^1(\Omega) \times L^2(\Omega)$ *solution of the optimal control problem* (9.37)–(9.35).

Proof. We use Theorem 9.4. Let us check assumptions (1) – (4).

(1) and (2) directly follow from (9.39).

(3) Let $\{(y_k, u_k)\} \subset V \times \mathcal{U}_{ad}$, with $G(y_k, u_k) = 0$, such that $y_k \rightharpoonup y$ in V and $u_k \rightharpoonup u$ in $L^2(\Omega)$. We have to show that (y, u) is feasible. Clearly $(y, u) \in V \times \mathcal{U}_{ad}$ since $\mathcal{U}_{ad}$ is weakly closed. We prove that $G(y, u) = 0$. For each $\varphi \in V$ we have:

$$\begin{aligned}\langle G(y_k, u_k) - G(y, u), \varphi\rangle_{V^*,V} &= \langle A(y_k - y) + B[\Phi(y_k) - \Phi(y) - (u_k - u)], \varphi\rangle_{V^*,V} \\ &= \int_\Omega \nabla(y_k - y)\nabla\varphi + \int_\Omega (y_k^3 - y^3)\varphi - \int_\Omega (u_k - u)\varphi.\end{aligned}$$

Clearly both the first and the third terms on the right tend to zero thanks to the weak convergence of (y_k, u_k) to (y, u). For the second term we observe that, as $\{y_k\}$ is bounded in $H^2(\Omega)$, by Rellich Theorem A.13, $\{y_k\}$ (not only a subsequence, in this case) converges to y (strongly) in V, and hence in $L^6(\Omega)$, too. Therefore we can write, using Holder inequality with exponents $1/6, 1/3, 1/2$:

$$\left|\int_\Omega (y_k^3 - y^3)\varphi\right| \le \|y_k - y\|_{L^6(\Omega)} \left\|y_k^2 - y_k y + y^2\right\|_{L^3(\Omega)} \|\varphi\|_{L^2(\Omega)} \le C\|y_k - y\|_{L^6(\Omega)} \|\varphi\|_{L^2(\Omega)}$$

so that $0 = G(y_k, u_k) \rightharpoonup G(y, u)$ in V^*.

(4) F is clearly continuous and therefore also sequentially weakly continuous in $L^6(\Omega) \times \mathcal{U}_{ad}$.

Finally, by Theorem 9.4, we can conclude that there exists an optimal pair $(\hat{y}, \hat{u})$, solution of the optimal control problem (9.35)–(9.37). □

First order optimality conditions. We can recover a system of first-order optimality conditions for the OCP (9.35)–(9.37) by applying the result of Theorem 9.6. We can show the following

Theorem 9.8. *Let* $d \le 3$. *Let* $(\hat{y}, \hat{u})$ *be an optimal solution. There exists* $\hat{p} \in H^1(\Omega)$ *such that* $\hat{y}, \hat{u}$ *and* $\hat{p}$ *fulfill the following system:*

$$\begin{cases} -\Delta\hat{y} + \hat{y} + \hat{y}^3 = \hat{u} & \text{in } \Omega, \quad \dfrac{\partial\hat{y}}{\partial n} = 0 \text{ on } \Gamma \\ -\Delta\hat{p} + \hat{p} + 3\hat{y}^2\hat{p} = -(\hat{y} - z_d)^5 & \text{in } \Omega, \quad \dfrac{\partial\hat{p}}{\partial n} = 0, \text{ on } \Gamma \\ (\beta\hat{u} - \hat{p}, v - \hat{u})_{L^2(\Omega)} \ge 0 & \forall v \in \mathcal{U}_{ad}. \end{cases}$$

Proof. We check assumptions $(i), (ii), (iii)$ of Theorem 9.6.

(i) $F : V \times L^2(\Omega) \to \mathbb{R}$ and $G : V \times L^2(\Omega) \to V^*$ are continuously $\mathcal{F}$-differentiable; indeed, we have:

$$\begin{aligned} &F'_y(y, u)\psi = \left((y - z_d)^5, \psi\right)_{L^2(\Omega)} && \forall\psi \in V \\ &F'_u(y, u)v = (u, v)_{L^2(\Omega)} && \forall v \in L^2(\Omega) \\ &G'_y(y, u)\varphi = A\varphi + B[\Phi'(y)\varphi] && \forall\varphi \in V \\ &G'_u(y, u)v = -Bv && \forall v \in L^2(\Omega). \end{aligned} \tag{9.44}$$

Note that $(y - z_d)^5 \in L^{6/5}(\Omega) \subset V^*$.

(ii) We have already checked that, for each $u \in L^2(\Omega)$, the state equation $G(y, u) = 0$ in Z defines a $\mathcal{G}$−differentiable control-to-state map $u \mapsto y(u)$.

(iii) The partial derivative $G'_y(y, u)$ is a continuous isomorphism between V and V^* for every $(y, u) \in V \times L^2(\Omega)$. In fact, let $f \in V^*$ and consider the problem

$$G'_y(y, u)\varphi = A\varphi + B[\Phi'(y)\varphi] = f \quad \text{in } V^*. \tag{9.45}$$

In weak form, we have

$$a(\varphi,\eta)=\int_\Omega\left[\nabla\varphi\cdot\nabla\eta+\varphi\eta+3y^2\varphi\eta\right]=\langle f,\eta\rangle_*\qquad\forall\eta\in V.$$

The bilinear form a is clearly coercive in V and symmetric. It is also continuous, since, by Hölder inequality with exponents $\frac{1}{3},\frac{1}{6},\frac{1}{6}$, we have

$$\left|\int_\Omega y^2\varphi\eta\right|\leq\left\|y^2\right\|_{L^3(\Omega)}\|\varphi\|_{L^6(\Omega)}\|\varphi\|_{L^6(\Omega)}\leq c\|y\|^2_{L^6(\Omega)}\|\varphi\|_V\|\eta\|_V\,.$$

From Lax-Milgram Theorem, problem (9.45) has a unique solution $\varphi\in V$ and $G_y(y,u)$ is a continuous isomorphism between V and V^*.

To write the optimality condition, introduce the Lagrangian

$$\mathcal{L}(y,u;p)=\frac{1}{6}\|y-z_d\|^6_{L^6(\Omega)}+\frac{\beta}{2}\|u\|^2_{L^2(\Omega)}+\langle p,Ay+B[\Phi(y)-u]\rangle_{V,V^*}\,.\tag{9.46}$$

Recalling (9.26) we deduce that the adjoint equation is

$$\mathcal{L}'_y(\hat{y},\hat{u};\hat{p})\psi=\left((\hat{y}-z_d)^5,\psi\right)_{L^2(\Omega)}+\left\langle\hat{p},A\psi+B[\Phi'(\hat{y})\psi]\right\rangle_{V,V^*}=0\quad\forall\psi\in V$$

or

$$a(\hat{p},\psi)+\left(3\hat{y}^2\hat{p},\psi\right)_{L^2(\Omega)}=\left((\hat{y}-z_d)^5,\psi\right)_{L^2(\Omega)}\qquad\forall\psi\in V,$$

which is the weak form of the Neumann problem

$$\begin{cases}-\Delta\hat{p}+\hat{p}+3\hat{y}^2\hat{p}=-(\hat{y}-z_d)^5 & \text{in }\Omega,\\ \dfrac{\partial\hat{p}}{\partial n}=0 & \text{on }\Gamma.\end{cases}$$

Finally, the variational inequality

$$\mathcal{L}_u(\hat{y},\hat{u};\hat{p})(v-\hat{u})\geq0\qquad\forall v\in\mathcal{U}_{ad}$$

gives

$$(\beta\hat{u}-\hat{p},v-\hat{u})_{L^2(\Omega)}\geq0\qquad\forall v\in\mathcal{U}_{ad}.$$

□

Remark 9.14. Thanks to Lemma 9.1, we can equivalently express the optimality conditions of Theorem 9.8 as follows:

$$\begin{cases}-\Delta\hat{y}+\hat{y}+\hat{y}^3=\hat{u} & \text{in }\Omega, & \dfrac{\partial\hat{y}}{\partial n}=0\text{ on }\Gamma\\ -\Delta\hat{p}+\hat{p}+3\hat{y}^2\hat{p}=-(\hat{y}-z_d)^5 & \text{in }\Omega, & \dfrac{\partial\hat{p}}{\partial n}=0\text{ on }\Gamma\\ \beta\hat{u}-\hat{p}+\hat{\lambda}_b-\hat{\lambda}_a=0 & \text{a.e. in }\Omega & \\ \hat{u}\geq a,\quad\hat{\lambda}_a\geq0, & \hat{\lambda}_a(\hat{u}-a)=0, & \\ \hat{u}\leq b,\quad\hat{\lambda}_b\geq0, & \hat{\lambda}_b(b-\hat{u})=0. & \end{cases}$$

•

*Second derivatives and second order condition**. Let $X=V\cap C\left(\overline{\Omega}\right)$. From Examples A.17, A.18, A.19 we deduce that $F:X\times L^2(\Omega)\to\mathbb{R}$, $G:X\times L^2(\Omega)\to L^2(\Omega)$, and hence (Remark 9.6) also the control-to-state map $u\mapsto y(u)$, are twice continuously $\mathcal{F}-$differentiable. To calculate $J''(u)$ we use formula (9.32). We have, for the cost functional,

$$F''_{yy}(y,u)[\varphi_1,\varphi_2]=5\int_\Omega(y-z_d)^4\varphi_1\varphi_2,\qquad F''_{uu}(y,u)[v_1,v_2]=\beta\int_\Omega v_1v_2$$

and

$$F''_{yu}(y,u)\left[\varphi_1, v_2\right] = 0, \qquad F''_{uy}(y,u)\left[\varphi_1, v_2\right] = 0$$

for any $\varphi_1, \varphi_2 \in X$ and any $v_1, v_2 \in L^2(\Omega)$.

For the state operator, we find

$$G''_{yy}(y,u)[\varphi_1, \varphi_2] = B[\Phi''(y)][\varphi_1, \varphi_2] \in V^*$$

or, more explicitly, if $\psi \in V$,

$$G''_{yy}(y,u)[\varphi_1, \varphi_2]\psi = 6\int_\Omega y\varphi_1\varphi_2\psi.$$

Finally,

$$G''_{yu}(y,u) = 0, \quad G''_{uy}(y,u) = 0, \quad G''_{uu}(y,u) = 0.$$

Hence, from (9.32),

$$J''(u)\left[v_1, v_2\right] = 5\int_\Omega \left(\hat{y} - z_d\right)^4 z_1 z_2 + 6\int_\Omega \hat{y} z_1 z_2 \hat{p} + \beta\int_\Omega v_1 v_2$$

where $z_j = z_j\left(v_j\right) = y'\left(u\right)v_j$, $j = 1, 2$.

Remark 9.15. In this case, condition (9.33) becomes

$$5\int_\Omega \left(\hat{y} - z_d\right)^4 \hat{z}^2 + 6\int_\Omega \hat{y} z^2 \hat{p} + \beta\int_\Omega v^2 \geq \alpha\int_\Omega v^2 \tag{9.47}$$

for any $v \in L^2(\Omega)$ and $\hat{z} = \hat{z}(v)$ satisfying the linearized equation

$$\begin{cases} -\Delta\hat{z} + \hat{z} + 3\hat{y}^2\hat{z} = v & \text{in } \Omega \\ \dfrac{\partial\hat{z}}{\partial n} = 0 & \text{on } \Gamma. \end{cases}$$

If $\beta > 0$ and

$$\int_\Omega [5\left(\hat{y} - z_d\right)^4 + 6\hat{y}\hat{p}]\hat{z}^2 \geq 0$$

then (9.47) is satisfied for $\alpha = \beta$. The case $\beta = 0$ is essentially more difficult since (9.47) is hardly satisfied. For a discussion of this case see, e.g., [59]. •

9.5 Least squares approximation of a reaction coefficient

We now consider the identification problem mentioned in Sect. 9.1. Let us denote by $Q_T = \Omega \times (0, T)$ the space-time cylinder, where Ω is a bounded, Lipschitz domain in $\mathbb{R}^d$, and take $z_d \in L^2(\Omega)$, $f \in L^2\left(0, T; H^{-1}\left(\Omega\right)\right)$, $g \in L^2\left(\Omega\right)$. We consider the following OCP:

$$F\left(y\right) = \frac{1}{2}\int_\Omega \left(y\left(\mathbf{x}, T\right) - z_d(\mathbf{x})\right)^2 d\mathbf{x} \to \min \tag{9.48}$$

subject to the state system

$$\begin{cases} y_t - \Delta y + u(\mathbf{x})\, y = f(\mathbf{x},t) & \text{in } Q_T = \Omega \times (0,T) \\ y = 0 & \text{on } \Gamma \times (0,T) \\ y(\mathbf{x},0) = g(\mathbf{x}) & \text{in } \Omega \end{cases} \tag{9.49}$$

and to the box constraint

$$u \in \mathcal{U}_{ad} = \{u \in L^\infty(\Omega) : a \le u(\mathbf{x}) \le b \text{ a.e. in } \Omega\}. \tag{9.50}$$

The above problem can be interpreted as an *identification* problem: we want to identify the coefficient u such that, at time T, $y(u)$ is the best approximation (in a least squares sense) of the given target (or measurement) z_d. The coefficient u thus will play the role of a control function.

Analysis of the state problem. Since $u \in L^\infty(\Omega)$ we already know that (9.49) is well posed in the space $Y = H^1(0,T; H_0^1(\Omega), H^{-1}(\Omega))$; see Sect. A.6.3 for the definition of the functional space $H^1(0,T;V,V^*)$. In our case $V = H_0^1(\Omega)$.

Precisely, the following proposition holds, where $u^- = \max\{-u, 0\}$ is the negative part of u.

Proposition 9.6. *For every $u \in L^\infty(\Omega)$, problem (9.49) has a unique solution $y(u) \in Y$. Moreover*

$$\|y(u)\|_Y^2 \le C_0 \left(\|f\|^2_{L^2(0,T;H^{-1}(\Omega))} + \|g\|^2_{L^2(\Omega)} \right) \tag{9.51}$$

where $C_0 = C_0(\mathcal{M}, \|u^-\|_{L^\infty(\Omega)}, T)$, $\mathcal{M} = 1 + C_P^2 \|u\|_{L^\infty(\Omega)}$ and C_P is the Poincaré constant.

Remark 9.16. By Proposition 9.6, for fixed $u \in L^\infty(\Omega)$, the map $(f,g) \mapsto y(u)$ realizes a continuous isomorphism between $Z = L^2(0,T;H^{-1}(\Omega)) \times L^2(\Omega)$ and Y. Note that if $f \in L^2(Q_T)$, then $\|f\|_{L^2(0,T;H^{-1}(\Omega))} \le C_P \|f\|_{L^2(Q_T)}$. •

Proof. We check the assumptions of Theorem 7.1 for the bilinear form

$$B(\varphi,\psi) = \int_\Omega (\nabla\varphi \cdot \nabla\psi + u\varphi\psi)$$

in $H_0^1(\Omega)$. We can write

$$\begin{aligned} \left| \int_\Omega (\nabla\varphi \cdot \nabla\psi + u\varphi\psi) \right| &\le \|\varphi\|_{H_0^1(\Omega)} \|\psi\|_{H_0^1(\Omega)} + \|u\|_{L^\infty(\Omega)} \|\varphi\|_{L^2(\Omega)} \|\psi\|_{L^2(\Omega)} \\ &\le \left(1 + C_P^2 \|u\|_{L^\infty(\Omega)}\right) \|\varphi\|_{H_0^1(\Omega)} \|\psi\|_{H_0^1(\Omega)} \end{aligned}$$

so that B is continuous in $H_0^1(\Omega)$ with continuity constant $\mathcal{M} = 1 + C_P^2 \|u\|_{L^\infty(\Omega)}$. Moreover, we have $u = u^+ - u^- \ge -u^-$ a.e. in Ω and therefore

$$B(\varphi,\varphi) = \int_\Omega (|\nabla\varphi|^2 + u\varphi^2) \ge \int_\Omega (|\nabla\varphi|^2 - u^-\varphi^2) \ge \|\varphi\|^2_{H_0^1(\Omega)} - \|u^-\|_{L^\infty(\Omega)} \|\varphi\|^2_{L^2(\Omega)}.$$

Thus $B(\varphi,\varphi) + \|u^-\|_{L^\infty(\Omega)} \|\varphi\|^2_{L^2(\Omega)}$ is coercive in $H_0^1(\Omega)$ with coercivity constant $\alpha = 1$. Existence, uniqueness and stability estimates now follow directly from Theorem 7.1. □

Differentiability of the control-to-state map. To prove that the control-to-state map is $\mathcal{F}$−differentiable we must show that, for every $u \in L^\infty(\Omega)$, there exists a bounded, linear operator $y'(u) : L^\infty(\Omega) \to Y$, such that

$$y(u+v) - y(u) - y'(u)\, v = R(u,v)$$

with

$$\frac{\|R(u,v)\|_Y}{\|v\|_{L^\infty(\Omega)}} \to 0 \quad \text{as } \|v\|_{L^\infty(\Omega)} \to 0.$$

To find the expression of $y'(u)$, we formally take the derivative with respect to u along the direction v of the three equations in (9.49). Then $y'(u)$ is the linear map that associates to $v \in L^2(\Omega)$ the weak solution $z = y'(u)\,v$ of the problem

$$\begin{cases} z_t - \Delta z + uz = -vy(u) & \text{in } Q_T \\ z = 0 & \text{on } \Gamma \times (0,T) \\ z(\mathbf{x},0) = 0 & \text{in } \Omega. \end{cases} \tag{9.52}$$

Note that from Proposition 9.6 we get (see Remark 9.16)

$$\begin{aligned} \|z\|_Y^2 &\le C_0 \|vy(u)\|^2_{L^2(0,T;H^{-1}(\Omega))} \le C_0 C_P \|v\|^2_{L^\infty(\Omega)} \|y(u)\|^2_{L^2(Q_T)} \\ &\le C_1 \|v\|^2_{L^\infty(\Omega)} \left(\|f\|^2_{L^2(0,T;H^{-1}(\Omega))} + \|g\|^2_{L^2(\Omega)}\right) \end{aligned} \tag{9.53}$$

with $C_1 = C_0^2 C_P^2$, so that $y'(u) \in \mathcal{L}(L^\infty(\Omega), Y)$ with

$$\|y'(u)\|^2_{\mathcal{L}(L^\infty(\Omega),Y)} \le C_1 \left(\|f\|^2_{L^2(0,T;H^{-1}(\Omega))} + \|g\|^2_{L^2(\Omega)}\right).$$

Rigorously, the following proposition holds:

Proposition 9.7. *The map $y = y(u) \in C^1(L^\infty(\Omega); Y)$ and, for every $v \in L^\infty(\Omega)$,*

$$y'(u)\,v = z \tag{9.54}$$

where z is the solution to problem (9.52).

Proof. We use the definition of $\mathcal{F}-$differentiability. Let $u,\ u+v \in L^\infty(\Omega)$ with corresponding solutions $y(u+v)$, $y(u)$. Let z be the unique solution to problem (9.52). Using the weak formulations of the equations satisfied by $y(u+v)$, $y(u)$ and z, we obtain that

$$R = R(u,v) = y(u+v) - y(u) - z$$

satisfies the equation

$$\langle R_t, \varphi\rangle_* + \int_\Omega [\nabla R \cdot \nabla\varphi + (u+v)R\varphi] = -\int_\Omega vz\varphi \qquad \forall \varphi \in H_0^1(\Omega),\ \text{a.e. in } (0,T)$$

with $R(0) = 0$. Therefore,

$$\begin{aligned} \|R\|_Y^2 &\le C_0 \|vz\|^2_{L^2(0,T;H^{-1}(\Omega))} \le C_0 C_P \|v\|^2_{L^\infty(\Omega)} \|z\|^2_{L^2(\Omega)} \\ &\le C_1 \|v\|^4_{L^\infty(\Omega)} \left(\|f\|^2_{L^2(0,T;H^{-1}(\Omega))} + \|g\|^2_{L^2(\Omega)}\right) \end{aligned}$$

from which we deduce that $\|R\|_Y / \|v\|_{L^\infty(\Omega)} \to 0$ as $\|v\|_{L^\infty(\Omega)} \to 0$ and therefore $u \mapsto y(u)$ is $\mathcal{F}-$differentiable in $L^\infty(\Omega)$ and (9.54) holds. □

Existence of an optimal control. The following result holds.

Proposition 9.8. *There exists at least an optimal solution $(\hat{y}, \hat{u})$ of the optimal control problem* (9.48)–(9.50).

Proof. We use Theorem 9.4 and Remark 9.4. Here we have: $V = Y, \mathcal{U} = L^\infty(\Omega), Z = L^2\left(0, T; H^{-1}(\Omega)\right) \times L^2(\Omega)$ and $G : Y \times L^2(\Omega) \to Z$ is given by

$$G(y, u) = \begin{pmatrix} y_t - \Delta y + uy - f \\ y(0) - g \end{pmatrix}.$$

We check assumptions (1) – (4) in that theorem.

(1) It is trivially satisfied, since $F(y) \geq 0$.

(2) Let $\{(y_k, u_k)\} \subset Y \times \mathcal{U}_{ad}$ be a (feasible) *minimizing* sequence: $F(y_k) \to \mu = \inf F$, $G(y_k, u_k) = 0$. Since $\mathcal{U}_{ad}$ is a bounded set in $L^\infty(\Omega)$, we have $\|u_k\|_{L^\infty(\Omega)} \leq M = \max\{|a|, |b|\}$. From Proposition 9.6 we deduce that

$$\|y_k\|_Y \leq C_0 \left(\|f\|^2_{L^2(0,T;H^{-1}(\Omega))} + \|g\|^2_{L^2(\Omega)}\right) = M_0$$

so that $\{(y_k, u_k)\}$ is bounded.

(3) The set of feasible points is weakly* sequentially closed in $Y \times \mathcal{U}$. Indeed, let $(y_n, u_n) \in Y \times \mathcal{U}_{ad}$ with $u_n \overset{*}{\rightharpoonup} u$ (weakly*) in $L^\infty(\Omega)$, $y_n \rightharpoonup y$ in Y and $G(y_n, u_n) = 0$. Since $\mathcal{U}_{ad}$ is weakly* sequentially closed (see Example 9.3) we infer that $(y, u) \in Y \times \mathcal{U}_{ad}$. Moreover,

$$\int_0^T \left(\langle \partial_t y_n, \varphi \rangle_* + (\nabla y_n, \nabla \varphi)_{L^2(\Omega)}\right) dt \to \int_0^T \left(\langle \partial_t y, \varphi \rangle_* + (\nabla y, \nabla \varphi)_{L^2(\Omega)}\right) dt \tag{9.55}$$

for each $\varphi \in L^2\left(0, T; H_0^1(\Omega)\right)$. We now check that

$$\int_0^T \int_\Omega u_n y_n \varphi \, d\mathbf{x}dt \to \int_0^T uy\varphi \, d\mathbf{x}dt \tag{9.56}$$

for each $\varphi \in L^2\left(0, T; H_0^1(\Omega)\right)$, and

$$y_n(0) \rightharpoonup y(0) \qquad \text{in } L^2(\Omega). \tag{9.57}$$

Passing if necessary to a subsequence, by Theorem A.19(c) we may assume that $y_n \to y$ (strongly) in $L^2(Q_T)$. Now, for $\varphi \in L^2\left(0, T; H_0^1(\Omega)\right)$, write

$$\int_0^T \int_\Omega (u_n y_n - uy)\varphi \, d\mathbf{x}dt = \int_0^T \int_\Omega y\varphi(u_n - u) d\mathbf{x}dt + \int_0^T \int_\Omega (y_n - y) u_n \varphi \, d\mathbf{x}dt.$$

The first integral on the right tends to zero thanks to the Lebesgue dominated convergence (Theorem A.8) since[4] $y\varphi \in L^1(Q_T)$ and $u_n \overset{*}{\rightharpoonup} u$ in $L^\infty(\Omega)$. For the second term we have

$$\left|\int_0^T \int_\Omega (y_n - y) u_n \varphi \, d\mathbf{x}dt\right| \leq \|u_n\|_{L^\infty(\Omega)} \|\varphi\|_{L^2(Q_T)} \|y_n - y\|_{L^2(Q_T)} \leq M \|\varphi\|_{L^2(Q_T)} \|y_n - y\|_{L^2(Q_T)}$$

and therefore also this term converges to zero.
Finally, using the integration by parts formula (A.61) with $\varphi = w\psi$, $w \in H_0^1(\Omega)$, $\psi \in C^1[0, T]$, $\psi(T) = 0$, $\psi(0) = 1$, we deduce

$$\int_0^T \langle \partial_t y_n, \varphi \rangle_* \, dt = -\int_0^T (y_n, \psi')_{L^2(\Omega)} w dt - (y_n(0), w)_{L^2(\Omega)},$$

$$\int_0^T \langle \partial_t y, \varphi \rangle_* \, dt = -\int_0^T (y, \psi')_{L^2(\Omega)} w dt - (y(0), w)_{L^2(\Omega)}.$$

Since $y_n \rightharpoonup y$ in Y, it follows that $y_n(0) \rightharpoonup y(0)$ in $L^2(\Omega)$. From (9.55), (9.56), (9.57) we conclude that $G(y, u) = 0$ and that (3) holds.

(4) Being F continuous in Y, it is also sequentially weakly lower semicontinuous in Y.

Applying Theorem 9.4 and Remark 9.4 we infer the existence of an optimal solution $(\hat{y}, \hat{u})$. □

---

[4] We have that $w_n(t) = \int_\Omega y(\mathbf{x}, t)\varphi(\mathbf{x}, t)(u_n(\mathbf{x}) - u(\mathbf{x}))\, d\mathbf{x} \to 0$ a.e. in $(0, T)$ as $n \to \infty$ and $|w_n(t)| \leq 2M \int_\Omega |y(\mathbf{x}, t)\varphi(\mathbf{x}, t)|\, d\mathbf{x} \in L^1(0, T)$.

Optimality conditions. To write the necessary optimality conditions, we use Theorem 9.6, whose assumptions have to be verified. Since $F : Y \to \mathbb{R}$ is a quadratic functional and $G : Y \times \mathcal{U} \to Z$ is a bilinear mapping, assumption (i) of Theorem 9.6 holds with:

$$\begin{aligned}
&F'(y) : Y \to \mathbb{R}, && F'(y)\,\psi = (y(T) - z_d, \psi(T))_{L^2(\Omega)} && \forall \psi \in Y \\
&G'_y(y,u) : Y \to Z, && G'_y(y,u)\,\varphi = \begin{pmatrix} \varphi_t - \Delta\varphi + u\varphi \\ \varphi(0) \end{pmatrix} && \forall \varphi \in Y \\
&G'_u(y,u) : \mathcal{U} \to Z && G'_u(y,u)\,v = \begin{pmatrix} vy \\ 0 \end{pmatrix} && \forall v \in L^\infty(\Omega).
\end{aligned} \tag{9.58}$$

Regarding assumption (ii), we already checked that, for each $u \in \mathcal{U}$, the state equation $G(y,u) = 0$ in Z defines a unique continuously $\mathcal{G}$–differentiable (actually $\mathcal{F}$–differentiable) control-to-state map $u \mapsto y(u)$.

Finally, concerning assumption (iii), we have that the partial derivative $G'_y(y,u) \in \mathcal{L}(Y,Z)$ is a continuous isomorphism between Y and Z, for every $u \in \mathcal{U}$. In fact, for each $z = (f,g) \in Z$ the equation $G'_y(y,u)\,\varphi = z$ is nothing but problem (9.49) for φ and the assertion comes from Remark 9.16. Hence, by applying Theorem 9.6 we derive the following result.

Theorem 9.9. *Let $(\hat{u}, \hat{y})$ be an optimal solution of the OCP (9.48), (9.49), (9.50). Then there exists a unique multiplier $\hat{p} \in Y$ such that, $\hat{u}, \hat{y}, \hat{p}$ fulfill (in a weak sense) the following system of optimality conditions:*

- *the state equation*

$$\begin{cases} \hat{y}_t - \Delta\hat{y} + \hat{u}\hat{y} = f & \text{in } Q_T \\ \hat{y} = 0 & \text{on } \Gamma \times (0,T) \\ \hat{y}(0) = g & \text{in } \Omega, \end{cases} \tag{9.59}$$

- *the adjoint equation*

$$\begin{cases} -\hat{p}_t - \Delta\hat{p} + \hat{u}\hat{p} = 0 & \text{in } Q_T \\ \hat{p} = 0 & \text{on } \Gamma \times (0,T) \\ \hat{p}(T) = z_d - \hat{y}(T) & \text{in } \Omega, \end{cases} \tag{9.60}$$

- *and the variational inequality*

$$\int_0^T (\hat{y}(t)\,\hat{p}(t), v - \hat{u})_{L^2(\Omega)}\, dt \geq 0 \qquad \forall v \in \mathcal{U}_{ad}. \tag{9.61}$$

Proof. Let $z^* = (p,q) \in Z^* = L^2\left(0,T; H_0^1(\Omega)\right) \times L^2(\Omega)$ be a multiplier and introduce the Lagrangian function

$$\begin{aligned}
\mathcal{L}(y,u,p,q) &= F(y) + \langle z^*, G(y,u)\rangle_{Z^*,Z} \\
&= \frac{1}{2}\|y(T) - z_d\|^2_{L^2(\Omega)} + \int_0^T \left\{ \langle y_t(t), p(t)\rangle_* + (\nabla y(t), \nabla p(t))_{L^2(\Omega)} + uy(t)\,p(t) \right\} dt \\
&\qquad + (y(0) - z_d, q)_{L^2(\Omega)}.
\end{aligned}$$

From (9.58), we obtain the following system of optimality conditions for $\hat{u}$, $\hat{y}$ and $(\hat{p}, \hat{q})$, where $\hat{\mathcal{L}}'_{(\cdot)} = \mathcal{L}'_{(\cdot)}(\hat{y}, \hat{u}, \hat{p}, \hat{q})$:

- $\forall \eta \in L^2\left(0,T; H_0^1(\Omega)\right), \forall w \in L^2(\Omega),$

$$\begin{aligned}\hat{\mathcal{L}}'_p\eta &= \int_0^T \left\{ \langle \hat{y}_t(t), \eta(t)\rangle_* + (\nabla \hat{y}(t), \nabla \eta(t))_{L^2(\Omega)} + \hat{u}\hat{y}(t)\eta(t) \right\} dt = 0 \\ \hat{\mathcal{L}}'_q w &= (y(0) - z_d, w)_{L^2(\Omega)} = 0.\end{aligned} \tag{9.62}$$

System (10.35), is equivalent to the state system (9.49) in weak form;

- $\forall v \in \mathcal{U}_{ad}$,

$$\hat{\mathcal{L}}'_u(v - \hat{u}) = \int_0^T (y(t)\hat{p}(t), v - \hat{u})_{L^2(\Omega)}\, dt \geq 0 \tag{9.63}$$

which is the variational inequality (9.61) expressing the optimality condition;

- $\forall \psi \in Y$,

$$\begin{aligned}\hat{\mathcal{L}}'_y\psi = \int_0^T & \left\{ \langle \psi_t(t), \hat{p}(t)\rangle_* + (\nabla \psi(t), \nabla \hat{p}(t))_{L^2(\Omega)} + \hat{u}\hat{p}(t)\psi(t) \right\} dt \\ & + (\hat{y}(T) - z_d, \psi(T))_{L^2(\Omega)} + (\psi(0), \hat{q})_{L^2(\Omega)} = 0\end{aligned} \tag{9.64}$$

which is nothing but the adjoint equation. To better express it, first note that a solution $\hat{p} \in L^2\left(0, T; H_0^1(\Omega)\right)$ of (9.64) actually belongs to Y (see Exercise 8). Then, integrating by parts, we get

$$\int_0^T \langle \psi_t(t), \hat{p}(t)\rangle_*\, dt = \int_0^T - \langle \hat{p}_t(t), \psi(t)\rangle_*\, dt + (\hat{p}(T), \psi(T))_{L^2(\Omega)} - (\psi(0), \hat{p}(0))_{L^2(\Omega)}$$

so that the adjoint equation takes the form

$$\begin{aligned}\int_0^T & \left\{ - \langle \hat{p}_t(t), \psi(t)\rangle_* + (\nabla \psi(t), \nabla \hat{p}(t))_{L^2(\Omega)} + \hat{u}\hat{p}(t)\psi(t) \right\} dt \\ & + (\hat{y}(T) - z_d + \hat{p}(T), \psi(T))_{L^2(\Omega)} + (\psi(0), \hat{q} - \hat{p}(0))_{L^2(\Omega)} = 0.\end{aligned}$$

Choosing first $\psi \in C\left(0, T; H_0^1(\Omega)\right)$, we get

$$\int_0^T \left\{ - \langle \hat{p}_t(t), \psi(t)\rangle_* + (\nabla \psi(t), \nabla \hat{p}(t))_{L^2(\Omega)} + \hat{u}\hat{p}(t)\psi(t) \right\} dt = 0.$$

Going back to $\psi \in Y$, we infer

$$\hat{p}(T) = -(\hat{y}(T) - z_d) \text{ and } \hat{q} = \hat{p}(0).$$

Since $C\left(0, T; H_0^1(\Omega)\right)$ is dense in $L^2\left(0, T; H_0^1(\Omega)\right)$, we deduce that $\hat{p}$ is the (unique) weak solution of problem (9.60) and that $\hat{q} = \hat{p}(0)$. The proof is then complete. □

The gradient of the reduced functional $J(u) = F(y(u))$ is the element in $L^2(\Omega)$ given by

$$\nabla J(u) = \int_0^T y(t)\hat{p}(t)\, dt.$$

Second derivative of $J(u)$. It is easy to check that F and G are twice continuously differentiable. We have, recalling (9.58)

$$F''(y)[\psi_1, \psi_2] = (\psi_1(T), \psi_2(T))_{L^2(\Omega)} \qquad \forall \psi_1, \psi_2 \in Y$$

$$G''_{yy}(y, u)[\psi_1, \psi_2] = G''_{uu}(y, u)[\psi_1, \psi_2] = \begin{pmatrix} 0 \\ 0 \end{pmatrix} \qquad \forall \psi_1, \psi_2 \in Y$$

$$G''_{uy}(y, u)[v_1, v_2] = G''_{uy}(y, u)[v_1, v_2] = \begin{pmatrix} v_1 v_2 \\ 0 \end{pmatrix} \qquad \forall v_1, v_2 \in L^\infty(\Omega).$$

Thus, using (9.32), we find, setting $z_j = z_j(v_j) = y'(u)v_j$, $j = 1, 2$,

$$J''(y(u), u)[v_1, v_2] = \int_\Omega z_1(T) z_2(T) + \int_{Q_T} p v_1 v_1 dt.$$

9.6 Numerical Approximation of Nonlinear OCPs

In this section we provide an overview of the most common strategies used to approximate a nonlinear OCP. A broad variety of methods can be found in literature, depending on *(i)* the type of nonlinearity of the state problem, and *(ii)* the presence of control and/or state constraints. Far from being exhaustive, here we present some basic ideas on how to extend the numerical methods introduced in Chapters 6 and 8 to the case of nonlinear OCPs. In particular, we consider iterative methods (based on descent directions) in Sect. 9.6.1, and the sequential quadratic programming method as main representative of all-at-once methods in Sect. 9.6.2. For the sake of simplicity, we focus on unconstrained OCPs, postponing to Sect. 9.8 some hints to handle control constraints. The interested reader can refer, e.g., to [140] for a more in-depth review of algorithmic approaches for solving PDE-constrained optimization problems in the nonlinear case, as well as to, e.g., [39, 150, 117].

9.6.1 Iterative Methods

We can introduce *iterative (or black-box, or reduced space) methods* for the numerical solution of nonlinear OCPs as we already did in the case of linear-quadratic OCPs in Sect. 6.3. We recall that in an iterative method we set $y = y(u)$ and the control variable u is the only optimization variable. No matter if an *optimize-then-discretize* approach or a *discretize-then-optimize* approach is employed, iterative methods can be set up by applying classical algorithms for numerical optimization to the discrete counterpart of problem (9.17).

For the sake of convenience, here we rely on the *optimize-then-discretize* approach to exploit the system of first-order optimality conditions (9.24) we have already derived. This latter is a nonlinear system because of the nonlinearity of the state problem – the adjoint problem is indeed always linear.

Let us denote by $\mathbf{y} \in \mathbb{R}^{N_y}$, $\mathbf{u} \in \mathbb{R}^{N_u}$ and $\mathbf{p} \in \mathbb{R}^{N_p}$ the discrete state, control and adjoint vectors, respectively; let us assume that the same space is used to discretize both state and adjoint variables, so that $N_p = N_y$. At the optimum, $\mathbf{y}$, $\mathbf{u}$ and $\mathbf{p}$ must fulfill the following system of optimality conditions:

$$\begin{cases} \mathbf{G}\left(\mathbf{y},\mathbf{u}\right) = \mathbf{0} & \text{(state equation)} \\ \left(\mathbf{G}'_y\left(\mathbf{y},\mathbf{u}\right)\right)^\top \mathbf{p} = -\mathbf{F}'_y(\mathbf{y},\mathbf{u}) & \text{(adjoint equation)} \\ \mathbf{F}'_u(\mathbf{y},\mathbf{u}) + \left(\mathbf{G}'_u(\mathbf{y},\mathbf{u})\right)^\top \mathbf{p} = \mathbf{0} & \text{(optimality condition)} \end{cases} \tag{9.65}$$

where:

1. $\mathbf{G}(\mathbf{y},\mathbf{u}) \in \mathbb{R}^{N_y}$ is the vector representing the residual of the discrete state system;
2. $\mathbf{G}'_y\left(\mathbf{y},\mathbf{u}\right) \in \mathbb{R}^{N_y \times N_y}$ and $\mathbf{G}'_u\left(\mathbf{y},\mathbf{u}\right) \in \mathbb{R}^{N_y \times N_u}$ are two matrices, representing the discrete counterpart of the partial derivatives of G with respect to y and u, respectively;
3. $\mathbf{F}'_y\left(\mathbf{y},\mathbf{u}\right) \in \mathbb{R}^{N_y}$ and $\mathbf{F}'_u\left(\mathbf{y},\mathbf{u}\right) \in \mathbb{R}^{N_u}$ are two vectors, representing the discrete counterpart of the partial derivatives of F with respect to y and u, respectively.

Below we detail the construction of system (9.65), the solution of the nonlinear state system, and the use of a descent method for the sake of numerical optimization. A system similar to (9.65) would be obtained by following the *discretize-then-optimize* approach as well.

Solving the state equation

Because of the nonlinearity, the numerical approximation of the state equation now yileds a nonlinear algebraic system that can be solved by using traditional methods, such as the Newton method or the fixed point (or Picard) iterations method. Assuming hereon that $W = V^*$, from the nonlinear state equation

$$G(y, u) = 0 \qquad \text{in } V^*, \tag{9.66}$$

we first retrieve its weak formulation: given $u \in \mathcal{U}$, find $y = y(u) \in V$ such that

$$g(y, \varphi; u) = 0 \qquad \forall \varphi \in V, \tag{9.67}$$

where the variational form $g(\cdot, u; \cdot) : V \times V \to \mathbb{R}$ is defined as

$$g(y, \varphi; u) = \langle G(y, u), \varphi \rangle_{V^*,V} \qquad \forall y, \varphi \in V.$$

Moreover, given $\bar{y} \in V$, $\varphi \in V$, we denote the partial Fréchet derivative of $g(\cdot, \varphi; u)$ with respect to y at $\bar{y} \in V$, evaluated in the direction z, by

$$g'_y[\bar{y}](z, \varphi; u) = \langle G'_y(\bar{y}, u)z, \varphi \rangle_{V^*,V} \qquad \forall z \in V$$

where $G'_y : V \to \mathcal{L}(V, V^*)$.

For the numerical approximation of (9.67), let us introduce two suitable finite-dimensional subspaces $V_h \subset V$ and $\mathcal{U}_h \subset \mathcal{U}$ of dimension N_y and N_u, respectively. The Galerkin problem reads as follows: given $u_h \in \mathcal{U}_h$, find $y_h = y_h(u_h) \in V_h$ such that

$$g(y_h, \varphi; u_h) = 0 \qquad \forall \varphi \in V_h. \tag{9.68}$$

The discrete problem (9.68) is equivalent to the solution of a nonlinear system of N_y equations. Indeed, denoting by $\{\varphi_j\}_{j=1}^{N_y}$, $\{\psi_j\}_{j=1}^{N_u}$ a basis for V_h, $\mathcal{U}_h$, respectively, we set

$$y_h = \sum_{j=1}^{N_y} y_{h,j}\varphi_j, \qquad u_h = \sum_{j=1}^{N_u} u_{h,j}\psi_j$$

and denote by $\mathbf{y}$, $\mathbf{u}$ the vectors having as components the unknown coefficients $y_{h,j}$ or $u_{h,j}$, respectively. Then, (9.68) is equivalent to: given $\mathbf{u} \in \mathbb{R}^{N_u}$, find $\mathbf{y} = \mathbf{y}(\mathbf{u}) \in \mathbb{R}^{N_y}$ such that

$$\mathbf{G}(\mathbf{y}, \mathbf{u}) = \mathbf{0}, \tag{9.69}$$

where the *residual vector* $\mathbf{G}(\cdot, \mathbf{u}) \in \mathbb{R}^{N_y}$ is defined as

$$\big(\mathbf{G}(\mathbf{y}, \mathbf{u})\big)_i = g(y_h, \varphi_i; u_h), \qquad i = 1, \ldots, N_y.$$

The nonlinear system (9.69) can be solved through the Newton method. Starting from an initial guess $\mathbf{y}^{(0)} \in \mathbb{R}^{N_y}$, at each step $n = 0, 1, \ldots$ we seek $\delta\mathbf{y} \in \mathbb{R}^{N_y}$ as the solution of

$$\mathbf{G}'_y(\mathbf{y}^{(n)}, \mathbf{u})\, \delta\mathbf{y} = -\mathbf{G}(\mathbf{y}^{(n)}, \mathbf{u}). \tag{9.70}$$

The latter is a linear system; the *Jacobian matrix* $\mathbf{G}'_y(\bar{\mathbf{y}}, \mathbf{u}) \in \mathbb{R}^{N_y \times N_y}$ is defined as

$$\big(\mathbf{G}'_y(\bar{\mathbf{y}}, \mathbf{u})\big)_{ij} = g'_y[\bar{y}_h](\varphi_j, \varphi_i; u_h), \qquad i, j = 1, \ldots, N_y \tag{9.71}$$

is nonsingular provided the assumptions of the Kantorovich theorem (see Theorem B.7) are fulfilled. The Newton method for the solution of (9.69) is reported in Algorithm 9.1.

Algorithm 9.1 Newton method for algebraic nonlinear problems

Input: $\mathbf{u} \in \mathbb{R}^{N_u}$, tolerance $\delta_N > 0$, max number of iterations N_{max}, $\mathbf{y}^{(0)} \in \mathbb{R}^{N_y}$
Output: $\mathbf{y} = \mathbf{y}(\mathbf{u})$

```
1: n = 0
2: while n ≤ N_max and ||G(y^(n), u)|| > δ_N
3:     solve G'_y(y^(n), u) δy = −G(y^(n), u)
4:     set y^(n+1) = y^(n) + δy
5:     n = n + 1
6: end while
7: y(u) = y^(n)
```

An alternative method to solve nonlinear system (9.69) is to rely on the so-called fixed point (or Picard) iterations method. In this case, starting from an initial guess $\mathbf{y}^{(0)} \in \mathbb{R}^{N_y}$, at each step we seek $\mathbf{y}^{(n+1)} \in \mathbb{R}^{N_y}$ as the solution of the following linear system,

$$\mathbf{G}'_y(\mathbf{y}^{(n)}, \mathbf{u})\, \mathbf{y}^{(n+1)} = \mathbf{0}, \qquad n = 0, 1, \ldots .$$

until a stopping criterion similar to the one used in Algorithm 9.1 is satisfied.

Example 9.5. For the problem of Sect. 9.4 we have $V = H_0^1(\Omega)$ and $V^* = H^{-1}(\Omega)$, and

$$G(y, u) = -\Delta y + y + y^3 - u$$

so that

$$\langle G(y, u), \varphi \rangle_{H^{-1}(\Omega), H^1(\Omega)} = g(y, \varphi; u) = \int_\Omega \nabla y \cdot \nabla \varphi \, d\mathbf{x} + \int_\Omega (y + y^3)\varphi \, d\mathbf{x} - \int_\Omega u\varphi \, d\mathbf{x};$$

as a consequence,

$$g'_y[\bar{y}](v, \varphi; u) = \int_\Omega \nabla v \cdot \nabla \varphi \, d\mathbf{x} + \int_\Omega (v + 3\bar{y}^2 v)\varphi \, d\mathbf{x}. \tag{9.72}$$

Note that, given $u \in \mathcal{U}$, for any $\bar{y} \in V$ this bilinear form is symmetric and coercive. The problem to be solved at each Newton's step is the discrete counterpart of the following boundary value problem: given $y^{(n)} \in V$, find $\delta y \in V$ such that

$$a(\delta y, \varphi) + \int_\Omega 3(y^{(n)})^2 \delta y \, \varphi d\mathbf{x} = -a(y^{(n)}, \varphi) - \int_\Omega (y^{(n)})^3 \varphi d\mathbf{x} + (f + u, \varphi)_{L^2(\Omega)} \quad \forall \varphi \in V.$$

Summing $\int_\Omega 3(y^{(n)})^2 y^{(n)} \varphi d\mathbf{x}$ to both sides, and denoting by $y^{(n+1)} = y^{(n)} + \delta y$ we finally obtain: given $y^{(n)} \in V$, find $y^{(n+1)} \in V$ such that

$$a(y^{(n+1)}, \varphi) + \int_\Omega 3(y^{(n)})^2 y^{(n+1)} \varphi d\mathbf{x} = 2 \int_\Omega (y^{(n)})^3 \varphi d\mathbf{x} + (f + u, \varphi)_{L^2(\Omega)} \quad \forall \varphi \in V.$$

Alternatively, if we rely on the fixed point (Picard) iterations method, at each step we would need to solve the following problem: given $y^{(n)} \in V$, find $y^{(n+1)} \in V$ such that

$$a(y^{(n+1)}, \varphi) + \int_\Omega 3(y^{(n)})^2 y^{(n+1)} \varphi d\mathbf{x} = (f + u, \varphi)_{L^2(\Omega)} \quad \forall \varphi \in V.$$

•

Solving the adjoint equation and updating the control

Given the state $\mathbf{y}$ and the control $\mathbf{u}$, the adjoint equation

$$(\mathbf{G}'_y(\mathbf{y},\mathbf{u}))^\top \mathbf{p} = -\mathbf{F}'_y(\mathbf{y},\mathbf{u}) \tag{9.73}$$

is a linear system. The matrix at the left-hand side is the transpose of the Jacobian matrix defined in (9.71), while the right-hand side has components given by

$$(\mathbf{F}'_y(\mathbf{y},\mathbf{u}))_i = F'_y(y_h,u_h)\varphi_i, \qquad i = 1,\dots,N_y.$$

Finally, since we are dealing with unconstrained OCPs, the optimality condition turns to an algebraic equation, which involves the discrete control and adjoint vectors $\mathbf{u}$ and $\mathbf{p}$,

$$\mathbf{F}'_u(\mathbf{y},\mathbf{u}) + (\mathbf{G}'_u(\mathbf{y},\mathbf{u}))^\top \mathbf{p} = \mathbf{0}. \tag{9.74}$$

In the equation above we denote by:

- the matrix $\mathbf{G}'_u(\mathbf{y},\mathbf{u}) \in \mathbb{R}^{N_y \times N_u}$, whose components are defined as

  $$\left(\mathbf{G}'_u(\mathbf{y},\bar{\mathbf{u}})\right)_{ij} = g'_u[\bar{u}_h](y_h;\varphi_i,\psi_j), \qquad i = 1,\dots,N_y,\ j = 1,\dots,N_u;$$

  here, we denote the partial Fréchet derivative of $g(y,\cdot;\bar{u})$, with respect to u at $\bar{u} \in \mathcal{U}$, by

  $$g'_u[\bar{u}](y;w,\psi) = \langle G'_u(y,\bar{u})w,\psi\rangle_{V^*,V} \qquad \forall w \in \mathcal{U}, \psi \in V,$$

  where $G'_u : \mathcal{U} \to \mathcal{L}(\mathcal{U}, V^*)$.
- $\mathbf{F}'_u(\mathbf{y},\mathbf{u}) \in \mathbb{R}^{N_u}$ is a vector, with components

  $$(\mathbf{F}'_u(\mathbf{y},\mathbf{u}))_i = F'_u(y_h,u_h)\psi_i, \qquad i = 1,\dots,N_u.$$

When a descent method is employed, given an initial guess $\mathbf{u}_0 \in \mathbb{R}^{N_u}$, we iteratively find a sequence of discrete control vectors $\{\mathbf{u}_k\}$ such that

$$\mathbf{u}_{k+1} = \mathbf{u}_k + \tau_k\, \mathbf{d}_k, \qquad k = 0,1,\dots$$

where $\mathbf{d}_k \in \mathbb{R}^{N_u}$ represents a descent direction and $\tau_k > 0$ is a suitable stepsize, ensuring that

$$J_h(\mathbf{u}_k + \tau_k \mathbf{d}_k) < J_h(\mathbf{u}_k);$$

here we denote by J_h the discrete counterpart of the reduced cost functional. The descent direction $\mathbf{d}_k$ can be calculated according to one of the methods described in Sect. 6.3; the solution of the nonlinear state system is then embedded into the optimization loop, each step requiring the solution of (at least) a state problem and an adjoint problem.

The *steepest descent* method uses the direction

$$\mathbf{d}_k = -\nabla J_h(\mathbf{u}_k)$$

at each step; note that evaluating the reduced gradient

$$\nabla J_h(\bar{\mathbf{u}}) = \mathbf{F}'_u(\mathbf{y},\bar{\mathbf{u}}) + (\mathbf{G}'_u(\mathbf{y},\bar{\mathbf{u}}))^\top \mathbf{p}$$

at an arbitrary point $\bar{\mathbf{u}}$ requires to solve both the state and the adjoint problem.

The *Newton* method for numerical optimization[5] uses the direction

$$\mathbf{d}_k = -(\mathbb{H}(\mathbf{u}_k))^{-1}\nabla J_h(\mathbf{u}_k) \tag{9.75}$$

[5] Choosing (9.75) as descent direction is nothing but the result of the application of the Newton method to the nonlinear system $\nabla J_h(\mathbf{u}) = \mathbf{0}$.

at each step, where $\mathbb{H}(\mathbf{u}_k) \in \mathbb{R}^{N_u \times N_u}$ is the Hessian of the reduced cost functional (also referred to as *reduced Hessian* matrix). In other words, at each Netwon's step the descent direction is obtained by solving the linear system

$$\mathbb{H}(\mathbf{u}_k)\mathbf{d}_k = -\nabla J_h(\mathbf{u}_k), \tag{9.76}$$

whose dimension is that of the control space $\mathcal{U}_h$. The right-hand side $-\nabla J_h(\mathbf{u}_k)$ must be calculated at each Newton step, requiring the solution of a (nonlinear) state problem and of an adjoint problem. In the case of a nonlinear state system (and/or a general nonlinear cost functional) we refrain from explicitly calculating the Hessian matrix $\mathbb{H}$ because of the large computational effort that would be required. The evaluation of the matrix-vector product $\mathbb{H}(\mathbf{u})\mathbf{d}$ requires two additional problems to be solved, a linearized state problem and an adjoint problem, see Remark 9.11. Note that the same holds true in the linear-quadratic case, see Sect. 6.3.1. To avoid the difficulties of evaluating the action of the exact Hessian in (9.76), it is also possible to rely on quasi-Newton methods.

Remark 9.17. Summarizing, for the outer iteration of the Newton method one has to solve the state problem to obtain $\mathbf{y}_k = \mathbf{y}_k(\mathbf{u}_k)$, and one adjoint problem to obtain $\mathbf{z}_k = \mathbf{z}_k(\mathbf{u}_k)$. An inner loop is then necessary to solve (9.76) iteratively, for every k; at every inner iteration loop, a linearized state problem and an adjoint problem are to be solved. •

Remark 9.18. In the case of time-dependent nonlinear OCPs, iterative methods can be exploited in a very similar way. In these cases, the state problem yields a nonlinear system of ODEs to be solved once a spatial discretization has been operated. The adjoint equation will be a time-dependent, backward in time, linear system of ODEs. •

9.6.2 All-at-once Methods: Sequential Quadratic Programming

In contrast to iterative methods, all-at-once methods treat the control and state variables as independent optimization variables. Here we limit ourselves to introduce one of the most popular strategies to address nonlinear OCPs in the framework of all-at-once methods, namely the *sequential quadratic programming* (SQP) method. First, we focus on the unconstrained case; some options to handle control constraints will be illustrated at a later stage.

As already remarked in Sect. 3.5.2 for a general finite-dimensional optimization problem, SQP generates, at each step, a linear-quadratic optimization problem, in which a quadratic approximation of the cost functional is minimized, subject to a linearization of the constraints.

Here we show how this method can be applied to solve a nonlinear OCP under the form (9.15), in the unconstrained case $\mathcal{U}_{ad} = \mathcal{U}$. Based on the results of Sect. 9.3.4, we first formulate the SQP method directly at the continuous level, to unveil the structure of the linear-quadratic problem generated at each step. Later on, we turn to the finite dimensional problem that arises from the numerical approximation of this linear-quadratic problem.

This *optimize-then-discretize* approach allows us to better highlight the connection between the second-order optimality conditions stated for a general nonlinear OCP and the discrete problem to be solved. In the unconstrained case $\mathcal{U}_{ad} = \mathcal{U}$, the SQP method can be obtained by applying the Newton method to solve the nonlinear system of optimality conditions (9.26); for this reason, the SQP method is also referred to as *Lagrange-Newton method.*

Rearranging the equations, the system (9.26) can be rewritten as (see Remark 9.9):

$$\begin{cases} F'_y(\hat{y},\hat{u}) + G'_y\,(\hat{y},\hat{u})^*\,\hat{p} = 0 \\ F'_u\,(\hat{y},\hat{u}) + G'_u\,(\hat{y},\hat{u})^*\,\hat{p} = 0 \\ G\,(\hat{y},\hat{u}) = 0, \end{cases} \tag{9.77}$$

or, in terms of the Lagrangian functional $\mathcal{L}\,(y,u,p) = F\,(y,u) + \langle p, G\,(y,u)\rangle_{W^*,W}$,

$$\begin{cases} \mathcal{L}'_{(y,u)}\,(\hat{y},\hat{u},\hat{p}) = 0 \\ G\,(\hat{y},\hat{u}) = 0, \end{cases} \tag{9.78}$$

where

$$\mathcal{L}'_{(y,u)}\,(y,u,p) = \begin{pmatrix} \mathcal{L}'_y(y,u,p) \\ \mathcal{L}'_u(y,u,p) \end{pmatrix}.$$

Here (y,u,p) are treated as independent variables and, similarly to the all-at-once methods introduced in Sect. 6.6, both y and u are optimization variables.

By applying the Newton method to the nonlinear system above, we obtain the following Newton iteration: given (y_0,u_0,p_0), until convergence solve

$$\begin{bmatrix} \mathcal{L}''_{yy}(y_k,u_k,p_k) & \mathcal{L}''_{yu}(y_k,u_k,p_k) & G'_y\,(y_k,u_k)^* \\ \mathcal{L}''_{uy}(y_k,u_k,p_k) & \mathcal{L}''_{uu}(y_k,u_k,p_k) & G'_u\,(y_k,u_k)^* \\ G'_y\,(y_k,u_k) & G'_u\,(y_k,u_k) & 0 \end{bmatrix} \begin{bmatrix} \delta y \\ \delta u \\ \delta p \end{bmatrix} = - \begin{bmatrix} F'_y(y_k,u_k) + G'_y\,(y_k,u_k)^*\,p_k \\ F'_u\,(y_k,u_k) + G'_u\,(y_k,u_k)^*\,p_k \\ G\,(y_k,u_k) \end{bmatrix} \tag{9.79}$$

and set

$$y_{k+1} = y_k + \delta y, \quad u_{k+1} = u_k + \delta u, \quad p_{k+1} = p_k + \delta p.$$

To stop the iterations, we can use, for instance, a criterion based on the difference of two successive iterates, that is, we iterate until

$$\sqrt{\|\delta y\|^2_V + \|\delta u\|^2_{\mathcal{U}} + \|\delta p\|^2_V} < \varepsilon,$$

where $\varepsilon > 0$ is a prescribed tolerance. System (9.79) is well-posed provided the second order sufficient condition (9.33) is verified. Indeed, under this assumption, the Newton method generates a unique sequence of iterates converging quadratically to $(\hat{y},\hat{u},\hat{p})$; see, e.g., [10, 11].

We can derive the expression of the algebraic system corresponding to the Newton step (9.79) by introducing two suitable finite-dimensional subspaces $V_h \subset V$ and $\mathcal{U}_h \subset \mathcal{U}$ of dimension N_y and N_u, respectively, and denoting by $\{\varphi_j\}_{j=1}^{N_y}$, $\{\psi_j\}_{j=1}^{N_u}$ a basis for V_h, $\mathcal{U}_h$, respectively; we assume that V_h is used to discretize both the state and the adjoint variable. The discretization of those terms involving the state operator G and the first derivative of G and F has already been considered in Sect. 9.6.1; regarding instead the other terms, let us recall that the second derivatives $\mathcal{L}''_{yy}$, $\mathcal{L}''_{yu}$, $\mathcal{L}''_{uy}$, $\mathcal{L}''_{uu}$ can be identified with suitable continuous bilinear forms, so that we can define the following matrices:

$$(\mathbb{H}_{yy}(\mathbf{y},\mathbf{u},\mathbf{p}))_{ij} = \mathcal{L}''_{yy}(y_h,u_h,p_h)[\varphi_i,\varphi_j], \qquad i,j = 1,\ldots,N_y,$$

$$(\mathbb{H}_{yu}(\mathbf{y},\mathbf{u},\mathbf{p}))_{ij} = \mathcal{L}''_{yu}(y_h,u_h,p_h)[\varphi_i,\psi_j], \qquad i = 1,\ldots,N_y, \quad j = 1,\ldots,N_u,$$

$$(\mathbb{H}_{yu}(\mathbf{y},\mathbf{u},\mathbf{p}))_{ij} = \mathcal{L}''_{uy}(y_h,u_h,p_h)[\psi_i,\varphi_j], \qquad i = 1,\ldots,N_u, \quad j = 1,\ldots,N_y,$$

$$(\mathbb{H}_{uu}(\mathbf{y},\mathbf{u},\mathbf{p}))_{ij} = \mathcal{L}''_{uu}(y_h,u_h,p_h)[\psi_i,\psi_j], \qquad i,j = 1,\ldots,N_u,$$

corresponding to the discretization of the Hessian of the Lagrangian function.

Hence, given $\mathbf{y}_0 \in \mathbb{R}^{N_y}$, $\mathbf{u}_0 \in \mathbb{R}^{N_u}$, $\mathbf{p}_0 \in \mathbb{R}^{N_y}$, at each step we solve the linear system:

$$\begin{bmatrix} \mathbb{H}_{yy}(\mathbf{y}_k,\mathbf{u}_k,\mathbf{p}_k) & \mathbb{H}_{yu}(\mathbf{y}_k,\mathbf{u}_k,\mathbf{p}_k) & (\mathbf{G}'_y(\mathbf{y}_k,\mathbf{u}_k))^\top \\ \mathbb{H}_{uy}(\mathbf{y}_k,\mathbf{u}_k,\mathbf{p}_k) & \mathbb{H}_{uu}(\mathbf{y}_k,\mathbf{u}_k,\mathbf{p}_k) & (\mathbf{G}'_u(\mathbf{y}_k,\mathbf{u}_k))^\top \\ \mathbf{G}'_y(\mathbf{y}_k,\mathbf{u}_k) & \mathbf{G}'_u(\mathbf{y}_k,\mathbf{u}_k) & 0 \end{bmatrix} \begin{bmatrix} \delta\mathbf{y} \\ \delta\mathbf{u} \\ \delta\mathbf{p} \end{bmatrix}$$
$$= - \begin{bmatrix} \mathbf{F}'_y(\mathbf{y}_k,\mathbf{u}_k) + (\mathbf{G}'_y(\mathbf{y}_k,\mathbf{u}_k))^\top \mathbf{p}_k \\ \mathbf{F}'_u(\mathbf{y}_k,\mathbf{u}_k) + (\mathbf{G}'_u(\mathbf{y}_k,\mathbf{u}_k))^\top \mathbf{p}_k \\ \mathbf{G}(\mathbf{y}_k,\mathbf{u}_k) \end{bmatrix}, \tag{9.80}$$

and then set

$$\begin{bmatrix} \mathbf{y}_{k+1} \\ \mathbf{u}_{k+1} \\ \mathbf{p}_{k+1} \end{bmatrix} = \begin{bmatrix} \mathbf{y}_k \\ \mathbf{u}_k \\ \mathbf{p}_k \end{bmatrix} + \begin{bmatrix} \delta\mathbf{y} \\ \delta\mathbf{u} \\ \delta\mathbf{p} \end{bmatrix},$$

until convergence. We summarize the resulting procedure in Algorithm 9.2; the norms of the increments $\delta\mathbf{y}$, $\delta\mathbf{u}$, $\delta\mathbf{p}$ are those inherited by the norms of V and $\mathcal{U}$.

Alternatively, the *discretize-then-optimize* approach would also yield a similar system to be solved at each step; indeed, we highlight the formal analogy between (9.80) and system (4.32), which had been obtained in Sect. 4.2.1 in the case of an algebraic OCP.

Let us now highlight the structure of the linear-quadratic problem that is indeed generated, and solved, at each step. Introducing the abridged notation

$$\mathcal{L}''_{(y,u)} = \begin{bmatrix} \mathcal{L}''_{yy} & \mathcal{L}''_{yu} \\ \mathcal{L}''_{uy} & \mathcal{L}''_{uu} \end{bmatrix},$$

system (9.79) corresponds to the (necessary and sufficient) optimality conditions of the following linear-quadratic problem:

$$\frac{1}{2}\mathcal{L}''_{(y,u)}(y_k,u_k,p_k)[(\delta_y,\delta_u)^2] + \mathcal{L}'_{(y,u)}(y_k,u_k,p_k)(\delta_y,\delta_u) \to \min \tag{9.81}$$

subject to the linearized equation

$$G'_y(y_k,u_k)\,\delta_y + G'_u(y_k,u_k)\,\delta_u + G(y_k,u_k) = 0 \qquad \text{in } W. \tag{9.82}$$

Similarly, system (9.80) corresponds to the (necessary and sufficient) optimality conditions for the discrete linear-quadratic problem

$$\begin{gathered} \frac{1}{2}\begin{bmatrix} \delta\mathbf{y} \\ \delta\mathbf{u} \end{bmatrix}^\top \mathbb{H}(\mathbf{y}_k,\mathbf{u}_k,\mathbf{p}_k) \begin{bmatrix} \delta\mathbf{y} \\ \delta\mathbf{u} \end{bmatrix} + \begin{bmatrix} \mathbf{G}'_y(\mathbf{y}_k,\mathbf{u}_k) \\ \mathbf{G}'_u(\mathbf{y}_k,\mathbf{u}_k) \end{bmatrix}^\top \begin{bmatrix} \delta\mathbf{y} \\ \delta\mathbf{u} \end{bmatrix} \to \min_{(\delta\mathbf{y},\delta\mathbf{u})^\top} \\ \text{s.t.} \\ \mathbf{G}'_y(\mathbf{y}_k,\mathbf{u}_k)\delta\mathbf{y} + \mathbf{G}'_u(\mathbf{y}_k,\mathbf{u}_k)\delta\mathbf{u} + \mathbf{g}(\mathbf{y}_k,\mathbf{u}_k) = \mathbf{0}. \end{gathered} \tag{9.83}$$

Algorithm 9.2 SQP method for unconstrained nonlinear OCPs

Input: initial guess $\mathbf{y}_0$, $\mathbf{u}_0$, $\mathbf{p}_0$

1: set $k = 0$, $\delta = 1$
2: **while** $\delta > \varepsilon$
3: solve the system (9.80)
4: set $\mathbf{y}_{k+1} = \mathbf{y}_k + \delta\mathbf{y}$, $\mathbf{u}_{k+1} = \mathbf{u}_k + \delta\mathbf{u}$, $\mathbf{p}_{k+1} = \mathbf{p}_k + \delta\mathbf{p}$
5: set $\delta = \sqrt{|\delta\mathbf{y}|^2 + |\delta\mathbf{u}|^2 + |\delta\mathbf{p}|^2}$
6: set k = k+1
7: **end while**

In the equation above, we have denoted by

$$\mathbb{H}(\mathbf{y}_k,\mathbf{u}_k,\mathbf{p}_k)=\begin{bmatrix}\mathbb{H}_{yy}(\mathbf{y}_k,\mathbf{u}_k,\mathbf{p}_k) & \mathbb{H}_{yu}(\mathbf{y}_k,\mathbf{u}_k,\mathbf{p}_k)\\ \mathbb{H}_{uy}(\mathbf{y}_k,\mathbf{u}_k,\mathbf{p}_k) & \mathbb{H}_{uu}(\mathbf{y}_k,\mathbf{u}_k,\mathbf{p}_k)\end{bmatrix}.$$

Note that, in contrast to the Newton method, the iterates $(\mathbf{y}_k,\mathbf{u}_k)$ generated by the SQP method are infeasible for the nonlinear state equation, that is, the SQP method generates control/state pairs that satisfy the state equation only in the limit.

Remark 9.19. In the linear-quadratic case, the SQP method yields the saddle-point system (6.70) to solve. Indeed, let us assume to apply the Algorithm 9.2 to a linear-quadratic OCP whose discrete version (6.23) can be formulated as a quadratic programming problem with equality constraint, as done in Sect. 6.6,

$$\frac{1}{2}\mathbf{x}^T\mathbb{J}\mathbf{x}-\mathbf{h}^T\mathbf{x}\to\min\qquad\text{subject to}\qquad \mathbf{x}\in\mathbb{R}^{N_y+N_u},\ \mathbb{D}\mathbf{x}=\mathbf{f}. \tag{9.84}$$

Here

$$\mathbf{x}=\begin{bmatrix}\mathbf{y}\\ \mathbf{u}\end{bmatrix},\quad \mathbb{J}=\begin{bmatrix}\mathbb{M} & 0\\ 0 & \beta\mathbb{N}\end{bmatrix},\quad \mathbf{h}=\begin{bmatrix}\mathbb{M}\mathbf{z}_d\\ \mathbf{0}\end{bmatrix},\quad \mathbb{D}=[\mathbb{A}\quad -\mathbb{M}].$$

In this case, the Newton step (9.80) would read

$$\begin{bmatrix}\mathbb{J} & \mathbb{D}^\top\\ \mathbb{D} & 0\end{bmatrix}\begin{bmatrix}\delta\mathbf{x}\\ \delta\mathbf{p}\end{bmatrix}=\begin{bmatrix}\mathbb{J}\mathbf{x}_k-\mathbf{h}+\mathbb{D}\mathbf{p}_k\\ \mathbb{D}\mathbf{x}_k+\mathbf{f}\end{bmatrix}$$

that is,

$$\begin{bmatrix}\mathbb{M} & 0 & \mathbb{A}^\top\\ 0 & \beta\mathbb{N} & -\mathbb{M}^\top\\ \mathbb{A} & -\mathbb{M} & 0\end{bmatrix}\begin{bmatrix}\delta\mathbf{y}\\ \delta\mathbf{u}\\ \delta\mathbf{p}\end{bmatrix}=\begin{bmatrix}\mathbb{M}\mathbf{y}_k-\mathbb{M}\mathbf{z}_d+\mathbb{A}^\top\mathbf{p}_k\\ \beta\mathbb{N}\mathbf{u}_k-\mathbb{M}^\top\mathbf{p}_k\\ \mathbb{A}\mathbf{y}_k-\mathbb{M}\mathbf{u}_k+\mathbf{f}\end{bmatrix}.$$

Taking $\mathbf{y}_0=\mathbf{0}$, $\mathbf{u}_0=\mathbf{0}$, $\mathbf{p}_0=\mathbf{0}$, the previous system is nothing but the saddle point system (6.70) stemming from the KKT optimality conditions in the linear-quadratic case. •

Example 9.6. We detail the procedure exploiting the nonlinear OCP introduced in Sect. 9.4. In this case, the Lagrangian functional (9.46) can be rewritten as

$$\mathcal{L}(y,u;p)=\frac{1}{6}\|y-z_d\|^6_{L^6(\Omega)}+\frac{\beta}{2}\|u\|^2_{L^2(\Omega)}+\int_\Omega\nabla p\cdot\nabla y+\int_\Omega y^3p+\int_\Omega yp-\int_\Omega up.$$

Consequently, for any $(\psi,v),(\eta,w)\in V\times\mathcal{U}$,

$$\begin{aligned}\mathcal{L}'_{(y,u)}(y,u,p)\begin{pmatrix}\psi\\ v\end{pmatrix}&=((y-z_d)^5,\psi)_{L^2(\Omega)}+\int_\Omega\nabla p\cdot\nabla\psi+3\int_\Omega y^2\psi p+\int_\Omega\psi p\\ &\quad+\beta(u,v)_{L^2(\Omega)}-\int_\Omega vp,\end{aligned}$$

$$\mathcal{L}''_{(y,u)}(y,u,p)\left[\begin{pmatrix}\psi\\ v\end{pmatrix},\begin{pmatrix}\eta\\ w\end{pmatrix}\right]=5\int_\Omega(y-z_d)^4\psi\eta+6\int_\Omega yp\psi\eta+\beta\,(v,w)_{L^2(\Omega)}.$$

On the other hand, we can rewrite the relations in (9.44) as

$$\begin{aligned}&F'_y(y,u)\,\psi=((y-z_d)^5,\psi)_{L^2(\Omega)} && \forall\psi\in V\\ &F'_u(y,u)\,v=(u,v)_{L^2(\Omega)} && \forall v\in\mathcal{U}=L^2(\Omega)\\ &\langle G'_y(y,u)\,\varphi,\psi\rangle_{V^*,V}=\int_\Omega\nabla\varphi\cdot\nabla\psi+3\int_\Omega y^2\varphi\psi && \forall\varphi,\psi\in V\\ &\langle G'_u(y,u)\,v,w\rangle_{\mathcal{U}^*,\mathcal{U}}=-\int_\Omega vw && \forall v,w\in\mathcal{U}=L^2(\Omega).\end{aligned} \tag{9.85}$$

According to (9.79), at each Newton step we must solve the following system:

$$\begin{cases} 6\int_\Omega y_k p_k \delta y \varphi + 5\int_\Omega (y_k - z_d)^4 \delta y \varphi + \int_\Omega \nabla\varphi\cdot\nabla\delta p + 3\int_\Omega y_k^2 \delta p \varphi \\ \qquad\qquad = -\int_\Omega \nabla p_k \cdot \nabla\varphi - 3\int_\Omega y_k^2 \varphi p_k - \int_\Omega (y_k - z_d)^5 \varphi \\ \beta\int_\Omega \delta u v - \int_\Omega v\delta p = -\beta\int_\Omega u_k v + \int_\Omega p_k v \\ \int_\Omega \nabla\delta y\cdot\nabla\psi + 3\int_\Omega y_k^2 \delta y \psi + \int_\Omega \delta u \psi = -\int_\Omega \nabla y_k \cdot\nabla\psi - \int_\Omega y_k^3 \psi + \int_\Omega u_k \psi \end{cases}$$

for all $\varphi \in V$, $v \in \mathcal{U}$, $\psi \in V$. The discrete counterpart of the system above results as follows:

$$\begin{bmatrix} 5\mathbb{M}[(\mathbf{y}_k - \mathbf{z}_d)^4] + 6\mathbb{M}[\mathbf{y}_k\mathbf{p}_k] & 0 & \mathbb{A}^\top + 3\mathbb{M}[\mathbf{y}_k^2]^\top \\ 0 & \beta\mathbb{N} & -\mathbb{B}^\top \\ \mathbb{A} + 3\mathbb{M}(\mathbf{y}_k^2) & -\mathbb{B} & 0 \end{bmatrix} \begin{bmatrix} \delta\mathbf{y} \\ \delta\mathbf{u} \\ \delta\mathbf{p} \end{bmatrix}$$
$$= -\begin{bmatrix} \mathbb{M}(\mathbf{y}_k - \mathbf{z}_d)^5 + (\mathbb{A} + 3\mathbb{M}[\mathbf{y}_k^2])^\top \mathbf{p}_k \\ \beta\mathbb{N}\mathbf{u}_k - \mathbb{B}^\top \mathbf{p}_k \\ \mathbb{A}\mathbf{y}_k + \mathbb{M}\mathbf{y}_k^3 - \mathbb{M}\mathbf{u}_k \end{bmatrix}, \tag{9.86}$$

where we have defined:

- the *stiffness* matrix $\mathbb{A} \in \mathbb{R}^{N_y \times N_y}$: $(\mathbb{A})_{ij} = a(\varphi_j, \varphi_i)$, $i, j = 1, \dots, N_y$;
- the *control* matrix $\mathbb{B} \in \mathbb{R}^{N_y \times N_u}$: $(\mathbb{B})_{ij} = (\psi_j, \varphi_i)_{L^2(\Omega)}$, $i = 1, \dots, N_y$, $j = 1, \dots, N_u$;
- the mass matrix $\mathbb{M} \in \mathbb{R}^{N_y \times N_y}$: $(\mathbb{M})_{ij} = (\varphi_j, \varphi_i)_{L^2(\Omega)}$, $i, j = 1, \dots, N_y$;
- the (control) mass matrix $\mathbb{N} \in \mathbb{R}^{N_u \times N_u}$: $(\mathbb{N})_{ij} = (\psi_j, \psi_i)_{L^2(\Omega)}$, $i, j = 1, \dots, N_u$;
- the *weighted* mass matrix $\mathbb{M}[\mathbf{q}] \in \mathbb{R}^{N_y \times N_y}$: $(\mathbb{M}[\mathbf{q}])_{ij} = \int_\Omega q_h \varphi_j \varphi_i$, $i, j = 1, \dots, N_y$, having exploited the equivalence between $q_h \in V_h$ and the vector $\mathbf{q} = (q_{h,1}, \dots, q_{h,N_y})^\top \in \mathbb{R}^{N_y}$. Note that a further approximation is made by taking $y_h^2 \approx \sum_j y_{h,j}^2 \varphi_j$ when assembling the matrix $\mathbb{M}[\mathbf{y}_k^2]$). •

Remark 9.20. Both the Newton and SQP methods presented above require the repeated solution of linear systems (9.76) and (9.80). These saddle point systems are inherently large scale and ill-conditioned. Hence, their efficient solution requires the use of preconditioned iterative solvers. The design of efficient preconditioner is usually carried out through general strategies for preconditioning saddle-point systems, indeed very similar to the ones introduced in Sect. 6.6.1; see also, e.g., [146, 212]. •

Remark 9.21. Solving the nonlinear state problem at every step on an iterative optimization method, like the Newton method, might seem disadvantageous compared to the SQP method (see Algorithm 9.2), which only requires linearized equations to be solved. However, this gap can be closed for time-dependent problems, if the nonlinearities are treated by considering a semi-implicit method for time discretization. Solving the nonlinear state problem in this way indeed becomes as expensive as solving the linearized state problem. Further details about SQP methods in the case of time-dependent problems can be found, e.g., in [146]. •

Remark 9.22. Other *all-at-once* algorithms are avaialble. For instance, *augmented Lagrangian methods* or *penalty methods* embed the PDE constraint in the cost functional, yielding to a family of unconstrained problems under the form

$$\tilde{J}(y, u) + \frac{c}{2}\|e(y, u)\|^2_{V^*} \to \min \quad \text{over } (y, u) \in V \times \mathcal{U}$$

where the augmentation parameter c has to be iteratively updated. See, e.g., [151, 269] for further details. •

9.7 Numerical Examples

In this section we show some numerical results obtained for the distributed control problem of a semilinear state equation introduced in Sect. 9.4; further results dealing with more advanced OCPs (both stationary and time-independent) will be the object of Chapter 10.

We deal with two differen cost functionals to be minimized, and will use either an iterative (steepest descent) method, or the SQP method. For the sake of simplicity, we address the case of an unconstrained OCP – with $\mathcal{U}_{ad} = L^2(\Omega)$ – and homogeneous Dirichlet boundary conditions, that is, we consider the following state problem:

$$\begin{cases} -\Delta y + y + \sigma y^3 = f + u & \text{in } \Omega = (0,1)^2 \\ y = 0 & \text{on } \Gamma \end{cases} \tag{9.87}$$

instead of (9.37). In particular:

- in *Case 1* we consider $f = x + y$, the following cost functional,

$$F(y,u) = \frac{1}{2}\|y - z_d\|^2_{L^2(\Omega)} + \frac{\beta}{2}\|u\|^2_{L^2(\Omega)}, \tag{9.88}$$

  with the target function
$$z_d = x_1 x_2 (1 - x_1)(1 - x_2);$$
- in *Case 2* we consider $f = 10^2(x + y)$, the cost functional (9.35), that is,

$$F(y,u) = \frac{1}{6}\|y - z_d\|^6_{L^6(\Omega)} + \frac{\beta}{2}\|u\|^2_{L^2(\Omega)}.$$

  with the target function

$$z_d = 10^3 x_1 x_2 \left(\frac{1}{4} - x_1^2\right)\left(\frac{1}{4} - x_2^2\right)(1 - x_1)(1 - x_2).$$

Regarding Case 1, we discretize in space the system of optimality conditions applying the Galerkin-FE method using a finite-dimensional space $V_h \subset V = H^1_0(\Omega)$ of dimension $\dim(V_h) = N_y = \dim(\mathcal{U}_h) = N_u = 11750$, using linear finite elements, and a mesh with 23894 triangular elements and 12148 vertices.

We consider the steepest descent method, exploiting the Newton method to solve the state problem at each iteration, and stopping the iteration when

$$|\nabla J_h(\mathbf{u}_k)| \,/\, |\nabla J_h(\mathbf{u}_0)| < \text{tol}$$

for tol $= 5 \cdot 10^{-2}$; we consider a fixed step length $\tau_k = \tau = 5$ and different values of the regularization parameter $\beta = 10^{-2}, 10^{-3}, \ldots, 10^{-6}$, to assess its impact on the numerical optimization procedure. First, we take $\sigma = 1$ as the coefficient of the nonlinear term; then, we compare the results obtained with $\beta = 10^{-2}$ and $\beta = 10^{-6}$ for different values of $\sigma = 1, 10, 10^2$.

Numerical results for the case $\sigma = 1$ are reported in Fig. 9.1, where we display the optimal state, adjoint and control variables obtained for the two values $\beta = 10^{-2}$ and $\beta = 10^{-6}$, and in Table 9.1. As expected, the regularization parameter has a strong impact also in the case of a nonlinear OCP, in terms of number of iterations (ranging from 51 to 1312) required to achieve convergence. We highlight that alternative step length (adaptive) selection strategies, relying, e.g., on inexact line searches or on the Armijio strategy, might yield convergence of the algorithm in a (remarkably) smaller number of iterations. Moreover, the norm of the optimal control shows a mild increase as β decreases. Numerical results for different values of $\sigma = 1, 10, 10^2$ are reported in Tables 9.2–9.3, showing that, for the case at hand, the regularization parameter has a stronger impact than the magnitude of the nonlinear term regarding the convergence of the iterative method – the number of iterations is almost constant when $\beta = 10^{-2}$, and mildly affected by σ when $\beta = 10^{-6}$. The overall computational cost is however larger in the case of higher values of σ, since the Newton method requires more steps to converge, at each solve of the state system, when $\sigma = 10^2$ (7, on average), compared to the case $\sigma = 1$ (3, on average), for the same tolerance $\delta_N = 10^{-10}$.

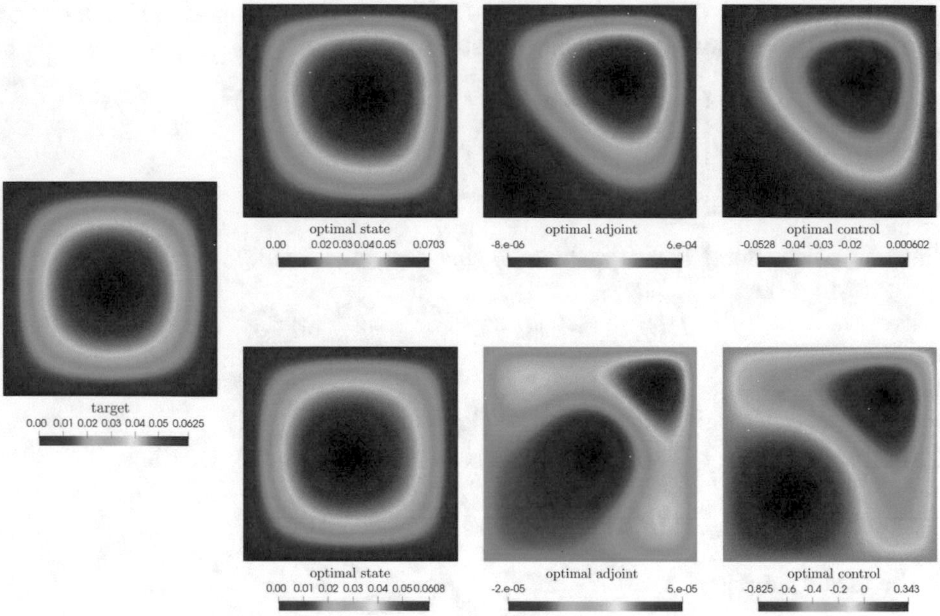

Fig. 9.1 Case 1. Target z_d, optimal state $\hat{y}$, adjoint $\hat{p}$ and control $\hat{u}$ for $\beta = 10^{-2}$ (top) and $\beta = 10^{-6}$ (bottom)

β	$J_h(\hat{u})$	# iter	$\|\hat{u}\|_{L^2(\Omega)}$	$\|\hat{y} - z_d\|_{L^2(\Omega)}$
10^{-2}	$3.2385 \cdot 10^{-5}$	51	0.02403	0.00543104
10^{-3}	$1.9712 \cdot 10^{-5}$	298	0.11072	0.00368553
10^{-4}	$6.2104 \cdot 10^{-6}$	916	0.24485	0.00179243
10^{-5}	$1.9645 \cdot 10^{-6}$	1252	0.30237	0.00122774
10^{-6}	$1.3985 \cdot 10^{-6}$	1312	0.31098	0.00116199

Table 9.1 Case 1. Values of the cost functional for the optimal control $J_h(\hat{u})$, number of iterations required by the steepest descent method to converge, norm of the optimal control $\|\hat{u}\|_{L^2(\Omega)}$, value of the cost $\|\hat{y} - z_d\|_{L^2(\Omega)}$, for different values of the regularization parameter β, in the case $\sigma = 1$

σ	$J_h(\hat{u})$	# iter	$\|\hat{u}\|_{L^2(\Omega)}$	$\|\hat{y} - z_d\|_{L^2(\Omega)}$
1	$3.23854 \cdot 10^{-5}$	51	0.0240383	0.00543104
10	$3.21316 \cdot 10^{-5}$	51	0.0237927	0.00541306
100	$2.97785 \cdot 10^{-5}$	52	0.0215791	0.00523935

Table 9.2 Case 1. Values of the cost functional for the optimal control $J_h(\hat{u})$, number of iterations required by the steepest descent method to converge, norm of the optimal control $\|\hat{u}\|_{L^2(\Omega)}$, value of the cost $\|\hat{y} - z_d\|_{L^2(\Omega)}$, for different values of the coefficient σ, in the case $\beta = 10^{-2}$

Regarding Case 2, we use the SQP method, solving a sequence of quadratic problems according to Algorithm 9.2, and following the derivation carried out in Example 9.6. For the case at hand, system (9.86) is obtained by applying on the Galerkin-FE method, using a finite-

σ	$J_h(\hat{u})$	# iter	$\|\hat{u}\|_{L^2(\Omega)}$	$\|\hat{y} - z_d\|_{L^2(\Omega)}$
1	$1.39858 \cdot 10^{-6}$	1312	0.310982	0.00116199
10	$1.38473 \cdot 10^{-6}$	1326	0.311496	0.00115595
100	$1.24759 \cdot 10^{-6}$	1482	0.317031	0.00109423

Table 9.3 Case 1. Values of the cost functional for the optimal control $J_h(\hat{u})$, number of iterations required by the steepest descent method to converge, norm of the optimal control $\|\hat{u}\|_{L^2(\Omega)}$, value of the cost $\|\hat{y}-z_d\|_{L^2(\Omega)}$, for different values of the coefficient σ, in the case $\beta = 10^{-6}$

dimensional space $V_h \subset V = H_0^1(\Omega)$ of dimension $\dim(V_h) = N_y = \dim(\mathcal{U}_h) = N_u = 8065$, using linear finite elements, and a mesh with 16384 triangular elements and 8321 vertices.

Choosing a tolerance $\varepsilon = 10^{-12}$, the SQP method requires from 12 to 26 iterations to converge, when different values of the regularization parameter $\beta = 10^{-3}, 10^{-6}, 10^{-9}$ are considered. Numerical results are reported in Fig. 9.2, where we display the optimal state, the adjoint and control variables obtained for the two values $\beta = 10^3$ and $\beta = 10^{-9}$, and Table 9.4. Hence, the regularization parameter has only a mild effect on the SQP convergence, making the solution of the problem feasible also for very small values of β.

Despite its implementation might be more involved than the one associated to the use of iterative methods (especially in those cases where already built-in subroutines are not available), the SQP method reveals to be a more efficient strategy from a computational standpoint.

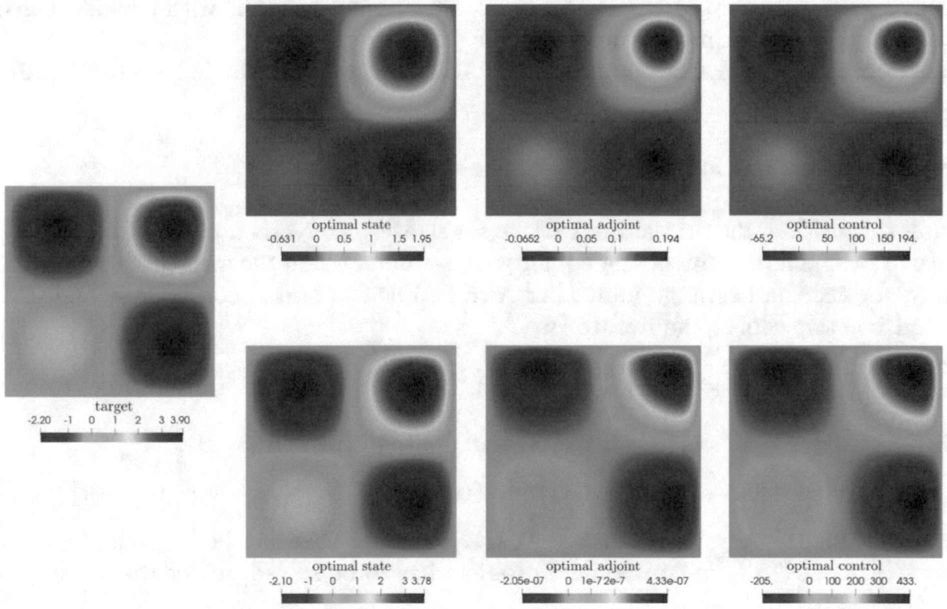

Fig. 9.2 Case 2. Target z_d, optimal state $\hat{y}$, adjoint $\hat{p}$ and control $\hat{u}$ for $\beta = 10^{-3}$ (top) and $\beta = 10^{-9}$ (bottom)

β	# iter
10^{-3}	12
10^{-6}	19
10^{-9}	26

Table 9.4 Case 2. Number of iterations required by the SQP method to converge for different values of the regularization parameter

9.8 Numerical Treatment of Control Constraints*

Special attention must be paid to the treatment of control constraints when solving nonlinear OCPs numerically, although similar strategies to those introduced for linear-quadratic OCPs can be employed.

Let us now consider the following constrained OCP,

$$\begin{gathered}\mathbf{F}(\mathbf{y},\mathbf{u}) \to \min \\ \text{subject to} \\ (\mathbf{y},\mathbf{u}) \in \mathbb{R}^{N_y} \times \mathcal{U}_{ad}, \quad \mathbf{G}(\mathbf{y},\mathbf{u}) = \mathbf{0},\end{gathered} \tag{9.89}$$

and focus on the case of *box constraints*, which in discrete form can be expressed as

$$\mathcal{U}_{ad} = \{\mathbf{u} \in \mathbb{R}^{N_u} \;:\; u_{a,i} \le u_i \le u_{b,i},\ i = 1,\dots,N_u\} \subset \mathbb{R}^{N_u}.$$

As in the linear-quadratic case, a projection method can be used to project the new iterate – updated using the descent direction computed for the unconstrained problem – onto the set of admissible controls, as we did for linear-quadratic OCPs in Sect. 6.4.

Alternatively, a primal-dual active set (PDAS) strategy, similar to the one introduced in Sect. 6.8, can be used in combination with the Newton method or the SQP method to handle control constraints, at least in two possible different ways [140]:

- by solving, at each step of the PDAS strategy, a nonlinear OCP with equality constraints (option 1), thus requiring two nested loops;
- by combining the linearization and the PDAS strategy into a single loop (option 2).

Option 1: two nested loops

Following this option, the primal-dual active set strategy converts a problem like (9.89) into a sequence of problems only involving equality constraints. Using the same notation of Sect. 6.8, we can introduce the Lagrange multiplier $\boldsymbol{\lambda}$ to handle the bound constraints, and define the active and inactive sets at the iterate $(\mathbf{u}_k, \boldsymbol{\lambda}_k)$ as

$$\begin{aligned}\mathcal{A}_k^+ &= \{i \in \{1,\dots,N_u\} \;:\; (\boldsymbol{\lambda}_k)_i + c(\mathbf{u}_k - \mathbf{b})_i > 0\},\\ \mathcal{A}_k^- &= \{i \in \{1,\dots,N_u\} \;:\; (\boldsymbol{\lambda}_k)_i + c(\mathbf{u}_k - \mathbf{a})_i < 0\},\\ \mathcal{I}_k &= \{i \in \{1,\dots,N_u\} \;:\; (\boldsymbol{\lambda}_k)_i + c(\mathbf{u}_k - \mathbf{b})_i \le 0 \le (\boldsymbol{\lambda}_k)_i + c(\mathbf{u}_k - \mathbf{a})_i\},\end{aligned}$$

for some $c > 0$. At every iteration $k = 0, 1, \dots$ of the PDAS method, given the active and inactive sets $\mathcal{A}_k^+$, $\mathcal{A}_k^-$, $\mathcal{I}_n$, a constrained problem has to be solved, under the form

$$\begin{gathered}\mathbf{F}(\mathbf{y}_{k+1},\mathbf{u}_{k+1}) \to \min \\ \text{subject to} \\ \mathbf{G}(\mathbf{y}_{k+1},\mathbf{u}_{k+1}) = \mathbf{0}, \quad \mathbf{u}_{k+1} = \mathbf{a} \text{ in } \mathcal{A}_k^-, \quad \mathbf{u}_{k+1} = \mathbf{b} \text{ in } \mathcal{A}_k^+.\end{gathered} \tag{9.90}$$

Inequalities are thus removed at the expense of an *outer* iteration, then an iterative (e.g., Newton) method or the SQP method can be used as an *inner* iteration, to solve problem (9.90) at each step $k = 0, 1, \dots$, see Algorithm 9.3. The resulting methods are referred to as PDAS-Newton or PDAS-SQP methods, respectively. Note that in this case the active/inactive sets depend on the solution of the nonlinear state problem.

Algorithm 9.3 PDAS method for constrained nonlinear OCPs

Input: initial guess $\mathbf{u}_0$, $\boldsymbol{\lambda}_0$, $c \in \mathbb{R}_+$, k_{max}

Initialize active and inactive sets:

$$\begin{aligned}
\mathcal{A}_0^+ &= \{i \in \{1,\ldots,N_u\} \ : \ (\boldsymbol{\lambda}_0)_i + c(\mathbf{u}_0 - \mathbf{b})_i > 0\},\\
\mathcal{A}_0^- &= \{i \in \{1,\ldots,N_u\} \ : \ (\boldsymbol{\lambda}_0)_i + c(\mathbf{u}_0 - \mathbf{a})_i < 0\},\\
\mathcal{I}_0 &= \{i \in \{1,\ldots,N_u\} \ : \ (\boldsymbol{\lambda}_0)_i + c(\mathbf{u}_0 - \mathbf{b})_i \le 0 \le (\boldsymbol{\lambda}_0)_i + c(\mathbf{u}_0 - \mathbf{a})_i\}
\end{aligned}$$

set $k = 0$ and `converged = false`

while `converged = false` and $k < k_{max}$

  solve problem (9.90) for $\mathbf{u}_{k+1}$ and associated Lagrange multipliers $\boldsymbol{\lambda}_a$, $\boldsymbol{\lambda}_b$

  set $\boldsymbol{\lambda}_{k+1} = \boldsymbol{\lambda}_b - \boldsymbol{\lambda}_a$

  given $(\mathbf{u}_{k+1}, \boldsymbol{\lambda}_{k+1})$, update active and inactive sets:

$$\begin{aligned}
\mathcal{A}_{k+1}^+ &= \{i \in \{1,\ldots,N_u\} \ : \ (\boldsymbol{\lambda}_{k+1})_i + c(\mathbf{u}_{k+1} - \mathbf{b})_i > 0\},\\
\mathcal{A}_{k+1}^- &= \{i \in \{1,\ldots,N_u\} \ : \ (\boldsymbol{\lambda}_{k+1})_i + c(\mathbf{u}_{k+1} - \mathbf{a})_i < 0\},\\
\mathcal{I}_{k+1} &= \{i \in \{1,\ldots,N_u\} \ : \ (\boldsymbol{\lambda}_{k+1})_i + c(\mathbf{u}_{k+1} - \mathbf{b})_i \le 0 \le (\boldsymbol{\lambda}_{k+1})_i + c(\mathbf{u}_{k+1} - \mathbf{a})_i\}
\end{aligned}$$

  check convergence:

  if $\mathcal{A}_{k+1}^+ = \mathcal{A}_k^+$, $\mathcal{A}_{k+1}^- = \mathcal{A}_k^-$ and $\mathcal{I}_{k+1} = \mathcal{I}_k$

    `converged = true`

  else

    k = k +1;

  end if

end while

Option 2: a single loop

An alternative option is the one in which we update the active and inactive sets iteratively, at each Newton or SQP step, thus combining the linearization and PDAS into one single loop. In this case, the active/inactive sets depend on the solution of a linearized state problem at each step.

In the case a Newton method is used to solve the optimization problem (in the control variable), the Newton update at iteration k can be obtained as the solution to

$$\begin{bmatrix} \mathbb{H}(\mathbf{u}_k) & \mathbb{I} \\ c\mathbb{I}_{\mathcal{A}_k} & -\mathbb{I}_{\mathcal{I}_k} \end{bmatrix} \begin{bmatrix} \delta\mathbf{u} \\ \delta\boldsymbol{\lambda} \end{bmatrix} = - \begin{bmatrix} \nabla J_h(\mathbf{u}_k) + \boldsymbol{\lambda}_k \\ c\mathbb{I}^+_{\mathcal{A}_k}(\mathbf{u}_k - \mathbf{b}) + c\mathbb{I}^-_{\mathcal{A}_k}(\mathbf{u}_k - \mathbf{a}) - \mathbb{I}_{\mathcal{I}_k}\boldsymbol{\lambda}_k \end{bmatrix}, \tag{9.91}$$

where:

- the matrices $\mathbb{I}_{\mathcal{A}_k}, \mathbb{I}_{\mathcal{I}_k} \in \mathbb{R}^{N_u \times N_u}$ are diagonal, with components

$$(\mathbb{I}_{\mathcal{A}_k})_{jj} = \begin{cases} 1 & \text{if } j \in \mathcal{A}_k^+ \cup \mathcal{A}_k^- \\ 0 & \text{otherwise,} \end{cases} \qquad (\mathbb{I}_{\mathcal{I}_k})_{jj} = \begin{cases} 1 & \text{if } j \in \mathcal{I}_k \\ 0 & \text{otherwise;} \end{cases}$$

- the matrices $\mathbb{I}^+_{\mathcal{A}_k}, \mathbb{I}^-_{\mathcal{A}_k} \in \mathbb{R}^{N_u \times N_u}$ are diagonal, too, with components

$$(\mathbb{I}^+_{\mathcal{A}_k})_{jj} = \begin{cases} 1 & \text{if } j \in \mathcal{A}_k^+ \\ 0 & \text{otherwise,} \end{cases} \qquad (\mathbb{I}^-_{\mathcal{A}_k})_{jj} = \begin{cases} 1 & \text{if } j \in \mathcal{A}_k^- \\ 0 & \text{otherwise.} \end{cases}$$

The expression of the system above can be obtained by rewriting the system (6.99) in the case of a general nonlinear problem, and performing block elimination for the variable $\mathbf{u}$; this interpretation is motivated by reformulating the complementarity condition by relying on a (regularized) semismooth Newton method, see Remark 6.10. In practice, it is more convenient to convert (9.91) into the following symmetric form, as we did in (6.101):

$$\begin{bmatrix} \mathbb{H}(\mathbf{u}_k) & \tilde{\mathbb{P}}_{\mathcal{A}_k} \\ \tilde{\mathbb{P}}_{\mathcal{A}_k}^\top & 0 \end{bmatrix} \begin{bmatrix} \delta\mathbf{u} \\ \delta\boldsymbol{\lambda}_{\mathcal{A}_k} \end{bmatrix} = - \begin{bmatrix} \nabla J_h(\mathbf{u}_k) + \mathbb{I}_{\mathcal{A}_k}\boldsymbol{\lambda}_k \\ \mathbb{P}_{\mathcal{A}_k^+}(\mathbf{u}_k - \mathbf{b}) + \mathbb{P}_{\mathcal{A}_k^-}(\mathbf{u}_k - \mathbf{a}) \end{bmatrix}, \tag{9.92}$$

where $\delta\boldsymbol{\lambda}_{\mathcal{A}_k}$ denotes the restriction of $\delta\boldsymbol{\lambda}$ to the active set, and $\boldsymbol{\lambda}_{k+1}$ is set to zero on the inactive set. Here $\tilde{\mathbb{P}}_{\mathcal{A}_k} \in \mathbb{R}^{(N_y+N_u)\times N_{\mathcal{A}_k}}$ is a rectangular matrix, $\tilde{\mathbb{P}}_{\mathcal{A}_k} = \begin{bmatrix} 0 & \mathbb{P}_{\mathcal{A}_k} \end{bmatrix}$; $\mathbb{P}_{\mathcal{A}_k} \in \mathbb{R}^{N_u \times N_{\mathcal{A}_k}}$ is also rectangular, consisting of those rows of the matrix $\mathbb{I}_{\mathcal{A}_k}$ which correspond to the active indexes. In the same way, $\mathbb{P}_{\mathcal{A}_k^\pm} \in \mathbb{R}^{N_u \times N_{\mathcal{A}_k^\pm}}$ are rectangular matrices consisting of those rows of $\mathbb{I}_{\mathcal{A}_k}$ which correspond to the indexes of $\mathcal{A}_k^-$ and $\mathcal{A}_k^+$, respectively. We summarize the resulting procedure in Algorithm 9.4; see, e.g., [141] for further details.

Differently than the Newton method, a SQP method yields a quadratic programming problem to be solved at each step, similarly to (9.83); given the current iterate $(\mathbf{y}_n, \mathbf{u}_n, \mathbf{p}_n)$,

$$\begin{gathered} \frac{1}{2}\begin{bmatrix} \mathbf{y}-\mathbf{y}_n \\ \mathbf{u}-\mathbf{u}_n \end{bmatrix}^\top \mathbb{H}(\mathbf{y}_n, \mathbf{u}_n, \mathbf{p}_n) \begin{bmatrix} \mathbf{y}-\mathbf{y}_n \\ \mathbf{u}-\mathbf{u}_n \end{bmatrix} + \begin{bmatrix} \mathbf{G}_y'(\mathbf{y}_n, \mathbf{u}_n) \\ \mathbf{G}_u'(\mathbf{y}_n, \mathbf{u}_n) \end{bmatrix}^\top \begin{bmatrix} \mathbf{y}-\mathbf{y}_n \\ \mathbf{u}-\mathbf{u}_n \end{bmatrix} \to \min \\ \text{s.t.} \\ \mathbf{G}_y'(\mathbf{y}_n, \mathbf{u}_n)(\mathbf{y}-\mathbf{y}_n) + \mathbf{G}_u'(\mathbf{y}_n, \mathbf{u}_n)(\mathbf{u}-\mathbf{u}_n) + \mathbf{g}(\mathbf{y}_n, \mathbf{u}_n) = \mathbf{0} \\ \text{and } \ \mathbf{a} \le \mathbf{u} \le \mathbf{b}. \end{gathered} \tag{9.93}$$

Problem (9.93) is a constrained linear-quadratic problem like the one addressed in Sect. 6.8, hence it can be solved by means of the PDAS method. In this problem, the Hessian of the Lagrangian $\mathbb{H}(\mathbf{y}_n, \mathbf{u}_n, \mathbf{p}_n)$ is evaluated at the current iterate $(\mathbf{y}_n, \mathbf{u}_n, \mathbf{p}_n)$, whereas the solution

Algorithm 9.4 Newton method with PDAS strategy for constrained nonlinear OCPs

Input: initial guess $\mathbf{u}_0$, $\boldsymbol{\lambda}_0$, $c \in \mathbb{R}_+$, k_{max}

Initialize active and inactive sets:

$$\begin{aligned} \mathcal{A}_0^+ &= \{i \in \{1,\dots,N_h\} \ : \ (\boldsymbol{\lambda}_0)_i + c(\mathbf{u}_0 - \mathbf{b})_i > 0\}, \\ \mathcal{A}_0^- &= \{i \in \{1,\dots,N_h\} \ : \ (\boldsymbol{\lambda}_0)_i + c(\mathbf{u}_0 - \mathbf{a})_i < 0\}, \\ \mathcal{I}_0 &= \{i \in \{1,\dots,N_h\} \ : \ (\boldsymbol{\lambda}_0)_i + c(\mathbf{u}_0 - \mathbf{b})_i \le 0 \le (\boldsymbol{\lambda}_0)_i + c(\mathbf{u}_0 - \mathbf{a})_i\} \end{aligned}$$

set $k = 0$ and `converged = false`

while `converged = false` and $k < k_{max}$

  evaluate the reduced gradient $\nabla J_h(\mathbf{u}_k)$

  solve the Newton system (9.91)

  set $\mathbf{u}_{k+1} = \mathbf{u}_k + \delta\mathbf{u}$

  set $(\boldsymbol{\lambda}_{k+1})_j = (\boldsymbol{\lambda}_k)_j + \delta\boldsymbol{\lambda}_{\mathcal{A}_k}$ if $j \in \mathcal{A}_k = \mathcal{A}_k^+ \cup \mathcal{A}_k^-$, and $(\boldsymbol{\lambda}_{k+1})_j = 0$ if $j \in \mathcal{I}_k$

  given $(\mathbf{u}^{k+1}, \boldsymbol{\lambda}^{k+1})$, update active and inactive sets:

$$\begin{aligned} \mathcal{A}_{k+1}^+ &= \{i \in \{1,\dots,N_h\} \ : \ (\boldsymbol{\lambda}^{k+1})_i + c(\mathbf{u}^{k+1} - \mathbf{b})_i > 0\}, \\ \mathcal{A}_{k+1}^- &= \{i \in \{1,\dots,N_h\} \ : \ (\boldsymbol{\lambda}^{k+1})_i + c(\mathbf{u}^{k+1} - \mathbf{a})_i < 0\}, \\ \mathcal{I}_{k+1} &= \{i \in \{1,\dots,N_h\} \in \Omega \ : \ (\boldsymbol{\lambda}^{k+1})_i + c(\mathbf{u}^{k+1} - \mathbf{b})_i \le 0 \le (\boldsymbol{\lambda}^{k+1})_i + c(\mathbf{u}^{k+1} - \mathbf{a})_i\} \end{aligned}$$

  if $\mathcal{A}_{k+1}^+ = \mathcal{A}_k^+$, $\mathcal{A}_{k+1}^- = \mathcal{A}_k^-$ and $\mathcal{I}_{k+1} = \mathcal{I}_k$

    `converged = true`

  else

    k = k+1;

  end if

end while

$\mathbf{y}, \mathbf{u}$ and the adjoint state $\mathbf{p}$ – playing the role of Lagrange multiplier associated to the equality constraint in (9.93) – yield the successive iterate $(\mathbf{y}_{n+1}, \mathbf{u}_{n+1}, \mathbf{p}_{n+1})$.

Because of the presence of inequality constraints in problem (9.93), the corresponding optimality system is nonlinear, and cannot be solved at one go. It is then possible to adopt a PDAS strategy, denoting by $k = 0, 1, \ldots$ the iterations with respect to this further loop, by $\mathcal{A}_k^+$, $\mathcal{A}_k^-$ the active sets, and by $\mathcal{I}_k$ the inactive set. At a given iterate $(\mathbf{y}_{n+1,k}, \mathbf{u}_{n+1,k}, \mathbf{p}_{n+1,k})$, we thus need to find the active and inactive sets $\mathcal{A}_k^+$, $\mathcal{A}_k^-$ and $\mathcal{I}_k$ and then solve the following problem:

$$
\begin{bmatrix}
\mathbb{H}_{yy} & \mathbb{H}_{yu} & (\mathbf{G}_y')^\top & 0 \\
\mathbb{H}_{uy} & \mathbb{H}_{uu} & (\mathbf{G}_u')^\top & \tilde{\mathbb{P}}_{\mathcal{A}_k} \\
\mathbf{G}_y' & \mathbf{G}_u' & 0 & 0 \\
0 & \tilde{\mathbb{P}}_{\mathcal{A}_k}^\top & 0 & 0
\end{bmatrix}
\begin{bmatrix}
\mathbf{y}_{n+1,k+1} - \mathbf{y}_n \\
\mathbf{u}_{n+1,k+1} - \mathbf{u}_n \\
\mathbf{p}_{n+1,k+1} \\
\boldsymbol{\lambda}_{k+1,\mathcal{A}_k}
\end{bmatrix}
= -
\begin{bmatrix}
\mathbf{F}_y'(\mathbf{y}_n, \mathbf{u}_n) \\
\mathbf{F}_u'(\mathbf{y}_n, \mathbf{u}_n) \\
\mathbf{G}(\mathbf{y}_n, \mathbf{u}_n) \\
\mathbb{P}_{\mathcal{A}_n^+}(\mathbf{u}_n - \mathbf{b}) + \mathbb{P}_{\mathcal{A}_n^-}(\mathbf{u}_n - \mathbf{a})
\end{bmatrix}
\tag{9.94}
$$

where $\boldsymbol{\lambda}_{k+1,\mathcal{A}_k}$ denotes the restriction of $\boldsymbol{\lambda}_{k+1}$ to the active set; all matrix coefficients are evaluated at $(\mathbf{y}_n, \mathbf{u}_n, \mathbf{p}_n)$. The nonlinear OCP is then linearized around each solution $\mathbf{y}_{n+1}, \mathbf{u}_{n+1}$ resulting from the application of the PDAS method. The SQP method with PDAS strategy for constrained nonlinear OCPs is summarized in Algorithm 9.5.

Note that in this case the PDAS strategy is applied to a linearized problem. The main difference between Algorithm 9.4 and 9.5 is that the iterates of the former satisfy the nonlinear state system, while the iterates of the latter satisfy the linearized state system as it can be seen from the third equation in (9.94). The nonlinear state system is satisified only in the limit.

Algorithm 9.5 SQP method with PDAS strategy for constrained nonlinear OCPs

---

Input: initial guess $\mathbf{y}_0$, $\mathbf{u}_0$, $\mathbf{p}_0$, $\boldsymbol{\lambda}_0$, $c \in \mathbb{R}_+$, k_{max}

1: set $n = 0$, $\delta = 1$
2: Initialize active and inactive sets:

$$
\begin{aligned}
\mathcal{A}_0^+ &= \{i \in \{1, \ldots, N_u\} \; : \; (\boldsymbol{\lambda}^0)_i + c(\mathbf{u}^0 - \mathbf{b})_i > 0\}, \\
\mathcal{A}_0^- &= \{i \in \{1, \ldots, N_u\} \; : \; (\boldsymbol{\lambda}^0)_i + c(\mathbf{u}^0 - \mathbf{a})_i < 0\}, \\
\mathcal{I}_0 &= \{i \in \{1, \ldots, N_u\} \; : \; (\boldsymbol{\lambda}^0)_i + c(\mathbf{u}^0 - \mathbf{b})_i \leq 0 \leq (\boldsymbol{\lambda}^0)_i + c(\mathbf{u}^0 - \mathbf{a})_i\}
\end{aligned}
$$

3: **while** $\delta > \varepsilon$
4: 　　set $k = 0$ and converged = false
5: 　　**while** converged = false and $k < k_{max}$
6: 　　　　solve the system (9.94)
7: 　　　　set $(\boldsymbol{\lambda}_{k+1})_j = 0$ if $j \in \mathcal{I}_k$
8: 　　　　**if** $\mathcal{A}_{k+1}^+ = \mathcal{A}_k^+$, $\mathcal{A}_{k+1}^- = \mathcal{A}_k^-$ and $\mathcal{I}_{k+1} = \mathcal{I}_k$
9: 　　　　　　converged = true
10: 　　　　**else**
11: 　　　　　　$k = k+1$;
12: 　　　　**end if**
13: 　　**end while**
14: 　　set $\mathbf{y}_{n+1} = \mathbf{y}_{n+1,k}$, $\mathbf{u}_{n+1} = \mathbf{u}_{n+1,k}$, $\mathbf{p}_{n+1} = \mathbf{p}_{n+1,k}$
15: 　　set $\delta = \sqrt{|\mathbf{y}_{n+1} - \mathbf{y}_n|^2 + |\mathbf{u}_{n+1} - \mathbf{u}_n|^2 + |\mathbf{p}_{n+1} = \mathbf{p}_n|^2}$
16: 　　set $k = k+1$
17: **end while**

9.9 Exercises

1. [*Minty lemma*] Under the assumptions of Theorem 9.2, show that, if J is a convex functional, condition (9.13) is equivalent to

$$\langle J'_G(v), v-\hat{u}\rangle_* \geq 0 \qquad \forall v \in \mathcal{U}_{ad}. \tag{9.95}$$

[*Solution:* To show that (9.95) implies (9.13) consider the function $g(t) = J(\hat{u}+t(v-\hat{u}))$. Prove that g is differentiable in $[0,1]$, for any $\hat{u}, v$ chosen in $\mathcal{U}_{ad}$. Moreover, since J is convex,

$$g'(t) = \langle J'_G(\hat{u}+t(v-\hat{u})), v-\hat{u}\rangle_*$$

is a monotone, nondecreasing function; hence, g' is continuous and in particular we have

$$\langle J'_G(\hat{u}+t(v-\hat{u})), v-\hat{u}\rangle \to \langle J'_G(\hat{u}), v-\hat{u}\rangle_* \quad \text{when} \quad t \to 0.$$

Taking $v = \hat{u}+t(\hat{u}-v)$, $t \in (0,1)$ in (9.95) and dividing by t, we obtain

$$\langle J'_G(\hat{u}+t(\hat{u}-v)), \hat{u}-v\rangle \geq 0$$

from which, taking the limit for $t \to 0$, we obtain (9.13). We use (A.69) to show that

$$\langle J'_G(v), v-\hat{u}\rangle_* \geq \langle J'_G(\hat{u}), v-\hat{u}\rangle_* \geq 0 \qquad \forall v \in \mathcal{U}_{ad}$$

and, finally, to show that (9.13) implies (9.95)].

2. If $\mathcal{U}_{ad}$ is a convex cone with vertex at $v=0$, relation (9.13) is equivalent to the system

$$\begin{cases} \langle J'_G(\hat{u}), v\rangle_* \geq 0 \;\; \forall v \in \mathcal{U}_{ad} \\ \langle J'_G(\hat{u}), \hat{u}\rangle_* = 0. \end{cases} \tag{9.96}$$

[*Hint:* To show that (9.96) implies (9.13) it is sufficient to subtract the two equations in (9.96). To show that the opposite is true, let $v = 2\hat{u}+w$; then $v \in \mathcal{U}_{ad}$, since $\mathcal{U}_{ad}$ is a convex cone. Insert v into (9.13) to get $\langle J'_G(\hat{u}), \hat{u}\rangle_* \geq 0$. Insert $v=0$ into (9.13) to get $\langle J'_G(\hat{u}), -\hat{u}\rangle_* \geq 0$. Use (9.13) to get the first equation in (9.96)].

3. Let us consider $\mathcal{U} = L^2(\Omega)$, with Ω a bounded domain and

$$K = \left\{ u \in L^2(\Omega);\; \frac{1}{|\Omega|}\int_\Omega u = c \right\},$$

where c is a prescribed constant. Prove that

$$P_K(u) = u - \frac{1}{|\Omega|}\int_\Omega (u-c).$$

4. Show that the control-to-state map $u \mapsto y(u)$ of problem (9.1) is $\mathcal{G}$–differentiable.
[*Hint:* Write in weak form the equation satisfied by

$$w(s;u,v) = \frac{y(u+sv)-y(u)}{s}$$

for s and v fixed. Show that, for u, v fixed, $w(s; u, v) \rightharpoonup w_0(u; v)$ in $H^1(\Omega)$, as $s \to 0$, strongly in $L^2(\Omega)$. Then pass to the limit as $s \to 0$ into the equation].

5. Show that the operator $\Phi : L^6(\Omega) \to L^2(\Omega)$, given by

$$\Phi : y \mapsto y^3,$$

is $\mathcal{F}-$differentiable at every point and

$$\Phi'(y)\, h = 3y^2 h \qquad \forall u, h \in L^6(\Omega).$$

[*Hint:* Use Hölder inequality with exponent $p = 3$, $q = 3/2$ to write

$$\left\|(y+h)^3 - y^3 - 3y^2 h\right\|_{L^2(\Omega)} = \left\|3yh^2 + h^3\right\|_{L^2(\Omega)} \leq 3\,\|y\|_{L^6(\Omega)}\,\|h\|^2_{L^6(\Omega)} + \|h\|^3_{L^6(\Omega)}\,.$$

and then apply the definition of $\mathcal{F}-$differentiability.]

6. Show that the solution map $u \mapsto y(u)$ of the semilinear problem (9.1) is $\mathcal{F}-$differentiable and that $y'(u)$ is the linear map that associates to $v \in L^2(\Omega)$ the weak solution $z(u, v) = y'(u)\, v$ of the problem

$$\begin{cases} -\Delta z + z + 3y^2(u)\, z = v & \text{in } \Omega \\ \dfrac{\partial z}{\partial n} = 0 & \text{on } \Gamma. \end{cases} \tag{9.97}$$

[*Hint:* The equation satisfied by $h(u, v) = y(u+v) - y(u)$ is

$$-\Delta h + h + y^3(u+v) - y^3(u) = v$$

with $\frac{\partial h}{\partial n} = 0$ on Γ. From Exercise 5

$$y^3(u+v) - y^3(u) = 3y^2(u)\, h + E(u, v)$$

where $\|E\|_{L^2(\Omega)} = o(\|h\|_{L^6(\Omega)}) = o(\|h\|_{H^1(\Omega)})$, since $d \leq 3$. Thus $h = z + e$ where z is the solution of (9.97) and e solves

$$\begin{cases} -\Delta e + e + 3y^2(u)\, e = -E(u, v) & \text{in } \Omega \\ \dfrac{\partial e}{\partial n} = 0 & \text{on } \Gamma. \end{cases} \tag{9.98}$$

From Remark 9.13, $\|h\|_{H^1(\Omega)} \leq \|v\|_{L^2(\Omega)}$. Thus

$$\|e\|_{H^1(\Omega)} \leq \|E(u, v)\|_{L^2(\Omega)} = o(\|h\|_{H^1(\Omega)}) = o(\|v\|_{L^2(\Omega)})$$

which shows the $\mathcal{F}-$differentiability of $u \mapsto y(u)$].

7. Analyze the following control problem:

$$J(y, u) = \frac{1}{6}\,\|y - z_d\|^6_{L^6(\Omega)} + \frac{\beta}{2}\,\|u\|^2_{L^2(\Omega)} \to \min$$

subject to

$$u \in \mathcal{U}_{ad} = \left\{u \in L^2(\Omega) : a \leq u(\mathbf{x}) \leq b, \text{ a.e. in } \Omega\right\},$$

with semilinear state system (provided Ω is a smooth domain)

$$\begin{cases} -\Delta y + y^3 = u & \text{in } \Omega \\ y = 0 & \text{on } \Gamma. \end{cases}$$

8. Show that the weak solution $\hat{p} \in L^2\left(0,T; H_0^1\left(\Omega\right)\right)$ of (9.64) actually belongs to Y. [*Hint:* After choosing a test function of the form $\psi\left(\mathbf{x},t\right) = \varphi\left(t\right) v\left(\mathbf{x}\right)$ with $\varphi \in C_0^\infty\left(0,T\right), v \in H_0^1\left(\Omega\right)$, write the equation in the form

$$\left(\int_0^T \hat{p}\left(t\right) \varphi'\left(t\right) dt, v\right)_{L^2(\Omega)} = -\int_0^T \left\langle f\left(t\right) \varphi\left(t\right), v\right\rangle_* dt$$

where $f\left(t\right) v = \left(\nabla \hat{p}\left(t\right), \nabla v\right)_{L^2(\Omega)} + \hat{u}\hat{p}\left(t\right) v$ for a.e. $t \in \left(0,T\right)$. Check that $f \in L^2\left(0,T; H^{-1}\left(\Omega\right)\right)$ and deduce that

$$\int_0^T \hat{p}\left(t\right) \varphi'\left(t\right) dt = -\int_0^T f\left(t\right) \varphi\left(t\right) dt$$

which shows that $\hat{p}_t = f$ in $\mathcal{D}'\left(0,T; H^{-1}\left(\Omega\right)\right)$.]

Chapter 10
Advanced Selected Applications

In this chapter we consider few, selected OCPs set in the context of fluid dynamics or cardiac electrophysiology mathematical models. Nonlinear state systems, more involved cost functionals, and additional constraints required to ensure control feasibility, make these OCPs more challenging than those addressed so far, both for their analysis and for their numerical approximation. We propose a theoretical framework for the well-posedness analysis of (some of) them, and suitable adaptations of the numerical approaches previously introduced. Moreover, we indicate alternative approaches available from literature.

10.1 Optimal Control of Steady Navier-Stokes Flows

The first application we consider concerns the steady Navier-Stokes equations. Optimal control of fluid flows is crucial to many applications in engineering and has been extensively analyzed. Examples include the maximization of lift, or the minimization of drag or resistance forces, turbulent kynetic energy, vorticity or wall shear stresses. Early contributions date back to the 90's and are addressed by Temam [1], in a series of articles by Gunzburger ([119, 127, 129, 130], as well as by Ghattas [103], Heinkenschloss [137] and many other authors; see, e.g., [144, 164, 77].

In this section we address a mixed boundary value problem for the incompressible steady Navier-Stokes equations, both in two and three dimensions, where we prescribe homogeneous Neumann conditions (on the normal stress) on a portion of the boundary, homogeneous Dirichlet conditions on the velocity field on another boundary portion, and a *velocity control* on the remaining portion of the domain boundary. We consider the minimization of quadratic cost functionals involving either the square of the L^2 norm of the vorticity, or a velocity tracking-type term, on the whole computational domain. For the problem at hand, we introduce an ad-hoc saddle point formulation of the state problem that entails the presence of both the velocity and the pressure fields. This formulation is more apt for numerical approximation than the one with a single field $\boldsymbol{v}$ that is obtained after projecting the Navier-Stokes equations onto a divergence free space. For the control problem, the saddle point formulation enables us to derive a system of first-order optimality conditions, and to carry out a rigorous well-posedness analysis.

Similar problems have been studied in [128, 124] for the case of full Dirichlet boundary conditions, in [127] for the case of viscous drag reduction, and in [119, 98] in the case of boundary Neumann control. We do not report all the proofs of the corresponding theoretical results; the interested reader can however refer to our paper [190].

A. Manzoni et al., *Optimal Control of Partial Differential Equations*,
Applied Mathematical Sciences 207, https://doi.org/10.1007/978-3-030-77226-0_10

10.1.1 Problem Formulation

In this section we consider a state system given by the velocity-pressure formulation of the Navier-Stokes equations for modeling steady, incompressible and viscous fluid flows. We analyze a minimization problem for quadratic functionals of the velocity field through a velocity control operated on a portion of the inflow boundary. For a concrete example we denote by $\Omega \subset \mathbb{R}^d$, $d = 2, 3$ the domain shown in Figure 10.1, occupied by the fluid and by $\Gamma = \partial\Omega$ its boundary. The inflow boundary $\Gamma_1 = \Gamma_c \cup \Gamma_{in}$ is composed by Γ_c, the control boundary, and by Γ_{in}, where a Dirichlet data is assigned; Γ_{out} and Γ_w are the outflow and the wall boundary portions, where zero stress and zero velocity are prescribed, respectively.

Let

$$X = H^1_{\Gamma_w}(\Omega)^d = \left\{ \boldsymbol{v} \in H^1(\Omega)^d : \boldsymbol{v} = \boldsymbol{0} \text{ on } \Gamma_w \right\}$$

denote the state space, endowed with the norm $\|\boldsymbol{v}\|_X = \|\nabla \boldsymbol{v}\|_0$ where

$$\|\nabla \boldsymbol{v}\|_0 = \left(\sum_{i,j=1}^{d} \int_\Omega \left(\frac{\partial v_i}{\partial x_j} \right)^2 dx \right)^{1/2}.$$

Our goal is to minimize either the vorticity functional

$$\tilde{J}_1(\boldsymbol{v}, \boldsymbol{u}) = \frac{1}{2} \int_\Omega |\nabla \times \boldsymbol{v}|^2 \, d\mathbf{x} + \frac{\tau}{2} \int_{\Gamma_c} |\nabla \boldsymbol{u}|^2 \, d\sigma \tag{10.1}$$

or a velocity tracking-type cost functional

$$\tilde{J}_2(\boldsymbol{v}, \boldsymbol{u}) = \frac{1}{2} \int_\Omega |\boldsymbol{v} - \boldsymbol{z}_d|^2 \, d\mathbf{x} + \frac{\tau}{2} \int_{\Gamma_c} |\nabla \boldsymbol{u}|^2 \, d\sigma \tag{10.2}$$

where $\tau \geq 0$, $\boldsymbol{u} \in \mathcal{U} = H^1_0(\Gamma_c)^d$ is a control variable and $(\boldsymbol{v}, \pi) = (\boldsymbol{v}(\boldsymbol{u}), \pi(\boldsymbol{u})) \in X \times L^2(\Omega)$ is a weak solution of the following system:

$$\begin{cases} -\nu \Delta \boldsymbol{v} + (\boldsymbol{v} \cdot \nabla)\boldsymbol{v} + \nabla \pi = \boldsymbol{0} & \text{in } \Omega \\ \operatorname{div} \boldsymbol{v} = 0 & \text{in } \Omega \\ \boldsymbol{v} = \boldsymbol{v}_{in} & \text{on } \Gamma_{in} \\ \boldsymbol{v} = \boldsymbol{u} & \text{on } \Gamma_c \\ \boldsymbol{v} = \boldsymbol{0} & \text{on } \Gamma_w \\ \pi \mathbf{n} - \nu \dfrac{\partial \boldsymbol{v}}{\partial n} = \boldsymbol{0} & \text{on } \Gamma_{out}, \end{cases} \tag{10.3}$$

where $\boldsymbol{v}$ and π denote the fluid velocity and pressure, respectively, ν is a given kynematic viscosity, and $\mathbf{n}$ is the unit outward normal direction. This configuration may e.g. refer to an arterial bypass where the lower branch is the hosting artery (that is partially or totally occluded) while the upper branch is the bypass. $\boldsymbol{v}_{in}$ indicates the residual blood flow velocity

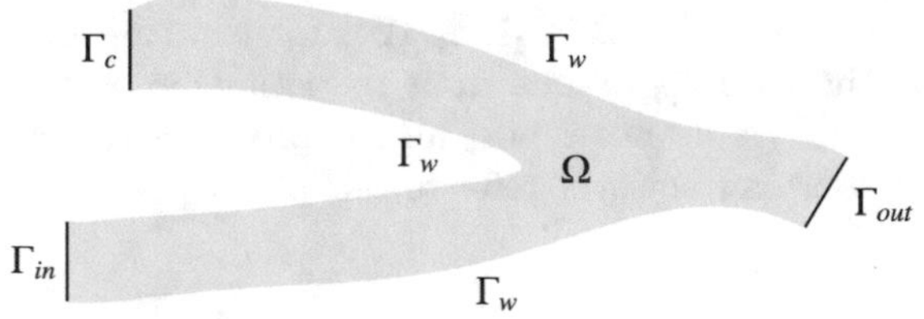

Fig. 10.1 The domain and its boundary portions

in the occluded vessel, while the control function $\boldsymbol{u}$ expresses the flow field at the bypass inflow section. See Sect. 10.1.6 for more. In the case of $\tilde{J}_2$, a flow matching problem is solved, aiming at driving the velocity field toward a target velocity profile $\boldsymbol{z}_d \in X$; this latter can be given, e.g., by solving the Stokes flow problem in the same domain, if the goal is to minimize recirculation and vortex generation to regularize the flow. As usual, if $\tau > 0$ the second term in (10.1) or (10.2) (a penalization term) helps in the well-posedness analysis and prevents using "too large" controls in the minimization of the cost functional. Concerning the choice of the norm in the penalization term $\|\boldsymbol{u}\|^2_{H^1_0(\Gamma_c)^d}$, see Sect. 5.13.2.

We handle the Dirichlet boundary conditions on Γ_1 (involving the control function) by a Lagrange multiplier approach. Let e and b be the following bilinear forms

$$e(\boldsymbol{v},\boldsymbol{w}) = \int_\Omega \nu \nabla \boldsymbol{v} : \nabla \boldsymbol{w}\, dx, \qquad b(\boldsymbol{v},q) = -\int_\Omega q \operatorname{div} \boldsymbol{v}\, dx \tag{10.4}$$

and c be the trilinear form

$$c(\boldsymbol{v},\boldsymbol{w},\boldsymbol{z}) = \int_\Omega (\boldsymbol{v}\cdot\nabla)\boldsymbol{w}\cdot\boldsymbol{z}\, dx = \sum_{i,j=1}^d \int_\Omega v_i \frac{\partial w_j}{\partial x_i} z_j\, dx. \tag{10.5}$$

Using Hölder inequality, and the Sobolev inequality

$$\|\boldsymbol{v}\|_{L^4(\Omega)} \le C_S \|\nabla \boldsymbol{v}\|_0 = C_S \|\boldsymbol{v}\|_X ,$$

we find that

$$|c(\boldsymbol{v},\boldsymbol{w},\boldsymbol{z})| \le \|\boldsymbol{v}\|_{L^4(\Omega)} \|\nabla \boldsymbol{w}\|_0 \|\boldsymbol{z}\|_{L^4(\Omega)} \le C_S^2 \|\boldsymbol{v}\|_X \|\boldsymbol{w}\|_X \|\boldsymbol{z}\|_X \quad \forall \boldsymbol{v},\boldsymbol{w},\boldsymbol{z} \in X. \tag{10.6}$$

Let us now define

$$\boldsymbol{g} = \boldsymbol{g}(\boldsymbol{u}) = \begin{cases} \boldsymbol{u} & \text{on } \Gamma_c \\ \boldsymbol{0} & \text{on } \Gamma_w \\ \boldsymbol{v}_{in} & \text{on } \Gamma_{in}. \end{cases}$$

Assuming that $\boldsymbol{v}_{in} \in H^{1/2}_{00}(\Gamma_{in})^d$, we have that $\boldsymbol{g}|_{\Gamma_1} \in H^{1/2}_{00}(\Gamma_1)^d$, since $\boldsymbol{u} \in H^1_0(\Gamma_c)^d \subset H^{1/2}_{00}(\Gamma_c)^d$ with continuous injection, that is, for some positive constant c_0,

$$\|\boldsymbol{u}\|_{H^{1/2}_{00}(\Gamma_c)^d} \le c_0 \|\boldsymbol{u}\|_{H^1_0(\Gamma_c)^d} . \tag{10.7}$$

We introduce the notation $\mathcal{T}_1 = H^{1/2}_{00}(\Gamma_1)^d$ and denote by $\langle\cdot,\cdot\rangle_{\Gamma_1}$ the duality pairing between $\mathcal{T}_1$ and its dual $\mathcal{T}_1^*$. In the same way, we set $\mathcal{T}_c = H^{1/2}_{00}(\Gamma_c)^d$, and denote by $\langle\cdot,\cdot\rangle_{\Gamma_c}$ the duality pairing between $\mathcal{T}_c$ and its dual $\mathcal{T}_c^*$.

10.1.2 Analysis of the State Problem

Let $V = X \times L^2(\Omega) \times \mathcal{T}_1^*$. As done in Sect. 5.13.2, to analyze the well-posedness of problem (10.3), we use a saddle-point formulation, introducing a new variable $\boldsymbol{t} \in \mathcal{T}_1$, representing the *boundary stress* $\pi\mathbf{n} - \nu\frac{\partial \boldsymbol{v}}{\partial n}$ on Γ_1, and appending a multiplier $\boldsymbol{\lambda} \in \mathcal{T}_1^*$ to the Dirichlet boundary conditions: *find* $(\boldsymbol{v},\pi,\boldsymbol{t}) \in V$ *such that*[1]

$$\begin{cases} e(\boldsymbol{v},\boldsymbol{\varphi}) + c(\boldsymbol{v},\boldsymbol{v},\boldsymbol{\varphi}) + b(\boldsymbol{\varphi},\pi) + \langle \boldsymbol{t},\boldsymbol{\varphi}\rangle_{\Gamma_1} = 0 & \forall \boldsymbol{\varphi} \in X \\ b(\boldsymbol{v},\phi) = 0 & \forall \phi \in L^2(\Omega) \\ \langle \boldsymbol{\lambda},\boldsymbol{v}\rangle_{\Gamma_1} = \langle \boldsymbol{\lambda},\boldsymbol{g}\rangle_{\Gamma_1} & \forall \boldsymbol{\lambda} \in \mathcal{T}_1^*. \end{cases} \tag{10.8}$$

[1] Observe that if $\boldsymbol{\phi} \in X$, then $\boldsymbol{\phi}|_{\Gamma_1} \in H^{1/2}_{00}(\Gamma_1)^d$.

Note that a formal integration by parts in the first equation returns

$$\boldsymbol{t} = \pi\mathbf{n} - \nu\frac{\partial \boldsymbol{v}}{\partial n} \qquad \text{on } \Gamma_1, \tag{10.9}$$

and

$$\mathbf{0} = \pi\mathbf{n} - \nu\frac{\partial \boldsymbol{v}}{\partial n} \qquad \text{on } \Gamma_{out}.$$

Thus, the first equation in (10.8) enforces in a natural way the Neumann condition on Γ_{out}. By grouping together the last two equations in (10.8), we obtain the following equivalent formulation: *find* $\boldsymbol{v} \in X$, $(\pi, \boldsymbol{t}) \in Q = L^2(\Omega) \times \mathcal{T}_1^*$ *such that*

$$\begin{cases} e(\boldsymbol{v}, \boldsymbol{\varphi}) + c\,(\boldsymbol{v}, \boldsymbol{v}, \boldsymbol{\varphi}) + b(\boldsymbol{\varphi}, \pi) + \langle \boldsymbol{t}, \boldsymbol{\varphi}\rangle_{\Gamma_1} = 0 & \forall \boldsymbol{\varphi} \in X \\ b(\boldsymbol{v}, \phi) + \langle \boldsymbol{\lambda}, \boldsymbol{v}\rangle_{\Gamma_1} = \langle \boldsymbol{\lambda}, \boldsymbol{g}\rangle_{\Gamma_1} & \forall (\phi, \boldsymbol{\lambda}) \in Q. \end{cases} \tag{10.10}$$

Upon defining the bilinear form $\tilde{b} : X \times Q \to \mathbb{R}$ as

$$\tilde{b}(\boldsymbol{v}, (\phi, \boldsymbol{\lambda})) = b(\boldsymbol{v}, \phi) + \langle \boldsymbol{\lambda}, \boldsymbol{v}\rangle_{\Gamma_1} \tag{10.11}$$

and the norm

$$\|(\phi, \boldsymbol{\lambda})\|_Q = \sqrt{\|\phi\|^2_{L^2(\Omega)} + \|\boldsymbol{\lambda}\|^2_{\mathcal{T}_1^*}},$$

we finally reformulate problem (10.10) as: *find* $\boldsymbol{v} \in X, (\pi, \boldsymbol{t}) \in Q$ *such that*

$$\begin{cases} e(\boldsymbol{v}, \boldsymbol{\varphi}) + c\,(\boldsymbol{v}, \boldsymbol{v}, \boldsymbol{\varphi}) + \tilde{b}(\boldsymbol{\varphi}, (\pi, \boldsymbol{t})) = 0 & \forall \boldsymbol{\varphi} \in X \\ \tilde{b}(\boldsymbol{v}, (\phi, \boldsymbol{\lambda})) = \langle G, (\phi, \boldsymbol{\lambda})\rangle_{Q^*,Q} & \forall (\phi, \boldsymbol{\lambda}) \in Q, \end{cases} \tag{10.12}$$

with $G \in Q^*$ *given by*

$$\langle G, (\phi, \boldsymbol{\lambda})\rangle_{Q^*,Q} = \langle \boldsymbol{\lambda}, \boldsymbol{g}\rangle_{\Gamma_1}. \tag{10.13}$$

As in Lemma 5.2, the inf-sup condition holds for the bilinear form $\tilde{b}$, that is,

Lemma 10.1. *There exists a constant* $\tilde{\beta} > 0$ *such that*

$$\inf_{0 \neq (\phi, \boldsymbol{\lambda}) \in Q} \sup_{0 \neq \boldsymbol{v} \in X} \frac{\tilde{b}(\boldsymbol{v}, (\phi, \boldsymbol{\lambda}))}{\|\boldsymbol{v}\|_X \|(\phi, \boldsymbol{\lambda})\|_Q} \geq \tilde{\beta}.$$

The following result is proved in [190, Thm. 3.3].

Theorem 10.1. *There exists* $\eta = \eta(\nu, \tilde{\beta}, C_S) > 0$ *such that, if* $\|\mathbf{g}(\mathbf{u})\|_{\mathcal{T}_1} \leq \eta$, *then problem (10.12) has a unique solution* $(\boldsymbol{v}, \pi, \boldsymbol{t}) = (\boldsymbol{v}(\boldsymbol{u}), \pi(\boldsymbol{u}), \boldsymbol{t}(\boldsymbol{u})) \in X \times L^2(\Omega) \times \mathcal{T}_1^*$. *Moreover,*

$$\|\boldsymbol{v}\|_X \leq 5\tilde{\beta}^{-1}\|\boldsymbol{g}(\boldsymbol{u})\|_{\mathcal{T}_1}$$

and

$$\|\pi\|_{L^2(\Omega)} + \|\boldsymbol{t}\|_{\mathcal{T}_1^*} \leq C_1\left(\|\boldsymbol{g}(\boldsymbol{u})\|_{\mathcal{T}_1} + \|\boldsymbol{g}(\boldsymbol{u})\|^2_{\mathcal{T}_1}\right) \tag{10.14}$$

where C_1 *depends only on* ν, $\tilde{\beta}, C_S$. *Furthermore, for each* $\boldsymbol{u}_1$, $\boldsymbol{u}_2 \in \mathcal{U}_{ad}$,

$$\|\boldsymbol{v}(\boldsymbol{u}_1) - \boldsymbol{v}(\boldsymbol{u}_2)\|_X \leq C_L\|\boldsymbol{u}_1 - \boldsymbol{u}_2\|_{\mathcal{T}_c}, \tag{10.15}$$

with C_L *depending only on* ν, $\tilde{\beta}, C_S$.

Remark 10.1. A possible choice of η is (see [190])

$$\eta = \frac{\tilde{\beta}}{60C_S^2}\min\{\nu, \nu^{-1}\}. \tag{10.16}$$

•

Remark 10.2. Theorem 10.1 also holds if a volume force $\boldsymbol{f} \in X^*$, $\boldsymbol{f} \neq \mathbf{0}$ is considered at the right hand side of the momentum equation in (10.3), such that $\|\boldsymbol{f}\|_{X^*}$ is small enough. In this case, the unique solution $(\boldsymbol{v}, \pi, \boldsymbol{t})$ satisfies the estimates

$$\|\boldsymbol{v}\|_X \leq \tilde{C}_1 \left(\|\boldsymbol{f}\|_{X^*} + \|\boldsymbol{g}(\boldsymbol{u})\|_{\mathcal{T}_1}\right), \tag{10.17}$$

$$\|\pi\|_{L^2(\Omega)} + \|\boldsymbol{t}\|_{\mathcal{T}_1^*} \leq \tilde{C}_2 \left(\|\boldsymbol{f}\|_{X^*} + \|\boldsymbol{g}(\boldsymbol{u})\|_{\mathcal{T}_1} + \|\boldsymbol{g}(\boldsymbol{u})\|_{\mathcal{T}_1}^2\right) \tag{10.18}$$

where $\tilde{C}_1$ and $\tilde{C}_2$ only depend on the constants ν, $\tilde{\beta}, C_S$. •

10.1.3 Existence of an Optimal Control

We are now ready for investigating the well-posedness analysis of the optimal control problem

$$\tilde{J}(\boldsymbol{v}, \boldsymbol{u}) \to \min \quad \text{subject to} \quad (10.8), \quad \boldsymbol{v} \in X, \quad (\pi, \boldsymbol{t}) \in Q, \quad \boldsymbol{u} \in \mathcal{U}_{ad}. \tag{10.19}$$

Here $\tilde{J}$ is given by either (10.1) or (10.2) and

$$\mathcal{U}_{ad} = \{\boldsymbol{u} : \ c_0\|\boldsymbol{u}\|_{H_0^1(\Gamma_c)^d} \leq \eta/2\},$$

is a closed, convex set in $H_0^1(\Gamma_c)^d$, where η denotes the constant defined in (10.16). Note that the definition of $\mathcal{U}_{ad}$ depends on the positive constant c_0 in (10.7). If we assume that $\|\boldsymbol{v}_{in}\|_{\mathcal{T}_1} \leq \eta/2$ then

$$\|\boldsymbol{g}(\boldsymbol{u})\|_{\mathcal{T}_1} \leq \|\boldsymbol{v}_{in}\|_{\mathcal{T}_1} + c_0\|\boldsymbol{u}\|_{H_0^1(\Gamma_c)^d} \leq \eta$$

and by Theorem 10.1 there exists a unique solution $\boldsymbol{v}(\boldsymbol{u}) \in X$, $(\pi(\boldsymbol{u}), \boldsymbol{t}(\boldsymbol{u})) \in Q$, with $\|\boldsymbol{v}(\boldsymbol{u})\|_X \leq 5\tilde{\beta}^{-1}\|\boldsymbol{g}(\boldsymbol{u})\|_{\mathcal{T}_1} \leq 5\tilde{\beta}^{-1}\eta$. The following result holds.

Theorem 10.2. *If $\|\boldsymbol{v}_{in}\|_{\mathcal{T}_1} \leq \eta/2$, there exists an optimal control $\hat{\mathbf{u}} \in \mathcal{U}_{ad}$ and a corresponding optimal state $\hat{\boldsymbol{v}} \in X$, $(\hat{\pi}, \hat{\boldsymbol{t}}) \in Q$.*

Proof. Define the set of feasible points by

$$\mathcal{F} = \{(\boldsymbol{v}, \pi, \boldsymbol{t}, \boldsymbol{u}) \in S_\delta \times Q \times \mathcal{U}_{ad} \text{ such that (10.8) is satisfied}\},$$

where $S_\delta = \{\boldsymbol{v} \in X : \|\boldsymbol{v}\|_X \leq \delta\}$, $\delta = 5\tilde{\beta}^{-1}\eta$. The proof follows from Theorem 9.4:

(1) $\tilde{J}(\boldsymbol{v}, \boldsymbol{u}) \geq 0$ and $\mathcal{F}$ is nonempty by Theorem 10.1.
(2) Let $\{\boldsymbol{v}_k, \pi_k, \boldsymbol{t}_k, \boldsymbol{u}_k\} \in \mathcal{F}$ be a minimizing sequence of *tildeJ* for our OCP (10.19). Then $\{\boldsymbol{v}_k, \pi_k, \boldsymbol{t}_k, \boldsymbol{u}_k\}$ is bounded. Indeed $\{\boldsymbol{v}_k, \boldsymbol{u}_k\} \subset S_\delta \times \mathcal{U}_{ad}$ and therefore is bounded in $X \times H_0^1(\Gamma_1)^d$. From (10.14),

$$\|\pi_k\|_{L^2(\Omega)} + \|\boldsymbol{t}_k\|_{\mathcal{T}_1^*} \leq C_1(\eta + \eta^2) \tag{10.20}$$

and hence also $\{\pi_k, \boldsymbol{t}_k\}$ is bounded in Q.
(3) $\mathcal{F}$ is weakly sequentially closed. We only need to check that if $\{\boldsymbol{v}_k, \pi_k, \boldsymbol{t}_k, \boldsymbol{u}_k\}$ is a solution to

$$\begin{cases} e(\boldsymbol{v}_k, \boldsymbol{\varphi}) + c(\boldsymbol{v}_k, \boldsymbol{v}_k, \boldsymbol{\varphi}) + b(\boldsymbol{\varphi}, \pi_k) + \langle \boldsymbol{t}_k, \boldsymbol{\varphi}\rangle_{\Gamma_1} = 0 & \forall \boldsymbol{\varphi} \in X \\ b(\boldsymbol{v}_k, \phi) = 0 & \forall \phi \in L^2(\Omega) \\ \langle \boldsymbol{\lambda}, \boldsymbol{v}_k\rangle_{\Gamma_1} = \langle \boldsymbol{\lambda}, \boldsymbol{g}(\boldsymbol{u}_k)\rangle_{\Gamma_1} & \forall \boldsymbol{\lambda} \in \mathcal{T}_1^* \end{cases} \tag{10.21}$$

and $(\boldsymbol{v}_k, \pi_k, \boldsymbol{t}_k) \rightharpoonup (\boldsymbol{v}, \pi, \boldsymbol{t})$ in V, $\boldsymbol{u}_k \rightharpoonup \boldsymbol{u}$ in $H_0^1(\Gamma_c)^d$, then $(\boldsymbol{v}, \pi, \boldsymbol{t}, \boldsymbol{u})$ is a solution of (10.8). Since $\boldsymbol{v}_k \rightharpoonup \boldsymbol{v}$ in X and the trace operator $\boldsymbol{v} \mapsto \boldsymbol{v}_{|\Gamma_1}$ is continuous from X into $\mathcal{T}_1$, it follows that, for every $\boldsymbol{\varphi} \in X$, $\phi \in L^2(\Omega)$ and $\lambda \in \mathcal{T}_1^*$,

$$e(\boldsymbol{v}_k, \boldsymbol{\varphi}) \to e(\boldsymbol{v}, \boldsymbol{\varphi}), \qquad b(\boldsymbol{v}_k, \phi) \to b(\boldsymbol{v}, \phi) \quad \text{and} \quad \langle \boldsymbol{\lambda}, \boldsymbol{v}_k \rangle_{\Gamma_1} \to \langle \boldsymbol{\lambda}, \boldsymbol{v} \rangle_{\Gamma_1}.$$

Also, $\pi_k \rightharpoonup \pi$ in $L^2(\Omega)$ implies that $b(\boldsymbol{\varphi}, \pi_k) \to b(\boldsymbol{\varphi}, \pi)$. Moreover, the embedding of $H_0^1(\Gamma_1)^d$ into $\mathcal{T}_c$ is continuous and hence $\mathbf{u}_k \rightharpoonup \boldsymbol{u}$ in $H_0^1(\Gamma_c)^d$ implies that $\langle \boldsymbol{\lambda}, \boldsymbol{g}(\boldsymbol{u}_k) \rangle_{\Gamma_1} \to \langle \boldsymbol{\lambda}, \boldsymbol{g}(\boldsymbol{u}) \rangle_{\Gamma_1}$ and $\langle \boldsymbol{t}_k, \boldsymbol{\varphi} \rangle_{\Gamma_1} \to \langle \boldsymbol{t}, \boldsymbol{\varphi} \rangle_{\Gamma_1}$.
Let us write

$$c(\boldsymbol{v}_k, \boldsymbol{v}_k, \boldsymbol{\varphi}) - c(\boldsymbol{v}, \boldsymbol{v}, \boldsymbol{\varphi}) = c(\boldsymbol{v}_k - \boldsymbol{v}, \boldsymbol{v}_k, \boldsymbol{\varphi}) - c(\boldsymbol{v}, \boldsymbol{v} - \boldsymbol{v}_k, \boldsymbol{\varphi})$$

and recall that X is compactly embedded into $L^4(\Omega)^d$, so that $\boldsymbol{v}_k \rightharpoonup \boldsymbol{v}$ in X implies that $\boldsymbol{v}_k \to \boldsymbol{v}$ in $L^4(\Omega)^d$. Thus

$$\begin{aligned} |c(\boldsymbol{v}_k - \boldsymbol{v}, \boldsymbol{v}_k, \boldsymbol{\varphi})| &\le \|\mathbf{v}_k\|_X \|\boldsymbol{\varphi}\|_{L^4(\Omega)^d} \|\boldsymbol{v}_k - \boldsymbol{v}\|_{L^4(\Omega)^d} \\ &\le \delta \|\boldsymbol{\varphi}\|_{L^4(\Omega)^d} \|\boldsymbol{v}_k - \boldsymbol{v}\|_{L^4(\Omega)^d} \to 0. \end{aligned}$$

Finally, thanks to the weak convergence of $\{\boldsymbol{v}_k\}$ in X, we have

$$c(\boldsymbol{v}, \boldsymbol{v}_k - \boldsymbol{v}, \boldsymbol{\varphi}) \to 0.$$

Passing to the limit for $k \to \infty$ into (10.21), we deduce that $(\boldsymbol{v}, \pi, \boldsymbol{t}, \boldsymbol{u})$ is a solution of (10.8).

(4) J is continuous and convex and therefore also sequentially weakly lower semicontinuous. □

10.1.4 Differentiability of the Control-to-State Map

In order to derive first order optimality conditions, we need to show that the control-to-state map $\mathcal{S} : \boldsymbol{u} \mapsto (\boldsymbol{v}(\boldsymbol{u}), \pi(\boldsymbol{u}), \boldsymbol{t}(\boldsymbol{u}))$ is Fréchet differentiable in $\mathcal{U}_{ad}$. Precisely, we show the following result.

Theorem 10.3. *Let $\bar{\boldsymbol{u}} \in \mathcal{U}_{ad}, \boldsymbol{k} \in H_0^1(\Gamma_c)^d$, with $\bar{\boldsymbol{u}} + \boldsymbol{k} \in \mathcal{U}_{ad}$. The control-to-state map $\mathcal{S}$ is Fréchet-differentiable at $\bar{\boldsymbol{u}}$ and its derivative $\mathcal{S}'(\bar{\boldsymbol{u}})$ is given by*

$$\mathcal{S}'(\bar{\boldsymbol{u}})\boldsymbol{k} = (\boldsymbol{v}'(\bar{\boldsymbol{u}})\boldsymbol{k}, \pi'(\bar{\boldsymbol{u}})\boldsymbol{k}, \boldsymbol{t}'(\bar{\boldsymbol{u}})\boldsymbol{k}) = (\boldsymbol{w}, q, \boldsymbol{s})$$

where $(\boldsymbol{w}, q, \boldsymbol{s}) \in V$ is the solution to

$$\begin{cases} e(\boldsymbol{w}, \boldsymbol{\varphi}) + c(\boldsymbol{w}, \bar{\boldsymbol{v}}, \boldsymbol{\varphi}) + c(\bar{\boldsymbol{v}}, \boldsymbol{w}, \boldsymbol{\varphi}) + b(\boldsymbol{\varphi}, q) + \langle \boldsymbol{s}, \boldsymbol{\varphi} \rangle_{\Gamma_1} = \mathbf{0} & \forall \boldsymbol{\varphi} \in X \\ b(\boldsymbol{w}, \phi) = 0 & \forall \phi \in L^2(\Omega) \\ \langle \boldsymbol{\lambda}, \boldsymbol{w} \rangle_{\Gamma_1} = \langle \boldsymbol{\lambda}, \chi_{\Gamma_c} \boldsymbol{k} \rangle_{\Gamma_1} & \forall \boldsymbol{\lambda} \in \mathcal{T}_1^* \end{cases} \tag{10.22}$$

and $\bar{\boldsymbol{v}} = \boldsymbol{v}(\bar{\boldsymbol{u}})$.

Remark 10.3. Since $\|\bar{\boldsymbol{v}}\|_X \le 5\tilde{\beta}^{-1}\eta \le 12^{-1}C_S^{-2}\nu$ thanks to (10.16), the bilinear form

$$a(\boldsymbol{w}, \boldsymbol{\varphi}) = e(\boldsymbol{w}, \boldsymbol{\varphi}) + c(\boldsymbol{w}, \bar{\boldsymbol{v}}, \boldsymbol{\varphi}) + c(\bar{\boldsymbol{v}}, \boldsymbol{w}, \boldsymbol{\varphi}) \tag{10.23}$$

is $X-$coercive; hence, the linear problem (10.22) is well posed. In fact,

$$a(\boldsymbol{w}, \boldsymbol{w}) \ge (\nu - 2C_S^2 \|\bar{\boldsymbol{v}}\|_X) \|\boldsymbol{w}\|_X^2 \ge \frac{5}{6}\nu \|\boldsymbol{w}\|_X^2. \qquad \bullet$$

Proof. Let $\bar{\boldsymbol{u}} \in \mathcal{U}_{ad}$, $\boldsymbol{k} \in H_0^1(\Gamma_c)^d$ be such that $\bar{\boldsymbol{u}} + \boldsymbol{k} \in \mathcal{U}_{ad}$. Denote by $\mathcal{S}(\bar{\boldsymbol{u}} + \boldsymbol{k}) = (\boldsymbol{v}^{\boldsymbol{k}}, \pi^{\boldsymbol{k}}, \boldsymbol{t}^{\boldsymbol{k}})$ and $\mathcal{S}(\bar{\boldsymbol{u}}) = (\bar{\boldsymbol{v}}, \bar{\pi}, \bar{\boldsymbol{t}})$ the corresponding solutions in V to (10.8). Write

$$\begin{aligned} c\left(\boldsymbol{v}^{\boldsymbol{k}}, \boldsymbol{v}^{\boldsymbol{k}}, \boldsymbol{\varphi}\right) - c\left(\bar{\boldsymbol{v}}, \bar{\boldsymbol{v}}, \boldsymbol{\varphi}\right) &= c\left(\boldsymbol{v}^{\boldsymbol{k}}, \boldsymbol{v}^{\boldsymbol{k}} - \bar{\boldsymbol{v}}, \boldsymbol{\varphi}\right) - c\left(\bar{\boldsymbol{v}} - \boldsymbol{v}^{\boldsymbol{k}}, \bar{\boldsymbol{v}}, \boldsymbol{\varphi}\right) \\ &= -c\left(\bar{\boldsymbol{v}} - \boldsymbol{v}^{\boldsymbol{k}}, \bar{\boldsymbol{v}}, \boldsymbol{\varphi}\right) - c\left(\bar{\boldsymbol{v}}, \boldsymbol{v}^{\boldsymbol{k}} - \bar{\boldsymbol{v}}, \boldsymbol{\varphi}\right) + c\left(\bar{\boldsymbol{v}} - \boldsymbol{v}^{\boldsymbol{k}}, \bar{\boldsymbol{v}} - \boldsymbol{v}^{\boldsymbol{k}}, \boldsymbol{\varphi}\right). \end{aligned}$$

Then the difference $(\boldsymbol{d}, p, \boldsymbol{r}) = (\boldsymbol{v}^{\boldsymbol{k}} - \bar{\boldsymbol{v}}, \pi^{\boldsymbol{k}} - \bar{\pi}, \boldsymbol{t}^{\boldsymbol{k}} - \bar{\boldsymbol{t}}) = \mathcal{S}\left(\bar{\boldsymbol{u}} + \boldsymbol{k}\right) - \mathcal{S}\left(\bar{\boldsymbol{u}}\right)$ satisfies the system

$$\begin{cases} e(\boldsymbol{d}, \boldsymbol{\varphi}) - c\left(\boldsymbol{d}, \bar{\boldsymbol{v}}, \boldsymbol{\varphi}\right) - c\left(\bar{\boldsymbol{v}}, \boldsymbol{d}, \boldsymbol{\varphi}\right) + b(\boldsymbol{\varphi}, p) + \langle \boldsymbol{r}, \boldsymbol{\varphi}\rangle_{\Gamma_1} = -c\left(\boldsymbol{d}, \boldsymbol{d}, \boldsymbol{\varphi}\right) \\ b(\boldsymbol{d}, \phi) = 0 \\ \langle \boldsymbol{\lambda}, \boldsymbol{d}\rangle_{\Gamma_1} = \langle \boldsymbol{\lambda}, \chi_{\Gamma_c} \boldsymbol{k}\rangle_{\Gamma_1} \end{cases} \tag{10.24}$$

for all $(\boldsymbol{\varphi}, \phi, \boldsymbol{\lambda}) \in V$. Therefore, the triple

$$(\boldsymbol{d} - \boldsymbol{w}, p - q, \boldsymbol{r} - \boldsymbol{s}) = \mathcal{S}\left(\bar{\boldsymbol{u}} + \boldsymbol{k}\right) - \mathcal{S}\left(\bar{\boldsymbol{u}}\right) - \mathcal{S}'\left(\bar{\boldsymbol{u}}\right) \boldsymbol{k}$$

is a solution to the problem

$$\begin{cases} e(\boldsymbol{d} - \boldsymbol{w}, \boldsymbol{\varphi}) + c\left(\boldsymbol{d} - \boldsymbol{w}, \bar{\boldsymbol{v}}, \boldsymbol{\varphi}\right) + c\left(\bar{\boldsymbol{v}}, \boldsymbol{d} - \boldsymbol{w}, \boldsymbol{\varphi}\right) + b(\boldsymbol{\varphi}, p - q) + \langle \boldsymbol{r} - \boldsymbol{s}, \boldsymbol{\varphi}\rangle_{\Gamma_1} = -c\left(\boldsymbol{d}, \boldsymbol{d}, \boldsymbol{\varphi}\right) \\ b(\boldsymbol{d} - \boldsymbol{w}, \phi) = 0 \\ \langle \boldsymbol{\lambda}, \boldsymbol{d} - \boldsymbol{w}\rangle_{\Gamma_1} = 0 \end{cases} \tag{10.25}$$

for all $(\boldsymbol{\varphi}, \phi, \boldsymbol{\lambda}) \in V$. From estimates (10.17), (10.18) we obtain

$$\|\boldsymbol{d} - \boldsymbol{w}\|_X \leq C \, \|\boldsymbol{f}\|_{X^*} \tag{10.26}$$

and

$$\|p - q\|_{L^2(\Omega)} + \|\boldsymbol{r} - \boldsymbol{s}\|_{\mathcal{T}_1^*} \leq C \, \|\boldsymbol{f}\|_{X^*} \tag{10.27}$$

where C depends only on ν, $\tilde{\beta}, c_0, C_S$ and $\boldsymbol{f}$ is the functional $\boldsymbol{\varphi} \longmapsto -c\left(\boldsymbol{d}, \boldsymbol{d}, \boldsymbol{\varphi}\right)$.

Using (10.15) we obtain

$$\left|c\left(\boldsymbol{d}, \boldsymbol{d}, \boldsymbol{\varphi}\right)\right| \leq C_S^2 \, \|\boldsymbol{d}\|_X^2 \, \|\boldsymbol{\varphi}\|_X \leq C_S^2 C_L \, \|\boldsymbol{k}\|_{\mathcal{T}_c}^2 \, \|\boldsymbol{\varphi}\|_X$$

from which

$$\|\boldsymbol{f}\|_{X^*} \leq C_S^2 C_L \, \|\boldsymbol{k}\|_{\mathcal{T}_c}^2 . \tag{10.28}$$

From (10.27) and (10.28) we have

$$\left\|\mathcal{S}\left(\bar{\boldsymbol{u}} + \boldsymbol{k}\right) - \mathcal{S}\left(\bar{\boldsymbol{u}}\right) - \mathcal{S}'\left(\bar{\boldsymbol{u}}\right) \boldsymbol{k}\right\|_V \leq C_3 \, \|\boldsymbol{k}\|_{\mathcal{T}_c}^2 \leq c_0 C_3 \, \|\boldsymbol{k}\|_{H_0^1(\Gamma_c)^d}^2$$

where C_3 only depends on the constants ν, $\tilde{\beta}$, C_S. □

Remark 10.4. From the previous proof, it is clear that $\mathcal{S}$ is differentiable also with respect to the $\mathcal{T}_c$-norm. •

10.1.5 First Order Optimality Conditions

We are now ready to write the first order optimality conditions for our OCP (10.19). The following result holds.

Theorem 10.4. *For the optimal control problem* (10.19), *let* $\hat{\boldsymbol{u}}$ *be an optimal control and* $(\hat{\boldsymbol{v}}, \hat{\pi}, \hat{\boldsymbol{t}})$ *the corresponding optimal state. There exists a unique multiplier* $(\hat{\boldsymbol{z}}, \hat{q}, \hat{\boldsymbol{s}}) \in V$ *such that* $\hat{\boldsymbol{u}}$ *and* $(\hat{\boldsymbol{z}}, \hat{q}, \hat{\boldsymbol{s}})$ *satisfy the following optimality system:*

- state problem (10.8)*;*
- adjoint problem*: for every* $(\boldsymbol{\psi}, \phi, \boldsymbol{\lambda}) \in V$,

$$\begin{cases} e(\boldsymbol{\psi}, \hat{\boldsymbol{z}}) + c(\hat{\boldsymbol{v}}, \boldsymbol{\psi}, \hat{\boldsymbol{z}}) + c(\boldsymbol{\psi}, \hat{\boldsymbol{v}}, \hat{\boldsymbol{z}}) + b(\boldsymbol{\psi}, \hat{q}) + \langle \hat{\boldsymbol{s}}, \boldsymbol{\psi} \rangle_{\Gamma_1} = \tilde{J}'_{\boldsymbol{v}}(\hat{\boldsymbol{v}}, \hat{\boldsymbol{u}})\,\boldsymbol{\psi} \\ b(\hat{\boldsymbol{z}}, \phi) = 0 \\ \langle \boldsymbol{\lambda}, \hat{\boldsymbol{z}} \rangle_{\Gamma_1} = 0; \end{cases} \tag{10.29}$$

- variational inequality*: for every* $\boldsymbol{k} \in \mathcal{U}_{ad}$,

$$J'(\hat{\boldsymbol{u}})(\boldsymbol{k} - \hat{\boldsymbol{u}}) = (\tau \hat{\boldsymbol{u}}, \boldsymbol{k} - \hat{\boldsymbol{u}})_{H_0^1(\Gamma_c)^d} + \langle \hat{\boldsymbol{s}}, \boldsymbol{k} - \hat{\boldsymbol{u}} \rangle_{\Gamma_c} \geq 0 \tag{10.30}$$

where $J(\boldsymbol{u}) = \tilde{J}(\boldsymbol{v}(\boldsymbol{u}), \boldsymbol{u})$ *is the reduced cost functional.*

Proof. Let us recall (see Sect. 5.13.2) the definition of the Stokes solver $R : V^* \to V$: given $(\mathbf{F}, h, \mathbf{G}) \in V^*$,

$$R(\mathbf{F}, h, \mathbf{G}) = (\boldsymbol{v}, \pi, \boldsymbol{t}) \in V \tag{10.31}$$

is the unique solution of the problem

$$\begin{cases} e(\boldsymbol{v}, \boldsymbol{\varphi}) + b(\boldsymbol{\varphi}, \pi) + \langle \mathbf{t}, \boldsymbol{\varphi} \rangle_{\Gamma_1} = \langle \mathbf{F}, \boldsymbol{\varphi} \rangle_{X^*, X} & \forall \boldsymbol{\varphi} \in X \\ b(\boldsymbol{v}, \phi) = (h, \phi)_{L^2(\Omega)} & \forall \phi \in L^2(\Omega) \\ \langle \boldsymbol{\lambda}, \boldsymbol{v} \rangle_{\Gamma_1} = \langle \boldsymbol{\lambda}, \mathbf{G} \rangle_{\Gamma_1} & \forall \boldsymbol{\lambda} \in \mathcal{T}_1^*. \end{cases} \tag{10.32}$$

Moreover, define the nonlinear operator $N : V \times \mathcal{T}_c \to V^*$ by

$$N(\boldsymbol{v}, \pi, \boldsymbol{t}, \boldsymbol{u}) = ((\boldsymbol{v} \cdot \nabla)\boldsymbol{v}, -h, -\boldsymbol{g}(\boldsymbol{u})).$$

The nonlinear component of N is a continuous bilinear operator and therefore N is Fréchet differentiable[2], with

$$N'_{(\boldsymbol{v}, \pi, \boldsymbol{t})}(\boldsymbol{v}, \pi, \boldsymbol{t}, \boldsymbol{u})(\boldsymbol{\psi}, q, \boldsymbol{r}) = ((\boldsymbol{v} \cdot \nabla)\boldsymbol{\psi} + (\boldsymbol{\psi} \cdot \nabla)\boldsymbol{v}, 0, \mathbf{0}) \tag{10.33}$$

and

$$N'_{\boldsymbol{u}}(\boldsymbol{v}, \pi, \boldsymbol{t}, \boldsymbol{u})\,\boldsymbol{k} = (\mathbf{0}, 0, -\boldsymbol{g}_0(\boldsymbol{k})) \tag{10.34}$$

where

$$\boldsymbol{g}_0(\boldsymbol{k}) = \begin{cases} \boldsymbol{k} & \text{on } \Gamma_c \\ \mathbf{0} & \text{on } \Gamma_w \cup \Gamma_{in}. \end{cases}$$

In terms of R and N, the state problem can be written as

$$\mathcal{G}(\boldsymbol{v}, \pi, \boldsymbol{t}, \boldsymbol{u}) = R^{-1}(\mathbf{v}, \pi, \boldsymbol{t}) + N(\boldsymbol{v}, \pi, \boldsymbol{t}, \boldsymbol{u}) = \mathbf{0} \tag{10.35}$$

with $\mathcal{G} : V \times \mathcal{T}_1 \to V^*$.

In Lemma 10.2 below we prove that:

(a) the adjoint operator $N'_{(\boldsymbol{v}, \pi, \boldsymbol{t})}(\boldsymbol{v}, \pi, \boldsymbol{t}, \boldsymbol{u})^* : V \to V^*$ is given by

$$N'_{(\boldsymbol{v}, \pi, \boldsymbol{t})}(\boldsymbol{v}, \pi, \boldsymbol{t}, \boldsymbol{u})^*(\boldsymbol{z}, q, \boldsymbol{s}) = (-(\boldsymbol{v} \cdot \nabla)\boldsymbol{z} + (\nabla \boldsymbol{v})^\top \boldsymbol{z} + [(\boldsymbol{v} \cdot \mathbf{n})\boldsymbol{z}]_{|\Gamma_1}, 0, \mathbf{0}) \tag{10.36}$$

where $\mathbf{n}$ is the exterior unit normal to Γ_c;

(b) the adjoint operator $N'_{\boldsymbol{u}}(\boldsymbol{v}, \pi, \boldsymbol{t}, \boldsymbol{u})^* : V \to \mathcal{T}_c^*$ is given by

$$N'_{\boldsymbol{u}}(\boldsymbol{v}, \pi, \boldsymbol{t}, \boldsymbol{u})^*(\boldsymbol{z}, q, \boldsymbol{s}) = -\boldsymbol{s}. \tag{10.37}$$

By Theorem 9.5, the existence of a unique multiplier $(\hat{\boldsymbol{z}}, \hat{q}, \hat{\mathbf{s}}) \in V$ follows from the properties $(i), (ii), (iii)$ listed below.

(i) $\tilde{J}$ and $\mathcal{G}$ are Fréchet differentiable, with

[2] Since $H_0^1(\Gamma_c)$ is continuously embedded into $H_{00}^{1/2}(\Gamma_c)$ it follows that N is differentiable also as an operator from $H_0^1(\Gamma_c)^d$ into V^*, and expressions of the two differentials coincide.

$$\tilde{J}'_{\boldsymbol{v}}(\boldsymbol{v},\boldsymbol{u})\,\boldsymbol{w} = \begin{cases} (\nabla\times\boldsymbol{v},\nabla\times\boldsymbol{w})_{L^2(\Omega)^d} & \text{in the case of } \tilde{J}_1(\boldsymbol{v},\boldsymbol{u}) \\ (\boldsymbol{v}-\boldsymbol{z}_d,\boldsymbol{w})_{L^2(\Omega)^d} & \text{in the case of } \tilde{J}_2(\boldsymbol{v},\boldsymbol{u}) \end{cases} \qquad \forall \boldsymbol{w}\in X$$

$$\tilde{J}'_{\boldsymbol{u}}(\boldsymbol{v},\boldsymbol{u})\,\boldsymbol{k} = \tau\,(\boldsymbol{u},\boldsymbol{k})_{H^1_0(\Gamma_c)^d} \qquad \forall \boldsymbol{k}\in H^1_0(\Gamma_c)^d$$

and $\tilde{J}'_\pi(\boldsymbol{v},\boldsymbol{u})=0$, $\tilde{J}'_{\boldsymbol{t}}(\boldsymbol{v},\boldsymbol{u})=\mathbf{0}$. Moreover,

$$\mathcal{G}'(\boldsymbol{v},\pi,\boldsymbol{t},\boldsymbol{u})(\boldsymbol{w},q,\boldsymbol{r},\boldsymbol{k}) = R^{-1}(\boldsymbol{w},q,\boldsymbol{r}) + N'(\boldsymbol{v},\pi,\boldsymbol{t},\boldsymbol{u})(\boldsymbol{w},q,\boldsymbol{r},\boldsymbol{k}).$$

(*ii*) From Theorem 10.3 we know that, for every $\boldsymbol{u}\in\mathcal{U}_{ad}$, the state equation (10.35) defines a unique *control-to-state map* $\mathcal{S}(\boldsymbol{u})$ which is Fréchet-differentiable.

(*iii*) For every $\bar{\boldsymbol{u}}\in\mathcal{U}_{ad}$, the partial derivative $\mathcal{G}'_{(\boldsymbol{v},q,\boldsymbol{s})}(\bar{\boldsymbol{v}},\bar{\pi},\bar{\boldsymbol{t}},\bar{\boldsymbol{u}})$, with $(\bar{\boldsymbol{v}},\bar{\pi},\bar{\mathbf{t}}) = (\mathbf{v}(\bar{\boldsymbol{u}}),\pi(\bar{\boldsymbol{u}}),\boldsymbol{t}(\bar{\boldsymbol{u}}))$ is a continuous isomorphism between V and V^*. Indeed, given $(\mathbf{F},h,\boldsymbol{g})\in V^*$, the equation

$$\mathcal{G}'_{(\boldsymbol{v},\pi,\boldsymbol{t})}(\bar{\boldsymbol{v}},\bar{\pi},\bar{\mathbf{t}},\bar{\mathbf{u}})(\boldsymbol{\psi},\eta,\mathbf{r}) = R^{-1}(\boldsymbol{\psi},\eta,\mathbf{r}) + N'_{(\boldsymbol{v},\pi,\boldsymbol{t})}(\bar{\boldsymbol{v}},\bar{\pi},\bar{\boldsymbol{t}},\bar{\boldsymbol{u}})(\boldsymbol{\psi},\eta,\boldsymbol{r}) = (\mathbf{F},h,\boldsymbol{g}), \tag{10.38}$$

in weak form reads:

$$\begin{cases} e(\boldsymbol{\psi},\boldsymbol{\varphi}) + c(\boldsymbol{\psi},\bar{\boldsymbol{v}},\boldsymbol{\varphi}) + c(\bar{\boldsymbol{v}},\boldsymbol{\psi},\boldsymbol{\varphi}) + b(\boldsymbol{\varphi},\eta) + \langle\boldsymbol{r},\boldsymbol{\varphi}\rangle_{\Gamma_1} = \langle\mathbf{F},\boldsymbol{\varphi}\rangle_{X^*,X} & \forall\boldsymbol{\varphi}\in X \\ b(\boldsymbol{\psi},\phi) = (h,\phi)_{L^2(\Omega)} & \forall\phi\in L^2(\Omega) \\ \langle\boldsymbol{\lambda},\boldsymbol{\psi}\rangle_{\Gamma_1} = \langle\boldsymbol{\lambda},\boldsymbol{g}\rangle_{\Gamma_1} & \forall\boldsymbol{\lambda}\in\mathcal{T}_1^*. \end{cases} \tag{10.39}$$

Since $\bar{\boldsymbol{u}}\in\mathcal{U}_{ad}$, the bilinear form

$$a(\boldsymbol{\psi},\boldsymbol{\varphi}) = e(\boldsymbol{\psi},\boldsymbol{\varphi}) + c(\boldsymbol{\psi},\bar{\boldsymbol{v}},\boldsymbol{\varphi}) + c(\bar{\boldsymbol{v}},\boldsymbol{\psi},\boldsymbol{\varphi})$$

is $X-$coercive, therefore problem (10.39) is well posed and $\mathcal{G}'_{(\boldsymbol{v},q,\boldsymbol{s})}(\bar{\boldsymbol{v}},\bar{\pi},\bar{\boldsymbol{t}},\bar{\boldsymbol{u}})$ is a continuous isomorphism between V and V^*.

Introducing the Lagrangian functional

$$\mathcal{L}((\boldsymbol{v},\pi,\boldsymbol{t}),\boldsymbol{u},(\boldsymbol{z},q,\boldsymbol{s})) = \tilde{J}(\boldsymbol{v},\boldsymbol{u}) - \langle\mathcal{G}(\boldsymbol{v},\pi,\boldsymbol{t},\boldsymbol{u}),(\boldsymbol{z},q,\boldsymbol{s})\rangle_{V^*,V},$$

where $(\boldsymbol{z},q,\boldsymbol{s})$ is a multiplier, the first order optimality conditions are obtained by differentiating $\mathcal{L}$. Precisely, the adjoint problem reads $(\hat{\mathcal{L}}'_{(\cdot)} = \mathcal{L}'_{(\cdot)}((\hat{\boldsymbol{v}},\hat{\pi},\hat{\boldsymbol{t}}),\hat{\boldsymbol{u}},(\hat{\boldsymbol{z}},\hat{q},\hat{\boldsymbol{s}})))$:

$$\hat{\mathcal{L}}'_{(\boldsymbol{v},\pi,\boldsymbol{t})}(\boldsymbol{\psi},\eta,\boldsymbol{r}) = \langle\tilde{J}'_{(\boldsymbol{v},\pi,\boldsymbol{t})}(\hat{\boldsymbol{v}},\hat{\boldsymbol{u}}),(\boldsymbol{\psi},\eta,\boldsymbol{r})\rangle_{V^*,V} - \langle\mathcal{G}'_{(\boldsymbol{v},\pi,\boldsymbol{t})}(\hat{\boldsymbol{v}},\hat{\pi},\hat{\boldsymbol{t}},\hat{\boldsymbol{u}})(\boldsymbol{\psi},\eta,\boldsymbol{r}),(\hat{\boldsymbol{z}},\hat{q},\hat{\boldsymbol{s}})\rangle_{V^*,V} = 0$$

or, equivalently, upon introducing the adjoint operator $\mathcal{G}'_{(\boldsymbol{v},\pi,\boldsymbol{t})}(\hat{\boldsymbol{v}},\hat{\pi},\hat{\boldsymbol{t}},\hat{\boldsymbol{u}})^*$,

$$\left\langle\tilde{J}'_{(\boldsymbol{v},\pi,\boldsymbol{t})}(\hat{\boldsymbol{v}},\hat{\boldsymbol{u}}) - \mathcal{G}'_{(\boldsymbol{v},\pi,\boldsymbol{t})}(\hat{\boldsymbol{v}},\hat{\pi},\hat{\boldsymbol{t}},\hat{\boldsymbol{u}})^*(\hat{\boldsymbol{z}},\hat{q},\hat{\boldsymbol{s}}),(\boldsymbol{\psi},\eta,\boldsymbol{r})\right\rangle_{V^*,V} = 0 \tag{10.40}$$

for every $(\boldsymbol{\psi},\eta,\boldsymbol{r})\in V$. Since R is self-adjoint, we have (see the proof of Lemma 10.2)

$$\begin{aligned} &\left\langle\mathcal{G}'_{(\boldsymbol{v},\pi,\boldsymbol{t})}(\hat{\boldsymbol{v}},\hat{\pi},\hat{\boldsymbol{t}},\hat{\boldsymbol{u}})^*(\hat{\boldsymbol{z}},\hat{q},\hat{\boldsymbol{s}}),(\boldsymbol{\psi},\eta,\boldsymbol{r})\right\rangle_{V^*,V} \\ &= \left\langle R^{-1}(\hat{\boldsymbol{z}},\hat{q},\hat{\boldsymbol{s}}) + N'_{(\boldsymbol{v},\pi,\boldsymbol{t})}(\hat{\boldsymbol{v}},\hat{\pi},\hat{\boldsymbol{t}},\hat{\boldsymbol{u}})^*(\hat{\boldsymbol{z}},\hat{q},\hat{\boldsymbol{s}}),(\boldsymbol{\psi},\eta,\boldsymbol{r})\right\rangle_{V^*,V} \\ &= \left\langle R^{-1}(\hat{\boldsymbol{z}},\hat{q},\hat{\boldsymbol{s}}),(\boldsymbol{\psi},\eta,\boldsymbol{r})\right\rangle_{V^*,V} + c(\hat{\boldsymbol{v}},\boldsymbol{\psi},\hat{\boldsymbol{z}}) + c(\boldsymbol{\psi},\hat{\boldsymbol{v}},\hat{\boldsymbol{z}}) + (\eta,0)_{L^2(\Omega)} + \langle\boldsymbol{r},\mathbf{0}\rangle_{\Gamma_1} = 0 \end{aligned}$$

and therefore (10.40) corresponds to the following system:

$$\begin{cases} e(\boldsymbol{\psi},\hat{\boldsymbol{z}}) + c(\hat{\boldsymbol{v}},\boldsymbol{\psi},\hat{\boldsymbol{z}}) + c(\boldsymbol{\psi},\hat{\boldsymbol{v}},\hat{\boldsymbol{z}}) + b(\boldsymbol{\psi},\hat{q}) + \langle\hat{\boldsymbol{s}},\boldsymbol{\psi}\rangle_{\Gamma_1} = \tilde{J}'_{\boldsymbol{v}}(\boldsymbol{v},\boldsymbol{u})\,\boldsymbol{\psi} \\ b(\hat{\boldsymbol{z}},\eta) = 0 \\ \langle\boldsymbol{r},\hat{\boldsymbol{z}}\rangle_{\Gamma_1} = 0 \end{cases} \tag{10.41}$$

for every $(\boldsymbol{\psi},\eta,\boldsymbol{r})\in V$, which is (10.29). The variational inequality reads, recalling (10.37),

$$\begin{aligned} \hat{\mathcal{L}}'_{\boldsymbol{u}}(\boldsymbol{k}-\hat{\boldsymbol{u}}) &= \langle\tilde{J}'_{\boldsymbol{u}}(\hat{\boldsymbol{v}},\hat{\boldsymbol{u}}),\boldsymbol{k}-\hat{\boldsymbol{u}}\rangle_{\mathcal{U}^*,\mathcal{U}} + \left\langle N'_{\boldsymbol{u}}(\hat{\boldsymbol{v}},\hat{\pi},\hat{\boldsymbol{t}},\hat{\boldsymbol{u}})^*(\hat{\boldsymbol{z}},\hat{q},\hat{\boldsymbol{s}}),\boldsymbol{k}-\hat{\boldsymbol{u}}\right\rangle_{V^*,V} \\ &= (\tau\hat{\boldsymbol{u}},\boldsymbol{k}-\hat{\boldsymbol{u}})_{H^1_0(\Gamma_c)^2} + \langle\hat{\boldsymbol{s}},\boldsymbol{k}-\hat{\boldsymbol{u}}\rangle_{\Gamma_c} \geq 0 \qquad \forall\boldsymbol{k}\in\mathcal{U}_{ad} \end{aligned}$$

which is (10.30). □

As in Remark 10.3, since $\hat{\boldsymbol{u}} \in \mathcal{U}_{ad}$, the following bilinear form (the adjoint of the bilinear form (10.23))

$$a^*(\boldsymbol{z}, \boldsymbol{\psi}) = e(\boldsymbol{\psi}, \boldsymbol{z}) + c(\boldsymbol{\psi}, \hat{\boldsymbol{v}}, \boldsymbol{z}) + c(\hat{\boldsymbol{v}}, \boldsymbol{\psi}, \boldsymbol{z})$$

is X-coercive and the linear problem (10.29) is well posed. To write the adjoint system in strong form, note that

$$c(\boldsymbol{\psi}, \hat{\boldsymbol{v}}, \hat{\boldsymbol{z}}) = \int_\Omega (\boldsymbol{\psi} \cdot \nabla)\hat{\boldsymbol{v}} \cdot \hat{\boldsymbol{z}}\, d\mathbf{x} = \sum_{i,j=1}^{d} \int_\Omega \psi_i \frac{\partial \hat{v}_j}{\partial x_i} \hat{z}_j\, d\mathbf{x} = \int_\Omega (\nabla \hat{\boldsymbol{v}})^\top \hat{\boldsymbol{z}} \cdot \boldsymbol{\psi}\, d\mathbf{x} \tag{10.42}$$

and

$$\begin{aligned} c(\hat{\boldsymbol{v}}, \boldsymbol{\psi}, \hat{\boldsymbol{z}}) &= \int_\Omega (\hat{\boldsymbol{v}} \cdot \nabla)\boldsymbol{\psi} \cdot \hat{\boldsymbol{z}}\, d\mathbf{x} \\ &= \int_\Gamma (\hat{\boldsymbol{v}} \cdot \mathbf{n})(\hat{\boldsymbol{z}} \cdot \boldsymbol{\psi})\, d\sigma - \int_\Omega (\hat{\boldsymbol{v}} \cdot \nabla)\hat{\boldsymbol{z}} \cdot \boldsymbol{\psi}\, d\mathbf{x} - \int_\Omega \mathrm{div}\hat{\boldsymbol{v}}\, (\hat{\boldsymbol{z}} \cdot \boldsymbol{\psi})\, d\mathbf{x} \\ &= \int_{\Gamma_{\text{out}}} (\hat{\boldsymbol{v}} \cdot \mathbf{n})(\hat{\boldsymbol{z}} \cdot \boldsymbol{\psi})\, d\sigma - c(\hat{\boldsymbol{v}}, \hat{\boldsymbol{z}}, \boldsymbol{\psi}) \end{aligned} \tag{10.43}$$

since div $\hat{\boldsymbol{v}} = 0$, $\hat{\boldsymbol{z}} = \mathbf{0}$ on $\Gamma \setminus \Gamma_{\text{out}}$.

Moreover, in the case of the vorticity functional $\tilde{J}_1$,

$$\int_\Omega (\nabla \times \hat{\boldsymbol{v}}) \cdot (\nabla \times \boldsymbol{\psi})\, d\mathbf{x} = \int_\Omega (\nabla \times \nabla \times \hat{\boldsymbol{v}}) \cdot \boldsymbol{\psi}\, d\mathbf{x} + \int_\Gamma (\nabla\hat{\boldsymbol{v}} - (\nabla\hat{\boldsymbol{v}})^\perp)\mathbf{n} \cdot \boldsymbol{\psi}\, d\sigma. \tag{10.44}$$

Therefore, the adjoint system in strong form reads, setting the rotation tensor $\hat{\mathbf{W}} = \nabla\hat{\boldsymbol{v}} - (\nabla\hat{\boldsymbol{v}})^\perp$,

$$\begin{cases} -\nu \Delta \hat{\boldsymbol{z}} - (\hat{\boldsymbol{v}} \cdot \nabla)\hat{\boldsymbol{z}} + (\nabla\hat{\boldsymbol{v}})^\top \hat{\boldsymbol{z}} + \nabla \hat{q} = \nabla \times \nabla \times \hat{\boldsymbol{v}} & \text{in } \Omega \\ \mathrm{div}\hat{\boldsymbol{z}} = 0 & \text{in } \Omega \\ \hat{\boldsymbol{z}} = \mathbf{0} & \text{on } \Gamma_0 \cup \Gamma_1 \\ \hat{q}\mathbf{n} - \nu \dfrac{\partial \hat{\boldsymbol{z}}}{\partial n} - (\hat{\boldsymbol{v}} \cdot \mathbf{n})\, \hat{\boldsymbol{z}} + \hat{\mathbf{W}}\mathbf{n} = \mathbf{0} & \text{on } \Gamma_{out}. \end{cases} \tag{10.45}$$

We can also deduce that, regarding the multiplier,

$$\hat{\boldsymbol{s}} = \hat{q}\mathbf{n} - \nu \frac{\partial \hat{\boldsymbol{z}}}{\partial n} + \hat{\mathbf{W}}\mathbf{n} \quad \text{on } \Gamma_1. \tag{10.46}$$

In the case of the cost functional $\tilde{J}_2$, the strong form of the adjoint problem reads:

$$\begin{cases} -\nu \Delta \hat{\boldsymbol{z}} - (\hat{\boldsymbol{v}} \cdot \nabla)\hat{\mathbf{z}} + (\nabla\hat{\boldsymbol{v}})^\top \hat{\boldsymbol{z}} + \nabla \hat{q} = \hat{\boldsymbol{v}} - \mathbf{z}_d & \text{in } \Omega \\ \mathrm{div}\hat{\boldsymbol{z}} = 0 & \text{in } \Omega \\ \hat{\boldsymbol{z}} = \mathbf{0} & \text{on } \Gamma_0 \cup \Gamma_1 \\ \hat{q}\mathbf{n} - \nu \dfrac{\partial \hat{\boldsymbol{z}}}{\partial n} - (\hat{\mathbf{v}} \cdot \mathbf{n})\, \hat{\boldsymbol{z}} = \mathbf{0} & \text{on } \Gamma_{out}. \end{cases} \tag{10.47}$$

In this case, we find

$$\hat{\boldsymbol{s}} = \hat{q}\mathbf{n} - \nu \frac{\partial \hat{\boldsymbol{z}}}{\partial n} \qquad \text{on } \Gamma_1. \tag{10.48}$$

Remark 10.5. From (10.30) we deduce that the derivative of the reduced functional $J(\boldsymbol{u}) = \tilde{J}(\boldsymbol{v}(\boldsymbol{u}), \boldsymbol{u})$ at $\boldsymbol{u} = \hat{\boldsymbol{u}} \in \mathcal{U}_{ad}$ is given by

$$J'(\hat{\boldsymbol{u}})\boldsymbol{k} = (\tau\hat{\boldsymbol{u}}, \boldsymbol{k})_{H_0^1(\Gamma_c)^d} + \langle \hat{\boldsymbol{s}}, \boldsymbol{k} \rangle_{\Gamma_c} \qquad \forall \boldsymbol{k} \in \mathcal{U}_{ad}.$$

Introducing the embedding operator $B : \mathcal{T}_c^* \to H^{-1}(\Gamma_c)^d$ defined by

$$\langle B\boldsymbol{s}, \boldsymbol{k}\rangle_{H^{-1}(\Gamma_c)^d, H_0^1(\Gamma_c)^d} = \langle \boldsymbol{s}, \boldsymbol{k}\rangle_{\Gamma_c} \quad \forall \boldsymbol{k} \in H_0^1(\Gamma_c)^d,$$

we find that

$$\nabla J(\hat{\boldsymbol{u}}) = \tau\hat{\boldsymbol{u}} + \hat{\boldsymbol{w}}$$

where $\hat{\boldsymbol{w}} \in H_0^1(\Gamma_c)^d$ is the unique weak solution to

$$-\Delta\hat{\boldsymbol{w}} = B\hat{\boldsymbol{s}} \quad \text{in } \Gamma_c. \tag{10.49}$$

Thus, the variational inequality (10.30) can be written in the form

$$(\tau\hat{\boldsymbol{u}} + \hat{\boldsymbol{w}}, \boldsymbol{k} - \hat{\boldsymbol{u}})_{H_0^1(\Gamma_c)^d} \geq 0 \quad \forall \boldsymbol{k} \in \mathcal{U}_{ad}. \tag{10.50}$$

If $B\hat{\boldsymbol{s}} \neq \boldsymbol{0}$ and $\tau = 0$, (10.50) gives

$$\hat{\boldsymbol{u}} = -\frac{\eta}{2c_0}\frac{\hat{\boldsymbol{w}}}{\|\hat{\boldsymbol{w}}\|_{H_0^1(\Gamma_c)^d}}. \tag{10.51}$$

If $B\hat{\boldsymbol{s}} \neq \boldsymbol{0}$ and $\tau > 0$, we have $\hat{\boldsymbol{u}} = -\hat{\boldsymbol{w}}/\tau$ if $\hat{\boldsymbol{w}}/\tau \in \mathcal{U}_{ad}$, otherwise $\hat{\boldsymbol{u}}$ is given by (10.51). •

Lemma 10.2. *The following expressions hold:*
(a) $N'_{(\boldsymbol{v},\pi,\boldsymbol{t})}(\boldsymbol{v}, \pi, \boldsymbol{t}, \boldsymbol{u})^* : V \to V^*$ *is given by*

$$N'_{(\boldsymbol{v},\pi,\boldsymbol{t})}(\boldsymbol{v}, \pi, \boldsymbol{t}, \boldsymbol{u})^*(\boldsymbol{z}, q, \boldsymbol{s}) = (-(\boldsymbol{v}\cdot\nabla)\mathbf{z} + (\nabla\boldsymbol{v})^\top\boldsymbol{z} + [((\boldsymbol{v}\cdot\mathbf{n})]_{|\Gamma_1}, 0, \boldsymbol{0}). \tag{10.52}$$

where $\mathbf{n}$ *is the exterior unit normal to* Γ_c*;*
(b) $N'_{\boldsymbol{u}}(\boldsymbol{v}, \pi, \boldsymbol{t}, \boldsymbol{u})^* : V \to \mathcal{T}_c^*$ *is given by*

$$N'_{\boldsymbol{u}}(\boldsymbol{v}, \pi, \boldsymbol{t}, \boldsymbol{u})^*(\boldsymbol{z}, q, \boldsymbol{s}) = -\boldsymbol{s}. \tag{10.53}$$

Proof. (*a*) We check (10.52). Using (10.42), (10.43) and recalling (10.33), we can write

$$\begin{aligned}
&\left\langle N'_{(\boldsymbol{v},\pi,\boldsymbol{t})}(\boldsymbol{v}, \pi, \boldsymbol{t}, \boldsymbol{u})(\boldsymbol{\psi}, \eta, \mathbf{r}), (\boldsymbol{z}, q, \boldsymbol{s})\right\rangle_{V^*,V} \\
&= c(\boldsymbol{v}, \boldsymbol{\psi}, \boldsymbol{z}) + c(\boldsymbol{\psi}, \boldsymbol{v}, \boldsymbol{z}) = \int_\Omega (\boldsymbol{v}\cdot\nabla)\boldsymbol{\psi}\cdot\boldsymbol{z}\, d\mathbf{x} + \int_\Omega (\boldsymbol{\psi}\cdot\nabla)\boldsymbol{v}\cdot\boldsymbol{z}\, d\mathbf{x} \\
&= \int_\Omega \{-(\boldsymbol{v}\cdot\nabla)\boldsymbol{z}\cdot\boldsymbol{\psi} + (\nabla\boldsymbol{v})^\top\boldsymbol{z}\cdot\boldsymbol{\psi}\}\, d\mathbf{x} + \int_{\Gamma_1} (\boldsymbol{v}\cdot\mathbf{n})(\boldsymbol{z}\cdot\boldsymbol{\psi})\, d\sigma + (\eta, 0)_{L^2(\Omega)} + \langle \boldsymbol{r}, \boldsymbol{0}\rangle_{\Gamma_1} \\
&= \left\langle N'_{(\boldsymbol{v},\pi,\boldsymbol{t})}(\boldsymbol{v}, \pi, \boldsymbol{t}, \boldsymbol{u})^*(\boldsymbol{z}, q, \boldsymbol{s}), (\boldsymbol{\psi}, \eta, \mathbf{r})\right\rangle_{V^*,V}
\end{aligned} \tag{10.54}$$

for every $(\boldsymbol{z}, q, \boldsymbol{s}) \in V$ and every $(\boldsymbol{\psi}, \eta, \mathbf{r}) \in V^*$.

(*b*) We check (10.53). We have, using (10.34),

$$\langle N'_{\boldsymbol{u}}(\boldsymbol{v}, \pi, \boldsymbol{t}, \boldsymbol{u})\boldsymbol{k}, (\boldsymbol{z}, q, \boldsymbol{s})\rangle_{V^*,V} = \langle \boldsymbol{s}, -\boldsymbol{k}\rangle_{\Gamma_c} = \langle -\boldsymbol{s}, \boldsymbol{k}\rangle_{\Gamma_c} = \langle N'_{\boldsymbol{u}}(\boldsymbol{v}, \pi, \boldsymbol{t}, \boldsymbol{u})^*(\boldsymbol{z}, q, \boldsymbol{s}), \boldsymbol{k}\rangle_{\Gamma_c}$$

for every $(\boldsymbol{z}, q, \boldsymbol{s}) \in V$ and $\boldsymbol{k} \in \mathcal{T}_c$. □

10.1.6 Numerical Approximation

For the numerical solution of the OCP (10.19) we use an iterative method for the minimization of the cost functional, and we exploit the Galerkin finite element method for the numerical discretization of the state equation, the adjoint equation and the optimality equation. In this way we can take advantage of the system of first-order (necessary) optimality conditions we have derived in the previous sections. Several alternatives have been proposed in literature; see, e.g., [103, 137, 149, 217] for the case of sequential quadratic programming methods applied to boundary control of steady Navier-Stokes flows, [81] for an augmented Lagrangian method, and [38] for inexact Newton methods.

Numerical discretization of state and adjoint equations

The state system consists of the steady Navier-Stokes equations, while the adjoint system is a steady Oseen problem [90]. Following the saddle-point approach described so far, for both these problems Dirichlet boundary conditions are enforced in weak form, by means of the Lagrange multipliers $\mathbf{t}$ and $\mathbf{s}$, respectively. These two latter variables are indeed related with the boundary stress over the Dirichlet boundary, see, e.g., equations (10.9) and (10.46)–(10.48). For the numerical approximation, however, a direct imposition of the (essential) Dirichlet conditions is more convenient, by introducing appropriate functional spaces and suitable lifting terms. This option allows us to avoid the explicit computation of the Lagrange multipliers $\mathbf{t}$ and $\mathbf{s}$.

We therefore introduce two subspaces $X_h \subset X$ and $Q_h \subset L^2(\Omega)$, of dimension $N_V, N_Q < +\infty$, respectively, and let us denote by $\boldsymbol{v}_h \in X_h$ and $\pi_h \in Q_h$ the FE approximations for the velocity and the pressure fields [224]. Similarly, $\boldsymbol{z}_h \in X_h$ and $q_h \in Q_h$ will denote the FE approximations for the adjoint velocity and pressure fields, respectively. We also denote by $\boldsymbol{u}_h \in \mathcal{U}_h \cap \mathcal{U}_{ad}$ the FE approximation of the control variable, where $\mathcal{U}_h \subset \mathcal{U}$ is a subspace of the control space $\mathcal{U} = H_0^1(\Gamma_c)^d$ of dimension $N_U < +\infty$.

The Galerkin-FE approximation of the state system (10.8) thus reads as follows: we seek $(\boldsymbol{v}_h, \pi_h) \in X_h \times Q_h$ such that

$$\begin{cases} \tilde{e}(\boldsymbol{v}_h, \boldsymbol{\varphi}_h) + c(\boldsymbol{v}_h, \boldsymbol{v}_h, \boldsymbol{\varphi}_h) + b(\boldsymbol{\varphi}_h, \pi_h) = F_1 \boldsymbol{\varphi}_h & \forall \boldsymbol{\varphi}_h \in X_h \\ b(\boldsymbol{v}_h, \phi_h) = F_2 \phi_h & \forall \phi_h \in Q_h, \end{cases} \tag{10.55}$$

where $\boldsymbol{v}_h^D = \boldsymbol{v}_h^D(\boldsymbol{u}_h) \in X_h$ is a discrete (lifting) function interpolating the Dirichlet data on the boundary,

$$\tilde{e}(\boldsymbol{v}_h, \boldsymbol{\varphi}_h) = e(\boldsymbol{v}_h, \boldsymbol{\varphi}_h) + c(\boldsymbol{v}_h^D, \boldsymbol{v}_h, \boldsymbol{\varphi}_h) + c(\boldsymbol{v}_h, \boldsymbol{v}_h^D, \boldsymbol{\varphi}_h),$$
$$F_1 \boldsymbol{\varphi}_h = -e(\boldsymbol{v}_h^D, \boldsymbol{\varphi}_h) - c(\boldsymbol{v}_h^D, \boldsymbol{v}_h^D, \boldsymbol{\varphi}_h), \qquad F_2 \phi_h = -b(\boldsymbol{v}_h^D, \phi_h).$$

All the previous forms are continuous over the discrete spaces X_h and Q_h. The approximation stability is ensured by imposing that the coercivity and inf-sup conditions are still valid at the discrete level. In particular, the coercivity of e over X_h is inherited from that over X; on the other hand, we require that b is inf-sup stable over $X_h \times Q_h$, so that the following discrete inf-sup condition holds [53]:

$$\exists\, \tilde{\beta}_h > 0 : \inf_{q \in Q_h} \sup_{\boldsymbol{w} \in X_h} \frac{b(q, \boldsymbol{w})}{\|\boldsymbol{w}\|_V \|q\|_Q} \geq \tilde{\beta}_h. \tag{10.56}$$

This last property is ensured e.g., by choosing $X_h \times Q_h$ as the space of Taylor-Hood $\mathbb{P}_2 - \mathbb{P}_1$ finite elements for the velocity and the pressure, respectively (in both the state and the adjoint problem); however, this choice is not restrictive — the whole construction keeps holding for other choices of the discrete spaces, too [113].

We now derive the matrix formulation corresponding to the Galerkin-FE approximation (10.55). Let us denote by $\{\boldsymbol{\phi}_i^{\mathbf{v}}\}_{i=1}^{N_v}$ and $\{\phi_i^p\}_{i=1}^{N_p}$ the Lagrangian basis of the FE spaces X_h and Q_h, respectively, so that we can express the FE velocity and pressure as

$$\boldsymbol{v}_h = \sum_{i=1}^{N_v} v_{h,i}\boldsymbol{\phi}_i^v, \qquad \pi_h = \sum_{i=1}^{N_p} \pi_{h,i}\phi_i^p. \tag{10.57}$$

We remark that the solution to (10.55) is vanishing on the whole Dirichlet boundary, so that the corresponding velocity approximation fulfilling the boundary conditions is given by $\boldsymbol{v}_h + \boldsymbol{v}_h^D$. Thus, by denoting with $\mathbf{v} \in \mathbb{R}^{N_v}$ and $\boldsymbol{\pi} \in \mathbb{R}^{N_p}$ the vectors of the degrees of freedom appearing in (10.57), problem (10.55) can be rewritten as:

$$\begin{bmatrix} \mathbb{E} + \mathbb{C}(\mathbf{v}) & \mathbb{B}^\top \\ \mathbb{B} & 0 \end{bmatrix} \begin{bmatrix} \mathbf{v} \\ \boldsymbol{\pi} \end{bmatrix} = \begin{bmatrix} \mathbf{f}_1 \\ \mathbf{f}_2 \end{bmatrix} \tag{10.58}$$

where, for $1 \le i, j \le N_v$ and $1 \le k \le N_p$,

$$\begin{gathered}(\mathbb{E})_{ij} = \tilde{e}(\boldsymbol{\phi}_j^v, \boldsymbol{\phi}_i^v), \qquad (\mathbb{B})_{ki} = b(\phi_k^p, \boldsymbol{\phi}_i^v), \qquad (\mathbb{C}(\mathbf{w}))_{ij} = \sum_{m=1}^{N_v} w_{h,m} c\left(\boldsymbol{\phi}_m^v, \boldsymbol{\phi}_j^v, \boldsymbol{\phi}_i^v\right), \\ (\mathbf{f}_1)_i = F_1 \boldsymbol{\phi}_i^v, \qquad (\mathbf{f}_2)_k = F_2 \phi_k^p. \end{gathered} \tag{10.59}$$

We solve the nonlinear saddle-point problem (10.58) by the *Picard iteration* method, see, e.g., [90, Sect. 8.2] as its ball of convergence is larger than Newton's method (see e.g., [90], Chapter 7.2 and references therein). Moreover, if a *small data* condition (like, e.g., (5.5) for the case at hand) is satisfied, this fixed-point method is *globally* convergent (that is, for every initial guess). Starting from an initial guess $(\mathbf{v}^{(0)}, \mathbf{p}^{(0)})$, for $n \ge 1$ we solve

$$\begin{bmatrix} \mathbb{E} + \mathbb{C}(\mathbf{v}^{(n-1)}) & \mathbb{B}^\top \\ \mathbb{B} & 0 \end{bmatrix} \begin{bmatrix} \mathbf{v}^{(n)} \\ \boldsymbol{\pi}^{(n)} \end{bmatrix} = \begin{bmatrix} \mathbf{f}_1 \\ \mathbf{f}_2 \end{bmatrix}, \tag{10.60}$$

to obtain $(\mathbf{v}^{(n)}, \boldsymbol{\pi}^{(n)})$, until $\|\mathbf{v}^{(n)} - \mathbf{v}^{(n-1)}\|_V \le \varepsilon_{\text{tol}}^{NS}$, given a small tolerance $\varepsilon_{\text{tol}}^{NS} > 0$. As initial guess, we take the Stokes solution of (10.58). Each Oseen system (10.60) is solved by means of a *sparse* LU factorization.

Remark 10.6. Using instead Newton's method, a different *linearized* problem has to be solved at each iteration. Starting from an initial guess $(\mathbf{v}^{(0)}, \boldsymbol{\pi}^{(0)})$, for $n \ge 1$ we solve

$$\begin{bmatrix} \mathbb{E} + \mathbb{C}(\mathbf{v}^{(n-1)}) + \mathbb{N}(\mathbf{v}^{(n-1)}) & \mathbb{B}^\top \\ \mathbb{B} & 0 \end{bmatrix} \begin{bmatrix} \mathbf{v}^{(n)} \\ \boldsymbol{\pi}^{(n)} \end{bmatrix} = \begin{bmatrix} \mathbf{f}_1 + \mathbf{f}_N \\ \mathbf{f}_2 \end{bmatrix}, \tag{10.61}$$

where

$$(\mathbb{N}(\mathbf{w}))_{ij} = \sum_{m=1}^{N_v} w_{h,m} c\left(\boldsymbol{\phi}_j^v, \boldsymbol{\phi}_m^v, \boldsymbol{\phi}_i^v\right), \qquad (\mathbf{f}_N)_i = \sum_{l,m=1}^{N_v} w_{h,l} w_{h,m} c\left(\boldsymbol{\phi}_l^v, \boldsymbol{\phi}_m^v, \boldsymbol{\phi}_i^v\right).$$

The additional arrays appearing in the linear system above arise from the residual vector and the Jacobian matrix. See, for comparison, problem (9.70) that was derived from a generic nonlinear system. •

Regarding instead the numerical approximation of the adjoint variables, the adjoint system (10.41) is linear with respect to both the adjoint velocity and pressure. Moreover, it depends on the state variables only through the term $c(\hat{\boldsymbol{v}}, \boldsymbol{\psi}, \hat{\boldsymbol{z}}) + c(\boldsymbol{\psi}, \hat{\boldsymbol{v}}, \hat{\boldsymbol{z}})$ and the right-hand side of (10.41). Hence, the FE arrays corresponding to these two terms must be reassembled at each step during the optimization procedure. Given the approximation $\boldsymbol{v}_h = \boldsymbol{v}_h(\boldsymbol{u}_h)$ of the (state) velocity and $\boldsymbol{u}_h$ of the control function, the Galerkin-FE approximation of the adjoint system (10.41) thus reads as follows: we seek $(\boldsymbol{z}_h, q_h) \in X_h \times Q_h$ such that

$$\begin{cases} e(\boldsymbol{\varphi}_h, \boldsymbol{z}_h) + c(\boldsymbol{v}_h, \boldsymbol{z}_h, \boldsymbol{\varphi}_h) + c(\boldsymbol{z}_h, \boldsymbol{v}_h, \boldsymbol{\varphi}_h) + b(\boldsymbol{\varphi}_h, q_h) = G_1(\boldsymbol{v}_h, \boldsymbol{\varphi}_h) & \forall \boldsymbol{\varphi}_h \in X_h \\ b(\boldsymbol{z}_h, \phi_h) = 0 & \forall \phi_h \in Q_h, \end{cases} \tag{10.62}$$

where

$$G_1(\boldsymbol{v}_h, \boldsymbol{\varphi}_h) = \tilde{J}'_{\boldsymbol{v}}(\boldsymbol{v}_h, \boldsymbol{u}_h)\,\boldsymbol{\varphi}_h.$$

Problem (10.62) can be rewritten algebraically as:

$$\begin{bmatrix} \mathbb{E}^\top + \mathbb{D}(\mathbf{v}) & \mathbb{B}^\top \\ \mathbb{B} & 0 \end{bmatrix} \begin{bmatrix} \mathbf{z} \\ \mathbf{q} \end{bmatrix} = \begin{bmatrix} \mathbf{g}_1(\mathbf{v}) \\ \mathbf{0} \end{bmatrix} \tag{10.63}$$

where, for $1 \le i, j \le N_v$ and $1 \le k \le N_p$,

$$(\mathbb{D}(\mathbf{w}))_{ij} = \textstyle\sum_{m=1}^{N_v} w_{h,m} \left(c(\boldsymbol{\phi}^v_m, \boldsymbol{\phi}^v_j, \boldsymbol{\phi}^v_i) + c(\boldsymbol{\phi}^v_j, \boldsymbol{\phi}^v_m, \boldsymbol{\phi}^v_i) \right), \qquad (\mathbf{g}_1(\mathbf{v}))_i = G_1(\mathbf{v}, \boldsymbol{\phi}^v_i). \tag{10.64}$$

Numerical optimization

We now describe the numerical optimization method exploited for the solution of our OCP. We use an iterative method exploiting the gradient $\nabla J(\boldsymbol{u}_h)$ to iteratively update the control until a suitable convergence criterion is fulfilled; a similar approach can be found, e.g., in [77]. In the simplest case of the steepest descent method, starting from an initial guess $\boldsymbol{u}_h^{(0)}$, we iteratively generate a sequence

$$\boldsymbol{u}_h^{(k+1)} = P_{\mathcal{U}_{ad}} \left(\boldsymbol{u}_h^{(k)} - \lambda_k \nabla J(\boldsymbol{u}_h^{(k)}) \right), \qquad k \ge 0,$$

where $\lambda_k > 0$ is a step size, until, e.g., a suitable stopping criterion is fulfilled. As usual, $P_{\mathcal{U}_{ad}}$ denotes the projection onto $\mathcal{U}_h \cap \mathcal{U}_{ad}$, to be evaluated similarly to what we have done in (10.51). As stopping criterion, we can require either

$$\|\nabla J(\boldsymbol{u}_h^{(k)})\|_{L^2(\Omega)^d} < \varepsilon$$

or

$$|J(\boldsymbol{u}_h^{(k)}) - J(\boldsymbol{u}_h^{(k-1)})| < \varepsilon \tag{10.65}$$

for a given tolerance $\varepsilon > 0$ (not necessarily the same).

At each step of the steepest descent method, we solve the state Navier-Stokes system (10.55), from which we obtain $\boldsymbol{v}^{(k)}$, and the value of the cost functional $\tilde{J}(\boldsymbol{v}^{(k)}, \boldsymbol{u}^{(k)})$. Then, we solve the adjoint system (10.62) ($\boldsymbol{v}^{(k)}$ and $\boldsymbol{u}^{(k)}$ being given) and obtain $(\boldsymbol{z}^{(k)}, q^{(k)})$. From these variables, we determine the boundary stress in weak from,

$$\boldsymbol{s}_h^{(k)} = q_h^{(k)}\mathbf{n} - \nu \frac{\partial \boldsymbol{z}_h^{(k)}}{\partial n} \qquad \text{on } \Gamma_c,$$

and approximate (by the Galerkin finite element method) the solution of the following problem,

$$\int_{\Gamma_c} \nabla \boldsymbol{w}_h^{(k)} : \nabla \boldsymbol{\varphi}_h d\sigma = \int_{\Gamma_c} \boldsymbol{s}_h^{(k)} \cdot \boldsymbol{\varphi}_h d\sigma \qquad \forall \boldsymbol{\varphi}_h \in H_0^1(\Gamma_c)^d$$

in order to evaluate the gradient

$$\nabla J(\boldsymbol{u}_h^{(k)}) = \tau \boldsymbol{u}_h^{(k)} + \boldsymbol{w}_h^{(k)}.$$

At each step of the steepest descent method, the initial guess of the Newton method is given by the state velocity computed at the previous step of the optimization procedure, for the sake of computational efficiency.

Numerical results

In this section we present the numerical results related with an OCP for the optimal control of blood flows through a bypass graft. This latter usually provides blood flow through an alternative bridging path in order to overcome critically occluded arteries; one of the most dangerous cases is related to coronary arteries, which supply the oxygen-rich blood perfusion to the heart muscle. The design of the connection between the graft and the host arterial vessels is a critical factor in avoiding post-operative recurrence of the occlusion, since fluid dynamic phenomena such as recirculation, oscillating or untypical high/low shear rates, and stagnation areas, can cause the growth of another stenosis downstream the arterial-graft connection. Hence, a typical design of a bypass graft aims at minimizing some cost functionals related to haemodynamic quantities [83, 199]. However, a rigorous model for blood circulation should take into account *(i)* the flow unsteadiness, and *(ii)* the arterial wall deformability. Optimal control (and optimal design) problem require the repeated simulation of these flow equations (and the evaluation of the cost functional to be minimized); to reduce the computational burden, we adopt steady incompressible Navier-Stokes equations for laminar Newtonian flows.

We focus on the minimization of the tracking-type functional (10.2) in order to drive the blood velocity towards a specified velocity target state $\mathbf{z}_d$, featuring a regular pattern; this latter is given by the solution of the Stokes system in the same domain. Other possible cost functionals for the problem at hand have been employed, e.g., in [164, 169]; the effect of uncertainty (e.g., affecting either the residual flow across Γ_{in}, or the geometrical configuration of the bypass) has been explored in [169, 170, 238].

Here a two-dimensional geometrical configuration is considered, although the whole theoretical analysis and the computational pipeline are perfectly valid for three-dimensional problems as well. For the case at hand, we consider a fluid flow in a tract of blood vessel where a bypass is inserted. As a matter of fact, the whole bypass graft is not simulated; its action is represented via a velocity control $\mathbf{u}$ acting on the boundary $\Gamma_c \subset \Gamma$, the interface where the final portion of the graft and the host vessel meet (see Figure 10.1). We consider the case in which a residual blood flow is present in the host artery, hence $\boldsymbol{v}_{in} \neq \mathbf{0}$ on Γ_{in}. In particular, we set

$$\boldsymbol{v}_{in} = \left(a \exp\left(-\frac{(x_2 - \bar{x}_2)^2}{2b^2} \right), 0 \right)$$

with $a = 50$ and $b = 10^{-2}$, to model a (limited) blood flow across the occluded artery vessel; here $\bar{x}_2$ denotes the x_2-coordinate of the midpoint of the (vertical) boundary segment Γ_{in} – for this reason, the residual flow is assumed to have only a horizontal component.

We choose $\rho = 1$, $\bar{v} = 25 cm\, s^{-1}$, $\bar{d} = |\Gamma_c| \approx 0.5cm$ from which, defining the Reynolds number $Re = \rho \, \bar{v} \, \bar{d}/\mu$, which is set equal to $Re = 200$, the dynamic viscosity coefficient μ can be deduced. We first compute the solution to the Stokes system, to determine the target velocity profile $\boldsymbol{v}_d$; we set the relaxation parameter $\lambda_k = \lambda = 0.1$ and the regularization parameter $\tau = 0.05$. We consider the stopping criterion based on the difference of two successive iterates (10.65), for which we select $\varepsilon = 0.01$; the tolerance for the stopping criterion of the Newton method is instead given by $\varepsilon_{\text{tol}}^{NS} = 10^{-5}$. For our computations we use a mesh with 27233 triangular elements and 14048 vertices.

In Figure 10.2 we report the velocity (both its magnitude and the streamlines) and pressure fields corresponding to the initial control function $\boldsymbol{u}_h^{(0)}$ and the optimal control $\hat{\boldsymbol{u}}_h$ obtained when the steepest descent algorithm stops, after $N_{it} = 151$ iterations. In this case, the cost functional is reduced of about 37%, decreasing from $J(\boldsymbol{u}_h^{(0)}) = 289.796$ to $J(\hat{\boldsymbol{u}}_h) = 183.565$.

A reduction in the magnitude of the adjoint velocity field, which can be observed by comparing the plots displayed in Figure 10.3, also shows that the derivative of the cost functional evaluated at $\boldsymbol{u}_h = \hat{\boldsymbol{u}}_h$ is smaller than the corresponding quantity calculated on the initial

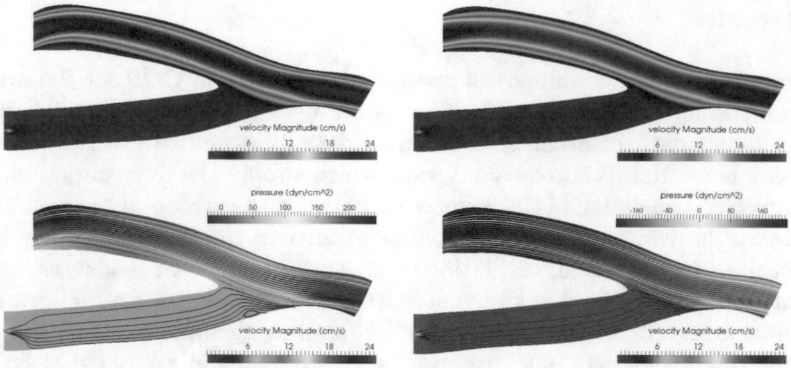

Fig. 10.2 State velocity magnitude (top) and pressure with velocity streamlines (bottom) for the initial control (left) and the optimal control (right). Case $\boldsymbol{v}_{in} \neq \mathbf{0}$.

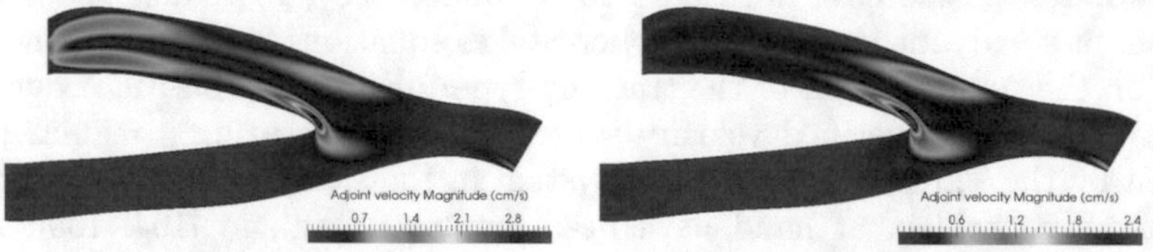

Fig. 10.3 Adjoint velocity magnitude for the initial (left) and the optimal (right) control. Case $\boldsymbol{v}_{in} \neq \mathbf{0}$.

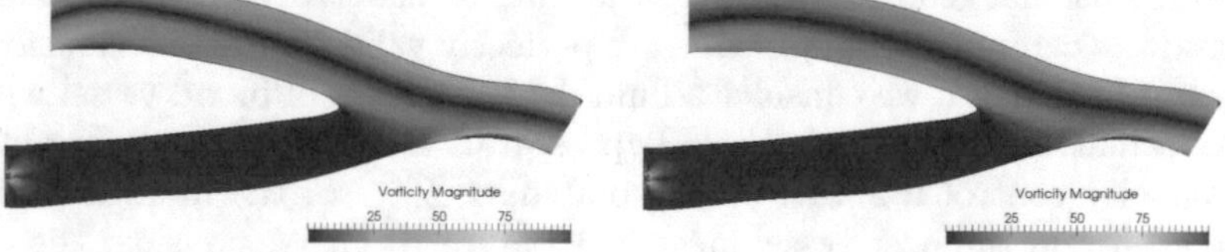

Fig. 10.4 Vorticity magnitude for the initial (left) and the optimal (right) control. Case $\boldsymbol{v}_{in} \neq \mathbf{0}$.

guess $\boldsymbol{u}_h = \boldsymbol{u}_h^{(0)}$. The magnitude of the vorticity is reported in Figure 10.4: a global decrease of this quantity is indeed obtained also by minimizing the tracking-type cost functional, as it results by comparing the vorticity magnitude for the initial and the optimal control.

10.2 Time optimal Control in Cardiac Electrophysiology

In this section we consider an optimal time control problem, first introduced and analyzed in [161, 163], arising in the mathematical modeling of cardiac electrophysiology.

10.2.1 Problem Formulation

The rhytmic contraction of the heart is triggered by an electrical stimulus (see [221, 68, 220]). Disturbances and/or anomalies in the formation/propagation of electrical impulses may increase the heart activity and lead to *fibrillations*. A common therapy to restore a normal cardiac rhythm consists in delivering a defibrillation pulse through implantable cardioverter defribrillators, which acts as a control and has to be carefully operated, based on accurate measurements.

The state system. For the mathematical setting of this problem we first need to write a model for electrical propagation in the myocardium (the cardiac muscle). Let Ω be a bounded domain in $\mathbb{R}^d$, $d = 2, 3$, representing a layer of heart muscle tissue, and let t_f be the *terminal time* of the control horizon $I_f = (0, t_f)$, that will be part of our main cost functional. The state variables are $v = v(\mathbf{x}, t)$, the *transmembrane voltage*, and $w = w(\mathbf{x}, t)$, a *gating or recovery ionic variable*, both defined in the cylinder $Q_{t_f} = \Omega \times I_f$. The following reaction diffusion system, known as *the monodomain equations*, models the electrophysiological activity of the heart (see [68])

$$\begin{cases} v_t - \operatorname{div}(\boldsymbol{\sigma}\nabla v) + I_{ion}(v, w) = I_e & \text{in } Q_{t_f} \\ w_t + g(v, w) = 0 & \text{in } Q_{t_f} \\ \boldsymbol{\sigma}\nabla v \cdot \mathbf{n} = 0 & \text{on } \Gamma \times I_f \\ v(0) = v_0, \; w(0) = w_0 & \text{in } \Omega. \end{cases} \tag{10.66}$$

Here $\Gamma = \partial\Omega$, $\boldsymbol{\sigma} : \Omega \to \mathbb{R}^{d\times d}$ is a time-independent *symmetric and uniformly elliptic* $d \times d$ matrix (the *intracellular conductivity tensor*). Then, $\boldsymbol{\sigma}^\top = \boldsymbol{\sigma}$ and there exists $\Lambda > \sigma_0 > 0$ such that

$$\sigma_0 |\boldsymbol{\xi}|^2 \le \boldsymbol{\xi}^\top \boldsymbol{\sigma}(\mathbf{x}) \boldsymbol{\xi} \le \Lambda |\xi|^2 \qquad \text{for all } \boldsymbol{\xi} \in \mathbb{R}^d \text{ and a.e. } \mathbf{x} \in \Omega. \tag{10.67}$$

The functions I_{ion} and g represent ionic activities in the myocardium, given by

$$\begin{aligned} I_{ion}(v, w) &= \eta_0 v \left(1 - \frac{v}{v_{th}}\right)\left(1 - \frac{v}{v_{pk}}\right) + \eta_1 v w = R(v) + \eta_1 v w \\ g(v, w) &= \eta_2 \left(\eta_3 w - \frac{v}{v_{pk}}\right), \end{aligned} \tag{10.68}$$

respectively, where $\eta_j > 0$, $j = 0, 1, 2, 3$, $v_{th} > 0$ is a *threshold potential* and $v_{pk} > v_{th}$ is the *peak potential.* I_e is an extracellular stimulation current which depends on the defibrillation pulse to be controlled. The current is induced by external devices consisting of electrode plates Ω_j, $j = 1, \ldots, N$, placed on the cardiac tissue. Each plate Ω_j delivers an independent pulse $u_j = u_j(t)$, so that I_e may be modeled as

$$I_e(\mathbf{u}; \mathbf{x}, t) = \sum_{j=0}^{N} u_j(t) \chi_{\Omega_j}(\mathbf{x}) \chi_{[0,t_f]}(t), \tag{10.69}$$

where χ_{Ω_j} and $\chi_{[0,t_f]}$ denote the characteristic function of Ω_j and $[0, t_f]$, respectively, and $\mathbf{u}(t) = (u_1(t), \ldots, u_N(t))$ is the control vector, that we assume to belong to $\mathcal{U} = L^2(0, t_f)^N$.

The cost functional. As we have already mentioned, the duration time t_f of the electrical pulse has to be controlled to achieve a successful defribrillation. Aiming at a low energy pulse and a small final value of the transmembrane voltage, the cost functional can be set as follows,

$$J(v, w, \mathbf{u}, t_f) = \int_0^{t_f} \left(\kappa + \frac{\alpha}{2} |\mathbf{u}(t)|^2\right) dt + \frac{\mu}{2} \|v(t_f)\|^2_{L^2(\Omega)} \tag{10.70}$$

subject to the pointwise constraints

$$\mathbf{u} \in \mathcal{U}_{ad} = \{\mathbf{u} \in \mathcal{U} : |u_k(t)| \le M_j, \; k = 1, \ldots, N, \; 0 \le t \le t_f\}; \tag{10.71}$$

κ, α, μ in (10.70) denote *positive* constants. Our *time optimal control problem* reads therefore

$$J(v, w, \mathbf{u}, t_f) \to \min \tag{10.72}$$

under the state system (10.66) and the constraint $\mathbf{u} \in \mathcal{U}_{ad}$.

The existence of a minimizer is shown in Section 10.2.3. Under a suitable assumption we can ensure the optimal time t_f to be positive. Then, our main goal is to derive first order necessary optimality conditions. Our presentation follows the approach of [161], framing it in the general theoretical setting introduced in Chapter 9.

10.2.2 Analysis of the Monodomain System

The analysis of existence, uniqueness and regularity for the solutions of the state system (10.66) requires some advanced tools from Functional Analysis. We will only state the main results and sketch some of the proofs, referring to the original paper [161] for a complete coverage.

Let $V = H^1(\Omega)$, $L^p = L^p(\Omega), 1 \leq p \leq \infty$ where no ambiguity on the domain occurs. Recall that, by Sobolev Embedding Theorem A.15, $V \hookrightarrow L^6(\Omega)$, $d = 2, 3$. We denote by $\langle \cdot, \cdot \rangle_*$ the duality pairing between V^* and V.

Weak solutions. Weak solutions to the monodomain equations are represented by pairs (v, w) with[3]

$$v \in Y = L^2(I_f; V) \cap L^4(Q_{t_f}),\ v_t \in Y^* = L^2(I_f; V^*) \oplus L^{4/3}(Q_{t_f})$$
$$w \in H^1(I_f; L^2) \text{ or, equivalently, } w \in L^2(Q_{t_f}),\ w_t \in L^2(Q_{t_f}),$$

such that

$$\int_0^{t_f} \{\langle v_t, \varphi_1 \rangle_* + (\boldsymbol{\sigma} \nabla v, \nabla \varphi_1)_{L^2} + (I_{ion}(v,w), \varphi_1)_{L^2}\}\, dt = \int_0^{t_f} \langle I_e(\mathbf{u}), \varphi_1 \rangle_* dt, \quad (10.73)$$

$$\int_0^{t_f} \{\langle w_t, \varphi_2 \rangle_* + (g(v,w), \varphi_2)_{L^2}\}\, dt = 0, \quad (10.74)$$

for every $\varphi_1 \in L^2(I_f; V) \cap L^4(Q_{t_f})$, $\varphi_2 \in L^2(Q_{t_f})$, and

$$v(0) = v_0,\ w(0) = w_0. \quad (10.75)$$

Remark 10.7. The choice of the space $L^4(Q_{t_f})$ for both the solution v and the test functions φ_1 guarantees that the integral $\int_0^{t_f} (I_{ion}(v,w), \varphi_1)_{L^2}\, dt$ is well defined; see Exercise 3. Since, in particular, $v \in L^2(I_f; V)$, $v_t \in L^2(I_f; V^*)$, we have $v \in C(\overline{I}_f; L^2)$ so that the initial conditions (10.75) are well defined, too. •

Remark 10.8. (Solution of the inhibitor equation). Equation (10.74), known as the *inhibitor equation*, is equivalent to the pointwise form

$$w_t + g(v,w) = 0 \quad \text{a.e. in } Q_{t_f}. \quad (10.76)$$

•

Given the specific choice of g in (10.68), equation (10.76) is linear with respect to w and it can be solved in closed form in terms of v, via the *variation of constant formula*, yielding

$$w(t) = e^{-\eta_2 \eta_3 t} w_0 + \frac{\eta_2}{v_{pk}} \int_0^t e^{-\eta_2 \eta_3 (t-s)} v(s)\, ds. \quad (10.77)$$

[3] The space $Y^* = L^2(I_f; V^*) \oplus L^{4/3}(Q_{t_f}) = \{f = f_1 + f_2 : f_1 \in L^2(I_f; V^*), f_2 \in L^{4/3}(Q_{t_f})\}$ with norm

$$\|f\|_Z = \inf\{\|f_1\|_{L^2(I_f;V^*)} + \|f_2\|_{L^4(Q_{t_f})} : f = f_1 + f_2\}$$

is the dual of $Y = L^2(I_f; V) \cap L^4(Q_{t_f})$ endowed with the norm

$$\|v\|_Y = \max\{\|v\|_{L^2(I_f;V)}, \|v\|_{L^4(Q_{t_f})}\}.$$

From this formula we can derive that $w \in H^1(I_f; L^2) \cap C(\overline{I}_f; L^2)$ and, for every $t \in [0, t_f]$,

$$\|w(t)\|_{L^2} \leq \left\{ \|w_0\|_{L^2} + \sqrt{\frac{\eta_2}{2v_{pk}^2\eta_3}} \|v\|_{L^2(Q_{t_f})} \right\}. \tag{10.78}$$

The following result, true in dimension $d = 2, 3$, establishes existence and uniqueness of a weak solution to problem (10.66) under mild assumptions.

Theorem 10.5 (Weak solutions). *Let $v_0, w_0 \in L^2(\Omega)$ and $I_e \in L^2(I_f; V^*)$. Then there exists a unique weak solution (v, w) to problem (10.66). Moreover there exists a constant C, independent of v_0, w_0 and I_e, such that*

$$\|v\|_Y + \|v_t\|_{Y^*} + \|v\|_{C(\overline{I}_f;L^2)} + \|w\|_{H^1(I_f;L^2)} \leq C\left(1 + \|v_0\|_{L^2} + \|w_0\|_{L^2} + \|I_e\|_{L^2(\overline{I}_f;V^*)}\right). \tag{10.79}$$

Proof. See [46] for the existence and the estimate (10.79). Uniqueness can be proved as follows. Assume $(v_1, w_1), (v_2, w_2)$ are two solutions to (10.66) sharing the same initial condition and the same right hand side. Let $h = v_1 - v_2$, $k = w_1 - w_2$. Then for (h, k) we get the equations

$$\int_0^{t_f} \left\{ \langle h_t, \varphi_1 \rangle_* + (\boldsymbol{\sigma}\nabla h, \nabla\varphi_1)_{L^2} + (R(v_1) - R(v_2) + \eta_1(v_1 w_1 - v_2 w_2), \varphi_1)_{L^2} \right\} ds = 0 \tag{10.80}$$

and

$$k(t) = \frac{\eta_2}{v_{pk}} \int_0^t e^{-\eta_2\eta_3(t-s)} h(s)\, ds \tag{10.81}$$

with $h(0) = k(0) = 0$. Observe now that

$$R'(v) = \eta_0 \left[1 - 2\left(\frac{1}{v_{th}} + \frac{1}{v_{pk}}\right) v + \frac{3}{v_{th}v_{pk}} v^2 \right] \geq R'\left(\frac{v_{th} + v_{pk}}{3}\right) = -c_0. \tag{10.82}$$

having set $c_0 = \eta_0(v_{th}^2 + v_{pk}^2 - v_{th}v_{pk})/v_{th}v_{pk} > 0$. Therefore

$$(R(v_1) - R(v_2), \varphi_1)_{L^2} = \int_0^1 (R'(\theta v_1 + (1-\theta) v_2) h, \varphi_1)_{L^2} d\theta \geq -c_0 (h, \varphi_1)_{L^2}.$$

Now we test equation (10.80) against $\varphi_1 = \chi_{(0,t)} h$, $0 < t < t_f$. By using (10.67), observing that $v_1 w_1 - v_2 w_2 = h w_1 + v_2 k$ and integrating the first term, we get

$$\|h(t)\|_{L^2}^2 + \sigma_0 \int_0^t \|h(s)\|_V^2\, ds \leq c_0 \int_0^t \|h(s)\|_{L^2}^2\, ds + \eta_1 \int_0^t |(hw_1, h)_{L^2}| + \left|(v_2 k, h)_{L^2}\right| ds.$$

Hölder inequality and (10.79) give

$$\left|(hw_1, h)_{L^2}\right| \leq \|h\|_{L^4}^2 \|w_1\|_{L^2} \leq M \|h\|_{L^4}^2$$

where $M = \sqrt{\frac{\eta_2}{2v_{pk}^2\eta_3}}$. Now, we first use Hölder inequality then Young inequality[4] with $p = 4, q = 4/3$, $a = \varepsilon^{-3/4} C_S^{3/2} \|h\|_{L^2}^{1/2}$, $b = \varepsilon^{3/4} \|h\|_V^{3/2}$, $\varepsilon > 0$, and the Sobolev embedding $V \hookrightarrow L^6(\Omega)$. We get:

$$\|h\|_{L^4}^2 \leq \|h\|_{L^2}^{1/2} \|h\|_{L^6}^{3/2} \leq C_S^{3/2} \|h\|_{L^2}^{1/2} \|h\|_V^{3/2} \leq \frac{C_S^6}{4\varepsilon^3} \|h\|_{L^2}^2 + \frac{3\varepsilon}{4} \|h\|_V^2. \tag{10.83}$$

[4] Young inequality states that

$$ab \leq \frac{a^p}{p} + \frac{a^q}{q}, \qquad \frac{1}{p} + \frac{1}{q} = 1,\ p, q > 1 \qquad \forall a, b \geq 0.$$

On the other hand, using (10.79),

$$\left|(v_2k,h)_{L^2}\right| \le \|v_2\|_{L^4}\|h\|_{L^4}\|k\|_{L^2} \le \frac{1}{2}\|h\|_{L^4}^2 + \frac{1}{2}\|v_2\|_{L^4}^2\|k\|_{L^2}^2 \tag{10.84}$$

$$\le \frac{1}{2}C^2\|I_e\|_{L^2(\bar{I}_f;V^*}^2\|k\|_{L^2}^2 + \frac{C_S^6}{8\varepsilon^3}\|h\|_{L^2}^2 + \frac{3\varepsilon}{8}\|h\|_V^2 . \tag{10.85}$$

By collecting the previous estimates we can write

$$\|h(t)\|_{L^2}^2 + \sigma_0\int_0^t \|h(s)\|_V^2\,ds \le C_1\int_0^t \left\{\left(1+\varepsilon^{-3}\right)\|h(s)\|_{L^2}^2 + \varepsilon\|h\|_V^2 + \|k\|_{L^2}^2\right\}ds$$

where C_1 is independent of h, k and ε. Choosing $\varepsilon = 1/2C_1$, the term containing the $V-$norm of h can be absorbed into the left-hand side so that

$$\|h(t)\|_{L^2}^2 \le C_2\int_0^t \left\{\|h(s)\|_{L^2}^2 + \|k(s)\|_{L^2}^2\right\}ds$$

with C_2 independent of h, k. For k, (10.78) gives

$$\|k(t)\|_{L^2}^2 \le C^2\|h(t)\|_{L^2}^2 . \tag{10.86}$$

Thus we deduce the estimate

$$\|h(t)\|_{L^2}^2 + \|k(t)\|_{L^2}^2 \le (C_2 + C^2)\int_0^t \left\{\|h(s)\|_{L^2}^2 + \|k(s)\|_{L^2}^2\right\}ds$$

for every $t \in \left[0, t_f\right]$. From Gronwall Lemma (see Lemma A.5 in Appendix A) we deduce that $h = k = 0$. □

Strong solutions and regularity. Due to the cubic nonlinearity of $R(v)$, to derive the optimality conditions for our control problem, we need more regularity of the solution than the one provided in Theorem 10.5. In fact, in the end we shall need to work with *bounded* solutions. Notice that from (10.69) and (10.71) one has $I_e \in L^\infty(Q_{t_f})$.

Since $\boldsymbol{\sigma}$ is only bounded, we cannot use the standard results (see, e.g., [118]) to infer additional regularity of the weak solution. Thus we introduce an ad-hoc Hilbert space, tailored on the operator $\mathcal{A} : V \to V^*$, defined through the formula

$$\langle \mathcal{A}v, \varphi\rangle_* = (\boldsymbol{\sigma}\nabla v, \nabla\varphi)_{L^2} \qquad \text{for all } \varphi \in V, \tag{10.87}$$

with domain

$$D_{\mathcal{A}} = \left\{v \in V : \mathcal{A}v \in L^2\right\}. \tag{10.88}$$

$D_{\mathcal{A}}$ is a Hilbert space[5] when endowed with the *graph norm* $\|v\|_{D_{\mathcal{A}}}^2 = \|v\|_{L^2}^2 + \|\mathcal{A}v\|_{L^2}^2$. In dimension $d \le 3$, we have the embedding $D_{\mathcal{A}} \hookrightarrow C^{0,\beta}\left(\overline{\Omega}\right)$ for some $\beta > 0$; see, e.g., [114].

Accordingly, looking at the equation $v_t = \mathcal{A}v + I_{ion}(v,w) + I_e(\mathbf{u})$, it is natural to introduce the time dependent Sobolev space

$$W(I_f) = L^2(I_f; D_{\mathcal{A}}) \cap H^1\left(I_f; L^2\right)$$

with norm given by

$$\|v\|_{W(I_f)} = \left(\int_0^{t_f}\left(\|v(s)\|_{D_{\mathcal{A}}}^2 + \|v_t(s)\|_{L^2}^2\right)ds\right)^{1/2} .$$

In the next lemma we collect some embeddings we are going to use later on.

[5] For smooth $\boldsymbol{\sigma}$ and Ω, $\mathcal{D}_{\mathcal{A}} = \left\{v \in H^2(\Omega) : \boldsymbol{\sigma}\nabla v \cdot \mathbf{n} = 0 \text{ on } \Gamma\right\}$ and $\mathcal{A}v = -\mathrm{div}(\boldsymbol{\sigma}\nabla v)$.

Lemma 10.3. *The following continuous embeddings hold in dimension $d \le 3$:*

1. $W(I_f) \hookrightarrow C(\overline{I}_f; V) \hookrightarrow L^\infty(I_f; L^6) \hookrightarrow L^6(Q_{t_f})$;
2. $L^2(I_f; D_{\mathcal{A}}) \hookrightarrow L^2(I_f; L^\infty)$;
3. $H^1(I_f; L^2) \hookrightarrow C^{0,1/2}(\overline{I}_f; L^2) \hookrightarrow L^\infty(I_f; L^2)$.

Proof. For 1, see [176]; 2 is a consequence of the embedding $D_{\mathcal{A}} \hookrightarrow C^{0,\beta}(\overline{\Omega})$. For 3, see Exercise 4. □

We are now ready to state the following regularity result (see, e.g., [161] for the proof).

Theorem 10.6. *(Regularity of weak solutions). Let (v, w) be a weak solution to* (10.66).

a) If $v_0 \in V$ and $w_0 \in L^3$, then $(v, w) \in W(I_f) \times H^1(I_f; L^3)$ and

$$\|v\|_{W(I_f)} + \|w\|_{H^1(I_f;L^3)} \le C\left(1 + \|v_0\|_V + \|w_0\|_{L^3} + \|I_e\|_{L^2(Q_{t_f})}\right). \tag{10.89}$$

b) If moreover $v_0 \in V \cap L^\infty$ and $w_0 \in L^\infty$, then $(v, w) \in L^\infty(Q_{t_f}) \times L^\infty(Q_{t_f})$ with norms bounded by a constant C depending only on $t_f, \|v_0\|_V, \|w_0\|_{L^\infty}, \|w_0\|_{L^\infty}$ and $\|I_e(\mathbf{u})\|_{L^\infty(Q_{t_f})}$.

Remark 10.9. If $v \in W(I_f)$, $w \in H^1(I_f; L^2)$ a weak solution (v, w) to (10.66) is called a *strong solution.* In this case the equation

$$v_t + \mathcal{A}v + I_{ion}(v, w) = I_e(\mathbf{u})$$

makes sense a.e. in Q_{t_f}. Also, by Lemma 10.3, one can check that $I_{ion}(v, w) \in L^2(Q_{t_f})$ and therefore for strong solutions we can choose test functions $\varphi_1 \in V \cap L^2(Q_{t_f})$. Moreover, we have $\langle v_t, \varphi_1\rangle_* = (v_t, \varphi_1)_{L^2}$. •

10.2.3 Existence of an Optimal Control $(\hat{\mathbf{u}}, \hat{t}_f)$

We now prove that our optimal control problem (10.72) has at least one solution. We work under the assumptions of Theorem 10.6.

Theorem 10.7. *There exists at least one optimal solution $(\hat{t}_f, \hat{\mathbf{u}})$ to (10.72) with corresponding optimal state $(\hat{v}, \hat{w})$ and $\hat{t}_f \ge 0$.*

Proof. Let $\{t_{fk}, \mathbf{u}_k; v_k, w_k\}$ be a minimizing sequence. Up to a subsequence we may assume that $t_{fk} \to \hat{t}_f$, as $k \to \infty$. Let $t_{\max} = \max_k\{t_{fk}\}$ and extend $\mathbf{u}_k$ to the interval $I_{\max} = (0, t_{\max})$, maintaining $\mathbf{u}_k \in U_{ad}$. Accordingly, we extend v_k, w_k to the interval $I_{\max}$ by solving the monodomain system with the extended control $\mathbf{u}_k$.

To show existence, set $V_{\max} = W(I_{\max}) \times H^1(I_{\max}; L^2)$ and $\mathcal{U}_{\max} = \bar{I}_{\max} \times L^2(I_{\max}; \mathbb{R}^N)$. From (10.89), $\{t_{fk}, \mathbf{u}_k; v_k, w_k\}$ is bounded in $\mathcal{U}_{\max} \times V_{\max}$ and (again up to a subsequence) weakly convergent to $\{\hat{t}_f, \hat{\mathbf{u}}; \hat{v}, \hat{w}\}$ in $\mathcal{U}_{\max} \times V_{\max}$.

Now, $\mathcal{U}_{ad}$ is closed and convex in $L^2(I_{\max}; \mathbb{R}^N)$ and therefore weakly sequentially closed. Moreover, it is not difficult to show (see Exercise 5) that we can pass to the limit into the weak formulation (10.73), (10.74) and conclude that $\hat{v}, \hat{w}$ are the solutions corresponding to the control $\hat{\mathbf{u}}$ in $Q_{t_{\max}}$.

To show the optimality of $\{\hat{t}_f, \hat{\mathbf{u}}; \hat{v}, \hat{w}\}$ we observe that F is weakly sequentially lower semicontinuous. Indeed, since the map

$$\mathbf{u} \mapsto \|\mathbf{u}\|^2_{L^2(I_{\max};\mathbb{R}^N)}$$

is weakly sequentially lower semicontinuous (being a convex and continuous functional), we have only to check that $\|v_k(t_{fk})\|_{L^2} \to \|\hat{v}(\hat{t}_f)\|_{L^2}$. Indeed, we show that $\|v_k(t_{fk}) - \hat{v}(t_f)\|_{L^2} \to 0$. Write

$$\|v_k(t_{fk}) - \hat{v}(\hat{t}_f)\|_{L^2} \le \|v_k(t_{fk}) - v_k(\hat{t}_f)\|_{L^2} + \|v_k(\hat{t}_f) - \hat{v}(\hat{t}_f)\|_{L^2}.$$

By the embedding 3 in Lemma 10.3 we deduce

$$\|v_k(t_{fk}) - v_k(\hat{t}_f)\|_{L^2} \le C|t_{fk} - \hat{t}_f|^{1/2}$$

with C independent of k. Therefore $\|v_k(t_{fk}) - v_k(\hat{t}_f)\|_{L^2} \to 0$ as $k \to \infty$.

On the other hand, since $W(I_{\max}) \hookrightarrow C(\overline{I}_{\max}; V)$ (Lemma 10.3), we have

$$\|v(\hat{t}_f)\|_V \le C\|v\|_{W(I_{\max})}.$$

Hence the point evaluation in V at $\hat{t}_f$ is a bounded linear operator on $W(I_{\max})$. Thus, since $v_k \rightharpoonup v$ in $W(I_{\max})$, we infer that $v_k(\hat{t}_f) \rightharpoonup v(\hat{t}_f)$ in V. Since V is compactly embedded into L^2 we conclude that $v_k(\hat{t}_f) \to \hat{v}(\hat{t}_f)$ in L^2. Thus

$$J(\hat{t}_f, \hat{\mathbf{u}}; \hat{v}, \hat{w}) \le \liminf J(t_{fk}, \mathbf{u}_k; v_k, w_k) = \inf J(t_f, \mathbf{u}; v, w). \qquad \square$$

We emphasize that an optimal time $\hat{t}_f$ should be positive. To this purpose we make the following assumption, reasonable since we expect the optimal potential v to be small at the final time. Let $\delta > 0$ be a small number.

Assumption[6] **H:** *There exists $\overline{v}, \overline{w}, \overline{\mathbf{u}}$, with $\overline{\mathbf{u}} \in U_{ad}$, $\overline{t}_f > 0$ and such that $(\overline{v}, \overline{w})$ solves the monodomain system and fulfills the terminal condition $\|\overline{v}(\overline{t}_f)\|_{L^2} \le \delta$.*

Define

$$\overline{f} = \int_0^{\overline{t}_f} \left(\kappa + \frac{\alpha}{2}|\overline{\mathbf{u}}(t)|^2\right) dt.$$

The following result holds.

Theorem 10.8. *Let assumption **H** hold and let $\delta_1 > \delta$. If $\mu \ge 2\overline{f}/(\delta_1^2 - \delta^2)$ any optimal solution $(\hat{v}_\mu, \hat{w}_\mu, \hat{\mathbf{u}}_\mu, \hat{t}_{f_\mu},)$ satisfies the condition $\|\hat{v}_\mu(\hat{t}_{f_\mu})\|_{L^2} \le \delta_1$. If $\delta_1 < \|v_0\|_{L^2}$, then $\hat{t}_{f_\mu} > 0$.*

Proof. Define

$$\hat{f}_\mu = \int_0^{\hat{t}_{f\mu}} \left(\kappa + \frac{\alpha}{2}|\hat{\mathbf{u}}_\mu(t)|^2\right) dt$$

and observe that $\overline{f}$ is independent of μ. By optimality, for all $\mu > 0$,

$$\hat{f}_\mu + \frac{\mu}{2}\|\hat{v}_\mu(\hat{t}_{f_\mu})\|_{L^2} \le \overline{f} + \frac{\mu}{2}\|\overline{v}(\overline{t}_f)\|_{L^2} \le \overline{f} + \frac{\mu}{2}\delta^2.$$

Since $\hat{f}_\mu > 0$, we get

$$\|\hat{v}_\mu(\hat{t}_{f_\mu})\|_{L^2} \le \frac{2}{\mu}\overline{f} + \delta^2$$

so that if $\mu \ge 2\overline{f}/(\delta_1^2 - \delta^2)$, $\|\hat{v}_\mu(\hat{t}_{f_\mu})\|_{L^2} \le \delta_1$. $\square$

[6] At present, conditions on data and controls that ensure the validity of assumption **H** are not known. See however [161] for a discussion on the large time behavior of the solution of the monodomain system.

10.2.4 Reduction to a Control Problem with Fixed End Time

To derive optimality conditions we reformulate our time optimal control problem as a control problem with fixed time horizon $I_T = (0, T)$, thanks to a suitable transformation. This is convenient for implementing numerical approximation algorithms, too.

To this purpose we introduce a new control variable ν, subject to the constraint

$$\nu \in N_{ad} = \{\nu \in L^\infty(0,T) : \text{ess inf}\, \nu > 0\};$$

N_{ad} is an open subset of $L^\infty(0,T)$. Then we set

$$t = \theta(\tau) = \int_0^\tau \nu(r)\, dr$$

so that

$$t_f = \theta(T) = \int_{I_T} \nu(r)\, dr.$$

Since θ is strictly monotonically increasing and Lipschitz continuous, it is a bijective mapping from $[0, T]$ onto $[0, t_f]$.

Let now (v, w) be a strong solution (in the sense of Remark 10.9) to the monodomain systems corresponding to a control $\mathbf{u}$. If we set

$$y = v \circ \theta, \; z = w \circ \theta, \; \mathbf{c} = \mathbf{u} \circ \theta,$$

we have

$$y_\tau = \nu v_t \circ \theta, \; z_\tau = \nu z_t \circ \theta.$$

Thus $(y, z) \in W(I_T) \times H^1(I_T; L^2)$ and (y, z) is a *strong* solution to the problem

$$\int_{I_T} \{(y_t, \varphi_1)_{L^2} + \nu(\boldsymbol{\sigma}\nabla y, \nabla\varphi_1)_{L^2} + \nu(I_{ion}(y, z), \varphi_1)_{L^2}\}\, d\tau = \int_{I_T} \nu(\tilde{I}_e(\mathbf{c}), \varphi_1)_{L^2} d\tau \quad (10.90)$$

$$\int_{I_T} \{(z_t, \varphi_2)_{L^2} + \nu(g(y, z), \varphi_2)_{L^2}\}\, d\tau = 0, \quad (10.91)$$

where

$$\tilde{I}_e(\mathbf{c}; \mathbf{x}, \tau) = \sum_{j=0}^{N} c_j(\tau)\, \chi_{\Omega_j}(\mathbf{x})\, \chi_{[0,T]}(\tau), \quad (10.92)$$

for every $\varphi_1 \in L^2(I_T; V) \cap L^2(Q_T)$, $\varphi_2 \in L^2(Q_T)$, and

$$y(0) = v_0, \; z(0) = w_0. \quad (10.93)$$

Vice versa, for any strong solution (y, z) to system (10.90), (10.91) with initial conditions (10.93) and controls $(\nu, \mathbf{c})$, the inverse transformation

$$v = y \circ \theta^{-1}, \; w = z \circ \theta^{-1}$$

yields a strong solution to the original monodomain system, corresponding to the control $\mathbf{u} = \mathbf{c} \circ \theta^{-1}$. Indeed, for $\nu \in N_{ad}$ we have

$$\partial_t \theta^{-1} = \frac{1}{\nu \circ \theta^{-1}} \in L^\infty(I_T).$$

In terms of the new variables, the optimal control problem (10.72) reads as follows:

$$\tilde{J}(y,z,\mathbf{c},\nu) = \int_{I_T} \nu\left(\kappa + \frac{\alpha}{2}|\mathbf{c}|^2\right)d\tau + \frac{\mu}{2}\|y(T)\|^2_{L^2(\Omega)} \to \min \tag{10.94}$$

over $\mathbf{c} \in \mathcal{U}_{ad}, \nu \in N_{ad}$ and $(y,z) \in W(I_T) \times H^1(I;L^2)$, subject to (10.90), (10.91) and (10.93).

Remark 10.10. The two problems (10.72) and (10.94) are equivalent in the following sense. If $(y,z,\mathbf{c},\nu)$ is feasible for (10.94), then $v = y \circ \theta^{-1}$, $w = z \circ \theta^{-1}, \mathbf{u} = \mathbf{c} \circ \theta^{-1}, t_f = \theta(T)$ is feasible for (10.72) and $\tilde{J}(y,z,\mathbf{c},\nu) = J(v,w,\mathbf{u},t_f)$. Vice versa, if $(v,w,\mathbf{u},t_f)$ is feasible for (10.72) and $t_f > 0$, then for every $\nu \in N_{ad}$ such that $\theta(T) = t_f$, the quadruplet $y = v \circ \theta$, $z = w \circ \theta$, $\mathbf{c} = \mathbf{u} \circ \theta, \nu$ is feasible for (10.94) and $J(v,w,\mathbf{u},t_f) = \tilde{J}(y,z,\mathbf{c},\nu)$. •

10.2.5 First Order Optimality Conditions

The main result of this section provides a system of optimality conditions for our fixed end time OCP. Since we need the boundedness of strong solutions to the state system, from now on it is understood that the initial conditions v_0 and $w_0 \in L^\infty$ are chosen in $V \cap L^\infty$ and L^∞, respectively. For any adjoint state $(p,q) \in W(I_T) \times H^1(I_T;L^2)$ we introduce the Hamiltonian function (of time τ)

$$H(y,z,\mathbf{c},p,q) = \left(\kappa + \frac{\alpha}{2}|\mathbf{c}|^2\right) + (\tilde{I}_e(\mathbf{c}),p)_{L^2} - (\boldsymbol{\sigma}\nabla y, \nabla p)_{L^2} + (I_{ion}(y,z),p)_{L^2} - (g(y,z),q)_{L^2}.$$

The following result holds.

Theorem 10.9 (First order optimality conditions). *Let $\hat{y},\hat{z},\hat{\mathbf{c}},\hat{\nu} \in W(I_T) \times H^1(I_T;L^2) \times \mathcal{U}_{ad} \times N_{ad}$ be an optimal solution of (10.94). Then there exists a unique adjoint state $(\hat{p},\hat{q}) \in W(I_T) \times H^1(I_T;L^2)$ such that the following optimality conditions are fulfilled:*

a) adjoint system (in strong form):

$$\begin{cases} \hat{p}_t + \hat{\nu}\mathrm{div}(\boldsymbol{\sigma}\nabla\hat{p}) - \hat{\nu}I'_{ion,y}(\hat{y},\hat{z})\hat{p} - \hat{\nu}g'_y(\hat{y},\hat{z})\hat{q} = 0 & \text{in } Q_T \\ \hat{q}_t - \hat{\nu}I'_{ion,z}(\hat{y},\hat{z})\hat{p} - \hat{\nu}g'_z(\hat{y},\hat{z})\hat{q} = 0 & \text{in } Q_T \\ \boldsymbol{\sigma}\nabla\hat{p}\cdot\mathbf{n} = 0 & \text{on } \Gamma \times I_T \\ \hat{p}(T) = \mu\hat{y}(T),\ \hat{q}(T) = 0 & \text{in } \Omega. \end{cases} \tag{10.95}$$

b) Regularity of the optimal control. For $j = 1,\dots,N$

$$\hat{c}_j(\tau) = Pr_{\mathcal{U}_{ad}}\left(-\frac{1}{\alpha}(\chi_{\Omega_j},\hat{p}(\tau))_{L^2}\right). \tag{10.96}$$

In particular, $\hat{\mathbf{c}} \in H^1(I_T;\mathbb{R}^N)$.

c) The Hamiltonian function $\tau \mapsto H(\hat{y}(\tau),\hat{z}(\tau),\hat{\mathbf{c}}(\tau),\hat{\nu}(\tau),\hat{p}(\tau),\hat{q}(\tau))$ satisfies the following transversality condition

$$H(\hat{y}(\tau),\hat{z}(\tau),\hat{\mathbf{c}}(\tau),\hat{p}(\tau),\hat{q}(\tau)) = 0 \quad \text{for every } \tau \in [0,T]. \tag{10.97}$$

Proof. In order to apply the theory developed in Sect. 9.3, in particular Theorem 9.6, we check assumptions $(i),(ii),(iii)$ of Theorem 9.6.

The functional $\tilde{J}$ is linear in ν and quadratic with respect to $\mathbf{c}$ and y. Since $\nu \in L^{\infty}(I_T)$, the Fréchet continuous differentiability of $\tilde{J}$ follows from a direct computation. The rest of the assumptions are proved in Lemma 10.4 and Theorem 10.10 below.

We now write the optimality conditions in the form of system (9.26). Let us introduce the adjoint state $\mathbf{p} = (p, q) \in W(I_T) \times H^1(I; L^2)$ and the Lagrangian functional, expressed in terms of the Hamiltonian function H, by

$$\mathcal{L}(y, z, \mathbf{c}, \nu, p, q) = \int_{I_T} \{\nu H(y, z, \mathbf{c}, p, q) - (y_t, p)_{L^2} - (z_t, q)_{L^2}\} \, d\tau + \frac{\mu}{2} \|y(T)\|_{L^2}^2 .$$

The adjoint system is obtained from the vanishing of $\mathcal{L}'_y(\hat{y}, \hat{z}, \hat{\mathbf{c}}, \hat{\nu}, \hat{p}, \hat{q})$ and $\mathcal{L}'_z(\hat{y}, \hat{z}, \hat{\mathbf{c}}, \hat{\nu}, \hat{p}, \hat{q})$. This yields, for all $\varphi_1, \varphi_2 \in W(I_T) \times H^1(I; L^2)$, with $\varphi_1(0) = \varphi_2(0) = 0$,

$$\begin{aligned}
&\mathcal{L}'_y(\hat{y}, \hat{z}, \hat{\mathbf{c}}, \hat{\nu}, \hat{p}, \hat{q})\varphi_1 \\
&= \int_{I_T} \left\{-\hat{\nu}(\boldsymbol{\sigma}\nabla\varphi_1, \nabla\hat{p})_{L^2} - \hat{\nu}(I'_{ion,y}(\hat{y}, \hat{z})\varphi_1, \hat{p})_{L^2} - \hat{\nu}\left(g'_y(\hat{y}, \hat{z})\varphi_1, \hat{q}\right)_{L^2} - (\partial_t\varphi_1, \hat{p})_{L^2}\right\} d\tau \\
&\quad + \mu(\hat{y}(T), \varphi_1(T))_{L^2} \qquad (10.98) \\
&= \int_{I_T} \left\{-\hat{\nu}(\boldsymbol{\sigma}\nabla\hat{p}, \nabla\varphi_1)_{L^2} - \hat{\nu}(I'_{ion,y}(\hat{y}, \hat{z})\hat{p}, \varphi_1)_{L^2} - \hat{\nu}\left(g'_y(\hat{y}, \hat{z})\hat{q}, \varphi_2\right)_{L^2} + (\hat{p}_t, \varphi_1)_{L^2}\right\} d\tau \\
&\quad + (\mu\hat{y}(T) - \hat{p}(T), \varphi_1(T))_{L^2} . \qquad (10.99)
\end{aligned}$$

where the last equality follows by an integration by parts.

Analogously,

$$\mathcal{L}'_z(\hat{y}, \hat{z}, \hat{\mathbf{c}}, \hat{\nu}, \hat{p}, \hat{q})\varphi_2 = \int_{I_T} \left\{-\hat{\nu}\left(I'_{ion,z}(\hat{y}, \hat{z})\varphi_2, \hat{p}\right)_{L^2} - \hat{\nu}\left(g'_z(\hat{y}, \hat{z})\varphi_2, \hat{q}\right)_{L^2} - (\partial_t\varphi_2, \hat{q})_{L^2}\right\} d\tau$$

$$= \int_{I_T} \left\{-\hat{\nu}\left(I'_{ion,z}(\hat{y}, \hat{z})\hat{p}, \varphi_2\right)_{L^2} - \hat{\nu}\left(g'_z(\hat{y}, \hat{z})\hat{q}, \varphi_2,\right)_{L^2} + (\hat{q}_t, \varphi_2)_{L^2}\right\} d\tau - (\hat{q}(T), \varphi_2(T))_{L^2} .$$

The adjoint system

$$\begin{cases} \mathcal{L}'_y(\hat{y}, \hat{z}, \hat{\mathbf{c}}, \hat{\nu}, \hat{p}, \hat{q})\varphi_1 = 0 \\ \mathcal{L}'_z(\hat{y}, \hat{z}, \hat{\mathbf{c}}, \hat{\nu}, \hat{p}, \hat{q})\varphi_2 = 0 \end{cases}$$

for all $\varphi_1, \varphi_2 \in W(I_T) \times H^1(I; L^2)$, with $\varphi_1(0) = \varphi_2(0) = 0$, is the weak form of system (10.95).

Moreover, we have the variational inequalities

$$\mathcal{L}'_\nu(\hat{y}, \hat{z}, \hat{\mathbf{c}}, \hat{\nu}, \hat{p}, \hat{q})(\nu - \hat{\nu}) \geq 0 \text{ for all } \nu \in N_{ad} \qquad (10.100)$$

and, for $j = 1, ..., N$,

$$\mathcal{L}'_{c_j}(\hat{y}, \hat{z}, \hat{\mathbf{c}}, \hat{\nu}, \hat{p}, \hat{q})(\xi_j - \hat{c}_j) \geq 0 \text{ for all } \xi \in U_{ad}. \qquad (10.101)$$

Since N_{ad} is open in $L^{\infty}(I_T)$, (10.100) reduces to the equation

$$\mathcal{L}'_\nu(\hat{y}, \hat{z}, \hat{\mathbf{c}}, \hat{\nu}, \hat{p}, \hat{q})\nu = \int_{I_T} \nu H(\hat{y}, \hat{z}, \hat{\mathbf{c}}, \hat{p}, \hat{q}) \, d\tau = 0 \text{ for all } \nu \in N_{ad}$$

which implies the *transversality condition*

$$H(\hat{y}(\tau), \hat{z}(\tau), \hat{\mathbf{c}}(\tau), \hat{p}(\tau), \hat{q}(\tau)) = 0 \quad \text{for a. e. } \tau \in I_T. \qquad (10.102)$$

Since

$$\mathcal{L}'_{c_j}(\hat{y}, \hat{z}, \hat{\mathbf{c}}, \hat{\nu}, \hat{p}, \hat{q})\xi_j = \int_{I_T} \hat{\nu}\left\{\alpha\hat{c}_j\xi_j + \xi_j(\chi_{\Omega_j}, \hat{p})_{L^2}\right\} \qquad (10.103)$$

(10.101) writes

$$\int_{I_T} \hat{\nu}\left(\alpha\hat{c}_j + (\chi_{\Omega_j}, \hat{p})_{L^2}\right)(\xi_j - \hat{c}_j) \, d\tau \geq 0 \text{ for all } \xi \in U_{ad}. \qquad (10.104)$$

As we know, formula (10.104) expresses each component $\hat{c}_j$ of the optimal control vector as the projection

$$\hat{c}_j(\tau) = \mathrm{Pr}_{U_{ad}}\left(-\frac{1}{\alpha}(\chi_{\Omega_j}, \hat{p}(\tau))_{L^2}\right) = \max\left\{-M_j, \min\left\{-\frac{1}{\alpha}\int_{\Omega_j}\hat{p}(\mathbf{x},\tau)\,d\mathbf{x}, M_j\right\}\right\}$$
$$= \min\left\{M_j, \max\left\{-M_j, -\frac{1}{\alpha}\int_{\Omega_j}\hat{p}(\mathbf{x},\tau)\,d\mathbf{x}\right\}\right\}.$$

From the last formula we deduce that $\hat{\mathbf{c}} \in H^1(I_T; \mathbb{R}^N)$ and therefore that $\hat{\mathbf{c}}$ is continuous in $[0,T]$. It follows, from Lemma 10.3, that the Hamiltonian function $\tau \mapsto H(\hat{y}(\tau), \hat{z}(\tau), \hat{\mathbf{c}}(\tau), \hat{\nu}(\tau), \hat{p}(\tau), \hat{q}(\tau))$ is continuous in $[0,T]$ and hence the transversality condition 10.102 actually holds for *every* $\tau \in [0,T]$. □

In the next lemma we prove that the control-to-state map $S : (\mathbf{c}, \nu) \mapsto (y, z)$ is continuously Fréchet differentiable.

Lemma 10.4 (differentability of the control-to-state map). *Let*

$$S : U_{ad} \times N_{ad} \subset \mathcal{U} \times L^\infty(I_T) \to W(I_T) \times H^1(I_T; L^2)$$

be the control-to-state map. Then, for each $(\bar{\mathbf{c}}, \bar{\nu}) \in U_{ad} \times N_{ad}$ there exists a neighborhood of $(\bar{\mathbf{c}}, \bar{\nu})$ in $U_{ad} \times N_{ad}$ where $(y,z) = S(\mathbf{c}, \nu)$ is continuously Fréchet differentiable.

Proof. We use the Implicit Function Theorem A.23 for the state equation that we write in operator form as

$$G(y, z, \mathbf{c}, \nu) = 0 \tag{10.105}$$

where $G : W(I_T) \times H^1(I; L^2) \times \mathcal{U} \times L^\infty(I_T) \to L^2(Q_T) \times L^2(Q_T) \times V \times L^3$ is defined by

$$G(y, z, \mathbf{c}, \nu) = \begin{pmatrix} y_t + \nu \mathcal{A} y + \nu I_{ion}(y,z) - \nu \tilde{I}_e(\mathbf{c}) \\ z_t + \nu g(y,z) \\ y(0) - v_0 \\ z(0) - w_0 \end{pmatrix}$$

where $\mathcal{A}$ is defined in (10.87), (10.88). Note that the first equation in (10.105) incorporates the homogeneous Neumann boundary condition. We check that G is continuously Fréchet differentiable by inspecting its first two components.

Since $\mathcal{A} : W(I_T) \to L^2(Q_T)$ and $g : W(I_T) \times H^1(I; L^2) \to L^2(Q_T)$ are linear and bounded, and $\mathbf{c} \to \tilde{I}_e(\mathbf{c})$ is quadratic, both these operators are continuously Fréchet differentiable.

Write (see 10.68)

$$I_{ion}(y,z) = \eta_0 v \left(1 - \frac{y}{v_{th}}\right)\left(1 - \frac{y}{v_{pk}}\right) + \eta_1 y z = R(y) + \eta_1 y z.$$

Since $W(I_T) \hookrightarrow L^6(Q_T)$, the differentiability of the cubic operator $R : L^6(Q_T) \to L^2(Q_T)$ follows by direct computations as done in the Example A.14 of Sect. A.7.3 in the Appendix A. By property *d*), Sect. A.7.1, it follows that R is also differentiable when regarded as an operator from $W(I_T)$ into $L^2(Q_T)$.

For the bilinear operator $(y,z) \to b(y,z) = yz$ we have $b'_y(y,z)k = kz$ and $b'_z(y,z)h = yh$ and

$$b(y+k, z+h) - b(y,z) - b'_y(y,z)k - b'_z(y,z)h = hk.$$

Using the embeddings $L^2(I_T; D_A) \hookrightarrow L^2(I_T; L^\infty)$ and $H^1(I_T; L^2) \hookrightarrow L^\infty(I_T; L^2)$ in Lemma 10.3, we find

$$\|hk\|^2_{L^2(Q_T)} = \int_{I_T} d\tau \int_\Omega h^2(\mathbf{x},\tau)\, k(\mathbf{x},\tau)^2\, d\mathbf{x} \le \int_{I_T} \|h(\tau)\|^2_{L^\infty} d\tau \int_\Omega k(\mathbf{x},\tau)^2\, d\mathbf{x}$$
$$\le \|h\|^2_{L^2(I_T;L^\infty)} \|k\|^2_{L^\infty(I_T;L^2)} \le C \|h\|^2_{W(I_T)} \|k\|^2_{H^1(I_T;L^2)}$$

from which the differentiability of b follows. Finally, since $\nu \in L^\infty(I_T)$, the differentiability of G easily follows.

The Implicit Function Theorem requires that, for any fixed $(\bar{\mathbf{c}}, \bar{\nu})$ and $(\bar{y}, \bar{z}) = S(\bar{\mathbf{c}}, \bar{\nu})$, the linear operator

$$G'_{(y,z)}(\bar{y}, \bar{z}, \bar{\mathbf{c}}, \bar{\nu}) : W(I_T) \times H^1(I; L^2) \to L^2(Q_T) \times L^2(Q_T) \times V \times L^3$$

defined by

$$G'_{(y,z)}(\bar{y}, \bar{z}, \bar{\mathbf{c}}, \bar{\nu})(h, k) = \begin{pmatrix} h_t + \bar{\nu}\mathcal{A}'(\bar{y})h + \bar{\nu}I'_{ion,y}(\bar{y}, \bar{z})h + \bar{\nu}I'_{ion,z}(\bar{y}, \bar{z})k \\ k_t + \bar{\nu}g(h, k) \\ h(0) \\ k(0) \end{pmatrix},$$

has a bounded inverse. Thus we have to check that the linear system

$$\begin{cases} h_t + \bar{\nu}\mathcal{A}'(\bar{y})h + \bar{\nu}I'_{ion,y}(\bar{y}, \bar{z})h + \bar{\nu}I'_{ion,z}(\bar{y}, \bar{z})k = f_1(\mathbf{x},t) & \text{in } Q_T \\ k_t + \bar{\nu}g(h, k) = f_2(\mathbf{x},t) & \text{in } Q_T \\ \boldsymbol{\sigma}\nabla h \cdot \mathbf{n} = 0 & \text{on } \Gamma \times I_T \\ h(0) = h_0, k(0) = k_0 & \text{in } \Omega \end{cases} \tag{10.106}$$

has a unique solution for every $(f_1, f_2) \in L^2(Q_T) \times L^2(Q_T)$ and every $(h_0, k_0) \in V \times L^3$. Since $\bar{\nu} \in L^\infty(I_T)$ and $(\bar{y}, \bar{z}) \in L^\infty(Q_T) \times L^\infty(Q_T)$, recalling Remark 10.8 and, in particular, formula (10.77), we can rewrite (10.106) in the form of system (10.107). Applying Theorem 10.10 we infer that $G'_{(y,z)}(\bar{y}, \bar{z}, \bar{\mathbf{c}}, \bar{\nu})$ has bounded inverse. Finally, applying the Implicit Function Theorem we conclude the proof. □

Our next result states the well-posedness of the integro-differential equation obtained by linearization of the monodomain system. Consider the system

$$\begin{cases} \psi_t - \operatorname{div}(\boldsymbol{\sigma}\nabla\psi) + a\psi + bE_\delta(\psi) = f & \text{in } Q_T \\ \boldsymbol{\sigma}\nabla\psi \cdot \mathbf{n} = 0 & \text{on } \Gamma \times I_T \\ \psi(0) = 0 & \text{in } \Omega, \end{cases} \tag{10.107}$$

where $a, b \in L^\infty(Q_T), f \in L^2(Q_T)$ and

$$E_\delta(\psi)(\mathbf{x}, t) = \int_0^t e^{-\delta(t-s)}\psi(\mathbf{x},s)\, ds.$$

A function ψ is a *strong solution* of (10.107) if $\psi \in W(I_T)$, $\psi(0) = 0$ and

$$\int_0^T ((\psi_t, \varphi)_{L^2} + (\boldsymbol{\sigma}\nabla\psi, \nabla\varphi)_{L^2} + (a\psi + bE_\delta(\psi), \varphi)_{L^2})\, dt = \int_0^T (f, \varphi)_{L^2}\, dt \tag{10.108}$$

for every $\varphi \in L^2(I_T; V)$. We want to prove the following theorem:

Theorem 10.10 (linearization). *There exists a unique strong solution to equation (10.108). Moreover,*

$$\|\psi\|_{W(I_T)} \leq C\|f\|_{L^2(Q_T)} \tag{10.109}$$

with C independent of ψ and f.

Proof. First we replace ψ by the new variable $\bar{\psi} = e^{-\gamma t}\psi$, $\gamma > 0$. The equation for $\bar{\psi}$ becomes

$$\bar{\psi}_t - \operatorname{div}(\boldsymbol{\sigma}\nabla\bar{\psi}) + (a + \gamma)\bar{\psi} + bE_{\delta+\gamma}(\bar{\psi}) = e^{-\gamma t}f. \tag{10.110}$$

We use the Contraction Mapping Theorem A.21 in $L^2(Q_T)$. Fix $\psi_0 \in L^2(Q_T)$, choose $\gamma > 1 + \|a\|_{L^\infty(Q_T)}$ and solve the equation

$$\bar{\psi}_t - \operatorname{div}(\boldsymbol{\sigma}\nabla\bar{\psi}) + (a + \gamma)\bar{\psi} = -bE_{\delta+\gamma}(\psi_0) + e^{-\gamma t}f \tag{10.111}$$

with

$$\bar{\psi}(0) = 0 \text{ and } \boldsymbol{\sigma}\nabla\bar{\psi} \cdot \mathbf{n} = 0 \text{ on } \Gamma \times I_f. \tag{10.112}$$

In weak form,

$$\int_0^T \left(\bar{\psi}_t, \varphi\right)_{L^2} + \left(\boldsymbol{\sigma}\nabla\bar{\psi}, \nabla\varphi\right)_{L^2} + \left((a+\gamma)\,\bar{\psi}, \varphi\right)_{L^2} dt = \int_0^T \left(e^{-\gamma t} f - bE_{\delta+\gamma}\left(\psi_0\right), \varphi\right)_{L^2} dt \tag{10.113}$$

or, equivalently,

$$\int_0^T \left(\bar{\psi}_t + \mathcal{A}\bar{\psi} + (a+\gamma)\,\bar{\psi}, \varphi\right)_{L^2} dt = \int_0^T \left(e^{-\gamma t} f - bE_{\delta+\gamma}\left(\psi_0\right), \varphi\right)_{L^2} dt \tag{10.114}$$

for all $\varphi \in L^2\left(I_f; V\right)$. We have, by Cauchy-Schwarz inequality,

$$\left\|E_{\delta+\gamma}\left(\psi_0\right)\right\|^2_{L^2(Q_T)} = \int_{Q_T} \left(\int_0^t e^{-(\delta+\gamma)(t-s)} \psi_0\left(s, \mathbf{x}\right) ds\right)^2 d\mathbf{x}dt \leq \frac{T}{2(\delta+\gamma)} \left\|\psi_0\right\|^2_{L^2(Q_T)}. \tag{10.115}$$

Thus Theorem 7.1 and Remark 7.2 give existence and uniqueness of a weak solution $\bar{\psi} \in L^\infty\left(I_T; V\right)$, with $\bar{\psi}_t \in L^2\left(I_T; L^2\right)$.

In particular, testing (10.113) with $\varphi = \bar{\psi}$, since $\bar{\psi}\left(0\right) = 0$, $a + \gamma > 1$,

$$\left(\boldsymbol{\sigma}\nabla\bar{\psi}, \nabla\bar{\psi}\right)_{L^2} \geq c_0 \left\|\nabla\bar{\psi}\right\|^2_{L^2} \text{ and } \int_0^T \left(\bar{\psi}_t, \bar{\psi}\right)_{L^2} dt = \frac{1}{2} \left\|\bar{\psi}\left(T\right)\right\|^2_{L^2},$$

we can write

$$\begin{aligned} \left\|\bar{\psi}\right\|_{L^2(Q_T)} &\leq \left\|f\right\|_{L^2(Q_T)} + \left\|b\right\|_{L^\infty(Q_T)} \left\|E_{\delta+\gamma}\left(\psi_0\right)\right\|_{L^2(Q_T)} \\ &\leq \left\|f\right\|_{L^2(Q_T)} + \sqrt{\frac{T}{2(\delta+\gamma)}} \left\|b\right\|_{L^\infty(Q_T)} \left\|\psi_0\right\|_{L^2(Q_T)}. \end{aligned} \tag{10.116}$$

In this way we have defined a map $S : L^2(Q_T) \to L^2(Q_T)$, $\psi_0 \mapsto \bar{\psi} = S\left(\psi_0\right)$, where $S\left(\psi_0\right)$ is the solution to system (10.111), (10.112). From (10.115) and (10.116) we have

$$\left\|S\left(\psi_0\right) - S\left(\psi_1\right)\right\|_{L^2(Q_T)} \leq \sqrt{\frac{T}{2(\delta+\gamma)}} \left\|b\right\|_{L^\infty(Q_T)} \left\|\psi_0 - \psi_1\right\|_{L^2(Q_T)} \leq \frac{1}{2} \left\|\psi_0 - \psi_1\right\|^2_{L^2(Q_T)}$$

provided γ is large enough to have, say $\sqrt{\frac{T}{2(\delta+\gamma)}} \left\|b\right\|_{L^\infty(Q_T)} \leq \frac{1}{2}$.

Therefore S is a strict contraction in $L^2(Q_f)$ and has a unique fixed point $\bar{\psi}^*$. Notice that, from (10.116) we deduce

$$\left\|\bar{\psi}^*\right\|_{L^2(Q_T)} \leq 2 \left\|f\right\|_{L^2(Q_T)}. \tag{10.117}$$

Together with the parabolic estimate in Remark 7.2, this gives

$$\left\|\bar{\psi}^*_t\right\|^2_{L^2(Q_T)} \leq C_0 \left\|f\right\|^2_{L^2(Q_T)}.$$

Finally, from equation (10.114), we infer $\mathcal{A}\bar{\psi}^* \in L^2\left(I_f; L^2\right)$ and

$$\left\|\mathcal{A}\bar{\psi}^*\right\|_{W(I_T)} \leq C_1 \left\|f\right\|_{L^2(Q_T)}.$$

The constants C_0, C_1 depend, in general, only on T, c_0, Λ, δ, $\left\|a\right\|_{L^\infty(Q_T)}$ and $\left\|b\right\|_{L^\infty(Q_T)}$. Going back to $\psi^* = e^{\gamma t}\bar{\psi}^*$ we deduce that $\psi^* \in W\left(I_T\right)$ and (10.109) holds. □

Remark 10.11. The adjoint system admits a unique solution. Indeed equation (10.95) shares the same structure of problem (10.107), which is well-posed according to Theorem 10.10. •

10.2.6 Numerical Approximation

We show some numerical results obtained on a simplified version of the optimal time control problem (10.72), inspired by the problem addressed by Kunisch et al. in [163]. This problem is posed on a fixed time horizon $(0, t_f)$ at the end of which defibrillation must be achieved, while the defibrillation pulse (to be controlled) is applied on the first part $(0, \bar{T})$ of the time interval, with $\bar{T} \in (0, t_f)$ being part of the optimization process. Moreover, to reach an effective defibrillation at time $t = t_f$, we aim at bringing as much tissue to the resting state as possible.

Hence, in this section we assume that $\bar{T}$ is also fixed; as a result, we deal with control functions acting on $(0, t_f)$, with the assumption that $\mathbf{u}(t) = \mathbf{u}(\bar{T})$ in the interval $(\bar{T}, t_f)$. As control function, we assume that similarly to (10.69) the current is induced through N electrode plates Ω_j, $j = 1, \ldots, N$, each one delivering a pulse $u_j = u_j(t)$. Hence, we minimize the cost functional

$$\tilde{J}(v, w, \mathbf{u}) = \frac{\mu}{2} \int_\Omega |v(t_f)|^2 d\mathbf{x} + \frac{\alpha}{2} \int_0^{t_f} |\mathbf{u}(t)|^2 dt \tag{10.118}$$

under the constraints $\mathbf{u} \in \mathcal{U}_{ad}$, where

$$\mathcal{U}_{ad} = \left\{ \mathbf{u} \in \mathcal{U} : |u_k(t)| \le M_j,\ k = 1, \ldots, N,\ 0 \le t \le \bar{T},\ \mathbf{u} = \mathbf{u}(\bar{T}),\ \bar{T} \le t \le t_f \right\}. \tag{10.119}$$

The first-order optimality conditions for the OCP (10.66), (10.119), (10.118), can be recovered similarly to the time-distributed control problem of Sect. 7.5.2; see Exercise 6 for further details. In particular, the adjoint problem is given by the following backward in time, linearized monodomain system:

$$\begin{cases} p_t + \operatorname{div}(\boldsymbol{\sigma}\nabla p) - (I_{ion})'_v(v, w)\, p - g'_v(v, w) q = 0 & \text{in } Q_{t_f} \\ q_t - (I_{ion})'_w(v, w)\, p - g'_w(v, w) q = 0 & \text{in } Q_{t_f} \\ \boldsymbol{\sigma}\nabla p \cdot \mathbf{n} = 0 & \text{on } \Gamma \times I_f \\ p(t_f) = \mu v(t_f),\ w(t_f) = 0 & \text{in } \Omega, \end{cases} \tag{10.120}$$

whereas the gradient of the reduced cost functional $J(\mathbf{u}) = \tilde{J}(v(\mathbf{u}), w(\mathbf{u}), \mathbf{u})$, according to the formula (10.103) (read in terms of the original time variable) is provided by

$$(\nabla J(\mathbf{u}))_j = \alpha u_j + \int_{\Omega_j} p, \qquad j = 1, \ldots, N. \tag{10.121}$$

For the numerical solution of the OCP (10.66), (10.118), (10.71), we use an iterative method for the minimization of the cost functional and the Galerkin finite element method for the numerical discretization of both the state and the adjoint equation. Numerical results obtained for problem (10.72) can be found, e.g., in [161, 163], where a trust region semismooth Newton method has been employed for the stabilization of a reentry wave in minimal time and with minimal energy input.

Numerical discretization of state and adjoint equations

We discuss a possible strategy to solve the state problem, which can also be exploited to approximate the adjoint problem. This latter indeed shares the same structure of the state problem. It is a linear, backward in time, PDE, whereas the state problem is a time-dependent nonlinear PDE (the monodomain equation) coupled to a system of ODEs. The ODE system has to be solved at each node of the spatial discretization used to approximate the monodomain equation.

We apply the Galerkin-FE method on two finite-dimensional spaces $V_h \subset V = H^1(\Omega)$, $W_h \subset W = L^2(\Omega)$, of (usually very large) dimension $\dim(V_h) = \dim(W_h) = N_h$; as usual, h denotes a parameter related to the mesh size of the computational grid. The Galerkin-FE approximation of the state system (10.73)–(10.75) thus reads as follows: given $\mathbf{u} \in \mathcal{U}_{ad}$, for each $t \in (0, t_f)$, we seek $(v_h, w_h) = (v_h(t), w_h(t)) \in V_h \times W_h$ such that

$$\begin{cases} \displaystyle\int_\Omega (\partial_t v_h + I_{ion}(v_h, w_h))\,\varphi + \int_\Omega \boldsymbol{\sigma}\nabla v_h \cdot \nabla\varphi = \int_\Omega I_e(\mathbf{u})\varphi & \forall \varphi \in H^1(\Omega), \\ \displaystyle\int_\Omega (\partial_t w_h + g(v_h, w_h))\,\psi = 0 & \forall \psi \in L^2(\Omega), \\ v_h(0) = v_{h0}, \quad w_h(0) = w_{h0}. \end{cases} \tag{10.122}$$

Here we denote by $v_{h0} \in V_h$, $w_{h0} \in W_h$ two suitable approximations of the initial conditions.

We can now derive the matrix formulation corresponding to the Galerkin-FE approximation (10.122). Let us denote by $\{\varphi_i\}_{i=1}^{N_h}$ the Lagrangian basis of the FE space V_h, so that we can express the FE approximation to $v(\mathbf{x}, t)$ as

$$v_h(\mathbf{x}, t) = \sum_{i=1}^{N_h} v_{h,i}(t)\varphi_i(\mathbf{x}); \tag{10.123}$$

as a matter of fact, at each $t \in (0, t_f)$, $\mathbf{v}(t) \in \mathbb{R}^{N_h}$ will denote the vector of the degrees of freedom appearing in (10.123). Similarly, we denote by $\mathbf{w}(t) \in \mathbb{R}^{N_h}$, for each $t \in (0, t_f)$, the vector of the nodal values of $w_h(\mathbf{x}, t)$. Formally speaking, $w_h(\mathbf{x}, t)$ cannot be expressed as a function of V_h; nevertheless, for the sake of algebraic calculation, its vector representation has indeed the same dimension of $\mathbf{v}(t) \in \mathbb{R}^{N_h}$. Problem (10.122) can thus be rewritten as: given $\mathbf{v}(0) = \mathbf{v}_0$, $\mathbf{w}(0) = \mathbf{w}_0$, solve

$$\begin{cases} \mathbb{M}\dot{\mathbf{v}}(t) + \mathbb{A}\mathbf{v}(t) + \mathbf{I}_{ion}(\mathbf{v}(t), \mathbf{w}(t)) = \mathbf{I}_e(\mathbf{u}) \\ \dot{\mathbf{w}}(t) + \mathbf{g}(\mathbf{v}(t), \mathbf{w}(t)) = \mathbf{0}, \end{cases} \qquad t \in (0, t_f). \tag{10.124}$$

Here $\mathbb{M}_h \in \mathbb{R}^{N_h \times N_h}$ and $\mathbb{A}_h \in \mathbb{R}^{N_h \times N_h}$ denote the mass matrix and the stiffness matrix,

$$(\mathbb{M})_{ij} = \int_\Omega \varphi_i\varphi_j, \qquad (\mathbb{A})_{ij} = \int_\Omega \boldsymbol{\sigma}\nabla\varphi_i \cdot \nabla\varphi_j,$$

respectively, for $1 \le i, j \le N_h$; the source term $\mathbf{I}_e \in \mathbb{R}^{N_h}$ is given by

$$(\mathbf{I}_e(\mathbf{u}))_j = \int_\Omega I_e(\mathbf{u})\varphi_j$$

while the nonlinear terms $\mathbf{I}_{ion}, \mathbf{g} \in \mathbb{R}^{N_h}$ are given by

$$(\mathbf{I}_{ion}(\mathbf{v}, \mathbf{w}))_j = \int_\Omega I_{ion}(\mathbf{v}, \mathbf{w})\varphi_j, \quad (\mathbf{g}(\mathbf{v}, \mathbf{w}))_j = \int_\Omega g(\mathbf{v}, \mathbf{w})\varphi_j,$$

respectively. Regarding the treatment of nonlinear terms and time discretization, we use a semi-implicit, first order, one-step scheme [67].

Given a uniform partition of $(0, t_f)$ in subintervals (t^n, t^{n+1}) of length $\Delta t = t^{n+1} - t^n$, $n = 0, 1, \ldots, N_t - 1$, with $N_t\Delta t = t_f$, at each time-step t^{n+1} the nonlinear vector $\mathbf{I}_{ion,h}$ is evaluated around the solution already computed at time $t^{(n)}$. This decouples the PDE from the ODE leading to a linear system to be solved at each time step. Moreover, a ionic current interpolation strategy is used to evaluate the ionic current term, so that only the nodal values are used to build a (piecewise linear) interpolant of the ionic current.

The control vector $\mathbf{u} = \mathbf{u}(t)$ is piecewise constant in time, with values $\mathbf{u}^n \in \mathbb{R}^N$ at each time step t^n, $n = 1, \ldots, N_t$; hence, we have

$$\mathbf{u}(t) = \sum_{n=1}^{N_t} \mathbf{u}^n \chi_{(t^{n-1},t^n]}(t);$$

its discrete counterpart is a vector $\mathbf{u}^n \in \mathbb{R}^N$, $n = 1, \ldots, N_t$.

In conclusion, given $\{\mathbf{u}^n\}_{n=1}^{N_t}$, solving the state problem requires to find $\mathbf{v}^{n+1}$ and $\mathbf{w}^{n+1}$ such that $\mathbf{v}(0) = \mathbf{v}_0$, $\mathbf{w}(0) = \mathbf{w}_0$ and, for $n = 0, \ldots, N_t - 1$,

$$\begin{cases} \dfrac{\mathbf{w}^{n+1} - \mathbf{w}^n}{\Delta t} + \mathbf{g}(\mathbf{v}^n, \mathbf{w}^{n+1}) = \mathbf{0}, \\ \mathbb{M}\dfrac{\mathbf{v}^{n+1} - \mathbf{v}^n}{\Delta t} + \mathbb{A}\mathbf{v}^{n+1} + \mathbf{I}_{ion}(\mathbf{v}^n, \mathbf{w}^{n+1}) - \mathbf{I}_e(\mathbf{u}^{n+1}) = \mathbf{0}. \end{cases} \tag{10.125}$$

The major computational costs are due to assembling the terms $\mathbf{I}_{ion}$ and $\mathbf{g}$ at each time step and by the solution of the linear system $(10.125)_2$. On its turn, the time step Δt is required to be sufficiently small to ensure the stability (and the convergence) of the method. Once the state problem has been solved, the cost functional (10.118) can be evaluated as

$$\tilde{J}(v_h, w_h, \mathbf{u}) = \frac{\mu}{2}(\mathbf{v}^{N_t})^\top \mathbb{M}(\mathbf{v}^{N_t}) + \frac{\alpha}{2}\Delta t \sum_{n=1}^{N_t-1} \left(|\mathbf{u}^n|^2 + (|\mathbf{u}^{n+1}|^2\right).$$

Numerical optimization

Regarding the minimization of the cost functional $J_h(\mathbf{u}) = \tilde{J}(v_h(\mathbf{u}), w_h(\mathbf{u}), \mathbf{u})$, an iterative method with projection is employed exploiting a limited-memory BFGS (LM-BFGS) to compute the descent direction, through a suitable approximation of the Hessian matrix of the cost functional. A backtracking algorithm is exploited to update the step length τ_k so that it fulfills the Wolfe conditions, according to Algorithm 3.3; in particular, $\sigma = 0.5$ and $\rho = 0.2$ have been chosen. The box constraints on the control, $\mathbf{u} \in \mathcal{U}_{ad}$, have been taken into account by projecting, at each optimization step, the updated control on the set of admissible controls. As stopping criterion, we require that

$$|\nabla J_h(\mathbf{u}_k)|/|\nabla J_h(\mathbf{u}_0)| < \varepsilon, \tag{10.126}$$

for a given tolerance $\varepsilon > 0$. At each step of the BFGS method, we solve the state monodomain system (10.66), from which we obtain the state variables and the value of the cost functional. Then, we solve the adjoint problem (10.120), similarly to the state problem, and finally we evaluate the gradient of the cost functional. The control is then updated according to the computed descent direction, and then projecting the new iterate onto $\mathcal{U}_{ad}$.

Numerical results

We now present the numerical results related with the optimal electrical defibrillation problem. A mesh with 16,384 triangular elements and 8,385 vertices has been built over the domain $\Omega = (0, 2) \times (0, 0.8)$ reported in Fig. 10.5. On the time interval $(0, 70)$ (in milliseconds) a partition of $N_t = 1750$ time intervals of length $\Delta t = 0.04$ has been introduced.

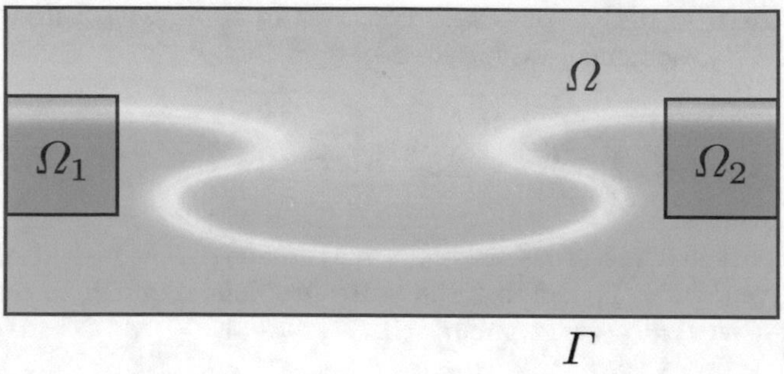

Fig. 10.5 The domain Ω and the subdomains $\Omega_{1,2}$ where the control acts simulating an electrical defibrillation. On the background, the initial condition on the trasmembrane potential v_0

The initial condition (v_0, w_0) describes a reentry wave of the so-called *figure of eight* type, and is constructed by the usual S1-S2 protocol (see, e.g., [5]) as follows [163]. Start by exciting the lower edge of the domain by setting $v(\mathbf{x}, 0) = 101$ if $\{(x_1, x_2) : x_2 < 1/160\}$ and $v(\mathbf{x}, 0) = 0$ otherwise, with $w(\mathbf{x}, 0) = 0$. Then compute the uncontrolled solution until $t = 130$ ms using a fixed step size $\Delta t = 0.1$ ms. The solution describes a planar wave front travelling from the bottom up. As soon as the centre gets excitable again, a second stimulus is based on a circle around the midpoint with radius 0.3 for 2 ms, that is, by setting $I_e = 200\chi_{\Omega_{S2}}(\mathbf{x})\chi_{[130,132]}(t)$ with $\Omega_{S2} = B_{0.3}(1, 0.4)$. We carry on the simulation without any further stimulus up to $T = 217$ ms; the resulting $v(\mathbf{x}, T)$ and $w(\mathbf{x}, T)$ are kept as initial condition (v_0, w_0) of the state problem.

Regarding the state (and the adjoint) problem, the choice of parameters is inspired by [69], where further details on the adimensionalization procedure used in this example can also be found. In particular, we take $\boldsymbol{\sigma} = diag(3 \cdot 10^{-3}, 3.1525 \cdot 10^{-4})$ and

$$\eta_0 = 1.5, \quad \eta_1 = 4.4, \quad \eta_2 = 0.012, \quad \eta_3 = 1.0, \quad v_{th} = 13, \quad v_{pk} = 100$$

in the definition of the functions I_{ion} and g. We consider a symmetric case, in which it is possible to defibrillate with just one control pulse, that is, $N = 1$, acting on the control domain $\Omega_1 \cup \Omega_2 = (0, 0.25) \times (0.3, 0.55) \cup (1.75, 2) \times (0.3, 0.55)$ (see Fig. 10.5).

Regarding the optimal control problem, we set $\bar{T} = 34.12$ ms, $t_f = 64$ ms, $\alpha = 10^{-3}$, $\mu = 10^3$ and the initial control to a constant value, $u = -50$; for the bound constraints, we set $M_1 = 100$. The LM-BFGS method takes 13 steps to fulfill the stopping criterion (10.126) with the tolerance set to $\varepsilon = 10^{-2}$; the value of the cost functional decreases from the initial value

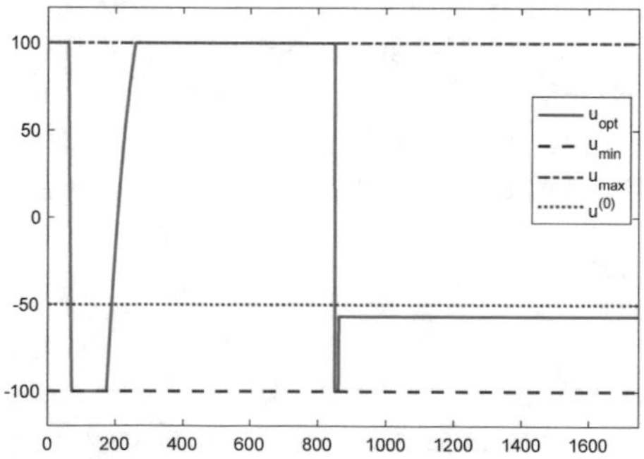

Fig. 10.6 Optimal control, initial guess, control constraints as functions of time t^n, $n = 1, \ldots, 1750$. Time $\bar{T} = 34.12$ ms corresponds to $n = 853$

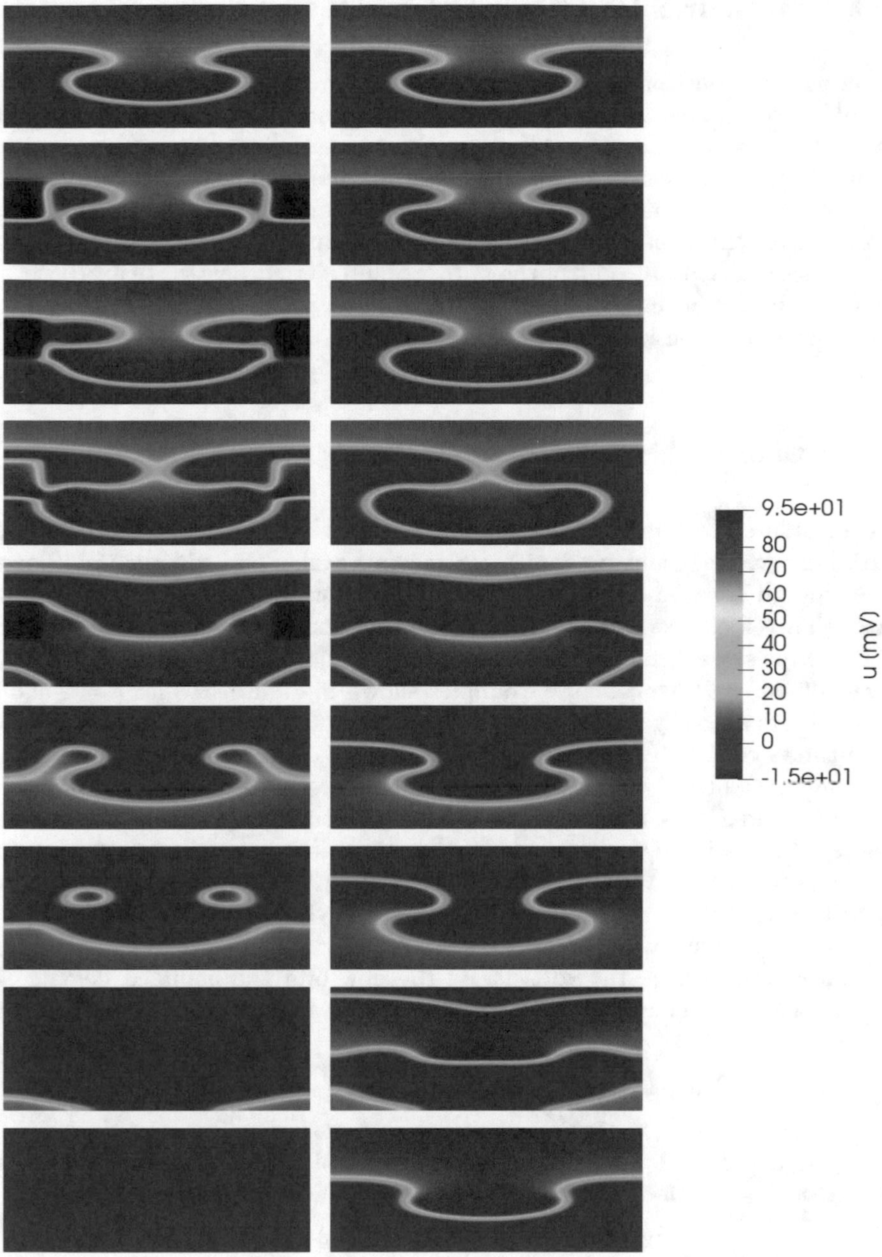

Fig. 10.7 From top to bottom: transmembrane potential v at time instants $t = 0, 50, 100, 250, 500, 1000, 1100, 1500, 1750$ ms with optimal control (left) and without control (right)

$J^{(0)} = 1.5775 \cdot 10^{10}$ to $\hat{J} = 5.4094 \cdot 10^3$, thus being reduced by a factor of about 99%. The profile of the computed optimal control is reported in Fig. 10.6, while a comparison between the uncontrolled solution (in terms of transmembrane potential) and the optimal solution is shown in Fig. 10.7, where the suppression of the reentry wave through the action of the control is clearly visible. Indeed, the defibrillation at the final time $t = t_f$ is effective since almost the whole tissue is at the resting state (left column); conversely, without defibrillation, the reentry wave is still present (right column).

10.3 Optimal Dirichlet Control of Unsteady Navier-Stokes Flows

In this section we consider an optimal flow control problem inspired by the problem of flow regularization considered in [32]. We study a mixed initial-boundary value problem for the incompressible unsteady Navier-Stokes equations in two dimensions, where we prescribe homogeneous Neumann conditions on a portion of the boundary, Dirichlet conditions (both homogeneous and non-homogeneous) on the velocity on another boundary portion, and a velocity control on the remaining portion of the domain boundary. The goal is to operate a dynamic control (that is, to control the flow in time) so that flow vorticity (both in space and time) is minimized. Similar problems, dealing however with Dirichlet conditions on the whole boundary, have been considered, e.g., in [122, 123, 97, 158].

10.3.1 Problem Formulation

We first formulate the state system, given by the unsteady Navier-Stokes equations, in primitive variables (velocity and pressure), for incompressible and viscous fluid flows. For this state system, we consider a minimization problem for quadratic functionals of the velocity field through a velocity control operated dynamically on a portion of the inflow boundary. For concrete examples, we denote by $\Omega \subset \mathbb{R}^d$ the domain occupied by the fluid and by $\Gamma = \partial\Omega$ its boundary. We refer to the concrete examples shown in Figure 10.8 for $d = 2$ and $d = 3$. The time interval is denoted by $(0, T]$, for a given final time $T > 0$; we also denote by $Q = \Omega \times (0, T]$ the space-time cylinder. The inflow boundary $\Gamma_1 = \Gamma_c \cup \Gamma_{in}$ is composed by Γ_c, the control boundary (a segment if $d = 2$, a disc if $d = 3$) and by Γ_{in} the inlet where Dirichlet data are assigned; Γ_{out} and Γ_w are the outflow and the wall boundary portions, where zero stress and zero velocity are assigned, respectively. Unless explicitly stated, notation conforms to those introduced in Sect. 10.1. We set an inflow parabolic profile on Γ_{in} – simulating, e.g., a fully developed flow through a planar duct if $d = 2$, or a cylinder if $d = 3$ – setting its inclination by prescribing the value of the angle $\eta \in (-\pi/2, \pi/2)$.

Our goal is to minimize the vorticity of the flow over the bounded domain $\Omega \subset \mathbb{R}^d$, by controlling a Dirichlet inflow $\boldsymbol{u} = \boldsymbol{u}(\mathbf{x}, t)$. Hence, we want to minimize the cost functional

$$\tilde{J}(\boldsymbol{v}, \boldsymbol{u}) = \frac{1}{2} \int_{Q_T} \xi |\nabla \times \boldsymbol{v}|^2 \, d\mathbf{x}dt + \frac{\tau}{2} \int_0^T \int_{\Gamma_c} |\nabla \boldsymbol{u}|^2 \, d\sigma dt \tag{10.127}$$

where $\tau \geq 0$, $\boldsymbol{u} \in \mathcal{U} = L^2(0, T; H_0^1(\Gamma_c)^d)$ is a control variable and $(\boldsymbol{v}, \pi) = (\boldsymbol{v}(\boldsymbol{u}), \pi(\boldsymbol{u})) \in X \times Q$ is a weak solution of the following unsteady Navier-Stokes system:

$$\begin{cases} \boldsymbol{v}_t - \nu \Delta \boldsymbol{v} + (\boldsymbol{v} \cdot \nabla)\boldsymbol{v} + \nabla \pi = \mathbf{0} & \text{in } Q_T \\ \operatorname{div} \boldsymbol{v} = 0 & \text{in } Q_T \\ \boldsymbol{v} = \boldsymbol{v}_{in} & \text{on } S_{in} \;\; = \Gamma_{in} \times (0, T) \\ \boldsymbol{v} = \boldsymbol{u} & \text{on } S_c \;\;\; = \Gamma_c \times (0, T) \\ \boldsymbol{v} = \mathbf{0} & \text{on } S_w \;\; = \Gamma_w \times (0, T) \\ \pi \mathbf{n} - \nu \dfrac{\partial \boldsymbol{v}}{\partial n} = \mathbf{0} & \text{on } S_{out} = \Gamma_{out} \times (0, T) \\ \boldsymbol{v}(\mathbf{x}, 0) = \boldsymbol{h} & \text{in } \Omega \end{cases} \tag{10.128}$$

where $\boldsymbol{v}$ and π denote the fluid velocity and pressure, respectively. Here $\xi = \xi(\mathbf{x})$ is a weight function emphasizing the effect of vorticity far from the boundaries Γ_c and Γ_{in}.

Moreover, let

$$\mathcal{U}_{ad} = \{\boldsymbol{u} \in L^2\left(0, T, H_0^1(\Gamma_c)^d\right) : \|\boldsymbol{u}(t)\|_{H_0^1(\Gamma_c)^d} \le \eta \text{ a.e. } t \in (0,T)\}$$

be the set of admissible controls, for a given constant $\eta > 0$. Let us now define

$$\boldsymbol{g}(t) = \boldsymbol{g}(t;\boldsymbol{u}) = \begin{cases} \boldsymbol{u}(t) & \text{on } \Gamma_c \\ \boldsymbol{0} & \text{on } \Gamma_w \\ \boldsymbol{v}_{in}(t) & \text{on } \Gamma_{in}. \end{cases}$$

Assuming that $\boldsymbol{v}_{in} \in L^2(0,T,H_{00}^{1/2}(\Gamma_{in})^d)$, we have that $\boldsymbol{g}|_{\Gamma_1} \in L^2(0,T,\mathcal{T}_1)$, being $\mathcal{T}_1 = H_{00}^{1/2}(\Gamma_1)^d$, since $\boldsymbol{u} \in L^2(0,T,H_0^1(\Gamma_c)^d)$ and $H_0^1(\Gamma_c)^d \subset H_{00}^{1/2}(\Gamma_c)^d$ with continuous injection.

If η, v_{in} and h are sufficiently small, the well-posedness of a suitable weak formulation of problem (10.128) can be inferred following what shown in, e.g., [160].

To write the optimality conditions, we take advantage of the setting already introduced in Sect. 7.6 (regarding the optimal control of unsteady Stokes equations) and Sect. 10.1 (concerning the case of steady Navier-Stokes equations). In particular, to handle the inhomogeneous Dirichlet conditions, we rewrite the state problem into a parabolic saddle point formulation, introducing a new variable $\boldsymbol{s} \in \mathcal{T}_1$, representing the boundary stress $\pi\mathbf{n} - \nu\frac{\partial \boldsymbol{v}}{\partial n}$ on Γ_1, and appending a multiplier $\boldsymbol{s} \in \mathcal{T}_1^*$ to the Dirichlet boundary conditions. This exempts us to use a lifting for the Dirichlet data and divergence free spaces for the velocity. Indeed, we choose

$$X = \left\{\varphi \in H^1(\Omega)^d : \varphi_{|\Gamma_1} \in H_{00}^{1/2}(\Gamma_1)^d\right\}$$

as velocity space and $L^2(\Omega)$ as pressure space, respectively. Moreover, we assume that $\boldsymbol{h} \in X$ is such that $\text{div}\boldsymbol{h} = 0$ in Ω, and set $H = L^2(\Omega)^d$.

The saddle point formulation of the problem is obtained by: multiplying the first equation in (10.128) by a test function $\boldsymbol{\varphi} \in X$, integrating over Ω, integrating by parts and taking into account the boundary conditions, then by multiplying the other equations by test functions $\phi \in Q$, $\boldsymbol{\lambda} \in \mathcal{T}_1^*$, respectively, and integrating over Ω. Thus, we find the following formulation: *find* $\boldsymbol{v} \in L^2(0,T;X) \cap H^1(0,T;H)$, $\pi \in L^2(0,T;L^2(\Omega))$ *and* $\boldsymbol{s} \in L^2(0,T;\mathcal{T}_1^*)$ *such that, for a.e.* $t \in (0,T)$,

$$\begin{cases} (\boldsymbol{v}_t(t), \boldsymbol{\varphi})_H + e(\boldsymbol{v}(t), \boldsymbol{\varphi}) + c(\boldsymbol{v}(t), \boldsymbol{v}(t), \boldsymbol{\varphi}) + b(\pi(t), \boldsymbol{\varphi}) + \langle \boldsymbol{s}(t), \boldsymbol{\varphi}\rangle_{\Gamma_1} = 0 & \forall \boldsymbol{\varphi} \in X \\ b(\boldsymbol{v}(t), \phi) = 0 & \forall \phi \in L^2(\Omega) \\ \langle \boldsymbol{\lambda}, \boldsymbol{v}(t)\rangle_{\Gamma_1} = \langle \boldsymbol{\lambda}, \boldsymbol{g}(t)\rangle_{\Gamma_1} & \forall \boldsymbol{\lambda} \in \mathcal{T}_1 \\ (\boldsymbol{v}(0), \boldsymbol{\varphi})_H = (\boldsymbol{h}, \boldsymbol{\varphi})_H & \forall \boldsymbol{\varphi} \in X. \end{cases} \quad (10.129)$$

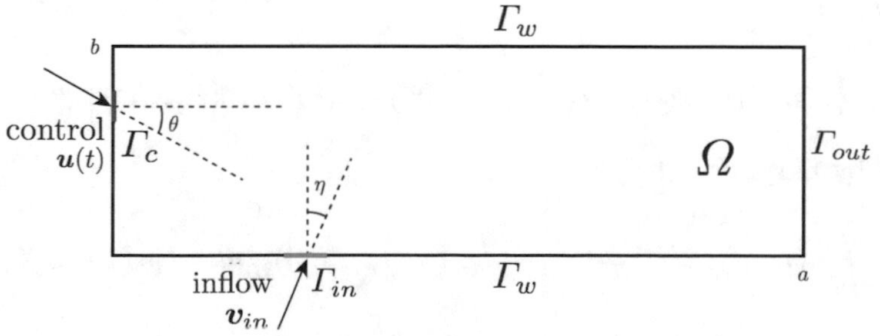

Fig. 10.8 The domain and its boundary portions

By grouping together the second and the third equation in (10.129), defining $Q = L^2(\Omega) \times \mathcal{T}_1^*$ and introducing the bilinear form $\tilde{b} : X \times Q \to \mathbb{R}$ as in (10.11), we obtain the following equivalent formulation of (10.129): *find* $\boldsymbol{v} \in L^2(0,T;X) \cap H^1(0,T;H)$, $(\pi, \boldsymbol{s}) \in L^2(0,T,Q) \times L^2(0,T;\mathcal{T}_1^*)$ *such that, for a.e.* $t \in (0,T)$,

$$\begin{cases} (\boldsymbol{v}_t(t), \boldsymbol{\varphi})_H + e(\boldsymbol{v}(t), \boldsymbol{\varphi}) + c(\boldsymbol{v}(t), \boldsymbol{v}(t), \boldsymbol{\varphi}) + \tilde{b}(\boldsymbol{\varphi}, (\pi(t), \boldsymbol{s}(t))) = 0 & \forall \boldsymbol{\varphi} \in X \\ \tilde{b}(\boldsymbol{v}(t), (\phi, \boldsymbol{\lambda})) = \langle G(t), (\phi, \boldsymbol{\lambda}) \rangle_{Q^*,Q} & \forall (\phi, \boldsymbol{\lambda}) \in Q \times \mathcal{T}_1^* \\ (\boldsymbol{v}(0), \boldsymbol{\varphi})_H = (\boldsymbol{h}, \boldsymbol{\varphi})_H & \forall \boldsymbol{\varphi} \in X, \end{cases} \tag{10.130}$$

with $G \in Q^*$ defined, for a.e. $t \in (0,T)$, by

$$\langle G(t), (\phi, \boldsymbol{\lambda}) \rangle_{Q^*,Q} = \langle \boldsymbol{\lambda}, \boldsymbol{g}(t) \rangle_{\Gamma_1}.$$

10.3.2 Optimality Conditions

Following what we did in Sect. 7.6 in the case of unsteady Stokes equations, and proceeding formally, we can recover the system of first-order necessary optimality conditions in weak form, introducing the Lagrangian $\mathcal{L} = \mathcal{L}((\boldsymbol{v}, \pi, \boldsymbol{s}), \boldsymbol{u}, (\boldsymbol{z}, q, \boldsymbol{\lambda}))$ given by

$$\mathcal{L}((\boldsymbol{v}, \pi, \boldsymbol{s}), \boldsymbol{u}, (\boldsymbol{z}, q, \boldsymbol{\lambda})) = \frac{1}{2} \int_{Q_T} \xi |\nabla \times \boldsymbol{v}|^2 \, d\mathbf{x} dt + \frac{\tau}{2} \int_0^T \int_{\Gamma_c} |\nabla \boldsymbol{u}|^2 \, d\sigma dt$$

$$- \int_0^T \Big\{ (\boldsymbol{v}_t(t), \boldsymbol{z}(t)) + e(\boldsymbol{v}(t), \boldsymbol{z}(t)) + c(\boldsymbol{v}(t), \boldsymbol{v}(t), \boldsymbol{z}(t)) + \tilde{b}(\boldsymbol{z}(t), (\pi(t), \boldsymbol{s}(t)) \Big\} dt$$

$$- \int_0^T \Big\{ \tilde{b}(\boldsymbol{v}(t), (q(t), \boldsymbol{\lambda}(t))) - \langle G(t), (q(t), \boldsymbol{\lambda}(t)) \rangle_{Q^*,Q} \Big\} dt,$$

where $\boldsymbol{z} \in L^2(0,T;X) \cap H^1(0,T;H)$ and $(\pi, \boldsymbol{\lambda}) \in L^2(0,T;L^2(\Omega)) \times L^2(0,T;\mathcal{T}_1^*)$ are the adjoint variables (or Lagrange multipliers) related to the constraints represented by the equations in the state problem (10.129).

We now differentiate the Lagrangian with respect to the state variables $\boldsymbol{v}$, π, $\boldsymbol{s}$ and set each derivative, evaluated at the optimal state and adjoint variables, equal to zero. We put $\hat{\mathcal{L}}'_{(\cdot)} = \mathcal{L}'_{(\cdot)}((\hat{\boldsymbol{v}}, \hat{\pi}, \hat{\boldsymbol{s}}), \hat{\boldsymbol{u}}, (\hat{\boldsymbol{z}}, \hat{q}, \hat{\boldsymbol{\lambda}}))$.

- $\hat{\mathcal{L}}'_{\boldsymbol{v}} \boldsymbol{w} = 0$ for all $\boldsymbol{w} \in L^2(0,T;X) \cap H^1(0,T;H)$, $\boldsymbol{w}(0) = \mathbf{0}$, gives

$$\hat{\mathcal{L}}'_{\boldsymbol{v}} \boldsymbol{w} = \int_0^T \int_\Omega \xi (\nabla \times \hat{\boldsymbol{v}}(t)) \cdot (\nabla \times \boldsymbol{w}(t)) d\mathbf{x} dt - \int_0^T \{ (\boldsymbol{w}_t(t), \hat{\boldsymbol{z}}(t))_H + e(\boldsymbol{w}(t), \hat{\boldsymbol{z}}(t) \} dt$$

$$- \int_0^T \Big\{ c(\boldsymbol{w}, \hat{\boldsymbol{v}}(t), \hat{\boldsymbol{z}}(t)) + c(\hat{\boldsymbol{v}}(t), \boldsymbol{w}, \hat{\boldsymbol{z}}(t)) + \tilde{b}(\boldsymbol{w}(t), (\hat{q}(t), \hat{\boldsymbol{\lambda}}(t)) \Big\} dt = 0. \tag{10.131}$$

Integrating by parts we get

$$\int_0^T (\boldsymbol{w}_t(t), \hat{\boldsymbol{z}}(t))_H \, dt = - \int_0^T (\boldsymbol{w}(t), \hat{\boldsymbol{z}}_t(t))_H \, dt + (\boldsymbol{w}(T), \hat{\boldsymbol{z}}(T))_H.$$

Choosing $\boldsymbol{w}(t) = a(t)\boldsymbol{\psi}$ with $a \in C^1([0,T])$, $a(0) = 0$, $\boldsymbol{\psi} \in X$, as a test function in (10.131), we derive the following equation, for a.e. $t \in (0,T)$:

$$-\left(\hat{\boldsymbol{z}}_t(t), \boldsymbol{\psi}\right)_H + e(\boldsymbol{\psi}, \hat{\boldsymbol{z}}(t)) + c(\boldsymbol{\psi}, \hat{\boldsymbol{v}}(t), \hat{\boldsymbol{z}}(t)) + c(\hat{\boldsymbol{v}}(t), \boldsymbol{\psi}, \hat{\boldsymbol{z}}(t)) + b(\boldsymbol{\psi}, \hat{q}) + \langle \hat{\boldsymbol{\lambda}}, \boldsymbol{\psi}\rangle_{\Gamma_1} = \int_\Omega \xi(\nabla \times \hat{\boldsymbol{v}}(t)) \cdot (\nabla \times \boldsymbol{\psi}) \qquad \forall \boldsymbol{\psi} \in X, \tag{10.132}$$

and the final condition

$$\hat{\boldsymbol{z}}(T) = 0. \tag{10.133}$$

- $\hat{\mathcal{L}}'_\pi r = 0$ for all $r \in L^2(Q_T)$ yields

$$\hat{\mathcal{L}}'_\pi r = \int_0^T \left(r(t), \operatorname{div}\hat{\boldsymbol{z}}(t)\right)_{L^2(\Omega)dt} = 0$$

which implies

$$\operatorname{div}\hat{\boldsymbol{z}} = 0 \quad \text{in } Q_T. \tag{10.134}$$

- $\hat{\mathcal{L}}'_{\boldsymbol{s}}\boldsymbol{\mu} = 0$ for all $\boldsymbol{\mu} \in L^2(0,T;\mathcal{T}_1^*)$ gives

$$\langle \boldsymbol{\mu}(t), \hat{\boldsymbol{z}}(t)\rangle_{\Gamma_1} = 0$$

which implies

$$\hat{\boldsymbol{z}} = \mathbf{0} \text{ on } \Gamma_1 \times (0,T). \tag{10.135}$$

Using (10.42), (10.43), and (10.44), and defining the rotation tensor $\nabla\hat{\boldsymbol{v}}(t) - (\nabla\hat{\boldsymbol{v}}(t))^\perp = \hat{\mathbf{W}}(t)$, we recognize that system (10.132), (10.133), (10.134) and (10.135) represents the saddle point formulation of the following adjoint problem:

$$\begin{cases} -\hat{\boldsymbol{z}}_t - \nu\,\Delta\hat{\boldsymbol{z}} - (\hat{\boldsymbol{v}}\cdot\nabla)\hat{\boldsymbol{z}} + (\nabla\hat{\boldsymbol{v}})^\top\hat{\boldsymbol{z}} + \nabla\hat{q} = \nabla\times(\xi\nabla\times\hat{\boldsymbol{v}}) & \text{in } Q_T \\ \operatorname{div}\hat{\boldsymbol{z}} = 0 & \text{in } Q_T \\ \hat{\boldsymbol{z}} = \mathbf{0} & \text{on } (\Gamma_0\cup\Gamma_1)\times(0,T) \\ \hat{q}\mathbf{n} - \nu\dfrac{\partial\hat{\boldsymbol{z}}}{\partial n} - (\hat{\boldsymbol{v}}\cdot\mathbf{n})\,\hat{\boldsymbol{z}}\xi + \hat{\mathbf{W}}\mathbf{n} = \mathbf{0} & \text{on } \Gamma_{out}\times(0,T) \\ \hat{\boldsymbol{z}}(\mathbf{x},T) = \mathbf{0} & \text{in } \Omega. \end{cases} \tag{10.136}$$

This latter is a backward in time Navier-Stokes problem with final condition, homogeneous Dirichlet/Neumann conditions and source term (for the momentum equation) equal to $\nabla\times(\xi\nabla\times\hat{\boldsymbol{v}}(t))$. We can also deduce that, regarding the multiplier,

$$\hat{\boldsymbol{s}}(t) = \hat{q}(t)\mathbf{n} - \nu\frac{\partial\hat{\boldsymbol{z}}(t)}{\partial n} + \xi\hat{\mathbf{W}}(\mathbf{t})\mathbf{n} \quad \text{on } \Gamma_1. \tag{10.137}$$

- We now differentiate the Lagrangian with respect to the control variable and set $\hat{\mathcal{L}}'_{\boldsymbol{u}}(\boldsymbol{k}-\hat{\boldsymbol{u}}) \geq 0$ for all $\boldsymbol{k} \in \mathcal{U}_{ad}$, obtaining the following variational inequality

$$\hat{\mathcal{L}}'_{\boldsymbol{u}}(\boldsymbol{k}-\hat{\boldsymbol{u}}) = \int_0^T (\tau\hat{\boldsymbol{u}}, \boldsymbol{k}-\hat{\boldsymbol{u}})_{H_0^1(\Gamma_c)^2}\,dt + \int_0^T \langle\hat{\boldsymbol{s}}, \boldsymbol{k}-\hat{\boldsymbol{u}}\rangle_{\Gamma_c}dt \geq 0 \qquad \forall\boldsymbol{k}\in\mathcal{U}_{ad}. \tag{10.138}$$

Then, introducing the embedding operator $B: \mathcal{T}_c^* \to H^{-1}(\Gamma_c)^d$ defined by

$$\langle B\boldsymbol{s}(t), \boldsymbol{k}\rangle_{H^{-1}(\Gamma_c)^d, H_0^1(\Gamma_c)^d} = \langle\boldsymbol{s}(t), \boldsymbol{k}\rangle_{\Gamma_c} \quad \forall\boldsymbol{k}\in H_0^1(\Gamma_c)^d, \text{ a.e. } t\in(0,T)$$

we find that

$$\nabla J(\hat{\boldsymbol{u}}) = \tau\hat{\boldsymbol{u}} + \hat{\boldsymbol{w}}$$

where, for a.e. $t\in(0,T)$, $\hat{\boldsymbol{w}}(t)\in H_0^1(\Gamma_c)^d$, is the unique weak solution to

$$-\Delta\hat{\boldsymbol{w}}(t) = B\hat{\boldsymbol{s}}(t) \quad \text{on } \Gamma_c. \tag{10.139}$$

Finally, by requiring that the derivatives of the Lagrangian with respect to the adjoint variables vanish, we obtain the state system (10.129).

10.3.3 Dynamic Boundary Action

To simplify the problem discussed so far, we can assume that the flow is controlled by suction and blowing on a part of the boundary, so that boundary control takes place through a parameterized, prescribed velocity distribution on the control boundary Γ_c. As we will see below, this also enables to turn the constrained OCP of the previous section into a more favorable unconstrained OCP.

Practically, the control flow is meant to enhance the flow emission from the outflow Γ_{out}. More specifically, following [32] we assume that the function $\boldsymbol{u}(t)$ is defined as

$$\boldsymbol{u}(t) = \boldsymbol{\psi}(\theta)u(t)$$

where:

- $\boldsymbol{\psi} : \mathcal{I}_{ad} \to \mathbb{R}^d$ is a (chosen) one-parameter family of velocity distributions on Γ_c, where $\mathcal{I}_{ad} \subset \mathbb{R}$ is a closed interval. For the case at hand, θ is a control quantity, representing the angle of incidence for an injected flow featuring a fixed, parabolic velocity profile on a flat part Γ_c of the boundary; in particular, $\boldsymbol{\psi}(\theta) \in H_0^1(\Gamma_c)^d$;
- $u : (0,T) \to \mathbb{R}$ is a control function, representing the (time-varying) amplitude of the injected flow.

We can also specify the choice of the cost functional (10.127), by considering

$$\tilde{J}(\boldsymbol{v}, u, \theta) = \frac{1}{2}\int_{Q_T} \xi |\nabla \times \boldsymbol{v}|^2 \, d\mathbf{x}dt + \frac{\tau}{2}\int_0^T (u(t))^2 dt + \frac{1}{\delta}\varphi(\theta) \tag{10.140}$$

where, in addition to the vorticity term and to the second (penalization) term, we handle the box constraint on the control variable θ (which must belong to $\mathcal{I}_{ad}$) by introducing a convex function $\varphi \in C^1(\mathbb{R})$ such that $\varphi = 0$ on $\mathcal{I}_{ad}$ and $\varphi > 0$ otherwise, and a second penalization parameter $\delta > 0$, typically small. Intuitively, φ forces θ to take values in $\mathcal{I}_{ad}$. In this way, instead of

$$\mathcal{U}_{ad} = \{(u, \theta) \, : \, u \in L^2(0,T), \; \theta \in \mathcal{I}_{ad}\}$$

we can take as space of admissible controls the (convex) set

$$\mathcal{U}_{ad} = \mathcal{U} = \{(u, \theta) \, : \, u \in L^2(0,T), \; \theta \in \mathbb{R}\}.$$

Hence, the OCP reads as follows:

$$\begin{gathered} \tilde{J}(\boldsymbol{v}, u, \theta) \to \min \\ \text{subject to (10.130)} \\ \boldsymbol{v} \in L^2(0,T;X) \cap H^1(0,T;H), \;\; (\pi, \boldsymbol{s}) \in L^2(0,T,Q) \times L^2(0,T;\mathcal{T}_1^*) \\ (u, \theta) \in \mathcal{U} = L^2(0,T) \times \mathbb{R}, \end{gathered} \tag{10.141}$$

where $\tilde{J}$ is given by (10.140). It is possible to show (see Exercise 7) the following result.

Theorem 10.11. *Let $(\hat{u}, \hat{\theta})$ be an optimal control and $(\hat{\boldsymbol{v}}, \hat{\pi}, \hat{\boldsymbol{s}})$ be the corresponding optimal state. The optimality system is given by*

- *the* state problem (10.130)*;*
- *the* adjoint problem*: for a.e. $t \in (0, T)$*

$$\begin{aligned} &-\left(\hat{\boldsymbol{z}}_t\left(t\right), \boldsymbol{\psi}\right)_H + e(\boldsymbol{\psi}, \hat{\boldsymbol{z}}(t)) + c\left(\boldsymbol{\psi}, \hat{\boldsymbol{v}}(t), \hat{\boldsymbol{z}}(t)\right) + c\left(\hat{\boldsymbol{v}}(t), \boldsymbol{\psi}, \hat{\boldsymbol{z}}(t)\right) \\ &\quad + \tilde{b}(\boldsymbol{\psi}, (\hat{q}(t), \hat{\boldsymbol{\lambda}}(t))) + \langle \hat{\boldsymbol{\lambda}}, \boldsymbol{\psi} \rangle_{\Gamma_1} = \textstyle\int_\Omega \xi(\nabla \times \hat{\boldsymbol{v}}(t)) \cdot (\nabla \times \boldsymbol{\psi}) \qquad \forall \boldsymbol{\psi} \in X \\ &\quad \tilde{b}(\hat{\boldsymbol{z}}(t), (\phi, \boldsymbol{\mu})) = 0 \qquad \forall (\phi, \boldsymbol{\mu}) \in Q \times \mathcal{T}_1^* \\ &\quad \left(\hat{\boldsymbol{z}}\left(T\right), \boldsymbol{\psi}\right)_H = (\boldsymbol{0}, \boldsymbol{\psi}) \qquad \forall \boldsymbol{\psi} \in X; \end{aligned} \tag{10.142}$$

- *the* Euler equations*:*

$$J'_u(\hat{u}, \hat{\theta})k = \int_0^T \tau\, \hat{u}\, k\, dt + \int_0^T k\,\langle \hat{\boldsymbol{s}}, \boldsymbol{\psi}(\hat{\theta}) \rangle_{\Gamma_c} dt = 0 \qquad \forall k \in L^2(0, T), \tag{10.143}$$

$$J'_\theta(\hat{u}, \hat{\theta}) = \frac{1}{\delta} \varphi'(\hat{\theta}) + \int_0^T \hat{u} \langle \hat{\boldsymbol{s}}, \nabla \boldsymbol{\psi}(\hat{\theta}) \rangle_{\Gamma_c} dt = 0, \tag{10.144}$$

where $J(u, \theta) = \tilde{J}(\boldsymbol{v}(u, \theta), u, \theta)$ is the reduced cost functional and

$$\nabla \boldsymbol{\psi} : \mathbb{R} \to \mathbb{R}^d, \qquad \nabla \boldsymbol{\psi}(\theta) = \left(\frac{\partial \psi_1}{\partial \theta}, \ldots, \frac{\partial \psi_d}{\partial \theta} \right).$$

10.3.4 Numerical Approximation

Numerical results are shown for different values of two characteristic parameters, namely the inflow inclination η and the Reynolds number $Re = d_1 U/\nu$, being U the maximum inflow velocity, ν the kynematic viscosity of the fluid, and d_1 the length of the inflow nozzle. In particular, for moderate Reynolds numbers we include stabilization terms in the solver for the state problem, and modify the system of optimality conditions accordingly.

Numerical discretization of state and adjoint equations

The state system consists of the unsteady Navier-Stokes equations, while the adjoint system is given by an unsteady Oseen problem (backward in time). Similarly to the steady case of Sect. 10.1.6, we rewrite both equations by introducing appropriate functional spaces and suitable lifting terms to take into account non-homogeneous boundary conditions; unless explicitly stated, all the notations required to set the algebraic problems is the same as the one introduced in Sect. 10.1.6.

The Galerkin-FE approximation of the state system (10.128) thus reads as follows: for each $t \in (0, T)$, we seek $(\boldsymbol{v}_h, \pi_h) = (\boldsymbol{v}_h(t), \pi_h(t)) \in X_h \times Q_h$ such that

$$\begin{cases} \displaystyle\int_\Omega \partial_t \boldsymbol{v}_h \cdot \boldsymbol{\varphi}_h + \tilde{e}(\boldsymbol{v}_h, \boldsymbol{\varphi}_h) + c\left(\boldsymbol{v}_h, \boldsymbol{v}_h, \boldsymbol{\varphi}_h\right) + b(\boldsymbol{\varphi}_h, \pi_h) = F_1 \boldsymbol{\varphi}_h & \forall \boldsymbol{\varphi}_h \in X_h \\ b(\boldsymbol{v}_h, \phi_h) = F_2 \phi_h & \forall \phi_h \in Q_h, \end{cases} \tag{10.145}$$

such that $\boldsymbol{v}_h(0) = \boldsymbol{v}_{h0}$, where now the discrete (lifting) function interpolating the Dirichlet data $\boldsymbol{v}_h^D(t) = \boldsymbol{v}_h^D(u(t)) \in X_h$ for each $t \in (0,T)$, depends on time, too. Here we denote by $\boldsymbol{v}_{h0} \in X_h$ a suitable approximation of the initial condition.

The matrix formulation corresponding to the Galerkin-FE approximation (10.145) is then a nonlinear system of ODEs. Indeed, by expressing

$$\boldsymbol{v}_h(t) = \sum_{i=1}^{N_v} v_{h,i}(t)\boldsymbol{\phi}_i^v, \qquad \pi_h(t) = \sum_{i=1}^{N_p} \pi_{h,i}(t)\phi_i^p, \tag{10.146}$$

the system (10.145) can be equivalently rewritten as: given $\mathbf{v}(0) = \mathbf{v}_0$, solve

$$\begin{cases} \mathbb{M}\dot{\mathbf{v}}(t) + \mathbb{E}\mathbf{v}(t) + \mathbb{C}(\mathbf{v}(t))\mathbf{v}(t) + \mathbb{B}^\top \boldsymbol{\pi}(t) = \mathbf{f}_1(t) \\ \mathbb{B}\mathbf{v}(t) = \mathbf{f}_2(t), \end{cases} \qquad t \in (0,T). \tag{10.147}$$

Here $\mathbf{v}_0 \in \mathbb{R}^{N_v}$ is the vector representation (over the FE basis) of $\boldsymbol{v}_{h0}$ and $\mathbb{M}$ is the (velocity) mass matrix, whose elements are given by

$$(\mathbb{M})_{ij} = \int_\Omega \boldsymbol{\phi}_j^v \cdot \boldsymbol{\phi}_i^v, \qquad 1 \le i,j \le N_v;$$

all the other matrices and vectors are defined similarly to what we have done in (10.59). To discretize the ODEs system (10.147) in time, we use the backward (or implicit) Euler method (see Sect. B.3 in the Appendix B), by introducing a uniform partition of $(0,T)$ in subintervals (t^n, t^{n+1}) of length $\Delta t = t^{n+1} - t^n$, $n = 0, 1, \dots, N_t - 1$, with $N_t = T/\Delta t$, and discretizing the time derivative by a backward finite difference: for $n = 0, 1, \dots$, given $\mathbf{v}^0 = \mathbf{v}_0$, solve

$$\begin{cases} \mathbb{M}\dfrac{\mathbf{v}^{n+1} - \mathbf{v}^n}{\Delta t} + \mathbb{E}\mathbf{v}^{n+1} + \mathbb{C}(\mathbf{v}^{n+1})\mathbf{v}^{n+1} + \mathbb{B}^\top \mathbf{p}^{n+1} = \mathbf{f}_1^{n+1} \\ \mathbb{B}\mathbf{v}^{n+1} = \mathbf{f}_2^{n+1}. \end{cases}$$

Equivalently, we obtain a nonlinear system of equations to be solved at each time step, under the form

$$\begin{bmatrix} \dfrac{\mathbb{M}}{\Delta t} + \mathbb{E} + \mathbb{C}(\mathbf{v}^{n+1}) & \mathbb{B}^\top \\ \mathbb{B} & 0 \end{bmatrix} \begin{bmatrix} \mathbf{v}^{n+1} \\ \mathbf{p}^{n+1} \end{bmatrix} = \begin{bmatrix} \mathbf{f}_1^{n+1} + \dfrac{\mathbb{M}}{\Delta t}\mathbf{v}^n \\ \mathbf{f}_2^{n+1} \end{bmatrix}. \tag{10.148}$$

At each time step, a nonlinear system must then be solved; here we rely on the Newton method for small Reynolds numbers, and on the fixed point iteration method for moderate Reynolds numbers; these iterations are initialized with the solution computed at the previous time step. The resulting linear system to be solved at each (Newton or fixed point) step is solved with the GMRES method, employing the Cahouet-Chabard preconditioner (see, e.g., [90]). For moderate Reynolds numbers, we also introduce a streamline-upwind Petrov-Galerkin (SUPG) stabilization in the case of large Reynolds numbers.

Regarding instead the numerical approximation of the adjoint variables, the adjoint system (10.142) is backward in time, and is linear with respect to both the adjoint velocity and pressure. Given the approximation $\boldsymbol{v}_h = \boldsymbol{v}_h(t)$ of the (state) velocity, the Galerkin-FE approximation of the adjoint system (10.142) thus reads as follows: for each $t \in (0,T)$, we seek $(\boldsymbol{z}_h, q_h) \in X_h \times Q_h$ such that

$$\begin{cases} -\displaystyle\int_\Omega \partial_t \boldsymbol{z}_h \cdot \boldsymbol{\varphi}_h + e(\boldsymbol{\varphi}_h, \boldsymbol{z}_h) + c(\boldsymbol{v}_h, \boldsymbol{z}_h, \boldsymbol{\varphi}_h) + c(\boldsymbol{z}_h, \boldsymbol{v}_h, \boldsymbol{\varphi}_h) \\ \qquad\qquad + b(\boldsymbol{\varphi}_h, q_h) = G_1(\boldsymbol{v}_h, \boldsymbol{\varphi}_h) \qquad \forall \boldsymbol{\varphi}_h \in X_h \\ b(\boldsymbol{z}_h, \phi_h) = 0 \qquad \forall \phi_h \in Q_h, \end{cases} \tag{10.149}$$

and that $z_h(T) = \mathbf{0}$. The right-hand side of the first equation is related with the derivative of the cost functional with respect to the state velocity, and for the case at hand, it reads as

$$G_1(\boldsymbol{v}_h, \boldsymbol{\varphi}_h) = \int_\Omega (\nabla \times \boldsymbol{v}_h)(\nabla \times \boldsymbol{\varphi}_h).$$

Proceeding similarly to the state problem, the system (10.149) can be equivalently rewritten as: given $\mathbf{z}(T) = \mathbf{0}$, solve

$$\begin{cases} -\mathbb{M}\dot{\mathbf{z}}(t) + \mathbb{E}^\top \mathbf{z}(t) + \mathbb{D}(\mathbf{v}(t))\,\mathbf{z}(t) + \mathbb{B}^\top \mathbf{q}(t) = \mathbf{g}_1(t; \mathbf{v}) \\ \mathbb{B}\mathbf{z}(t) = \mathbf{0}, \end{cases} \qquad t \in (0, T), \tag{10.150}$$

where $\mathbb{D}$ and $\mathbf{g}_1(t; \mathbf{v})$ (this latter now depending on t) are defined as in (10.59).

To discretize the ODEs system (10.150) in time, we use the backward (or implicit) Euler method. Hence, for $n = N_t, N_t - 1, \ldots$, given $\mathbf{z}^{N_t} = \mathbf{0}$, we solve

$$\begin{cases} \mathbb{M}\dfrac{\mathbf{z}^n - \mathbf{z}^{n+1}}{\Delta t} + \mathbb{E}^\top \mathbf{v}^n + \mathbb{D}(\mathbf{v}^n)\mathbf{z}^n + \mathbb{B}^\top \mathbf{q}^n = \mathbf{g}_1^n \\ \mathbb{B}\mathbf{z}^n = \mathbf{0} \end{cases}$$

At each time step, a linear system of equations has to be solved, under the form

$$\begin{bmatrix} \dfrac{\mathbb{M}}{\Delta t} + \mathbb{E}^\top + \mathbb{D}(\mathbf{v}^n) & \mathbb{B}^\top \\ \mathbb{B}_h & 0 \end{bmatrix} \begin{bmatrix} \mathbf{z}^n \\ \mathbf{q}^n \end{bmatrix} = \begin{bmatrix} \mathbf{g}_1^n + \dfrac{\mathbb{M}}{\Delta t}\mathbf{z}^{n+1} \\ \mathbf{0} \end{bmatrix}. \tag{10.151}$$

In the case of the adjoint problem, a streamline diffusion stabilization is employed when dealing with moderate Reynolds numbers. This case then represents a further example of the *Optimize-then-Discretize* approach, since different discretization strategies are used to deal with the state and the adjoint problem, similarly to what we have shown in Sect. 6.2.4. Finally, the cost functional (10.140) can be evaluated as

$$\tilde{J}(\boldsymbol{v}_h, \mathbf{u}, \theta) = \frac{1}{2}\Delta t \sum_{n=1}^{N_t} \int_\Omega \xi |\nabla \times \boldsymbol{v}_h(t^n)|^2 + \frac{\mathcal{T}}{2}\Delta t \sum_{n=1}^{N_t} (u(t^n))^2 dt + \frac{1}{\delta}\varphi(\theta).$$

where $\mathbf{u} = (u(t^1), \ldots, u(t^{N_t}))$, then setting $J_h(\mathbf{u}, \theta) = \tilde{J}(\boldsymbol{v}_h, \mathbf{u}, \theta)$.

Numerical optimization

We rely on the steepest descent method to solve the optimization problem, with an adaptive choice of the step length τ^u and τ^θ at each iteration to update both u and θ. For the sake of simplicity, here we denote by τ the generic step length, since the strategy below is valid in both cases. First of all, too small values of τ would imply an extremely slow convergence of the algorithm, whereas too large values might cause the algorithm not to converge at all. The adaptive criterion – indeed very similar to the backtracking algorithm 3.3 – is set so that the cost functional always decreases from one iteration to another; in particular, we first solve the state equation and compute the initial value of the cost functional $J_h^{(0)}$; an initial value $\tau > 0$ of the step size is chosen. Then, as long as the stopping criterion is not satisfied, at each iteration k two controls are determined from $\mathbf{u}_k$ by performing a step of the gradient method by employing two different step lengths $\tau_1 = \alpha_1 \tau$ and $\tau_2 = \alpha_2 \tau$, with $\alpha_1 > 1$ and $\alpha_2 < 1$.

The corresponding state solutions, as well as the values $J_1^{(k+1)}$ and $J_2^{(k+1)}$ of the cost functional, are calculated. Then, if $J_h^{(k)} > J_1^{(k+1)}$ and $J_h^{(k)} > J_2^{(k+1)}$, either

- $J_1^{(k+1)} > J_2^{(k+1)}$, so that $J_h^{(k+1)} = J_2^{(k+1)}$ and $\tau_{k+1} = \alpha_2 \tau_k$, or
- $J_1^{(k+1)} < J_2^{(k+1)}$, so that $J_h^{(k+1)} = J_1^{(k+1)}$ and $\tau_{k+1} = \alpha_1 \tau_k$;

otherwise, if both $J_1^{(k+1)}$ and $J_2^{(k+1)}$ are larger than $J_h^{(k)}$, then we set $\tau_{k+1} = \tilde{\alpha}\tau_k$ with $\tilde{\alpha} < \alpha_2$. At each step of the steepest descent method, the values of both control variables v and θ are then updated.

Numerical results

We present the numerical results related with the OCP for unsteady flows assuming that the control takes place through a dynamic boundary action. A mesh with 19,523 triangular elements and 10,527 vertices has been built over the domain $\Omega = (0,a) \times (0,b)$ reported in Fig. 10.8, with $a = 2$ and $b = 0.6$. On the time interval $(0, 6.25)$ (in seconds) a partition of $N_t = 125$ time intervals of length $\Delta t = 0.05$ has been set. The Reynolds number is defined as $Re = d_1 U/\nu$, where

- U denotes the maximum value of the inflow (parabolic) profile;
- $d_1 = |\Gamma_{in}| = 0.125$ is the length of the inflow boundary portion;
- $\nu = 10^{-3}$ is the kynematic viscosity.

Two different cases have been considered: *(i)* $Re = 125$, and *(ii)* $Re = 375$, corresponding to the cases $U = 1$ and $U = 3$, respectively. The length of the control boundary portion is instead $d_c = |\Gamma_c| = 0.09$. Moreover, we have set $\boldsymbol{v}_0 = \mathbf{0}$ as initial condition, and defined the function $\xi(\mathbf{x}) = \xi_1(x_1)\xi_2(x_2)$ with

$$\xi_1(x_1) = \begin{cases} 2/5 & x_1 \le 0.35 \\ 6x_1 - 1.7 & x_1 \in (0.35, 0.45] \\ 1 & x_1 > 0.45 \end{cases}, \quad \xi_2(x_2) = \begin{cases} 0 & x_2 \le 0.1 \\ 2(x_2 - 0.1) & x_2 \in (0.1, 0.15] \\ 1 & x_2 > 0.15. \end{cases}$$

As initial step length we fix the same value for both τ^u and τ^θ in the case $Re = 125$, whereas we take $\tau^\theta = 0.1\tau^u$ in the case $Re = 375$. As stopping criterion for the steepest descent method, we consider

$$\frac{|J_h^{(k+1)} - J_h^{(k)}|}{|J_h^{(0)}|} < 10^{-6}$$

together with a maximum number of $N_{max} = 25$ iterations.

We consider as inflow inclination $\eta \in (0, \pi)$, focusing on the cases $\eta = \pi/6, \pi/4, \pi/3$. As initial guess we take either $u^{(0)}(t) = 1$ if $Re = 125$ or $u^{(0)}(t) = 3$ if $Re = 375$ and, under this assumption the initial control, the angle $\theta^{(0)}$ for which the cost functional is minimized. The decay of the cost functional, as well as the optimal control (in terms of angle θ and magnitude $u(t)$), are reported in Fig. 10.9 in the case $Re = 125$. The baseline (reported in blue) is given by the value of the cost functional J_h without boundary control. Since at $t = 0$ the fluid velocity is zero, vorticity induced by the inflow is rather low, so that the optimal control magnitude decreases in time in order to prevent an undesired increase of vorticity. Later on, the control magnitude becomes larger because of the presence of vorticity induced by the inflow. The optimal angle is, in all the three cases, smaller than the one of the initial control.

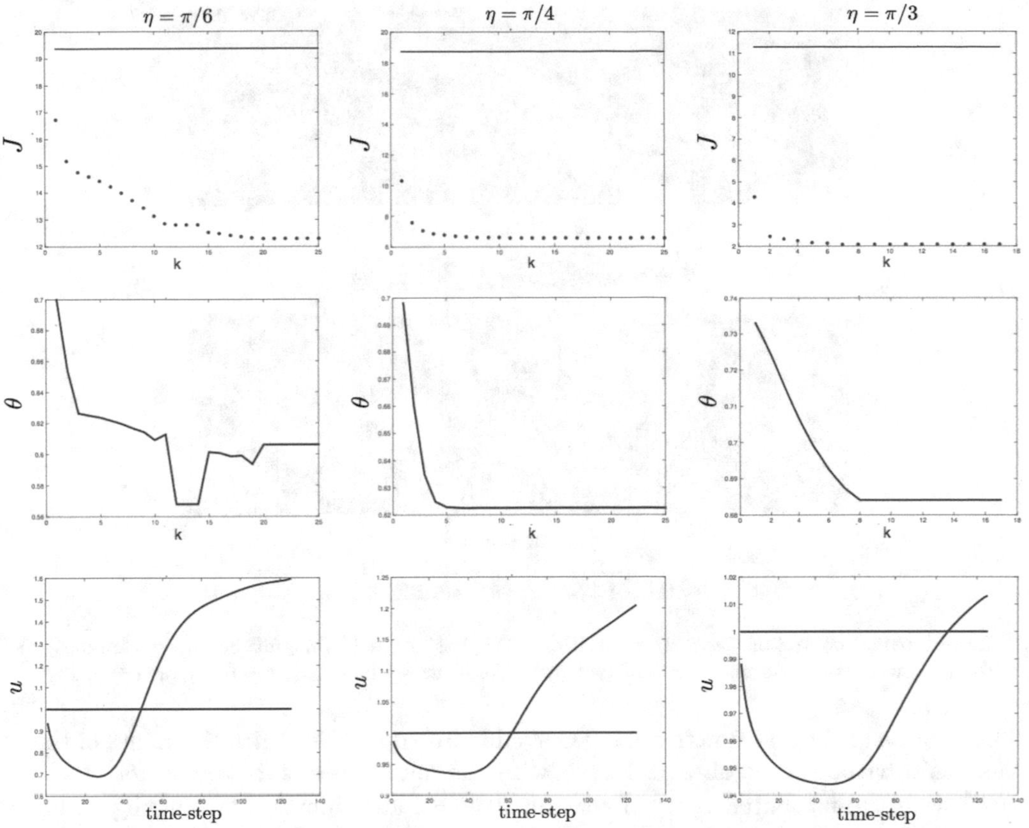

Fig. 10.9 Optimal control of unsteady flows, $Re = 125$. Top: decay of the cost functional during the steepest descent method; center: control angle θ at different iterations of the steepest descent method; bottom: optimal control velocity $\hat{u}(t)$ (in red) and initial guess $u^{(0)}(t)$ (in blue) as functions of time. From left to right: $\eta = \pi/6, \pi/4, \pi/3$

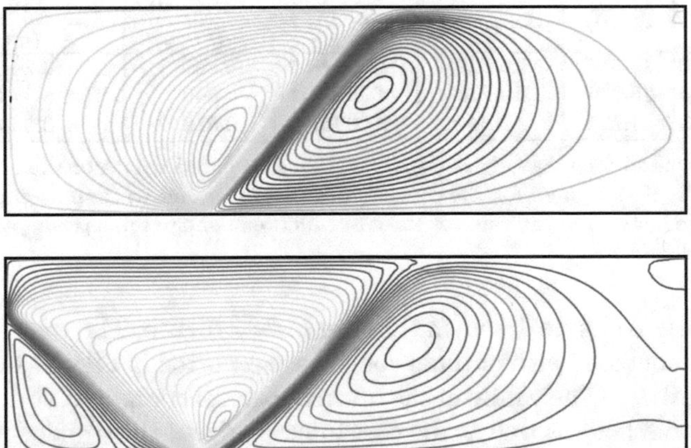

Fig. 10.10 Optimal control of unsteady flows, $Re = 125$. Streamlines of the velocity field without control (top) and with optimal control (bottom), in the case $\eta = \pi/4$, at the final time $t = 6.25$

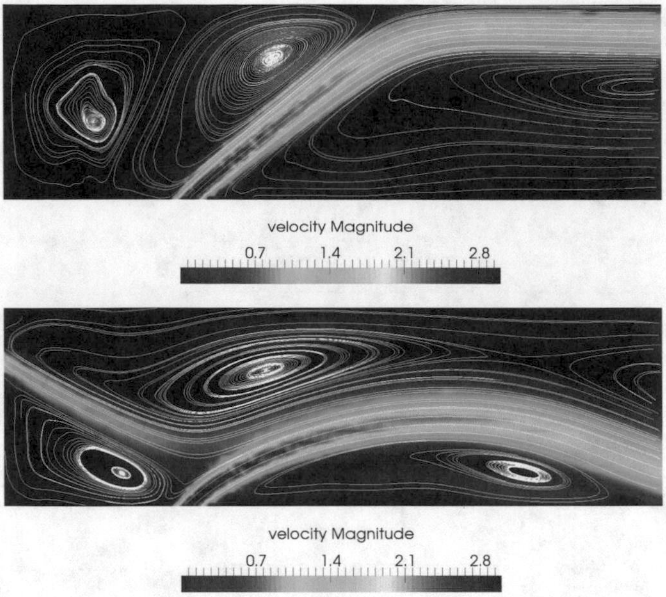

Fig. 10.11 Optimal control of unsteady flows, $Re = 375$. Velocity field (magnitude and streamlines) in the case without control (top) and with optimal control (bottom) at the final time $t = 6.25$, for $\eta = \pi/4$

The decrease in the cost functional is reported in Table 10.1, in which the values of the cost functional *(i)* without control, $J(0,0)$; *(ii)* with the initial guess for the control, $J(u_0,\ \theta_0)$; and *(iii)* with the computed optimal control, $J(\hat{u},\ \hat{\theta})$, are reported. We highlight that the larger the inflow inclination, the higher the decrease of the cost functional, compared to the case without control. In other words, vorticity is minimized by enhancing the outflow through Γ_{out}, and this effect is more evident as the inflow inclination η increases. In particular, we get a reduction in the value of the cost functional, compared to the case without control, ranging from 37% to 82% for different values of η. A detail of the streamlines of the velocity field without control and with optimal control are reported in Fig. 10.10 for the case $\eta = \pi/4$.

	$J(0,0)$	$J(u_0,\ \theta_0)$	$J(\hat{u},\ \hat{\theta})$	$1 - \dfrac{J(\hat{u},\ \hat{\theta})}{J(0,0)}$	$1 - \dfrac{J(\hat{u},\ \hat{\theta})}{J(u_0,\ \theta_0)}$
$\eta = \pi/6$	19.367	16.725	12.293	0.365	0.264
$\eta = \pi/4$	18.739	10.313	6.561	0.649	0.363
$\eta = \pi/3$	11.285	4.7244	2.0695	0.816	0.561

Table 10.1 Optimal control for unsteady flows, $Re = 125$. Computed values of the cost functional: without control, $J(0,0)$; with the initial guess for the control, $J(u^{(0)},\ \theta^{(0)})$; and with the computed optimal control, $J(\hat{u},\ \hat{\theta})$; and resulting reductions in the value of the cost functional due to numerical optimization, for different values of the inflow inclination η

Similar considerations can be drawn in the case $Re = 375$, too. The velocity field (magnitude and streamlines) without and with optimal control are reported in Fig. 10.11-10.12 for the cases $\eta = \pi/4$, $\pi/3$, respectively. The values of the optimal control computed at different iterations of the steepest descent method, as well as the cost functional and the step lengths, are reported in Figs. 10.13–10.14. We observe that the vorticity reduction during the optimization procedure is less pronounced in this case (the reduction in the cost functional is 7% against 37% for $\eta = \pi/4$, 21% against 56% for $\eta = \pi/3$). This might be due to an intrinsic difficulty in controlling a flow with three times larger magnitude. Moreover, the optimal angle tends to increase in the case $\eta = \pi/3$ (see Fig. 10.13, bottom) thus yielding a control flow directed

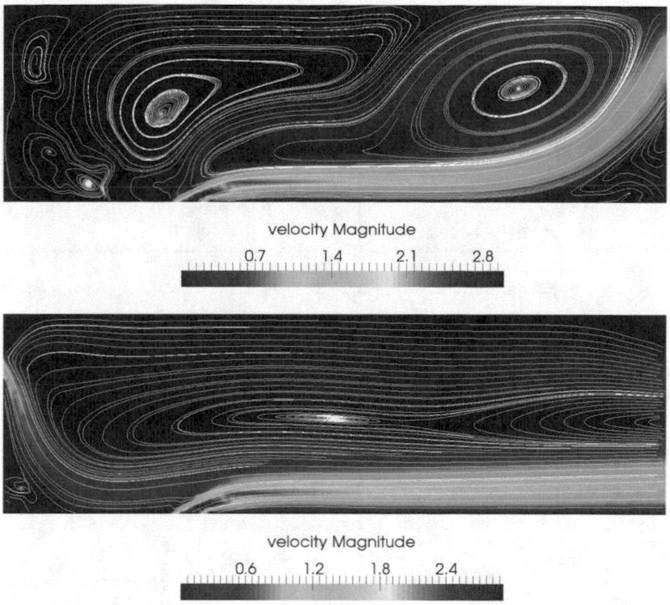

Fig. 10.12 Optimal control of unsteady flows, $Re = 375$. Velocity field (magnitude and streamlines) in the case without control (top) and with optimal control (bottom) at the final time $t = 6.25$, for $\eta = \pi/3$

towards the bottom. We also point out that the chance to select step lengths adaptively during steepest descent iterations plays a relevant role, as shown by the large variations the values of τ^u and τ^θ undergo in Fig. 10.14, bottom.

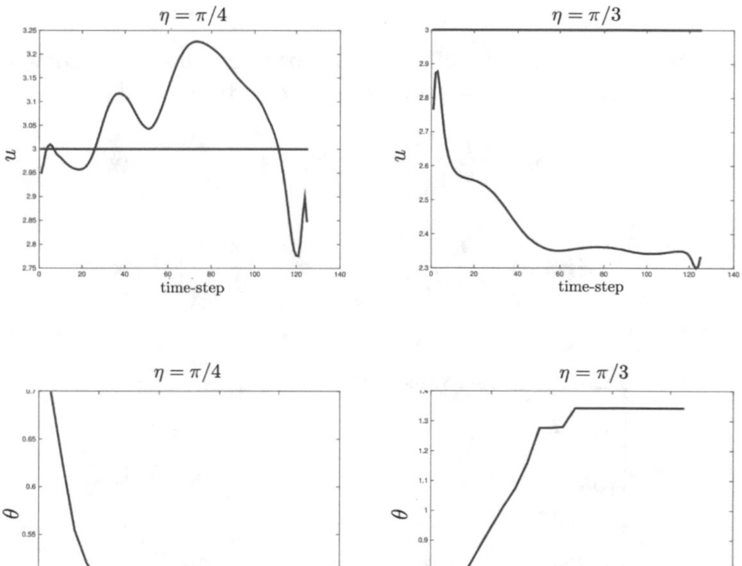

Fig. 10.13 Optimal control of unsteady flows, $Re = 375$. Top: optimal control velocity $\hat{u}(t)$ (in red) and initial guess $u^{(0)}(t)$ (in blue) versus the number of iterations; bottom: control angle θ at different iterations of the steepest descent method. Left: case $\eta = \pi/3$, right: $\eta = \pi/4$

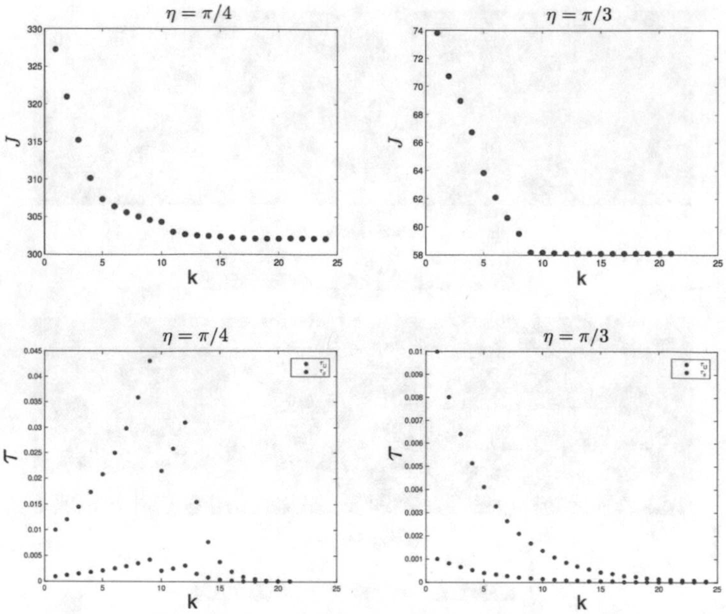

Fig. 10.14 Optimal control of unsteady flows, $Re = 375$. Top: optimal control velocity $\hat{u}(t)$ (in red) and initial guess $u^{(0)}(t)$ (in blue) as functions of time; bottom: control angle θ at different iterations of the steepest descent method. Left: case $\eta = \pi/3$, right: $\eta = \pi/4$

10.4 Exercises

1. *Distributed control of steady Navier-Stokes equations.* In the same framework of Sect. 10.1, analyze the minimization of each of the following cost functionals

$$J(\boldsymbol{v}, \boldsymbol{u}) = \frac{1}{2}\int_{\Omega} |\boldsymbol{v} - \boldsymbol{v}_d|^2 d\mathbf{x} + \frac{\alpha}{2}\int_{\Omega} |\boldsymbol{u}|^2 d\mathbf{x}$$

or

$$J(\boldsymbol{v}, \boldsymbol{u}) = \frac{1}{2}\int_{\Omega} |\nabla \times \boldsymbol{v}|^2 d\mathbf{x} + \frac{\alpha}{2}\int_{\Omega} |\boldsymbol{u}|^2 d\mathbf{x}$$

under the state system

$$\begin{cases} -\nu\, \Delta \boldsymbol{v} + (\boldsymbol{v}\cdot\nabla)\boldsymbol{v} + \nabla\pi = \boldsymbol{f} + \boldsymbol{u} & \text{in } \Omega \\ \operatorname{div} \boldsymbol{v} = 0 & \text{in } \Omega \\ \frac{\partial \boldsymbol{v}}{\partial n} - \pi\mathbf{n} = \boldsymbol{g} & \text{on } \Gamma_{in} \cup \Gamma_c \\ \boldsymbol{v} = \mathbf{0} & \text{on } \Gamma_w \cup \Gamma_{in} \\ -\pi\mathbf{n} + \nu\frac{\partial \boldsymbol{v}}{\partial n} = \mathbf{0} & \text{on } \Gamma_{out}. \end{cases} \tag{10.152}$$

   where $\boldsymbol{f} \in L^2(\Omega)^d$ $\boldsymbol{g} \in L^2(\Gamma_{in})^d$.
2. *Neumann control of steady Navier-Stokes equations.* In the same framework of Sect. 10.1, analyze the minimization of each of the following cost functionals

$$J(\boldsymbol{v},\boldsymbol{u})=\frac{1}{2}\int_{\Omega}|\boldsymbol{v}-\boldsymbol{v}_d|^2 d\mathbf{x}+\frac{\alpha}{2}\int_{\Omega}|\boldsymbol{u}|^2 d\mathbf{x}$$

or

$$J(\boldsymbol{v},\boldsymbol{u})=\frac{1}{2}\int_{\Omega}|\nabla\times\boldsymbol{v}|^2 d\mathbf{x}+\frac{\alpha}{2}\int_{\Omega}|\boldsymbol{u}|^2 d\mathbf{x}$$

under the state system

$$\begin{cases}-\nu\,\Delta\boldsymbol{v}+(\boldsymbol{v}\cdot\nabla)\boldsymbol{v}+\nabla\pi=\boldsymbol{f} & \text{in } \Omega\\ \operatorname{div}\boldsymbol{v}=0 & \text{in } \Omega\\ \frac{\partial\boldsymbol{v}}{\partial n}-p\mathbf{n}=\boldsymbol{u} & \text{on } \Gamma_c\\ \boldsymbol{v}=\mathbf{0} & \text{on } \Gamma_w\cup\Gamma_{in}\\ -\pi\mathbf{n}+\nu\frac{\partial\boldsymbol{v}}{\partial n}=\mathbf{0} & \text{on } \Gamma_{out}.\end{cases}\tag{10.153}$$

where $\mathbf{f}\in L^2(\Omega)$.

3. The Hölder inequality with three factors states that for every functions $f\in L^p(\Omega)$, $g\in L^q(\Omega)$, $h\in L^r(\Omega)$ such that $p,q,r\in(1,\infty)$ with $1/p+1/q+1/r=1$,

$$\int_{\Omega}|fgh|d\mathbf{x}\le\|f\|_{L^p(\Omega)}\|g\|_{L^q(\Omega)}\|h\|_{L^r(\Omega)}.$$

Using this inequality, show that the integral

$$\int_0^{t_f}(I_{ion}(v,w),\varphi_1)_H\,dt$$

appearing in (10.73) is well-defined.

4. Let V be a Hilbert space. Show that $H^1(0,T;V)\hookrightarrow C^{0,1/2}([0,T];V)$ where

$$C^{0,1/2}([0,T];V)=\left\{u\in C([0,T;V]):\sup_{s,t\in[0,T],t\neq s}|t-s|^{-1/2}\|u(t)-u(s)\|_V<\infty\right\},$$

is a Banach space with norm

$$\|u\|_{C^{0,1/2}([0,T];V)}=\max_{t\in[0,T]}\|u(t)\|_V+\sup_{s,t\in[0,T],t\neq s}|t-s|^{-1/2}\|u(t)-u(s)\|.$$

[*Hint.* By Theorem A.18, $H^1(0,T;V)\hookrightarrow C([0,T];V)$ and

$$u(t)-u(s)=\int_s^t u'(r)\,dr.$$

Use (A.56) and Schwarz inequality A.4.]

5. Complete the proof of Theorem 10.7. Consider the weak formulation (10.73), (10.74) of the monodomain problem and an admissible sequence $\{t_{fk},\mathbf{u}_k;v_k,w_k\}$, bounded in $\mathcal{U}_{\max}\times V_{\max}$ and weakly convergent to $\{\hat{t}_f,\hat{\mathbf{u}};\hat{v},\hat{w}\}$ in $\mathcal{U}_{\max}\times V_{\max}$. Show that we can pass to the limit into (10.73), (10.74) and conclude that $\hat{v},\hat{w}$ are the solutions corresponding to the control $\hat{\mathbf{u}}$ in $Q_{t_{\max}}$.

6. Show that, for the optimal control problem (10.66), (10.118), (10.71), the gradient of the cost functional (10.118) is given by (10.121), where the adjoint variables (p,q) solve the linearized monodomain system (10.120).

7. Check the optimality conditions for the OCP (10.141) in Theorem 10.11.

Chapter 11
Shape Optimization Problems

In this chapter we provide a short review on some of the most relevant analytical and numerical tools required to face shape optimization problems. Without pretending to be exhaustive on this broad and non-trivial topic, we nonetheless aim at *(i)* highlighting the specific difficulties intrinsic to shape optimization problems, *(ii)* deriving a system of (first-order, necessary) optimality conditions, and *(iii)* reviewing some numerical optimization algorithms already introduced for OCPs that can be exploited in this context too. For a more in depth presentation, the interested reader can refer to, e.g., [6, 79, 134, 139, 200, 215].

In Sect. 11.1 we introduce a general formulation of shape optimization problems. Domain deformations and shape derivatives of functions, which are essential tools in shape optimization, are introduced in Sect. 11.2. Some useful formulas to compute the shape derivative of functionals and solutions of boundary value problems are then provided in Sect. 11.3. Starting from a simple shape optimization model problem, the derivation of a system of first-order optimality conditions is presented in Sect. 11.4; in particular, the adjoint method introduced for OCPs is adapted to the case of shape optimization problems. Some techniques for the numerical approximation of shape optimization problems are then described in Sect. 11.5. Finally, we consider two applications in Sects. 11.6–11.7: the former is related to a minimum compliance problem for an elastic structure, while the latter deals with the optimal design of a body immersed in a fluid, subject to the minimization of its drag coefficient.

11.1 Formulation

In abstract form, a shape optimization problem (SOP) amounts to the minimization, over a set of admissible shapes $\mathcal{O}_{ad}$, of a cost functional $\tilde{J}(y, \Omega)$ that depends on the domain Ω and on the solution $y \in V$ of a state problem defined over Ω, that is:

$$\begin{aligned} &\tilde{J}(y, \Omega) \rightarrow \min \\ &\text{subject to} \\ &\mathcal{E}(y; \Omega) = 0, \quad y \in V, \quad \Omega \in \mathcal{O}_{ad}. \end{aligned} \tag{11.1}$$

Here $\tilde{J} : V \times \mathcal{O} \rightarrow \mathbb{R}$, $\mathcal{O}$ is a suitable set of shapes and $\mathcal{E}(y; \Omega) = 0$ represents the state problem. This formulation is analogue to the *full space* formulation of an OCP, see Sect. 1.8. As for OCPs, Ω denotes the spatial domain over which the state problem is set; the difference is that now the domain Ω is unknown, and represents in fact an optimization variable. To guarantee the existence of an optimal solution, $\mathcal{O}_{ad}$ has to be included in a larger set $\mathcal{O}$ fulfilling some regularity assumption.

A. Manzoni et al., *Optimal Control of Partial Differential Equations*,
Applied Mathematical Sciences 207, https://doi.org/10.1007/978-3-030-77226-0_11

When the state problem allows to express the state variable y as a function of Ω, that is, $y = y(\Omega) : \Omega \to \mathbb{R}$, problem (11.1) reduces to:

$$J(\Omega) = \tilde{J}(y(\Omega), \Omega) \to \min$$
$$\text{subject to} \qquad (11.2)$$
$$\Omega \in \mathcal{O}_{ad}.$$

This is a *reduced space* approach, see Sect. 1.8. As in the OC case, also a shape optimization problem is defined as soon as a state problem, a cost functional to be minimized, a set of admissible shapes where we seek the optimal solution, and further additional constraints (e.g. on the state) have been set.

The solution of a SOP requires, additionally to the techniques already developed for OCPs, flexible and computationally efficient tools for shape deformation and mesh motion, possibly avoiding remeshing.

Three families of shape optimization problems, and corresponding ways to define sets of admissible shapes, can be identified, in increasing order of complexity (see Fig. 11.1):

1. *parametric shape optimization.* The admissible shapes are described by means of a set of parameters $\{p_i\}$, $i = 1, \ldots, N$ having a physical meaning, e.g., thicknesses, diameters, angles, control points displacements, etc, or by a function. These parameters are in fact the optimization variables, and the shape optimization problem becomes
$$\min_{p_1,\ldots,p_N \in \mathcal{P}_{ad}} J(p_1, \ldots, p_N)$$
where $\mathcal{P}_{ad}$ is a set of admissible parameters. For this family of problems shape variations can be easily calculated. However, the variety of possible designs is severely restricted, and the use of this method implies an a priori knowledge of the desired optimal design;
2. *geometric shape optimization.* In this case the topology of the considered shapes is fixed, whereas the boundary Γ of Ω represents the optimization variable; in other words, all the domains generated during the design process are homeomorphic to the initial domain, so that formation or coalescence of *holes* in the domain are not allowed;
3. *topological shape optimization.* In the most general case, we deal with the optimization of a shape by acting on both the position of its free boundary and its topology, as for example the inclusion of holes.

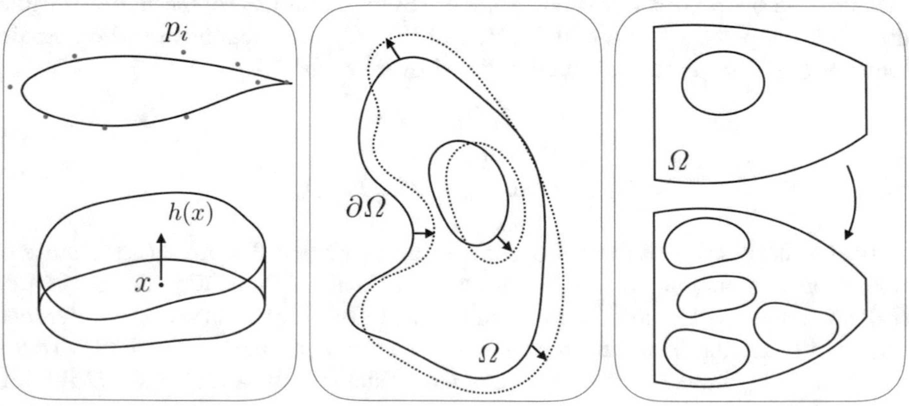

Fig. 11.1 Three families of shape optimization problems. Left: parametric SO (description of a wing profile by a non-uniform rational B-spline (NURBS), and of a plate through its thickness function $h = h(x)$). Center: geometric SO (optimization of a shape by free perturbations of its boundary). Right: topological SO.

Hereon we will restrict our discussion to geometric SOPs; for topological shape optimization the reader is referred, e.g., to [27, 61]. In particular, we pursue the same approach followed in Chapter 5: we first consider a model problem to address the main issues related with shape optimization, then we provide the mathematical and numerical tools required to address general shape optimization problems. The interested reader can refer, e.g., to [6, 79, 134, 250] for further details.

11.1.1 A Model Problem

Our initial illustrative example concerns the shape optimization of a two-dimensional elastic membrane (see Fig. 11.2). Let $\Omega \subset \mathbb{R}^2$ be its shape (that is, the region it occupies), and $\Gamma = \Gamma_D \cup \Gamma_N \cup \Gamma_C$ its boundary. Here Γ_C denotes the varying part of the boundary we want to act on, in order to, e.g., reach a target displacement $z_d \in L^2(\Omega)$. More precisely, the functional

$$J(\Omega) = \frac{1}{2}\int_{\Omega} |y(\Omega) - z_d|^2 d\mathbf{x} + \frac{\beta}{2}\int_{\Gamma_C} d\sigma \tag{11.3}$$

has to be minimized over all admissible shapes. Here $\beta \geq 0$ is a parameter that penalizes the length of the moving boundary Γ_C, while $y = y(\Omega) \in V = H^1_{0,\Gamma_D}(\Omega)$ denotes the vertical displacement of the membrane, that is, the solution of the following state problem:

$$\begin{cases} -\Delta y = f & \text{in } \Omega \\ y = 0 & \text{on } \Gamma_D \\ \dfrac{\partial y}{\partial n} = g & \text{on } \Gamma_N \\ \dfrac{\partial y}{\partial n} = 0 & \text{on } \Gamma_C. \end{cases}$$

The membrane is kept fixed on the boundary Γ_D, where its displacement is null. Moreover, $f, g : U \to \mathbb{R}$ are sufficiently smooth functions, and $U \subset \mathbb{R}^2$ denotes the so-called *universe* domain. This latter is assumed to be a closed subset of $\mathbb{R}^2$ and serves as the domain of definition for all data in order to make the underlying shape problem well posed. U is assumed to be sufficiently large to contain all the domains generated by an iterative process used to solve the shape optimization problem.

Another cost functional we might be interested to minimize is the *compliance*, that is, the work of external loads, which can be expressed as

$$J(\Omega) = \int_{\Omega} f y(\Omega) d\mathbf{x} + \int_{\Gamma_N} g y(\Omega) d\boldsymbol{\sigma}. \tag{11.4}$$

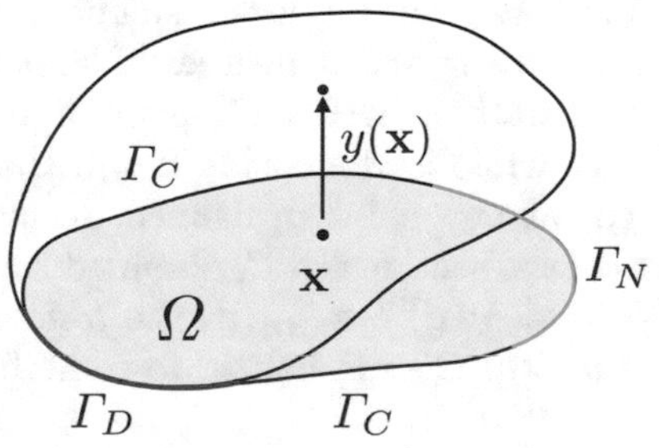

Fig. 11.2 Shape optimization of an elastic membrane: domain and boundaries.

This is equivalent to maximize the stiffness of the structure, which is equal and opposite in sign to the compliance (intuitively, the less the structure works, the higher its stiffness).

To determine the optimal shape we need to find a system of first-order optimality conditions, similarly to what we have done, when addressing OCPs. The main difficulty arises because, hereon, the control function is given by the domain Ω itself, and *derivation* with respect to the shape of a given quantity, such as the cost functional, is rather involved. This indeed requires:

- to introduce suitable notions of shape deformations; for instance, a suitable way to describe deformations of the shape Ω is to consider perturbations of the identity of the form $\Phi_t(\mathbf{x}) = \mathbf{x} + t\mathbf{V}(\mathbf{x})$, where $t \geq 0$ and $\mathbf{V} : U \to \mathbb{R}^d$ is a sufficiently smooth vector field. The unknown field $\mathbf{V}$ must also satisfy some physical constraints; for instance, $\mathbf{V} = \mathbf{0}$ on $\Gamma \setminus \Gamma_C$, since we can only act on Γ_C for the sake of optimal design;
- to define the shape derivatives $y'(\Omega; \mathbf{V})$ of the solution $y = y(\Omega)$ of the state problem and $J'(\Omega; \mathbf{V})$ of the shape functional $J(\Omega)$, respectively; both quantities are meant as *directional* derivatives, evaluated at Ω, along the direction $\mathbf{V}$;
- to extend the adjoint method and the Lagrange multipliers method to avoid the calculation of the shape derivative $y'(\Omega; \mathbf{V})$ when deriving $J(\Omega)$ with respect to the shape. Indeed, if we consider a cost functional like (11.3) (in the case $\beta = 0$ for simplicity), it is quite intuitive that according to the chain rule its shape derivative will be of the form

$$J'(\Omega; \mathbf{V}) = \int_{\Omega} (y(\Omega) - z_d) y'(\Omega; \mathbf{V}) d\mathbf{x} + \dots$$

and, as already done in the case of OCPs, we need to get rid of the term containing $y'(\Omega; \mathbf{V})$ to improve computational efficiency.

Proceeding in this way, in the case the cost functional is given by (11.4), we obtain[1]

$$J'(\Omega; \mathbf{V}) = -\int_{\Gamma_C} |\nabla y|^2 \mathbf{V} \cdot \mathbf{n} \, d\sigma.$$

Hence, in order to reduce the compliance (that is, to have $J'(\Omega; \mathbf{V}) \leq 0$), it is enough to enlarge the domain, that is, to choose $\mathbf{V} \cdot \mathbf{n} > 0$. This information will then be exploited by an optimization algorithm, by suitably extending *descent methods* to shape optimization problems.

11.2 Shape Functionals and Derivatives

Although shape optimization problems share a common structure with OCPs, both their analysis and numerical approximation involve further relevant difficulties. As previously mentioned, in this case we deal with the minimization of *shape functionals* $J : \mathcal{A} \to \mathbb{R}$, where $\mathcal{A}$ is a suitable family of admissible subsets of a (closed, smooth) *universe* domain $U \subseteq \mathbb{R}^d$. An element of $\mathcal{A}$ can be a *domain* Ω, with various degrees of smoothness, typically Lipschitz or of class C^k, its boundary Γ or a part of it. For C^2 plane domains the notion of curvature at any point of Γ is well defined, while in dimension $d = 3$ the surface Γ has two principal curvatures $\kappa_{\min}(\mathbf{p})$ and $\kappa_{\max}(\mathbf{p})$ at any point $\mathbf{p}$, defined, respectively, as the minimum and the maximum of the curvatures of the sections to Γ, obtained by planes normal to the tangent plane at $\mathbf{p}$. We denote by $\kappa(\mathbf{p}) = \kappa_{\min}(\mathbf{p}) + \kappa_{\max}(\mathbf{p})$ the *total curvature* of Γ at $\mathbf{p}$.

For each domain $\Omega \in \mathcal{A}$, let $y(\Omega) : \Omega \to \mathbb{R}$ belong to some function space, e.g. $C^k(\overline{\Omega})$ or $H^k(\Omega)$, $k \geq 1$. We call $y(\Omega)$ a *domain function*.

[1] As we will see in Sect. 11.4.2, the problem of minimizing the compliance is indeed self-adjoint, that is, the adjoint problem here coincides with the state problem. For this reason, $J'(\Omega; \mathbf{V})$ only depends on y.

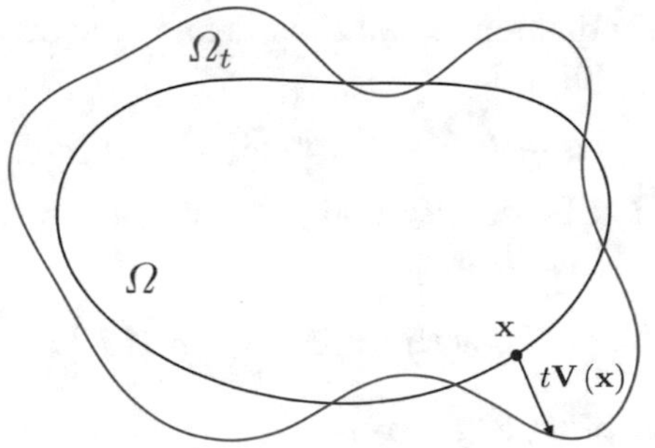

Fig. 11.3 Shape deformation of a two-dimensional domain obtained through the perturbation of identity map.

Correspondingly, the shape functional

$$J(\Omega) = \int_{\Omega} y(\Omega) = \int_{\Omega} y(\Omega)(\mathbf{x})\, d\mathbf{x}$$

takes the name of *domain functional.* For instance, if $y(\Omega) = 1$, then $J(\Omega)$ is the measure (area or volume) of Ω. On the other hand, the definition of $y(\Omega)$ could be more involved, e.g., it could represent the solution of an elliptic boundary value problem.

Analogously, *boundary functionals* can be given by

$$J(\Gamma) = \int_{\Gamma} z(\Gamma) = \int_{\Gamma} z(\Gamma)(\boldsymbol{\sigma})\, d\sigma$$

where $z(\Gamma) : \Gamma \to \mathbb{R}$ is a *boundary function* belonging to some space, e.g. $C^k(\Gamma)$ or $H^k(\Gamma)$, $k \geq 1$. Examples are $z(\Gamma) = 1$, in which case $J(\Gamma)$ represents the measure (length or surface area) of Γ, or $z(\Gamma) = \mathbf{V} \cdot \mathbf{n}$, where $\mathbf{n}$ is the outward unit normal to Γ and $\mathbf{V}$ a vector field.

To minimize a cost functional $J(\Omega)$ (or $J(\Gamma)$) with respect to Ω (or Γ), a suitable notion of derivative, called *shape derivative* must be called into play. This is a non-trivial issue that requires a notion of shape deformation (or perturbation). Here we describe the simplest way to define it, suitable for first order necessary conditions of optimality, restricting ourselves to dimension $d = 3$ for better clarity.

Our main focus is on the derivation of the necessary formulas to implement numerical approximation algorithms. We often proceed formally, referring to specialized texts for rigorous proofs. *Hereon, we omit the symbols* $d\mathbf{x}$ *and* $d\sigma$ *usually appearing in the integrals for ease of notation.*

11.2.1 Domain Deformations

Given a bounded domain $\Omega \subset U$, we introduce a one-parameter family of perturbations of the identity $\boldsymbol{\Phi}_t$, given by

$$\Phi_t(\mathbf{x}) = \mathbf{x} + t\mathbf{V}(\mathbf{x})$$

where $\mathbf{x} \in \Omega$, $t \in [0, \varepsilon]$, for some $\varepsilon > 0$ and $\mathbf{V} : U \to \mathbb{R}^3$ is a regular vector field, typically $\mathbf{V} \in C^k(U)^3$, $k \geq 1$, or $\mathbf{V} \in W^{1,\infty}(U)^3$ (e.g. locally Lipschitz continuous). We denote the gradient (or Jacobian matrix) of $\mathbf{V}$ the matrix whose elements are given by

$$(\nabla\mathbf{V})_{ij} = \frac{\partial V_i}{\partial x_j}, \qquad i, j = 1, \ldots, 3.$$

Starting from the reference domain Ω, we then define a deformed domain Ω_t as the image of Ω through the map Φ_t, $t \in [0, \varepsilon]$, i.e.

$$\Omega_t = \boldsymbol{\Phi}_t(\Omega) = \{\boldsymbol{\Phi}_t(\mathbf{x}) : \mathbf{x} \in \Omega\}.$$

Clearly $\Omega_0 = \Omega$ and $\boldsymbol{\Phi}_0 = \mathbf{I}$, the identity mapping, and, for ε small, $\Omega_t \subset U$. Similarly, we define a deformed boundary Γ_t of Ω_t as

$$\Gamma_t = \boldsymbol{\Phi}_t(\Gamma) = \{\boldsymbol{\Phi}_t(\boldsymbol{\sigma}) : \boldsymbol{\sigma} \in \Gamma\}.$$

The Jacobian matrix $\nabla\boldsymbol{\Phi}_t$ of $\boldsymbol{\Phi}_t$ is given by

$$\nabla\boldsymbol{\Phi}_t(\mathbf{x}) = \mathbf{I} + t\nabla\mathbf{V}(\mathbf{x})$$

so that

$$\frac{d}{dt}\nabla\boldsymbol{\Phi}_t(\mathbf{x})_{|t=0} = \nabla\mathbf{V}(\mathbf{x}) \text{ and } \det\nabla\boldsymbol{\Phi}_t(\mathbf{x}) = 1 + t\mathrm{div}\mathbf{V}(\mathbf{x}) + o(t) \qquad \text{as } t \to 0.$$

As a consequence, if ε is small enough, $\det\boldsymbol{\Phi}_t(\mathbf{x}) \geq C > 0$ and thanks to the Inverse Function Theorem, $\boldsymbol{\Phi}_t$ defines a family of bijective mappings (diffeomorphisms) between Ω and Ω_t. Moreover, Ω_t inherits the regularity properties of Ω and $\mathbf{V}$. Note also that

$$\frac{d}{dt}\det\nabla\boldsymbol{\Phi}_t(\mathbf{x})_{|t=0} = \mathrm{div}\mathbf{V}(\mathbf{x}). \tag{11.5}$$

Example 11.1. Of special interest are *normal deformations*, in which $\mathbf{V}(\mathbf{x})$ points in the direction of the outward unit normal $\mathbf{n}(\mathbf{x})$ at any point of Γ. In this case, the family of perturbations may take the form

$$\boldsymbol{\Phi}_t(\mathbf{x}) = \mathbf{x} + tv(\mathbf{x})\,\mathbf{n}(\mathbf{x})$$

where $v = v(\mathbf{x})$ is a smooth function. •

*11.2.2 Elements of Tangential (or Surface) Calculus**

We briefly introduce the *tangential* or *surface calculus* on the boundary Γ of a smooth bounded domain Ω. If $f \in C^1(\Gamma)$ is a scalar function defined on Γ, quantities like ∇f are meaningless, thus, to define tangential differential operators acting on f, we can use a C^1 extension $\tilde{f}$ of f to a tubular neighborhood $\mathcal{N}(\Gamma)$, where

$$\mathcal{N}(\Gamma) = \{\mathbf{x} \in U : \mathrm{dist}(\mathbf{x},\Gamma) \leq \delta\}$$

for some small δ. Analogous concepts are introduced for vector valued functions $\mathbf{w} \in C^1(\Gamma)^3$ with extension $\tilde{\mathbf{w}}$.

The signed distance

All the notions we are going to introduce are independent of the extension chosen for f. In particular an extension of objects defined on Γ can be realized in a canonical way by using the so called *signed distance function*, defined as

$$b(\mathbf{x}) = b(\Omega)(\mathbf{x}) = \begin{cases} \text{dist}(\mathbf{x};\Gamma) & \mathbf{x} \in \mathbb{R}^d \setminus \Omega \\ 0 & \mathbf{x} \in \Gamma \\ -\text{dist}(\mathbf{x};\Gamma) & \mathbf{x} \in \Omega \end{cases} \tag{11.6}$$

where

$$\text{dist}(\mathbf{x};\Gamma) = \inf_{\mathbf{y} \in \Gamma} |\mathbf{x} - \mathbf{y}| .$$

This function is globally Lipschitz, indeed[2]

$$|b(\mathbf{x}) - b(\mathbf{y})| \leq |\mathbf{x} - \mathbf{y}|$$

and if δ is small enough, for each point $\mathbf{x} \in \mathcal{N}(\Gamma)$ there exists a unique point $\mathbf{p}(\mathbf{x}) \in \Gamma$, the *projection* of $\mathbf{x}$ onto Γ, that realizes the distance of $\mathbf{x}$ from Γ:

$$|\mathbf{p}(\mathbf{x}) - \mathbf{x}| = \text{dist}(\mathbf{x};\Gamma) .$$

In this case, b is as smooth as Γ. Moreover[3]

$$|\nabla b| = 1 \qquad \text{in } \mathcal{N}(\Gamma)$$

and therefore, being Γ the zero-level set of b,

$$\nabla b = \mathbf{n} \qquad \text{on } \Gamma.$$

Thus, ∇b constitutes a unitary (canonical) extension to $\mathcal{N}(\Gamma)$ of the unit normal $\mathbf{n}$. Moreover, we have (see [79])

$$\Delta b = \kappa \quad \text{on } \Gamma. \tag{11.7}$$

We can write the projection operator $\mathbf{p} : \mathcal{N}(\Gamma) \to \Gamma$ in terms of the signed distance as

$$\mathbf{p}(\mathbf{x}) = \mathbf{x} - b(\mathbf{x}) \nabla b(\mathbf{x}) .$$

Definition 11.1. Given $f : \Gamma \to \mathbb{R}$, the *canonical or normal extension* of f to $\mathcal{N}(\Gamma)$ is the composite function $\tilde{f}(\mathbf{x}) = (f \circ \mathbf{p})(\mathbf{x}) = f(p(\mathbf{x}))$.

Remark 11.1. Clearly, $\mathbf{p}(\mathbf{y}) = \mathbf{p}(\mathbf{x})$ at every point $\mathbf{y}$ on the segment

$$[\mathbf{p}(\mathbf{x}), \mathbf{x}] = \{\mathbf{p}(\mathbf{x}) + t\mathbf{n}(\mathbf{x}),\ 0 \leq t \leq b(\mathbf{x})\} .$$

Along this segment, the canonical extension $\tilde{f}$ is constant and therefore $\frac{\partial \tilde{f}}{\partial n} = 0$ at $p(\mathbf{x})$. •

---

[2] Clearly, passing from a point $\mathbf{x}$ to a point $\mathbf{y}$, the distance cannot increase more than $|\mathbf{x} - \mathbf{y}|$.

[3] Let b be differentiable at $\mathbf{x} \in \mathcal{N}(\Gamma)$ and $\mathbf{z}$ be such that $\text{dist}(\mathbf{x}, \boldsymbol{\Gamma}) = |\mathbf{x} - \mathbf{z}|$. Then, for all $\mathbf{y}$ and any ε small, we have

$$b^2(\mathbf{x} + \varepsilon\mathbf{y}) = b^2(\mathbf{x}) + 2\varepsilon(\mathbf{x})\ \mathbf{y} \cdot \nabla b(\mathbf{x}) + o(\varepsilon)$$

$$b^2(\mathbf{x} + \varepsilon\mathbf{y}) \leq |\mathbf{x} + \varepsilon\mathbf{y} - \mathbf{z}|^2 = b^2(\mathbf{x}) + 2\varepsilon(\mathbf{y}, \mathbf{x} - \mathbf{z}) + o(\varepsilon) .$$

Hence, after dividing by $2\varepsilon > 0$ and letting $\varepsilon \to 0$, we get

$$b(\mathbf{x}) \nabla b(\mathbf{x}) \cdot \mathbf{y} \leq (\mathbf{y}, \mathbf{x} - \mathbf{z}) .$$

Since this inequality holds for every $\mathbf{y}$, it must be an equality, from which we infer $\nabla b(\mathbf{x}) = (\mathbf{x} - \mathbf{z})/b(\mathbf{x})$, whence $|\nabla b(\mathbf{x})| = 1$.

Tangential gradient and divergence

We now recall the definitions and some basic properties of the surface (or tangential) differential operators.

Tangential gradient. The tangential gradient ∇_Γ of f at a point $\mathbf{x} \in \Gamma$ is defined as the tangential component of $\nabla \tilde{f}(\mathbf{x})$, i.e. the projection of $\nabla \tilde{f}(\mathbf{x})$ onto the tangent plane to Γ at $\mathbf{x}$, that is

$$\nabla_\Gamma f = \nabla \tilde{f}_{|\Gamma} - (\nabla \tilde{f}_{|\Gamma} \cdot \mathbf{n})\mathbf{n} = \nabla \tilde{f}_{|\Gamma} - \frac{\partial \tilde{f}}{\partial n}\mathbf{n} \qquad \text{on } \Gamma. \tag{11.8}$$

Note that $\nabla_\Gamma f \cdot \mathbf{n} = 0$ and that the definition of $\nabla_\Gamma f$ is independent of the extension $\tilde{f}$ of f. Indeed, if $\tilde{f}_1$ is another extension of f, then $\tilde{g} = \tilde{f} - \tilde{f}_1 = 0$ on Γ and $\nabla \tilde{g}_{|\Gamma} = (\nabla \tilde{g}_{|\Gamma} \cdot \mathbf{n})\mathbf{n}$ on Γ. Therefore, $\nabla_\Gamma g = \mathbf{0}$ or $\nabla_\Gamma f = \nabla_\Gamma f_1$.

Remark 11.2. Thanks to Remark 11.1, using the canonical extension $\tilde{f} = f \circ \mathbf{p}$ we have $\frac{\partial \tilde{f}}{\partial n} = 0$ and therefore

$$\nabla_\Gamma f = \nabla \tilde{f}_{|\Gamma}. \tag{11.9}$$

•

The *tangential gradient* $\nabla_\Gamma \mathbf{w}$ of a vector valued function $\mathbf{w}$ is given by the following matrix, whose *rows* are the tangential gradients of the components of $\mathbf{w}$:

$$\nabla_\Gamma \mathbf{w} = \begin{bmatrix} \nabla_\Gamma w_1 \\ \nabla_\Gamma w_2 \\ \nabla_\Gamma w_3 \end{bmatrix}.$$

$\nabla_\Gamma \mathbf{w}$ can be written in the compact form[4]

$$\nabla_\Gamma \mathbf{w} = \nabla \tilde{\mathbf{w}}_{|\Gamma} - \frac{\partial \tilde{\mathbf{w}}}{\partial n} \otimes \mathbf{n}. \tag{11.10}$$

Note that $[\nabla_\Gamma \mathbf{w}]\, \mathbf{n} = \mathbf{0}$.

Tangential divergence. In analogy with the usual definition, the tangential divergence of $\mathbf{w}$ is defined as

$$\mathrm{div}_\Gamma \mathbf{w} = \mathrm{tr} \nabla_\Gamma \mathbf{w} = \mathrm{tr} \nabla \tilde{\mathbf{w}}_{|\Gamma} - \left[\nabla \tilde{\mathbf{w}}_{|\Gamma}\right] \mathbf{n} \cdot \mathbf{n} = \mathrm{div} \tilde{\mathbf{w}}_{|\Gamma} - \frac{\partial \tilde{\mathbf{w}}}{\partial n} \cdot \mathbf{n} \qquad \text{on } \Gamma. \tag{11.11}$$

A particularly important case is the tangential divergence of the unit normal vector $\mathbf{n}$.

Proposition 11.1. *Let κ be the total curvature of Γ. Then*

$$div_\Gamma \mathbf{n} = \kappa.$$

Proof. Let $\tilde{\mathbf{n}} = \nabla b(\Omega)$ be the canonical extension of $\mathbf{n}$, where $b(\Omega)$ is the signed distance function. We have the identity $|\tilde{\mathbf{n}}|^2 = 1$ in $\mathcal{N}(\Gamma)$. Taking the normal derivatives, we get

$$\frac{d\tilde{\mathbf{n}}}{dn} \cdot \mathbf{n} = 0 \qquad \text{on } \Gamma$$

[4] The symbol $\mathbf{w} \otimes \mathbf{z}$ denotes the tensor product between two vectors $\mathbf{w}$ and $\mathbf{z}$ is defined as the matrix

$$\mathbf{w} \otimes \mathbf{z} = (w_i z_j)_{i,j=1,2,3}\,.$$

Note that $tr(\mathbf{w} \otimes \mathbf{z}) = \mathbf{w} \cdot \mathbf{z}$.

whence, from (11.7),

$$\mathrm{div}_\Gamma \mathbf{n} = \mathrm{div}\tilde{\mathbf{n}}_{|\Gamma} = \mathrm{tr}\nabla\tilde{\mathbf{n}}_{|\Gamma} = \mathrm{tr}D^2 b_{|\Gamma} = \Delta b_{|\Gamma} = \kappa. \tag{11.12}$$

In particular $\Delta b\,(\Omega)$ constitutes an extension to $\mathcal{N}\,(\Gamma)$ of the total curvature κ. □

Example 11.2. For a surface ball $\Gamma_R = \left\{\mathbf{x} \in \mathbb{R}^3; |\mathbf{x}| = R\right\}$, we have $\kappa_{\min} = \kappa_{\max} = 1/R$ so that $\kappa = 2/R$. Let us check that, in this case, $\mathrm{div}_{\Gamma_R}\mathbf{n} = \kappa$. At any point $\boldsymbol{\sigma} \in \Gamma_R$, $\mathbf{n} = \boldsymbol{\sigma}/R$. The vector field $\tilde{\mathbf{n}} = \mathbf{x}/\left|\mathbf{x}\right|$ is a unitary extension of $\mathbf{n}$, and

$$\mathrm{div}\,\tilde{\mathbf{n}} = \mathrm{div}\frac{\mathbf{x}}{|\mathbf{x}|} = \frac{3}{|\mathbf{x}|} - \sum_{j=1}^{3} \frac{x_j^2}{|\mathbf{x}|^3}.$$

Therefore,

$$\mathrm{div}_{\Gamma_R}\mathbf{n} = \mathrm{div}\frac{\mathbf{x}}{|\mathbf{x}|}_{|\Gamma_R} = \frac{3}{R} - \frac{1}{R^3}\sum_{j=1}^{3} x_j^2 = \frac{2}{R} = \kappa.$$

•

Calculus rules and integration by parts formulas

The tangential operators share with their classical counterparts several calculus rules that we gather in the following proposition (see Exercise 2 for the proof).

Proposition 11.2. *Let $f \in C^1(\Gamma), \mathbf{w} \in C^1(\Gamma)^3$. The following rules hold:*

1.

$$\nabla_\Gamma(fg) = g\nabla_\Gamma f + f\nabla_\Gamma g \tag{11.13}$$

2.

$$\mathit{div}_\Gamma(f\mathbf{w}) = \nabla_\Gamma f \cdot \mathbf{w} + f\,\mathit{div}_\Gamma \mathbf{w}. \tag{11.14}$$

Let $\mathbf{w} \in C^1(\Gamma)$ and let

$$\mathbf{w}_\Gamma = \mathbf{w} - (\mathbf{w}\cdot\mathbf{n})\,\mathbf{n} \quad \text{on } \Gamma, \tag{11.15}$$

be its tangential component. The following proposition directly follows from the previous result (see Exercise 3 for the proof).

Proposition 11.3. *Let Ω be a C^2 domain and $f \in C^1(\Gamma), \mathbf{w} \in C^1(\Gamma)^3$. The following formulas hold on Γ:*

$$\mathit{div}_\Gamma(f\mathbf{n}) = f\kappa, \qquad \mathit{div}_\Gamma\mathbf{w} = \mathit{div}_\Gamma\mathbf{w}_\Gamma + \kappa\,(\mathbf{w}\cdot\mathbf{n}). \tag{11.16}$$

The following theorem provides some integration by parts formulas involving tangential operators.

Theorem 11.1. *Let Ω be a C^2 domain and $f \in C^1(\Gamma), \mathbf{w} \in C^1(\Gamma)^3$. Then:*
1. *if $\mathbf{w} \cdot \mathbf{n} = 0$,*
$$\int_\Gamma div_\Gamma \mathbf{w} = 0; \tag{11.17}$$
2.
$$\int_\Gamma div_\Gamma \mathbf{w} = \int_\Gamma \kappa \, \mathbf{w} \cdot \mathbf{n}; \tag{11.18}$$
3.
$$\int_\Gamma f \, div_\Gamma \mathbf{w} = \int_\Gamma \left[-\nabla_\Gamma f \cdot \mathbf{w} + \kappa f \, \mathbf{w} \cdot \mathbf{n} \right]. \tag{11.19}$$

Proof. (sketch). For (11.17), see [139]; (11.18) follows from (11.17) and the second formula in (11.16), since $\mathbf{w}_\Gamma \cdot \mathbf{n} = 0$. Finally, (11.19) follows from (11.18) and (11.14). □

Remark 11.3. If $\Gamma_0 \subset \Gamma$ has a smooth closed curve as its boundary $\partial\Gamma_0$ (e.g. a circle on a spherical surface), formula (11.18) must be modified as follows. Let τ_Γ be a unit tangent vector to Γ_0 and $\mathbf{n}_\Gamma = \tau_\Gamma \times \mathbf{n}$. Then
$$\int_\Gamma \mathrm{div}_\Gamma \mathbf{w} = \int_\Gamma \kappa \; \mathbf{w} \cdot \mathbf{n} + \int_{\Gamma_0} \mathbf{w} \cdot \mathbf{n}_\Gamma .$$
•

Tangential Laplacian. Let Ω be of class C^2 and $f \in C^2(\Gamma)$. By analogy with the classical Laplace operator, we define the tangential Laplacian (or Laplace-Beltrami operator) of f on Γ as
$$\Delta_\Gamma f = \mathrm{div}_\Gamma \nabla_\Gamma f. \tag{11.20}$$
Using (11.16) we have on Γ (see Exercise 4)
$$\Delta_\Gamma f = \Delta \tilde{f} - \left[D^2 \tilde{f} \right] \mathbf{n} \cdot \mathbf{n} - \kappa \frac{\partial \tilde{f}}{\partial n} = \Delta \tilde{f} - \kappa \frac{\partial \tilde{f}}{\partial n} - \frac{\partial^2 \tilde{f}}{\partial n^2} \tag{11.21}$$
where[5] $D^2 \tilde{f}$ denotes the Hessian matrix of $\tilde{f}$.

The following *Green formulas* are also very useful.

Proposition 11.4. *Let Ω be a C^2 domain and $f \in C^1(\Gamma)$, $g \in C^2(\Gamma)^3$. Then:*
1.
$$\int_\Gamma f \Delta_\Gamma g = - \int_\Gamma \nabla_\Gamma f \cdot \nabla_\Gamma g; \tag{11.22}$$
2.
$$\int_\Gamma \tilde{f} \; \Delta \tilde{g} = - \int_\Gamma \nabla \tilde{f} \cdot \nabla \tilde{g} + \int_\Gamma \left(\frac{\partial \tilde{f}}{\partial n} \frac{\partial \tilde{g}}{\partial n} + f \frac{\partial^2 \tilde{g}}{\partial n^2} + \kappa f \frac{\partial \tilde{g}}{\partial n} \right). \tag{11.23}$$

Proof. Formula (11.22) follows from (11.19), by taking $\mathbf{w} = \nabla_\Gamma g$. Regarding (11.23), on one hand we have, on Γ,
$$\nabla \tilde{f} \cdot \nabla \tilde{g} = \left(\nabla_\Gamma f + \frac{\partial \tilde{f}}{\partial n} \mathbf{n} \right) \cdot \left(\nabla_\Gamma g + \frac{\partial \tilde{g}}{\partial n} \mathbf{n} \right) = \nabla_\Gamma f \cdot \nabla_\Gamma g + \frac{\partial \tilde{f}}{\partial n} \frac{\partial \tilde{g}}{\partial n}. \tag{11.24}$$
On the other hand, (11.21) gives, on Γ,
$$\Delta_\Gamma g = \Delta \tilde{g} - \kappa \frac{\partial \tilde{g}}{\partial n} - \frac{\partial^2 \tilde{g}}{\partial n^2} \tag{11.25}$$
Inserting (11.24) and (11.25) into (11.22), (11.23) follows. □

[5] The notation $\frac{\partial^2 \tilde{f}}{\partial n^2}$ here stands for $\left[D^2 \tilde{f} \right] \mathbf{n} \cdot \mathbf{n}$ and not for $\frac{\partial}{\partial n} \left(\frac{\partial \tilde{f}}{\partial n} \right) = \mathbf{n} \cdot \nabla (\mathbf{n} \cdot \nabla f)$.

Remark 11.4. The above concepts and formulas make sense also for domain or boundary functions in Sobolev spaces provided the surface operators are intended in the sense of traces in these spaces. For instance, Theorem 11.1 still holds for $f \in H^2(\Omega)$, and $\mathbf{w} \in H^2(\Omega; \mathbb{R}^3)$. •

11.2.3 Shape Derivative of Functions

For domain and boundary functions, two kind of derivatives can be introduced, that reflect the *material* (or *Lagrangian*) and the *Eulerian* points of view.

Domain Functions

Let $y = y(\Omega)$ be a domain function and, recall, $\boldsymbol{\Phi}_t(\mathbf{x}) = \mathbf{x} + t\mathbf{V}(\mathbf{x})$, $\mathbf{x} \in \Omega$. Provided the following limit exists,

$$\dot{y}(\Omega; \mathbf{V})(\mathbf{x}) = \lim_{t\to 0} \frac{y(\Omega_t)(\boldsymbol{\Phi}_t(\mathbf{x})) - y(\Omega)(\mathbf{x})}{t} = \frac{d}{dt} y(\Omega_t)(\boldsymbol{\Phi}_t(\mathbf{x}))_{|t=0}, \tag{11.26}$$

it is called the *material derivative of* y (with respect to the family of deformations $\boldsymbol{\Phi}_t$) in the direction $\mathbf{V}$.

Remark 11.5. The limit (11.26) can be understood in several ways. It makes sense pointwise for $C^1(U)$ functions; if $\mathbf{V} \in C^k(U)$ and y is in $C^k(U)$ or $H^k(U)$, it could be intended in some suitable Sobolev norm, $k \geq 1$. Finally, for functions with weaker regularity, it can be understood in the corresponding weak sense. •

The material derivative is the derivative of $y(\Omega)$ with respect to a coordinate reference system that moves with the deformation $\boldsymbol{\Phi}_t$. This operator obeys the usual rules of calculus (product rule, chain rule,...) but *cannot* be interchanged with temporal or spatial derivatives.

If y does not depend on the geometry of Ω, e.g., if y can be seen as the restriction to Ω of a function $\hat{y}$ defined in U, the chain rule gives

$$\dot{y}(\Omega; \mathbf{V}) - \nabla y(\Omega) \cdot \mathbf{V} = 0. \tag{11.27}$$

In general, the left hand side of (11.27) encodes the sensitivity of $y(\Omega)$ with respect to domain deformations, but without the convective effect of moving the domain itself. Thus it makes sense to define the *shape* (or *Eulerian*) derivative of y in the direction $\mathbf{V}$ as

$$y'(\Omega; \mathbf{V}) = \dot{y}(\Omega; \mathbf{V}) - \nabla y(\Omega) \cdot \mathbf{V} \quad \text{in } \Omega. \tag{11.28}$$

This definition extends to vector valued functions $\mathbf{w}$ as follows

$$\mathbf{w}'(\Omega; \mathbf{V}) = \dot{\mathbf{w}}(\Omega; \mathbf{V}) - [\nabla \mathbf{w}(\Omega)]\,\mathbf{V} = \dot{\mathbf{w}}(\Omega; \mathbf{V}) - (\mathbf{V} \cdot \nabla)\,\mathbf{w} \quad \text{in } \Omega.$$

Example 11.3. Let $\Omega = B_R(\mathbf{0}) \backslash \overline{B}_1(\mathbf{0})$, $R \geq 2$, be a spherical shell in $\mathbb{R}^3$ and $y = y(\Omega)$ be the solution to the Dirichlet problem $\Delta y = 0$ in Ω, $y = 0$ on ∂B_1 and $y = 1$ on ∂B_R. It is easy to check that

$$y(\Omega)(\mathbf{x}) = \frac{R}{R-1}\left(1 - \frac{1}{|\mathbf{x}|}\right).$$

We want to keep the inner boundary fixed and move ∂B_R to ∂B_{R+vt}, along the normal $\mathbf{n}(\mathbf{x}) = \mathbf{x}/|\mathbf{x}|$, $t > 0$. One way to realize this deformation is to introduce a smooth and monotonically increasing function $\beta = \beta(s)$ such that $\beta(s) = 0$ for $s \leq 1$ and $\beta(s) = 1$ for

$s \geq 2$, and define

$$\mathbf{V}(\mathbf{x}) = v\beta(|\mathbf{x}|)\frac{\mathbf{x}}{|\mathbf{x}|}, \qquad v \geq 0.$$

Indeed, we have

$$|\boldsymbol{\Phi}_t(\mathbf{x})|^2 = \left|\mathbf{x}+tv\beta(|\mathbf{x}|)\frac{\mathbf{x}}{|\mathbf{x}|}\right|^2 = |\mathbf{x}|^2 + 2tv\beta(|\mathbf{x}|)|\mathbf{x}| + t^2v^2\beta^2(|\mathbf{x}|)$$

and for $|\mathbf{x}| = R$, $|\boldsymbol{\Phi}_t(\mathbf{x})| = R + tv$, since $R \geq 2$ and $\beta(R) = 1$.

Then $\Omega_t = B_{(1+vt)R}(\mathbf{0}) \backslash \overline{B}_1(\mathbf{0})$ and the solution of the Dirichlet problem $\Delta y = 0$ in Ω_t, $y = 0$ on ∂B_1 and $y = 1$ on ∂B_{R+vt} is given by

$$y(\Omega_t)(\mathbf{x}) = \frac{R+tv}{R+tv-1}\left(1 - \frac{1}{|\mathbf{x}|}\right).$$

Hence

$$y(\Omega_t)(\boldsymbol{\Phi}_t(\mathbf{x})) = \frac{R+tv}{R+tv-1}\left(1 - \frac{1}{|\boldsymbol{\Phi}_t(\mathbf{x})|}\right)$$

We have, after routine calculations,

$$\dot{y}(\Omega;\mathbf{V})(\mathbf{x}) = \frac{d}{dt}y(\Omega_t)(\boldsymbol{\Phi}_t(\mathbf{x}))_{|t=0} = -\frac{v}{(R-1)^2}\left(1 - \frac{1}{|\mathbf{x}|}\right) + \frac{R}{R-1}\frac{v\beta(|\mathbf{x}|)}{|\mathbf{x}|^2}.$$

Thus, since

$$\nabla y(\Omega)(\mathbf{x}) \cdot \mathbf{V}(\mathbf{x}) = \frac{R}{R-1}\frac{v\beta(|\mathbf{x}|)}{|\mathbf{x}|^2},$$

for the shape derivative we get

$$y'(\Omega;\mathbf{V})(\mathbf{x}) = \dot{y}(\Omega;\mathbf{V})(\mathbf{x}) - \nabla y(\Omega)(\mathbf{x}) \cdot \mathbf{V}(\mathbf{x}) = -\frac{v}{(R-1)^2}\left(1 - \frac{1}{|\mathbf{x}|}\right).$$

Note that $y' = y'(\Omega;\mathbf{V})$ solves the Dirichlet problem $\Delta y' = 0$ in Ω, $y' = 0$ on ∂B_1 and

$$y' = -\frac{v}{R(R-1)} = -\frac{\partial y}{\partial n}\mathbf{V} \cdot \mathbf{n} \quad \text{on } \partial B_R.$$

According to intuition, $y' < 0$, in Ω. We will see later a general formula for the shape derivation of a Dirichlet problem. •

Boundary Functions

For boundary functions $z(\Gamma)$, the definition of *material derivative* is analogous to (11.26):

$$\dot{z}(\Gamma;\mathbf{V})(\mathbf{x}) = \lim_{t\to 0}\frac{z(\Gamma_t)(\boldsymbol{\Phi}_t(\mathbf{x})) - z(\Gamma)(\mathbf{x})}{t} = \frac{d}{dt}z(\Gamma_t)(\boldsymbol{\Phi}_t(\mathbf{x}))_{|t=0}. \tag{11.29}$$

To define the *shape derivative*, since $\nabla z(\Gamma)$ does not make sense, we use the *canonical extension* $\tilde{z} = z \circ \mathbf{p}$ according to Definition (11.1), recalling (Remark 11.1) that $\nabla \tilde{z}_{|\Gamma} = \nabla_\Gamma z$:

$$z'(\Gamma;\mathbf{V}) = \dot{z}(\Gamma;\mathbf{V}) - \nabla_\Gamma z \cdot \mathbf{V} \quad \text{on } \Gamma. \tag{11.30}$$

Remark 11.6. If $z = z(\Gamma)$ is the restriction to the boundary of a domain function $y = y(\Omega)$ (e.g. in a Dirichlet problem), in general it is *not true* that $z'(\Gamma; \mathbf{V}) = y'(\Omega; \mathbf{V})_{|\Gamma}$. Indeed, assume that $y(\Omega)$ and $z(\Gamma)$ are restrictions to Ω and Γ, respectively, of the same function $\psi \in C^1(U)$. Then we have, since ψ is independent of Ω,

$$\dot{z}(\Gamma; \mathbf{V}) = \dot{y}(\Omega; \mathbf{V})_{|\Gamma} \tag{11.31}$$

since both quantities are equal to $\nabla\psi \cdot \mathbf{V}_{|\Gamma}$. However, from (11.28), we have $y'(\Omega; \mathbf{V}) = 0$, while

$$z'(\Gamma; \mathbf{V}) = \nabla\psi \cdot \mathbf{V} - \nabla_\Gamma \psi \cdot \mathbf{V} = \frac{\partial y}{\partial n} \mathbf{V} \cdot \mathbf{n} \qquad \text{on } \Gamma.$$

Formula actually holds (11.31) for any smooth extension $y(\Omega)$ of $z(\Gamma)$ and this leads to the following general formula

$$z'(\Gamma; \mathbf{V}) = y'(\Omega; \mathbf{V})_{|\Gamma} + \frac{\partial y}{\partial n} \mathbf{V} \cdot \mathbf{n}. \tag{11.32}$$

Clearly, if $y(\Omega)$ is the *canonical extension* of $z(\Gamma)$ in a tubular neighborhood of Γ, then $\frac{\partial y}{\partial n} = 0$ and $z'(\Gamma; \mathbf{V}) = y'(\Omega; \mathbf{V})_{|\Gamma}$. •

Basic Rules for Shape Derivatives

The following rules hold for shape derivatives; their proof can be obtained by direct verification using the definitions.

- *Product rule for shape derivatives.* Let $S = \Omega$ or $S = \Gamma$, with Γ smooth, and $u(S), v(S) : S \to \mathbb{R}$ be C^1 functions. Assume the shape derivatives $u'(S; \mathbf{V})$ and $v'(S; \mathbf{V})$ exist for any vector field $\mathbf{V}$. Then

$$(uv)'(S; \mathbf{V}) = u'(S; \mathbf{V})\, v(S; \mathbf{V}) + v'(S; \mathbf{V})\, u(S; \mathbf{V}). \tag{11.33}$$

- *Chain rule.* Let $f : \mathbb{R} \to \mathbb{R}$ be differentiable at $u(S)$. Then

$$(f \circ u)'(S; \mathbf{V}) = f'(u(S))\, u'(S; \mathbf{V}) \tag{11.34}$$

  where f' denotes the usual derivative of f.
- *Mixed shape-spatial derivative.* Since classical differentiation does not depend on the domain geometry, for *domain* functions it commutes with shape differentiation. Assume $y(\Omega), \mathbf{w}(\Omega)$ are smooth functions which are shape differentiable along the direction $\mathbf{V}$, with $\mathbf{V}$ also smooth. Then:

  1. $\nabla y(\Omega)$ and $\Delta y(\Omega)$ are also shape differentiable along the direction $\mathbf{V}$ and

$$(\nabla y)'(\Omega; \mathbf{V}) = \nabla y'(\Omega; \mathbf{V}), \quad (\Delta y)'(\Omega; \mathbf{V}) = \Delta y'(\Omega; \mathbf{V});$$

  2. $\nabla \mathbf{w}(\Omega)$ and $\mathrm{div}\mathbf{w}(\Omega)$ are also shape differentiable along the direction $\mathbf{V}$ and

$$(\nabla \mathbf{w})'(\Omega; \mathbf{V}) = \nabla \mathbf{w}'(\Omega; \mathbf{V}), \quad (\mathrm{div}\mathbf{w})'(\Omega; \mathbf{V}) = \mathrm{div}\mathbf{w}'(\Omega; \mathbf{V}).$$

We apply the formulas above to compute the shape derivative in the direction $\mathbf{V}$ of the signed distance $b(\Omega)$, as defined in (11.6) in a tubular neighborhood $\mathcal{N}(\Gamma)$, and of the normal unit vector $\mathbf{n}$.

Proposition 11.5. *The following formulas hold on Γ:*

$$b'(\Omega;\mathbf{V}) = -\mathbf{V}\cdot\mathbf{n}, \tag{11.35}$$

$$\mathbf{n}'(\Gamma;\mathbf{V}) = -\nabla_\Gamma(\mathbf{V}\cdot\mathbf{n}). \tag{11.36}$$

Proof. Since $b(\Omega_t) = 0$ on Γ_t, for t small, we have $b(\Omega_t)(\boldsymbol{\Phi}_t(\mathbf{x})) = 0$ on Γ whence $\dot{b}(\Omega;\mathbf{V}) = 0$. Thus

$$b'(\Omega;\mathbf{V}) = \dot{b}(\Omega;\mathbf{V}) - \mathbf{V}\cdot\nabla b(\Omega) = -\mathbf{V}\cdot\mathbf{n} \quad \text{on } \Gamma.$$

To prove (11.36), we use the canonical extension $\tilde{\mathbf{n}} = \nabla b(\Omega)$ of $\mathbf{n}$ in $\mathcal{N}(\Gamma)$. Taking the shape derivative of $|\tilde{\mathbf{n}}|^2 = 1$ we get, on Γ,

$$(\nabla b)'(\Omega;\mathbf{V})\cdot\mathbf{n} = \nabla b'(\Omega;\mathbf{V})\cdot\mathbf{n} = 0.$$

Thus $\nabla b'(\Omega;\mathbf{V})$ is orthogonal to $\mathbf{n}$ and therefore, using (11.35),

$$\nabla b'(\Omega;\mathbf{V})_{|\Gamma} = \nabla_\Gamma b'(\Omega;\mathbf{V})_{|\Gamma} = -\nabla_\Gamma(\mathbf{V}\cdot\mathbf{n}).$$ □

11.3 Shape Derivatives of Functionals and Solutions of Boundary Value Problems

In this section we provide the definition of shape derivative of a functional and several examples, in view of subsequent applications. Given a Banach space $\mathcal{V}$ of admissible vector fields (e.g. $C^1(U)$), we first define an Eulerian derivative as a *directional derivative*, along the direction of the domain perturbation, as follows.

Definition 11.2. Let $J : \mathcal{A} \to \mathbb{R}$ be a shape functional, $S \in \mathcal{A}$. The Eulerian derivative of J at S along the direction $\mathbf{V}$ is defined as the limit

$$J'(S;\mathbf{V}) = \lim_{t\to 0} \frac{J(S_t) - J(S)}{t}$$

where, as usual $S_t = \boldsymbol{\Phi}_t(S)$. We call J *shape differentiable* at S if $J'(S;\mathbf{V})$ exists for every $\mathbf{V} \in \mathcal{V}$ and the map $\mathcal{G}(S) : \mathbf{V} \to J'(S;\mathbf{V})$ is linear and continuous in $\mathcal{V}$. In this case $\mathcal{G}(S)$ takes the name of *shape derivative of* J at S.

A classical result due to Hadamard, called *Structure Theorem* (see, e.g., [201]), states that, *under appropriate smoothness conditions*, the shape derivative of a differentiable functional $J(\Omega)$ at Ω, in the direction of $\boldsymbol{V}$, only depends on the normal trace $\boldsymbol{V}\cdot\mathbf{n}$ on Γ.

Theorem 11.2. *Let $\Omega \in \mathcal{A}$ be an admissible domain and $J : \mathcal{A} \to \mathbb{R}$ be shape differentiable at Ω; denote by $\mathcal{G}(\Omega)$ the shape derivative of J at Ω. Then:*

1. the support of $\mathcal{G}(\Omega)$ is contained in Γ;
2. if Ω is at least of class C^1, there exists a distribution $G(\Gamma) \in \mathcal{D}'(\mathbb{R}^d;\mathbb{R}^d)$ such that

$$J'(\Omega;\mathbf{V}) = \langle G(\Gamma), \mathbf{V}|_\Gamma\cdot\mathbf{n}\rangle \qquad \forall\, \mathbf{V} \in \mathcal{D}(\mathbb{R}^d;\mathbb{R}^d).$$

3. if $G(\Gamma) \in L^1(\Gamma)$ then

$$J'(\Omega;\mathbf{V}) = \int_\Gamma G(\Gamma)\mathbf{V}\cdot\mathbf{n}. \tag{11.37}$$

The function $G(\Gamma)$ is often referred as the *shape gradient* of $J(\Omega)$, a scalar field which depends on J and on the current shape S. Indeed, the shape derivative is concentrated on Γ. We emphasize that formula (11.37) requires the domain Ω to be at least C^1. Indeed, it is not valid, e.g., for piecewise smooth domains (see, e.g., [250]).

Heuristically, at first order, a vector field $\boldsymbol{V}$ tangential to Γ only yields a pure transport of the shape Ω, and it is somehow natural to expect that $J'(\Omega;\mathbf{V}) = 0$.

We note that thanks to (11.37) we can evaluate a descent direction at Ω in a straighforward way (at least formally): letting $\mathbf{V} = -G(\Gamma)\,\mathbf{n}$, for a small enough descent step $t > 0$, one has

$$J(\Omega_t) = J(\Omega) - t\int_\Gamma (G(\Gamma))^2 + o(t) < J(\Omega).$$

11.3.1 Domain Functionals

We calculate the shape derivative of some significant domain functionals. We start with integrands that *do not depend* on Ω.

Proposition 11.6. *[250, Prop. 2.45] Let $\Omega \subset U$ be a bounded Lipschitz domain. If $y \in H^1(U)$ and $\mathbf{V} \in W^{1,\infty}(U)^3$, the functional*

$$J(\Omega) = \int_\Omega y$$

is shape differentiable and its shape derivative along the direction $\mathbf{V}$ is given by

$$J'(\Omega;\mathbf{V}) = \int_\Omega div(y\mathbf{V}) = \int_\Gamma y\mathbf{V}\cdot\mathbf{n}. \tag{11.38}$$

Proof. The classical formula for the change of variables in volume integrals gives

$$J(\Omega_t) = \int_{\Omega_t} y = \int_\Omega y\circ\mathbf{\Phi}_t \det\nabla\mathbf{\Phi}_t$$

(note that $\det\nabla\mathbf{\Phi}_t(\mathbf{x}) > 0$ for t small). Using

$$\left.\frac{d}{dt}(y\circ\mathbf{\Phi}_t)\right|_{t=0} = \nabla y\cdot\mathbf{V}$$

and (see (11.5))

$$\left.\frac{d}{dt}(\det\nabla\mathbf{\Phi}_t)\right|_{t=0} = \mathrm{div}\mathbf{V},$$

we have[6], using first the chain and product rules and then the divergence theorem,

$$\begin{aligned} J'(\Omega_t;\mathbf{V}) &= \left.\frac{d}{dt}\int_\Omega y\circ\mathbf{\Phi}_t\det\nabla\mathbf{\Phi}_t\right|_{t=0} = \int_\Omega \left.\frac{d}{dt}(y\circ\mathbf{\Phi}_t\det\nabla\mathbf{\Phi}_t)\right|_{t=0} \\ &= \int_\Omega (\nabla y\cdot\mathbf{V} + y\mathrm{div}\mathbf{V}) = \int_\Omega \mathrm{div}(y\mathbf{V}) = \int_\Gamma y\mathbf{V}\cdot\mathbf{n}. \end{aligned}$$

□

Note that (11.38) is in accordance with the Structure Theorem 11.2, provided we take $G(\Gamma) = y_{|\Gamma}$.

We now consider the case $y = y(\Omega)$. We have the following result.

[6] It is possible to exchange integral and $t-$derivative.

Proposition 11.7. *[250, Sect. 2.31] Let $\Omega \subset U$ be a bounded Lipschitz admissible domain and $\mathbf{V} \in W^{1,\infty}(U)^3$. Let $y = y(\Omega)$ be so that the material derivative $\dot{y}(\Omega; \mathbf{V})$ and the shape derivative $y'(\Omega; \mathbf{V})$ exist in $L^1(\Omega)$. Then, the functional*

$$J(\Omega) = \int_\Omega y(\Omega)$$

is shape differentiable and

$$J'(\Omega; \mathbf{V}) = \int_\Omega y'(\Omega; \mathbf{V}) + \int_\Omega div(y(\Omega)\mathbf{V}) \tag{11.39}$$

$$= \int_\Omega y'(\Omega; \mathbf{V}) + \int_\Gamma y(\Omega)\mathbf{V} \cdot \mathbf{n} \tag{11.40}$$

Proof. We have, using the definitions of material and shape derivative for $y(\Omega)$,

$$\begin{aligned} J'(\Omega; \mathbf{V}) &= \frac{d}{dt} \int_\Omega y(\Omega_t) \circ \mathbf{\Phi}_t \det \nabla \mathbf{\Phi}_t \Big|_{t=0} = \int_\Omega \frac{d}{dt} \left(y(\Omega_t) \circ \mathbf{\Phi}_t \det \nabla \mathbf{\Phi}_t \right) \Big|_{t=0} \\ &= \int_\Omega \left(\dot{y}(\Omega; \mathbf{V}) + y(\Omega) \operatorname{div} \mathbf{V} \right) = \int_\Omega \left(\dot{y}(\Omega; \mathbf{V}) - \nabla y(\Omega) \cdot \mathbf{V} + \operatorname{div}(y(\Omega)\mathbf{V}) \right) \\ &= \int_\Omega \left(y'(\Omega; \mathbf{V}) + \operatorname{div}(y(\Omega)\mathbf{V}) \right) = \int_\Omega y'(\Omega; \mathbf{V}) + \int_\Gamma y(\Omega)\mathbf{V} \cdot \mathbf{n}. \end{aligned}$$

□

When y is a solution of a boundary value problem, also formula (11.40) can be put in the form (11.37) by introducing a suitable adjoint problem, as we shall see later.

11.3.2 Boundary Functionals

We now consider integrals on $\Gamma = \partial\Omega$, starting again with integrands independent of Ω. We shall need the following change of variables formula for boundary integrals, where $\Gamma_t = \partial\Omega_t$:

$$\int_{\Gamma_t} f(\boldsymbol{\sigma}) = \int_\Gamma f(\mathbf{\Phi}_t(\boldsymbol{\sigma})) \det \nabla \mathbf{\Phi}_t(\boldsymbol{\sigma}) \left| (\nabla \mathbf{\Phi}_t(\boldsymbol{\sigma}))^{-\top} \mathbf{n}(\boldsymbol{\sigma}) \right|, \tag{11.41}$$

where $(\nabla \mathbf{\Phi}_t)^{-\top}$ is the inverse of the transpose of $\nabla \mathbf{\Phi}_t$ and $\mathbf{n}(\boldsymbol{\sigma})$ is the outward unit normal to Γ at $\boldsymbol{\sigma}$. For the proof of (11.41) see, e.g., [139]. The next lemma shows some preliminary calculations.

Lemma 11.1. *Let $\boldsymbol{\gamma}(t) = (\nabla \mathbf{\Phi}_t)^{-\top} \mathbf{n} = (\mathbf{I} + t(\nabla \mathbf{V})^\top)^{-1}\mathbf{n}$. Then*

$$\boldsymbol{\gamma}'(0) = -(\nabla \mathbf{V})^\top \mathbf{n} \quad and \quad \frac{d}{dt} |\boldsymbol{\gamma}(t)|_{|t=0} = -(\nabla \mathbf{V})^\top \mathbf{n} \cdot \mathbf{n} = -(\nabla \mathbf{V})\mathbf{n} \cdot \mathbf{n} \tag{11.42}$$

Proof. For any differentiable invertible matrix $M(t)$, we have

$$\frac{d}{dt} M^{-1}(t) = -M^{-1}(t) \frac{d}{dt} M(t) M^{-1}(t),$$

see Exercise 9. Therefore

$$\boldsymbol{\gamma}'(t) = -(\nabla \mathbf{\Phi}_t)^{-\top} \frac{d}{dt} (\nabla \mathbf{\Phi}_t)^\top \ (\nabla \mathbf{\Phi}_t)^{-\top} \mathbf{n}$$

and, since $\boldsymbol{\gamma}(0) = \mathbf{n}$ and $(\nabla \mathbf{\Phi}_0)^\top = \mathbf{I}$, evaluating at $t = 0$ we get $\boldsymbol{\gamma}'(0) = -(\nabla \mathbf{V})^\top \mathbf{n}$. Finally,

$$\frac{d}{dt}|\gamma(t)| = \frac{d}{dt}\sqrt{\sum_{j=1}^{d}\gamma_j(t)^2} = \frac{1}{|\gamma(t)|}\gamma(t)\cdot\gamma'(t)$$

and

$$\frac{d}{dt}|\gamma(t)|_{|t=0} = \frac{1}{|\gamma(0)|}\gamma'(0)\cdot\gamma(0) = -(\nabla\mathbf{V})\,\mathbf{n}\cdot\mathbf{n}.$$

□

Proposition 11.8. *[250, Prop. 2.50] Let $z \in H^2(U)$, $\mathbf{V} \in C^1(U)^3$ and $\Omega \subset U$ be a C^2 domain. Then the functional*

$$J(\Gamma) = \int_\Gamma z$$

is shape differentiable and its shape derivative is

$$J'(\Gamma;\mathbf{V}) = \int_\Gamma (\nabla z\cdot\mathbf{V} + z\,div_\Gamma\mathbf{V}) = \int_\Gamma\left(\frac{\partial z}{\partial n} + \kappa z\right)(\mathbf{V}\cdot\mathbf{n}). \tag{11.43}$$

Proof. We have

$$J'(\Gamma;\mathbf{V}) = \frac{d}{dt}\int_{\Gamma_t} z\Big|_{t=0} = \int_\Gamma \frac{d}{dt}\left(z\circ\mathbf{\Phi}_t \det\nabla\mathbf{\Phi}_t\left|(\nabla\mathbf{\Phi}_t)^{-\top}\mathbf{n}\right|\right)\Big|_{t=0}.$$

From Lemma 11.1, letting $\boldsymbol{\gamma}(t) = (\nabla\mathbf{\Phi}_t)^{-\top}\mathbf{n} = \left(\mathbf{I} + t(\nabla\mathbf{V})^\top\right)^{-1}\mathbf{n}$, we have $\gamma(0) = \mathbf{n}$ and

$$\boldsymbol{\gamma}'(0) = -(\nabla\mathbf{V})^\top\mathbf{n}, \quad \frac{d}{dt}|\gamma(t)|_{|t=0} = (\nabla\mathbf{V})\,\mathbf{n}\cdot\mathbf{n}.$$

Thus, using the chain and product rules,

$$\int_\Gamma \frac{d}{dt}\left(z\circ\mathbf{\Phi}_t\det\nabla\mathbf{\Phi}_t\,|\gamma(t)|\right)\Big|_{t=0} = \int_\Gamma\left(\nabla z\cdot\mathbf{V} + z\mathrm{div}\mathbf{V} - z\ (\nabla\mathbf{V})\,\mathbf{n}\cdot\mathbf{n}\right).$$

Since, on Γ,

$$\mathrm{div}_\Gamma\mathbf{V} = \mathrm{div}\mathbf{V} - (\nabla\mathbf{V})\,\mathbf{n}\cdot\mathbf{n},$$

we can write

$$J'(\Gamma;\mathbf{V}) = \int_\Gamma(\nabla z\cdot\mathbf{V} + z\mathrm{div}_\Gamma\mathbf{V}).$$

Finally, using formula (11.19), we get

$$\int_\Gamma z\mathrm{div}_\Gamma\mathbf{V} = \int_\Gamma[-\nabla_\Gamma z\cdot\mathbf{V} + \kappa z\ \mathbf{V}\cdot\mathbf{n}];$$

recall that κ denotes the total curvature of Γ. Since $\nabla z - \nabla_\Gamma z = \frac{\partial z}{\partial n}\mathbf{n}$, we find

$$J'(\Gamma;\mathbf{V}) = \int_\Gamma(\nabla z\cdot\mathbf{V} - \nabla_\Gamma z\cdot\mathbf{V} + \kappa z\ \mathbf{V}\cdot\mathbf{n}) = \int_\Gamma\left(\frac{\partial z}{\partial n} + \kappa z\right)\mathbf{V}\cdot\mathbf{n}.$$

□

Remark 11.7. Also (11.43) is coherent with the structure theorem, letting $G(\Gamma) = \left(\frac{\partial z}{\partial n} + \kappa z\right)_\Gamma$. In particular, let $J(\Gamma) = |\Gamma|$, the surface area of Γ. We have

$$J'(\Gamma;\mathbf{V}) = \int_\Gamma \mathrm{div}_\Gamma\mathbf{V} = \int_\Gamma \kappa\ \mathbf{V}\cdot\mathbf{n}.$$

•

Example 11.4. The presence of the curvature in (11.43) is not surprising, as the following simple case shows. Let $\Gamma_R = \partial B_R$, where B_R is the ball of radius R centered at the origin, and $z \in C^1(\mathbb{R}^3)$ independent of Γ_R. When R changes, Γ_R expands (or contracts) along the normal direction $\mathbf{V}(\mathbf{x}) = \mathbf{x}/|\mathbf{x}|$. Consider the integral

$$J(\Gamma_R) = \int_{\Gamma_R} z(\omega)\,d\omega.$$

Putting $\omega = R\sigma$, we have $d\omega = R^2 d\sigma$ and

$$J(\Gamma_R) = \int_{\Gamma_1} z(R\boldsymbol{\sigma})\, R^2 d\sigma.$$

Taking the derivative with respect to R (i.e. the shape derivative), we obtain

$$J'(\Gamma_R) = \int_{\Gamma_1} \nabla z(R\boldsymbol{\sigma}) \cdot \boldsymbol{\sigma}\; R^2 d\sigma + 2z(R\boldsymbol{\sigma})\, R d\sigma.$$

Going back to ω, we find

$$J'(\Gamma_R) = \int_{\Gamma_R} \left(\nabla z(\omega) \cdot \frac{\omega}{R} + \frac{2}{R} z(\omega) \right) d\omega.$$

Since $\frac{\omega}{R} = \mathbf{V}(\omega) = \mathbf{n}(\omega)$, $\mathbf{V}(\omega) \cdot \mathbf{n}(\omega) = 1$ and $2/R = \kappa(\Gamma_R)$, we finally can write

$$J'(\Gamma_R) = \int_{\Gamma_R} \left(\frac{\partial z}{\partial n} + \kappa(\Gamma_R)\, z \right) \mathbf{V} \cdot \mathbf{n}\; d\omega.$$

•

Consider now the case $z = z(\Gamma)$. We have the following result.

Proposition 11.9. *[250, Sect. 2.33] Let $z = z(\Gamma)$ be so that the material derivative $\dot{z}(\Gamma; \mathbf{V})$ and the shape derivative $z'(\Gamma; \mathbf{V})$ exist and both belong to $L^1(\Gamma)$. Then we have*

$$J'(\Gamma; \mathbf{V}) = \int_{\Gamma} z'(\Gamma; \mathbf{V}) + \int_{\Gamma} \kappa z(\Gamma) \mathbf{V} \cdot \mathbf{n}. \tag{11.44}$$

If $z(\Gamma) = y(\Omega)|_{\Gamma}$, then

$$J'(\Gamma; \mathbf{V}) = \int_{\Gamma} y'(\Omega; \mathbf{V})|_{\Gamma} + \int_{\Gamma} \left(\frac{\partial y(\Omega)}{\partial n} + \kappa y(\Omega) \right) \mathbf{V} \cdot \mathbf{n}. \tag{11.45}$$

Proof. We have

$$\begin{aligned} J'(\Gamma; \mathbf{V}) &= \frac{d}{dt} \int_{\Gamma_t} z(\Gamma_t) \circ \boldsymbol{\Phi}_t \det \nabla \boldsymbol{\Phi}_t \left| (\nabla \boldsymbol{\Phi}_t)^{-\top} \mathbf{n} \right| \Big|_{t=0} \\ &= \int_{\Gamma} \frac{d}{dt} \left(z(\Gamma_t) \circ \boldsymbol{\Phi}_t \det \nabla \boldsymbol{\Phi}_t \left| (\nabla \boldsymbol{\Phi}_t)^{-\top} \mathbf{n} \right| \right) \Big|_{t=0} . \end{aligned}$$

Proceeding as in Proposition 11.8, we end up with

$$J'(\Gamma; \mathbf{V}) = \int_{\Gamma} (\dot{z}(\Gamma; \mathbf{V}) + z \operatorname{div}_{\Gamma} \mathbf{V}) = \int_{\Gamma} (\dot{z}(\Gamma; \mathbf{V}) - \nabla_{\Gamma} z \cdot \mathbf{V} + \kappa z\; \mathbf{V} \cdot \mathbf{n}).$$

In the first case $\dot{z}(\Gamma; \mathbf{V}) = z'(\Gamma; \mathbf{V}) + \nabla_{\Gamma} z(\Gamma) \cdot \mathbf{V}$, therefore

$$J'(\Gamma; \mathbf{V}) = \int_{\Gamma} \left(z'(\Gamma; \mathbf{V}) + \kappa z\; \mathbf{V} \cdot \mathbf{n} \right). \tag{11.46}$$

If $z(\Gamma) = y(\Omega)|_{\Gamma}$, then by (11.32) $z'(\Gamma; \mathbf{V}) = y'(\Omega; \mathbf{V})_{|\Gamma} + \frac{\partial y(\Omega)}{\partial n} \mathbf{V} \cdot \mathbf{n}$, and a substitution into (11.46) gives (11.45). □

11.3.3 Chain Rules

To use Propositions 11.7 and 11.9 for functionals of the type

$$J_0(\Omega) = \int_\Omega F_0(y(\Omega)) \quad \text{or} \quad J_1(\Gamma) = \int_\Gamma F_1(z(\Gamma))$$

where $F_0, F_1 \in C^1(\mathbb{R})$ and $\Omega, \Gamma \in \mathcal{A}$, we need to compute the shape derivative of $y(\Omega)$ and $z(\Gamma)$. As a matter of fact, under the usual smoothness assumptions, the chain rule holds for the shape derivatives, giving

$$J_0'(\Omega; \mathbf{V}) = \int_\Omega \frac{dF_0}{dy}(y(\Omega))\, y'(\Omega; \mathbf{V}) + \int_\Gamma F_0(y(\Omega))\, \mathbf{V} \cdot \mathbf{n} \tag{11.47}$$

and

$$J_1'(\Gamma; \mathbf{V}) = \int_\Gamma \frac{dF_1}{dz}(z(\Gamma))\ z'(\Gamma; \mathbf{V}) + \int_\Gamma \kappa F_1(z(\Gamma))\, \mathbf{V} \cdot \mathbf{n}$$

or, if $z(\Gamma) = y(\Omega)|_\Gamma$,

$$J_1'(\Gamma; \mathbf{V}) = \int_\Gamma \frac{dF_1}{dz}(z(\Gamma))\ [y'(\Gamma; \mathbf{V}) + \frac{\partial y(\Omega)}{\partial n} \mathbf{V} \cdot \mathbf{n}] + \int_\Gamma \kappa F_1(z(\Gamma))\, \mathbf{V} \cdot \mathbf{n}.$$

In the following sections we consider the shape derivative of solutions to simple elliptic boundary value problems.

11.3.4 Shape Derivative for the Solution of a Dirichlet Problem

In this section we consider the solution $y = y(\Omega)$ to the problem

$$\begin{cases} -\Delta y = f & \text{in } \Omega \\ y = g & \text{on } \Gamma \end{cases} \tag{11.48}$$

where $\Omega \in \mathcal{A}$ is of class C^2, $f \in H^1(U), g \in H^2(U)$. Thus f, g are functions defined on the universe domain U and do not depend on Ω. The following theorem holds.

Theorem 11.3. *Let $\mathbf{V} \in W^{1,\infty}(U)$. Then $y \in H^2(\Omega)$ and it is shape differentiable along the direction $\mathbf{V}$. The shape derivative along the direction $\mathbf{V}$ of y is the unique variational solution $y' = y'(\Omega; \mathbf{V}) \in H^1(\Omega)$ to the problem*

$$\begin{cases} -\Delta y' = 0 & \text{in } \Omega \\ y' = -\dfrac{\partial (y-g)}{\partial n} \mathbf{V} \cdot \mathbf{n} & \text{on } \Gamma. \end{cases} \tag{11.49}$$

Proof. For the proof of the shape differentiablility of y see [139] or [6]. We now prove (11.49) starting with $-\Delta y' = 0$ in a weak sense. If A is an open set compactly supported in Ω, then

$$\int_A \nabla y \cdot \nabla \varphi = \int_A f\varphi \qquad \forall \varphi \in C_0^\infty(A).$$

Observe that neither the domain of integration nor the test function depend on Ω_t, for t small enough. Thus, we can take the shape derivative and find

$$\int_A \nabla y' \cdot \nabla \varphi = 0 \qquad \forall \varphi \in C_0^\infty(A).$$

Since A is arbitrary, we obtain the equation $-\Delta y' = 0$ in Ω, in a weak sense.

To get the boundary condition, we write $y = g$ in the weak form

$$\int_{\Gamma} (y - g)\psi = 0 \qquad \forall \psi \in C_0^{\infty}(U).$$

Taking the shape derivative we find, using (11.45) and $y = g$ on Γ,

$$\int_{\Gamma} y'\psi + \int_{\Gamma} \left(\kappa (y - g) \psi + \frac{\partial ((y - g) \psi)}{\partial n} \right) \mathbf{V} \cdot \mathbf{n} = \int_{\Gamma} y'\psi + \int_{\Gamma} \frac{\partial (y - g)}{\partial n} \psi \mathbf{V} \cdot \mathbf{n} = 0,$$

which yields the boundary condition $y' = -\frac{\partial (y-g)}{\partial n} \mathbf{V} \cdot \mathbf{n}$ by the arbitrariness of ψ. □

11.3.5 Shape Derivative for the Solution of a Neumann Problem

In this section we compute the shape derivative of the solution $y = y(\Omega)$ to the problem

$$\begin{cases} -\Delta y + y = f & \text{in } \Omega \\ \dfrac{\partial y}{\partial n} = g & \text{on } \Gamma \end{cases} \tag{11.50}$$

where $\Omega \in \mathcal{A}$ is of class C^2, $f \in H^1(U), g \in H^2(U)$. The following theorem holds.

Theorem 11.4. *Let $\mathbf{V} \in W^{1,\infty}(U)$. Then $y \in H^2(\Omega)$ is shape differentiable along the direction $\mathbf{V}$ and the shape derivative along the direction $\mathbf{V}$ is the unique solution $y' = y'(\Omega; \mathbf{V}) \in H^1(\Omega)$ of the variational problem*

$$\int_{\Omega} (\nabla y' \cdot \nabla v + y'v) = -\int_{\Gamma} (\nabla_{\Gamma} y \cdot \nabla_{\Gamma} v + yv) \mathbf{V} \cdot \mathbf{n} + \int_{\Gamma} \left(f + \kappa g + \frac{\partial g}{\partial n} \right) v \, \mathbf{V} \cdot \mathbf{n} \tag{11.51}$$

for every $v \in C_0^2(U)$. Moreover, if Ω is of class C^3, then $y \in H^3(\Omega)$ and $y' \in H^2(\Omega)$ is the unique solution of

$$\begin{cases} -\Delta y' + y' = 0 & \text{in } \Omega \\ \dfrac{\partial y'}{\partial n} = \left(\dfrac{\partial g}{\partial n} - \dfrac{\partial^2 y}{\partial n^2} \right) \mathbf{V} \cdot \mathbf{n} + \nabla_{\Gamma} y \cdot \nabla_{\Gamma} (\mathbf{V} \cdot \mathbf{n}) & \text{on } \Gamma. \end{cases} \tag{11.52}$$

Proof. For the proof of the shape differentiablility of y see [139]. We now prove (11.51) starting with the weak formulation of problem (11.50):

$$\int_{\Omega} (\nabla y \cdot \nabla v + yv) = \int_{\Omega} fv + \int_{\Gamma} gv$$

for each $v \in C_0^2(U)$.

Taking the shape derivative we get, using (11.40) and (11.45),

$$\int_{\Omega} (\nabla y' \cdot \nabla v + y'v) + \int_{\Gamma} (\nabla y \cdot \nabla v + yv) \mathbf{V} \cdot \mathbf{n} = \int_{\Gamma} \left(fv + \kappa gv + \frac{\partial (gv)}{\partial n} \right) \mathbf{V} \cdot \mathbf{n}, \tag{11.53}$$

since $v'(\Omega; \mathbf{V}) = (gv)'(\Omega; \mathbf{V})_{\Gamma} = 0$. From (11.24) we get, on Γ,

$$\nabla y \cdot \nabla v = \nabla_{\Gamma} y \cdot \nabla_{\Gamma} v + \frac{\partial y}{\partial n} \frac{\partial v}{\partial n} = \nabla_{\Gamma} y \cdot \nabla_{\Gamma} v + g \frac{\partial v}{\partial n}.$$

Inserting this expression into (11.53) gives (11.51). Choosing $v \in C_0^2(\Omega)$ we obtain that y' is a weak solution to

$$-\Delta y' + y' = 0 \quad \text{in } \Omega.$$

If Ω is of class C^3, thanks to the elliptic regularity, we have $y \in H^3(\Omega)$ and $y' \in H^2(\Omega)$; hence, we can integrate by parts and get

$$\int_\Omega (\nabla y' \cdot \nabla v + y'v) = \int_\Gamma \frac{\partial y'}{\partial n} v + \int_\Omega (-\Delta y' + y')v = \int_\Gamma \frac{\partial y'}{\partial n} v. \tag{11.54}$$

Moreover, using (11.22),

$$\begin{aligned}\int_\Gamma \nabla_\Gamma y \cdot \nabla_\Gamma v \, \mathbf{V} \cdot \mathbf{n} &= \int_\Gamma \nabla_\Gamma y \cdot \nabla_\Gamma (v\mathbf{V} \cdot \mathbf{n}) - \int_\Gamma v \nabla_\Gamma y \cdot \nabla_\Gamma (\mathbf{V} \cdot \mathbf{n}) \\ &= -\int_\Gamma v \Delta_\Gamma y \, \mathbf{V} \cdot \mathbf{n} - \int_\Gamma v \nabla_\Gamma y \cdot \nabla_\Gamma (\mathbf{V} \cdot \mathbf{n}).\end{aligned}$$

On the other hand, (11.21) gives, on Γ,

$$-\Delta_\Gamma y = -\Delta y + \kappa \frac{\partial y}{\partial n} + \frac{\partial^2 y}{\partial n^2} = -y + f + \kappa g + \frac{\partial^2 y}{\partial n^2}. \tag{11.55}$$

Thus

$$\int_\Gamma (\nabla y \cdot \nabla v + yv) \mathbf{V} \cdot \mathbf{n} = \int_\Gamma v \left(f + \kappa g + \frac{\partial^2 y}{\partial n^2} \right) \mathbf{V} \cdot \mathbf{n} - \int_\Gamma v \nabla_\Gamma y \cdot \nabla_\Gamma (\mathbf{V} \cdot \mathbf{n}). \tag{11.56}$$

Inserting (11.54), (11.56) into (11.53) we obtain, after simple adjustments,

$$\int_\Gamma \frac{\partial y'}{\partial n} v = \int_\Gamma \left(\frac{\partial (vg)}{\partial n} - v \frac{\partial^2 y}{\partial n^2} \right) \mathbf{V} \cdot \mathbf{n} + \int_\Gamma v \nabla_\Gamma y \cdot \nabla_\Gamma (\mathbf{V} \cdot \mathbf{n}).$$

Choose now v such that $\frac{\partial v}{\partial n} = 0$. Then

$$\int_\Gamma \frac{\partial y'}{\partial n} v = \int_\Gamma v \left(\frac{\partial g}{\partial n} - \frac{\partial^2 y}{\partial n^2} \right) \mathbf{V} \cdot \mathbf{n} + \int_\Gamma v \nabla_\Gamma y \cdot \nabla_\Gamma (\mathbf{V} \cdot \mathbf{n}).$$

Observing that the set of restrictions to Γ of functions in $C_0^2(U)$ with vanishing normal derivative on Γ is dense[7] in $L^2(\Gamma)$ we deduce the Neumann condition

$$\frac{\partial y'}{\partial n} = \left(\frac{\partial g}{\partial n} - \frac{\partial^2 y}{\partial^2 n} \right) \mathbf{V} \cdot \mathbf{n} + \nabla_\Gamma y \cdot \nabla_\Gamma (\mathbf{V} \cdot \mathbf{n}).$$ □

11.4 Gradient and First-Order Necessary Optimality Conditions

In this section we derive a system of first-order necessary optimality conditions for a shape optimization problem. We first present the formal calculation of the shape derivative of a functional depending on the solution of a simple boundary value problem on a smooth domain, then we introduce a general framework exploiting the Lagrange multipliers approach to derive a system of optimality conditions.

11.4.1 A Model Problem

Let Ω be a C^2 domain and $\mathcal{A}_\Phi$ be the set of shapes obtained by deforming Ω through the perturbation of the identity map $\Phi_t(\mathbf{x}) = \mathbf{x} + t\mathbf{V}(\mathbf{x})$, $\mathbf{V} \in W^{1,\infty}(U)$. For a fixed constant m_0, define

$$\mathcal{O}_{ad} = \{\Omega \in \mathcal{A}_\Phi \ : \ |\Omega| = m_0\}.$$

[7] The space $C^2(\Gamma)$ is dense in $L^2(\Gamma)$ and every $v \in C^2(\Gamma)$ has an extension (see Definition 11.1) in $C_0^2(U)$ such that $\frac{\partial v}{\partial n} = 0$ on Γ.

Consider the following shape optimization problem:

$$J(\Omega) = \frac{1}{2} \int_{\Omega} (y(\Omega) - z_d)^2 \to \min \tag{11.57}$$

subject to $\Omega \in \mathcal{O}_{ad}$ and to

$$\begin{cases} -\Delta y(\Omega) = f & \text{in } \Omega \\ y(\Omega) = 0 & \text{on } \Gamma. \end{cases} \tag{11.58}$$

Under the assumptions of Sect. 11.3.4 and using formula (11.40) we can write, setting for simplicity $y = y(\Omega)$,

$$J'(\Omega; \mathbf{V}) = \int_{\Omega} (y - z_d) y' + \int_{\Gamma} (y - z_d)^2 \, \mathbf{V} \cdot \mathbf{n} \tag{11.59}$$

where the shape derivative $y' = y'(\Omega; \mathbf{V})$ of y is the solution to the Dirichlet problem

$$\begin{cases} -\Delta y' = 0 & \text{in } \Omega \\ y' = -\dfrac{\partial y}{\partial n} \mathbf{V} \cdot \mathbf{n} & \text{on } \Gamma. \end{cases}$$

At an optimal shape $\hat{\Omega}$, we require that

$$J'(\hat{\Omega}; \mathbf{V}) = 0$$

for all admissible vector fields $\mathbf{V}$.

Through an *adjoint method*, as in optimal control theory, it is possible to calculate $J'(\Omega; \mathbf{V})$ in terms of $\mathbf{V}$, determining also the gradient of $J(\Omega)$, without the need of calculating the shape derivative y'. Let us introduce the adjoint state $p = p(\Omega)$ as the solution of the following adjoint problem

$$\begin{cases} -\Delta p(\Omega) = y(\Omega) - z_d & \text{in } \Omega \\ p(\Omega) = 0 & \text{on } \Gamma. \end{cases} \tag{11.60}$$

By multiplying the adjoint equation by y' and integrating over Ω we get

$$-\int_{\Omega} y' \Delta p = \int_{\Omega} y'(y - z_d).$$

Integrating by parts twice, we get

$$-\int_{\Gamma} y' \frac{\partial p}{\partial n} = \int_{\Omega} y'(y - z_d)$$

being $\Delta y' = 0$ in Ω and $p = 0$ on Γ. Finally, since $y' = -\frac{\partial y}{\partial n} \mathbf{V} \cdot \mathbf{n}$ on Γ, we end up with the following expression for the shape derivative of J,

$$J'(\Omega; \mathbf{V}) = \int_{\Gamma} \left(\frac{\partial y}{\partial n} \frac{\partial p}{\partial n} + (y - z_d)^2 \right) \mathbf{V} \cdot \mathbf{n}. \tag{11.61}$$

Note that both y and p belong to $H^2(\Omega)$ by elliptic regularity, so that (11.61) makes sense. We note that the equation above can be expressed under the form (11.37), where the shape gradient of J is given by

$$G(\Gamma) = (y - z_d)^2 + \frac{\partial y}{\partial n} \frac{\partial p}{\partial n}.$$

On the other hand, differentiating the constraint

$$\int_\Omega 1 = |\Omega| = m_0$$

along the direction $\mathbf{V}$, we get

$$\int_\Gamma \mathbf{V} \cdot \mathbf{n} = 0. \tag{11.62}$$

Summarizing, the optimality conditions for the OCP (11.57), (11.58) are, other than the state equation itself, the adjoint equation (11.60) and

$$J'(\Omega; \mathbf{V}) = \int_\Gamma G(\Gamma) \mathbf{V} \cdot \mathbf{n} = 0$$

for all $\mathbf{V}$ satisfying (11.62).

11.4.2 The Lagrange Multipliers Approach

We have seen that, by introducing the adjoint variable, we can express $J'(\Omega; \mathbf{V})$ as a function of y and p, without involving the shape derivative $y'(\Omega; \mathbf{V})$. The goal is to find, in general, an expression of the form

$$J'(\Omega; \mathbf{V}) = \int_\Gamma g(y,p) \mathbf{V} \cdot \mathbf{n}$$

where $g(y,p)$ is the *gradient* of J. From a numerical standpoint, y and p are two variables independent of the field $\mathbf{V}$ and can be calculated prior to the choice of the direction $\mathbf{V}$ along which we move the boundary. Two questions arise:

- How to obtain the system of optimality conditions in a simpler way?
- How to compute the numerical solution of the problem?

A concrete approach is based on the *Lagrange multipliers* method – in this case also referred to as *Céa's method* – allowing to determine the adjoint problem and the gradient in a more natural way. We start with Neumann boundary conditions. As we shall shortly see, the case of Dirichlet boundary conditions is sligthly more delicate.

Neumann boundary conditions

Consider the problem

$$J(\Omega) = \int_\Omega C(y(\Omega)) \to \min$$

where $C = C(\cdot)$ is a smooth function, subject to $\Omega \in A_\Phi$ and to

$$\begin{cases} -\Delta y(\Omega) + y(\Omega) = f & \text{in } \Omega \\ \dfrac{\partial y}{\partial n}(\Omega) = g & \text{on } \Gamma. \end{cases} \tag{11.63}$$

The weak formulation of (11.63) is

$$\int_\Omega (\nabla y \cdot \nabla \psi + y\psi - f\psi) = \int_\Gamma g\psi \qquad \forall \psi \in H^1(\Omega).$$

Under the assumptions of Sect. 11.3.5, i.e., $f \in H^1(\mathbb{R}^d)$, $g \in H^2(\mathbb{R}^d)$, we have $y = y(\Omega) \in H^2(\Omega)$ and it is shape differentiable (Theorem 11.4).

We introduce the Lagrangian functional

$$\mathcal{L}(\Omega, z, p) = \int_\Omega C(z) - \int_\Omega (\nabla z \cdot \nabla p + zp - fp) + \int_\Gamma gp$$

which is defined for any shape $\Omega \in \mathcal{O}_{ad}$, and for any z and $p \in H^1(\mathbb{R}^d)$, so that the variables Ω, z and p are *independent*.

In general, the Lagrangian functional is given by the difference between the cost functional $J(\cdot)$ and either the weak or the strong form of the state problem. Observe that evaluating $\mathcal{L}$ at $z = y(\Omega)$ gives[8]

$$\mathcal{L}(\Omega, y(\Omega), p) = \int_\Omega C(y(\Omega)) = J(\Omega).$$

As in the OCPs case, the vanishing of the partial derivatives of $\mathcal{L}$ with respect to z and p and the shape derivative of $\mathcal{L}$ yield, at the optimum, the system of optimality conditions.

Recovering the state equation. We have

$$\mathcal{L}'_p(\Omega, z, p)\,\psi = -\int_\Omega (\nabla z \cdot \nabla \psi + z\psi - f\psi) + \int_\Gamma g\psi \quad \forall \psi \in H^1(\mathbb{R}^d).$$

We require that, *at the optimum* $z = \hat{y} = y(\hat{\Omega})$, with $\hat{\Omega}$ optimal shape and $\hat{\Gamma} = \partial\hat{\Omega}$,

$$-\int_{\hat{\Omega}} (\nabla \hat{y} \cdot \nabla \psi + \hat{y}\psi - f\psi) + \int_{\hat{\Gamma}} g\psi = 0 \qquad \forall \psi \in H^1(\mathbb{R}^d).$$

Choosing in particular[9] $\psi \in \mathcal{D}(\overline{\Omega})$, we recover the weak formulation of the *state problem*.

The adjoint problem. The partial derivative of $\mathcal{L}$ with respect to z, evaluated at ξ, is

$$\mathcal{L}'_z(\Omega, z, p)\xi = \int_\Omega C'(z)\xi - \int_\Omega (\nabla \xi \cdot \nabla p + \xi p) \qquad \forall \xi \in H^1(\mathbb{R}^d).$$

We choose $p = \hat{p}$ such that, *at* $\hat{\Omega}, \hat{y}$,

$$\mathcal{L}'_z(\hat{\Omega}, \hat{y}, \hat{p})\xi = 0 \qquad \forall \xi \in H^1(\mathbb{R}^d). \tag{11.64}$$

Choosing again $\xi \in \mathcal{D}(\overline{\Omega})$, (11.64) gives the weak formulation of the *adjoint problem*

$$\begin{cases} -\Delta \hat{p} + \hat{p} = C'(\hat{y}) & \text{in } \hat{\Omega} \\ \dfrac{\partial \hat{p}}{\partial n} = 0 & \text{on } \hat{\Gamma}. \end{cases} \tag{11.65}$$

Note that, by elliptic regularity, $\hat{p} \in H^2(\Omega)$.

The shape derivative $J'(\hat{\Omega}; \mathbf{V})$. Since

$$J(\Omega) = \mathcal{L}(\Omega, y(\Omega), p) \quad \forall p \in H^1(\mathbb{R}^d),$$

we have, using the chain rule,

$$J'(\Omega; \mathbf{V}) = \mathcal{L}'_\Omega(\Omega, y(\Omega), p)(\mathbf{V}) + \mathcal{L}'_z(\Omega, y(\Omega), p)y'(\Omega, \mathbf{V}),$$

---

[8] Recall that there exists a continuous extension operator $E : H^1(\Omega) \to H^1(\mathbb{R}^d)$.

[9] Recall that $\mathcal{D}(\overline{\Omega})$ is the set of restrictions to $\overline{\Omega}$ of functons in $C_0^\infty(\mathbb{R}^d)$, which is dense into $H^1(\Omega)$, if Ω is a bounded Lipschitz domain.

where $\mathcal{L}'_\Omega(\Omega, z, p)(\mathbf{V})$ denotes the shape derivative of $\mathcal{L}$ along $\mathbf{V}$, evaluated at Ω, z, p. At the optimum, since $\mathcal{L}'_z(\hat{\Omega}, \hat{y}, \hat{p}) = 0$, we get

$$J'(\hat{\Omega}; \mathbf{V}) = \mathcal{L}'_\Omega(\hat{\Omega}, y(\hat{\Omega}), \hat{p})(\mathbf{V}). \tag{11.66}$$

To calculate $\mathcal{L}'_\Omega(\Omega, z, p)(\mathbf{V})$, $p \in H^2(\mathbb{R}^d)$, we use formulas (11.40) and (11.45), since the integrands are fixed functions; we find

$$\mathcal{L}'_\Omega(\Omega, z, p)(\mathbf{V}) = \int_\Gamma \left(C(z) - \nabla z \cdot \nabla p - zp + fp + \frac{\partial(gp)}{\partial n} + \kappa gp \right) \mathbf{V} \cdot \mathbf{n}.$$

Evaluating this derivative at $\hat{\Omega}, \hat{y}$ and $p = \hat{p}$, we finally find, since $\frac{\partial \hat{p}}{\partial n} = 0$ on $\hat{\Gamma}$,

$$J'(\hat{\Omega}; \mathbf{V}) = \int_{\hat{\Gamma}} \left(C(\hat{y}) - \nabla \hat{y} \cdot \nabla \hat{p} + (f - \hat{y} + \frac{\partial g}{\partial n} + \kappa g)\hat{p} \right) \mathbf{V} \cdot \mathbf{n}.$$

Thus the gradient of J is given by

$$G(\Gamma) = C(\hat{y}) - \nabla \hat{y} \cdot \nabla \hat{p} + (f - \hat{y} + \frac{\partial g}{\partial n} + \kappa g)\hat{p}.$$

Dirichlet boundary conditions

Consider now the problem

$$J(\Omega) = \int_\Omega C(y(\Omega)) \to \min$$

where $C = C(\cdot)$ is a smooth function, subject to $\Omega \in A_\Phi$ and to

$$\begin{cases} -\Delta y(\Omega) + y(\Omega) = f & \text{in } \Omega \\ \qquad\quad y(\Omega) = g & \text{on } \Gamma, \end{cases} \tag{11.67}$$

with $f \in H^1(\mathbb{R}^d)$, $g \in H^2(\mathbb{R}^d)$. A weak formulation of (11.67) is: find $w = y - g \in H_0^1(\Omega)$ such that

$$\int_\Omega (\nabla w \cdot \nabla \varphi + w\varphi) = \int_\Omega (f\varphi - \nabla g \cdot \nabla \varphi - g\varphi) \qquad \forall \varphi \in H_0^1(\Omega).$$

By elliptic regularity, $y = y(\Omega) \in H^2(\Omega)$ and it is shape differentiable by Theorem 11.2. It would seem natural to introduce the Lagrangian functional

$$\mathcal{L}(\Omega, w, p) = \int_\Omega C(w + g) + \int_\Omega (f - gp - \nabla w \cdot \nabla p - wp);$$

however, $\mathcal{L}(\Omega, w, p)$ should be defined for any shape $\Omega \in A_\Phi$, and here $w, p \in H_0^1(\Omega)$. Unfortunately, in this case Ω, w and p are not independent; indeed w and p vanish on Γ so that they are tied to Ω. This means that the shape derivative of this $\mathcal{L}$ does not give the correct result.

To avoid this inconvenient, multiply the differential equation by $\varphi \in H^1(\mathbb{R}^d)$, integrate over Ω to get, after an integration by parts,

$$\int_\Omega (\nabla y \cdot \nabla \varphi + y\varphi) + \int_\Gamma \frac{\partial y}{\partial n}\varphi = \int_\Omega f\varphi \quad \forall \varphi \in H^1(\mathbb{R}^d).$$

To take care of the Dirichlet conditions, we introduce a second Lagrange multiplier λ and, for $\Omega \in A_\Phi$, $z \in H^2(\mathbb{R}^d), p, \lambda \in H^1(\mathbb{R}^d)$, we define the Lagrangian functional

$$\mathcal{L}(\Omega, z, p, \lambda) = \int_{\Omega} C(z) + \int_{\Omega} (f - \nabla z \cdot \nabla p - zp) + \int_{\Gamma} \frac{\partial z}{\partial n} p + \int_{\Gamma} \lambda (z - g) \tag{11.68}$$

Since z, q and λ belong to spaces independent of Ω, *all the arguments in* (11.68) *are independent.* Note again that evaluating $\mathcal{L}$ at $z = y(\Omega)$ gives

$$\mathcal{L}(\Omega, y(\Omega), p, \lambda) = \int_{\Omega} C(y(\Omega)) = J(\Omega) \quad \forall p, \lambda \in H^1\left(\mathbb{R}^d\right).$$

Recovering the state problem. The vanishing of the partial derivatives $\mathcal{L}'_p$ and $\mathcal{L}'_\lambda$ along $\psi \in H^2(\mathbb{R}^d)$ and $\mu \in H^1(\mathbb{R}^d)$ give, at the optimum $z = \hat{y}$, $\Omega = \hat{\Omega}$,

$$\int_{\hat{\Omega}} (f - \nabla \hat{y} \cdot \nabla \psi - \hat{y}\psi) + \int_{\hat{\Gamma}} \frac{\partial \hat{y}}{\partial n} \psi = 0 \quad \forall \psi \in H^2(\mathbb{R}^d) \tag{11.69}$$

and

$$\int_{\hat{\Gamma}} \mu (\hat{y} - g)\, d\sigma = 0 \quad \forall \mu \in H^1(\mathbb{R}^d). \tag{11.70}$$

Choosing both ψ and μ in $\mathcal{D}(\overline{\Omega})$, counter-integrating by parts in (11.69), we recover the state equation $\Delta \hat{y} - \hat{y} = f$ in $\hat{\Omega}$ and, from (11.70), the boundary condition $\hat{y} = g$ on $\hat{\Gamma}$.

*The adjoint problem.*The partial derivative of $\mathcal{L}$ with respect to z, evaluated at $\xi \in H^2(\mathbb{R}^d)$, is

$$\mathcal{L}'_z(\Omega, z, p, \lambda)\xi = \int_{\Omega} C'(z)\xi - \int_{\Omega} (\nabla \xi \cdot \nabla p - \xi p) + \int_{\Gamma} \frac{\partial \xi}{\partial n} p + \int_{\Gamma} \lambda \xi.$$

We choose $p = \hat{p}$, $\lambda = \hat{\lambda}$ such that, *at the optimum* $\hat{\Omega}, \hat{y}$,

$$\int_{\hat{\Omega}} C'(\hat{y})\xi - \int_{\hat{\Omega}} (\nabla \xi \cdot \nabla \hat{p} - \xi \hat{p}) + \int_{\hat{\Gamma}} \frac{\partial \xi}{\partial n} \hat{p} + \int_{\hat{\Gamma}} \hat{\lambda} \xi = 0 \qquad \forall \xi \in H^2(\mathbb{R}^d). \tag{11.71}$$

To identify $\hat{p}$, choose $\xi \in \mathcal{D}(\Omega)$, arbitrary. We deduce

$$\int_{\hat{\Omega}} C'(\hat{y})\xi - \int_{\hat{\Omega}} (\nabla \xi \cdot \nabla \hat{p} - \xi \hat{p}) = 0 \qquad \forall \xi \in \mathcal{D}(\Omega). \tag{11.72}$$

Choosing now $\xi \in \mathcal{D}\left(\overline{\Omega}\right)$, $\xi = 0$ on $\hat{\Gamma}$ and arbitrary $\frac{\partial \xi}{\partial n}$, we obtain

$$\hat{p} = 0 \quad \text{on } \hat{\Gamma}. \tag{11.73}$$

Thus $\hat{p} \in H_0^1(\Omega)$ and it satisfies the adjoint equation (11.72) in weak form.

Note that elliptic regularity gives that $\hat{p} \in H^2(\Omega)$. Thus, the *adjoint problem* for the multiplier $\hat{p}$ in strong form is

$$\begin{cases} -\Delta \hat{p} + \hat{p} = C'(\hat{y}) & \text{in } \hat{\Omega} \\ \hat{p} = 0 & \text{on } \hat{\Gamma}. \end{cases} \tag{11.74}$$

To identify $\hat{\lambda}$, choose $\xi \in \mathcal{D}\left(\overline{\Omega}\right)$, arbitrary, in (11.71). Taking into account (11.74), we deduce, after an integration by parts of the first term,

$$\int_{\hat{\Gamma}} \left(-\frac{\partial \hat{p}}{\partial n} + \hat{\lambda} \right) \xi = 0 \qquad \forall \xi \in \mathcal{D}\left(\overline{\Omega}\right)$$

from which

$$\hat{\lambda} = \frac{\partial \hat{p}}{\partial n} \qquad \text{on } \hat{\Gamma}. \tag{11.75}$$

The shape derivative $J'(\hat{\Omega}; \mathbf{V})$. From

$$J(\Omega) = \mathcal{L}(\Omega, y(\Omega), p, \lambda) \qquad \forall p, \lambda \in H^1(\mathbb{R}^d),$$

we have

$$J'(\Omega; \mathbf{V}) = \mathcal{L}'_\Omega(\Omega, y(\Omega), p, \lambda)(\mathbf{V}) + \mathcal{L}'_z(\Omega, y(\Omega), p, \lambda) y'(\Omega, \mathbf{V}).$$

Since $\mathcal{L}'_z(\hat{\Omega}, \hat{y}, \hat{p}, \hat{\lambda}) = 0$, we get

$$J'(\hat{\Omega}; \mathbf{V}) = \mathcal{L}'_\Omega(\hat{\Omega}, \hat{y}, \hat{p}, \hat{\lambda})(\mathbf{V}).$$

Formulas (12.35) and (12.40) give, being z, p and λ independent of Ω,

$$\mathcal{L}'_\Omega(\Omega, z, p, \lambda)(\mathbf{V}) = \int_{\hat{\Gamma}} \left(C(z) + (f + \Delta z - z)p + \frac{\partial((z-g)\lambda)}{\partial n} + \kappa(z-g)\lambda p \right) \mathbf{V} \cdot \mathbf{n}.$$

Evaluating this derivative at $(\hat{\Omega}, \hat{y}, \hat{p}, \hat{\lambda})$ and using that $\hat{y} = g$, $\hat{\lambda} = \frac{\partial \hat{p}}{\partial n}$ and $\hat{p} = 0$ on $\hat{\Gamma}$, we find

$$J'(\hat{\Omega}; \mathbf{V}) = \int_{\hat{\Gamma}} \left(C(\hat{y}) + \frac{\partial(\hat{y} - g)}{\partial n} \frac{\partial \hat{p}}{\partial n} \right) \mathbf{V} \cdot \mathbf{n}.$$

Thus the gradient of J is given by

$$G(\hat{\Gamma}) = C(\hat{y}) + \frac{\partial(\hat{y} - g)}{\partial n} \frac{\partial \hat{p}}{\partial n}.$$

Other types of functionals and boundary conditions can be considered, as shown below.

Mixed Dirichlet-Neumann boundary conditions

We consider the optimization of the shape of an elastic membrane (see Fig. 11.2), whose displacement y is the solution to the following mixed problem:

$$\begin{cases} -\Delta y = f & \text{in } \Omega \\ y = 0 & \text{on } \Gamma_D \\ \dfrac{\partial y}{\partial n} = g & \text{on } \Gamma_N \\ \dfrac{\partial y}{\partial n} = 0 & \text{on } \Gamma_C. \end{cases} \tag{11.76}$$

Only the part Γ_C is free to move, while Γ_D and Γ_N are kept fixed. Accordingly, we assume that $\mathbf{V} = \mathbf{0}$ on Γ_D and Γ_N. Moreover, $f \in H^1(\mathbb{R}^d)$ and $g \in H^2(\mathbb{R}^d)$. The goal is to minimize the compliance of the structure, which can be expressed as

$$J(\Omega) = \int_\Omega f y(\Omega) + \int_{\Gamma_N} g y(\Omega). \tag{11.77}$$

Let us introduce the following Lagrangian functional,

$$\mathcal{L}(\Omega, z, p, \lambda) = \int_{\Omega} fz + \int_{\Gamma_N} gz + \int_{\Omega} fp + \int_{\Gamma_N} gp - a(z,p) + \int_{\Gamma_D} \lambda z,$$

where $z, p, \lambda \in H^1(\mathbb{R}^d)$, and

$$a(z,p) = \int_{\Omega} \nabla z \cdot \nabla p.$$

The state equation and the corresponding Dirichlet boundary condition can be recovered as in the previous section, requiring that, respectively,

$$\mathcal{L}'_p(\hat{\Omega}, \hat{y}, \hat{p}, \hat{\lambda})\varphi = \int_{\hat{\Omega}} f\varphi + \int_{\hat{\Gamma}_N} g\varphi - a(\hat{y}, \varphi) = 0 \quad \forall \varphi \in H^1(\mathbb{R}^d)$$
$$\mathcal{L}'_\lambda(\hat{\Omega}, \hat{y}, \hat{p}, \hat{\lambda})\mu = \int_{\hat{\Gamma}_D} \mu\hat{y} = 0 \quad \forall \mu \in H^1(\mathbb{R}^d).$$

The partial derivative of $\mathcal{L}$ with respect to z evaluated at ξ is

$$\mathcal{L}'_z(\Omega, z, p, \lambda)\xi = \int_{\Omega} f\xi + \int_{\Gamma_N} g\xi - a(\xi, p) + \int_{\Gamma_D} \lambda\xi.$$

We require that, *at the optimum*,

$$\mathcal{L}'_z(\hat{\Omega}, \hat{y}, \hat{p}, \hat{\lambda})\xi = \int_{\hat{\Omega}} f\xi + \int_{\hat{\Gamma}_N} g\xi - a(\xi, \hat{p}) + \int_{\hat{\Gamma}_D} \hat{\lambda}\xi = 0 \quad \forall \xi \in H^1(\mathbb{R}^d). \tag{11.78}$$

Choose $\xi \in \mathcal{D}\left(\overline{\Omega}\right)$, $\xi = 0$ on $\hat{\Gamma}_D$. We find

$$\int_{\hat{\Omega}} f\xi + \int_{\hat{\Gamma}_N} g\xi - a(\xi, \hat{p}) = 0 \qquad \forall \xi \in \mathcal{D}\left(\overline{\Omega}\right), \ \ \xi = 0 \text{ on } \hat{\Gamma}_D.$$

Thus, we obtain that $\hat{p} = \hat{y}$, that is, the adjoint problem is equivalent to the state problem. In other words, the initial minimization problem is self-adjoint. Moreover, integrating by parts in (11.78) and taking into account the adjoint problem (in strong form), we (formally) find, for each $\xi \in H^1(\mathbb{R}^d)$,

$$\int_{\hat{\Omega}} f\xi + \int_{\hat{\Gamma}_N} g\xi + \int_{\hat{\Omega}} \Delta\hat{p}\xi - \int_{\partial\hat{\Omega}} \frac{\partial \hat{p}}{\partial n}\xi + \int_{\hat{\Gamma}_D} \hat{\lambda}\xi = \int_{\hat{\Gamma}D} \left(\hat{\lambda} - \frac{\partial \hat{p}}{\partial n}\right)\xi = 0.$$

Hence

$$\hat{\lambda} = \frac{\partial \hat{p}}{\partial n} = \frac{\partial \hat{y}}{\partial n} \qquad \text{on } \hat{\Gamma}_D.$$

Since the initial minimization problem is self-adjoint, to compute the shape derivative of J, we can refer to a reduced Lagrangian function by setting $p = z$, that is

$$\mathcal{L}(\Omega, z, \lambda) = 2\int_{\Omega} (fz - |\nabla z|^2) + 2\int_{\Gamma_N} gz + \int_{\Gamma_D} \lambda z.$$

We have that

$$J(\Omega) = \mathcal{L}(\Omega, y(\Omega), \lambda) \qquad \forall \lambda \in H^1(\mathbb{R}^d)$$

and, as in the previous section,

$$J'(\hat{\Omega}; \mathbf{V}) = \mathcal{L}'_\Omega(\hat{\Omega}, \hat{y}, \hat{\lambda})(\mathbf{V}).$$

Since f, z, g, λ are independent of Ω,

$$
\begin{aligned}
\mathcal{L}'_{\Omega}(\Omega, z, \lambda)(\mathbf{V}) &= \int_{\Gamma} (2fz - |\nabla z|^2)\mathbf{V}\cdot\mathbf{n} + 2\int_{\Gamma_N} \left(\frac{\partial(gz)}{\partial n} + \kappa g z\right)\mathbf{V}\cdot\mathbf{n} \\
&\quad + \int_{\Gamma_D} \left(\frac{\partial(\lambda z)}{\partial n} + \kappa\lambda z\right)\mathbf{V}\cdot\mathbf{n} \\
&= \int_{\hat{\Gamma}_C} (2fz - |\nabla z|^2)\mathbf{V}\cdot\mathbf{n}
\end{aligned}
$$

since $\mathbf{V} = \mathbf{0}$ on $\Gamma_D \cup \Gamma_N$. Therefore,

$$
J'(\hat{\Omega}; \mathbf{V}) = \int_{\hat{\Gamma}_C} (2f\hat{y} - |\nabla \hat{y}|^2)\, \mathbf{V}\cdot\mathbf{n}.
$$

In absence of external forces, that is if $f = 0$,

$$
J'(\hat{\Omega}; \mathbf{V}) = -\int_{\hat{\Gamma}_C} |\nabla \hat{y}|^2 \mathbf{V}\cdot\mathbf{n}.
$$

We can conclude that, in order to reduce the compliance, that is, to have $J'(\hat{\Omega}; \mathbf{V}) \leq 0$, a possible way is to enlarge the domain, that is, to choose $\mathbf{V}\cdot\mathbf{n} \leq 0$.

11.5 Numerical Approximation of Shape Optimization Problems

The numerical approximation of shape optimization problems poses additional challenges than for OCPs, because of the need to handle shape deformations, to enhance shape regularity, and possibly to avoid remeshing. In this respect, a key issue is the way shape deformations are expressed as functions of the *shape gradient* of the cost functional. In this section we illustrate a basic technique relying on the *gradient* (or *steepest descent*) *method*, and point out further possible extensions available in literature.

Starting from the expression

$$
J'(\Omega; \mathbf{V}) = \int_{\Gamma} G(\Gamma)\mathbf{V}\cdot\mathbf{n},
$$

the idea of the *steepest descent* (or *gradient*) *method* is to deform, at each step $k = 0, 1, \ldots$, the shape of the domain $\Omega^{(k)}$ into a new shape configuration

$$
\Omega^{(k+1)} = (I + \tau_k \mathbf{V}^{(k)})(\Omega^{(k)})
$$

where τ_k is a pseudo time-step and $\mathbf{V}^{(k)}$ is the descent direction at step k. In particular, if we set

$$
\mathbf{V}^{(k)} = \begin{cases} -G(\Gamma^{(k)})\mathbf{n}^{(k)} & \text{on } \Gamma_C^{(k)} \\ 0 & \text{on } \Gamma^{(k)} \setminus \Gamma_C^{(k)} \end{cases} \tag{11.79}
$$

where $\Gamma_C^{(k)}$ denotes the part of the boundary (at step k) that we can modify, we obtain that

$$
J(\Omega^{(k+1)}) < J(\Omega^{(k)})
$$

since

$$
J'(\Omega^{(k+1)}; \mathbf{V}^{(k)}) = \int_{\Gamma_C^{(k)}} G(\Gamma^{(k)})(-G(\Gamma^{(k)})\mathbf{n}^{(k)}\cdot\mathbf{n}^{(k)}) = -\int_{\Gamma_C^{(k)}} |G(\Gamma^{(k)})|^2 < 0.
$$

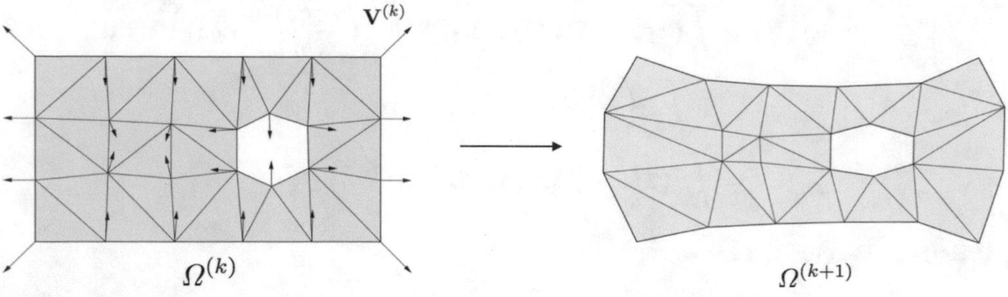

Fig. 11.4 Shape deformation through local boundary variation

In the case at each step of the gradient method the shape is deformed through a local displacement of its boundary, we obtain the so-called *local boundary variation* method. This technique assumes shapes are described through boundary parametrization, so that boundary portions are defined analytically.

The corresponding numerical algorithm can be obtained by adapting the idea of the descent method of Sect. 6.3, see Algorithm 11.1. In this algorithm, the numerical approximation of both state and adjoint problems is performed through the finite element method, yielding the discrete state $\mathbf{y} = \mathbf{y}(\Omega)$ and adjoint state $\mathbf{p} = \mathbf{p}(\Omega)$, respectively.

Algorithm 11.1 Gradient method for shape optimization, local boundary variation

Input: initial shape $\Omega^{(0)}$
compute $\mathbf{y}_0 = \mathbf{y}(\Omega^{(0)})$ by solving the state equation in $\Omega^{(0)}$
compute $\mathbf{p}_0 = \mathbf{p}(\Omega^{(0)})$ by solving the adjoint equation in $\Omega^{(0)}$
compute $J(\Omega^{(0)})$ and the shape gradient $G(\Gamma^{(0)})$
set $k = 0$
while $|J'(\Omega^{(k)}; \mathbf{V}^{(k)})|/|J'(\Omega^{(0)}; \mathbf{V}^{(0)})| > \text{tol}$ or $|J(\Omega^{(k)}) - J_h(\Omega^{(k-1)})| > \text{tol}$
    compute a descent direction $\boldsymbol{V}^{(k)}$ fulfilling (11.79)
    compute a step length τ_k with a line search routine
    set $\Omega^{(k+1)} = (I + \tau_k \mathbf{V}^{(k)})(\Omega^{(k)})$
    compute $\mathbf{y}_{k+1} = \mathbf{y}(\Omega^{(k+1)})$ by solving the state equation on $\Omega^{(k+1)}$
    compute $\mathbf{p}_{k+1} = \mathbf{p}(\Omega^{(k+1)})$ by solving the adjoint equation on $\Omega^{(k+1)}$
    compute $J(\Omega^{(k+1)})$ and $G(\Gamma^{(k+1)})$
    $k = k + 1$
end while

An update of the computational mesh is required at each step k. For the sake of computational efficiency, we would like to avoid remeshing at each step. Mesh motion, obtained by relocating each mesh node $\mathbf{x}_h^{(k)}$ to $\mathbf{x}_h^{(k)} + \tau_k \mathbf{V}^{(k)}(\mathbf{x}_h^{(k)})$, is a more feasible option, however two crucial questions arise since:

- the mesh quality must be preserved at each step, avoiding numerical artifacts, and
- the shape gradient only provides information on the free boundary Γ_C of the domain shape which we can act on during the iterative design process.

Hence, we need to extend the definition of $\mathbf{V}^{(k)}$ inside the domain, and to regularize it at the same time. Indeed, if the step size τ_k can be properly tuned to avoid large deformations (often breaking mesh validity), to enhance the regularity of the mesh displacement we can exploit a harmonic (or solid) extension mesh moving technique, by evaluating e.g. an *harmonic extension* of the shape gradient of the cost functional at each optimization step. This operation can be performed in (at least) two possible ways:

1. by defining $\mathbf{V}^{(k)}$ as the solution of the problem

$$\begin{cases} -\Delta \mathbf{V}^{(k)} = \mathbf{0} & \text{in } \Omega^{(k)} \\ \mathbf{V}^{(k)} = -G(\Gamma^{(k)})\mathbf{n}^{(k)} & \text{on } \Gamma_C^{(k)} \\ \mathbf{V}^{(k)} = \mathbf{0} & \text{on } \Gamma^{(k)} \setminus \Gamma_C^{(k)}. \end{cases}$$

   It is straightforward to check that $\mathbf{V}^{(k)}$ is a descent direction for J. However, a possible low regularity of $G(\Gamma^{(k)})$ may yield numerical oscillations of the boundary during the optimization procedure. The following option enables to gain a further degree of regularity;
2. by defining $\mathbf{V}^{(k)}$ as the solution of the problem

$$\begin{cases} -\Delta \mathbf{V}^{(k)} = \mathbf{0} & \text{in } \Omega^{(k)} \\ \dfrac{\partial \mathbf{V}^{(k)}}{\partial \mathbf{n}^{(k)}} = -G(\Gamma^{(k)})\mathbf{n}^{(k)} & \text{on } \Gamma_C^{(k)} \\ \mathbf{V}^{(k)} = \mathbf{0} & \text{on } \Gamma^{(k)} \setminus \Gamma_C^{(k)}. \end{cases} \tag{11.80}$$

   In this case, from the weak form of the problem,

$$\int_{\Omega^{(k)}} \nabla \mathbf{V}^{(k)} \cdot \nabla \mathbf{w} = \int_{\Gamma_C^{(k)}} \frac{\partial \mathbf{V}^{(k)}}{\partial \mathbf{n}^{(k)}} \mathbf{w} = -\int_{\Gamma_C^{(k)}} G(\Gamma^{(k)}) \mathbf{w} \cdot \mathbf{n}^{(k)} \qquad \forall \mathbf{w} \in H^1_{\Gamma_D}(\Omega^{(k)})^d$$

   we obtain, inserting $\mathbf{w} = \mathbf{V}^{(k)}$,

$$J'(\Omega^{(k)}; \mathbf{V}^{(k)}) = \int_{\Gamma_C^{(k)}} G(\Gamma^{(k)}) \mathbf{V}^{(k)} \cdot \mathbf{n}^{(k)} = -\int_{\Omega^{(k)}} |\nabla \mathbf{V}^{(k)}|^2 < 0,$$

   hence $\mathbf{V}^{(k)}$ is indeed a descent direction.

11.5.1 Computational Aspects

The numerical approximation of the harmonic extension problem (11.80) can be carried out using the finite element method, on the same computational mesh already used for the discretization of both the state and the adjoint problems. Some comments are in order, to make this procedure computationally feasible.

Additional Geometric Constraints

Very often, geometric constraints are included in a shape optimization problem to fulfill specific properties depending on the problem at hand. For instance, the shape of a body immersed in a fluid is optimized in order to minimize drag forces, while requiring that the volume of the body remains constant, say,

$$m(\Omega) = \int_{\Omega} 1 = m_0 \tag{11.81}$$

for a given $m_0 > 0$. This is in fact a common constraint on the admissible open sets, as seen in Sect.11.4.1. We follow [9] to show how this constraint can be handled in the optimization algorithm. The same procedure can then be extended to treat other geometric constraints.

Recall that the shape derivative of $m(\Omega)$ in the direction $\mathbf{V}$ is given by

$$m'(\Omega;\mathbf{V}) = \int_\Gamma \mathbf{V}\cdot\mathbf{n}.$$

Hence, to fulfill (11.81), we can minimize the modified cost functional

$$\tilde{J}(\Omega) = J(\Omega) + \xi\,(m(\Omega) - m_0),$$

where $\xi \in \mathbb{R}$ is a Lagrange multiplier; the descent direction is then computed by evaluating the shape derivative of $\tilde{J}(\Omega)$.

From a numerical standpoint, the value of the Lagrange multiplier is updated at each step so that at convergence the domain shape fulfills the volume constraint. Since mesh motion is an expensive task, we prefer not to enforce the volume constraint exactly at each step; rather, we increase the value of the Lagrange multiplier if the current shape volume is larger than the target volume, or we decrease it otherwise. To avoid numerical oscillations in the value of the shape volume, the constraint is relaxed by taking into account the value of the Lagrange multiplier computed by assuming that the optimality condition

$$J'(\Omega;\mathbf{V}) + \xi m'(\Omega;\mathbf{V}) = 0$$

is satisfied at least by averaging on the value of $\mathbf{V}$ on the boundary, so that

$$\xi = -\frac{\int_\Gamma G(\Gamma)}{\int_\Gamma 1}.$$

Finally, the Lagrange multiplier is updated at each iteration according to the formula

$$\xi^{(k+1)} = \frac{\xi^{(k)} + \xi}{2} + \varepsilon_l(m - m_0)$$

where $\varepsilon_l > 0$ is a sufficiently small costant, fixed a priori.

Mesh Handling

From a numerical standpoint, shape regularization can be achieved by using two meshes to represent the computational domain during the optimization process. We denote by $\mathcal{T}_{h,fine}^{(k)}$ the (fine) mesh used to perform the discretization of the state and the adjoint problems, as well as for the solution of the harmonic extension problem to determine $\mathbf{V}^{(k)}$, and by $\mathcal{T}_{h,coarse}^{(k)}$ a coarser mesh used to perform shape deformation, after the displacement field $\mathbf{V}^{(k)}$ has been interpolated over $\mathcal{T}_{h,coarse}^{(k)}$. The updated mesh $\mathcal{T}_{h,coarse}^{(k+1)}$ is generated by moving the nodes of the coarse mesh; then, the fine mesh $\mathcal{T}_{h,fine}^{(k+1)}$ is built by adapting the coarse one.

Despite the flexibility of local boundary variations, alternative approaches are often employed. Indeed, at each step of the optimization procedure local boundary variation requires the evaluation of a further differential problem and the need to handle mesh structures. Although remeshing is avoided, the whole procedure is rather expensive in any case. Suitable *shape parametrizations* are introduced to prescribe admissible shape deformations by a set of geometrical parameters (or *design variables*). In this way, a geometrical SOP is cast in the (simpler) family of parametric SOPs. For instance, shapes can be described in terms of a set of control points – this is. e.g., the case when using Computer Aided Design (CAD) – exploiting given families of polynomial (or rational) functions, such as Bézier curves, B-splines or NURBS. In these cases, the position of control points can be updated at each step of the optimization procedure, according to the information provided by the shape gradient; see, e.g., [60] for a derivation of such a procedure.

Another popular approach relies on the so-called *volume-based representations* [99] which allow to perform deformations whatever the complexity of the shape. In particular, by means of maps relying on such a representation, geometry can be parametrized independently from PDE models, discrete formulations and computational meshes. Volume based representations operate on a control volume, regardless of the object to be deformed, contained into the volume. Parametric maps are thus defined by introducing a set of control points over the control volume and considering their displacements (which actually induce a shape deformation) as geometrical parameters, rather than geometrical properties directly related with the shape itself. In this way, shifting a control point causes a deformation of the embedding space – the parametric map is defined inside the whole control volume – and thereby induces a global modification of each shape located inside this volume. Two possible techniques leading to free shape representations are *free-form deformations* [243, 173, 82, 12] and *radial basis functions interpolation* [152, 239, 272]. See, e.g., [17, 189, 188] for their application to the numerical approximation of shape optimization problems.

Another possible – and indeed often used – approach to represent shapes relies on the *level set* method [245], in which a bounded domain is given by the zero level set of a suitable function $\phi : \mathbb{R}^d \to \mathbb{R}$ and its deformations are determined through a level set advection equation, which is indeed related to the Hamilton-Jacobi equation, see, e.g., [8, 7].

11.6 Minimizing the Compliance of an Elastic Structure

In this section we consider a problem arising in structural mechanics, namely the optimal design of a solid structure in order to minimize its compliance, under suitable physical constraints, e.g., on the total mass of the structure. The compliance represents the work of external loads, and its minimization is equivalent to the maximization of the stiffness of the structure.

We consider a structure occupying the domain $\Omega \subset \mathbb{R}^d$, which is fixed on a portion of its boundary $\Gamma_D \subset \partial\Omega$ and submitted to a surface load g, applied on Γ_N. A third portion, Γ_C, represents the free boundary portion, and has to be optimally designed; we assume that Γ_C is neither fixed, nor subject to surface loads. Moreover, we assume that $\Gamma = \Gamma_D \cup \Gamma_N \cup \Gamma_C$, with $\Gamma_D, \Gamma_N, \Gamma_C$ pairwise disjoint. The linear elastic deformation of an isotropic solid occupying the domain $\Omega \subset \mathbb{R}^d$ is described in terms of the stress tensor $\boldsymbol{\sigma} : \mathbb{R}^d \to \mathbb{R}^{d\times d}$, the strain tensor $\boldsymbol{\varepsilon} : \mathbb{R}^d \to \mathbb{R}^{d\times d}$, the applied body force $\mathbf{f} : \mathbb{R}^d \to \mathbb{R}^d$ and the displacement field $\boldsymbol{y} : \mathbb{R}^d \to \mathbb{R}^d$, this latter being the unknown of the problem. The governing equations consist of an equation stating the equilibrium of forces

$$-\mathrm{div}(\boldsymbol{\sigma}) = \boldsymbol{f} \quad \text{in } \Omega,$$

the strain-displacement relation

$$\boldsymbol{\varepsilon}(\boldsymbol{y}) = \frac{1}{2}(\nabla \boldsymbol{y} + (\nabla \boldsymbol{y})^\top)$$

and the constitutive law, which in the linear isotropic case takes the form

$$\boldsymbol{\sigma} = 2\mu\boldsymbol{\varepsilon}(\boldsymbol{y}) + \lambda(\mathrm{div}(\boldsymbol{y}))\boldsymbol{I}.$$

Here μ and λ are the Lamé coefficients, which can be expressed in terms of the Young modulus E and the Poisson coefficient ν as

$$\lambda = \frac{E\nu}{(1+\nu)(1-2\nu)}, \qquad \mu = \frac{E}{2(1+\nu)}.$$

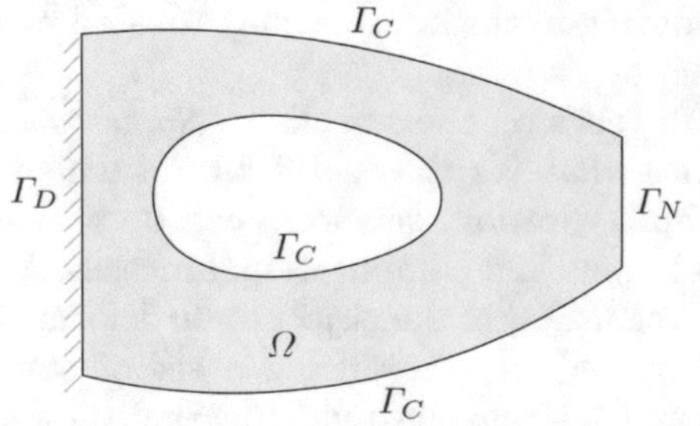

Fig. 11.5 Shape optimization of an elastic structure (in $d = 2$ dimensions): domain and boundaries.

The state equation can therefore be written as follows:

$$\begin{cases} -\mathrm{div}(\mu(\nabla \boldsymbol{y} + (\nabla \boldsymbol{y})^\top + \lambda(\mathrm{div}\boldsymbol{y})\boldsymbol{I}) = \boldsymbol{f} & \text{in } \Omega \\ \boldsymbol{y} = \boldsymbol{0} & \text{on } \Gamma_D \\ \boldsymbol{\sigma n} = \boldsymbol{g} & \text{on } \Gamma_N \\ \boldsymbol{\sigma n} = \boldsymbol{0} & \text{on } \Gamma_C. \end{cases} \tag{11.82}$$

Note that on the Dirichlet boundary Γ_D we impose a prescribed displacement, whereas on the boundary Γ_N we impose that the normal stress $\boldsymbol{\sigma n}$ equals a prescribed load vector $\boldsymbol{g}$. On the free boundary Γ_c we prescribe instead a homogeneous Neumann condition.

Our goal is to minimize the compliance

$$J(\Omega) = \int_\Omega \boldsymbol{f} \cdot \boldsymbol{y} + \int_{\Gamma_N} \boldsymbol{g} \cdot \boldsymbol{y}.$$

Moreover, we impose a constraint on the total mass of the structure, assuming that

$$\mathcal{O}_{ad} = \{\Omega \in \mathcal{A}_\Phi \ : \ \Gamma_D \cup \Gamma_N \subset \Gamma \text{ and } |\Omega| = m_0\}.$$

In order to ensure that the fixed portions $\Gamma_D \cup \Gamma_N$ are included in the boundary Γ of any admissible shape, it is sufficient to choose

$$\mathbf{V} = \mathbf{0} \qquad \text{on } \Gamma_D \cup \Gamma_N$$

in the parametrization of the vector fields $\mathbf{V}$.

The weak formulation of problem (11.82) reads: find $\boldsymbol{y} \in V = H^1_{\Gamma_D}(\Omega)^d$ such that

$$a(\boldsymbol{y}, \boldsymbol{v}) = F\boldsymbol{v} \qquad \forall \boldsymbol{v} \in V, \tag{11.83}$$

where we have defined the bilinear form $a\colon V \times V \to \mathbb{R}$ as

$$a(\boldsymbol{y}, \boldsymbol{v}) = \int_\Omega 2\mu\, \varepsilon(\boldsymbol{y}) : \varepsilon(\boldsymbol{v}) + \int_\Omega \lambda\, \mathrm{div}(\boldsymbol{y})\, \mathrm{div}(\boldsymbol{v}), \tag{11.84}$$

and the linear form $F\colon V \to \mathbb{R}$ as

$$F\boldsymbol{v} = \int_\Omega \boldsymbol{f} \cdot \boldsymbol{v} + \int_{\Gamma_N} \boldsymbol{g} \cdot \boldsymbol{v}. \tag{11.85}$$

The bilinear form (11.84) is symmetric and strongly coercive thanks to Korn's inequality (see e.g. [64]). Therefore problem (11.82) admits a unique solution $u \in V$ thanks to Lax-Milgram lemma. Note that $F\boldsymbol{y} = J(\Omega)$.

In order to make the choice of the state space independent of the domain, we can introduce a further variable to penalize the Dirichlet condition. Hence, the weak formulation (11.83) can be rewritten as: find $\boldsymbol{y} \in V = H^1(\Omega)^d, \boldsymbol{\eta} \in H^1(\Omega)^d$ such that

$$a(\boldsymbol{y}, \boldsymbol{v}) + \int_{\Gamma_D} \boldsymbol{\eta} \cdot \boldsymbol{v} = F\boldsymbol{v} \qquad \forall \boldsymbol{v} \in V. \tag{11.86}$$

Proceeding as in the case of problem (11.76)–(11.77) and exploiting the Céa's method, provided we consider the definition (11.84) of the bilinear form, we can obtain (see Exercise 10) the following expression for the shape derivative of the compliance $J(\Omega)$,

$$J'(\Omega; \mathbf{V}) = -\int_{\Gamma_C} \left(2\mu|\varepsilon(\boldsymbol{y})|^2 + \lambda\,|\mathrm{div}\ \boldsymbol{y}|^2\right) \mathbf{V} \cdot \mathbf{n}. \tag{11.87}$$

Also in this case, the problem is self-adjoint and the adjoint state $\boldsymbol{p}$ results from the solution of a problem equivalent to the state problem, that is, $\boldsymbol{p} = \boldsymbol{y}$.

11.6.1 Numerical Results

In this section we use the steepest descent method introduced in Sect. 11.5 for the numerical solution of the compliance minimization problem. We consider a solid made of a linear isotropic material with Young modulus $E = 25\,GPa$ and Poisson ratio $\nu = 0.45$, and initially occupying the two-dimensional domain $\Omega^{(0)}$ shown in Fig. 11.6.

There are no distributed forces ($\boldsymbol{f} = \mathbf{0}$) whereas the load vector acting on Γ_N is $\boldsymbol{g} = (0, -10)$. The state problem is discretized by means of $\mathbb{P}_1$ finite elements over a computational mesh made by 2435 vertices and 4640 elements. At each step of the steepest descent method, we solve the state problem and problem (11.80) to recover the displacement fields over the domain; the problem is self-adjoint, so that no further adjoint solutions are needed.

We further impose a geometric constraint on the shape area under the form (11.81), by considering three different cases:

(i) $m_0 = |\Omega^{(0)}|$ (which is equal to 99.6432 for the case at hand);
(ii) a slight reduction in the shape area, taking $m_0 = 95$;
(iii) a moderate reduction in the shape area, taking $m_0 = 75$.

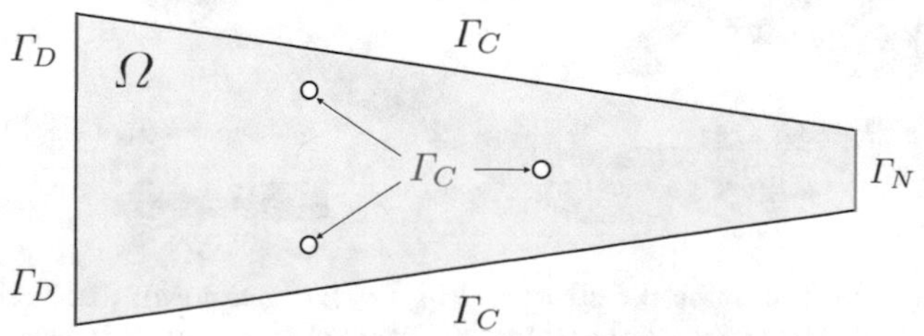

Fig. 11.6 Shape optimization of an elastic structure (in $d = 2$ dimensions): domain and boundaries

Shape regularization is achieved by introducing a coarser mesh to perform shape deformation, after the displacement field $\mathbf{V}^{(k)}$ has been interpolated. An adaptive choice of the step length τ_k is made at each step, depending on the value of $p_k = (\mathbf{V}^{(k-1)}, \mathbf{V}^{(k)})$: if $p_k < 0$ the current step size is decreased – to prevent a possible growth of the cost functional at that step – otherwise it is increased (proportionally to the value of $p_k > 0$). Furthermore, the current step is then decreased as soon as reversed triangles arise in the coarse mesh. Further details, and a possible implementation[10] of the procedure summarized so far, can be found, e.g., in [9].

The displacement magnitude computed on the initial shape and on the optimal shape obtained in the case (*ii*) are shown in Fig. 11.7; displacements' magnitude obtained in the cases (*i*) and (*iii*) are instead reported in Fig. 11.8. We note that in all cases the initial shape undergoes relevant shape variations, allowed by the large portions of free boundary Γ_C.

The values of the cost functional obtained during the optimization steps are reported in Fig. 11.9 for the cases (*i*) and (*iii*). We can clearly notice that the step size τ_k has been reduced several times to avoid the cost functional increased. In Tab. 11.1 we report the values of the cost functional and of the domain's area for the optimal shape; by means of the steepest descent method, a decrease in the cost functional in the range 45% - 58% has been reported. Finally, we report in Fig. 11.10 the values of the area of the domain obtained at each iteration, in all the three test cases, showing that the volume constraint imposed on the admissible shapes is accurately fulfilled.

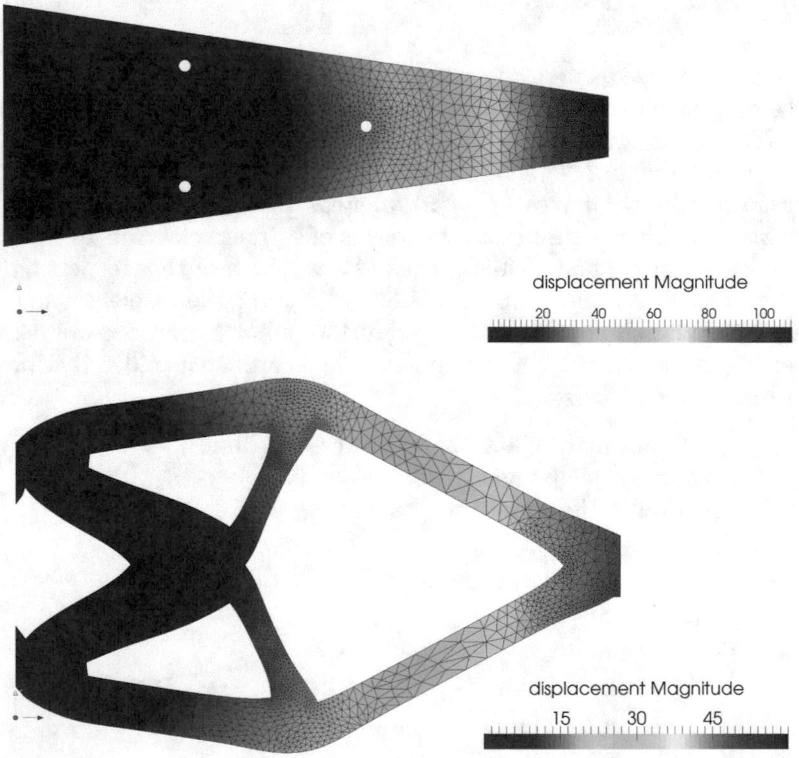

Fig. 11.7 Shape optimization of an elastic structure (in $d = 2$ dimensions): displacements and geometrical configurations at the initial step ($\Omega^{(0)}$, top) and at the optimum (bottom) in the case the final area is fixed to $m_0 = 95 < |\Omega^{(0)}|$

[10] The current implementation of the numerical procedure whose results are shown in this section is actually based on the algorithm proposed in [9].

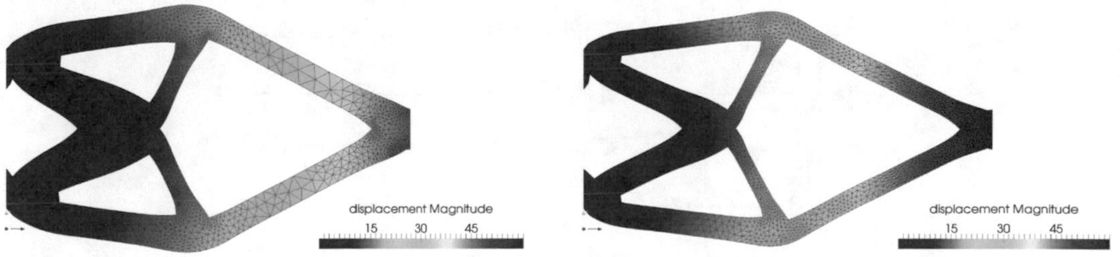

Fig. 11.8 Shape optimization of an elastic structure (in $d = 2$ dimensions): displacements and geometrical configurations at the optimum in the case the final area is fixed to $m_0 = |\Omega^{(0)}|$ (left) or to $m_0 = 75 < |\Omega^{(0)}|$ (right)

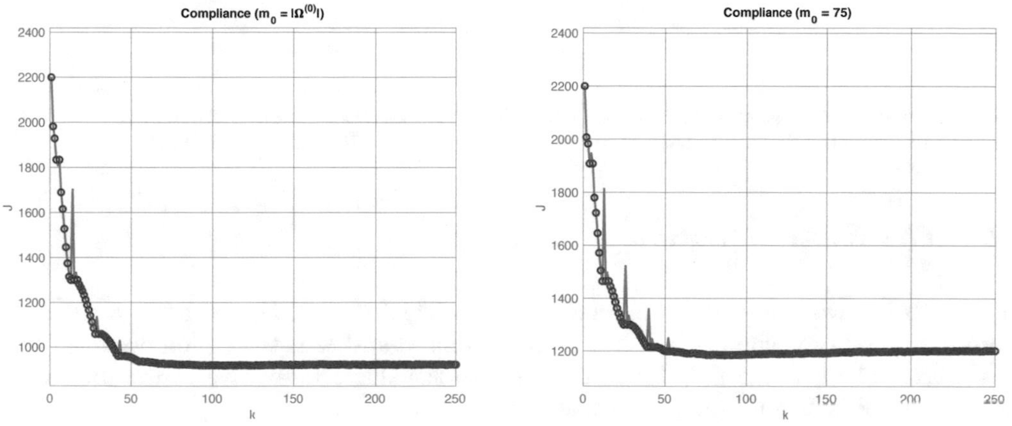

Fig. 11.9 Convergence history of the cost functional during the steepest descent method if the final area is fixed to $m_0 = |\Omega^{(0)}|$ (left) or to $m_0 = 75 < |\Omega^{(0)}|$ (right)

	$J(\hat{\Omega})$	$1 - J(\hat{\Omega})/J(\Omega^{(0)})$	m_0	$\lvert\hat{\Omega}\rvert$	$1 - \lvert\hat{\Omega}\rvert/m_0$
(i)	922.1	0.5807	99.6432	99.9175	-0.0028
(ii)	965.5	0.5610	95	95.4116	-0.0043
(iii)	1201.5	0.4537	75	75.8621	0.0115

Table 11.1 Cost functional $J(\hat{\Omega}$ and reduction with respect to the initial value $J(\Omega^{(0)}))$ shape areas (target m_0, optimal shape area $|\hat{\Omega}|$, and distance from the target m_0) for test cases (i), (ii) and (iii). The initial value of the compliance is $J(\Omega^{(0)}) = 2199.4$, whereas the area of the initial shape is $|\Omega^{(0)}| = 99.6432$

11.7 Drag Minimization in Navier-Stokes Flows

In this section we consider the drag T exerted on a body (or profile) by a viscous incompressible fluid flow travelling at a given uniform velocity γ, at the far field. Our purpose is to determine the optimal shape of the body in order to minimize the corresponding drag. We first carry out the analysis of this problem, following [26], and then present numerical results obtained for a two-dimensional profile. See also, e.g., [200, 173, 251, 51, 101, 100] for optimal shape design in fluid flows.

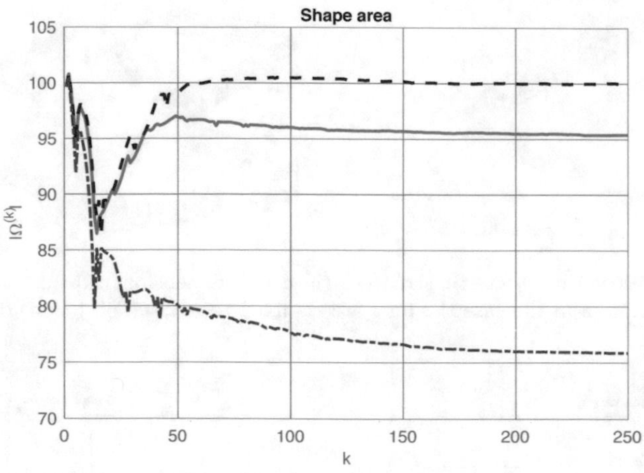

Fig. 11.10 Convergence history of the domain's area during the steepest descent method in the case the final area is fixed to $m_0 = |\Omega^{(0)}|$ (black, dashed), $m_0 = 95$ (blue) or $m_0 = 75$ (red, dash-dotted)

11.7.1 The State Equation

Let D and B be two bounded Lipschitz domains in $\mathbb{R}^2$, with $\overline{B} \subset D$, and let $\Omega = D \backslash \overline{B}$. Here B represents the body whose shape has to be optimized and Ω is the region occupied by the fluid, see Fig. 11.11. Since D is fixed, to compute the shape derivative of the drag we will choose vector fields $\mathbf{V} \in C^1(\mathbb{R}^n)$ such that

$$\mathbf{V} = \mathbf{0} \quad \text{on } \partial D.$$

The state problem can be stated in the following way. Let us define

$$Q = \left\{ q \in L^2(\Omega) : \int_\Omega q = 0 \right\}$$

and consider as universe domain $U = \mathbb{R}^2$. For the sake of simplicity, we consider a Dirichlet problem for the steady Navier-Stokes equations, and impose a velocity profile on the whole

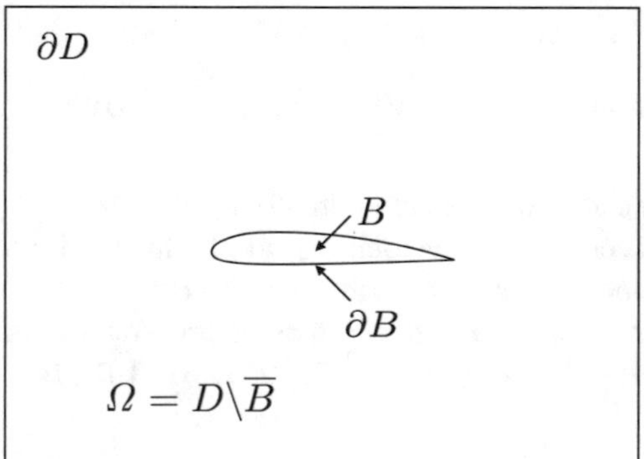

Fig. 11.11 Shape optimization of a body immersed in a fluid flow (in $d = 2$ dimensions): the domain and its boundaries.

boundary ∂D – rather than the more natural *zero flux* condition on the outflow boundary. Let $\gamma \in \mathbb{R}^2$ be a given constant vector. Then, choose $\boldsymbol{g} \in H^1(\mathbb{R}^2)^2$ such that

$$\mathrm{div}\boldsymbol{g} = 0 \quad \text{in } \Omega$$

and that

$$\boldsymbol{g} = \gamma \text{ in a neighborhood of } \partial D, \quad \boldsymbol{g} = \mathbf{0} \text{ in a neighborhood of } B. \tag{11.88}$$

Hence, we seek $(\boldsymbol{v}, \pi) \in H^1(\Omega)^2 \times Q$ such that $\boldsymbol{v} = \boldsymbol{g}$ on $\Gamma = \partial\Omega$ and

$$\begin{cases} -\nu\,\Delta\boldsymbol{v} + (\boldsymbol{v}\cdot\nabla)\boldsymbol{v} + \nabla\pi = \mathbf{0} & \text{in } \Omega \\ \mathrm{div}\boldsymbol{v} = 0 & \text{in } \Omega. \end{cases} \tag{11.89}$$

Thus, if B is well contained in D, e.g. if $\overline{B} \subset D_0$, for some $\overline{D}_0 \subset D$, the fluid velocity is constant far from B.

A weak formulation of problem (11.89) is obtained as follows: find $\boldsymbol{v} \in H^1(\Omega)^2$, $\boldsymbol{v} - \boldsymbol{g} \in H_0^1(\Omega)^2$ and $\pi \in Q$, such that

$$\begin{cases} e(\boldsymbol{v}, \boldsymbol{\varphi}) + c(\boldsymbol{v}, \boldsymbol{v}, \boldsymbol{\varphi}) + b(\boldsymbol{\varphi}, \pi) = 0 & \forall\boldsymbol{\varphi} \in H_0^1(\Omega)^2 \\ b(\boldsymbol{v}, q) = 0 & \forall q \in L^2(\Omega) \end{cases} \tag{11.90}$$

where the bilinear forms and the trilinear form are defined as in (10.4)–(10.5). If $|\gamma|$ is small enough, problem (11.90) is well posed. Precisely, the following result holds (for the proof see, e.g., [201, Theorem 2.1]; the $C^\infty-$ regularity can be found instead in, e.g., [165]).

Proposition 11.10. *There exists $\alpha > 0$ such that if $|\gamma| < \alpha\nu$, then (11.89) has a unique solution $(\boldsymbol{v}, \pi)$ with $\boldsymbol{v} \in H^1(\Omega)^2 \cap C^\infty(\Omega)$, $\pi \in L^2(\Omega) \cap C^\infty(\Omega)$, independent of the choice of the extension function $\mathbf{g}$ satisfying (11.88). Furthermore, if $\overline{D}_0 \subset D$ is given and $B \subset D_0$, for each $\varepsilon > 0$, α can be chosen to depend only on ε, D_0, D, in such a way that*

$$\|\boldsymbol{v}\|_{H^1(\Omega)} \leq \varepsilon\nu.$$

Finally, if Ω is a C^2 domain, then $(\boldsymbol{v}, \pi) \in H^2(\Omega)^2 \times H^1(\Omega)$.

11.7.2 The Drag Functional

If Ω is regular enough, thanks to Proposition 11.10, we can express the hydrodynamical drag as

$$T(\Omega) = -\gamma \cdot \int_{\partial B} (-\pi\mathbf{I} + 2\nu\boldsymbol{\sigma}(\boldsymbol{v})) \cdot \mathbf{n}$$

where

$$\boldsymbol{\sigma}(\boldsymbol{v}) = \frac{1}{2}\left(\nabla\boldsymbol{v} + (\nabla\boldsymbol{v})^\top\right).$$

By inserting a minus sign we conform to the convention that the force is positive when acting on the fluid. Exploiting the particular Dirichlet conditions of the state problem, it is possible to write an equivalent expression for the drag as a more comfortable volume integral, that makes sense also when Ω is a bounded and Lipschitz domain.

Lemma 11.2. *Let Ω be a C^2 domain and $(\boldsymbol{v}, \pi) \in H^2(\Omega)^2 \times H^1(\Omega)$ be a solution to* (11.90). *Then*

$$T(\Omega) = 2\nu \int_\Omega |\boldsymbol{\sigma}(\boldsymbol{v})|^2. \tag{11.91}$$

Proof. Since $\boldsymbol{v} - \boldsymbol{g} = \mathbf{0}$ on $\partial\Omega$ and $\boldsymbol{g} = \boldsymbol{\gamma}$ on ∂D, $\boldsymbol{g} = \mathbf{0}$ on ∂B, we can write

$$T(\Omega) = -\int_{\partial B} [\pi(\boldsymbol{v} - \boldsymbol{\gamma}) - 2\nu\boldsymbol{\sigma}(\boldsymbol{v})(\boldsymbol{v} - \boldsymbol{\gamma})] \cdot \mathbf{n} \;\; = \;\; -\int_\Omega \operatorname{div}[\pi(\boldsymbol{v} - \boldsymbol{\gamma}) - 2\nu\sigma(\boldsymbol{v})(\boldsymbol{v} - \boldsymbol{\gamma})]$$

thanks to Gauss formula. From $\operatorname{div}\boldsymbol{v} = 0$ and the symmetry of $\boldsymbol{\sigma}(\boldsymbol{v})$, we have

$$\begin{aligned} \operatorname{div}[\boldsymbol{\sigma}(\boldsymbol{v})(\boldsymbol{v} - \boldsymbol{\gamma})] &= \left(\operatorname{div}(\boldsymbol{\sigma}(\boldsymbol{v}))^\top\right) \cdot (\boldsymbol{v} - \boldsymbol{\gamma}) + (\boldsymbol{\sigma}(\boldsymbol{v}))^\top : \nabla\boldsymbol{v} \\ &= (\operatorname{div}\boldsymbol{\sigma}(\boldsymbol{v})) \cdot (\boldsymbol{v} - \boldsymbol{\gamma}) + \boldsymbol{\sigma}(\boldsymbol{v}) : \nabla\boldsymbol{v} \;\; = \;\; \frac{1}{2}\Delta\boldsymbol{v} \cdot (\boldsymbol{v} - \boldsymbol{\gamma}) + \sigma(\boldsymbol{v}) : \nabla\boldsymbol{v}. \end{aligned}$$

Therefore

$$T(\Omega) = \int_\Omega [(\nabla\pi - \nu\Delta\boldsymbol{v} \cdot (\boldsymbol{v} - \boldsymbol{\gamma}) + 2\nu\boldsymbol{\sigma}(\boldsymbol{v}) : \nabla\boldsymbol{v}].$$

From the Navier Stokes momentum equation, we get

$$(\nabla\pi - \nu\Delta\boldsymbol{v}) \cdot (\boldsymbol{v} - \boldsymbol{\gamma}) = (\boldsymbol{v} \cdot \nabla)\boldsymbol{v} \cdot (\boldsymbol{v} - \boldsymbol{\gamma}) = \frac{1}{2}\operatorname{div}(|\boldsymbol{v} - \boldsymbol{\gamma}|^2\,\boldsymbol{v}).$$

Moreover (see Exercise 11)

$$\boldsymbol{\sigma}(\boldsymbol{v}) : \nabla\boldsymbol{v} = \sigma(\boldsymbol{v}) : \boldsymbol{\sigma}(\boldsymbol{v}) = |\boldsymbol{\sigma}(\boldsymbol{v})|^2.$$

Thus

$$\begin{aligned} T(\Omega) &= \int_\Omega \left[\frac{1}{2}\operatorname{div}(|\mathbf{y} - \gamma|^2\,\mathbf{y}) + 2\nu\,|\boldsymbol{\sigma}(\boldsymbol{v})|^2\right] \\ &= \frac{1}{2}\int_{\partial\Omega} |\boldsymbol{v} - \boldsymbol{\gamma}|^2\,\boldsymbol{v} \cdot \mathbf{n} + 2\nu\int_\Omega |\boldsymbol{\sigma}(\boldsymbol{v})|^2 = 2\nu\int_\Omega |\boldsymbol{\sigma}(\boldsymbol{v})|^2. \end{aligned}$$

□

The expression (11.91) makes sense when Ω is a bounded Lipschitz domain and for $\boldsymbol{v} \in H^1(\Omega)^2$. Therefore from now on we adopt

$$T(\Omega) = 2\nu \int_\Omega |\sigma(\boldsymbol{v})|^2$$

as cost functional.

11.7.3 Shape Derivative of the State

The problem on the deformed domain $\Omega_t = \boldsymbol{\Phi}_t(\Omega)$ can be stated in the following form. Given $\mathbf{g} \in H^1(\mathbb{R}^2)^2$, find $\boldsymbol{v}_t \in H^1(\Omega_t)^2$, $\pi_t \in L^2(\Omega_t)$ such that

$$\boldsymbol{v}_t - \mathbf{g} \in H_0^1(\Omega_t)^2, \qquad \int_\Omega \pi_t \circ \boldsymbol{\Phi}_t = 0 \tag{11.92}$$

and[11]

$$\begin{cases} -\nu\Delta\boldsymbol{v}_t + (\boldsymbol{v}_t \cdot \nabla)\boldsymbol{v}_t + \nabla\pi_t = \mathbf{0} & \text{in } \Omega_t \\ \operatorname{div}\boldsymbol{v}_t = 0 & \text{in } \Omega_t. \end{cases} \tag{11.93}$$

[11] Here, the subscript t **does not** mean *derivative with respect to t*.

If t is small enough, $\mathbf{g} = \mathbf{0}$ in a neighborhood of $B_t = \mathbf{\Phi}_t(B)$. Notice that the normalization condition for the pressure in (11.92) is slightly different from

$$\int_{\Omega_t} \pi_t = \int_{\Omega} \pi_t \circ \mathbf{\Phi}_t \det \mathbf{\Phi}_t = 0,$$

but somewhat more convenient for the calculations.

From [26, Theorem 6], we deduce that the map $t \mapsto (\boldsymbol{v}_t, \pi_t) \circ \mathbf{\Phi}_t$ is differentiable at $t = 0$, and its derivative at 0, denoted by $(\dot{\boldsymbol{v}}, \dot{\pi})$, belongs to $H_0^1(\Omega)^2 \times Q$. Since for the shape derivatives $\boldsymbol{v}'$ and π' along $\mathbf{V}$ we have

$$\boldsymbol{v}' = \dot{\boldsymbol{v}} - (\mathbf{V} \cdot \nabla) \boldsymbol{v}, \qquad \pi' = \dot{\pi} - \mathbf{V} \cdot \nabla \pi,$$

then

$$\boldsymbol{v}' + (\mathbf{V} \cdot \nabla) \boldsymbol{v} \in H_0^1(\Omega)^2, \qquad \pi' + \mathbf{V} \cdot \nabla \pi \in Q$$

and, from the state equation (11.89),

$$\begin{cases} -\nu \Delta \boldsymbol{v}' + (\boldsymbol{v}' \cdot \nabla) \boldsymbol{v} + (\boldsymbol{v} \cdot \nabla) \boldsymbol{v}' + \nabla \pi' = \mathbf{0} & \text{in } \Omega \\ \operatorname{div} \boldsymbol{v}' = 0 & \text{in } \Omega. \end{cases} \tag{11.94}$$

The above formulas hold when Ω is a bounded Lipschitz domain. If Ω is a C^2 domain, then, since $\dot{\boldsymbol{v}} = \mathbf{0}$ and $\boldsymbol{v}$ is constant on Γ, we have on Γ,

$$\boldsymbol{v}' = -(\mathbf{V} \cdot \nabla) \boldsymbol{v}$$

and, using (11.10),

$$\mathbf{0} = \nabla_\Gamma \boldsymbol{v} = \nabla \boldsymbol{v}_{|\Gamma} - \frac{\partial \boldsymbol{v}}{\partial n} \otimes \mathbf{n}.$$

Thus, on Γ,

$$[(\mathbf{V} \cdot \nabla) \boldsymbol{v}]_j = \sum_{i=1}^{2} V_i \frac{\partial v_j}{\partial n} n_i = (\mathbf{V} \cdot \mathbf{n}) \frac{\partial v_j}{\partial n}, \qquad j = 1, 2$$

and therefore

$$\boldsymbol{v}' = -\frac{\partial \boldsymbol{v}}{\partial n} \mathbf{V} \cdot \mathbf{n} \quad \text{on } \Gamma.$$

11.7.4 Shape Derivative and Gradient of $T(\Omega)$

In this section we compute shape derivative and gradient of the cost functional. Recall that our vector fields $\mathbf{V}$ vanish on ∂D.

Theorem 11.5. *Let Ω be a bounded Lipschitz domain. Then*

$$\boldsymbol{\sigma}(\boldsymbol{v}) : \boldsymbol{\sigma}(\boldsymbol{v}') + \frac{1}{2} div\left(|\boldsymbol{\sigma}(\boldsymbol{v})|^2 \mathbf{V}\right) \in L^1(\Omega)$$

and

$$T'(\Omega; \mathbf{V}) = 4\nu \int_{\Omega} \left[\boldsymbol{\sigma}(\boldsymbol{v}) : \boldsymbol{\sigma}(\boldsymbol{v}') + \frac{1}{2} div\left(|\boldsymbol{\sigma}(\boldsymbol{v})|^2 \mathbf{V}\right)\right] \tag{11.95}$$

where $\boldsymbol{v}' = \boldsymbol{v}'(\Omega; \mathbf{V})$. If Ω is a C^2 domain, then

$$T'(\Omega; \mathbf{V}) = -4\nu \int_{\Omega} \Delta \boldsymbol{v} \cdot \boldsymbol{v}' - 2\nu \int_{\partial B} \left|\frac{\partial \boldsymbol{v}}{\partial n}\right|^2 (\mathbf{V} \cdot \mathbf{n}). \tag{11.96}$$

Proof. We check only formulas (11.95) and (11.96). Formula (11.95) follows from the chain rule and (11.39). Let now Ω be a C^2 domain. To derive (11.96) observe that, since $\boldsymbol{v}' = -\mathbf{V} \cdot \mathbf{n} \frac{\partial \boldsymbol{v}}{\partial n}$ on $\partial\Omega$ and $\mathbf{V} = \mathbf{0}$ on ∂D, we have

$$\begin{aligned}\int_{\Omega} \boldsymbol{\sigma}(\boldsymbol{v}) : \boldsymbol{\sigma}(\boldsymbol{v}') &= \int_{\Omega} \boldsymbol{\sigma}(\boldsymbol{v}) : \nabla \boldsymbol{v}' = -\int_{\Omega} (\mathrm{div}\boldsymbol{\sigma}(\boldsymbol{v})) \cdot \boldsymbol{v}' + \int_{\partial B} \boldsymbol{\sigma}(\boldsymbol{v}) \mathbf{n} \cdot \boldsymbol{v}' \\ &= -\int_{\Omega} \Delta \boldsymbol{v} \cdot \boldsymbol{v}' - \int_{\partial B} \boldsymbol{\sigma}(\boldsymbol{v}) \mathbf{n} \cdot \frac{\partial \boldsymbol{v}}{\partial n} (\mathbf{V} \cdot \mathbf{n}).\end{aligned}$$

Now, $\boldsymbol{v}$ is constant on ∂B and $\mathrm{div}\boldsymbol{v} = 0$. This implies that (see Exercise 8) $(\nabla \boldsymbol{v})^\top \mathbf{n} = \mathbf{0}$ and therefore $\boldsymbol{\sigma}(\boldsymbol{v}) \mathbf{n} = \frac{\partial \boldsymbol{v}}{\partial n}$. Hence

$$\int_{\Omega} \boldsymbol{\sigma}(\boldsymbol{v}) : \boldsymbol{\sigma}(\boldsymbol{v}') = -\int_{\Omega} \Delta \boldsymbol{v} \cdot \boldsymbol{v}' - \int_{\partial B} \left|\frac{\partial \boldsymbol{v}}{\partial n}\right|^2 (\mathbf{V} \cdot \mathbf{n}). \tag{11.97}$$

On the other hand,

$$\frac{1}{2} \int_{\Omega} \mathrm{div}(|\boldsymbol{\sigma}(\boldsymbol{v})|^2 \mathbf{V}) = \frac{1}{2} \int_{\Gamma} |\boldsymbol{\sigma}(\boldsymbol{v})|^2 \mathbf{V} \cdot \mathbf{n} = \frac{1}{2} \int_{\partial B} \boldsymbol{\sigma}(\boldsymbol{v}) : \nabla \boldsymbol{v} \, (\mathbf{V} \cdot \mathbf{n}) \tag{11.98}$$

Since $\boldsymbol{v} = \mathbf{0}$ on ∂B, we have $\nabla_{\partial B} \boldsymbol{v} = 0$ and from (11.10) we get

$$\nabla \boldsymbol{v}_{|\partial B} = (\nabla \boldsymbol{v})_{|\partial B} \, \mathbf{n} \otimes \mathbf{n} = \frac{\partial \boldsymbol{v}}{\partial n} \otimes \mathbf{n}$$

Therefore, on ∂B,

$$\nabla \boldsymbol{v} : \nabla \boldsymbol{v} = \sum_{i,j=1}^{2} \frac{\partial v_i}{\partial n} n_j \frac{\partial v_i}{\partial x_j} = \left|\frac{\partial \boldsymbol{v}}{\partial n}\right|^2$$

while

$$\nabla \boldsymbol{v} : (\nabla \boldsymbol{v})^\top = \sum_{i,j=1}^{2} \frac{\partial v_i}{\partial n} n_j \frac{\partial v_j}{\partial x_i} = (\nabla \boldsymbol{v}) \mathbf{n} \cdot (\nabla \boldsymbol{v})^\top \mathbf{n} = 0.$$

Thus, from (11.98), we get

$$\frac{1}{2} \int_{\Omega} \mathrm{div}(|\boldsymbol{\sigma}(\boldsymbol{v})|^2 \mathbf{V}) = \frac{1}{2} \int_{\partial B} \left|\frac{\partial \boldsymbol{v}}{\partial n}\right|^2 (\mathbf{V} \cdot \mathbf{n}) \tag{11.99}$$

and finally (11.96), follows from (11.97) and (A.61). □

Formulas (11.95) or (11.96) require to solve problem (11.94) for each direction $\mathbf{V}$, because of the presence of the shape derivative $\boldsymbol{v}'$. As we did for OCPs and in the model problem in Sect. 11.4.1, it is possible to introduce an adjoint problem for a multiplier $(\boldsymbol{z}, q)$ to get rid of $\boldsymbol{v}'$. We do it in the case of C^2 domains and formula (11.96). Precisely, the following result holds (recall Proposition 11.10).

Theorem 11.6. *Let Ω be a C^2 domain, $\mathbf{V} \in W^{2,\infty}(\mathbb{R}^2)$ and $(\boldsymbol{v}, \pi) \in H^2(\Omega)^2 \times H^1(\Omega)$ be the solution of problem (11.89). Let $(\boldsymbol{z}, q)$ with $\boldsymbol{z} \in H^2(\Omega)^2 \cap H_0^1(\Omega)^2$ and $q \in H^1(\Omega) \cap Q$, be the unique solution to the adjoint problem*

$$\begin{cases} -\nu\Delta \boldsymbol{z} + (\nabla \boldsymbol{v})^\top \boldsymbol{z} - (\boldsymbol{v}\cdot\nabla)\boldsymbol{z} + \nabla q = -2\nu\Delta\boldsymbol{v} & \text{in } \Omega \\ \operatorname{div} \mathbf{z} = 0 & \text{in } \Omega. \end{cases} \tag{11.100}$$

Then

$$T'(\Omega; \mathbf{V}) = 2\nu \int_{\partial B} (\mathbf{V}\cdot\mathbf{n}) \left(\frac{\partial \boldsymbol{z}}{\partial n} - \frac{\partial \boldsymbol{v}}{\partial n}\right) \cdot \frac{\partial \boldsymbol{v}}{\partial n} \tag{11.101}$$

and hence the gradient of T is given by

$$G(\Gamma) = 2\nu \left(\frac{\partial \boldsymbol{z}}{\partial n} - \frac{\partial \boldsymbol{v}}{\partial n}\right) \cdot \frac{\partial \boldsymbol{v}}{\partial n}.$$

Proof. Using the first equation of the adjoint problem we can write

$$T'(\Omega; \mathbf{V}) = 2\int_\Omega (-\nu\Delta\boldsymbol{z} + (\nabla\boldsymbol{v})^\top \boldsymbol{z} - (\boldsymbol{v}\cdot\nabla)\boldsymbol{z} + \nabla q)\cdot\boldsymbol{v}' - 2\nu \int_{\partial B} \left|\frac{\partial \boldsymbol{v}}{\partial n}\right|^2 (\mathbf{V}\cdot\mathbf{n}). \tag{11.102}$$

We have, since $\boldsymbol{v}' = -\frac{\partial \boldsymbol{v}}{\partial n}(\mathbf{V}\cdot\mathbf{n})$ on ∂B, $\mathbf{V} = \mathbf{0}$ on ∂D,

$$\begin{aligned} -\nu\int_\Omega \Delta\boldsymbol{z}\cdot\boldsymbol{v}' &= -\nu\int_{\partial B} \frac{\partial \boldsymbol{z}}{\partial n}\cdot\boldsymbol{v}' + \nu\int_\Omega \nabla\boldsymbol{z} : \nabla\boldsymbol{v}' \\ &= \nu\int_{\partial B} \frac{\partial \boldsymbol{z}}{\partial n}\cdot\frac{\partial \boldsymbol{v}}{\partial n}(\mathbf{V}\cdot\mathbf{n}) - \nu\int_\Omega \boldsymbol{z}\Delta\boldsymbol{v}'. \end{aligned}$$

Moreover, since $\boldsymbol{v} = \mathbf{0}$ on ∂B, $\boldsymbol{v}' = \mathbf{0}$ on ∂D, $\operatorname{div}\boldsymbol{v} = \mathbf{0}$, $\operatorname{div}\boldsymbol{v}' = \mathbf{0}$ in Ω,

$$\int_\Omega ((\nabla\boldsymbol{v})^\top\boldsymbol{z} - (\boldsymbol{v}\cdot\nabla)\boldsymbol{z} + \nabla q)\cdot\boldsymbol{v}' = \int_\Omega ((\boldsymbol{v}'\cdot\nabla)\boldsymbol{v}\cdot\boldsymbol{z} + (\boldsymbol{v}\cdot\nabla)\boldsymbol{v}'\cdot\boldsymbol{z}).$$

Thus, from the first equation in (11.94)

$$\begin{aligned} \int_\Omega (-\nu\Delta\boldsymbol{z} + (\nabla\boldsymbol{v})^\top\boldsymbol{z} - (\boldsymbol{v}\cdot\nabla)\boldsymbol{z} + \nabla q)\cdot\boldsymbol{v}' &= \int_\Omega -\nu\nabla\pi'\cdot\boldsymbol{v}' + \nu\int_{\partial B} \frac{\partial \boldsymbol{z}}{\partial n}\cdot\frac{\partial \boldsymbol{v}}{\partial n}(\mathbf{V}\cdot\mathbf{n}) \\ &= \nu\int_{\partial B} \frac{\partial \boldsymbol{z}}{\partial n}\cdot\frac{\partial \boldsymbol{v}}{\partial n}(\mathbf{V}\cdot\mathbf{n}) \end{aligned}$$

and finally, from (11.102), we get (11.101). □

Remark 11.8. To derive the adjoint equation (11.100), observe that the first equation of (11.94) in weak form reads ($\boldsymbol{v}$ is given)

$$\tilde{e}(\boldsymbol{v}', \boldsymbol{\varphi}) + b(\boldsymbol{\varphi}, \pi') = 0 \quad \forall \boldsymbol{\varphi} \in H_0^1(\Omega)^2$$

where

$$\tilde{e}(\boldsymbol{v}', \boldsymbol{\varphi}) = \int_\Omega [\nu\nabla\boldsymbol{v}' : \nabla\boldsymbol{\varphi} + (\boldsymbol{v}'\cdot\nabla)\boldsymbol{v}\cdot\boldsymbol{\varphi} + (\boldsymbol{v}\cdot\nabla)\boldsymbol{v}'\cdot\boldsymbol{\varphi}]$$

and $b(\cdot,\cdot)$ is defined as in (10.4). The adjoint bilinear form $e^*(\boldsymbol{z}, \boldsymbol{\varphi}) = e(\boldsymbol{\varphi}, \boldsymbol{z})$ is given by (see the calculations in Sect. 10.1)

$$\tilde{e}^*(\boldsymbol{z},\boldsymbol{\varphi}) = \int_\Omega [\nu\nabla\boldsymbol{z} : \nabla\boldsymbol{\varphi} + (\boldsymbol{\varphi}\cdot\nabla)\,\boldsymbol{v}\cdot\boldsymbol{z} + (\boldsymbol{v}\cdot\nabla)\,\boldsymbol{\varphi}\cdot\boldsymbol{z}]$$

$$= \int_\Omega [-\nu\nabla\boldsymbol{z} : \nabla\boldsymbol{\varphi} + (\nabla\boldsymbol{v})^\top\boldsymbol{z}\cdot\boldsymbol{\varphi} - (\boldsymbol{v}\cdot\nabla)\,\boldsymbol{z}\cdot\boldsymbol{\varphi}]$$

where we have used $\boldsymbol{z} = \mathbf{0}$ on Γ and $\mathrm{div}\boldsymbol{z} = 0$ in Ω. The first equation in the adjoint problem (11.100) is the strong formulation of

$$e^*(\boldsymbol{z},\boldsymbol{\varphi}) + b(\boldsymbol{\varphi}, q) = -2\nu\,(\Delta\boldsymbol{v}\cdot\boldsymbol{\varphi}) \qquad \forall\boldsymbol{\varphi}\in H_0^1(\Omega)^2 .$$

•

11.7.5 Numerical Results

In this section we report some numerical results for the drag minimization problem for two-dimensional Navier-Stokes flows. In particular, we investigate the impact of the Reynolds number on the optimal shape, as well as the robustness of the steepest descent method with respect to the initial shape to be optimized.

First, we consider as initial shape the bluff body already introduced in Sects. 1.6–1.7, and impose that the area occupied by the body B (or, equivalently, by the fluid in the domain $\Omega = D\setminus\bar{B}$) remains constant. Here, $D = (-a,a)\times(-b,b)$ with $a = 0.5$, $b = 0.4$. We discretize both the state and the adjoint problems using Taylor-Hood $\mathbb{P}_2/\mathbb{P}_1$ finite elements for velocity and pressure, respectively, and adopt the same regularization procedure on the mesh introduced in Sect. **??**. The initial mesh is made of 2872 vertices and 1436 elements, whereas the mesh corresponding to the optimal shape (after iterative shape regularization and mesh refinement) is made of about 6000 vertices and 3000 elements (slight differences are obtained from case to case). The far field velocity has a parabolic profile $\boldsymbol{\gamma}(\mathbf{x}) = V_\gamma g(x_2)\mathbf{e}_1$, vanishing on the upper and the lower portion of the boundary[12] ∂D, since $g(x_2) = 1-(y/b)^2$; note that the maximum is reached for $g(0) = 1$.

As shape functional to minimize, we consider a slight variation of the drag functional $T(\Omega)$, namely, we take the *drag coefficient* $C_D(\Omega)$, a dimensionless quantity commonly used to quantify the resistance of an object immersed in a fluid. For the two-dimensional case at hand, this is given by

$$J(\Omega) = \frac{T(\Omega)}{q_\gamma\, d} = \frac{T(\Omega)}{\frac{1}{2}\rho V_\gamma^2\, d} = C_D(\Omega)$$

where ρ is the density of the fluid, d is a characteristic linear dimension of the body, $\mathbf{e}_\infty$ is the unit vector directed as the incoming flow $\boldsymbol{\gamma} = V_\gamma\mathbf{e}_1$, and $q_\gamma = \frac{1}{2}\rho V_\gamma^2$. In this case, the Reynolds number is defined as

$$Re = \frac{\rho V_\gamma d}{\nu}.$$

Numerical optimization is performed following the procedure introduced in Sect. 11.6.1. In particular, shape regularization is achieved by introducing a coarser mesh to perform shape deformation, after the computed displacement field has been interpolated over it, and an adaptive choice of the step length τ_k is made at each iteration of the steepest descent method. This latter is then stopped as soon as

$$\frac{|J'(\Omega^{(k)};\mathbf{V}^{(k)})|}{|J'(\Omega^{(0)};\mathbf{V}^{(0)})|} < \mathrm{tol};$$

[12] For the sake of numerical calculations, here we have opted for the more usual choice of imposing a zero-stress Neumann condition at the outflow; as a matter of fact, only slight variations are required in the expression of the adjoint problem and the shape gradient of the cost functional.

in our case, tol $= 5 \cdot 10^{-3}$. Three cases with increasing Reynolds number – $Re = 10$, $Re = 100$, and $Re = 500$ – have been considered; the values of the cost functional and of the domain's area for the optimal shape, compared to those of the initial shape, are reported in Table 11.2. In these three cases, a decrease in the drag coefficient ranging from 14% to 29% has been observed. The computational meshes for the initial shape, as well as for the optimal shapes obtained in these three cases, are reported in Fig. 11.12. As expected, the larger the Reynolds number, the higher the tapering of the optimal shape of the profile.

	$J(\hat{\Omega})$	$J(\Omega^{(0)})$	$1 - J(\hat{\Omega})/J(\Omega^{(0)})$	$\|J'(\Omega^{(k)}; \mathbf{V}^{(k)})\|/\|J'(\Omega^{(0)}; \mathbf{V}^{(0)})\|$	$1 - \|\hat{\Omega}\|/\|\Omega^{(0)}\|$
$Re = 10$	3.9984	4.6516	0.1404	0.0022	-0.0005
$Re = 100$	0.5540	0.7012	0.2099	0.0014	-0.0013
$Re = 500$	0.2144	0.3036	0.2938	0.0055	0.0003

Table 11.2 Values of the cost functional for the optimal shape $J(\hat{\Omega})$ and the initial shape $J(\Omega^{(0)})$, and corresponding reduction; ratio between shape derivatives $|J'(\Omega^{(k)}; \mathbf{V}^{(k)})|/|J'(\Omega^{(0)}; \mathbf{V}^{(0)})|$; discrepancy between the target area $|\Omega^{(0)}| = 0.79107$ (of the domain with initial shape) and the area of the domain $|\hat{\Omega}|$ corresponding to the optimal shape, for different Reynolds numbers.

Computed velocity and pressure fields for both the initial and the optimal shape, as well as the adjoint velocity and pressure fields, are reported in Figs. 11.13 and 11.14, respectively, for the case $Re = 10$. In Fig. 11.15 we report instead the computed velocity and pressure fields for the initial and the optimal shape in the case $Re = 500$. In this latter case, the vortex cores at the rear of the optimal profile are almost completely suppressed, as we notice by comparing the streamlines of the velocity field for the initial and the optimal shape.

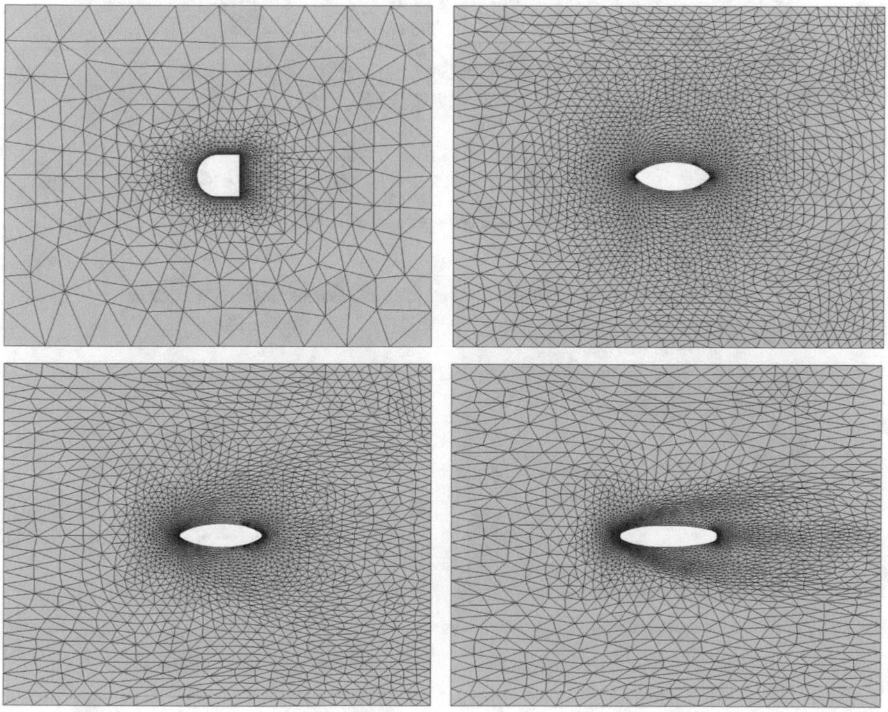

Fig. 11.12 From top, left to bottom, right: initial shape $\Omega^{(0)}$ and optimal shapes $\hat{\Omega}$ for different Reynolds numbers ($Re = 10$, top right; $Re = 100$, bottom left; $Re = 500$, bottom right) and corresponding computational meshes

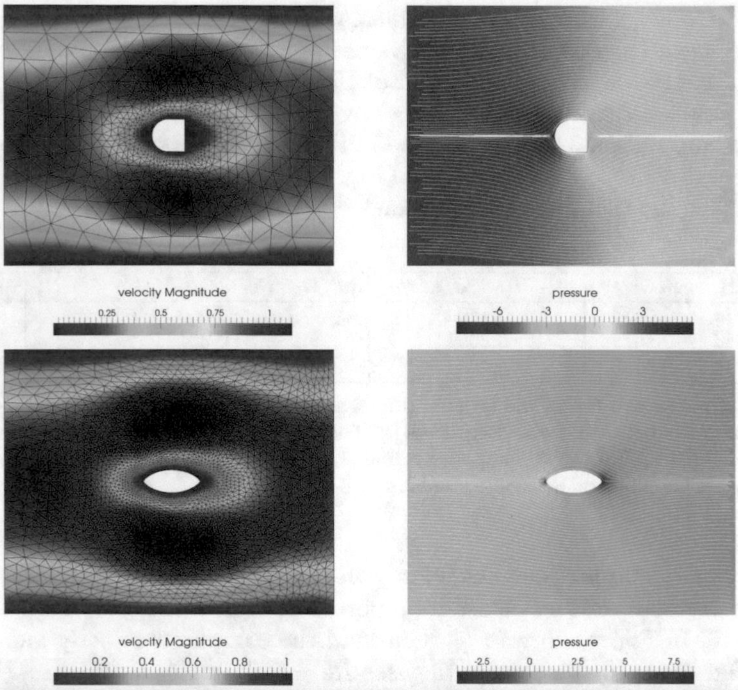

Fig. 11.13 Case $Re = 10$: initial (top) and optimal (bottom) velocity field magnitude (left) and pressure field with velocity streamlines (right)

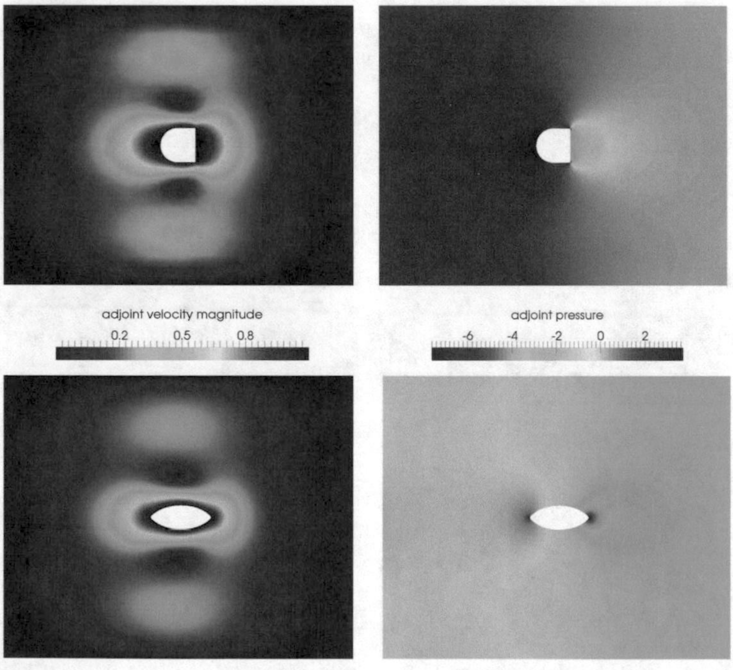

Fig. 11.14 Case $Re = 10$: initial (top) and optimal (bottom) adjoint velocity field magnitude (left) and adjoint pressure field

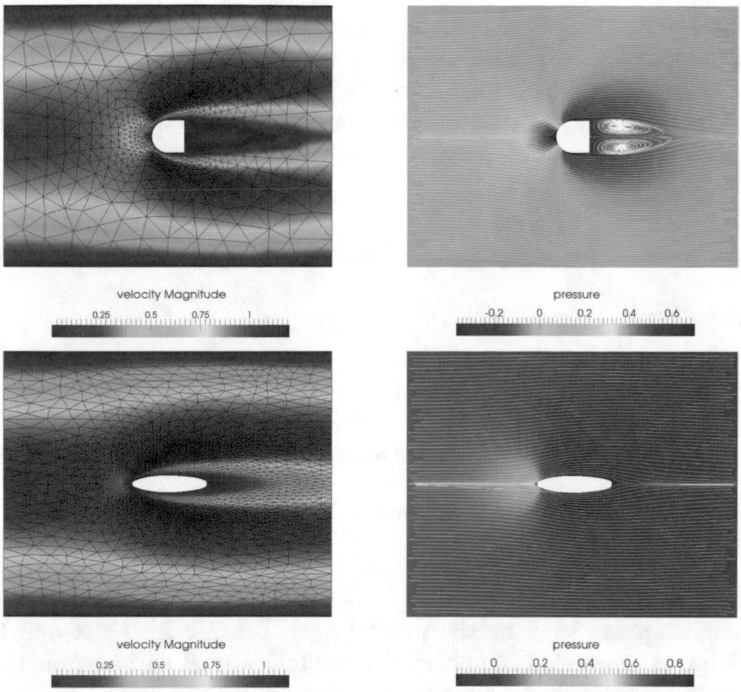

Fig. 11.15 Case $Re = 500$: initial (top) and optimal (bottom) velocity field magnitude (left) and pressure field with velocity streamlines (right)

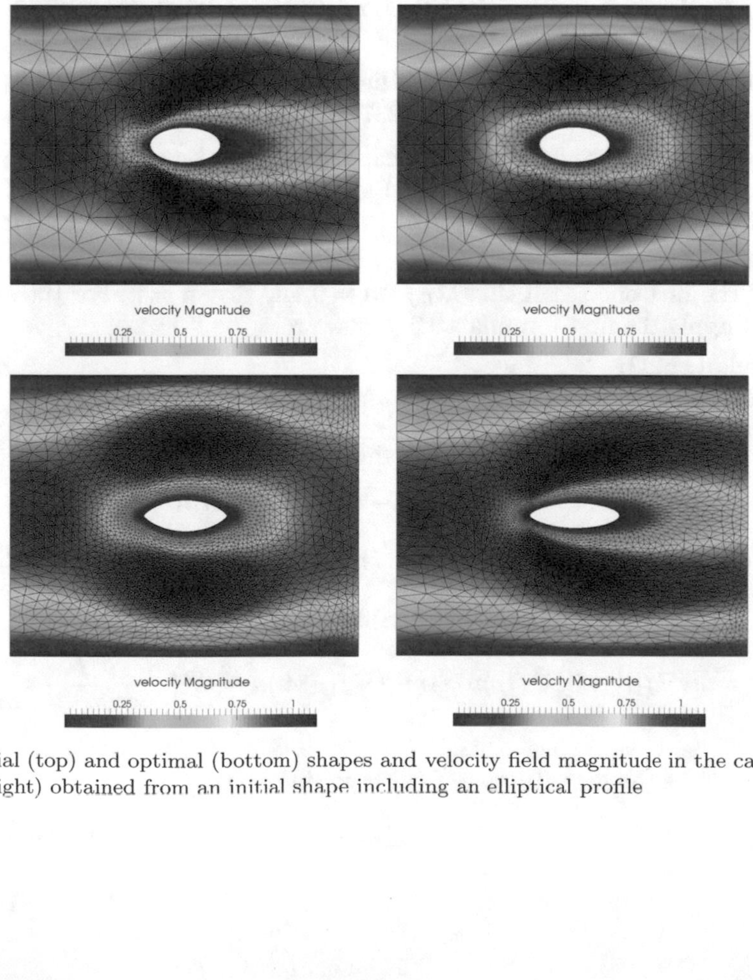

Fig. 11.16 Initial (top) and optimal (bottom) shapes and velocity field magnitude in the case $Re = 10$ (left) and $Re = 100$ (right) obtained from an initial shape including an elliptical profile

Finally, we show in Fig. 11.16 the optimal shape obtained in the case $Re = 10$ and $Re = 100$ starting from a different initial shape, namely, a domain in which the immersed body has an elliptical shape. Comparing the optimal shapes with the one reported in Fig. 11.12 (top, right and bottom, left, respectively) we can observe that the steepest descent method is robust with respect to the choice of the initial shape – in fact, the computed optimal shapes are indeed very similar.

11.8 Exercises

1. *Tangential gradient on the unit sphere.* Let

$$\mathbf{x} = \mathbf{x}(\theta, \psi) = (\cos\theta \sin\psi, \sin\theta \sin\psi, \cos\psi) \qquad 0 \le \theta \le 2\pi, 0 \le \psi \le \pi$$

be the parametrization of the unit sphere S. The vectors

$$\mathbf{x}_\theta(\theta, \psi) = -\sin\theta\sin\psi\mathbf{i} + \cos\theta\sin\psi\mathbf{j}, \ \ \mathbf{x}_\psi(\theta, \psi) = \cos\theta\cos\psi\mathbf{i} + \sin\theta\cos\psi\mathbf{j} - \sin\psi\mathbf{k}$$

span the tangent plane to S at the point (θ, ψ). Let $f : \mathbb{R}^3 \to \mathbb{R}$ and let $\tilde{f}(r, \theta, \psi) = f(r\mathbf{x}(\theta, \psi))$ be its expression in spherical coordinates (r, θ, ψ). Assume that $\tilde{f}$ is a differentiable function. Check that, in terms of the tangent basis $\mathbf{x}_\theta, \mathbf{x}_\psi$,

$$(\nabla_S f \circ \mathbf{x})(\theta, \psi) = \frac{1}{(\sin\psi)^2}\tilde{f}_\theta(1, \theta, \psi)\,\mathbf{x}_\theta(\theta, \psi) + \tilde{f}_\psi(1, \theta, \psi)\,\mathbf{x}_\psi(\theta, \psi). \tag{11.103}$$

[*Hint.* Observe that $\boldsymbol{\nu}(\theta, \psi) = \mathbf{x}(\theta, \psi)$ is the unit (exterior) normal to S at (θ, ψ). In terms of the basis $\boldsymbol{\nu}, \mathbf{x}_\theta, \mathbf{x}_\psi$, one has (see, e.g., [237] p. 674) $\nabla\tilde{f} = \tilde{f}_r \nu + \frac{1}{(\sin\psi)^2}\tilde{f}_\theta \mathbf{x}_\theta + \tilde{f}_\psi \mathbf{x}_\psi \dots$.]

2. Check the rules (11.13) and (11.14) for shape derivatives by using the definitions.
[*Hint.* Apply the definition of tangential gradient to fg to get (11.13). For the second formula, apply (11.13) to each row of the matrix $\nabla_\Gamma(f\mathbf{w})$ and take the trace.]
3. Prove formulas (11.16).
[*Hint.* For the first one recall that $\nabla_\Gamma f \cdot \mathbf{n} = 0$ and $\mathrm{div}_\Gamma \mathbf{n} = \kappa$. For the second one, use (11.15) and apply the first formula with $f = \mathbf{w} \cdot \mathbf{n}$.]
4. Show formula (11.21).
[*Hint.* First, use definition (11.20) to obtain

$$\begin{aligned}\Delta_\Gamma f = \mathrm{div}_\Gamma \nabla_\Gamma f &= \mathrm{div}_\Gamma(\nabla\tilde{f}|_\Gamma - (\nabla\tilde{f}|_\Gamma \cdot \mathbf{n})\mathbf{n}) \\ &= \mathrm{div}_\Gamma(\nabla\tilde{f}|_\Gamma) - \mathrm{div}_\Gamma(\nabla\tilde{f}|_\Gamma \cdot \mathbf{n})\mathbf{n}.\end{aligned}$$

Then, use the first relation in (11.16) and the definition (11.12) to conclude that

$$\begin{aligned}&\mathrm{div}_\Gamma(\nabla\tilde{f}|_\Gamma) - \mathrm{div}_\Gamma(\nabla\tilde{f}|_\Gamma \cdot \mathbf{n})\mathbf{n} \\ &= \mathrm{div}\nabla\tilde{f}|_\Gamma - ([D^2\tilde{f}|_\Gamma]\mathbf{n}) \cdot \mathbf{n} - (\nabla\tilde{f}|_\Gamma \cdot \mathbf{n})\kappa = \Delta\tilde{f}|_\Gamma - \frac{\partial^2 \tilde{f}}{\partial n^2} - \kappa\frac{\partial f}{\partial n}.]\end{aligned}$$

5. Let $\mathbf{f} : \Gamma \to \mathbb{R}^3$ be the identity map, $\mathbf{f}(\mathbf{x}) = \mathbf{x}$. Show that

$$-\Delta_\Gamma \mathbf{f} = \kappa\mathbf{n}.$$

6. Show that, if Ω is a smooth domain,

$$\nabla_{\Gamma}\mathbf{n} = D^2 b_{|\Gamma}.$$

where $b = b(\Omega)$ is the signed distance.

7. Let Ω be a smooth domain with boundary Γ and $\mathbf{w}$ be a smooth vector field in $\mathbb{R}^3$. Derive the formula

$$\int_{\Gamma} \mathrm{div}_{\Gamma}\mathbf{w} = \int_{\Gamma} \kappa\,\mathbf{w}\cdot\mathbf{n}$$

by taking the shape derivative on both members of the classical Gauss formula

$$\int_{\Omega} \mathrm{div}\mathbf{w} = \int_{\Gamma} \mathbf{w}\cdot\mathbf{n}.$$

[*Hint.* Use the extension $\tilde{\mathbf{n}} = \nabla b$, where $b = b(\Omega)$ is the signed distance to Ω. Use that $D^2 b_{|\Gamma}\mathbf{n} = \mathbf{0}$].]

8. Let $\mathbf{y} = (y_1, y_2)$ be constant on ∂B and $\mathrm{div}\,\mathbf{y} = 0$. Show that $(\nabla\mathbf{y})^{\top}\mathbf{n} = \mathbf{0}$ on ∂B. [*Hint.* Since $\frac{\partial y_1}{\partial x_1} + \frac{\partial y_2}{\partial x_2} = 0$, the first component of $(\nabla\mathbf{y})^{\top}\mathbf{n}$ is given by

$$\left[(\nabla\mathbf{y})^{\top}\mathbf{n}\right]_1 = \frac{\partial y_1}{\partial x_1}n_1 + \frac{\partial y_2}{\partial x_1}n_2 = -\frac{\partial y_2}{\partial x_2}n_1 + \frac{\partial y_2}{\partial x_1}n_2 = \nabla y_2\cdot\boldsymbol{\tau}$$

where $\boldsymbol{\tau} = (n_2, -n_1)$ is a tangential vector on ∂B. Since y_2 is constant on ∂B, $\nabla y_2\cdot\tau = 0$. By a similar argument we conclude that also $\left[(\nabla\mathbf{y})^{\top}\mathbf{n}\right]_2 = 0$.]

9. Let $M = M(t)$ be a square non singular matrix with differentiable entries. Check the formula

$$\frac{d}{dt}M^{-1}(t) = -M^{-1}(t)\left(\frac{d}{dt}M(t)\right)M^{-1}(t).$$

[*Hint.* Observe that $0 = (\frac{d}{dt}M^{-1}(t))M(t) = \ldots$.]

10. By adapting the calculations done for problem (11.76)–(11.77), show formula (11.87).

11. Recalling that $\boldsymbol{\sigma}(\boldsymbol{v}) = \frac{1}{2}\left(\nabla\boldsymbol{v} + (\nabla\boldsymbol{v})^{\top}\right)$, show that

$$\boldsymbol{\sigma}(\boldsymbol{v}) : \nabla\boldsymbol{v} = \boldsymbol{\sigma}(\boldsymbol{v}) : \boldsymbol{\sigma}(\boldsymbol{v}) = |\boldsymbol{\sigma}(\boldsymbol{v})|^2.$$

Appendix A
Toolbox of Functional Analysis

In this Appendix we provide many definitions and results related to Banach and Hilbert spaces, which form a useful toolbox for the analysis of OCPs governed by PDEs. Further details can be found on, e.g., [237, 52, 274, 65].

A.1 Metric, Banach and Hilbert Spaces

A.1.1 Metric Spaces

A *metric space* is a set M endowed with a distance $d : M \times M \to \mathbb{R}$, satisfying the following three properties, for every x, y, $z \in M$:

$$
\begin{array}{ll}
D_1 : d(x,y) \geq 0, d(x,y) = 0 \text{ if and only if } x = y & (\textit{positivity}) \\
D_2 : d(x,y) = d(y,x) & (\textit{symmetry}) \\
D_3 : d(x,z) \leq d(x,y) + d(y,z) & (\textit{triangular inequality}).
\end{array}
$$

A metric space *does not need to be a linear (vector)* space. A sequence $\{x_n\} \subset M$ *converges* to x in M, and we write $x_m \to x$ in M, if $d(x_m, x) \to 0$ as $m \to \infty$.

The limit is unique: if $d(x_m, x) \to 0$ and $d(x_m, y) \to 0$ as $m \to \infty$, then, from the triangular inequality, we have

$$d(x,y) \leq d(x_m, x) + d(x_m, y) \to 0 \quad \text{as} \quad m \to \infty$$

which implies $x = y$. We call

$$B_r(x) = \{y \in M : d(x,y) < r\}$$

the *open* ball of radius r, centered at x. As in the finite dimensional space $\mathbb{R}^d$, we can define a *topology* in M, induced by the metric. Let $A \subseteq M$; a point $x \in A$ is:

- an interior point if there exists a ball $B_r(x) \subset A$;
- a boundary point if any ball $B_r(x)$ contains both points of A and of its complement $M \setminus A$; the set ∂A of boundary points of A is called *boundary* of A;
- x is a limit point of A if there exists a sequence $\{x_k\}_{k\geq 0} \subset A$ such that $x_k \to x$.

A is *open* if every point in A is an interior point; the set $overline A = A \cup \partial A$ is the *closure* of A; A is *closed* if $\overline{A} = A$. A set is closed if and only if it contains all its limit points. In other words, A is closed if and only if A is sequentially closed, that is if, for every sequence $\{x_k\}_{k\geq 0} \subset A$ such that $x_k \to x$, then $x \in A$.

A. Manzoni et al., *Optimal Control of Partial Differential Equations*,
Applied Mathematical Sciences 207, https://doi.org/10.1007/978-3-030-77226-0

An open set A is *connected* if for every couple of points $x, y \in A$ there exists a regular curve joining them entirely contained in A. In the case of $M = \mathbb{R}^d$, by *domain* we mean an open connected set, usually denoted by the letter Ω. Moreover, we say that A is bounded if it is contained in some ball $B_r(0)$.

A sequence $\{x_m\} \subset M$ is a *Cauchy* (or *fundamental*) sequence if $d(x_m, x_k) \to 0$ as $m, k \to \infty$. If $\{x_m\}$ is convergent then it is a Cauchy sequence. The converse in not true, in general. A metric space in which every Cauchy sequence converges is called *complete.*

We say that $l \in \mathbb{R} \cup \{-\infty\} \cup \{+\infty\}$ is a limit point of $\{x_m\}$ if there exists a subsequence $\{x_{m_k}\}$ such that $x_{m_k} \to l$ as $k \to \infty$. Denoting by E the set of limit points of $\{x_m\}$, we define

$$\liminf_{m\to\infty} x_m = \inf E, \qquad \limsup_{m\to\infty} x_m = \sup E.$$

A.1.2 Banach Spaces

Let X be a *linear space* over the scalar field $\mathbb{R}$. A *norm* in X is a real function

$$\|\cdot\| : X \to \mathbb{R} \tag{A.1}$$

such that, for each scalar λ and every x, $y \in X$, the following properties hold:

$N_1 : \|x\| \geq 0; \|x\| = 0$ if and only if $x = 0$ (*positivity*)
$N_2 : \|\lambda x\| = |\lambda| \, \|x\|$ (*homogeneity*)
$N_3 : \|x + y\| \leq \|x\| + \|y\|$ (*triangular inequality*).

A *normed space* is a linear space X endowed with a norm $\|\cdot\|$. A norm induces a *distance* given by $d(x, y) = \|x - y\|$, which makes X a metric space. A normed space which is complete with respect to this induced distance is called a *Banach space.*

Proposition A.1. *Every norm in a linear space X is continuous in X.*

Definition A.1. Two norms $\|\cdot\|_1$ and $\|\cdot\|_2$, defined in the same space X, are *equivalent* if there exist two positive numbers c_1, c_2 such that

$$c_1 \|x\|_2 \leq \|x\|_1 \leq c_2 \|x\|_2 \qquad \text{for every } x \in X.$$

Definition A.2. A seminorm on V is a function $|\cdot| : X \to \mathbb{R}$ fulfilling properties N_2, N_3, whereas property N_1 is replaced by $|x| \geq 0 \; \forall x \in X$.

Example A.1. Let $\Omega \subset \mathbb{R}^d$ be a *bounded open set.*

1. *The space*
$$C\left(\overline{\Omega}\right) = \left\{u : \overline{\Omega} \to \mathbb{R},\ u \text{ continuous}\right\}$$
is a Banach space if endowed with the norm (called the *maximum norm*)
$$\|v\|_{C(\overline{\Omega})} = \sup_{\overline{\Omega}} v.$$

2. *The spaces* $C^k\left(\overline{\Omega}\right)$, $k \geq 0$ *integer.* The symbol $C^k\left(\overline{\Omega}\right)$ denotes the set of functions whose derivatives, up to the order k included, are continuous in Ω and can be extended continuously up to $\Gamma = \partial\Omega$. To denote a derivative of order m, it is convenient to use the symbol

$$D^\alpha = \frac{\partial^{\alpha_1}}{\partial x_1^{\alpha_1}} \cdots \frac{\partial^{\alpha_n}}{\partial x_n^{\alpha_n}},$$

where $\alpha = (\alpha_1, ..., \alpha_n)$ is an $n -$ *uple* of nonnegative integers (a *multi-index*), of *length* $|\alpha| = \alpha_1 + \ldots + \alpha_n = m$. Endowed with the norm (*maximum norm of order* k)

$$\|f\|_{C^k(\overline{\Omega})} = \|f\|_{C(\overline{\Omega})} + \sum_{|\alpha|=1}^{k} \|D^\alpha f\|_{C(\overline{\Omega})},$$

$C^k\left(\overline{\Omega}\right)$ is a Banach space. Note that $C^0(\overline{\Omega}) \equiv C(\overline{\Omega})$.

3. *The spaces* $C^{0,\alpha}\left(\overline{\Omega}\right)$, $0 \leq \alpha \leq 1$. A function f is *Hölder continuous in* Ω *with exponent* α if

$$\sup_{\substack{\mathbf{x},\mathbf{y}\in\Omega \\ \mathbf{x}\neq\mathbf{y}}} \frac{|f(\mathbf{x}) - f(\mathbf{y})|}{|\mathbf{x} - \mathbf{y}|^\alpha} = C_H(f;\Omega) < \infty. \tag{A.2}$$

The quotient in the left hand side of (A.2) represents an "incremental quotient of order α". The number $C_H(f;\Omega)$ is called the *Hölder constant of* f in Ω. If $\alpha = 1$, f is *Lipschitz continuous.* Typical examples of Hölder continuous functions in $\mathbb{R}^d$ with exponent α are the powers $f(\mathbf{x}) = |\mathbf{x}|^\alpha$.
We use the symbol $C^{0,\alpha}\left(\overline{\Omega}\right)$ to denote the set of functions of $C\left(\overline{\Omega}\right)$, Hölder continuous in Ω with exponent α. Endowed with the norm

$$\|f\|_{C^{0,\alpha}(\overline{\Omega})} = \|f\|_{C(\overline{\Omega})} + C_H(f;\Omega),$$

$C^{0,\alpha}\left(\overline{\Omega}\right)$ is a Banach space.

4. *The spaces* $C^{k,\alpha}(\overline{\Omega})$, $k \geq 1$ integer, $0 \leq \alpha \leq 1$. $C^{k,\alpha}(\overline{\Omega})$ denotes the set of functions of $C^k(\overline{\Omega})$, whose derivatives up to order k included, are Hölder continuous in Ω with exponent α. Endowed with the norm

$$\|f\|_{C^{k,\alpha}(\overline{\Omega})} = \|f\|_{C^k(\overline{\Omega})} + \sum_{|\beta|=k} C_H(D^\beta f;\Omega)$$

$C^{k,\alpha}(\overline{\Omega})$ becomes a Banach space.

•

Throughout the book we denote by

$$X^d = \underbrace{X \times X \times \ldots \times X}_{\text{d times}}$$

the (Cartesian) product space for, say, vector-valued functions. For instance, $\boldsymbol{v} \in C(\Omega)^d = C(\Omega;\mathbb{R}^d)$ denotes a vector field whose components $v_1, \ldots, v_d$ belong to $C(\Omega)$.

A.1.3 Hilbert Spaces

Let X be a linear space over $\mathbb{R}$. An *inner* or *scalar product* in X is a function

$$(\cdot,\cdot) : X \times X \to \mathbb{R}$$

with the following three properties. For every x, y, $z \in X$ and every scalar λ, $\mu \in \mathbb{R}$:

$$\begin{aligned} &I_1 : (x,x) \geq 0 \text{ and } (x,x) = 0 \text{ if and only if } x = 0 && (\textit{positivity}) \\ &I_2 : (x,y) = (y,x) && (\textit{symmetry}) \\ &I_3 : (\lambda x + \mu y, z) = \lambda\,(x,z) + \mu\,(y,z) && (\textit{bilinearity}). \end{aligned}$$

A linear space endowed with an inner product is called an *inner product space*. Property I_3 shows that the inner product is linear with respect to its first argument. From I_2, the same is true for the second argument as well. Then, we say that $(\cdot,\cdot)$ constitutes a *symmetric bilinear* form in X. When different inner product spaces are involved it may be necessary the use of notations like $(\cdot,\cdot)_X$, to avoid confusion.

An inner product induces a norm, given by

$$\|x\| = \sqrt{(x,x)}. \tag{A.3}$$

In fact, properties N_1 and N_2 in the definition of norm are immediate, while the triangular inequality is a consequence of the two properties stated in the following

Theorem A.1. *Let $x, y \in X$. Then:*

1. Schwarz inequality*:*

$$|(x,y)| \leq \|x\|\,\|y\|, \tag{A.4}$$

where equality holds in (A.4) if and only if x and y are linearly dependent;

2. Parallelogram law*:*

$$\|x+y\|^2 + \|x-y\|^2 = 2\,\|x\|^2 + 2\,\|y\|^2.$$

The parallelogram law generalizes an elementary result in Euclidean plane geometry: *in a parallelogram, the sum of the squares of the sides length equals the sum of the squares of the diagonals length.*

The Schwarz inequality implies the continuity of the inner product; in fact,

$$|(w,z)-(x,y)| = |(w-x,z)+(x,z-y)| \leq \|w-x\|\,\|z\| + \|x\|\,\|z-y\|$$

and if $w \to x$, $z \to y$, then $(w,z) \to (x,y)$.

Definition A.3. Let H be an inner product space. We say that H is a *Hilbert space* if it is complete with respect to the norm (A.3), induced by the inner product.

Two Hilbert spaces H_1 and H_2 are *isometric* if there exists a linear map $L : H_1 \to H_2$, called *isometry*, which preserves the norm, that is

$$\|x\|_{H_1} = \|Lx\|_{H_2} \qquad \forall x \in H_1. \tag{A.5}$$

An isometry preserves also the inner product since from

$$\|x-y\|^2_{H_1} = \|L(x-y)\|^2_{H_2} = \|Lx - Ly\|^2_{H_2}$$

we get

$$\|x\|_{H_1}^2 - 2(x,y)_{H_1} + \|y\|_{H_1}^2 = \|Lx\|_{H_2}^2 - 2(Lx,Ly)_{H_2} + \|Ly\|_{H_2}^2$$

and from (A.5) we infer

$$(x,y)_{H_1} = (Lx,Ly)_{H_2} \qquad \forall x,y \in H_1.$$

It is easy to show that every real linear space of dimension n is isometric to $\mathbb{R}^d$.

A.2 Linear Operators and Duality

In this section we focus on linear operators and functionals, on the concepts of dual space, dual and adjoint operators.

A.2.1 Bounded Linear Operators

Given two normed spaces X and Y, a linear operator from X to Y is a map $L : X \to Y$ such that, for all $\alpha, \beta \in \mathbb{R}$ and for all $x, y \in X$

$$L(\alpha x + \beta y) = \alpha Lx + \beta Ly.$$

If L is linear, when no confusion arises, we shall write Lx instead of $L(x)$. Moreover, we say that L is a bounded operator if there exists a constant $C > 0$ such that

$$\|Lx\|_Y \le C\|x\|_X \qquad \forall x \in X, \tag{A.6}$$

or, equivalently,

$$\sup_{\|x\|_X=1} \|Lx\|_Y = K < +\infty. \tag{A.7}$$

Indeed, if $x \neq 0$, using the linearity of L we can express (A.6) under the form

$$\left\| L\left(\frac{x}{\|x\|_X}\right)\right\|_Y \le C,$$

which is equivalent to (A.7) since $x/\|x\|_X$ is a unit vector in X; clearly, $K \le C$.

Proposition A.2. *A linear operator $L : X \to Y$ is bounded if and only if it is continuous.*

We denote by $\mathcal{L}(X,Y)$ the set of linear and bounded operators from X to Y. In particular, $\mathcal{L}(X,Y)$ is a vector space, over which we can define the following norm

$$\|L\|_{\mathcal{L}(X,Y)} = \sup_{\|x\|_X=1} \|Lx\|_Y, \tag{A.8}$$

which is the number K appearing in (A.7). Thus, for every $L \in \mathcal{L}(X,Y)$, we have

$$\|Lx\|_Y \le \|L\|_{\mathcal{L}(X,Y)}\|x\|_X.$$

Furthermore, endowed with the norm (A.8), $\mathcal{L}(X,Y)$ is a Banach space.

Example A.2. Let $\mathbf{A} \in \mathbb{R}^{m\times n}$; then, the map $L : \mathbf{x} \to \mathbf{A}\mathbf{x}$ is a linear operator from $\mathbb{R}^n$ into $\mathbb{R}^m$. Moreover, since

$$\|\mathbf{A}\mathbf{x}\|^2 = \mathbf{A}\mathbf{x} \cdot \mathbf{A}\mathbf{x} = \mathbf{A}^\top\mathbf{A}\mathbf{x} \cdot \mathbf{x}$$

and $\mathbf{A}^\top\mathbf{A}$ is symmetric and nonnegative, from Linear Algebra we have that

$$\sup_{\|\mathbf{x}\|=1} \mathbf{A}^\top\mathbf{A}\mathbf{x} \cdot \mathbf{x} = \sigma_M$$

where σ_M is the maximum eigenvalue of $\mathbf{A}^\top\mathbf{A}$, that is to say, the maximum singular value of $\mathbf{A}$. Thus, $\|L\| = \sqrt{\sigma_M}$. •

Example A.3. Let V and H be Hilbert spaces, with $V \subset H$. Considering an element in V as an element of H, we define the operator $I_{V\to H} : V \to H$,

$$I_{V\to H}(u) = u$$

which is called *embedding of V into H*. This is a linear operator, and it is also bounded if there exists a constant C such that $\|u\|_H \le C\|u\|_V$ for every $u \in V$. In this case, we say that V is *continuously embedded in* H and we write $V \hookrightarrow H$. •

A.2.2 Functionals, Dual space and Riesz Theorem

If X is a normed space, a linear operator $L : X \to \mathbb{R}$ is called (linear) functional. The space $\mathcal{L}(X, \mathbb{R})$ of linear and bounded (or continuous) functionals over X is a Banach space, called *dual space* of X and is denoted by X^*. This space is endowed with the norm

$$\|L\|_{X^*} = \sup_{\|x\|_X=1} |Lx| \qquad \forall L \in X^*.$$

To denote the action of an element $L \in X^*$ on $v \in X$ we use the notations Lv or $\langle L, v\rangle_{X^*,X}$, called *duality pairing*. The following important theorem, known as the *Riesz Representation Theorem*, gives a characterization of the dual space of a Hilbert space.

Theorem A.2 (Riesz). *Let H be a Hilbert space with inner product $(\cdot,\cdot)$. The dual space H^* is isometric to H. Precisely, for every $u \in H$, the linear functional L_u defined by*

$$L_u v = (u, v)$$

belongs to H^* with norm $\|L_u\|_{H^*} = \|u\|_H$. Conversely, for every $L \in H^*$ there exists a unique $u_L \in H$ such that

$$Lv = (u_L, v) \qquad \forall v \in H.$$

Moreover, $\|L\|_{H^*} = \|u_L\|$.

See, e.g., [237] for the proof. Since it is linear, one-to-one and it preserves norms, the Riesz map $\mathcal{R}_H : H^* \to H$, given by $L \longmapsto u_L$ is an isometry – also referred to as Riesz canonical isometry – and we say that u_L is the *Riesz element associated with* L, with respect to the scalar product $(\cdot,\cdot)$. Moreover, H^* is a Hilbert space, when endowed with the inner product

$$(L_1, L_2)_{H^*} = (u_{L_1}, u_{L_2}). \tag{A.9}$$

Since X^* is a normed space (actually a Banach space) we can consider its dual $(X^*)^* = X^{**}$, called the *bidual* of X, a Banach space, normed by

$$\|x^{**}\|_{X^{**}} = \sup_{\|L\|_{X^*}=1} |\langle x^{**}, L\rangle_{X^{**},X^*}|.$$

Between X and X^{**} there exists a *canonical injection* $J : X \to X^{**}$ defined by

$$\langle Jx, L\rangle_{X^{**},X^*} = \langle L, x\rangle_{X^*,X} \qquad \forall L \in X^*.$$

Clearly Jx belongs to X^{**} since $|\langle Jx, L\rangle_{X^{**},X^*}| \le \|x\|_X \|L\|_{X^*}$ and (by the Hahn-Banach Theorem, see, e.g., [235]) $\|Jx\|_{X^{**}} = \|x\|_X$; thus J is an *isometry*. In general $J(X) \subset X^{**}$.

Definition A.4. When $J(X) = X^{**}$ we say that X is *reflexive.*

Thus, if X is reflexive we can identify X with X^{**} *through the canonical isometry.* Theorem A.2 allows the identification of a Hilbert space with its dual. Applying Riesz Theorem to H^*, we conclude that *every Hilbert space is reflexive.*

A.2.3 Hilbert Triplets

In the analysis of boundary value problems, we often meet a pair of Hilbert spaces V, H with the following properties:

1. $V \hookrightarrow H$, i.e. V is *continuously embedded in* H. Recall that this simply means that the embedding operator $I_{V\hookrightarrow H}$, from V into H, introduced in Example A.3, is continuous or, equivalently that there exists a number c such that
$$\|u\|_H \le c\,\|u\|_V \qquad \forall u \in V; \tag{A.10}$$
2. V *is dense in* H.

Using Riesz Representation Theorem A.2, we may identify H with H^*. Also, we may *continuously embed* H into V^*, so that any element in H can be thought as an element of V^*, with action given by

$$\langle u, v\rangle_{V^*,V} = (u, v)_H \qquad \forall v \in V. \tag{A.11}$$

Finally, V (and therefore also H) *is dense in* V^*. Summarizing, we have

$$V \hookrightarrow H \hookrightarrow V^*$$

with *dense embeddings.* We call $\{V, H, V^*\}$ a *Hilbert triplet.*

Remark A.1. Given the identification of H with H^*, the scalar product in H achieves a privileged role. As a consequence, a further identification of V with V^* through the Riesz map $\mathcal{R}_V^{-1}$ is forbidden, since it would give rise to a nonsense, unless $H = V$. •

A.2.4 Adjoint Operators

The basic relation that characterizes the definition of the transpose $\mathbf{A}^\top$ of a matrix $\mathbf{A}$,

$$(\mathbf{Ax}, \mathbf{y})_{\mathbb{R}^m} = (\mathbf{x}, \mathbf{A}^\top\mathbf{y})_{\mathbb{R}^n} \qquad \forall \mathbf{x} \in \mathbb{R}^n, \forall \mathbf{y} \in \mathbb{R}^m,$$

can be generalized to define the adjoint of a bounded linear operator.

Let X, Y be Banach spaces. For an operator $T \in \mathcal{L}(X, Y)$ we can define the *adjoint* or *dual operator* $T^* \in \mathcal{L}(Y^*, X^*)$ by the relation

$$\langle T^* L, x\rangle_{X^*,X} = \langle L, Tx\rangle_{Y^*,Y} \tag{A.12}$$

for all $L \in Y^*$, $x \in X$. It holds $\|T^*\|_{\mathcal{L}(Y^*,X^*)} = \|T\|_{\mathcal{L}(X,Y)}$. We say that L is *selfadjoint* if $X = Y$ and $T^* = T$.

Example A.4. Let $\{V, H, V^*\}$ be a Hilbert triplet and $T = I_{H \hookrightarrow V^*}$ the embedding of H into V^*. Then, $T^* : V \to H^* = H$ is given by

$$(T^* v, h)_H = \langle v, Th\rangle_{V,V^*} = (v, h)_H \qquad \forall h \in H,\ \forall v \in V.$$

Thus $T^* = I_{V \hookrightarrow H}$, the embedding of V into H. •

A.3 Compactness and Weak Convergence

A.3.1 Compactness

We briefly review the notion of compactness. Let X be a normed space. The general definition of compact set involves open coverings: an *open covering* of $E \subseteq X$ is a family of *open* sets whose union contains E.

Definition A.5. We say that $E \subseteq X$ is compact if from every open covering of E it is possible to extract a finite subcovering of E.

It is somewhat more convenient to work with *pre-compact* sets as well, that is sets whose *closure* is compact. In finite dimensional spaces the characterization of pre-compact sets is well known: $E \subset \mathbb{R}^d$ is pre-compact if and only if E is bounded.

What about infinitely many dimensions? Let us introduce a characterization of precompact sets in normed spaces in terms of convergent sequences, much more comfortable to use. First, let us agree that a subset E of a normed space X is *sequentially precompact* (resp. *compact*), if for every sequence $\{x_k\} \subset E$ there exists a subsequence $\{x_{k_s}\}$, convergent in X (resp. in E). The following result holds.

Theorem A.3. *Let X be a normed space and $E \subset X$. Then E is precompact (compact) if and only if it is sequentially precompact (compact).*

While a compact set is always *closed and bounded*, the following example exhibits a closed and bounded set which is *not* compact.

Example A.5. Consider the Hilbert space

$$l^2 = \left\{ \mathbf{x} = \{x_k\}_{k=1}^{\infty} : \sum_{k=1}^{\infty} x_k^2 < \infty, x_k \in \mathbb{R} \right\}$$

endowed with

$$(\mathbf{x}, \mathbf{y}) = \sum_{k=1}^{\infty} x_k y_k \qquad \text{and} \qquad \|\mathbf{x}\|^2 = \sum_{k=1}^{\infty} x_k^2.$$

Let $E = \{\mathbf{e}^k\}_{k\geq 1}$, where $\mathbf{e}^1 = \{1, 0, 0, ...\}$, $\mathbf{e}^2 = \{0, 1, 0, ...\}$, etc.. Observe that E constitutes an orthonormal basis in l^2. Then, E is closed and bounded in l^2. However, E *is not sequentially compact.* Indeed, $\|\mathbf{e}^j - \mathbf{e}^k\| = \sqrt{2}$, if $j \neq k$, and therefore no subsequence of E can be convergent. •

Thus, in infinite dimensions, closed and bounded does not imply compact. Actually, in a normed space, this can only happen in the finite-dimensional case. In fact, the following theorem holds.

Theorem A.4. *Let B be a Banach space. B is finite-dimensional if and only if the unit ball $\{\mathbf{x} : \|\mathbf{x}\| \leq 1\}$ is compact.*

A.3.2 Weak Convergence, Reflexivity and Weak Sequential Compactness

We have seen that compactness in a normed space is equivalent to sequential compactness. In the applications, this translates into a very strong requirement for approximating sequences. Luckily, in normed spaces there are other notions of convergence, much more flexible, which turn out to be better suited to our purposes.

Let X be a Banach space with norm $\|\cdot\|$ and X^* be its dual space.

Definition A.6. We say that a sequence $\{u_k\} \subset X$ converges weakly to $u \in X$, and we write $u_k \rightharpoonup u$, if

$$Lu_k \to Lu \qquad \forall L \in X^*. \tag{A.13}$$

If X is a Hilbert space with inner product $(\cdot, \cdot)$, then (A.13) is equivalent to

$$(u_k, v) \to (u, v) \qquad \forall v \in \mathcal{U},$$

thanks to Riesz Representation Theorem. Since by definition any $L \in X^*$ is a continuous functional in X, it follows that if $u_k \to u$ then $Lu_k \to Lu$ and therefore

$$u_k \to u \qquad \text{implies} \qquad u_k \rightharpoonup u.$$

The converse is not true in general, as the following example shows. This justifies the denomination of *weak* convergence.

Example A.6. Let $X = L^2(0, 2\pi)$ and $x_k(t) = \frac{1}{\sqrt{\pi}} \sin kt$, $t \in (0, 2\pi)$; see Sect. A.5.1 for the definition of the $L^2(a, b)$ space. Then,

$$\lim_{k\to 0} \int_0^{2\pi} \sin kt\, \varphi(t) dt = 0 \qquad \forall \varphi \in X$$

so that $x_k \rightharpoonup 0$ in X. On the other hand, $x_k \nrightarrow 0$ strongly in X since

$$\|x_k\|_X^2 = \frac{1}{\pi} \int_0^{2\pi} (\sin kt)^2 dt = 1. \qquad \bullet$$

The following properties are consequences of the Hahn-Banach and the Banach-Steinhaus theorems in Functional Analysis (see, e.g., [235]).

Proposition A.3. *Let* $\{u_k\} \subset X$, $u_k \rightharpoonup u$. *Then:*

(*a*) *the weak limit is unique: if* $u_k \rightharpoonup v$ *then* $v = u$;
(*b*) $\{u_k\}$ *is bounded;*
(*c*) $\|u\| \leq \liminf \|u_k\|$;
(*d*) *if* $\|L_k - L\|_{X^*} \to 0$, *then* $L_k u_k \to Lu$.

We have observed that the norm in a normed space is strongly continuous. Property (*c*) in Proposition (A.3) expresses the fact that the norm is only *sequentially lower semicontinuous* with respect to weak convergence. The following fundamental result states the connection of weak convergence with reflexivity.

Proposition A.4. *Let* X *be a reflexive Banach space. Then every bounded sequence* $\{u_n\} \subset X$ *contains a weakly convergent subsequence.*

Remark A.2. Thanks to the Riesz representation theorem A.2, every Hilbert space H is reflexive. Hence, usual control spaces such as those introduced in Chapter 5 are all reflexive, since they are Hilbert spaces. In particular, Proposition A.4 applies to any Hilbert space. •

A.3.3 Compact Operators

By definition, every operator in $\mathcal{L}(X,Y)$ transforms bounded sets in X into bounded sets in Y. The subclass of operators that transform *bounded sets* into *precompact* sets is particularly important.

Definition A.7. Let X and Y be Banach spaces, and $L \in \mathcal{L}(X,Y)$. We say that L is *compact* if, for every bounded $E \subset X$, the image $L(X)$ is pre-compact in Y.

From Theorem A.4, the identity operator $I : X \to X$ is compact if and only if $\dim X < \infty$. Also, any bounded operator with finite dimensional range is compact. The following proposition is also very useful.

Proposition A.5. *Let* $L : X \to Y$ *be compact. Then:*

(*a*) $L^* : X \to Y$ *is compact;*
(*b*) *if* $G \in \mathcal{L}(Y,Z)$ *or* $G \in \mathcal{L}(Z,X)$, *the operators* $G \circ L$ *and* $L \circ G$ *are compact.*

Remark A.3. Linear compact operators convert weak convergence into strong convergence. If X is reflexive, also the converse is true. Precisely, let $T \in L(X,Y)$, X, Y be Banach spaces, and X reflexive. Then T is compact if and only if

$$\forall \{x_k\} \subset X, \ x_k \rightharpoonup x \ \text{ implies } \ Tx_k \to Tx.$$

•

A.3.4 Convexity and Weak Closure

We have seen in Sect. A.3.2 that a strongly convergent sequence is also weakly convergent: however, not always a property holding in the strong sense also holds in the weak one. For instance, a (strongly) closed set in a Banach space X is not necessarily weakly sequentially closed. The sequence $\{x_k\}$ of Example A.6 is contained in the unit sphere S_1 of $L^2(0, 2\pi)$, which is a closed set in this space. However $x_k \rightharpoonup 0 \notin S_1$, and therefore S_1 is not weakly sequentially closed.

Here *convexity* comes into play: as soon as the (strongly) closed set is also *convex*, it is also weakly sequentially closed, according to the following

Proposition A.6. *Let $K \subseteq X$ be a closed convex set. Then K is weakly sequentially closed.*

Thus, a convex set is closed if and only if it is weakly sequentially closed.

A.3.5 Weak* Convergence, Separability and Weak* Sequential Compactness

In several problems the natural space where we seek a control variable is not reflexive. This is for instance the case in which a control $u \geq 0$ acts as a reaction coefficient or a potential in a state equation like

$$-\Delta y + uy = f \qquad \text{in } \Omega,$$

or as a diffusion coefficient in

$$-\text{div}\,(u\nabla y) = f \qquad \text{in } \Omega.$$

In these situations $u \in L^\infty(\Omega)$, which is not a reflexive Banach space, see Sect. A.5.1.

When reflexivity is lacking, weak convergence is not appropriate to recover compactness. In fact, the dual X^* of a Banach space can be endowed with another type of convergence, called *weak** convergence.

Definition A.8. We say that a sequence $\{L_k\} \subset X^*$ converges weakly* to $L \in X^*$, and we write $L_k \overset{*}{\rightharpoonup} L$, if

$$L_k u \to Lu \qquad \forall u \in X. \tag{A.14}$$

Remark A.4. If X is reflexive, the weak* convergence is nothing else than weak convergence in X^*. •

For the weak* convergence an analog of Proposition A.4 holds.

Proposition A.7. *Let X be a Banach space and $\{L_k\} \subset X^*$ such that $L_k \overset{*}{\rightharpoonup} L$. Then:*

(a) *the weak* limit is unique: if $L_k \overset{*}{\rightharpoonup} G$ then $G = L$;*
(b) *$\{L_k\}$ is bounded in X^*;*
(c) *if $\|u_k - u\|_X \to 0$, then $L_k u_k \to Lu$.*

The relevance of weak* convergence relies in the following fundamental result that states a connection with the separability of X. We recall that X is *separable* if there exists a countable subset D which is *dense* in X, that is $D \subseteq X$ and $\overline{D} = X$.

Theorem A.5 (Banach-Alaoglu). *If X is separable, any bounded sequence in X^* contains a weakly* convergent subsequence.*

A.4 Abstract Variational Problems

We recall in this section some results essential for the correct variational formulation of a broad variety of boundary value problems, including several examples of state problems for optimal control problems discussed along this book.

In particular, we briefly review (strongly) coercive and saddle-point problems, presenting the main well-posedness results; see, e.g., [224, 64, 121, 237, 218, 41] for their proofs.

In this context, a key role is played by *bilinear forms.* Given two linear spaces V, W, a *bilinear form* in $V \times W$ is a map

$$a : V \times W \to \mathbb{R}$$

satisfying the following properties:

(i) for every $y \in W$, the function $x \mapsto a(x, y)$ is linear in V;
(ii) for every $x \in V$, the function $y \mapsto a(x, y)$ is linear in W.

When $V = W$, we simply say that a is a *bilinear form* in V.

A.4.1 Coercive Problems

Let V be a Hilbert space, a be a bilinear form in V and $F \in V^*$ (the dual space of V). Consider now the following problem, called *abstract variational* problem:

find $y \in V$ such that

$$a(y, v) = \langle F, v \rangle_{V^*,V} \qquad \forall v \in V. \tag{P_1}$$

The fundamental result is provided by the following

Theorem A.6 (Lax-Milgram). *Let V be a real Hilbert space endowed with inner product $(\cdot,\cdot)$ and norm $\|\cdot\|$, and $a(\cdot,\cdot)$ a bilinear form in V which is:*
(i) continuous, *i.e. there exists a constant M such that*

$$|a(v, w)| \le M \|v\| \|w\| \qquad \forall v, w \in V;$$

(ii) $V-$coercive, *i.e. there exists a constant $\alpha > 0$ such that*

$$a(v, v) \ge \alpha \|v\|^2 \qquad \forall v \in V. \tag{A.15}$$

Then, problem (P_1) has a unique solution $y \in V$. Moreover, y satisfies the stability estimate

$$\|y\| \le \frac{1}{\alpha} \|F\|_{V^*}. \tag{A.16}$$

If a is also symmetric, then it induces in V the inner product

$$(v,w)_a = a(v,w).$$

In this case, existence, uniqueness and stability for problem (P_1) follow directly from Riesz Representation Theorem A.2. Moreover (see Sect. 5.1.2), by defining the quadratic functional

$$J(v) = \frac{1}{2}a(v,v) - \langle F, v\rangle_{V^*,V}, \tag{A.17}$$

the following property holds.

Proposition A.8. *Let V be a real Hilbert space, $a: V \times V \to \mathbb{R}$ a continuous, symmetric and positive bilinear form on $V \times V$, and $F \in V^*$. Then, u is the (unique) solution of (P_1) if and only if*

$$J(u) = \min_{v\in V} J(v). \tag{A.18}$$

A.4.2 Saddle-Point Problems

Mixed variational problems or *saddle point problems* do not fall in the class of *coercive* problems, thus requiring a different treatment[1]. Two remarkable cases, addressed in this book, are Stokes equations and the optimality system arising from unconstrained linear-quadratic OCPs. For a more in-depth reading we refer e.g. to [41].

Given two Hilbert spaces X and Q along with their duals X^* and Q^*, respectively, the bilinear forms $e\colon X\times X \to \mathbb{R}$, $b\colon X \times Q \to \mathbb{R}$ and $F_1 \in X^*$, $F_2 \in Q^*$, we consider the following mixed variational problem:

find $(x,p) \in X \times Q$ such that

$$\begin{cases} e(x,w) + b(w,p) = \langle F_1, w\rangle_{W^*,W} & \forall w \in X \\ b(x,q) = \langle F_2, q\rangle_{Q^*,Q} & \forall q \in Q. \end{cases} \tag{P_2}$$

The following theorem [53, 41] establishes sufficient conditions for the saddle-point problem (P_2) to be well posed. Let us denote by

$$X_0 = \{w \in X : b(w,q) = 0 \quad \forall q \in Q\} \subset X.$$

Theorem A.7. *Assume that the following conditions hold:*

1. $e(\cdot,\cdot)$ *is* continuous, *i.e. there exists $M_e > 0$ such that*

$$|e(x,w)| \le M_e \|x\|_X \|w\|_X \qquad \forall x, w \in X; \tag{A.19}$$

2. $e(\cdot,\cdot)$ *is* weakly coercive *on X_0, i.e. there exists $\alpha_0 > 0$ such that*

$$\inf_{x\in X_0} \sup_{w\in X_0} \frac{e(x,w)}{\|x\|_X \|w\|_X} \ge \alpha_0, \qquad \inf_{w\in X_0} \sup_{x\in X_0} e(x,w) > 0; \tag{A.20}$$

[1] These are special instances of *weakly coercive* problems, which can be analyzed according to the Nečas Theorem, see, e.g., [203] for further details. This may happen if, for instance, *(i)* in (P_1) $a(\cdot,\cdot)$ fails to be coercive, or *(ii)* $y \in V$ but we ask the variational equation to hold for every $\varphi \in W \neq V$, with $F \in W^*$.

3. $b(\cdot,\cdot)$ *is* continuous, *i.e. there exists* $M_b > 0$ *such that*

$$|b(w,q)| \le M_b \|w\|_X \|q\|_Q \qquad \forall w \in X,\ q \in Q; \tag{A.21}$$

4. *the bilinear form* $b(\cdot,\cdot)$ *satisfies the* inf-sup condition, i.e. there exists $\beta_0^s > 0$ *such that*

$$\inf_{q\in Q} \sup_{w\in X} \frac{b(w,q)}{\|w\|_X \|q\|_Q} \ge \beta_0^s. \tag{A.22}$$

Then, there exists a unique solution $(x,p) \in X \times Q$ *to the mixed variational problem* (P_2). *Moreover, the following stability estimates hold:*

$$\|x\|_X \le \frac{1}{\alpha}\Big[\|f_1\|_{X^*} + \frac{\alpha_0 + M_e}{\beta_0^s}\|F_2\|_{Q^*}\Big], \tag{A.23a}$$

$$\|p\|_Q \le \frac{1}{\beta_0^s}\Big[\Big(1 + \frac{M_e}{\alpha_0}\Big)\|F_1\|_{X^*} + \frac{M_e(\alpha_0 + M_e)}{\alpha_0 + \beta_0^s}\|F_2\|_{Q^*}\Big]. \tag{A.23b}$$

Remark A.5. Note that if $e(\cdot,\cdot)$ is symmetric, the second condition in (A.20) automatically follows from the first one. •

We have seen that for symmetric bilinear forms the variational problem (P_1) is equivalent to a minimization problem for the quadratic functional J defined in (A.17). Similarly, the following result states that the mixed variational problem (P_2) is equivalent to a *constrained minimization problem* for the same quadratic functional, provided $e(\cdot,\cdot)$ is a symmetric and positive bilinear form, in the following sense.

Given a linear mapping $\mathcal{L} : X \times Q \to \mathbb{R}$, (x,p) is called a *saddle point* of $\mathcal{L}$ if

$$\mathcal{L}(x,q) \le \mathcal{L}(x,p) \le \mathcal{L}(w,p) \qquad \forall (w,q) \in X \times Q,$$

or equivalently,

$$\inf_{w\in X} \sup_{q\in Q} \mathcal{L}(w,q) = \sup_{q\in Q} \mathcal{L}(x,q) = \mathcal{L}(x,p) = \inf_{w\in X} \mathcal{L}(w,p) = \sup_{q\in Q} \inf_{w\in X} \mathcal{L}(w,q).$$

Proposition A.9. *Let* X *and* Q *be two Hilbert spaces and consider the functionals* $F_1 \in X^*$ *and* $F_2 \in Q^*$, *as well as the bilinear forms* $e(\cdot,\cdot)$ *and* $b(\cdot,\cdot)$ *on* $X \times X$ *and* $X \times Q$, *respectively. Assume that* $d : X \times X \to \mathbb{R}$ *is symmetric and positive. Then* $(x,p) \in X \times Q$ *is a solution to problem* (P_2) *if and only if* (x,p) *is a saddle point of*

$$\mathcal{L}(w,q) = \frac{1}{2}e(w,w) + b(w,q) - \langle F_1, w\rangle_{X^*,X} - \langle F_2, q\rangle_{Q^*,Q}, \tag{A.24}$$

that is, if and only if x *is a minimizer of*

$$J(w) = \frac{1}{2}e(w,w) - \langle F_1, w\rangle_{X^*,X}$$

over X *under the constraint* $b(x,q) = \langle F_2, q\rangle_{Q^*,Q} \ \ \forall q \in Q$. *Equivalently,*

$$x = \arg\min_{w\in X} J(w) \quad \textit{subject to} \quad b(x,q) = \langle F_2, q\rangle_{Q^*,Q} \quad \forall q \in Q. \tag{A.25}$$

If, additionally, the assumptions of Theorem A.7 are fulfilled, problem (P_2) *has a unique solution, which is the unique saddle point of* $\mathcal{L}$.

Hence, (P_2) can be seen as a set of first-order necessary (and sufficient) optimality conditions for the constrained minimization problem (A.25).

A.5 Sobolev spaces

Sobolev spaces provide the natural functional setting for the PDEs that we consider in this textbook. After introducing their definition, we recall the main results concerning these spaces. It is understood that by a *measure* $|E|$ of a set $E \subseteq \mathbb{R}^d$ we mean the Lebesgue measure defined over the ($\sigma-$algebra of the) Lebesgue measurable sets. Accordingly, if Ω is a measurable set, a function $u : \Omega \to \mathbb{R}$, is *measurable* if the set $\{\mathbf{x} \in \Omega : u(\mathbf{x}) > t\}$ is measurable $\forall t \in \mathbb{R}$. For a survey on measure theory we refer, e.g., to [234, 280].

A.5.1 Lebesgue Spaces

Let Ω be an *open set* in $\mathbb{R}^d$ and $p \in \mathbb{R}$, $p \geq 1$. We denote by $L^p(\Omega)$ the set of measurable functions $u : \Omega \to \mathbb{R}$ such that $|u|^p$ is Lebesgue integrable[2] in Ω. Identifying two functions u and v when they are *equal a.e.*[3] in Ω, $L^p(\Omega)$ becomes a Banach space when equipped with the norm (*integral norm of order p*)

$$\|u\|_{L^p(\Omega)} = \left(\int_\Omega |u|^p \right)^{1/p}.$$

$L^2(\Omega)$ is a Hilbert space when endowed with the inner product

$$(u, v)_{L^2(\Omega)} = \int_\Omega uv.$$

The identification of two functions equal a.e. amounts to saying that an element of $L^p(\Omega)$ is not a single function but, actually, an equivalence class of functions, different from one another only on subsets of zero measure. When dealing with a *function* in $L^p(\Omega)$, for most practical purposes, one may refer to the more convenient representative of the class. However, as we shall see later, some care is due, especially when pointwise properties are involved.

The symbol $L^\infty(\Omega)$ denotes the set of *essentially bounded* functions in Ω. We say that $u : \Omega \to \mathbb{R}$ is essentially bounded if there exists M such that

$$|u(\mathbf{x})| \leq M \qquad \text{a.e. in } \Omega. \tag{A.26}$$

The infimum of all numbers M with the property (A.26) is called *essential supremum* of u, and denoted by

$$\|u\|_{L^\infty(\Omega)} = \operatorname*{ess\,sup}_\Omega |u|.$$

Note that the essential supremum may differ from the supremum. As an example, take the characteristic function of the rational numbers

$$\chi_{\mathbb{Q}}(x) = \begin{cases} 1 & x \in \mathbb{Q}, \\ 0 & x \in \mathbb{R} \setminus \mathbb{Q}. \end{cases} \tag{A.27}$$

Then $\sup_{\mathbb{R}} \chi_{\mathbb{Q}} = 1$, but $\|\chi_{\mathbb{Q}}\|_{L^\infty(\Omega)} = 0$, since $\mathbb{Q}$ has zero Lebesgue measure.

If we identify two functions when they are equal a.e., $\|u\|_{L^\infty(\Omega)}$ is a norm in $L^\infty(\Omega)$, and $L^\infty(\Omega)$ becomes a Banach space.

---

[2] A basic introduction on the construction of the Lebesgue integral can be found, e.g., in [237].

[3] A property is valid *almost everywhere* in a set Ω, *a.e.* in short, if it is true at all points in Ω, but for a subset of measure zero.

We list below some relevant results concerning the L^p spaces.

Lemma A.1 (Hölder inequality). *Let $p \in [1, \infty]$ and $q = p/(p-1)$ be its dual exponent (with $q = 1, \infty$ if $p = \infty, 1$, respectively). If $u \in L^p(\Omega)$, $v \in L^q(\Omega)$, then*

$$\int_\Omega |uv| \le \|u\|_{L^p(\Omega)} \|v\|_{L^q(\Omega)} \, . \tag{A.28}$$

A more general version of the Hölder inequality, involving three functions, states that if $p, q, r \in [1, \infty]$ with $1/p + 1/q + 1/r = 1$, and $u \in L^p(\Omega), v \in L^q(\Omega)$, and $w \in L^r(\Omega)$, then

$$\int_\Omega |uvw| \le \|u\|_{L^p(\Omega)} \|v\|_{L^q(\Omega)} \|w\|_{L^r(\Omega)} \, . \tag{A.29}$$

Note that, if Ω has *finite measure* and $1 \le p_1 < p_2 \le \infty$, from (A.28) we have, choosing $v = 1$ and taking $|u|^{p_1}$ as u , $p = p_2/p_1$ and $q = p_2/(p_2 - p_1)$,

$$\int_\Omega |u|^{p_1} \le |\Omega|^{1/q} \|u\|^{p_1}_{L^{p_2}(\Omega)}$$

where $|\Omega|$ denotes the Lebesgue measure of Ω, then $L^{p_2}(\Omega) \subset L^{p_1}(\Omega)$. If the measure of Ω is *infinite*, this inclusion is not true, in general; for instance, $f = 1$ belongs to $L^\infty(\mathbb{R})$ but not to $L^p(\mathbb{R})$ for $1 \le p < \infty$.

Lemma A.2. *For every $p \in [1, \infty]$, the dual of $L^p(\Omega)$ is identified with $L^q(\Omega)$, where $q = p/(p-1)$.*

This means that for $p \in (1, \infty)$ the space $L^p(\Omega)$ is reflexive. The spaces $L^p(\Omega)$, $1 \le p < \infty$ are all separable; $L^\infty(\Omega)$ is not.

Remark A.6. If X^* is separable, so is X, but clearly the converse is false. For instance, $L^\infty(\Omega)$ is the dual of $L^1(\Omega)$. •

Let $C^\infty(\Omega)$ be the set of infinitely differentiable functions in Ω and $C_0^\infty(\Omega)$ be the set of all functions $u \in C^\infty(\Omega)$ vanishing outside a compact subset of Ω. Then:

Lemma A.3. *$C_0^\infty(\Omega)$ is dense in $L^p(\Omega)$ for any $1 \le p < \infty$.*

As a consequence, we have the following useful result, where $L^1_{loc}(\Omega)$ denotes the set of functions integrable on every compact subset of Ω.

Lemma A.4. *If $u \in L^1_{loc}(\Omega)$ and*

$$\int_\Omega u\varphi = 0 \quad \text{for all } \varphi \in C_0^\infty(\Omega),$$

then $u = 0$ almost everywhere.

The following result is a cornerstone of the Lebesgue theory.

Theorem A.8 (Lebesgue dominated convergence). *Let $u_k : \Omega \to \mathbb{R}$, $k \geq 1$, be a sequence of measurable functions such that:*

(i) $u_k \to u$ *a.e. in* Ω;
(ii) there exists $g \in L^1(\Omega)$ *such that* $|u_k| \leq g$ *a.e. in* Ω $\forall k \geq 1$.

Then $u_k, u \in L^1(\Omega)$ *and* $\|u_k - u\|_{L^1(\Omega)} \to 0$ *as* $k \to \infty$.

A.5.2 Domain Regularity

A domain $\Omega \subset \mathbb{R}^d$ can be classified according to the degree of smoothness of its boundary $\Gamma = \partial\Omega$ (see Fig. A.1).

Definition A.9. We say that $\Omega \subset \mathbb{R}^d$ is a C^1−domain if for every point $\mathbf{p} \in \Gamma$ there exists a system of coordinates $(y_1, \ldots, y_{d-1}, y_d) = (\mathbf{y}', y_d)$ with origin at $\mathbf{p}$, a ball $B(\mathbf{p})$ and a function φ defined in a neighborhood $\mathcal{N}_{\mathbf{p}} \subset \mathbb{R}^{d-1}$ of $\mathbf{y}' = \mathbf{0}'$ such that

$$\varphi \in C^1(\mathcal{N}_{\mathbf{p}}), \ \varphi(\mathbf{0}') = 0$$

and

$$\Gamma \cap B(\mathbf{p}) = \{(\mathbf{y}', y_d) : y_d = \varphi(\mathbf{y}'), \ \mathbf{y}' \in \mathcal{N}_{\mathbf{p}}\},$$
$$\Omega \cap B(\mathbf{p}) = \{(\mathbf{y}', y_d) : y_d > \varphi(\mathbf{y}'), \ \mathbf{y}' \in \mathcal{N}_{\mathbf{p}}\}.$$

In other words, Γ locally coincides with the graph of a C^1−function, and Ω is locally placed on one side of its boundary.

The boundary of a C^1−domain does not have corners or edges and for every point $\mathbf{p} \in \Gamma$, a tangent straight line ($d = 2$) or plane ($d = 3$) or hyperplane ($d > 3$) is well defined, together with the *outward* normal unit vectors $\mathbf{n} = (n_1, \ldots, n_d)$. Moreover these vectors vary continuously on Γ. The couples $(\varphi, \mathcal{N}_{\mathbf{p}})$ appearing in the above definition are called *local charts.* If the functions φ are all C^k−functions, for some $k \geq 1$, Ω is said to be a C^k−*domain*; if this is true for every $k \geq 1$, Ω is a C^∞−*domain* (or *smooth* domain).

In several applications the relevant domains are rectangles, prisms, cones, cylinders or unions of them; moreover, polygonal domains obtained by *triangulation* procedures of smooth domains are very often employed in numerical approximation, see Sect. B.4. All these are *Lipschitz domains,* described by local charts where all the functions φ are Lipschitz continuous.

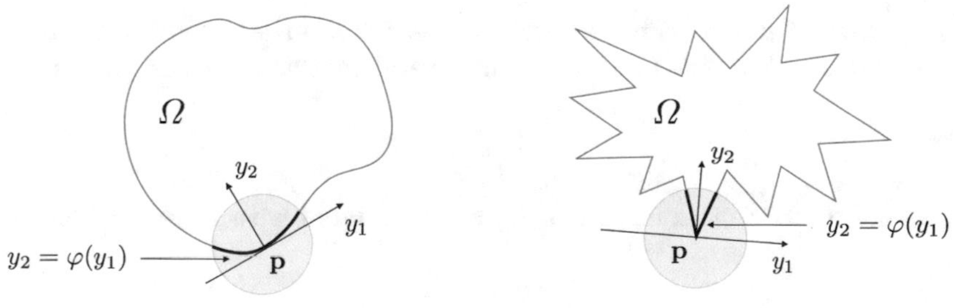

Fig. A.1 A C^1 domain (left) and a Lipschitz domain (right)

A.5.3 Integration by Parts and Weak Derivatives

The following integration by parts formula holds: let Ω be a bounded, Lipschitz domain in $\mathbb{R}^d$; if $u, v \in C^1(\overline{\Omega})$, then

$$\int_\Omega u_{x_j} v \, d\mathbf{x} = \int_\Gamma uv \, n_j d\sigma - \int_\Omega uv_{x_j} d\mathbf{x}, \qquad j = 1, \ldots, d \tag{A.30}$$

where $d\sigma$ denotes the surface area element; in terms of local charts, $d\sigma = \sqrt{1 + |\nabla\varphi|^2} d\mathbf{y}'$.

The notion of derivative in a weak sense (*weak derivative* or *distributional derivative*) relies on the fact that, using formula (A.30), for functions $u \in C^1(\overline{\Omega})$ the following relation holds

$$\int_\Omega u_{x_j} \varphi \, d\mathbf{x} = - \int_\Omega u\varphi_{x_j} d\mathbf{x} \qquad \forall \varphi \in C_0^\infty(\Omega). \tag{A.31}$$

Observing that the right hand side in (A.31) makes sense for $u \in L^1_{loc}(\Omega)$, we are led to the following definition.

Definition A.10. Let $\Omega \subseteq \mathbb{R}^d$ be open and $u \in L^1_{loc}(\Omega)$. A function $w \in L^1_{loc}(\Omega)$ is called the weak partial x_j–derivative of u, if

$$\int_\Omega w\varphi \, d\mathbf{x} = - \int_\Omega u\varphi_{x_j} d\mathbf{x} \qquad \forall \varphi \in C_0^\infty(\Omega), \qquad j = 1, \ldots, d. \tag{A.32}$$

We use the same symbols u_{x_j}, $\partial_{x_j} u$ or $\partial u / \partial x_j$ to denote the weak x_j–derivative of u. Note that w is uniquely determined thanks to Lemma A.4. Similarly, for any multi-index $\alpha = (\alpha_1, \ldots, \alpha_d)$, $|\alpha| > 0$, the $\alpha - th$ weak partial derivative of u, $w = D^\alpha u$ is defined by the relation

$$\int_\Omega w\varphi \, d\mathbf{x} = (-1)^{|\alpha|} \int_\Omega u D^\alpha \varphi d\mathbf{x} \qquad \forall \varphi \in C_0^\infty(\Omega). \tag{A.33}$$

A.5.4 The Space $H^1(\Omega)$

Let $\Omega \subseteq \mathbb{R}^d$ be an open set. The Sobolev space $H^1(\Omega)$ is the subspace of $L^2(\Omega)$ of the functions u having weak first partial derivatives which belong to $L^2(\Omega)$. Precisely:

Definition A.11. The Sobolev space $H^1(\Omega)$ is defined by

$$H^1(\Omega) = \left\{ u \in L^2(\Omega) : u \text{ has weak derivatives } u_{x_j} \in L^2(\Omega), j = 1, \ldots, d \right\}.$$

The weak gradient ∇u is a vector whose components are the weak first partial derivatives of u; then $u \in H^1(\Omega)$ if $\nabla u \in L^2(\Omega)^d$. Since in several contexts the Dirichlet integral

$$\int_\Omega |\nabla u|^2$$

represents an energy, the functions in $H^1(\Omega)$ are associated with *configurations having finite energy*.

Proposition A.10. *Endowed with the inner product*

$$(u,v)_{H^1(\Omega)} = \int_\Omega uv \; d\mathbf{x} + \int_\Omega \nabla u \cdot \nabla v \; d\mathbf{x}$$

$H^1(\Omega)$ *is a Hilbert space, continuously embedded in* $L^2(\Omega)$.

Example A.7. Let $\Omega = B_{1/2}(\mathbf{0}) = \{\mathbf{x} \in \mathbb{R}^2 : |\mathbf{x}| < 1/2\}$ and $u(\mathbf{x}) = (-\log|\mathbf{x}|)^a$, $\mathbf{x} \neq \mathbf{0}$. Using polar coordinates, we find

$$\int_{B_{1/2}(\mathbf{0})} u^2 d\mathbf{x} = 2\pi \int_0^{1/2} (-\log r)^{2a} \, r dr < \infty \qquad \forall a \in \mathbb{R},$$

so that $u \in L^2(B_{1/2}(\mathbf{0}))$ for every $a \in \mathbb{R}$. Moreover,

$$u_{x_i} = -a x_i |\mathbf{x}|^{-2} (-\log|\mathbf{x}|)^{a-1}, \; i = 1, 2$$

and therefore $|\nabla u| = |a(-\log|\mathbf{x}|)^{a-1}| \, |\mathbf{x}|^{-1}$. Still using polar coordinates,

$$\int_{B_{1/2}(\mathbf{0})} |\nabla u|^2 \, d\mathbf{x} = 2\pi a^2 \int_0^{1/2} |\log r|^{2a-2} \, r^{-1} dr.$$

This integral is finite only if $2 - 2a > 1$ or $a < 1/2$. In particular, ∇u represents the gradient of u in the weak sense. We conclude that $u \in H^1(B_1(\mathbf{0}))$ only if $a < 1/2$. We point out that when $a > 0$, u is *unbounded* near $\mathbf{0}$. •

As Example A.7 shows, a function in $H^1(\Omega)$ may be unbounded. In dimension $d = 1$ this cannot occur. In fact, the elements in $H^1(a,b)$ are continuous functions[4] in $[a,b]$.

Proposition A.11. *Let* $u \in L^2(a,b)$. *Then* $u \in H^1(a,b)$ *if and only if* u *is continuous in* $[a,b]$ *and there exists* $w \in L^2(a,b)$ *such that*

$$u(y) = u(x) + \int_x^y w(s) \, ds \qquad \forall x, y \in [a,b]. \tag{A.34}$$

Moreover $u' = w$ *(both a.e. and in the weak sense).*

Remark A.7. Since a function $u \in H^1(a,b)$ is continuous in $[a,b]$, the value $u(x_0)$ at every point $x_0 \in [a,b]$ makes sense. •

A.5.5 The Space $H_0^1(\Omega)$ and its Dual

We introduce an important subspace of $H^1(\Omega)$.

Definition A.12. We denote by $H_0^1(\Omega)$ the closure of $C_0^\infty(\Omega)$ in $H^1(\Omega)$.

Hence, $u \in H_0^1(\Omega)$ if and only if there exists a sequence $\{\varphi_k\} \subset C_0^\infty(\Omega)$ such that $\varphi_k \to u$ in $H^1(\Omega)$, that is, $\|\varphi_k - u\|_{L^2(\Omega)} \to 0$ and $\|\nabla\varphi_k - \nabla u\|_{L^2(\Omega)} \to 0$ as $k \to \infty$. Since the test functions in $C_0^\infty(\Omega)$ vanish on Γ, every $u \in H_0^1(\Omega)$ "inherits" this property and it is

[4] Rigorously: every *equivalence class in* $H^1(a,b)$ *has a representative continuous in* $[a,b]$.

reasonable to consider $H_0^1(\Omega)$ as the set of functions in $H^1(\Omega)$ vanishing on Γ. Clearly, $H_0^1(\Omega)$ is a Hilbert subspace of $H^1(\Omega)$.

In the applications of the Lax-Milgram theorem to boundary value problems, the dual of $H_0^1(\Omega)$ plays an important role and it is denoted by $H^{-1}(\Omega)$.

Theorem A.9. *$F \in H^{-1}(\Omega)$ if and only if there exists $f_0 \in L^2(\Omega)$ and $\mathbf{f} = (f_1, \ldots, f_d) \in L^2(\Omega)^d$ such that*

$$\langle F, v\rangle_{H^{-1},H_0^1} = (f_0, v)_{L^2(\Omega)} + (\mathbf{f}, \nabla v)_{L^2(\Omega)}\,; \tag{A.35}$$

moreover,

$$\|F\|_{H^{-1}(\Omega)} \leq \left\{\|f_0\|_{L^2(\Omega)} + \|\mathbf{f}\|_{L^2(\Omega)}\right\}. \tag{A.36}$$

Hence, the elements of $H^{-1}(\Omega)$ are represented by a linear combination of functions in $L^2(\Omega)$ and their weak first derivatives. In particular, $L^2(\Omega) \hookrightarrow H^{-1}(\Omega)$.

An important property of $H_0^1(\Omega)$, particularly useful in the analysis of boundary value problems, is the *Poincaré* inequality[5].

Theorem A.10. *Let $\Omega \subset \mathbb{R}^d$ be a bounded domain. Then, there exists a positive constant $C_P = C_P(d, diam(\Omega))$ (Poincaré constant) such that*

$$\|u\|_{L^2(\Omega)} \leq C_P \|\nabla u\|_{L^2(\Omega)^d} \qquad \forall u \in H_0^1(\Omega). \tag{A.37}$$

Inequality (A.37) implies that, if Ω is bounded, the norm $\|u\|_{H^1(\Omega)}$ is equivalent to $\|\nabla u\|_{L^2(\Omega)^d}$ in $H_0^1(\Omega)$. Indeed

$$\|u\|_{H^1(\Omega)} = \sqrt{\|u\|^2_{L^2(\Omega)} + \|\nabla u\|^2_{L^2(\Omega)^d}}$$

and from (A.37)

$$\|\nabla u\|_{L^2(\Omega)^d} \leq \|u\|_{H^1(\Omega)} \leq \sqrt{C_P^2 + 1}\, \|\nabla u\|_{L^2(\Omega)^d}.$$

In the space $H^1(\Omega)$, $\|\nabla u\|_{L^2(\Omega)^d}$ only defines a seminorm, denoted by

$$|u|_{H^1(\Omega)} = \|\nabla u\|_{L^2(\Omega)^d}.$$

Example A.8 (Weak convergence). We characterize the weak convergence in $L^2(\Omega)$ and $H_0^1(\Omega)$. We have that $\{u_k\} \subset L^2(\Omega)$ converges weakly to u if

$$\int_\Omega u_k \varphi \to \int_\Omega u\varphi \qquad \forall \varphi \in L^2(\Omega).$$

Similarly, $\{u_k\} \subset H_0^1(\Omega)$ converges weakly to u if

$$\int_\Omega \nabla u_k \cdot \nabla\varphi \to \int_\Omega \nabla u \cdot \nabla \varphi \qquad \forall \varphi \in H_0^1(\Omega).$$

•

Remark A.8. The optimal constant C_P in the Poincaré inequality is usually referred to as *the* Poincaré constant for the domain Ω. Its determination is related with the solution of an eigenvalue problem for the Laplace operator. Indeed, denoting by

$$\lambda_1 = \min_{u\in H_0^1(\Omega), u\neq 0} R(u), \qquad R(u) = \frac{\int_\Omega |\nabla u|^2 d\mathbf{x}}{\int_\Omega u^2 d\mathbf{x}},$$

[5] Recall that the diameter of a set Ω is given by $\operatorname{diam}(\Omega) = \sup_{\mathbf{x},\mathbf{y}\in\Omega} |\mathbf{x} - \mathbf{y}|$.

we have that

$$\|u\|_{L^2(\Omega)} \leq \frac{1}{\sqrt{\lambda_1}} \|\nabla u\|_{L^2(\Omega)^d}$$

and equality holds when u is an eigenvector corresponding to λ_1. $R(u)$ is usually called the Rayleigh quotient. Hence, $1/\sqrt{\lambda_1}$ is the best (i.e. the smallest) Poincaré constant for the domain Ω. See, e.g., [237, Chapter 8] for further details and related proofs. •

Another very useful result (see e.g. [276] for the proof) is provided by the following Gärding inequality, helpful when dealing with weakly coercive problems.

Proposition A.12. *Let $a : V \times V \to \mathbb{R}$ be a bilinear form in V, $H_0^1(\Omega) \subset V \subset H^1(\Omega)$, such that:*

1. *it fulfills the Gärding inequality: for some constants $\alpha > 0$ and $\lambda > 0$*

$$a(v,v) \geq \alpha \|v\|_V^2 - \lambda \|v\|_{L^2(\Omega)}^2 \qquad \forall v \in V; \tag{A.38}$$

2. *if $u \in V$ is such that $a(u,v) = 0$ for any $v \in V$, then $u = 0$.*

Then there exists a constant $\beta = \beta(\lambda) > 0$ such that

$$\sup_{w \in V} \frac{a(v,w)}{\|w\|_V} \geq \beta \|v\|_V \qquad \forall v \in V.$$

Indeed, a bilinear form fulfilling inequality (A.38) is often called *weakly coercive.*

A.5.6 The Spaces $H^m(\Omega)$, $m > 1$

Involving higher order derivatives, we may construct new Sobolev spaces. Let N be the number of multi-indexes $\alpha = (\alpha_1, \dots, \alpha_d)$, $\alpha_j \geq 0$, such that $|\alpha| = \sum_{i=1}^d \alpha_i \leq m$. We denote by $H^m(\Omega)$ the Sobolev space of the functions in $L^2(\Omega)$, whose weak derivatives up to order m included, are functions in $L^2(\Omega)$, that is (with slight abuse of notation)

$$H^m(\Omega) = \left\{ u \in L^2(\Omega) : D^\alpha u \in L^2(\Omega) \quad \forall \alpha : |\alpha| \leq m \right\}.$$

Proposition A.13. *Equipped with the inner product*

$$(u,v)_{H^m(\Omega)} = \sum_{|\alpha| \leq m} \int_\Omega D^\alpha u D^\alpha v \, d\mathbf{x}$$

$H^m(\Omega)$ is a Hilbert space, continuously embedded in $L^2(\Omega)$. Moreover,

$$\|u\|_{H^m(\Omega)} = \left(\|u\|_{L^2(\Omega)}^2 + \sum_{k=1}^m |u|_{H^k(\Omega)}^2 \right)^{1/2}. \tag{A.39}$$

where $|u|_{H^k(\Omega)}$ denotes the following seminorm in $H^k(\Omega)$, $k = 1, \dots, m$,

$$|u|_{H^k(\Omega)} = \sqrt{\sum_{|\alpha| = k} \|D^\alpha u\|_{L^2(\Omega)}^2}.$$

If $u \in H^m(\Omega)$, any derivative of u of order k belongs to $H^{m-k}(\Omega)$; more generally, if $|\alpha| = k \leq m$, then $D^\alpha u \in H^{m-k}(\Omega)$ and $H^m(\Omega) \hookrightarrow H^{m-k}(\Omega)$, $k \geq 1$.

Example A.9. Let $B_{1/2}(\mathbf{0}) \subset \mathbb{R}^3$ and consider $u(\mathbf{x}) = |\mathbf{x}|^{-a}$; then, $u \in H^1(B_{1/2}(\mathbf{0}))$ if $a < 1/2$. The second order derivatives of u are given by

$$u_{x_i x_j} = a(a+2)\, x_i x_j \,|\mathbf{x}|^{-a-4} - a\delta_{ij}\,|\mathbf{x}|^{-a-2}.$$

Then $|u_{x_i x_j}| \leq |a(a+2)|\,|\mathbf{x}|^{-a-2}$ so that $u_{x_i x_j} \in L^2(B_{1/2}(0))$ if $2a+4 < 3$, or $a < -\frac{1}{2}$. Thus $u \in H^2(B_{1/2}(\mathbf{0}))$ if $a < -1/2$. •

We denote by $H_0^m(\Omega)$ the closure of $C_0^\infty(\Omega)$ with respect to the norm (A.39). The elements in $H_0^m(\Omega)$ "vanish" on Γ together with all their derivatives of order up to $m-1$ included.

More in general, for $k \geq 0$ integer, and $1 \leq p \leq \infty$, we define the Sobolev spaces

$$W^{1,p}(\Omega) = \left\{ u \in L^p(\Omega) : \nabla u \in L^p(\Omega)^d \right\}.$$

A.5.7 Approximation by Smooth Functions. Traces

A function $u \in H^1(\Omega)$ is, in principle, defined only up to a set of zero measure. However, if Ω is a bounded Lipschitz domain, it is possible to approximate any element in $H^1(\Omega)$ by smooth functions in $C^\infty(\overline{\Omega})$, the set of infinitely differentiable functions, whose derivatives of any order can be continuously extended up to Γ. In other terms,

$$C^\infty(\overline{\Omega}) = \bigcap_{k=1}^{\infty} C^k(\overline{\Omega}).$$

Proposition A.14. *Let Ω be a bounded Lipschitz domain. Then $C^\infty(\overline{\Omega})$ is dense in $H^1(\Omega)$.*

The possibility of approximating any element $u \in H^1(\Omega)$ by smooth functions in $\overline{\Omega}$ represents a key tool for introducing the notion of *restriction of* u *on* Γ, denoted by $u_{|\Gamma}$. Such restriction is called the *trace of* u on Γ and it will be an element of $L^2(\Gamma)$. Since Γ is a set of zero measure, this notion is subtle and has to be addressed in a rather indirect way, by introducing a suitable trace operator, as shown in the following theorem.

Theorem A.11. *Let $\Omega \subset \mathbb{R}^d$ be a bounded, Lipschitz domain. Then there exists a linear operator (trace operator) $\tau_0 : H^1(\Omega) \to L^2(\Gamma)$ such that*

$$\tau_0 u = u_{|\Gamma} \quad \text{if} \quad u \in \mathcal{C}^\infty(\overline{\Omega})$$

and

$$\|\tau_0 u\|_{L^2(\Gamma)} \leq c_{tr}\, \|u\|_{H^1(\Omega)} \qquad \textit{(trace inequality)}, \tag{A.40}$$

for a suitable constant $c_{tr} = c_{tr}(\Omega, d)$; $\tau_0 u$ is called the trace of u on Γ.

It is not surprising that the kernel of τ_0 is precisely $H_0^1(\Omega)$, that is,

$$\tau_0 u = 0 \ \text{ if and only if } u \in H_0^1(\Omega).$$

For the trace $\tau_0 u$ we also use the notation $u_{|\Gamma}$ and, if there is no risk of confusion, inside a boundary integral we will simply write $\int_\Gamma u \, d\sigma$.

The operator $\tau_0 : H^1(\Omega) \to L^2(\Gamma)$ is not surjective. In fact, the image of τ_0 is strictly contained in $L^2(\Gamma)$. In other words, there are functions in $L^2(\Gamma)$ which are *not* traces of functions in $H^1(\Omega)$. We use the symbol $H^{1/2}(\Gamma)$ to denote Imτ_0, that is,

$$H^{1/2}(\Gamma) = \left\{\tau_0 u : u \in H^1(\Omega)\right\}. \tag{A.41}$$

$H^{1/2}(\Gamma)$ is a Hilbert space with norm given by

$$\|g\|_{H^{1/2}(\Gamma)} = \inf\left\{\|w\|_{H^1(\Omega)} : w \in H^1(\Omega),\ \tau_0 w = g\right\}. \tag{A.42}$$

This norm is equal to the smallest among the norms of all elements in $H^1(\Omega)$ sharing the same trace g on Γ and takes into account that the trace operator τ_0 is not injective, since we know that Ker $\tau_0 = H_0^1(\Omega)$.

In particular, the following *trace inequality* holds

$$\|\tau_0 u\|_{H^{1/2}(\Gamma)} \leq \|u\|_{H^1(\Omega)}, \tag{A.43}$$

which means that the trace operator τ_0 is continuous from $H^1(\Omega)$ onto $H^{1/2}(\Gamma)$.

We denote by $H^{-1/2}(\Gamma)$ the dual space of $H^{1/2}(\Gamma)$.

Remark A.9. It is possible to define a semi-norm in $H^{1/2}(\Gamma)$ thanks to the so-called Steklov-Poincaré operator. Indeed, for any $g \in H^{1/2}(\Gamma)$, there exists a unique function $u_g \in H^1(\Omega)$, called the harmonic extension of g, fulfilling

$$\begin{cases} -\Delta u_g = 0 & \text{in } \Omega \\ u_g = g & \text{on } \Gamma. \end{cases} \tag{A.44}$$

Exploiting the Green's identity, we can formally define the following semi-norm

$$|g|^2_{H^{1/2}(\Gamma)} = \int_\Gamma \frac{\partial u_g}{\partial n} u_g d\sigma = \int_\Omega |\nabla u_g|^2 d\mathbf{x}. \tag{A.45}$$

By introducing the Steklov-Poincaré operator $S : H^{1/2}(\Gamma) \to H^{-1/2}(\Gamma)$,

$$Sg = \frac{\partial u_g}{\partial n} \tag{A.46}$$

realizing the so-called *Dirichlet-to-Neumann* map [225] related to problem (A.44), we can rewrite (A.45) rigorously as

$$|g|^2_{H^{1/2}(\Gamma)} = \langle Sg, g\rangle_{H^{-1/2},H^{1/2}}.$$

As a consequence,

$$\|g\|^2_{H^{1/2}(\Gamma)} = \|g\|^2_{L^2(\Gamma)} + |g|^2_{H^{1/2}(\Gamma)}.$$

See, e.g., [209] for further details about applications in the field of OCPs. •

In a similar way, we may define the trace of $u \in H^1(\Omega)$ on a smooth subset $\Gamma_0 \subset \Gamma$.

Theorem A.12. *Assume that $\Omega \subset \mathbb{R}^d$ is a bounded, Lipschitz domain. Let Γ_0 be a smooth subset of Γ. Then there exists a trace operator $\tau_{\Gamma_0} : H^1(\Omega) \to L^2(\Gamma_0)$ such that*

$$\tau_{\Gamma_0} u = u_{|\Gamma_0} \ \textit{if}\ \ u \in C^\infty\left(\overline{\Omega}\right),$$

and

$$\|\tau_{\Gamma_0} u\|_{L^2(\Gamma_0)} \leq c_{tr}(\Omega, d)\, \|u\|_{H^1(\Omega)} \qquad \textit{(trace inequality)}. \tag{A.47}$$

The function $\tau_{\Gamma_0} u$ is called the *trace of* u *on* Γ_0, often denoted by $u_{|\Gamma_0}$. For the kernel of τ_{Γ_0}, a Hilbert subspace of $H^1(\Omega)$, we use the symbol $H^1_{\Gamma_0}(\Omega)$, that is

$$\tau_{\Gamma_0} u = 0 \text{ if and only if } u \in H^1_{\Gamma_0}(\Omega).$$

This space can be characterized in another way. Let V_{Γ_0} be the set of functions in $C^\infty(\overline{\Omega})$ vanishing in a neighborhood of $\overline{\Gamma}_0$. Then $H^1_{\Gamma_0}(\Omega)$ is the closure of V_{Γ_0} under the norm in $H^1(\Omega)$.

Proposition A.15. *The Poincaré inequality holds in* $H^1_{\Gamma_0}(\Omega)$. *Therefore,* $\|\nabla u\|_{L^2(\Omega)^d}$ *is a norm in* $H^1_{\Gamma_0}(\Omega)$.

We use the symbol $H^{1/2}(\Gamma_0)$ to denote the image of τ_{Γ_0}, that is,

$$H^{1/2}(\Gamma_0) = \left\{\tau_{\Gamma_0} u : u \in H^1(\Omega)\right\}. \tag{A.48}$$

$H^{1/2}(\Gamma_0)$ is a Hilbert space with induced norm given by

$$\|g\|_{H^{1/2}(\Gamma_0)} = \inf\left\{\|w\|_{H^1(\Omega)} : w \in H^1(\Omega),\ \tau_{\Gamma_0} w = g\right\}.$$

In particular,

$$\|\tau_{\Gamma_0} u\|_{H^{1/2}(\Gamma_0)} \leq \|u\|_{H^1(\Omega)} \tag{A.49}$$

which means that the trace operator τ_{Γ_0} is continuous from $H^1(\Omega)$ onto $H^{1/2}(\Gamma_0)$.

A.5.8 Compactness

Since

$$\|u\|_{L^2(\Omega)} \leq \|u\|_{H^1(\Omega)},$$

$H^1(\Omega)$ is *continuously embedded* in $L^2(\Omega)$ that is, if a sequence $\{u_k\}$ converges to u in $H^1(\Omega)$ it converges to u in $L^2(\Omega)$ as well.

If we assume that Ω is a bounded, Lipschitz domain, then the embedding of $H^1(\Omega)$ in $L^2(\Omega)$ is also *compact*. Thus, for any bounded sequence $\{u_k\} \subset H^1(\Omega)$, there exists a subsequence $\{u_{k_j}\}$ and $u \in H^1(\Omega)$ such that

$$u_{k_j} \to u \text{ in } L^2(\Omega) \quad \text{and} \quad u_{k_j} \rightharpoonup u \text{ in } H^1(\Omega).$$

Actually, only the former property follows from the compactness of the embedding; the latter – i.e., the fact that u_{k_j} converges weakly to u in $H^1(\Omega)$ – expresses indeed a general fact in every Hilbert space H, since every bounded subset $E \subset H$ is *sequentially weakly compact* (see Proposition A.4). We can state the following

Theorem A.13 (Rellich). *Let* Ω *be a bounded, Lipschitz domain. Then* $H^1(\Omega)$ *is compactly embedded in* $L^2(\Omega)$. *Moreover, under the same assumptions,* $H^{1/2}(\Gamma)$ *is dense and compactly embedded in* $L^2(\Gamma)$.

A.5.9 The space $H_{div}(\Omega)$

Another space which is indeed useful when dealing with incompressible fluid flows is $H_{div}(\Omega)$.

Definition A.13. Let $\Omega \subset \mathbb{R}^d$, $d \geq 2$, be a bounded Lipschitz domain. Then

$$H_{div}(\Omega) = \{\mathbf{u} \in L^2(\Omega)^d \,:\, \mathrm{div}\mathbf{u} \in L^2(\Omega)\}$$

is a Sobolev space, with norm

$$\|\mathbf{u}\|_{H_{div}(\Omega)} = \left(\|\mathbf{u}\|^2_{L^2(\Omega)} + \|\mathrm{div}\mathbf{u}\|^2_{L^2(\Omega)}\right)^{1/2}$$

and inner product

$$(\mathbf{u}, \mathbf{v})_{H_{div}(\Omega)} = \int_\Omega \mathbf{u} \cdot \mathbf{v}\, d\mathbf{x} + \int_\Omega \mathrm{div}\mathbf{u}\, \mathrm{div}\mathbf{v}\, d\mathbf{x}.$$

$H_{div}(\Omega)$ is a separable Hilbert space, continuously embedded into $L^2(\Omega)^d$. For each $\mathbf{u} \in H^1(\Omega)^d \subset H_{div}(\Omega)$, $d \geq 2$, we can define the trace of each component u_i on $\Gamma = \partial\Omega$ and, consequently, the so-called *normal trace* of $\mathbf{u}$,

$$\mathbf{u} \cdot \mathbf{n} = \sum_{i=1}^{d} u_i|_\Gamma n_i.$$

Since $C^\infty(\overline{\Omega})^d$ is dense also in $H_{div}(\Omega)$, the normal trace can be defined for vectors of $H_{div}(\Omega)$, according to the following

Theorem A.14. *Assume $\Omega \subset \mathbb{R}^d$ is a bounded, Lipschitz domain. Then there exists a linear operator (normal trace operator) $\tau : H_{div}(\Omega) \to H^{-1/2}(\Gamma)$ such that*

$$\tau\mathbf{u} = \mathbf{u} \cdot \mathbf{n}|_\Gamma \text{ if } \mathbf{u} \in C^\infty(\bar{\Omega})^d$$

and

$$\|\tau\mathbf{u}\|_{H^{-1/2}(\Gamma)} \leq c(\Omega, d)\|\mathbf{u}\|_{H_{div}(\Omega)}.$$

Moreover, for each $\mathbf{u} \in H_{div}(\Omega)$ and each $\varphi \in H^1(\Omega)$, the following generalized integration by parts formula holds

$$\int_\Omega \nabla\varphi \cdot \mathbf{u}\, d\mathbf{x} = -\int_\Omega \varphi\, div\mathbf{u}\, d\mathbf{x} + \langle \mathbf{u} \cdot \mathbf{n}, \varphi\rangle_{H^{-1/2}, H^{1/2}}. \tag{A.50}$$

A.5.10 The Space $H^{1/2}_{00}$ and Its Dual

The treatment of mixed problems for Stokes and Navier-Stokes equations requires the introduction of other trace spaces. In general, if Γ_0 is a smooth subset of Γ and $u \in H^{1/2}(\Gamma_0)$, the *trivial extension* $\tilde{u}$ of u to Γ, that is $\tilde{u} = u$ on Γ_0 and $\tilde{u} = 0$ on $\Gamma_1 = \Gamma\backslash\overline{\Gamma}_0$, does not belong[6] to $H^{1/2}(\Gamma)$. Thus, we introduce the new space

$$H^{1/2}_{00}(\Gamma_0) = \left\{u \in H^{1/2}(\Gamma_0) : \tilde{u} \in H^{1/2}(\Gamma)\right\}.$$

[6] Take for instance $\Omega = (-2, 2)^2 \subset \mathbb{R}^2$ and $\Gamma_0 = (-1, 1) \times \{0\}$. The characteristic function of Γ_0 does not belong to $H^{1/2}(\Gamma)$ (although it is not trivial to prove it) while clearly the function $v = 1$ on Γ_0 belongs to $H^{1/2}(\Gamma_0)$ since it is the trace of the function identically equal to 1 on Ω.

In other terms, $H_{00}^{1/2}(\Gamma_0)$ is the space of traces on Γ_0 of elements in $H_{\Gamma_1}^{1/2}(\Omega)$. $H_{00}^{1/2}(\Gamma_0)$ is a Hilbert space if endowed with the norm $\|u\|_{H_{00}^{1/2}(\Gamma_0)} = \|\tilde{u}\|_{H^{1/2}(\Gamma)}$. We denote its dual by $H_{00}^{-1/2}(\Gamma_0)$ and the duality between $H_{00}^{-1/2}(\Gamma_0)$ and $H_{00}^{1/2}(\Gamma_0)$ by the symbol $\langle\cdot,\cdot\rangle_{\Gamma_0}$.

It turns out that

$$H_{00}^{1/2}(\Gamma_0) \subset L^2(\Gamma_0) \subset H_{00}^{-1/2}(\Gamma_0)$$

is a Hilbert triplet, with $H_{00}^{1/2}(\Gamma_0)$ compactly embedded into $L^2(\Gamma_0)$. In particular, if $u \in L^2(\Gamma_0)$, $v \in H_{00}^{1/2}(\Gamma_0)$ then

$$\langle u, v\rangle_{\Gamma_0} = (u, v)_{L^2(\Gamma_0)}.$$

Using these spaces, it is possible to define the normal trace $\boldsymbol{u}\cdot\mathbf{n}$ of $\boldsymbol{u} \in H_{\text{div}}(\Omega)^d$ on Γ_0. Indeed, we can define $\boldsymbol{u}\cdot\mathbf{n}_{|\Gamma_0}$ as the restriction to Γ_0 of the trace $\boldsymbol{u}\cdot\mathbf{n}_{|\Gamma} \in H^{-1/2}(\Gamma)$. This restriction requires some care.

In fact, let $F \in H^{-1/2}(\Gamma)$. If F and v were functions in $L^2(\Gamma_0)$, the action of $F_{|\Gamma_0}$ on v would be defined by

$$F_{|\Gamma_0} v = \int_\Gamma F\,\tilde{v}.$$

The functional $F_{|\Gamma_0}$ acts on the trivial extension of v. Therefore, the restriction to Γ_0 of $F \in H^{-1/2}(\Gamma)$ must act on $H_{00}^{1/2}(\Gamma_0)$ and be an element of $H_{00}^{-1/2}(\Gamma_0)$. Its action is defined by the formula

$$\langle F_{|\Gamma_0}, v\rangle_{\Gamma_0} = \langle F, \tilde{v}\rangle_{H^{-1/2}(\Gamma),H^{1/2}(\Gamma)} \qquad \forall v \in H_{00}^{1/2}(\Gamma_0). \tag{A.51}$$

Going back to the normal trace of $\boldsymbol{u} \in H_{\text{div}}(\Omega)^d$ on Γ_0, we conclude that $\boldsymbol{u}\cdot\mathbf{n}_{|\Gamma_0}$ is well defined as an element in $H_{00}^{-1/2}(\Gamma_0)$. Moreover, the following integration by parts formula holds

$$\int_\Omega \nabla v\cdot\boldsymbol{u}\,d\mathbf{x} = -\int_\Omega v\ \text{div}\boldsymbol{u}\,d\mathbf{x} + \langle \boldsymbol{u}\cdot\mathbf{n}_{|\Gamma_0}, v\rangle_{\Gamma_0} \qquad \forall v \in H_{\Gamma_1}^1(\Omega).$$

A.5.11 Sobolev Embeddings

We know that the elements of $H^1(a,b)$ are actually continuous functions in $[a,b]$. In dimension $d \geq 2$, the functions in $H^1(\Omega)$ enjoy a better summability property, as indicated in the following theorem.

Theorem A.15. *Let $\Omega \subset \mathbb{R}^d$ be a bounded, Lipschitz domain. Then:*

1. *If $d > 2$, $H^1(\Omega) \hookrightarrow L^p(\Omega)$ for $2 \leq p \leq \frac{2d}{d-2}$.*
   Moreover, if $2 \leq p < \frac{2d}{d-2}$, the embedding of $H^1(\Omega)$ in $L^p(\Omega)$ is compact;
2. *if $d = 2$, $H^1(\Omega) \hookrightarrow L^p(\Omega)$ for $2 \leq p < \infty$, with compact embedding.*

In the above cases

$$\|u\|_{L^p(\Omega)} \leq c(d,p,\Omega)\,\|u\|_{H^1(\Omega)}.$$

For instance, if $d = 3$, $\frac{2d}{d-2} = 6$, hence

$$H^1(\Omega) \hookrightarrow L^6(\Omega) \qquad \text{and} \qquad \|u\|_{L^6(\Omega)} \leq c(\Omega)\,\|u\|_{H^1(\Omega)}.$$

Theorem A.15 shows that H^1-functions can enjoy better summability properties than just $L^2(\Omega)$. For H^m-functions, with $m > 1$, the following regularity theorem holds.

Theorem A.16. *Let Ω be a bounded, Lipschitz domain, and $m > d/2$. Then*

$$H^m(\Omega) \hookrightarrow C^{k,\alpha}\left(\overline{\Omega}\right), \tag{A.52}$$

for k integer, $0 \leq k < m - \frac{d}{2}$ and $\alpha = 1/2$ if d is odd, $\alpha \in (0,1)$ if d is even. The embedding in (A.52) is compact and

$$\|u\|_{C^{k,\alpha}(\overline{\Omega})} \leq c(d,m,\alpha,\Omega)\, \|u\|_{H^m(\Omega)}.$$

Theorem A.16 implies that, in dimension $d = 2$, two derivatives occur to get (Hölder) continuity

$$H^2(\Omega) \hookrightarrow C^{0,\alpha}\left(\overline{\Omega}\right), \qquad 0 < \alpha < 1.$$

In fact, in this case d is even, $m = 2$, and $m - \frac{d}{2} = 1$. Similarly

$$H^3(\Omega) \hookrightarrow C^{1,a}\left(\overline{\Omega}\right),\ 0 < \alpha < 1,$$

since $m - \frac{d}{2} = 2$.

In dimension $d = 3$ we have

$$H^2(\Omega) \subset C^{0,1/2}\left(\overline{\Omega}\right) \quad \text{and} \quad H^3(\Omega) \subset C^{1,1/2}\left(\overline{\Omega}\right),$$

since $m - \frac{d}{2} = \frac{1}{2}$ in the former case and $m - \frac{d}{2} = \frac{3}{2}$ in the latter.

Remark A.10. If $u \in H^m(\Omega)$ for any $m \geq 1$, then $u \in C^\infty\left(\overline{\Omega}\right)$. This kind of results is very useful in the regularity theory of boundary value problems. •

A.6 Functions with Values in Hilbert Spaces

Let V be a *separable* Hilbert space (e.g. $L^2(\Omega)$ or $H^1(\Omega)$). We consider functions $y : [0,T] \to V$ and introduce the main functional spaces we are going to use.

A.6.1 Spaces of Continuous Functions

We start by introducing the set $C([0,T];V)$ of the continuous functions $y : [0,T] \to V$, endowed with the norm

$$\|y\|_{C([0,T];V)} = \max_{0 \leq t \leq T} \|y(t)\|_V.$$

Clearly $C([0,T];V)$ is a Banach space. We say that $v : [0,T] \to V$ is the (strong) derivative v of y if

$$\left\| \frac{y(t+h) - y(t)}{h} - v(t) \right\|_V \to 0$$

for each $t \in [0,T]$. We write as usual y' or $\dot{y}$ for the derivative of y.

The symbol $C^1([0,T];V)$ denotes the Banach space of functions whose derivative exists and belongs to $C([0,T];V)$, endowed with the norm

$$\|y\|_{C^1([0,T];V)} = \|y\|_{C([0,T];V)} + \|y'\|_{C([0,T];V)}$$

Similarly we can define the spaces $C^k([0,T];V)$ and $C^\infty([0,T];V)$, while $C_0^\infty(0,T;V)$ denotes the subspace of $C^\infty([0,T];V)$ of the functions compactly supported in $(0,T)$.

A.6.2 Integrals and Spaces of Integrable Functions

The notions of measurability and integral (called *Bochner integral*) can be extended to these types of functions. First, we introduce the set of functions $s : [0,T] \to V$ which assume only a finite number of values; they are called *simple* and take the form

$$s(t) = \sum_{j=1}^{N} \chi_{E_j}(t)\, y_j, \qquad 0 \le t \le T, \tag{A.53}$$

where $y_1, \dots, y_N \in V$ and $E_1, \dots, E_N$ are Lebesgue measurable, mutually disjoint subsets of $[0,T]$. We say that $f : [0,T] \to V$ is *measurable* if there exists a sequence of simple functions $s_k : [0,T] \to V$ such that, as $k \to \infty$,

$$\|s_k(t) - f(t)\|_V \to 0 \qquad \text{a.e. in } [0,T].$$

It is not difficult to prove that, if f is measurable and $v \in V$, the (real) function $t \longmapsto (f(t), v)_V$ is Lebesgue measurable in $[0,T]$.

The notion of integral is defined first for simple functions. If s is given by (A.53), we define

$$\int_0^T s(t)\, dt = \sum_{j=1}^{N} |E_j|\, y_j$$

with the convention that $|E_j|\, y_j = 0$ if $y_j = 0$. We thus provide the following

Definition A.14. We say that $f : [0,T] \to V$ is summable in $[0,T]$ if it is measurable and there exists a sequence $s_k : [0,T] \to V$ of simple functions such that

$$\int_0^T \|s_k(t) - f(t)\|_V \, dt \to 0 \qquad \text{as } k \to +\infty. \tag{A.54}$$

If f is summable in $[0,T]$, we define the integral of f as follows

$$\int_0^T f(t)\, dt = \lim_{k \to +\infty} \int_0^T s_k(t)\, dt \qquad \text{as } k \to +\infty. \tag{A.55}$$

The limit (A.55) is well defined and does not depend on the choice of the approximating sequence $\{s_k\}$. Moreover, the following important theorem holds.

Theorem A.17 (Bochner). *A measurable function $f : [0,T] \to V$ is summable in $[0,T]$ if and only if the real function $t \longmapsto \|f(t)\|_V$ is summable in $[0,T]$. Moreover*

$$\left\| \int_0^T f(t)\, dt \right\|_V \le \int_0^T \|f(t)\|_V \, dt \tag{A.56}$$

and

$$\left(u, \int_0^T f(t)\, dt \right)_V = \int_0^T (u, f(t))_V \, dt, \qquad \forall u \in V. \tag{A.57}$$

The inequality (A.56) is well known in the case of real or complex functions. Thanks to the Riesz Representation Theorem, (A.57) shows that the action of any element of V^* commutes with the integrals.

Once the definition of integral has been given, we can introduce the space $L^p(0,T;V)$, $1 \le p \le \infty$, as the set of measurable functions $y : [0,T] \to V$ such that:

- if $1 \le p < \infty$,

$$\|y\|_{L^p(0,T;V)} = \left(\int_0^T \|y(t)\|_V^p \, dt \right)^{1/p} < \infty; \tag{A.58}$$

- if $p = \infty$

$$\|y\|_{L^\infty(0,T;V)} = \text{ess} \sup_{0 \le t \le T} \|y(t)\|_V < \infty.$$

Endowed with the above norms, $L^p(0,T;V)$ becomes a Banach space for $1 \le p \le \infty$. If $p = 2$, the norm (A.58) is induced by the inner product

$$(y,v)_{L^2(0,T;V)} = \int_0^T (y(t), v(t))_V \, dt$$

that makes $L^2(0,T;V)$ a Hilbert space.

If $y \in L^1(0,T;V)$, the function $t \longmapsto \int_0^t y(s)\,ds$ is continuous and

$$\frac{d}{dt} \int_0^t y(s)\,ds = y(t) \quad \text{a.e. in } (0,T).$$

Proposition A.16. *$C_0^\infty(0,T;V)$ is dense in $L^p(0,T;V)$, $1 \le p < \infty$.*

A.6.3 Sobolev Spaces Involving Time

To define *Sobolev spaces*, we need the notion of derivative in the sense of distributions for functions $y \in L^1_{loc}(0,T;V)$. We say that $y' \in L^1_{loc}(0,T;V)$ is the *derivative in the sense of distributions* of y if

$$\int_0^T \varphi(t)\, y'(t)\, dt = -\int_0^T \varphi'(t)\, y(t)\, dt \tag{A.59}$$

for every $\varphi \in C_0^\infty(0,T)$ or, equivalently, if

$$\int_0^T \varphi(t)\,(y'(t), v)_V \, dt = -\int_0^T \varphi'(t)\,(y(t), v)_V\, dt \qquad \forall v \in V. \tag{A.60}$$

We denote by $H^1(0,T;V)$ the Sobolev space of the functions $y \in L^2(0,T;V)$ such that $y' \in L^2(0,T;V)$. This is a Hilbert space with inner product

$$(y,\varphi)_{H^1(0,T;V)} = \int_0^T \{(y(t), \varphi(t))_V + (y'(t), \varphi'(t))_V\}\, dt.$$

Since functions in $H^1(a,b)$ are continuous in $[a,b]$, it makes sense to consider the value of y at any point of $[a,b]$. In a certain way, the functions in $H^1(0,T;V)$ depend only on the real variable t, so that the following theorem is not surprising.

Theorem A.18. *Let $u \in H^1(0,T;V)$. Then, $u \in C([0,T];V)$ and*

$$\max_{0\le t\le T} \|u(t)\|_V \le C(T)\|u\|_{H^1(0,T;V)}.$$

Moreover, the fundamental theorem of Calculus holds,

$$u(t) = u(s) + \int_s^t u'(r)\,dr, \qquad 0 \le s \le t \le T.$$

If V and W are separable Hilbert spaces with $V \hookrightarrow W$, it makes sense to define the Sobolev space

$$H^1(0,T;V,W) = \left\{ y \in L^2(0,T;V) : y' \in L^2(0,T;W) \right\},$$

where y' is intended in the sense of distributions in $\mathcal{C}_0^\infty(0,T;W)$, endowed with the scalar product

$$(y,\varphi)_{H^1(0,T;V,W)} = \int_0^T \{(y(t),\varphi(t))_V + (y'(t),\varphi'(t))_W\}\,dt.$$

and the norm

$$\|y\|_{H^1(0,T;V,W)} = \left(\|y\|^2_{L^2(0,T;V)} + \|y'\|^2_{L^2(0,T;W)}\right)^{1/2}.$$

This situation typically occurs in the applications to initial-boundary value problems. Given a Hilbert triplet (see Sect. A.2.3) $V \hookrightarrow H \hookrightarrow V^*$, with V and H separable, the natural functional setting is precisely the space $H^1(0,T;V,V^*)$. The following result (whose proof can be found, e.g., in [73, Volume 5, Chapter XVIII]) is of fundamental importance.

Theorem A.19. *Let (V,H,V^*) be a Hilbert triplet, with V and H separable. Then:*

(a) $C^\infty([0,T];V)$ is dense in $H^1(0,T;V,V^*)$;
(b) $H^1(0,T;V,V^*) \hookrightarrow C([0,T];H)$, that is

$$\|u\|_{C([0,T];H)} \le C(T)\|u\|_{H^1(0,T;V,V^*)};$$

*(c) (*Aubin Lemma*) $H^1(0,T;V,V^*)$ is compactly embedded into $L^2(0,T;H)$;*
(d) if $y,\varphi \in H^1(0,T;V,V^*)$, the following integration by parts formula holds,

$$\int_s^t \{\langle y'(r),\varphi(r)\rangle_* + \langle y(r),\varphi'(r)\rangle_*\}\,dr = (y(t),\varphi(t))_H - (y(s),\varphi(s))_H \qquad \text{(A.61)}$$

for all $s,t \in [0,T]$.

From (A.61) we infer that

$$\frac{d}{dt}(y(t),\varphi(t))_H = \langle y'(t),\varphi(t)\rangle_* + \langle y(t),\varphi'(t)\rangle_*$$

a.e. $t \in [0,T]$ and (letting $\varphi = y$)

$$\int_s^t \frac{d}{dt}\|y(r)\|_H^2\,dt = \|y(t)\|_H^2 - \|y(s)\|_H^2. \qquad \text{(A.62)}$$

A.6.4 The Gronwall Lemma

When dealing with initial-boundary value problems, the following result proves to be extremely useful in order to derive stability estimates (see, e.g., [224, Chap. 1] for the proof).

Lemma A.5 (Gronwall). *Let $A \in L^1(0,T)$ be a non-negative function, φ a continuous function on $[0,T]$. If g is non-decreasing and φ is such that*

$$\varphi(t) \le g(t) + \int_0^t A(\tau)\varphi(\tau)d\tau \qquad \forall t \in [0,T],$$

then

$$\varphi(t) \le g(t)\,\exp\left(\int_{t_0}^t A(\tau)d\tau\right) \qquad \forall t \in [t_0,T].$$

A.7 Differential Calculus in Banach Spaces

In order to write first order optimality conditions for OCPs in Banach spaces, we need some calculus in infinite dimensional vector spaces.

A.7.1 The Fréchet Derivative

We start by generalizing the notion of differential to operators between Banach spaces.

Definition A.15. Let X and Y be Banach spaces, and $U \subseteq X$ be a (non-empty) open subset of X. An operator $F : U \to Y$ is called Fréchet differentiable (briefly $\mathcal{F}-$differentiable) at $x_0 \in U$ if there exists an operator $L(x_0) \in \mathcal{L}(X,Y)$ such that if $h \in U$ and $x_0 + h \in U$,

$$F(x_0+h) - F(x_0) = L(x_0)h + R(h,x_0) \tag{A.63}$$

with $R(h,x_0) = o(\|h\|_X)$, that is

$$\lim_{h\to 0} \frac{\|R(h,x_0)\|_Y}{\|h\|_X} = 0. \tag{A.64}$$

$L(x_0)$ is called Fréchet derivative (or differential) of F at x_0, and we write $L(x_0) = F'(x_0)$.

The following properties hold:

a) *Uniqueness.* The Fréchet derivative $F'(x_0)$ is unique; indeed, if $L_1, L_2 \in \mathcal{L}(X,Y)$ and both satisfy (A.63), then we would have $(L_1 - L_2)h = o(\|h\|_X)$, which implies $L_1 = L_2$.

b) *$\mathcal{F}-$differentiability $\Longrightarrow$ continuity.* If F is $\mathcal{F}-$differentiable at x_0 then it is continuous at x_0; in fact, letting $h \to 0$ in (A.63) we get $F(x_0+h) - F(x_0) \to 0$.

c) *Linearity.* Fréchet differentiation is a linear operation: if $F, G : U \subseteq X \to Y$ are $\mathcal{F}-$differentiable at x_0, and $r, s \in \mathbb{R}$, then $rF + sG$ is $\mathcal{F}-$differentiable at x_0 with

$$(rF+sG)'(x_0) = rF'(x_0) + sG'(x_0).$$

d) *Different norms.* Let $\tilde{X} \hookrightarrow X$ and the operator $F : X \to Y$ be $\mathcal{F}-$differentiable at $x_0 \in \tilde{X}$. Then F is differentiable at x_0 when regarded as an operator from $\tilde{X}$ to Y and the analytical expression of the derivative does not change[7]. In particular, if F is differentiable with respect to a given norm, then it is also differentiable with respect to a *stronger* norm. If we introduce two equivalent norms over X or Y, a differentiable function with respect to one norm is still differentiable with respect to the other norm and its derivative does not change.

e) *Chain rule.* The chain rule for the derivation of the composition of functions still holds. Let X, Y, Z be Banach spaces, $F : U \subseteq X \to V \subseteq Y$ and $G : V \subseteq Y \to Z$, where $U$and V are open sets. If F is F-differentiable at x_0 and G is F-differentiable at $y_0 = F(x_0)$, then $G \circ F$ is F-differentiable at x_0 and

$$(G \circ F)'(x_0) = G'(y_0) \circ F'(x_0).$$

Example A.10. If F is constant, then $F' = 0$. If $F \in \mathcal{L}(X, Y)$, F is differentiable at any point and $F'(x) = F$ for every $x \in X$. •

Example A.11. Let X, Y, Z be Banach spaces. Introduce over $X \times Y$ the norm

$$\|(x, y)\|_{X \times Y} = (\|x\|^2 + \|y\|^2)^{1/2}$$

and let $F : X \times Y \to Z$ be a bilinear and continuous operator, that is $x \mapsto F(x, y)$ and $y \mapsto F(x, y)$ are linear for y and x respectively fixed, and there exists a number C such that

$$\|F(x, y)\|_Z \le C \|x\|_X \|y\|_Y \qquad \forall (x, y) \in X \times Y.$$

Then F is differentiable and $F'(x_0, y_0)$ is the linear map

$$(h, k) \longmapsto F(x_0, k) + F(h, y_0).$$

In fact,

$$\begin{aligned} &F(x_0 + h, y_0 + k) - F(x_0, y_0) \\ &= F(x_0 + h, y_0 + k) - F(x_0 + h, y_0) + F(x_0 + h, y_0) - F(x_0, y_0) \\ &= F(x_0 + h, k) + F(h, y_0) = F(x_0, k) + F(h, y_0) + F(h, k). \end{aligned}$$

Moreover, $F(h, k) = o(\| (h, k) \|_{X \times Y})$; indeed,

$$\frac{\|F(h, k)\|_Z}{\| (h, k) \|_{X \times Y}} \le C \frac{\|h\|_X \|k\|_Y}{(\|h\|^2 + \|k\|^2)^{1/2}} \le \frac{C}{2} \| (h, k) \|_{X \times Y},$$

which tends to 0 as $\| (h, k) \|_{X \times Y} \to 0$. •

If F is $\mathcal{F}-$differentiable at every point of U, the map

$$F' : U \to \mathcal{L}(X, Y)$$

given by $x \longmapsto F'(x)$ is called the *Fréchet derivative* of F. We say that F is *continuously* $\mathcal{F}-$differentiable and we write $F \in C^1(U; Y)$, if F' is continuous in U, that is if for any $x, x_0 \in U$ we have

$$\|F'(x) - F'(x_0)\|_{\mathcal{L}(X,Y)} \to 0 \quad \text{if} \quad \|x - x_0\|_X \to 0.$$

[7] Note that $\tilde{X} \hookrightarrow X$ means that $\|x\|_X \le c\|x\|_{\tilde{X}}$ $\forall x \in \tilde{X}$, for a given constant $c > 0$. If $L \in \mathcal{L}(X, Y)$, then $\|x\|_Y \le \|L\|_{\mathcal{L}(X,Y)} \|x\|_X \le c\|L\|_{\mathcal{L}(X,Y)} \|x\|_{\tilde{X}}$ $\forall x \in \tilde{X}$ and hence $L \in \mathcal{L}(X, Y)$ implies $L \in \mathcal{L}(\tilde{X}, Y)$. Then, it is possible to show that (A.63) holds with the same L, regarded as element of $\mathcal{L}(\tilde{X}, Y)$, and $R(h, x_0) = o(\|h\|_{\tilde{X}})$.

In the special case $Y = \mathbb{R}$, we have $\mathcal{L}(X, \mathbb{R}) = X^*$ and therefore if $J : U \subseteq X \to \mathbb{R}$ is $\mathcal{F}-$differentiable at any point of U, its Fréchet derivative is a map $J' : U \to X^*$. Thus, the action of $J'(x)$ on an element h of X is given by the duality

$$J'(x)\, h = \langle J'(x), h\rangle_{X^*,X},$$

and we can simply write $\langle J'(x), h\rangle_*$ if there is no risk of confusion.

If moreover X is a Hilbert space with inner product $(\cdot,\cdot)$, through the Riesz canonical isomorphism $\mathcal{R}$ (see Theorem A.2) we can identify X^* with X. The application $\mathcal{R} \circ J'$ deserves a special name and leads to the following definition.

Definition A.16. Let H be a Hilbert space and $J : U \subseteq H \to \mathbb{R}$ be $\mathcal{F}-$differentiable at any point of U. The application

$$\mathcal{R} \circ J' : U \subseteq H \to H$$

is called the *gradient* of J and is denoted by ∇J. Equivalently, ∇J is defined by the following relation

$$\langle J'(u), v\rangle_* = (\nabla J(u), v) \qquad \forall u \in U, \forall v \in H. \tag{A.65}$$

Example A.12. Quadratic functionals over a Hilbert space. Let H be a Hilbert space with inner product $(\cdot,\cdot)$ and norm $\|\cdot\|$. Let $a : H \times H \to \mathbb{R}$ be a bilinear, symmetric and continuous form on H (see Sect. A.4) and $J : H \to \mathbb{R}$ be the quadratic functional

$$J(u) = \frac{1}{2} a(u, u).$$

We want to show that $J \in C^1(H) = C^1(H; \mathbb{R})$, that

$$\langle J'(u), v\rangle_* = a(u, v) \qquad \forall u, v \in H \tag{A.66}$$

and to compute ∇J. Since a is bilinear and symmetric, we have

$$\begin{aligned} J(u+v) - J(u) &= \frac{1}{2} a(u+v, u+v) - \frac{1}{2} a(u, u) \\ &= a(u, v) + \frac{1}{2} a(v, v) = a(u, v) + o(\|v\|) \end{aligned}$$

as $a(v, v) \leq M \|v\|^2$. We can conclude that J is $\mathcal{F}$-differentiable and (A.66) holds. Moreover,

$$\langle J'(u+h) - J'(u), v\rangle_* = a(h, v)$$

and then, since a is continuous,

$$\|J'(u+h) - J'(u)\|_{H^*} = \sup_{\|v\|=1} |\langle J'(u+h) - J'(u), v\rangle_*| \leq M \|h\|$$

from which we deduce the (Lipschitz) continuity of J'.

Now we compute ∇J. From Riesz Representation Theorem A.2, there exists an element $u_a = \mathcal{R} \circ J'(u)$ such that we can rewrite (A.66) under the form

$$\langle J'(u), v\rangle_* = (u_a, v) \qquad \forall u, v \in H.$$

Therefore

$$\nabla J(u) = u_a.$$

In particular, let $z_0 \in H$ and consider the functional

$$J(u) = \frac{1}{2}\|u - z_0\|^2 = \frac{1}{2}\|u\|^2 - (z_0, u) + \frac{1}{2}\|z_0\|^2 .$$

Then $J \in C^1(H)$ and

$$\langle J'(u_0), v\rangle_* = (u_0 - z_0, v),$$

so that

$$\nabla J(u_0) = u_0 - z_0.$$

•

The *Fréchet partial derivatives* of an operator $F : U \subseteq X \times Y \to Z$,

$$F'_x \in \mathcal{L}(X, Z), \qquad F'_y \in \mathcal{L}(Y, Z)$$

at a point $(x_0, y_0) \in U$, can be defined by the following relations: for every $h \in X$, $k \in Y$,

$$F(x_0 + h, y_0) - F(x_0, y_0) = F'_x(x_0, y_0)\, h + o(\|h\|_X),$$

$$F(x_0, y_0 + k) - F(x_0, y_0) = F'_y(x_0, y_0)\, k + o(\|k\|_X).$$

As in the finite dimensional case, the following result holds:

Proposition A.17. *If $F'_x : U \to \mathcal{L}(X, Z)$ and $F'_y : U \to \mathcal{L}(Y, Z)$ are continuous at $(x_0, y_0) \in U$, then F is F-differentiable at (x_0, y_0) and*

$$F'(x_0, y_0)(h, k) = F'_x(x_0, y_0)h + F'_y(x_0, y_0)k \qquad \forall h \in X, \forall k \in Y.$$

A.7.2 The Gâteaux Derivative

A weaker notion of derivative is that of *Gâteaux derivative*, which generalizes the concept of *directional derivative.*

Definition A.17. Let X and Y be Banach spaces, and $U \subseteq X$ be an open subset of X. An operator $F : U \subseteq X \to Y$ is called Gâteaux differentiable (briefly $\mathcal{G}$−differentable) at $x_0 \in U$ if there exists an operator $A(x_0) \in \mathcal{L}(X, Y)$, such that, for any $h \in X$ and any $t \in \mathbb{R}$ such that $x_0 + th \in U$,

$$\lim_{t \to 0} \frac{F(x_0 + th) - F(x_0)}{t} = A(x_0)\, h.$$

$A(x_0)$ is called the Gâteaux derivative (or differential) of F at x_0 and we write $A(x_0) = F'_{\mathcal{G}}(x_0)$.

If F is $\mathcal{G}$−differentiable at any point in U, the application $F'_{\mathcal{G}} : U \to \mathcal{L}(X, Y)$ given by

$$F'_{\mathcal{G}} : x \mapsto F'_{\mathcal{G}}(x)$$

is called *Gâteaux derivative* of F. The following properties hold:

a) *uniqueness.* The Gâteaux derivative at x_0 is unique;

b) *commutativity with linear bounded functionals.* The Gâteaux differentiation commutes with any linear and continuous functional. In fact, if $y^* \in Y^*$ and F is $\mathcal{G}$−differentiable in a neighborhood of x_0, we can write

$$\frac{d}{dt}\langle y^*, F(x_0+th)\rangle = \lim_{s\to 0}\left\langle y^*, \frac{F(x_0+th+sh)-F(x_0+th)}{s}\right\rangle$$
$$= \langle y^*, F'_{\mathcal{G}}(x_0+th)h\rangle$$

and, in particular,

$$\frac{d}{dt}\langle y^*, F(x_0+th)\rangle_{|t=0} = \langle y^*, F'_{\mathcal{G}}(x_0)h\rangle;$$

c) *chain rule.* The chain rule holds in the following form: if F is $\mathcal{G}$–differentiable at x_0 and G is $\mathcal{F}$-differentiable at $y_0 = F(x_0)$, then $G \circ F$ is $\mathcal{G}$-differentiable at x_0 and

$$(G\circ F)'_{\mathcal{G}}(x_0) = G'(y_0)\circ F'_{\mathcal{G}}(x_0); \tag{A.67}$$

d) *$\mathcal{F}$ and $\mathcal{G}$–differentiability.* If F is $\mathcal{F}$-differentiable at x_0, then it is also $\mathcal{G}$-differentiable at x_0 and $F'(x_0) = F'_{\mathcal{G}}(x_0)$. Conversely, we have:

Proposition A.18. *Let $F : U \subseteq X \to Y$ be $\mathcal{G}$–differentiable in U. If $F'_{\mathcal{G}}$ is continuous at x_0, then F is $\mathcal{F}$-differentiable at x_0 and $F'(x_0) = F'_{\mathcal{G}}(x_0)$.*

This latter result is often employed to show the $\mathcal{F}$–differentiability of a given operator; in fact it is often simpler to compute the Gâteaux derivative and show that it is a continuous operator, than to check directly the $\mathcal{F}$-differentiability.

A.7.3 Derivative of Convex Functionals

When dealing with minimization problems, convex functionals play a relevant role. As already seen in the finite dimensional case, uniqueness of the minimizer is guaranteed for strictly convex functionals.

Let $J : U \subseteq X \to \mathbb{R}$ be a $\mathcal{G}$–differentiable functional in U. Then, $J'_{\mathcal{G}}(x) \in X^*$ for any $x \in U$, that is, $J'_{\mathcal{G}} : U \to X^*$. To denote the action of $J'_{\mathcal{G}}(x)$ (or of any other element of X^*) over $h \in X$, we will use the symbols

$$J'_{\mathcal{G}}(x)\,h \quad \text{or} \quad \langle J'_{\mathcal{G}}(x), h\rangle_* \quad \text{or} \quad \langle J'_{\mathcal{G}}(x), h\rangle_{X^*,X}$$

depending on the context. The following result (strictly related to OCPs, hence provided together with its proof) holds:

Theorem A.20. *Let U be an open convex set and $J : U \subseteq X \to \mathbb{R}$ be a $\mathcal{G}$–differentiable functional on U. The following properties are equivalent:*

1. *J is convex;*
2. *for any $x, y \in U$*
$$J(y) \geq J(x) + \langle J'_{\mathcal{G}}(x), y-x\rangle_*; \tag{A.68}$$
3. *$J'_{\mathcal{G}}$ is a monotone operator in the following sense*
$$\langle J'_{\mathcal{G}}(x) - J'_{\mathcal{G}}(y), x-y\rangle_* \geq 0 \qquad \forall x, y \in U. \tag{A.69}$$

Proof. (1) $\Rightarrow$ (2). Fix x, $y \in U$. Since J is convex, for all $t \in (0,1)$ we have

$$J(x+t(y-x)) - J(x) \leq t(J(y) - J(x)).$$

Dividing by t and letting $t \to 0^+$ we get

$$\langle J'_{\mathcal{G}}(x), (y-x)\rangle_* \leq J(y) - J(x). \tag{A.70}$$

(2) $\Rightarrow$ (3). Exchanging the roles of x and y in (A.70), we obtain

$$\langle J'_{\mathcal{G}}(y), (x-y)\rangle_* \leq J(x) - J(y). \tag{A.71}$$

By summing up (A.70) and (A.71) we deduce (A.69).

(3) $\Rightarrow$ (1). Set $g(t) = J(x+t(y-x))$, $t \in [0,1]$ and compute

$$g'(t) = \langle J'_{\mathcal{G}}(x+t(y-x)), (y-x)\rangle_*$$

Thus, if $t > s$,

$$\begin{aligned} g'(t) - g'(s) &= \langle J'_{\mathcal{G}}(x+t(y-x)) - J'_{\mathcal{G}}(x+s(y-x)), (y-x)\rangle_* \\ &= \frac{1}{t-s}\langle J'_{\mathcal{G}}(x+t(y-x)) - J'_{\mathcal{G}}(x+s(y-x)), (t-s)(y-x)\rangle_* \geq 0. \end{aligned}$$

It follows that g' is nondecreasing, therefore g is convex and hence, for $t \in (0,1)$,

$$g(t) \leq g(0) + t[g(1) - g(0)]$$

which is the convexity of J. □

We point out that when in (A.68) and (A.69) the inequality is replaced by a strict inequality for any $x, y \in U$, $x \neq y$, both these conditions are equivalent to the strict convexity of J.

Example A.13. We return to the case of quadratic functionals discussed in Sect. A.7.1. If the bilinear form a is *nonnegative*, that is $a(v,v) \geq 0$ $\forall v \in \mathcal{U}$, then the functional

$$J(u) = \frac{1}{2}a(u,u)$$

is *convex*. In fact, since $\langle J'(u), v\rangle_* = a(u,v)$, we have that

$$J(u+v) - J(u) = a(u,v) + \frac{1}{2}a(v,v) \geq a(u,v) \qquad \forall v \in \mathcal{U}.$$

If a is *positive*, that is, $a(v,v) > 0$ $\forall v \in \mathcal{U}$, $v \neq 0$, J is clearly *strictly convex*. •

Example A.14. Let $\Omega \subset \mathbb{R}^d$ be a bounded Lipschitz domain. The operator $\Phi : L^6(\Omega) \to L^2(\Omega)$ given by

$$\Phi : y \mapsto y^3$$

is $\mathcal{F}$–differentiable at every point and

$$\Phi'(y)h = 3y^2 h \qquad \forall y, h \in L^6(\Omega).$$

This follows by a straightforward application of the definition (A.15). By property *d)* in Sect. A.7.2, Φ is also $\mathcal{F}$–differentiable as an operator from $H^1(\Omega)$ into $L^2(\Omega)$, provided $d \leq 3$. Moreover $\Phi' : L^6(\Omega) \to \mathcal{L}(L^6(\Omega), L^2(\Omega))$ is continuous. In fact, using Hölder inequality (A.29) with exponents $p = q = r = 1/3$,

$$\begin{aligned} \|[\Phi'(y+z) - \Phi'(y)]h\|_{L^2(\Omega)} &= \|3(y^2 - z^2)h\|_{L^2(\Omega)} \\ &\leq 3\|y+z\|_{L^6(\Omega)}\|y-z\|_{L^6(\Omega)}\|h\|_{L^6(\Omega)} \end{aligned}$$

whence

$$
\begin{aligned}
\|\Phi'(y+z)-\Phi'(y)\|_{\mathcal{L}(L^6(\Omega),L^2(\Omega))} &= \sup_{\|h\|_{L^6(\Omega)}=1} \|[\Phi'(y+z)-\Phi'(y)]\,h\|_{L^2(\Omega)} \\
&\le 3\,\|y+z\|_{L^6(\Omega)}\,\|y-z\|_{L^6(\Omega)}
\end{aligned}
$$

and Φ' is (locally Lipschitz) continuous. •

Example A.15. Let $\Phi : L^p(0,1) \to L^p(0,1)$ be given by

$$\Phi : y \to \sin y.$$

Then Φ is $\mathcal{G}$−differentiable for every p, $1 \le p \le \infty$, with

$$\Phi'_{\mathcal{G}}(y)\,h = \cos(y)\,h \qquad \forall y, h \in L^p(\Omega).$$

In fact, we can show that

$$\frac{\sin(y(x)+th(x))-\sin(y(x))}{t} - \cos(y(x))\,h(x) = R(t;y,h)$$

with

$$R(t;y,h) = -h(x)\int_0^1 [\cos(y(x)+sth(x))-\cos(y(x))]\,ds.$$

Since $R(t;y,h) \to 0$ as $t \to 0$ and $|R(t;y,h)| \le 2\,|h| \in L^p(0,1)$, by the Lebesgue's dominated convergence Theorem A.8 we conclude that $\|R(t;y,h)\|_{L^p(0,1)} \to 0$ as $t \to 0$ and therefore is $\mathcal{G}$−differentiable with $\Phi'_{\mathcal{G}}(y)\,h = \cos(y)\,h$. •

Remark A.11. In general, if $1 \le p < \infty$, $\Phi : y \mapsto \sin y$ is *not* $\mathcal{F}$−differentiable, at any point. Indeed, it can be proved that an operator $\Phi : L^p(\Omega) \to L^p(\Omega)$ given by

$$\Phi : y \mapsto F(y)$$

is $\mathcal{F}$−differentiable if and only if $F(y) = cy$ for some constant c. •

Remark A.12. Maps of the form $\Phi : y \to f(\mathbf{x},y)$ are known as *Nemytskii operators* (see, e.g. [231, Section 10.3.4]). •

Example A.16. Let $F \in C^2(\mathbb{R})$. The operator $\Phi : L^\infty(\Omega) \to L^\infty(\Omega)$ given by

$$\Phi : y \to F(y)$$

is $\mathcal{F}$−differentiable with

$$\Phi'(y)\,h = F'(y)\,h, \tag{A.72}$$

and that $\Phi' : L^\infty(\Omega) \to \mathcal{L}(L^\infty(\Omega), L^\infty(\Omega))$ is (locally Lipschitz) continuous. Indeed, we can write, for a.e. $\mathbf{x} \in \Omega$,

$$
\begin{aligned}
&F(y(\mathbf{x})+h(\mathbf{x}))-F(y(\mathbf{x})) \\
&= \int_0^1 \frac{d}{dt}F(y(\mathbf{x})+th(\mathbf{x}))dt = \int_0^1 F'(y(\mathbf{x})+th(\mathbf{x}))h(\mathbf{x})\,dt \\
&= F'(y(\mathbf{x}))h(\mathbf{x}) + h(\mathbf{x})\int_0^1 [F'(y(\mathbf{x})+th(\mathbf{x}))-F'(y(\mathbf{x}))]\,dt \\
&= F'(y(\mathbf{x}))h(\mathbf{x}) + E(y(\mathbf{x}),h(\mathbf{x}))
\end{aligned}
$$

and

$$\begin{aligned} F'(y(\mathbf{x}) + th(\mathbf{x})) - F'(y(\mathbf{x})) &= \int_0^t \frac{d}{ds} F'(y(\mathbf{x}) + sh(\mathbf{x}))ds \\ &= \int_0^t F''(y(\mathbf{x}) + sh(\mathbf{x}))\, h(\mathbf{x})\, ds. \end{aligned}$$

Then,

$$\|E(y,h)\|_{L^\infty(\Omega)} \le \|h\|^2_{L^\infty(\Omega)} \|F''(y(\mathbf{x}) + sth(\mathbf{x}))\|_{L^\infty(\Omega)} \le C\|h\|^2_{L^\infty(\Omega)}$$

where $C = \sup_{s\in[0,1],\mathbf{x}\in\Omega} |F''(y(\mathbf{x}) + sh(\mathbf{x}))|$. Thus, we can deduce that Φ is $\mathcal{F}$-differentiable and that (A.72) holds. With the same method, it is possible to show that Φ' is (locally Lipschitz) continuous. •

*A.7.4 Second Order Derivatives**

As in Definition A.15, let $F : U \subseteq X \to Y$ be a $\mathcal{F}-$ differentiable operator, where U is an open subset of X. The mapping $y \mapsto F'(y)$ defines an operator from U into $\mathcal{L}(X, Y)$.

Definition A.18. The operator $F : U \subseteq X \to Y$ is twice differentiable at $y \in U$ if the operator $u \mapsto F'(y)$ is $\mathcal{F}-$differentiable at $y \in U$. $(F')'$ is called the second derivative of F and we can write $F''(y) = (F')'(y)$; moreover, we have

$$F''(y) \in \mathcal{L}(X, \mathcal{L}(X, Y)).$$

Thus, $F''(y)$ is a bounded linear operator from X into $\mathcal{L}(X, Y)$.

More explicitly, given $h \in X$, $F''(y)h$ represents a bounded linear operator in $\mathcal{L}(X, Y)$, whose action on another element $k \in X$ is given by $(F''(y)h)k \in Y$. Therefore, $F''(y)$ can be identified (see, e.g., [57, Sect. 5.1]) with a continuous bilinear form from $X \times X$ into Y, and its action can be denoted by

$$(h, k) \mapsto F''(y)[h, k]; \tag{A.73}$$

in particular, we set

$$F''(y)[h^2] = F''(y)[h, h].$$

With these notations, we have that

$$\|F''(y)\|_{\mathcal{L}(X,\mathcal{L}(X,Y))} = \sup_{\|h\|=1,\|k\|=1} \|F''(y)[h, k]\|_Y.$$

A result by Cartan (see [57, Thm. 5.1.1]) asserts that the continuous bilinear form (A.73) is symmetric, that is

$$(F''(y)h)k = (F''(y)k)h \quad \text{or} \quad F''(y)[h, k] = F''(y)[k, h].$$

In the particular case $Y = \mathbb{R}$, we have $F''(y) \in \mathcal{L}(X, X^*)$, so that $F''(y)h \in X^*$ and, for its action on $k \in X$, we may use the (hybrid) notation

$$\langle F''(y)h, k\rangle_{X^*,X}.$$

For twice $\mathcal{F}-$differentiable operators, the following Taylor expansion holds:

$$F(y+h) = F(y) + F'(y)h + \frac{1}{2}F''(y)[h^2] + \varepsilon(y,h)\|h\|_X^2 \tag{A.74}$$

with

$$\|\varepsilon(y,h)\|_Y \to 0 \text{ as } \|h\|_X \to 0.$$

If F is twice differentiable at every point $y \in U$, the mapping $y \mapsto F''(y)$ from U into $\mathcal{L}(X, \mathcal{L}(X,Y))$ is well defined. If this mapping is continuous, that is if

$$\|F''(y) - F''(y_0)\|_{\mathcal{L}(X,\mathcal{L}(X,Y))} \to 0 \text{ as } \|y - y_0\|_X \to 0,$$

we say that F'' is twice continuously $\mathcal{F}-$ differentiable.

Example A.17. We go back to Example A.14, and consider $\Phi : L^6(\Omega) \to L^2(\Omega) : y \mapsto y^3$. We have seen that

$$\Phi'(y) : L^6(\Omega) \to \mathcal{L}(L^6(\Omega), L^2(\Omega))$$

is given by the multiplication operator

$$h \mapsto 3y^2 h$$

We now show that Φ' is continuously $\mathcal{F}-$differentiable, with

$$\Phi''(y) : L^6(\Omega) \to \mathcal{L}(L^6(\Omega), \mathcal{L}(L^6(\Omega), L^2(\Omega))$$

given by

$$\Phi''(y)[h,k] = 6yhk.$$

Indeed,

$$\Phi'(y+h) - \Phi'(y) - 6yh = 3h^2,$$

to be intended as a multiplication operator from $L^6(\Omega)$ into $L^2(\Omega)$. By the three-factors Hölder inequality (A.29), with $p = q = r = 3$,

$$\|h^2\|_{\mathcal{L}(L^6(\Omega),L^2(\Omega))} = \sup_{\|k\|_{L^6(\Omega)}=1} \|h^2 k\|_{L^2(\Omega)} \leq \|h\|_{L^6(\Omega)}^2 ;$$

hence,

$$\frac{\|\Phi'(y+h) - \Phi'(y) - 6yh\|_{L^2(\Omega)}}{\|h\|_{L^6(\Omega)}} \to 0 \text{ as } \|h\|_{L^6(\Omega)} \to 0.$$

The (Lipschitz) continuity of Φ'' follows by the following relationship,

$$\begin{aligned}\|\Phi''(y) - \Phi''(z)\|_{\mathcal{L}(L^6(\Omega),\mathcal{L}(L^6(\Omega),L^2(\Omega))} &= 6 \sup_{\|h\|_{L^6(\Omega)}=\|k\|_{L^6(\Omega)}=1} \|(y-z)hk\|_{L^2(\Omega)} \\ &\leq 6\|y - z\|_{L^6(\Omega)}.\end{aligned}$$

•

Example A.18. The operator $\Phi : L^\infty(\Omega) \to L^\infty(\Omega)$ given by

$$u \mapsto f(u)$$

already considered in Example A.16 is twice continuously $\mathcal{F}-$differentiable, with

$$\Phi''(u) : L^\infty(\Omega) \to \mathcal{L}(L^\infty(\Omega), L^\infty(\Omega)^*)$$

given by

$$\Phi''(u)[h,k] = f''(u)hk.$$

•

Example A.19. The functional $F : L^\infty(\Omega) \to \mathbb{R}$ given by

$$F(y)=\frac{1}{6}\int_{\Omega}y^{6}d\mathbf{x}$$

is twice continuously $\mathcal{F}-$differentiable, with

$$F'(y)\,h=\int_{\Omega}y^{5}h\,d\mathbf{x} \tag{A.75}$$

and

$$F''(y)\,[h,k]=5\int_{\Omega}y^{4}hk\,d\mathbf{x}. \tag{A.76}$$

In fact, for $\|h\|_{L^{\infty}(\Omega)}$ small, we have

$$\begin{aligned}\left|F(y+h)-F(y)-\int_{\Omega}y^{5}hd\mathbf{x}\right| &\leq \int_{\Omega}\left|\frac{(y+h)^{6}-y^{6}}{6}-y^{5}h\right|d\mathbf{x}\\ &=\int_{\Omega}\left|\frac{5}{2}y^{4}h^{2}+\frac{10}{3}y^{3}h^{3}+\frac{5}{2}y^{2}h^{4}+6yh^{5}+h^{6}\right|d\mathbf{x}\\ &\leq C(\|y\|_{L^{\infty}(\Omega)})\,\|h\|^{2}_{L^{\infty}(\Omega)}\end{aligned}$$

so that F is $\mathcal{F}-$differentiable and (A.75) holds. Similarly, for $\|h\|_{L^{\infty}(\Omega)}$ small,

$$\begin{aligned}\left|F'(y+h)-F'(y)-5\int_{\Omega}y^{4}hd\mathbf{x}\right| &\leq \int_{\Omega}\left|(y+h)^{5}-y^{5}-5y^{4}h\right|d\mathbf{x}\\ &=\int_{\Omega}\left|\frac{5}{2}y^{4}h^{2}+\frac{10}{3}y^{3}h^{3}+\frac{5}{2}y^{2}h^{4}+6yh^{5}+h^{6}\right|d\mathbf{x}\\ &\leq C(\|y\|_{L^{\infty}(\Omega)})\,\|h\|^{2}_{L^{\infty}(\Omega)}\end{aligned}$$

and also (A.76) holds. Finally,

$$\begin{aligned}\|F''(y)-F''(z)\|_{\mathcal{L}(L^{\infty}(\Omega),L^{\infty}(\Omega)^{*})} &= \sup_{\|h\|_{L^{\infty}(\Omega)}=\|k\|_{L^{\infty}(\Omega)}=1} 5\left|\int_{\Omega}(y^{4}-z^{4})hkd\mathbf{x}\right|\\ &\leq C(\|y\|_{L^{\infty}(\Omega)},\|z\|_{L^{\infty}(\Omega)})\,\|y-z\|_{L^{\infty}(\Omega)}\end{aligned}$$

which shows that F'' is locally Lipschitz continuous. •

Second-order partial derivatives. Let $X=X_1\times X_2$ and $U=U_1\times U_2\subseteq X$, open, $F:U\subseteq X\rightarrow Y$ be a $\mathcal{F}-$differentiable operator. Then, there exist the partial derivatives $F_{x_1}:U\rightarrow\mathcal{L}(X_1,Y)$ and $F_{x_2}:U\rightarrow\mathcal{L}(X_2,Y)$. The second partial derivatives of F are defined as the partial derivatives of F_{x_1} and F_{x_2}. We have $(F_{x_ix_j}=\partial_{x_j}\partial_{x_i})$, for $h_1,k_1\in X_1$ and $k_1,k_2\in X_2$,

$$F''_{x_1x_1}:U\rightarrow\mathcal{L}(X_1,\mathcal{L}(X_1,Y)),\ F''_{x_1x_1}(u)\,[h_1,k_1]=\left(F''_{x_1x_1}(u)\,h_1\right)k_1$$

and

$$F''_{x_1x_2}:U\rightarrow\mathcal{L}(X_2,\mathcal{L}(X_1,Y)),\ F''_{x_1x_2}(u)\,[h_1,k_2]=\left(F''_{x_1x_2}(u)\,k_2\right)h_1,$$

while

$$F''_{x_2x_1}:U\rightarrow\mathcal{L}(X_1,\mathcal{L}(X_2,Y)),\ F''_{x_2x_1}(u)\,[h_1,k_2]=\left(F''_{x_2x_1}(u)\,h_1\right)k_2$$

and

$$F''_{x_2x_2}:U\rightarrow\mathcal{L}(X_2,\mathcal{L}(X_2,Y)),\ F''_{x_2x_2}(u)\,[h_2,k_2]=\left(F''_{x_2x_2}(u)\,h_2\right)k_2.$$

As in the finite dimensional case, if F is twice continuously $\mathcal{F}-$differentiable, the order of differentiation commutes,

$$F''_{x_2x_1}(u)[h_1,k_2] = F''_{x_1x_2}(u)[h_1,k_2].$$

A.8 Fixed Points and Implicit Function Theorems

Fixed point theorems provide a fundamental tool for solving nonlinear problems of the form

$$F(x) = x,$$

where $F : X \to X$ is an operator defined in some normed space X. We recall here two of these theorems. The first one is the contraction mapping theorem, that involves maps that contract distances (so called *contractions*).

Let X be a Banach space and $F : X \to X$; F is a *strict contraction* if there exists a number ρ, $0 < \rho < 1$, such that

$$\|F(x) - F(y)\| \leq \rho\|x - y\| \qquad \forall x, y \in X. \tag{A.77}$$

Clearly, every contraction is continuous. The inequality (A.77) states that F contracts the distance between two points at least by a percentage $\rho < 1$. This fact confers to a recursive sequence $x_{m+1} = F(x_m)$, $m \geq 0$, starting from any point $x_0 \in M$, a remarkable convergence behavior, particularly useful in numerical approximation methods. Precisely, the following result (also known as Banach-Caccioppoli fixed-point theorem) holds.

Theorem A.21 (Contraction Mapping Theorem). *Let X be a Banach space and $F : X \to X$ be a strict contraction. Then there exists a unique fixed point $x^* \in X$ of F. Moreover, for any choice of $x_0 \in X$, the recursive sequence*

$$x_{m+1} = F(x_m) \qquad \forall m \geq 0 \tag{A.78}$$

converges to x^*.

The second result is the Leray-Schauder theorem, that requires the compactness of F and the existence of a family of operators F_s, $0 \leq s \leq 1$, where $T_1 = T$ and T_0 is a compact operator which has a fixed point.

Theorem A.22 (Leray-Schauder). *Let X be a Banach space and $F : X \to X$ such that:*

i) F is continuous and compact;
ii) there exists M such that

$$\|x\| \leq M$$

for every solution (x, s) of the equation $x = sF(x)$ with $0 \leq s \leq 1$.

Then F has a fixed point.

Note that the Leray-Schauder Theorem yields no information on the uniqueness of the fixed point.

Another valuable tool for the analysis of nonlinear problems is the Implicit Function Theorem in Banach spaces. Recall that a linear operator $T : X \to Y$ is an *isomorphism* if it is bijective and both T and T^{-1} are bounded.

Theorem A.23 (Implicit Function). *Let X, Y and Z be three Banach spaces, $A \subseteq X \times Y$ an open set and $f : A \to Z$. Let us assume that:*

(i) $f \in C^1(A)$;
(ii) for a suitable $(x_0, y_0) \in A$, $f(x_0, y_0) = 0$ and $f_y(x_0, y_0)$ is a continuous isomorphism from Y onto Z.

Then:

(1) There exists a unique function $g : U \to Y$, defined in a neighborhood U of x_0, such that $g(x_0) = y_0$ and $f(x, g(x)) = 0$ in U;
(2) $g \in C^1(U)$ and the following formula holds,

$$g'(x) = -\left[f_y(x, g(x))\right]^{-1} \circ f_x(x, g(x)). \tag{A.79}$$

Remark A.13. In the above Implicit Function Theorem, an additional degree of regularity on f yields more regularity of the implicit function g, too. In particular, if $f \in C^2(A)$, then $g \in C^2(U)$. •

Appendix B
Toolbox of Numerical Analysis

In this Appendix we provide a short introduction to the numerical methods for the approximation of PDEs, as well as some basic matrix properties, used all along the book. The interested reader can refer, e.g., to [222, 224] for further details.

B.1 Basic Matrix Properties

We denote by $\mathbb{R}^{m\times n}$ the vector space of all $m \times n$ real matrices; the components of $\mathbb{A} \in \mathbb{R}^{m\times n}$ are denoted by $(\mathbb{A})_{ij} = a_{ij} \in \mathbb{R}$. A vector $\mathbf{x} \in \mathbb{R}^n$ is tacitly considered as a column vector, $\mathbf{x} = (x_1, \ldots, x_n)^\top$, being x_i its i-th component. Its p-norms are

$$\|\mathbf{x}\|_p = \left(\sum_{i=1}^{n} |x_i|^p\right)^{1/p}, \qquad p \geq 1, \qquad\qquad \|\mathbf{x}\|_\infty = \max_{1\leq i\leq n} |x_i|;$$

the Euclidean norm, corresponding to the case $p = 2$ is also denoted as $\|\mathbf{x}\|_2 = |\mathbf{x}|$. For any matrix $\mathbb{A}$, we denote its transpose by $\mathbb{A}^\top$; if $m = n$, $\mathbb{A}$ is a squared matrix. If $\mathbb{A} = \mathbb{A}^\top$, $\mathbb{A}$ is symmetric. If the inverse $\mathbb{A}^{-1}$ of $\mathbb{A}$ exists, $\mathbb{A}$ is said to be nonsingular – in this case, $\det \mathbb{A} \neq 0$.

B.1.1 Eigenvalues, Eigenvectors, Positive Definite Matrices

The eigenvalues of $\mathbb{A} \in \mathbb{R}^{n\times n}$ are the zeros of the characteristic polynomial $p(\lambda) = \det(\mathbb{A} - \lambda\mathbb{I})$. More precisely: if λ is an eigenvalue of $\mathbb{A}$, there exists $\mathbf{x} \in \mathbb{R}^n$, $\mathbf{x} \neq \mathbf{0}$, called *eigenvector* of $\mathbb{A}$ associated with λ, such that

$$\mathbb{A}\mathbf{x} = \lambda\mathbf{x}.$$

If $\mathbb{A} \in \mathbb{R}^{n\times n}$ has n independent eigenvectors $\mathbf{x}_1, \ldots, \mathbf{x}_n$ such that $\mathbb{A}\mathbf{x}_i = \lambda_i\mathbf{x}_i$, $i = 1, \ldots, n$, $\mathbb{A}$ is diagonalizable: by setting $\mathbb{X} = [\mathbf{x}_1 \mid \ldots \mid \mathbf{x}_n] \in \mathbb{R}^{n\times n}$,

$$\mathbb{X}^{-1}\mathbb{A}\mathbb{X} = \text{diag}(\lambda_1, \ldots, \lambda_n).$$

Such a decomposition always exists if $\mathbb{A}$ is symmetric; in this case all the eigenvalues are real, $\lambda_n(\mathbb{A}) \leq \ldots \leq \lambda_2(\mathbb{A}) \leq \lambda_1(\mathbb{A})$, and $\mathbb{X}$ is an orthogonal matrix – that is, its columns are orthonormal vectors: $\mathbf{x}_i^\top\mathbf{x}_j = \delta_{ij}$, where δ_{ij} denotes the Kronecker delta. The largest and the smallest eigenvalues of a symmetric matrix $\mathbb{A}$ satisfy

$$\lambda_1(\mathbb{A}) = \max_{\mathbf{x}\neq 0} \frac{\mathbf{x}^\top\mathbb{A}\mathbf{x}}{\mathbf{x}^\top\mathbf{x}}, \qquad\qquad \lambda_n(\mathbb{A}) = \min_{\mathbf{x}\neq 0} \frac{\mathbf{x}^\top\mathbb{A}\mathbf{x}}{\mathbf{x}^\top\mathbf{x}},$$

A. Manzoni et al., *Optimal Control of Partial Differential Equations*,
Applied Mathematical Sciences 207, https://doi.org/10.1007/978-3-030-77226-0

where the ratio being minimized or maximized is called *Rayleigh quotient*; moreover, we denote by $\rho(\mathbb{A}) = |\lambda_1(\mathbb{A})|$ the *spectral radius* of $\mathbb{A}$.

For any square matrix $\mathbb{A}$, the *Caley-Hamilton Theorem* states that $p(\mathbb{A}) = \mathbb{O}$, where $\mathbb{O}$ is the null matrix and $p(\cdot)$ denotes the characteristic polynomial of $\mathbb{A}$.

A matrix $\mathbb{A} \in \mathbb{R}^{n\times n}$ is *positive (negative) definite* if $\mathbf{x}^\top\mathbb{A}\mathbf{x} > 0$ (resp. < 0) for all $\mathbf{x} \in \mathbb{R}^n$, $\mathbf{x} \neq \mathbf{0}$, *positive (negative) semidefinite* is $\mathbf{x}^\top\mathbb{A}\mathbf{x} \geq 0$ (resp. ≤ 0) for all $\mathbf{x} \in \mathbb{R}^n$, $\mathbf{x} \neq \mathbf{0}$, and indefinite if we can find $\mathbf{x}, \mathbf{y} \in \mathbb{R}^n$ so that $(\mathbf{x}^\top\mathbb{A}\mathbf{x})(\mathbf{y}^\top\mathbb{A}\mathbf{y}) < 0$. Symmetric positive definite (SPD) matrices appear in many situations, and several methods for solving linear systems are tailored on this kind of matrices. A positive definite matrix is nonsingular; moreover, $\mathbb{A} \in \mathbb{R}^{n\times n}$ is positive definite if and only if its symmetric part $\frac{1}{2}(\mathbb{A} + \mathbb{A}^\top)$ has positive eigenvalues. In particular, an SPD matrix has positive eigenvalues. Moreover, if $\mathbb{A} \in \mathbb{R}^{n\times n}$ is SPD,

$$(\mathbf{x}, \mathbf{y})_{\mathbb{A}} = \mathbf{x}^\top\mathbb{A}\mathbf{y} = \mathbb{A}\mathbf{x}^\top\mathbf{y} = \sum_{i,j=1}^{n} a_{ij}x_i y_j \tag{B.1}$$

defines[1] another scalar product in $\mathbb{R}^n$, that induces the so-called *energy norm*, or $\mathbb{A}$-norm,

$$\|\mathbf{x}\|_{\mathbb{A}} = \sqrt{(\mathbf{x}, \mathbf{x})_{\mathbb{A}}}.$$

B.1.2 Singular Value Decomposition

Singular value decomposition is a diagonalization process involving left and right multiplication by orthogonal matrices. More precisely: if $\mathbb{A} \in \mathbb{R}^{m\times n}$ is a real matrix, there exist two orthogonal matrices $\mathbb{U} = [\boldsymbol{\zeta}_1 \mid \ldots \mid \boldsymbol{\zeta}_m] \in \mathbb{R}^{m\times m}$, $\mathbb{Z} = [\boldsymbol{\psi}_1 \mid \ldots \mid \boldsymbol{\psi}_n] \in \mathbb{R}^{n\times n}$ such that

$$\mathbb{A} = \mathbb{U}\Sigma\mathbb{Z}^\top, \qquad \text{with } \Sigma = \mathrm{diag}(\sigma_1, \ldots, \sigma_p) \in \mathbb{R}^{m\times n} \tag{B.2}$$

and $\sigma_1 \geq \sigma_2 \geq \ldots \geq \sigma_p \geq 0$, for $p = \min(m, n)$. The matrix factorization (B.2) is called singular value decomposition (SVD) of $\mathbb{A}$ and the numbers $\sigma_i = \sigma_i(\mathbb{A})$ are called *singular values* of $\mathbb{A}$. $\boldsymbol{\zeta}_1, \ldots, \boldsymbol{\zeta}_m$ are called *left singular vectors* of $\mathbb{A}$, whereas $\boldsymbol{\psi}_1, \ldots, \boldsymbol{\psi}_n$ are the *right singular vectors* of $\mathbb{A}$, as

$$\mathbb{A}\boldsymbol{\psi}_i = \sigma_i\boldsymbol{\zeta}_i, \qquad \mathbb{A}^\top\boldsymbol{\zeta}_j = \sigma_j\boldsymbol{\psi}_j, \qquad i, j = 1, \ldots, n.$$

Since $\mathbb{A}\mathbb{A}^\top$ and $\mathbb{A}^\top\mathbb{A}$ are SPD matrices, the left (resp. right) singular vectors of $\mathbb{A}$ are the eigenvectors of $\mathbb{A}\mathbb{A}^\top$ (resp. $\mathbb{A}^\top\mathbb{A}$). There is a close relationship between the SVD of $\mathbb{A}$ and the eigenvalues of $\mathbb{A}^\top\mathbb{A}$ and $\mathbb{A}\mathbb{A}^\top$; indeed

$$\sigma_i(\mathbb{A}) = \sqrt{\lambda_i(\mathbb{A}^\top\mathbb{A})}, \qquad i = 1, \ldots, p.$$

The largest and the smallest singular values are also denoted by

$$\sigma_{\max} = \max_{i=1,\ldots,p} \sigma_i, \qquad \sigma_{\min} = \min_{i=1,\ldots,p} \sigma_i.$$

Furthermore, if $\mathbb{A} \in \mathbb{R}^{n\times n}$ is symmetric, then $\sigma_i(\mathbb{A}) = |\lambda_i(\mathbb{A})|$, with $\lambda_1(\mathbb{A}) \geq \lambda_2(\mathbb{A}) \geq \ldots \geq \lambda_n(\mathbb{A})$ being the eigenvalues of $\mathbb{A}$. The most frequently employed matrix norms are the Frobenius norm

$$\|\mathbb{A}\|_F = \left(\sum_{i=1}^{m}\sum_{j=1}^{n} |a_{ij}|^2\right)^{1/2} \tag{B.3}$$

[1] Every inner product in $\mathbb{R}^n$ can be written in the form (B.1), with a suitable matrix $\mathbb{A}$.

and the p-norm, $p \in [1, \infty]$,

$$\|\mathbb{A}\|_p = \sup_{\mathbf{x} \neq \mathbf{0}} \frac{\|\mathbb{A}\mathbf{x}\|_p}{\|\mathbf{x}\|_p} = \sup_{\|\mathbf{x}\|_p = 1} \|\mathbb{A}\mathbf{x}\|_p.$$

This latter is also called the *natural* matrix norm, or matrix norm *induced* by the vector norm $\| \cdot \|_p$. Note that the 2-norm (also-called *spectral norm*) and the Frobenius norm can be expressed in terms of the singular values of $\mathbb{A}$: for any $\mathbb{A} \in \mathbb{R}^{m \times n}$,

$$\|\mathbb{A}\|_2 = \sigma_{\max}, \qquad \|\mathbb{A}\|_F = \sqrt{\sum_{i=1}^{p} \sigma_i^2}. \tag{B.4}$$

In particular, if $\mathbb{A}$ is symmetric then $\|\mathbb{A}\|_2 = \rho(\mathbb{A})$.

Finally, for any matrix p-norm the *condition number* of a matrix is defined as

$$\kappa_p(\mathbb{A}) = \|\mathbb{A}\|_p \|\mathbb{A}^{-1}\|_p.$$

If $\mathbb{A}$ has a large condition number, the solution of the linear system $\mathbb{A}\mathbf{x} = \mathbf{b}$ is more sensitive to perturbations both of the matrix coefficients a_{ij} and the right-hand side coefficients b_i.

If $\mathbb{A} \in \mathbb{R}^{n \times n}$ is nonsingular, by inverting (B.2) we obtain $\mathbb{A}^{-1} = \mathbb{Z}\Sigma^{-1}\mathbb{U}^\top$, with $\Sigma^{-1} = \text{diag}(\sigma_1^{-1}, \ldots, \sigma_n^{-1})$; σ_n^{-1} is the largest singular value of $\mathbb{A}^{-1}$, whence

$$\|\mathbb{A}^{-1}\|_2 = 1/\sigma_n, \qquad \kappa(\mathbb{A}) = \|\mathbb{A}\|_2 \|\mathbb{A}^{-1}\|_2 = \sigma_1/\sigma_n.$$

In particular, for any SPD matrix $\mathbb{A}$,

$$\kappa_2(\mathbb{A}) = \lambda_1(\mathbb{A})/\lambda_n(\mathbb{A}).$$

Further details related to matrix properties can be found in [222] and in specific monographs devoted to matrix computations like [106, 40].

B.1.3 Spectral Properties of Saddle-Point Systems

Although substantial progress has been made in the design of sparse direct solvers for large-scale linear systems of equations, very often preconditioned iterative solvers provide the only viable approach for the approximation of PDE problems, because of large dimension, exhagerate filling-in and conditioning issues that are inherently related with direct (factorization) methods.

Typical matrices requiring efficient preconditioners are those related to saddle-point problems, which are ubiquitous in PDE-constrained optimization. Indeed, quadratic programming for linear-quadratic OCPs and, more generally, optimization methods showing a faster-than-linear rate of convergence such as the SQP method, require to solve (once, or repeatedly) systems involving a *saddle point matrix* under the form

$$\begin{bmatrix} \mathbb{A} & \mathbb{B}_1^\top \\ \mathbb{B}_2 & -\mathbb{C} \end{bmatrix}.$$

Hereon, we will restrict ourselves to the case where $\mathbb{A}$ is symmetric, $\mathbb{B}_1 = \mathbb{B}_2$ and $\mathbb{C} = 0$, that is

$$\mathcal{K} = \begin{bmatrix} \mathbb{A} & \mathbb{B}^\top \\ \mathbb{B} & 0 \end{bmatrix}, \qquad \text{with } \mathbb{A} = \mathbb{A}^\top.$$

The saddle-point system

$$\begin{bmatrix} \mathbb{A} & \mathbb{B}^\top \\ \mathbb{B} & 0 \end{bmatrix} \begin{bmatrix} \mathbf{x} \\ \mathbf{p} \end{bmatrix} = \begin{bmatrix} \mathbf{f} \\ \mathbf{g} \end{bmatrix} \tag{B.5}$$

represents the KKT system for the finite dimensional optimization problem

$$\frac{1}{2}\mathbf{x}^\top \mathbb{A}\mathbf{x} - \mathbf{f}^\top \mathbf{x} \to \min_{\mathbf{x}\in\mathbb{R}^n} \qquad \text{s.t.} \quad \mathbb{B}\mathbf{x} = \mathbf{g}, \tag{B.6}$$

see Sect. 3.4.1. Throughout this section, we suppose that $\mathbb{A} \in \mathbb{R}^{n\times n}$ and $\mathbb{B} \in \mathbb{R}^{m\times n}$, with $m \leq n$. We recall that the matrix $\mathcal{K}$ in (B.5) is invertible if and only if $\mathbb{A} = \mathbb{A}^\top$ is positive semidefinite, $\mathbb{B}$ has full row rank m and $ker(\mathbb{A}) \cap ker(\mathbb{B}) = \{0\}$.

Saddle point matrices are indefinite. If $\mathbb{A} = \mathbb{A}^\top$ is positive definite on $ker(\mathbb{B})$, $\mathcal{K}$ has exactly n positive and m negative eigenvalues; moreover, problem (B.6) admits a unique solution. In particular, if $\mathbb{A} = \mathbb{A}^\top$ is positive definite on $\mathbb{R}^n$, the cost function in (B.6) is strictly convex, $\mathbb{A}$ is invertible and $\mathbb{S} = \mathbb{B}\mathbb{A}^{-1}\mathbb{B}$ denotes its Schur complement, which is positive definite, too.

Regarding instead the spectral properties of saddle point matrices, we have that if $\mathbb{A} = \mathbb{A}^\top$ is positive definite on $ker(\mathbb{B})$ and $\mathbb{A}$ is positive semidefinite, the spectrum of $\mathcal{K}$ is contained in the intervals $\mathcal{I}^- \cup \mathcal{I}^+$, where

$$\mathcal{I}^- = \left[\frac{1}{2}\left(\mu_n - \sqrt{\mu_n^2 + 4\sigma_1^2}\right), \frac{1}{2}\left(\mu_1 - \sqrt{\mu_1^2 + 4\sigma_m^2}\right)\right] \subset \mathbb{R}^-$$

$$\mathcal{I}^+ = \left[\mu_n, \frac{1}{2}\left(\mu_1 + \sqrt{\mu_1^2 + 4\sigma_1^2}\right)\right] \subset \mathbb{R}^+,$$

and $\mu_1 \geq \mu_2 \geq \ldots \geq \mu_n \geq 0$ are the eigenvalues of $\mathbb{A}$, whereas $\sigma_1 \geq \sigma_2 \geq \ldots \geq \sigma_m > 0$ denote the singular values of $\mathbb{B}$. In particular, the following estimate for the spectral condition number of $\mathcal{K}$ holds (see [110])

$$\kappa_2(\mathcal{K}) \leq \frac{\max\{\frac{1}{2}(\mu_1 + \sqrt{\mu_1^2 + 4\sigma_1^2}), -\frac{1}{2}(\mu_n - \sqrt{\mu_n^2 + 4\sigma_1^2})\}}{\min\{\gamma, -\frac{1}{2}(\mu_1 - \sqrt{\mu_1^2 + 4\sigma_m^2})}$$

where $\gamma > 0$ is the smallest positive root of the cubic equation

$$\mu^3 - \mu^2(\hat{\mu} + \mu_n) + \mu(\hat{\mu}\mu_n - M^2 - \sigma_m^2) + \hat{\mu}\sigma_m^2 = 0.$$

Here $M > 0$ is the continuity constant of the bilinear form $a(\cdot,\cdot)$ to which the matrix $\mathbb{A}$ is associated, whereas $\hat{\mu} = \lambda_n(\mathbb{Z}^\top \mathbb{A}\mathbb{Z})$ is the smallest eigenvalue of the reduced problem, and $\mathbb{Z}$ is a matrix whose columns span $ker(\mathbb{B})$. For a simple linear-quadratic OCP where the state system is given by the Laplace equation, the control is distributed and belongs to $\mathcal{U} = L^2(\Omega)$, the observation of the state is taken over $\Omega_{obs} \subset \Omega$ and $V_h, \mathcal{U}_h$ consist of piecewise linear and constant finite elements, respectively, such an estimate yields $\kappa_2(\mathcal{K}) \sim h^{-2}$ for problems in both $d = 2$ or $d = 3$ dimensions.

Remark B.1. In the case of more general PDE-constrained optimization problems,

$$J(\mathbf{x}) \to \min_{\mathbf{x}\in\mathbb{R}^n} \qquad \text{s.t.} \quad G(\mathbf{x}) = \mathbf{0}$$

the SQP method approximates the optimal solution by solving a sequence of problems under the form (B.5) with $\mathbb{A} = \nabla^2_{\mathbf{xx}} L(\mathbf{x}, \mathbf{p})$, the Hessian of the Lagrangian $L(\mathbf{x}, \mathbf{p}) = J(\mathbf{x}) + \mathbf{p}^\top G(\mathbf{x})$, and $\mathbb{B} = \nabla_{\mathbf{x}} G(\mathbf{x})$, the linearization of the constraints. The sufficient condition of Proposition (2.1) matches the assumptions on the block of the matrix $\mathcal{K}$ above – namely, we require that:

- $\hat{\mathbf{x}}$ is a local minimizer. In particular, the first-order necessary conditions $\nabla_{\mathbf{x}} L(\hat{\mathbf{x}}, \hat{\mathbf{p}}) = \mathbf{0}$ and $G(\hat{\mathbf{x}}) = \mathbf{0}$ hold;

- $\nabla_{\mathbf{x}} G(\hat{\mathbf{x}})$ has full rank, so that the constraint qualification (CQ) is satisfied at $\hat{\mathbf{x}}$;
- $\nabla^2_{\mathbf{xx}} L(\hat{\mathbf{x}}, \hat{\mathbf{p}})$ is positive definite on $ker(D_{\mathbf{x}} G(\hat{\mathbf{x}}))$.

By a continuity argument, we can expect these conditions to be fulfilled by the blocks appearing in matrices of the form

$$\begin{bmatrix} \nabla^2_{\mathbf{xx}} L(\mathbf{x}, \mathbf{p}) & (\nabla_{\mathbf{x}} G(\mathbf{x}))^\top \\ \nabla_{\mathbf{x}} G(\mathbf{x}) & 0 \end{bmatrix}$$

where $(\mathbf{x}, \mathbf{p})$ in a neighborhood of a point $(\hat{\mathbf{x}}, \hat{\mathbf{p}})$ which satisfies the second-order sufficient conditions. •

Since $\mathbb{A}$ is nonsingular, from the first equation of (B.5) we can formally obtain $\mathbf{x} = \mathbb{A}^{-1}(\mathbf{f} - \mathbb{B}^\top \mathbf{p})$ so that substituting this expression in the second equation of (B.5) yields the following *Schur complement* system

$$\mathbb{S}\mathbf{p} = \mathbf{g} + \mathbb{B}\mathbb{A}^{-1}\mathbf{f}, \qquad \text{with} \quad \mathbb{S} = \mathbb{B}\mathbb{A}^{-1}\mathbb{B}^\top.$$

A possible strategy for the solution of (B.5) consists in solving the above Schur complement system by an iterative method, such as the conjugate gradient (CG) method (as $\mathbb{A}$ is symmetric), the GMRES or the Bi-CGStab method. In all these cases, a convenient preconditioner is mandatory, see, e.g., [30].

Because of the potentially large size of the problem, the approach we pursue exploit multigrid solvers as part of block preconditioners employed within appropriate Krylov subspace iteration. Indeed, let us observe that the matrix $\mathcal{K}$ of system (B.5) can be expressed, by an LU factorization, as the product of two block triangular matrices

$$\mathcal{K} = \begin{bmatrix} \mathbb{A} & 0 \\ \mathbb{B} & -\mathbb{S} \end{bmatrix} \begin{bmatrix} \mathbb{I} & \mathbb{A}^{-1}\mathbb{B}^\top \\ 0 & \mathbb{I} \end{bmatrix}. \tag{B.7}$$

Similarly, $\mathcal{K}$ can be factorized through the following LDU decomposition,

$$\mathcal{K} = \begin{bmatrix} \mathbb{I} & 0 \\ \mathbb{B}\mathbb{A}^{-1} & \mathbb{I} \end{bmatrix} \begin{bmatrix} \mathbb{A} & 0 \\ 0 & -\mathbb{S} \end{bmatrix} \begin{bmatrix} \mathbb{I} & \mathbb{A}^{-1}\mathbb{B}^\top \\ 0 & \mathbb{I} \end{bmatrix}.$$

Most preconditioners are based on a combination of block factorizations with a suitable approximation of the Schur-complement matrix. For instance, each one of the two matrices

$$\mathcal{P}_{BD} = \begin{bmatrix} \mathbb{A} & 0 \\ 0 & -\mathbb{S} \end{bmatrix} \qquad \text{or} \qquad \mathcal{P}_{BT} = \begin{bmatrix} \mathbb{A} & 0 \\ \mathbb{B} & -\mathbb{S} \end{bmatrix}$$

provides an optimal preconditioner for $\mathbb{S}$, a *block diagonal* preconditioner ($\mathcal{P}_{BD}$), and a *block triangular* preconditioner ($\mathcal{P}_{BT}$).

If the saddle-point system (B.5) is preconditioned by $\mathcal{P}_{BD}$, then the resulting preconditioned system has only the three distinct eigenvalues 1, $1/2 \pm \sqrt{5}/2$. If the same system is preconditioned by $\mathcal{P}_{BT}$, then the resulting preconditioned system has only the two distinct eigenvalues ± 1 [271, 249, 248]. In both cases, a suitable Krylov subspace method (such as MINRES or GMRES in the symmetric and nonsymmetric case, respectively) will terminate with the correct solution after respectively three or two iterations. Indeed, the fewer the number of distinct eigenvalues or the number of eigenvalues, the faster the convergence.

Of course, computing $\mathbb{A}^{-1}$ and $\mathbb{S}^{-1}$ is impractical due to storage and CPU requirements; as a matter of fact, both these preconditioners are computationally expensive to apply. However, the result above show the way to preconditioning: if approximations to $\mathbb{A}$ and $\mathbb{S}$ are employed, the preconditioned matrix will show respectively three or two clusters of eigenvalues, so that Krylov subspace iterative convergence will still be very rapid. Therefore, all variants of block diagonal or triangular preconditioners contain a cheap approximation of $\mathbb{S}$. Hence, we can employ their approximants

$$\hat{\mathcal{P}}_{BD} = \begin{bmatrix} \hat{\mathbb{A}} & 0 \\ 0 & -\hat{\mathbb{S}} \end{bmatrix} \qquad \text{or} \qquad \hat{\mathcal{P}}_{BT} = \begin{bmatrix} \hat{\mathbb{A}} & \mathbb{B}^\top \\ 0 & -\hat{\mathbb{S}} \end{bmatrix}$$

where $\hat{\mathbb{A}}$ and $\hat{\mathbb{S}}$ are two inexpensive approximations of $\mathbb{A}$ and $\mathbb{S}$, respectively. For instance, the system $\mathbb{A}\mathbf{x} = \mathbf{f}$ is solved approximately, usually by a small number of iterations with an iterative method. On the other hand, the Schur complement is not formed, but approximated by a simple matrix; the way this approximation is done yields different block preconditioners.

The literature on this class of preconditioners is huge, also because of their role when dealing with Stokes and Navier-Stokes equations, see [88]. In the special case of the Stokes problem, the (properly scaled) pressure mass matrix is a cheap and spectrally equivalent approximation to the Schur-complement matrix. In the case of steady Navier-Stokes problem, popular methods are, e.g., the *pressure correction diffusion* [156], the *least squares commutator* [87] and the *augmented Lagrangian* [31] preconditioners; see, e.g., [230, 244, 136].

The LU factorization (B.7) in the case of Stokes (or linearized Navier-Stokes) equations also stands at the base of the so-called SIMPLE preconditioner [211], obtained by replacing $\mathbb{A}^{-1}$ in both factors by a triangular matrix $\mathbb{D}^{-1}$ (for instance, $\mathbb{D}$ could be the diagonal of $\mathbb{A}$). See, e.g., [244] or [218, Chap. 16] for possible generalizations of the SIMPLE preconditioner.

An in depth presentation of iterative methods and preconditioning techniques for sparse linear systems can be found, e.g., in [222, 236].

B.2 Numerical Approximation of Elliptic PDEs

In this book the Galerkin-finite element method (FEM) is used for the numerical approximation of OCPs governed by PDEs. In this section we introduce it for the case of scalar coercive second-order elliptic PDEs; parabolic PDEs are treated in Sect. B.3.

Since FE approximations are obtained by projecting the original problem upon a finite-dimensional subspace, their analysis follows from that of their infinite-dimensional counterpart, see Appendix A. Conditions to be fulfilled by discrete problems are in some cases (but not always!) automatically inherited from the ones on the original problems. For an in depth discussion we refer, e.g., to [218, 41].

B.2.1 Strongly Coercive Problems

We start from the approximation of a strongly coercive problem like (P_1), for which the assumption of the Lax-Milgram Theorem A.6 are fulfilled.

Let us denote by $V_h \subset V$ a family (depending on the real parameter $h > 0$) of finite-dimensional subspaces of V, such that dim $V_h = N_h$; $h > 0$ is a characteristic discretization parameter (most typically, the mesh size). The problem approximating (P_1) reads

find $y_h \in V_h$ such that

$$a(y_h, v_h) = f(v_h) \qquad \forall v_h \in V_h \qquad (P_1^h)$$

and is called *Galerkin problem*; y_h is called the *Galerkin approximation* of y. Since $V_h \subset V$, the exact solution y satisfies the weak problem (P_1) for each element $v = v_h \in V_h$,

$$a(y, v_h) = f(v_h) \qquad \forall v_h \in V_h. \tag{B.8}$$

Thanks to this property, the Galerkin problem is said *strongly consistent* (see, e.g., [218]). A remarkable consequence is that

$$a(y - y_h, v_h) = 0 \qquad \forall v_h \in V_h, \tag{B.9}$$

a property which goes under the name of *Galerkin orthogonality*. Indeed, should $a(\cdot,\cdot)$ be symmetric, (B.9) can be interpreted as the orthogonality – with respect to the scalar product $a(\cdot,\cdot)$ – of the approximation error $y - y_h$ from the subspace V_h.

A consequence of (B.9) is the optimality condition

$$\|y - y_h\|_a \le \|y - v_h\|_a \qquad \forall v_h \in V_h.$$

With these premises, the following result (of existence, uniqueness, stability and convergence) easily follows; see, e.g., [64, 224, 121]).

Theorem B.1. *Let the space V, the bilinear form $a(\cdot,\cdot)$ and $F \in V^*$ satisfy the hypotheses of the Lax-Milgram Theorem A.6. Let $V_h \subset V$ be a closed subspace. Hence, $a(\cdot,\cdot)$ is continuous on $V_h \times V_h$ with constant $M_h \le M$ and coercive on $V_h \times V_h$ with constant $\alpha_h \ge \alpha$. Then, for every $h > 0$, the approximate problem (P_1^h) has a unique solution $y_h \in V_h$, that satisfies the stability estimate*

$$\|y_h\|_V \le \frac{1}{\alpha_h}\|F\|_{V^*}. \tag{B.10}$$

Furthermore, the following optimal error inequality is satisfied

$$\|y - y_h\|_V \le \frac{M}{\alpha} \inf_{v_h \in V_h} \|y - v_h\|_V. \tag{B.11}$$

We remark that *optimality* here means that the error $\|y - y_h\|_V$ in a finite-dimensional approximation is bounded from above by the error of the best approximation out of the same finite-dimensional subspace V_h, multiplied by a constant independent of h. However, this is not sufficient to ensure the convergence $\|y - y_h\|_V \to 0$ when $h \to 0$: in fact, an additional property related to the approximability of the discrete spaces is required. As a matter of fact, thanks to (B.11) (also known as Céa Lemma), in order for the method to converge it will be sufficient to require that

$$\lim_{h\to 0} \inf_{v_h \in V_h} \|v - v_h\|_V = 0 \qquad \forall v \in V. \tag{B.12}$$

In that case, the Galerkin method is convergent and $\|y - y_h\|_V \to 0$ when $h \to 0$; V_h must therefore be chosen in order to guarantee the density property (B.12). In this way, by taking a sufficiently small h, it is possible to approximate y by y_h as accurately as desired.

The actual convergence rate will depend on the specific choice of the subspace V_h; see Sect. B.4. We underline that, in the finite element case, V_h is made of piecewise polynomials of degree $r \ge 1$, so that $\|y - y_h\|_V$ tends to zero as $O(h^r)$.

Remark B.2. Since $\inf_{v_h \in V_h} \|y - v_h\|_V \le \|y - y_h\|_V$, in the case where M/α has order 1 the error due to the Galerkin method can be identified with the best approximation error for y in V_h. On the other hand, both errors have in general the same infinitesimal order with respect to h. In the particular case where $a(\cdot,\cdot)$ is a symmetric bilinear form, then (B.11) can be improved by replacing M/α with $\sqrt{M/\alpha}$. •

B.2.2 Algebraic Form of (P_1^h)

Problem (P_1^h) can be turned into an algebraic linear system of equations. Indeed, if we denote by $\{\varphi^j\}_{j=1}^{N_h}$ a basis for the space V_h, then every $v_h \in V_h$ has a unique representation

$$v_h = \sum_{j=1}^{N_h} v_{h,j}\varphi_j, \qquad \text{with } \mathbf{v} = (v_{h,1},\ldots,v_{h,N_h})^\top \in \mathbb{R}^{N_h}. \tag{B.13}$$

By setting $y_h = \sum_{j=1}^{N_h} y_{h,j}\varphi_j$, and denoting by $\mathbf{y}$ the vector having as components the unknown coefficients $y_{h,j}$, (P_1^h) is equivalent to: find $\mathbf{y} \in \mathbb{R}^{N_h}$ such that

$$\sum_{j=1}^{N_h} a(\varphi_j,\varphi_i)y_{h,j} = F\varphi_i \qquad \forall i = 1,\ldots,N_h, \tag{B.14}$$

that is

$$\mathbb{A}\mathbf{y} = \mathbf{f}, \tag{B.15}$$

where $\mathbb{A} \in \mathbb{R}^{N_h \times N_h}$ is the *stiffness* matrix with elements $(\mathbb{A})_{ij} = a(\varphi_j,\varphi_i)$ and $\mathbf{f} \in \mathbb{R}^{N_h}$ the vector with components $(\mathbf{f})_i = F\varphi_i$.

Thanks to the coercivity of $a(\cdot,\cdot)$ over V_h, for every $h > 0$, the matrix $\mathbb{A}$ in (B.15) is positive definite (and thus nonsingular); moreover, if $a(\cdot,\cdot)$ is symmetric so is the matrix $\mathbb{A}$. Other properties, such as the condition number or the sparsity pattern of $\mathbb{A}$, depend on the chosen basis of V_h; for instance, basis functions with small support (like those in finite element approximations) reduce to zero the entries of $\mathbb{A}$ related to basis functions having non-intersecting supports. The numerical solution of (B.15) can be carried out using either direct methods, such as the LU (Cholesky in the symmetric case) factorization, or iterative methods, such as the GMRES or the conjugate gradient method in the symmetric case.

B.2.3 Saddle-Point Problems

We now turn to the approximation of a relevant class of weakly coercive problems, namely saddle-point problems. Such a problem can arise when dealing with OCPs governed by Stokes equations, like in the case addressed in Sect. 5.13. Even more importantly, whenever relying on a *full-space* (or *all-at-once*) approach to derive the system of optimality conditions for a linear-quadratic unconstrained OCP, we end up with a saddle-point problem to be solved.

Recalling the formulation provided in Sect. A.4.2, a saddle point problem reads as follows:

find $(x,p) \in X \times Q$ such that

$$\begin{cases} e(x,w) + b(w,p) = f_1(w) & \forall w \in X \\ b(x,q) = f_2(q) & \forall q \in Q. \end{cases} \tag{P_2}$$

Let $X_h \subset X$ and $Q_h \subset Q$ be two finite dimensional subspaces of X and Q, respectively. We consider the following variational problem:

find $(x_h, p_h) \in X_h \times Q_h$ such that

$$\begin{cases} e(x_h, w_h) + b(w_h, p_h) = f_1(w_h) & \forall w_h \in X_h \\ b(x_h, q_h) = f_2(q_h) & \forall q_h \in Q_h \end{cases} \qquad (P_2^h)$$

that represents a Galerkin approximation of (P_2). Let

$$X_0^h = \{w_h \in X_h \colon b(w_h, q_h) = 0 \quad \forall q_h \in Q_h\} \subset X_h.$$

The well-posedness of (P_2^h) follows by the discrete counterpart of Theorem A.7 (see, e.g., [41] for the proof).

Theorem B.2 (Brezzi). *Let the space X and Q, the bilinear forms $e(\cdot,\cdot)$, $b(\cdot,\cdot)$ and $F_1 \in X^*$ and $F_2 \in Q^*$ satisfy the hypotheses of Theorem A.7. Let $X_h \subset X$ and $Q_h \subset Q$ be two finite-dimensional subspaces. Then $e(\cdot,\cdot)$ and $b(\cdot,\cdot)$ are continuous on $X_h \times X_h$ and $X_h \times Q_h$, respectively. Assume that the bilinear form $e(\cdot,\cdot)$ satisfies the conditions*

$$\inf_{x_h \in X_0^h} \sup_{w_h \in X_0^h} \frac{e(x_h, w_h)}{\|x_h\|_X \|w_h\|_X} \geq \hat{\alpha} > 0,$$

$$\inf_{w_h \in X_0^h} \sup_{x_h \in X_0^h} \frac{e(x_h, w_h)}{\|x_h\|_X \|w_h\|_X} > 0.$$

Moreover suppose that the bilinear form $b(\cdot,\cdot)$ verifies the discrete inf-sup condition

$$\inf_{q_h \in Q_h} \sup_{w_h \in X_h} \frac{b(w_h, q_h)}{\|w_h\|_X \|q_h\|_Q} \geq \hat{\beta}^s, \qquad \text{(B.16)}$$

for a suitable constant $\hat{\beta}^s > 0$ independent of h. Then, for every $h > 0$, problem (P_2^h) has a unique solution $(x_h, p_h) \in X_h \times Q_h$, which satisfies the stability estimates

$$\|x_h\|_X \leq \frac{1}{\hat{\alpha}}\Big[\|F_1\|_{X^*} + \frac{\hat{\alpha} + M_d}{\hat{\beta}^s}\|F_2\|_{Q^*}\Big], \qquad \text{(B.17a)}$$

$$\|p_h\|_Q \leq \frac{1}{\hat{\beta}^s}\Big[\Big(1 + \frac{M_d}{\hat{\alpha}}\Big)\|F_1\|_{X^*} + \frac{M_d(\hat{\alpha} + M_d)}{\hat{\alpha} + \hat{\beta}^s}\|F_2\|_{Q^*}\Big]. \qquad \text{(B.17b)}$$

If $(x, p) \in X \times Q$ denotes the unique solution of (P_2), the following optimal error inequality holds

$$\|x - x_h\|_X + \|p - p_h\|_Q \leq C\Big(\inf_{w_h \in X_h} \|x - w_h\|_X + \inf_{q_h \in Q_h} \|p - q_h\|_Q\Big),$$

where $C = C(\hat{\alpha}, \hat{\beta}^s, M_d, M_b)$ is independent of h.

Condition (B.16) is also called Ladyzhenskaia-Babuška-Brezzi (LBB) inf-sup condition; see, e.g., [41]. In the case of mixed variational problems, well-posedness of the discrete approximation (P_2^h) is ensured provided the inf-sup conditions on both bilinear forms e and b are fulfilled over $X_h \subset X$ and $Q_h \subset Q$, respectively. As in the case of weakly coercive problems, these assumptions are not automatically verified if they are valid on X and Q. This is the reason why they should be explicitly required on the spaces X_h and Q_h.

B.2.4 Algebraic Form of (P_2^h)

If $\{\varphi_i\}_{i=1}^{N_h}$ and $\{\eta^i\}_{i=1}^{M_h}$ denote two bases for X_h and Q_h respectively, being $N_h^u = \dim V_h$ and $N_h^p = \dim Q_h$, then (P_2^h) is equivalent to the linear system

$$\begin{pmatrix} \mathbb{D}_h & \mathbb{B}_h^\top \\ \mathbb{B}_h & 0 \end{pmatrix} \begin{pmatrix} \mathbf{x}_h \\ \mathbf{p}_h \end{pmatrix} = \begin{pmatrix} \mathbf{f}_{1h} \\ \mathbf{f}_{2h} \end{pmatrix}, \tag{B.18}$$

where $x_h = \sum_{j=1}^{N_h^u} x_h^{(j)} \varphi_j$, $p_h = \sum_{k=1}^{N_h^p} p_h^{(k)} \eta^k$, $\mathbf{x}_h$ and $\mathbf{p}_h$ are the vectors whose components are the unknown coefficients $x_h^{(j)}$, $p_h^{(k)}$,

$$(\mathbb{D}_h)_{ij} = d(\varphi_j, \varphi_i), \quad (\mathbb{B}_h)_{kj} = b(\varphi_j, \eta^k), \quad i, j = 1, \ldots, N_h^u, \quad k = 1, \ldots, N_h^p$$

are the elements of the matrices $\mathbb{D}_h$ and $\mathbb{B}_h$, whereas $\mathbf{f}_{1h}$ and $\mathbf{f}_{2h}$ denote the vectors with components $(\mathbf{f}_{1h})_i = F_1 \varphi_i$ and $(\mathbf{f}_{2h})_k = F_2 \eta^k$, $i = 1, \ldots, N_h^u$, $k = 1, \ldots, N_h^p$.

From an algebraic standpoint, Theorem B.2 ensures that the matrix in (B.18) is nonsingular; in particular the inf-sup condition (B.16) is equivalent to the condition $\mathrm{Ker}(\mathbb{B}_h^\top) = \{\mathbf{0}\}$. Moreover, the matrix in (B.18) is always indefinite, see, e.g., [88, Chap. 6], for the proof.

B.3 Numerical Approximation of Parabolic PDEs

In this section we consider the numerical approximation of an initial-boundary value problem like (7.26) for uniformly parabolic linear equations, under the assumptions stated in Sect. 7.4. In particular, we consider the weak formulation (7.31) of problem (7.26), which we report here for the sake of convenience, with $f \in L^2(0, T; \Omega)$:

for a.e. $t \in (0, T)$, find $y(t) \in V$ such that

$$\begin{cases} (\dot{y}(t), v)_{L^2(\Omega)} + a(y(t), v; t) = (f(t), v)_{L^2(\Omega)} & \forall v \in V, \\ y(0) = g. \end{cases} \tag{B.19}$$

Here $a(\cdot, \cdot)$ is the bilinear form associated with an elliptic operator and f is the linear functional depending on the source term and on the boundary conditions; for ease of notation, we denote the time-derivative by $\dot{y}$.

The classical approach consists in discretizing in space by the *finite element method* and then in time by the *finite difference method*, limiting ourselves to *first-order* accurate methods in time. We set $t \in (0, T)$ and approximate the weak problem (B.19) at time t in Ω. Given a family of subspace of finite dimension $\{V_h\}_{h>0}$ of V, being $N_h = dim(V_h)$, for $t \in (0, T)$ we introduce the following *semi-discrete problem in space*:

for a.e. $t \in (0,T)$, find $y_h(t) \in V_h$ such that

$$\begin{cases} (\dot{y}_h(t), v_h)_{L^2(\Omega)} + a(y_h(t), v_h; t) = (f(t), v_h)_{L^2(\Omega)} & \forall v_h \in V_h, \\ y_h(0) = g_h. \end{cases} \tag{B.20}$$

Here g_h denotes a suitable approximation of g in V_h. Such a problem yields a coupled system of ODEs, whose solution $y_h(t)$ is an approximation of the exact solution for each $t \in (0,T)$.

Remark B.3. Assuming that $a(\cdot,\cdot)$ is a coercive bilinear form with coercivity constant α, and thanks to the Cauchy and the Young inequality, by applying the Gronwall lemma (see, e.g., [224, Sect. 1.4]) it is possible to show the stability of the semi-discretization in space, yielding

$$\|y_h(t)\|^2_{L^2(\Omega)} + 2\alpha \int_0^t \|y_h(s)\|^2_V ds \le e^t \left[\|u_{0h}\|^2_{L^2(\Omega)} + \int_0^t \|f(s)\|^2_{L^2(\Omega)} ds \right] \tag{B.21}$$

for all $t \in (0,T)$. In particular, this implies that $y_h \in L^2(0,T;V) \cap C^0(0,T;L^2(\Omega))$. •

Let us introduce a basis $\{\varphi_j\}$ of V_h, and write the solution of the Galerkin problem (B.20) as a linear combination on the basis,

$$y_h(\mathbf{x},t) = \sum_{j=1}^{N_h} y_{h,j}(t)\varphi_j(\mathbf{x}).$$

Problem (B.20) can thus be rewritten as

$$\int_\Omega \sum_{j=1}^{N_h} \dot{y}_{h,j}(t)\varphi_j\varphi_i + a\left(\sum_{j=1}^{N_h} y_{h,j}(t)\varphi_j, \varphi_i; t\right) = (f(t),\varphi_i)_{L^2(\Omega)}, \qquad i = 1,\dots,N_h,$$

that is,

$$\sum_{j=1}^{N_h} \dot{y}_{h,j}(t) \underbrace{\int_\Omega \varphi_j\varphi_i}_{m_{ij}} + \sum_{j=1}^{N_h} y_{h,j}(t)\, \underbrace{a(\varphi_j,\varphi_i;t)}_{a_{ij}(t)} = \underbrace{(f(t),\varphi_i)_{L^2(\Omega)}}_{f_i(t)}, \qquad i = 1,\dots,N_h. \tag{B.22}$$

Let $\mathbf{y} = (y_{h,1}(t), \dots, y_{h,N_h}(t))^\top$, $\mathbb{M} = (m_{ij})$ the mass matrix, $\mathbb{A}(t) = (a_{ij}(t))$ the stiffness matrix, and $\mathbf{f} = (f_{h,1}(t), \dots, f_{h,N_h}(t))^\top$ the vector of known terms; then, system (B.22) may be written in matrix form,

$$\mathbb{M}\dot{\mathbf{y}}(t) + \mathbb{A}(t)\mathbf{y}(t) = \mathbf{f}(t).$$

For the sake of simplicity, hereon we assume that $\mathbb{A}$ does not depend on t, that is, the elliptic operator is time-independent. To solve this system we can use, e.g., the θ-method, which consists in discretizing the time derivative by an incremental ratio and replacing the other terms by a linear combination of the values at times t^n and t^{n+1}, depending on $\theta \in [0,1]$:

$$\mathbb{M}\frac{\mathbf{y}^{n+1} - \mathbf{y}^n}{\Delta t} + \theta\mathbb{A}\mathbf{y}^{n+1} + (1-\theta)\mathbb{A}\mathbf{y}^n = \theta\mathbf{f}^{n+1} + (1-\theta)\mathbf{f}^n; \tag{B.23}$$

$\Delta t = t^{n+1} - t^n$, $n = 0,1,\dots$, denotes the discretization timestep, while the superscript n indicates that the considered quantity is evaluated at time t^n.

Some special cases of the θ-method are the following ones:

- if $\theta = 0$, we obtain the *forward (explicit) Euler* method, which is a first order method with respect to Δt;

- if $\theta = 1$, we have the *backward (implicit) Euler* method, which is also a first order method with respect to Δt;
- if $\theta = 1/2$, we have the *Crank-Nicolson (trapezoidal)* method, which is a second order method with respect to Δt. Note that $\theta = 1/2$ is the only value such that the corresponding method is of second order.

Remark B.4. Note that (B.23) is an implicit algebraic system of equations; however, in the special case $\theta = 0$, this system can be made explicit by diagonalizing the matrix $\mathbb{M}_h/\Delta t$ (achievable by *lumping* of the mass matrix). •

Regarding the stability analysis of the fully discretized problem, applying the θ-method to the Galerkin problem (B.20) we obtain

$$\begin{aligned}\left(\frac{y_h^{n+1} - y_h^n}{\Delta t}, v_h\right)_{L^2(\Omega)} &+ a\left(\theta y_h^{n+1} + (1-\theta) y_h^n, v_h\right) \\ &= \theta F^{n+1}(v_h) + (1-\theta) F^n(v_h) \qquad \forall v_h \in V_h,\end{aligned} \tag{B.24}$$

for all $n \geq 0$, with $y_h^0 = g_h$ and $F^n(v_h) = (f(t^n), v_h)_{L^2(\Omega)}$.

Let us first consider the case where $f = 0$ and $\theta = 1$ (implicit Euler method):

$$\left(\frac{y_h^{n+1} - y_h^n}{\Delta t}, v_h\right)_{L^2(\Omega)} + a(y_h^{n+1}, v_h) = 0 \qquad \forall v_h \in V_h.$$

Using again the coercivity of $a(\cdot,\cdot)$, the Cauchy-Schwarz and Young inequalities, and summing with respect to n from 0 to $m-1$, we find

$$\|y_h^m\|^2_{L^2(\Omega)} + 2\alpha\Delta t \sum_{n=0}^{m-1} \|y_h^{n+1}\|^2_V \leq \|g_h\|^2_{L^2(\Omega)}. \tag{B.25}$$

Finally, observing that $C\|y_h^{n+1}\|_V \geq \|y_h^{n+1}\|_{L^2(\Omega)}$ (where $C = 1$ if $V = H^1(\Omega)$, and $C = C_P$, the Poincaré constant, if $V = H^1_0(\Omega)$), we deduce from (B.25) that, for any chosen $\Delta t > 0$,

$$\lim_{n\to\infty} \|y_h^n\|_{L^2(\Omega)} = 0,$$

that is, the backward Euler method is *absolutely stable* with no hypothesis on the timestep Δt. This results can also be extended to the more general case where $f \neq 0$ by relying, e.g., on the discrete Gronwall lemma (see, e.g., [224, Sect. 1.4]) showing that

$$\|y_h^m\|^2_{L^2(\Omega)} + 2\alpha\Delta t \sum_{n=1}^{m} \|y_h^n\|^2_V \leq C(m)\left(\|g_h\|^2_{L^2(\Omega)} + \sum_{n=1}^{m} \Delta t \|f^n\|^2_{L^2(\Omega)}\right). \tag{B.26}$$

This relation resembles (B.21), provided we replace the integrals $\int_0^t \cdot\, ds$ by a composite numerical integration formula with timestep Δt.

In general, it is possible to show that:

- *if $\theta \geq 1/2$, the θ-method is unconditionally stable (i.e., it is stable for all Δt),*
- *if $\theta < 1/2$, the θ-method is only stable for*

$$\Delta t \leq \frac{2}{(1-2\theta)\kappa_2(\mathbb{A}_h)}.$$

where $\kappa_2(\mathbb{A}_h) \simeq C(\theta) h^{-2}$ and $C(\theta)$ denotes a positive constant dependent on θ. Hence, for $\theta < 1/2$, Δt cannot be chosen arbitrarily.

B.4 Finite Element Spaces and Interpolation Operator

In this section we recall the definition of the finite element spaces used throughout the book, as well as some relevant (more technical) tools required to derive error estimates for finite element approximations.

To define approximations of the space $H^1(\Omega)$ that depend on a parameter h, let us first introduce a *triangulation* of the domain Ω. For the sake of simplicity, most often we will consider domains $\Omega \subset \mathbb{R}^d$ ($d = 2, 3$) with polygonal shape, and meshes (or grids) $\mathcal{T}_h$ which represent their cover with non-overlapping triangles ($d = 2$) or tetrahedra ($d = 3$) K. We denote by h_K the diameter of K and define h to be the maximum value of h_K, $K \in \mathcal{T}_h$. Under this assumption, the discretized domain

$$\Omega_h = int\left(\bigcup_{K \in \mathcal{T}_h} K \right)$$

represented by the internal part of the union of the triangles of $\mathcal{T}_h$ coincides with Ω. Hence, from now on we will adopt the symbol Ω to denote without distinction both the computational domain and its (optional) approximation. Of course, the extension to domains with nonpolygonal shape is also possible, see e.g. [224] for a detailed definition. We also refer e.g. to [218] for a more detailed description of mesh generation, refinement and other essential features.

Finite Element Spaces

The most natural (and well-known) strategy to define a finite element space is to consider globally continuous functions that are polynomials of degree r on the single triangles (elements) of the triangulation $\mathcal{T}_h$, that is, to define

$$X_h^r = \left\{ v_h \in C^0(\overline{\Omega}) : v_h|_K \in \mathbb{P}_r \quad \forall K \in \mathcal{T}_h \right\}, \qquad r = 1, 2, \dots . \tag{B.27}$$

as the space of globally continuous functions that are polynomials of degree at most r on the single elements of the triangulation $\mathcal{T}_h$. Moreover, we can define

$$\overset{\circ}{X}{}_h^r = \{ v_h \in X_h^r : v_h|_{\partial\Omega} = 0 \} \tag{B.28}$$

and, more in general, $V_h = X_h^r \cap V$.

Remark B.5. We remark that the space of polynomials $\mathbb{P}_r$ in d variables is defined as follows

$$\mathbb{P}_r = \left\{ p(\mathbf{x}) = \sum_{\substack{0 \le i_1, \dots, i_d \\ i_1 + \dots + i_d \le d}} a_{i_1 \dots i_d} x_1^{i_1} \dots x_d^{i_d}, \quad a_{i_1 \dots i_d} \in \mathbb{R} \right\}; \tag{B.29}$$

$\mathbb{P}_r$ has dimension

$$N_r = \dim \mathbb{P}_r = \binom{r+d}{r} = \frac{1}{d!} \prod_{k=1}^{d} (r+k). \tag{B.30}$$

In particular, for $d = 2$ the spaces $\mathbb{P}_r$ have dimension $(r+1)(r+2)/2$, so that dim $\mathbb{P}_1 = 3$, dim $\mathbb{P}_2 = 6$ and dim $\mathbb{P}_3 = 10$, hence on every element of the grid $\mathcal{T}_h$ the generic function v_h is well defined whenever its value at 3, resp. 6, resp. 10 suitably chosen nodes, is known. •

The spaces X_h^r and $\overset{\circ}{X}{}_h^r$ are suitable for the approximation of $H^1(\Omega)$, resp. $H_0^1(\Omega)$, thanks to the following property (for its proof see, e.g., [224]):

Proposition B.1. *If $v \in C^0(\overline{\Omega})$ and $v|_K \in H^1(K)$ for each $K \in \mathcal{T}_h$, then $v \in H^1(\Omega)$.*

The spaces X_h^r are all subspaces of $H^1(\Omega)$, as they are constituted by continuous functions, differentiable inside the elements of the partition. The fact that the functions of X_h^r are locally (element-wise) polynomials will make the stiffness matrix sparse and easy to handle.

To this aim, let us now choose a basis $\{\varphi_i\}$ for the X_h^r space. First of all, it is convenient that the support of the generic basis function φ_i has non-empty intersection only with the support of a negligible number of other functions of the basis. In this way, many elements of the stiffness matrix will be null.

It is also convenient to deal with a *Lagrangian* basis: in that case, the coefficients of the expansion of a generic function $v_h \in X_h^r$ in the basis will be the values taken by v_h at carefully chosen points $\mathbf{N}_i$, with $i = 1, \ldots, N_h$ of the grid $\mathcal{T}_h$, which we call *nodes* (and which form, in general, a superset of the vertices of $\mathcal{T}_h$).

Thus, each function $v_h \in V_h$ is uniquely characterized by its values at the nodes $\mathbf{N}_i$, with $i = 1, \ldots, N_h$, excluding the boundary nodes where $v_h = 0$. Consequently, a basis for the space V_h can be the set of the characteristic Lagrangian functions $\varphi_j \in V_h$, $j = 1, \ldots, N_h$, such that

$$\varphi_j(\mathbf{N}_i) = \delta_{ij} = \begin{cases} 0 & i \neq j, \\ 1 & i = j, \end{cases} \qquad i, j = 1, \ldots, N_h. \tag{B.31}$$

In particular, if $r = 1$ the nodes are the vertices of the elements, with the exception of those belonging to the boundary of Ω, while the function φ_j is affine on each triangle and is equal to 1 at the node $\mathbf{N}_j$ and 0 at all the other nodes of the triangulation.

A function $v_h \in V_h$ can be expressed through a linear combination of the basis functions of V_h in the following way

$$v_h(\mathbf{x}) = \sum_{i=1}^{N_h} v_i \varphi_i(\mathbf{x}) \quad \forall\, \mathbf{x} \in \Omega, \qquad \text{with } v_i = v_h(\mathbf{N}_i) \tag{B.32}$$

so that, by expressing the discrete solution y_h in terms of the basis $\{\varphi_j\}$ via (B.32), we have that

$$y_j = y_h(\mathbf{N}_j),$$

that is, the unknowns of a linear systems like (B.15) are the nodal values of the FE solution. It is rather evident, since the *support* of the generic function with basis φ_i is only formed by the triangles having node $\mathbf{N}_i$ in common, that $\mathbb{A}$ is a sparse matrix. In particular, the number of non-null elements of $\mathbb{A}$ is of the order of N_h, as a_{ij} is different from zero only if $\mathbf{N}_j$ and $\mathbf{N}_i$ are nodes of the same triangle. $\mathbb{A}$ has not necessarily a definite structure (e.g. banded), as that will depend on how the nodes are numbered.

Interpolation Operator and Interpolation Error

A further ingredient required to obtain an estimate for the approximation error $\|y - y_h\|_V$ is the interpolation operator (and related interpolation error estimates). Here we recall the basic results; see, e.g. [218, 224] for their proof.

For each $v \in C^0(\overline{\Omega})$ we define *interpolant* of v in the space of X_h^1, determined by the grid $\mathcal{T}_h$, the function $\Pi_h^1 v$ such that

$$\Pi_h^1 v(\mathbf{N}_i) = v(\mathbf{N}_i) \text{ for each node } \mathbf{N}_i \text{ of } \mathcal{T}_h, \text{ for } i = 1, \ldots, N_h.$$

If $\{\varphi_i\}$ is the Lagrangian basis of the space X_h^1, then

$$\Pi_h^1 v(\mathbf{x}) = \sum_{i=1}^{N_h} v(\mathbf{N}_i)\varphi_i(\mathbf{x}).$$

The operator $\Pi_h^1 : C^0(\overline{\Omega}) \to X_h^1$, associating to a continuous function v its interpolant $\Pi_h^1 v$, is called *interpolation operator.*

Similarly, we can define an operator $\Pi_h^r : C^0(\overline{\Omega}) \to X_h^r$, for each integer $r \geq 1$. Having denoted by Π_K^r the local interpolation operator associated to a continuous function v the polynomial $\Pi_K^r v \in \mathbb{P}_r(K)$, interpolating v in the degrees of freedom of the element $K \in \mathcal{T}_h$, we define

$$\Pi_h^r v \in X_h^r \ : \quad \Pi_h^r v\big|_K = \Pi_K^r(v\big|_K) \qquad \forall K \in \mathcal{T}_h. \tag{B.33}$$

Let us denote by h_K the diameter of each triangle K, so that $h_K \leq h$ for all $K \in \mathcal{T}_h$. Then, for the interpolation operator the following *global* estimate holds:

Theorem B.3 (Global estimate for the interpolation error). *Let $\{\mathcal{T}_h\}_{h>0}$ be a family of regular grids of the domain Ω and let $m = 0, 1$ and $r \geq 1$. Then there exists a constant $C = C(r, m, \widehat{K}) > 0$ such that*

$$|v - \Pi_h^r v|_{H^m(\Omega)} \leq C \Big(\sum_{K\in\mathcal{T}_h} h_K^{2(r+1-m)} |v|^2_{H^{r+1}(K)} \Big)^{1/2} \qquad \forall v \in H^{r+1}(\Omega). \tag{B.34}$$

In particular, we obtain

$$|v - \Pi_h^r v|_{H^m(\Omega)} \leq C\, h^{r+1-m} |v|_{H^{r+1}(\Omega)} \qquad \forall v \in H^{r+1}(\Omega). \tag{B.35}$$

B.5 A Priori Error Estimation

We furthermore provide some results dealing with convergence of finite element approximations in the case of elliptic and parabolic PDEs. The interested reader may refer, e.g., to [224, 218] for a deeper analysis and further a priori error estimates.

B.5.1 Elliptic PDEs

The result of Theorem B.3 provides us with an estimate for the approximation error of the Galerkin method in the case of an elliptic equation. For the sake of simplicity we focus on the (strongly) coercive case described in Sect. B.2.1; similar results can be obtained for the other classes of problems we have considered so far.

Indeed, it is sufficient to apply (B.11) and Theorem B.3 (for $m = 1$) to obtain the following *a priori* error estimate:

Theorem B.4. *Let $y \in V$ be the exact solution of the variational problem (P_1) and y_h its approximate solution via the finite element method of degree r, i.e. the solution of problem (P_1^h) where $V_h = X_h^r \cap V$. If $y \in H^{r+1}(\Omega)$, then the following* a priori *error estimates hold:*

$$\|y - y_h\|_{H^1(\Omega)} \le \frac{M}{\alpha} C \Big(\sum_{K \in \mathcal{T}_h} h_K^{2r} |y|^2_{H^{r+1}(K)} \Big)^{1/2}, \tag{B.36}$$

$$\|y - y_h\|_{H^1(\Omega)} \le \frac{M}{\alpha} C h^r |y|_{H^{r+1}(\Omega)}, \tag{B.37}$$

C being a constant independent of h and y.

Note that (B.37) follows from (B.36). In order to increase the accuracy, two different strategies can therefore be pursued: *(i)* decreasing h, i.e. refining the grid; *(ii)* increasing r, i.e. using finite elements of higher degree. However, the latter approach can only be pursued if the solution y is regular enough. In general, we can say that if $y \in C^0(\bar{\Omega}) \cap H^{p+1}(\Omega)$ for some $p > 0$, then

$$\|y - y_h\|_{H^1(\Omega)} \le C h^s |y|_{H^{s+1}(\Omega)}, \quad s = \min\{r, p\}. \tag{B.38}$$

B.5.2 Parabolic PDEs

Regarding instead the convergence analysis of parabolic problems, we report an a priori estimate for the error between the semi-discrete solution $y_h(t^n)$ and the fully-discrete one y_h^n for any fixed h (see, e.g., [224, Chapter 11] and [262] for further details).

Theorem B.5. *Assume that the hypotheses for existence and uniqueness of the solution y to the initial-boundary value problem* (B.19) *are satisfied with $\lambda = 0$. When $0 \le \theta < 1/2$ assume, moreover, that the time-step restriction $\Delta t \le C(\theta) h^2$ is satisfied. Under the hypothesis that g, f and y are sufficiently regular, the functions y_h^n and $y_h(t)$ defined in* (B.24) *and* (B.20) *fulfill the following a priori error estimate holds: for all $n = 1, \ldots$*

$$\|y_h^n - y_h(t_n)\|^2_{L^2(\Omega)} + 2\alpha \Delta t \sum_{k=1}^{n} \|y_h^n - y_h(t_n)\|^2_V \le \tilde{C}(g, f, y)(\Delta t^{p(\theta)} + h^{2r})$$

if $V_h = X_h^r \cap V$, where $p(\theta) = 2$ if $\theta \neq 1/2$, $p(1/2) = 4$ and $\tilde{C}$ depends on its arguments but not on h and Δt.

B.6 Solution of Nonlinear PDEs

In this section we provide some basic notions about the numerical approximation of nonlinear PDEs through the Newton method. Let us consider a general problem under the form

$$G(y) = 0 \qquad \text{in } W, \tag{B.39}$$

where $G : V \to W$ and V, W are suitable Banach spaces. The weak formulation of problem (B.39) reads: find $y \in V$ such that

$$g(y; \varphi) = 0 \qquad \forall \varphi \in V, \tag{B.40}$$

where the variational form $g(\cdot;\cdot) : V \times V \to \mathbb{R}$ is defined as

$$g(y; \varphi) = \langle G(y), \varphi \rangle \qquad \forall y, \varphi \in V.$$

Moreover, we denote by

$$dg[\bar{y}](v, \varphi) = \langle G'_y(\bar{y})v, \varphi \rangle \qquad \forall v, \varphi \in V$$

the partial Fréchet derivative of $g(\bar{y}, \cdot)$ with respect to its first argument, evaluated at $\bar{y} \in V$.

For the numerical approximation of (B.40) we rely on the finite element method, introducing a finite-dimensional subspace $V_h \subset V$ of dimension $N_h = \dim(V_h)$: hence, we seek $y_h \in V_h$ such that

$$g(y_h; \varphi_h) = 0 \qquad \forall \varphi_h \in V_h. \tag{B.41}$$

For the discrete problem (B.41) to be well-posed we assume $dg[y_h](\cdot, \cdot)$ to be inf-sup stable, i.e. that there exists a constant $\beta_{0,h} > 0$ such that

$$\beta_h = \inf_{w_h \in V_h} \sup_{v_h \in V_h} \frac{dg[y_h](w_h, v_h)}{\|w_h\|_V \|v_h\|_V} \geq \beta_{0,h}, \tag{B.42}$$

and continuous, i.e. that there exists a positive constant $\gamma_{0,h} < \infty$ such that

$$\gamma_h = \sup_{v_h \in V_h} \sup_{w_h \in V_h} \frac{dg[y_h](w_h, v_h)}{\|w_h\|_V \|v_h\|_V} \leq \gamma_{0,h}. \tag{B.43}$$

A solution y_h to problem (B.41) is said to be *regular* if it fulfills (B.42)-(B.43).

Solving problem (B.41) requires nonlinear iterations with a linearized problem being solved at every step. Newton method is a very natural approach: given an initial guess $y_h^{(0)} \in V_h$, for $n = 0, 1, \ldots$ until convergence, we seek $\delta y_h \in V_h$ such that

$$dg[y_h^{(n)}](\delta y_h, \varphi_h) = -g(y_h^{(n)}; \varphi_h) \qquad \forall \varphi_h \in V_h, \tag{B.44}$$

and then set $y_h^{(n+1)} = y_h^{(n)} + \delta y_h$.

Newton method is quadratically convergent provided $dg[y_h](\cdot, \cdot)$ is locally Lipschitz continuous and $y_h^{(0)}$ is sufficiently close to y_h.

Theorem B.6. *Let y_h be a regular solution of (B.41) and $dg[y_h](\cdot, \cdot)$ locally Lipschitz continuous at y_h, i.e. there exist $\varepsilon > 0$ and $K_h > 0$ such that*

$$\|dg[y_h](\cdot, \cdot) - dg[v_h](\cdot, \cdot)\|_{\mathcal{L}(V_h, V_h')} \leq K_h \|y_h - v_h\|_V,$$

for all $v_h \in B_\varepsilon(y_h) = \{w_h \in V_h : \|y_h - w_h\|_V \leq \varepsilon\}$. Then there exists $\delta > 0$ such that if $\|y_h^{(0)} - y_h\| \leq \delta$, the Newton sequence $\{y_h^{(n)}\}$ is well-defined and converges to y_h. Furthermore, for some constant M with $M\delta < 1$, we have the error bound

$$\|y_h^{(n+1)} - y_h\|_V \leq M \|y_h^{(n)} - y_h\|_V^2, \qquad n \geq 0.$$

This local convergence theorem (see, e.g., [275, Chap. 5] for the proof) regretfully requires the existence of a solution u_h in advance. The Kantorovich theorem (see, e.g., [275, 65]) overcomes this drawback by establishing the existence of y_h simply on the basis of the knowledge of $g(y_h^{(0)}, \cdot)$ and $dg[y_h^{(0)}](\cdot, \cdot)$.

Theorem B.7 (Kantorovich). *Let $y_h^{(0)} \in V_h$ be a given initial guess such that:*

i) $\varepsilon = \|g(y_h^{(0)}; \cdot)\|_{V_h^*} < \infty$ and $dg[y_h^{(0)}](\cdot, \cdot)$ is continuous and inf-sup stable with

$$a^{-1} = \inf_{w_h \in V_h} \sup_{v_h \in V_h} \frac{dg[y_h^{(0)}](w_h, v_h)}{\|w_h\|_V \|v_h\|_V} > 0; \tag{B.45}$$

ii) there exists a positive constant K such that

$$\|dg[w_h](\cdot, \cdot) - dg[v_h](\cdot, \cdot)\|_{\mathcal{L}(V_h, V_h^*)} \leq K \|w_h - v_h\|_V \quad \forall v_h, w_h \in B_r(y_h^{(0)})$$

with $r = 1/(aK)$ and

$$0 < 2\varepsilon K a^2 \leq 1. \tag{B.46}$$

Define two positive numbers

$$r_- = \frac{1 - \sqrt{1 - 2\varepsilon K a^2}}{aK}, \qquad r_+ = \frac{1 + \sqrt{1 - 2\varepsilon K a^2}}{aK}.$$

Then

1. *problem* (B.41) *has a solution $y_h \in \overline{B}_{r_-}(y_h^{(0)})$. Moreover, if*

$$\|dg[w_h](\cdot, \cdot) - dg[v_h](\cdot, \cdot)\|_{\mathcal{L}(V_h, V_h^*)} \leq K \|w_h - v_h\|_V \quad \forall v_h, w_h \in B_{r_+}(y_h^{(0)}),$$

   this solution is unique in $\overline{B}_{r_+}(u_h^0)$;

2. *the Newton sequence $\{y_h^{(n)}\}$ converges to y_h and the following error bound holds for each $n \geq 0$*

$$\|y_h^{(n)} - y_h\|_V \leq \frac{r}{2^n} \left(\frac{r_-}{r} \right)^{2^n}.$$

References

1. F. Abergel and R. Temam. On some control problems in fluid mechanics. *Theoret. Comput. Fluid Dynamics*, 1:303–325, 1990.
2. V.I. Agoshkov, D. Ambrosi, V. Pennati, A. Quarteroni, and F. Saleri. Mathematical and numerical modelling of shallow water flow. *Comput. Mech.*, 11(5):280–299, 1993.
3. M. Ainsworth and J. T. Oden. A Posteriori *Error Estimation in Finite Element Analysis*. Wiley-Interscience, 2000.
4. V. Akcelik, G. Biros, O. Ghattas, J. Hill, D. Keyes, and B. Waanders. Parallel Algorithms for PDE-Constrained Optimization. In M.A. Heroux, P. Raghavan, and H.D. Simon, editors, *Parallel Processing for Scientific Computing*, Philadelphia, PA, USA, 2006. Society for Industrial and Applied Mathematics.
5. R. R. Aliev and A. V. Panfilov. A simple two-variable model of cardiac excitation. *Chaos Soliton Fract.*, 7(3):293–301, 1996.
6. G. Allaire. *Conception optimale de structures*. Springer Verlag, 2007. Vol. 58 Mathématiques et Applications.
7. G. Allaire, F. de Gournay, F. Jouve, and A.-M. Toader. Structural optimization using topological and shape sensitivity via a level-set method. *Control and Cybernetics*, 34(1):59–80, 2005.
8. G. Allaire, F. Jouve, and A.-M. Toader. Structural optimization using sensitivity analysis and a level-set method. *J. Comput. Phys.*, 194(1):363–393, 2004.
9. G. Allaire and O. Pantz. Structural optimization with `freefem++`. *Struct. Multidisc. Optim.*, 32(3):173–181, 2006.
10. W. Alt. The Lagrange-Newton method for infinite-dimensional optimization problems. *Numer. Funct. Anal. Optim.*, 11(3-4):201–224, 1990.
11. W. Alt and K. Malanowski. The Lagrange-Newton method for nonlinear optimal control problems. *Comput. Optim. Appl.*, 2(1):77–100, 1993.
12. E.I. Amoiralis and I.K. Nikolos. Freeform deformation versus B-spline representation in inverse airfoil design. *J. Comput. Inf. Sci. Engrg.*, 8(2), 2008.
13. J.S. Arora and Q. Wang. Review of formulations for structural and mechanical system optimization. *Struct. Multidisc. Optim.*, 30(4):251–272, 2005.
14. M. Asch, M. Bocquet, and M. Nodet. *Data Assimilation: Methods, Algorithms, and Applications*. SIAM, 2016.
15. K.J. Aström and R.M. Murray. *Feedback systems: an introduction for scientists and engineers*. Princeton University Press, 2010.
16. I. Babuška. The finite element method with lagrangian multipliers. *Numer. Math.*, 20:179–192, 1973.
17. F. Ballarin, A. Manzoni, G. Rozza, and S. Salsa. Shape optimization by free-form deformation: existence results and numerical solution for stokes flows. *J. Sci. Comput.*, 60(3):537–563, 2014.
18. H.T. Banks and K. Kunisch. The linear regulator problem for parabolic systems. *SIAM J. Control Optim.*, 22(5):684–698, 1984.
19. V. Barbu. Feedback control of time dependent stokes flows. In S.S. Sritharan, editor, *Optimal Control of Viscous Flow*, chapter 3, pages 63–77. Society for Industrial and Applied Mathematics, Philadelphia, PA, USA, 1998.
20. R.A. Bartlett, M. Heinkenschloss, D. Ridzal, and B. van Bloemen Waanders. Domain decomposition methods for advection dominated linear-quadratic elliptic optimal control problems. *Comput. Meth. Appl. Mech. Engrg.*, 195(44?-47):6428–6447, 2006.
21. G. Basile and G. Marro. *Controlled and Conditioned Invariants in Linear System Theory*. Prentice Hall Englewood Cliffs, NJ, 1992.

A. Manzoni et al., *Optimal Control of Partial Differential Equations*,
Applied Mathematical Sciences 207, https://doi.org/10.1007/978-3-030-77226-0

22. A. Battermann and M. Heinkenschloss. Preconditioners for karush-kuhn-tucker matrices arising in the optimal control of distributed systems. In *Control and Estimation of Distributed Parameter Systems*, pages 15–32. Springer, 1998.
23. R. Becker, H. Kapp, and R. Rannacher. Adaptive finite element methods for optimal control of partial differential equations: Basic concepts. *SIAM, J. Control Opt.*, 39(1):113–132, 2000.
24. R. Becker and R. Rannacher. A feed-back approach to error control in finite element method: Basic analysis and examples. *East - West J. Numer. Math.*, 4:237–264, 1996.
25. R. Becker and R. Rannacker. An optimal control approach to a posteriori error estimation in finite element methods. *Acta Numerica*, pages 1–102, 2001.
26. J.A. Bello, E. Fernandez-Cara, J. Lemoine, and J. Simon. The differentiability of the drag with respect to the variations of a Lipschitz domain in a Navier-Stokes flow. *SIAM J. Control. Optim.*, 35(2):626–640, 1997.
27. M.P. Bendsoe and O. Sigmund. *Topology optimization theory, methods and applications.* Springer-Verlag, Berlin Heidelberg, 2004.
28. P. Benner, S. Görner, and J. Saak. Numerical solution of optimal control problems for parabolic systems. In *Parallel Algorithms and Cluster Computing*, volume 52 of *Lecture Notes in Computational Science and Engineering*, pages 151–169. Springer, Berlin, Heidelberg, 2006.
29. M. Benzi, L. Ferragut, M. Pennacchio, and V. Simoncini. Solution of linear systems from an optimal control problem arising in wind simulation. *Numerical Linear Algebra with Applications*, 17(6):895–915, 2010.
30. M. Benzi, G.H. Golub, and J. Liesen. Numerical solution of saddle point problems. *Acta Numerica*, 14:1–137, 2005.
31. M. Benzi and M. A. Olshanskii. An augmented lagrangian-based approach to the oseen problem. *SIAM J. Sci. Comput.*, 28(6):2095–2113, 2006.
32. M. Berggren. Numerical solution of a flow-control problem: vorticity reduction by dynamic boundary action. *SIAM J. Sci. Comput.*, 19(3):829–860, 1998.
33. M. Berggren. Approximations of very weak solutions to boundary-value problems. *SIAM J. Numer. Anal.*, 42(2):860–877, 2004.
34. M. Bergounioux, K. Ito, and K. Kunisch. Primal-dual strategy for constrained optimal control problems. *SIAM J. Control Optim.*, 237(4):1176–1194, 1999.
35. M. Bergounioux and K. Kunisch. Primal-dual strategy for state-constrained optimal control problems. *Comput. Optim. Appl.*, 22(2):193–224, 2002.
36. D.P. Bertsekas. *Constrained optimization and Lagrange multiplier methods.* Academic Press, New York, 1982.
37. J.T. Betts. *Practical methods for optimal control and estimation using nonlinear programming.* SIAM, 2010.
38. G. Biros and O. Ghattas. Parallel lagrange–newton–krylov–schur methods for pde-constrained optimization. part ii: The lagrange–newton solver and its application to optimal control of steady viscous flows. *SIAM J. Sci. Comput.*, 27(2):714–739, 2005.
39. G. Biros and O. Ghattas. Parallel lagrange-newton-krylov-schur methods for pde- constrained optimization. part i: The krylov-schur solver. *SIAM J. Sci. Comput.*, 27:687–713, 2005.
40. A. Björck. *Numerical Methods in Matrix Computations*, volume 59 of *Texts in Applied Mathematics.* Springer International Publishing, Switzerland, 2015.
41. D. Boffi, F. Brezzi, and M. Fortin. *Mixed Finite Elements and Applications.* Springer-Verlag, Berlin-Heidelberg, 2013.
42. A. Borzì and K. Kunisch. A multigrid scheme for elliptic constrained optimal control problems. *Comput. Optim. Appl.*, 31:309–333, 2005.
43. A. Borzì and V. Schulz. Multigrid methods for PDE optimization. *SIAM Review*, 51(2):361–395, 2009.
44. A. Borzì and V. Schulz. *Computational optimization of systems governed by partial differential equations.* Society for Industrial and Applied Mathematics, Philadephia, PA, 2011.
45. M. Boulakia, M.A Fernández, J.F. Gerbeau, and N. Zemzemi. A coupled system of pdes and odes arising in electrocardiograms modeling. *Applied Math. Res. Exp.*, 28(2008(abn002)), 2008.
46. Y. Bourgault, Y. Coudiere, and C. Pierre. Existence and uniqueness of the solution for the bidomain model used in cardiac electrophysiology. *Nonlinear analysis: Real world applications*, 10(1):458–482, 2009.
47. J. Boyle, M. D. Mihajlović, and J. A. Scott. HSL MI20: an efficient AMG preconditioner, 2007.
48. D. Braess and P. Peisker. On the numerical solution of the Biharmonic equation and the role of squaring matrices for preconditioning. *IMA J. Numer. Anal.*, 6(4):393–404, 1986.
49. J. H. Bramble and J. E. Pasciak. A preconditioning technique for indefinite systems resulting from mixed approximations of elliptic problems. *Math. Comp.*, 50:1–17, 1988.
50. J.H. Bramble. *Multigrid Methods.* Pitman Research Notes in Mathematics Series, Essex, 1993.

51. C. Brandenburg, F. Lindemann, M. Ulbrich, and S. Ulbrich. A continuous adjoint approach to shape optimization for Navier Stokes flow. In K. Kunisch, J. Sprekels, G. Leugering, and F. Tröltzsch, editors, *Optimal Control of Coupled Systems of Partial Differential Equations*, volume 158 of *International Series of Numerical Mathematics*, pages 35–56. Birkhäuser Verlag, 2009.
52. H. Brezis. *Functional Analysis, Sobolev Spaces and Partial Differential Equations.* Springer-Verlag, New York, 2011.
53. F. Brezzi. On the existence, uniqueness, and approximation of saddle point problems arising from Lagrangian multipliers. *R.A.I.R.O., Anal. Numér.*, 2:129–151, 1974.
54. A. E. Bryson and Y.-C. Ho. *Applied optimal control: optimization, estimation and control.* CRC Press, 1975.
55. M. Cannon, C. Cullum, and E. Polak. *Theory and Optimal Control and Mathematical Programming.* McGraw-Hill, New York, 1970.
56. Y. Cao, S. Li, L. Petzold, and R. Serban. Adjoint sensitivity analysis for differential-algebraic equations: The adjoint dae system and its numerical solution. *SIAM J. Sci. Comput.*, 24(3):1076–1089, 2003.
57. H. Cartan. *Calcul Differentiel.* Hermann, Paris, 1967.
58. E. Casas and F. Tröltzsch. Error estimates for linear-quadratic elliptic control problems. In V. Barbu et al., editor, *Analysis and optimization of differential systems*, pages 89–100. Springer, 2003.
59. E. Casas and F. Tröltzsch. Second order optimality conditions and their role in pde control. *Jahresber. Dtsch. Math. Ver.*, 117(1):3–44, 2015.
60. J. Céa. Conception optimale ou identification de formes: calcul rapide de la dérivée directionelle de la fonction cout. *Mathematical Modelling and Numerical Analysis*, 20(3):371–402, 1986.
61. J. Céa, S. Garreau, P. Guillaume, and M. Masmoudi. The shape and topological optimizations connection. *Comput. Methods Appl. Mech. Engrg.*, 188:713–726, 2000.
62. N. Chamakuri, K. Kunisch, and G. Plank. Numerical solution for optimal control of the reaction-diffusion equations in cardiac electrophysiology. *Comput. Optim. Appl.*, 49:149–178, 2011.
63. K. Chrysafinos and S. Hou. Analysis and approximation of the evolutionary Stokes equations with inhomogeneous boundary and divergence data using a parabolic saddle point formulation. *ESAIM: Math. Model. Numer. Anal.*, 51:1501–1526, 2017.
64. P. G. Ciarlet. *The Finite Element Method for Elliptic Problems.* Classics in Applied Mathematics, 40. Society for Industrial and Applied Mathematics, Philadephia, PA, 2002.
65. P. G. Ciarlet. *Linear and nonlinear functional analysis with applications.* Society for Industrial and Applied Mathematics, Philadelphia, 2014.
66. P. Clément. Approximation by finite element functions using local regularization. *RAIRO, Anal. Num.*, 9:77–84, 1975.
67. P. Colli Franzone and L. F. Pavarino. A parallel solver for reaction–diffusion systems in computational electrocardiology. *Math. Mod. Meth. Appl. Sci.*, 14(06):883–911, 2004.
68. P. Colli Franzone, L.F. Pavarino, and S. Scacchi. *Mathematical cardiac electrophysiology*, volume 13 of *Modeling, Simulation and Applications (MS&A) Series.* Springer-Verlag Italia, Milano, 2014.
69. Piero Colli Franzone, Peter Deuflhard, Bodo Erdmann, Jens Lang, and Luca F Pavarino. Adaptivity in space and time for reaction-diffusion systems in electrocardiology. *SIAM J. Sci. Comput.*, 28(3):942–962, 2006.
70. A. R. Conn, N. I. M. Gould, and P. L. Toint. *Trust Region Methods.* Society for Industrial and Applied Mathematics, Philadelphia, PA, USA, 2000.
71. A. R. Conn, K. Scheinberg, and L. N. Vicente. *Introduction to Derivative-Free Optimization.* Society for Industrial and Applied Mathematics, Philadelphia, PA, USA, 2009.
72. J.-M. Coron. *Control and Nonlinearity.* American Mathematical Society, 2007.
73. R. Dautray and J. L. Lions. *Mathematical Analysis and Numerical Methods for Science and Technology.* Springer-Verlag, Berlin Heidelberg, 2000.
74. J.C. de los Reyes and K. Kunisch. A semi-smooth newton method for control constrained boundary optimal control of the navier-stokes equations. *Nonlinear Analysis*, 62(7):1289–1316, 2005.
75. K. Deckelnick, A. Günther, and M. Hinze. Finite element approximation of dirichlet boundary control for elliptic pdes on two-and three-dimensional curved domains. *SIAM J. Control Optim.*, 48(4):2798–2819, 2009.
76. K. Deckelnick and M. Hinze. Numerical analysis of a control and state constrained elliptic control problem with piecewise constant control approximations. In K. Kunisch, G. Of, and O. Steinbach, editors, *Numerical Mathematics and Advanced Applications*, pages 597–604. Springer, Berlin, Heidelberg, 2008.
77. L. Dedè. Optimal flow control for Navier-Stokes equations: Drag minimization. *Int. J. Numer. Meth. Fluids*, 55(4):347 – 366, 2007.
78. L. Dedè and A. Quarteroni. Optimal control and numerical adaptivity for advection–diffusion equations. *ESAIM Math. Modelling Numer. Anal.*, 39(5):1019–1040, 2005.
79. M.C. Delfour and J.P. Zolésio. *Shapes and geometries - analysis, differential calculus, and optimization.* Society for Industrial and Applied Mathematics, Philadephia, PA, 2nd edition, 2011.

80. J. Dennis and R. Schnabel. *Numerical Methods for Unconstrained Optimization and Nonlinear Equations.* Classics in Applied Mathematics. Society for Industrial and Applied Mathematics (SIAM), Philadelphia, PA, 1996.
81. M. Desai and K. Ito. Optimal controls of Navier–Stokes equations. *SIAM J. Control Optim.*, 32(5):1428–1446, 1994.
82. J.A. Désidéri, B. Abou El Majd, and A. Janka. Nested and self-adaptive Bézier parameterizations for shape optimization. *J. Comput. Phys.*, 224(1):117–131, 2007.
83. H. Do, A. A. Owida, and Y. S. Morsi. Numerical analysis of coronary artery bypass grafts: An over view. *Comput. Meth. Prog. Bio.*, 108:689–705, 2012.
84. H. S. Dollar, N. I. M. Gould, W. H. A. Schilders, and A. J. Wathen. Implicit-factorization preconditioning and iterative solvers for regularized saddle-point systems. *SIAM J. Matrix Anal. Appl.*, 28:170–189, 2006.
85. H. S. Dollar, N. I. M. Gould, W. H. A. Schilders, and A. J. Wathen. Using constraint preconditioners with regularized saddle-point problems. *Computational Optimization and Applications*, 36(2):249–270, 2007.
86. H.S. Dollar, N.I.M. Gould, M. Stoll, and A.J. Wathen. Preconditioning Saddle-Point Systems with Applications in Optimization. *SIAM J. Sci. Comput.*, 32:249–270, February 2010.
87. H. Elman, V. E. Howle, J. Shadid, R. Shuttleworth, and R. Tuminaro. Block preconditioners based on approximate commutators. *SIAM J. Sci. Comput.*, 27(5):1651–1668, 2006.
88. H. C. Elman, D. J. Silvester, and A.J. Wathen. *Finite Elements and Fast Iterative Solvers with Applications in Incompressible Fluid Dynamics.* Oxford University Press, New York, 2004.
89. H.C.. Elman and G.H. Golub. Inexact and preconditioned Uzawa algorithms for saddle point problems. *SIAM J. Numer. Anal.*, 6:1645–1661, 1994.
90. H.C. Elman, D.J. Silvester, and A.J. Wathen. *Finite Elements and Fast Iterative Solvers with Applications in Incompressible Fluid Dynamics.* Oxford Science Publications, Series in Numerical Mathematics and Scientific Computation, Clarendon Press, Oxford, 2005.
91. H.W. Engl, M. Hanke, and A. Neubauer. *Regularization of Inverse Problems.* Mathematics and its Applications. Kluwer Academic Publisher, Dordrecht, The Netherlands, 2000.
92. S. Ervedoza and E. Zuazua. *The Wave Equation: Control and Numerics*, pages 245–339. Springer Berlin Heidelberg, 2012.
93. S. Ervedoza and E. Zuazua. *Numerical Approximation of Exact Controls for Waves.* SpringerBriefs in Mathematics. Springer-Verlag New York, 2013.
94. R. S. Falk. Approximation of a class of optimal control problems with order of convergence estimates. *J. Math. Anal. Appl.*, 44:28–47, 1973.
95. R. Fletcher. *Practical methods of optimization.* Wiley-Interscience (John Wiley & Sons), New York, 2nd edition, 2001.
96. A. V. Fursikov. *Optimal Control of Distributed Systems. Theory and Applications.* American Mathematical Society, 1999.
97. A. V. Fursikov, M. D. Gunzburger, and L. S. Hou. Boundary value problems and optimal boundary control for the navier–stokes system: the two-dimensional case. *SIAM J. Control Optim.*, 36(3):852–894, 1998.
98. A.V. Fursikov and R. Rannacher. Optimal Neumann control for the two-dimensional steady-state Navier-Stokes equations. In A.V. Fursikov, G. P. Galdi, and V.V. Pukhnachev, editors, *New Directions in Mathematical Fluid Mechanics: The Alexander V. Kazhikhov Memorial Volume.* Birkhäuser Basel, Basel, 2010.
99. J. Gain and D. Bechmann. A survey of spatial deformation from a user-centered perspective. *ACM Trans. Graph.*, 27(4):107:1–107:21, 2008.
100. Z. Gao and Y. Ma. A new stabilized finite element method for shape optimization in the steady navier–stokes flow. *Appl. Num. Math.*, 60(8):816–832, 2010.
101. Z. Gao, Y. Ma, and H. Zhuang. Drag minimization for navier–stokes flow. *Numerical Methods for Partial Differential Equations*, 25(5):1149–1166, 2009.
102. T. Geveci. On the approximation of the solution of an optimal control problem governed by an elliptic equation. *R.A.I.R.O., Anal. Numér.*, 13(4):313–328, 1978.
103. O. Ghattas and J.H. Bark. Optimal control of two- and three-dimensional incompressible Navier-Stokes flows. *J. Comput. Phys.*, 136(2):231–244, 1997.
104. D. Gilbarg and N.S. Trudinger. *Elliptic Partial Differential Equations of Second Order.* Springer Verlag, Berlin, 1983.
105. V. Girault and P.-A. Raviart. *Finite element methods for Navier-Stokes equations: Theory and algorithms.* Springer-Verlag, Berlin and New York, 1986.
106. G.H. Golub and C.F. Van Loan. *Matrix computations (3rd ed.).* Johns Hopkins University Press, 1996.
107. W. Gong, M. Hinze, and Z. Zhou. Finite element method and a priori error estimates for dirichlet boundary control problems governed by parabolic pdes. *J. Sci. Comput.*, 66(3):941–967, 2016.

108. W. Gong and N. Yan. A posteriori error estimate for boundary control problems governed by the parabolic partial differential equations. *J. Comput. Math.*, pages 68–88, 2009.
109. Wei Gong, Michael Hinze, and Zhaojie Zhou. A priori error analysis for finite element approximation of parabolic optimal control problems with pointwise control. *SIAM J. Control Optim.*, 52(1):97–119, 2014.
110. N. Gould and V. Simoncini. Spectral analysis of saddle point matrices with indefinite leading blocks. *SIAM J. Matrix Anal. & Appl.*, 31(3):1152–1171, 2009.
111. N. I. M. Gould, M. E. Hribar, and J. Nocedal. On the solution of equality constrained quadratic programming problems arising in optimization. *SIAM J. Sci. Comput.*, 23:1376–1395, 2001.
112. T. Grätsch and K. J. Bathe. A posteriori error estimation techniques in practical finite element analysis. *Comput. and Struct.*, 83:235–265, 2005.
113. P.M. Gresho and R.L. Sani. *Incompressible Flow and the Finite Element Method: Advection-Diffusion and Isothermal Laminar Flow.* John Wiley & Sons, 1998.
114. J.A. Griepentrog and L. Recke. Linear elliptic boundary value problems with non-smooth data: Normal solvability on sobolev-campanato spaces. *Mathematische Nachrichten*, 225(1):39–74, 2001.
115. J. A Griepentrop. Linear elliptic boundary value problems with nonsmooth data: Campanato spaces of functionals. *Math. Nachr.*, 243:19–42, 2002.
116. R. Griesse and B. Vexler. Numerical sensitivity analysis for the quantity of interest in pde-constrained optimization. *SIAM J. Sci. Comput.*, 29:22–48, January 2007.
117. R. Griesse and S. Volkwein. A primal-dual active set strategy for optimal boundary control of a nonlinear reaction-diffusion system. *SIAM J. Control Optim.*, 44(2):467–494, 2005.
118. P. Grisvard. *Elliptic problems in nonsmooth domains.* SIAM, 2011.
119. M. Gunzburger, L. Hou, and T. P. Svobodny. Analysis and finite element approximation of optimal control problems for the stationary Navier-Stokes equations with distributed and Neumann controls. *Math. Comp.*, 57:123–151, 1991.
120. M. D. Gunzburger. Sensitivities, adjoints and flow optimization. *Int. J. Numer. Methods Fluids*, 31(1):53–78, 1999.
121. M. D. Gunzburger and P.B. Bochev. *Least-Squares Finite Element Methods.* Springer-Verlag, New York, 2009.
122. M. D. Gunzburger and S. Manservisi. The velocity tracking problem for navier–stokes flows with bounded distributed controls. *SIAM J. Control Optim.*, 37(6):1913–1945, 1999.
123. M. D. Gunzburger and S. Manservisi. Analysis and approximation of the velocity tracking problem for navier–stokes flows with distributed control. *SIAM J. Numer. Anal.*, 37(5):1481–1512, 2000.
124. M. D. Gunzburger and S. Manservisi. The velocity tracking problem for navier–stokes flows with boundary control. *SIAM J. Control Optim.*, 39(2):594–634, 2000.
125. Max D. Gunzburger and Steven L. Hou. Treating inhomogeneous essential boundary conditions in finite element methods and the calculation of boundary stresses. *SIAM J. Numer. Anal.*, 29(2):390–424, 1992.
126. M.D. Gunzburger and P.B. Bochev. Least-squares finite element methods for optimization and control problems for the Stokes equations. *Comput. Math. with Appl.*, 48(7–8):1035–1057, 2004.
127. M.D. Gunzburger, L. Hou, and T.P. Svobodny. Boundary velocity control of incompressible flow with an application to viscous drag reduction. *SIAM J. Control Optim.*, 30:167–181, 1992.
128. M.D. Gunzburger, L. S. Hou, and Th. P. Svobodny. Analysis and finite element approximation of optimal control problems for the stationary navier-stokes equations with dirichlet controls. *ESAIM: Math. Model. Numer. Anal.*, 25(6):711–748, 1991.
129. M.D. Gunzburger, L.S. Hou, and T.P. Svobodny. Optimal control problems for a class of nonlinear equations with an application to control of fluids. In S.S. Sritharan, editor, *Optimal Control of Viscous Flow*, pages 43–62. Society for Industrial and Applied Mathematics, 1998.
130. M.D. Gunzburger and H. Kim. Existence of an optimal solution of a problem for the stationary Navier-Stokes equations. *SIAM J. Control Optim.*, 36(3):895–909, 1998.
131. E. Haber and U. Ascher. Preconditioned all-at-once methods for large sparse parameter estimation problems. *Inverse Problems*, 17:1847–1864, 2000.
132. W. Hackbusch. *Multi-grid Methods and Applications.* Springer-Verlag, New York, 1985.
133. R. Haller-Dintelmann, C. Meyer, J. Rehberg, and A. Schiela. Hölder continuity and optimal control for nonsmooth elliptic problems. *Applied Mathematics & Optimization*, 60(3):397–428, 2009.
134. J. Haslinger and R. A. E. Mäkinen. *Introduction to Shape Optimization: Theory, Approximation, and Computation.* Society for Industrial and Applied Mathematics, Philadephia, PA, 2003.
135. J. Haslinger and P. Neittaanmäki. *Finite element approximation for optimal shape, material and topology design (2nd Ed).* Wiley & Sons, 2nd edition, 1996.
136. X. He and C. Vuik. Comparison of some preconditioners for the incompressible navier-stokes equations. *Numer. Math. Theor. Meth. Appl.*, 9(2):239–261, 2016.
137. M. Heinkenschloss. Formulation and analysis of a sequential quadratic programming method for the optimal dirichlet boundary control of Navier-Stokes flow. In *Optimal Control: Theory, Algorithms, and Applications*, pages 178–203. Springer US, 1998.

138. M. Heinkenschloss and H. Nguyen. Neumann-neumann domain decomposition preconditioners for linear-quadratic elliptic optimal control problems. *SIAM J. Sci. Comput.*, 28(3):1001–1028, 2006.
139. A. Henrot and M. Pierre. *Variation et optimisation de formes. Une analyse géométrique*, volume 48 of *Series Mathématiques et Applications*. Springer, 2005.
140. R. Herzog and K. Kunisch. Algorithms for pde-constrained optimization. *GAMM-Mitteilungen*, 33(2):163–176, 2010.
141. R. Herzog and E. Sachs. Preconditioned conjugate gradient method for optimal control problems with control and state constraints. *SIAM. J. Matrix Anal. Appl.*, 31:2291–2317, 2010.
142. M. Hintermüller and M. Hinze. Moreau-yosida regularization in state constrained elliptic control problems: Error estimates and parameter adjustment. *SIAM J. Numer. Anal.*, 47(3):1666–1683, 2009.
143. M. Hintermüller, K. Ito, and K. Kunisch. The primal-dual active set strategy as a semismooth newton method. *SIAM J. Optim.*, 13(3):865–888, 2002.
144. M. Hintermüller, K. Kunisch, Y. Spasov, and S. Volkwein. Dynamical systems-based optimal control of incompressible fluids. *Int. J. Numer. Meth. Fluids*, 46(4):345–359, 2004.
145. M. Hinze. A variational discretization concept in control constrained optimization: the linear-quadratic case. *Comput. Optim. Appl.*, 30(1):45–61, 2005.
146. M. Hinze and K. Kunisch. Second order methods for optimal control of time-dependent fluid flow. *SIAM J. Control Optim.*, 40(3):925–946, 2001.
147. M. Hinze and C. Meyer. Variational discretization of lavrentiev-regularized state constrained elliptic optimal control problems. *Comput. Optim. Appl.*, 46(3):487–510, 2010.
148. M. Hinze, R. Pinnau, M. Ulbrich, and S. Ulbrich. *Optimization with PDE Constraints*, volume 23 of *Mathematical Modelling: Theory and Applications*. Springer, 2009.
149. L. S. Hou and S. S. Ravindran. Numerical approximation of optimal flow control problems by a penalty method: Error estimates and numerical results. *SIAM J. Sci. Comput.*, 20(5):1753–1777, April 1999.
150. K. Ito and K. Kunisch. The primal-dual active set method for nonlinear optimal control problems with bilateral constraints. *SIAM J. Control Optim.*, 43(1):357–376, 2004.
151. K. Ito and K. Kunisch. *Lagrange multiplier approach to variational problems and applications*. Advances in design and control. Society for Industrial and Applied Mathematics, 2008.
152. S. Jakobsson and O. Amoignon. Mesh deformation using radial basis functions for gradient-based aerodynamic shape optimization. *Comput. Fluids*, 36:1119–1136, 2007.
153. A. Jameson. Aerodynamic design via control theory. *J. Sci. Comput.*, 3:233–260, 1988.
154. J. Kaipio and E. Somersalo. *Statistical and computational inverse problems*, volume 160 of *Applied Mathematical Sciences*. Springer Science+Business Media, Inc., 2005.
155. M. I. Kamien and N.L. Schwartz. *Dynamic Optimization: The Calculus of Variations and Optimal Control in Economics and Management*, volume 31 of *Advanced Textbooks in Economics*. North-Holland, Amsterdam, The Netherlands, 2nd edition, 1991.
156. D. Kay, D. Loghin, and A. Wathen. A preconditioner for the steady-state navier–stokes equations. *SIAM J. Sci. Comput.*, 24(1):237–256, 2002.
157. C. Keller, N. I. M. Gould, and A. J. Wathen. Constraint preconditioning for indefinite linear systems. *SIAM J. Matrix Anal. Appl.*, 21:1300–1317, 2000.
158. H. Kim. A boundary control problem for vorticity minimization in time-dependent 2D Navier-Stokes equations. *Korean J. Math.*, 23(2):293–312, 2006.
159. H. Kim and O.K. Kwon. On a vorticity minimization problem for the stationary 2D Stokes equations. *J. Korean Math. Soc.*, 43(1):45–63, 2006.
160. T. Kim and D. Cao. Non-stationary Navier-Stokes equations with mixed boundary conditions. *J. Math. Sci. Univ. Tokyo*, 24:159–194, 2017.
161. K. Kunisch, K. Pieper, and A. Rund. Time optimal control for a reaction diffusion system arising in cardiac electrophysiology: a monolithic approach. *ESAIM: Math. Model. Numer. Anal.*, 50(2):381–414, 2016.
162. K. Kunisch and A. Rösch. Primal-dual active set strategy for a general class of constrained optimal control problems. *SIAM J. Optim.*, 13(2):321–334, 2002.
163. K. Kunisch and A. Rund. Time optimal control of the monodomain model in cardiac electrophysiology. *IMA J. Appl. Math.*, 80(6):1664–1683, 2015.
164. K. Kunisch and B. Vexler. Optimal vortex reduction for instationary flows based on translation invariant cost functionals. *SIAM J. Control Optim*, 46(4):1368–1397, 2007.
165. O. Ladyzhenskaya. *The mathematical theory of viscous incompressible flow*. Gordon and Breach Science Publishers, New York, 2nd edition, 1969.
166. J. Lagarias, J. Reeds, M. Wright, and P. Wright. Convergence properties of the Nelder-Mead simplex method in low dimensions. *SIAM J. Optim.*, 9(1):112–147, 1999.
167. J. Larson, M. Menickelly, and S. M. Wild. Derivative-free optimization methods. *Acta Numerica*, 28:287–404, 2019.

168. I. Lasiecka and R. Triggiani. *Control Theory for Partial Differential Equations: Continuous and Approximation Theories. I: Abstract Parabolic Systems.* Cambridge University Press, 2010.
169. T. Lassila, A. Manzoni, A. Quarteroni, and G. Rozza. Boundary control and shape optimization for the robust design of bypass anastomoses under uncertainty. *ESAIM Math. Modelling Numer. Anal.*, 47(4):1107–1131, 2013.
170. T. Lassila, A. Manzoni, A. Quarteroni, and G. Rozza. A reduced computational and geometrical framework for inverse problems in haemodynamics. *Int. J. Numer. Methods Biomed. Engng.*, 29(7):741–776, 2013.
171. K. Law, A. Stuart, and K. Zygalakis. *Data Assimilation. A Mathematical Introduction*, volume 62 of *Texts in Applied Mathematics.* Springer International Publishing Switzerland, 2015.
172. E.B. Lee and L. Markus. *Foundations of Optimal Control Theory.* Krieger Pub. Co., 1986.
173. T. Lehnhäuser and M. Schäfer. A numerical approach for shape optimization of fluid flow domains. *Comput. Methods Appl. Mech. Engrg.*, 194:5221–5241, 2005.
174. R. Li, W. Liu, H. Ma, and T. Tang. Adaptive finite element approximation for distribuited elliptic optimal control problems. *SIAM J. Control Optim.*, pages 1321–1349, 2001.
175. J. L. Lions. Exact controllability, stabilization and perturbations for distributed systems. *SIAM Review*, 30(1):1–68, 1988.
176. J.-L. Lions and E. Magenes. *Non-Homogenous Boundary Value Problems and Applications.* Springer-Verlag, 1972.
177. J.L. Lions. *Optimal Control of Systems governed by Partial Differential Equations.* Springer-Verlag, Berlin Heidelberg, 1971.
178. J.L. Lions. *Some aspects of the optimal control of distributed parameter systems.* Society for Industrial and Applied Mathematics, Philadelphia, PA, USA, 1972.
179. W. Liu, H. Ma, T. Tang, and N. Yan. A posteriori error estimates for discontinuous galerkin time-stepping method for optimal control problems governed by parabolic equations. *SIAM J. Numer. Anal.*, 42(3):1032–1061, 2004.
180. W. Liu and N. Yan. A posteriori error estimates for some model boundary control problems. *J. Comput. Appl. Math.*, pages 159–173, 2000.
181. W. Liu and N. Yan. A posteriori error estimates for distribuited convex optimal control problems. *Adv. Comput. Math.*, 15:285–309, 2001.
182. W. Liu and N. Yan. A posteriori error estimates for control problems governed by stokes equations. *SIAM J. Numer. Anal.*, 40(5):1850–1869, 2002.
183. W. Liu and N. Yan. A posteriori error estimates for optimal control problems governed by parabolic equations. *Num. Math.*, 93(3):497–521, 2003.
184. D. Luenberger. *Introduction to Linear and Non Linear Programming.* Addison-Wesley, Reading, Massachusetts, 1973.
185. R. Luus. *Iterative Dynamic Programming.* Chapman & Hall, 2000.
186. J. Macki and A. Strauss. *Introduction to Optimal Control Theory.* Undergraduate Texts in Mathematics. Springer-Verlag, New York, 2nd edition, 1982.
187. A. Manzoni and S. Pagani. A certified rb method for PDE-constrained parametric optimization problems. *Commun. Appl. Ind. Math.*, 10(1):123–?152, 2019.
188. A. Manzoni, A. Quarteroni, and G. Rozza. Model reduction techniques for fast blood flow simulation in parametrized geometries. *Int. J. Numer. Methods Biomed. Engng.*, 28(6–7):604–625, 2012.
189. A. Manzoni, A. Quarteroni, and G. Rozza. Shape optimization of cardiovascular geometries by reduced basis methods and free-form deformation techniques. *Int. J. Numer. Methods Fluids*, 70(5):646–670, 2012.
190. A. Manzoni, A. Quarteroni, and S. Salsa. A saddle point approach to an optimal boundary control problem for steady navier-stokes equations. *Mathematics in Engineering*, 1(mine-01-02-252):252, 2019.
191. S. May, R. Rannacher, and B. Vexler. Error Analysis for a Finite Element Approximation of Elliptic Dirichlet Boundary Control Problems. *SIAM J. Control Optim.*, 51(3):2585–2611, 2013.
192. D. Meidner and B. Vexler. Adaptive space-time finite element methods for parabolic optimization problems. *SIAM J. Control Optim.*, 46(1):116–142, 2007.
193. D. Meidner and B. Vexler. A priori error estimates for space-time finite element discretization of parabolic optimal control problems Part i: Problems without control constraints. *SIAM J. Control Optim.*, 47(3):1150–1177, 2008.
194. D. Meidner and B. Vexler. A priori error estimates for space-time finite element discretization of parabolic optimal control problems part ii: problems with control constraints. *SIAM J. Control Optim.*, 47(3):1301–1329, 2008.
195. D. Meidner and B. Vexler. A priori error analysis of the petrov–galerkin crank–nicolson scheme for parabolic optimal control problems. *SIAM J. Control Optim.*, 49(5):2183–2211, 2011.
196. C. Meyer. Error estimates for the finite-element approximation of an elliptic control problem with pointwise state and control constraints. *Control and Cybernetics*, 37(1):51, 2008.

197. C. Meyer, U. Pruüfert, and F. Tröltzsch. On two numerical methods for state-constrained elliptic control problems. *Optim. Methods Softw.*, 22:871–899, 2007.
198. C. Meyer, A. Rösch, and F. Tröltzsch. Optimal control of PDEs with regularized pointwise state constraints. *Comput. Optim. Appl.*, 33(2):209–228, 2006.
199. F. Migliavacca and G. Dubini. Computational modeling of vascular anastomoses. *Biomech. Model. Mechanobiol.*, 3(4):235–250, 2005.
200. B. Mohammadi and O. Pironneau. *Applied shape optimization for fluids.* Oxford University Press, 2001.
201. F. Murat and J. Simon. *Sur le contrôle par un domaine géométrique*, volume Internal Report N. 76 015. Université Paris 6, 1976. Pré-publication du Laboratoire d'Analyse Numérique.
202. F.M. Murphy, G.H. Golub, and A.J. Wathen. A note on preconditioning for indefinite linear systems. *SIAM J. Sci. Comput.*, 21:1969–1972, December 1999.
203. J. Necas. *Les Methodes Directes en Theorie des Equations Elliptiques.* Masson, Paris, 1967.
204. F. Negri, A. Manzoni, and G. Rozza. Reduced basis approximation of parametrized optimal flow control problems for the Stokes equations. *Comput. & Math. with Appl.*, 69(4):319–336, 2015.
205. F. Negri, G. Rozza, A. Manzoni, and A. Quarteroni. Reduced basis method for parametrized elliptic optimal control problems. *SIAM J. Sci. Comput.*, 35(5):A2316–A2340, 2013.
206. J. Nelder and R. Mead. A simplex method for function minimization. *The Computer Journal*, 7:308–313, 1965.
207. J. Nocedal. Theory of algorithms for unconstrained optimization. *Acta Numerica*, 1:199–242, 1992.
208. J. Nocedal and S.J. Wright. *Numerical optimization.* Springer Series in Operations Research and Financial Engineering. Springer, New York, second edition, 2006.
209. G. Of, T.X. Phan, and O. Steinbach. An energy space finite element approach for elliptic dirichlet boundary control problems. *Num. Math.*, 129(4):723–748, 2015.
210. C. C. Paige and M. A. Saunders. Solution of sparse indefinite systems of linear equations. *SIAM J. Numer. Anal.*, 12:617–629, 1975.
211. S. V. Patankar. *Numerical Heat Transfer and Fluid Flow.* Hemisphere, Washington, 1980.
212. J. W. Pearson. Preconditioned iterative methods for navier–stokes control problems. *J. Comput. Phys.*, 292:194–207, 2015.
213. J. W. Pearson, M. Stoll, and A. J. Wathen. Regularization-robust preconditioners for time-dependent pde-constrained optimization problems. *SIAM Journal on Matrix Analysis and Applications*, 33(4):1126–1152, 2012.
214. J. W. Pearson and A. J. Wathen. A new approximation of the schur complement in preconditioners for PDE-constrained optimization. *Numer. Linear Algebr.*, 19:816?–829, 2012.
215. O. Pironneau. *Optimal shape design for elliptic systems.* Springer-Verlag, 1984.
216. L.S. Pontryagin, V.G. Boltyanskii, R.V. Gamkrelidze, and E.F. Mishchenko. *The Mathematical Theory of Optimal Processes.* Pergamon Press, New York, 1964.
217. E. E. Prudencio, R. Byrd, and X.-C. Cai. Parallel full space sqp lagrange–newton–krylov–schwarz algorithms for pde-constrained optimization problems. *SIAM J. Sci. Comput.*, 27(4):1305–1328, 2006.
218. A. Quarteroni. *Numerical Models for Differential Problems*, volume 16 of *Modeling, Simulation and Applications (MS&A).* Springer-Verlag Italia, Milano, 3rd edition, 2017.
219. A. Quarteroni, L. Dede', A. Manzoni, and C. Vergara. *Mathematical Modelling of the Human Cardiovascular System. Data, Numerical Approximation, Clinical Applications.* Cambridge Monographs on Applied and Computational Mathematics. Cambridge University Press, 2019.
220. A. Quarteroni, T. Lassila, S. Rossi, and R. Ruiz-Baier. Integrated heart – coupling multiscale and multiphysics models for the simulation of the cardiac function. *Comput. Methods Appl. Mech. Engrg.*, 314:345–407, 2017.
221. A. Quarteroni, A. Manzoni, and C. Vergara. The cardiovascular system: Mathematical modelling, numerical algorithms and clinical applications. *Acta Numerica*, 26:365–590, 2017.
222. A. Quarteroni, R. Sacco, and F. Saleri. *Numerical Mathematics*, volume 37 of *Texts in Applied Mathematics.* Springer, New York, 2000.
223. A. Quarteroni, F. Saleri, and P. Gervasio. *Scientific Computing with MATLAB and Octave*, volume 2 of *Texts in Computational Science and Engineering.* Springer-Verlag Berlin Heidelberg, 4 edition, 2014.
224. A. Quarteroni and A. Valli. *Numerical Approximation of Partial Differential Equations.* Springer-Verlag, Berlin-Heidelberg, 1994.
225. A. Quarteroni and A. Valli. *Domain Decomposition Methods for Partial Differential Equations.* Oxford Science Publications, 1999.
226. T. Rees, H. S. Dollar, and A. J. Wathen. Optimal solvers for PDE-constrained optimization. *SIAM J. Sci. Comput.*, 32:271–298, 2010.
227. T. Rees and M. Stoll. Block-triangular preconditioners for pde-constrained optimization. *Numer. Linear Algebra Appl.*, 17:977–996, 2010.
228. T. Rees, M. Stoll, and A.J. Wathen. All at once preconditioning in pde-constrained optimization. *Kybernetika*, 46(2):341–360, 2010.

229. T. Rees and A. J. Wathen. Preconditioning iterative methods for the optimal control of the Stokes equation. *SIAM J. Sci. Comput.*, 33(5):2903–2926, 2011.
230. M. ur Rehman, T. Geenen, C. Vuik, G. Segal, and S. P. MacLachlan. On iterative methods for the incompressible stokes problem. *Int. J. Numer. Meth. Fluids*, 65(10):1180–1200, 2011.
231. M. Renardy and R.C. Rogers. *An introduction to partial differential equations*, volume 13 of *Texts in Applied Mathematics*. Springer-Verlag New York, 2nd edition, 2004.
232. A. Rösch. Error estimates for linear-quadratic control problems with control constraints. *Optim. Meth. Software*, 21(1):121–134, 2006.
233. A. Rösch and B. Vexler. Optimal control of the stokes equations: A priori error analysis for finite element discretization with postprocessing. *SIAM J. Numer. Anal.*, 44(5):1903–1920, 2006.
234. W. Rudin. *Principles of Mathematical Analysis*. International Series in Pure and Applied Mathematics. McGraw-Hill, Inc., 3rd edition, 1976.
235. W. Rudin. *Real and complex analysis*. Tata McGraw-Hill Education, 1987.
236. Y. Saad. *Iterative Methods for Sparse Linear Systems*. Society for Industrial and Applied Mathematics, Philadelphia, second edition, 2003.
237. S. Salsa. *Partial Differential Equations in Action. From Modelling to Theory*, volume 99 of *Unitext*. Springer-Verlag Italia, Milano, 3rd edition, 2016.
238. S. Sankaran and A.L. Marsden. The impact of uncertainty on shape optimization of idealized bypass graft models in unsteady flow. *Phys. Fluids*, 22:121902, 2010.
239. R. Schaback and H. Wendland. Numerical techniques based on radial basis functions. In *Curve and Surface Fitting: Saint-Malo 1999*, pages 359–374. Vanderbilt University Press, 2000.
240. J. Schöberl, R. Simon, and W. Zulehner. A robust multigrid method for elliptic optimal control problems. *SIAM J. Numer. Anal.*, 49(4):1482–1503, 2011.
241. J. Schöberl and W. Zulehner. Symmetric Indefinite Preconditioners for Saddle Point Problems with Applications to PDE-Constrained Optimization Problems. *SIAM J. Matrix Anal. Appl.*, 29:752–773, 2007.
242. R. L. Scott and S. Zhang. Finite element interpolation of nonsmooth functions satisfying boundary conditions. *Math. Comp.*, (190):483?–493, 1990.
243. T.W. Sederberg and S.R. Parry. Free-form deformation of solid geometric models. *Comput. Graph.*, 20(4):151–160, 1986.
244. A. Segal, M. ur Rehman, and C. Vuik. Preconditioners for incompressible Navier-Stokes solvers. *Numer. Math. Theor. Meth. Appl.*, 3(3):245–275, 2010.
245. J.A. Sethian. *Level set methods and fast marching methods: evolving interfaces in computational geometry, fluid mechanics, computer vision, and materials science*, volume 3. Cambridge University Press, 1999.
246. L. F. Shampine and M. W. Reichelt. The `matlab` ODE suite. *SIAM J. Sci. Comput.*, 18:1–22, 1997.
247. G. Shultz, R. Schnabel, and R. Byrd. A family of trust region-based algorithms for unconstrained minimization with strong global convergence properties. *SIAM J. Numer. Anal.*, 22(1):47–67, 1985.
248. D. Silvester, H. Elman, D. Kay, and A. Wathen. Efficient preconditioning of the linearized navier–stokes equations for incompressible flow. *J. Comput. Appl. Math.*, 128(1):261–279, 2001.
249. D. Silvester and A. Wathen. Fast iterative solution of stabilised stokes systems part ii: using general block preconditioners. *SIAM J. Numer. Anal.*, 31(5):1352–1367, 1994.
250. J. Sokolowski and J.-P. Zolésio. *Introduction to shape optimization: shape sensitivity analysis*. Springer, 1992.
251. D. Srinath and S. Mittal. An adjoint method for shape optimization in unsteady viscous flows. *J. Comp. Phys.*, 229(6):1994–2008, 2010.
252. M. Stoll and A. Wathen. Preconditioning for partial differential equation constrained optimization with control constraints. *Numer. Linear Algebra Appl.*, 19(1):53–71, 2012.
253. M. Stoll and A. J. Wathen. Preconditioning for active set and projected gradient methods as semi-smooth newton methods for pde-constrained optimization with control constraints. Technical Report 25/2009, 2009.
254. M. Stoll and A. J. Wathen. All-at-once solution of time-dependent PDE-constrained optimization problems. Technical report, Oxford eprints archive, 2010.
255. M. Stoll and A. J. Wathen. All-at-once solution of time-dependent Stokes control. *J. Comput. Phys.*, (232):498–515, 2013.
256. A.M. Stuart. Inverse problems: a Bayesian perspective. *Acta Numerica*, 19(1):451–559, 2010.
257. T.J. Sullivan. *Introduction to Uncertainty Quantification*, volume 63 of *Texts in Applied Mathematics*. Springer International Publishing Switzerland, 2015.
258. W. Sun and Y.-X. Yuan. *Optimization theory and methods. Nonlinear programming*, volume 1 of *Springer Optimization and Its Applications*. Springer, New York, 2006.
259. J. Sundes, G.T. Lines, X. Cai, B.F. Nielsen, K.A. Mardal, and A. Tveito. *Computing the electrical activity in the heart*, volume 1 of *Monographs in Computational Science and Engineering Series*. Springer, 2006.

260. A. Tarantola. *Inverse Problem Theory and Methods for Model Parameter Estimation.* Society for Industrial and Applied Mathematics, Philadelphia, USA, 2004.
261. K. L. Teo, C. J. Goh, and K. H. Wong. *A Unified Computational Approach to Optimal Control Problems.* Longman Scientific & Technical, New York, 1991.
262. V Thomée. *Galerkin Finite Element Methods for Parabolic Problems*, volume 25 of *Springer Series in Computational Mathematics.* Springer Berlin Heidelberg, 2006.
263. F. Tröltzsch. *Optimal Control of Partial Differential equations: Theory, Methods, and Applications*, volume 112. American Mathematical Society, 2010.
264. F. Tröltzsch and I. Yousept. A regularization method for the numerical solution of elliptic boundary control problems with pointwise state constraints. *Comput. Optim. Appl.*, 42(1):43–66, 2009.
265. J.L. Troutman. *Variational Calculus and Optimal Control: Optimization with Elementary Convexity.* Undergraduate Texts in Mathematics. Springer, New York, 2nd edition, 1995.
266. R. Verfürth. *A posteriori Error Estimation Techniques for Finite Element Methods.* Oxford University Press, Oxford, 2013.
267. B. Vexler and W. Wollner. Adaptive finite elements for elliptic optimization problems with control constraints. *SIAM J. Control Optim.*, 47(1):509–534, 2008.
268. C.R. Vogel. *Computational Methods for Inverse Problems.* Frontiers in Applied Mathematics. Society for Industrial and Applied Mathematics, Philadephia, PA, 2002.
269. S. Volkwein. Mesh-independence for an augmented lagrangian-sqp method in hilbert spaces. *SIAM J. Control Optim.*, 38(3):767–785, 2000.
270. G. Walsh. *Methods of Optimization.* Wiley, 1975.
271. A. Wathen and D. Silvester. Fast iterative solution of stabilised stokes systems. part i: Using simple diagonal preconditioners. *SIAM J. Numer. Anal.*, 30(3):630–649, 1993.
272. H. Wendland. *Scattered Data Approximation.* Cambridge University Press, 2005.
273. M. Wolfe. *Numerical Methods for Unconstrained Optimization.* Van Nostrand Reinhold Company, New York, 1978.
274. K. Yosida. *Functional Analysis.* Springer-Verlag, Berlin, 1974.
275. E. Zeidler. *Nonlinear Functional Analysis and its Applications*, volume I: Fixed-Point Theorems. Springer-Verlag, New York, 1985.
276. E. Zeidler. *Nonlinear Functional Analysis and its Applications*, volume II/A: Linear Monotone Operators. Springer-Verlag, New York, 1990.
277. E. Zuazua. Propagation, observation, and control of waves approximated by finite difference methods. *SIAM Review*, 47 (2):197–243, 2005.
278. E. Zuazua. Controllability and observability of partial differential equations: Some results and open problems. In C.M. Dafermos and E. Feiteisl, editors, *Handbook of Differential Equations: Evolutionary Differential Equations*, volume 3, pages 527–621. Elsevier Science, 2006.
279. W. Zulehner. Nonstandard norms and robust estimates for saddle point problems. *SIAM. J. Matrix Anal. & Appl.*, 32(2):536–560, 2011.
280. R. Zygmund and R. Wheeden. *Measure and Integral.* Marcel Dekker, 1977.

Index

A. Manzoni et al., *Optimal Control of Partial Differential Equations*,
Applied Mathematical Sciences 207, https://doi.org/10.1007/978-3-030-77226-0

Printed in the United States
by Baker & Taylor Publisher Services